IN QUEST OF THE UNIVERSE

2ND EDITION

KARL F. KUHN

Eastern Kentucky University

JONES AND BARTLETT PUBLISHERS

Sudbury, Massachusetts

BOSTON   LONDON   SINGAPORE

Editorial, Sales, and Customer Service Offices
Jones and Bartlett Publishers
40 Tall Pine Drive
Sudbury, MA 01776
978-443-5000
info@jbpub.com
http://www.jbpub.com

Jones and Bartlett Publishers International
Barb House, Barb Mews
London W6 7PA
UK

Copyright © 1998 by Jones and Bartlett Publishers, Inc.

All rights reserved. No part of the material protected by this copyright notice may be reproduced or utilized in any form, electronic or mechanical, including photocopying, recording, or by any information storage and retrieval system, without written permission from the copyright owner.

PRODUCTION CREDITS
SPONSORING EDITOR Christopher W. Hyde
SR. DEVELOPMENTAL EDITOR Dean DeChambeau
PRODUCTION MANAGER Anne Spencer
COMPOSITION Carlisle Communications
COVER DESIGN Anne Spencer
TEXT DESIGN Diane Beasley
TECHNICAL ILLUSTRATIONS Randy Miyake, Miyake Illustration

Library of Congress Cataloging-in-Publication Data
Kuhn, Karl F.
    In quest of the universe, updated and enhanced web version / Karl. F. Kuhn.—2nd ed.
       p.   cm.
    Includes index.
    ISBN 0-7637-0605-1
     1. Astronomy I. Title
  QB45.K84
  520—dc20   93-26335
                                                     CIP

ABOUT THE COVER The cover is a composite of the W.M. Keck Telescope and the Hubble Deep Field image. Astronomers pointed the Hubble Space Telescope at one of the emptiest parts of the night sky and focused on an area the size of a dime seen from 75 feet to obtain this image. Only the few spiked points of light are stars; the remaining 1500 or so objects are galaxies, each containing billions of stars. Astronomers are using the W.M. Keck Telescope for follow-up surveys, observations, and comparison studies. (Keck photograph courtesy of Richard Wainscoat; Hubble Deep Field image courtesy of Robert Williams, the Hubble Deep Field Team, and NASA)

Printed in the United States of America
01 00 99 98 97   10 9 8 7 6 5 4 3 2 1

To my mother-in-law, Gladys. I got a bonus when your beautiful daughter took my hand thirty-three years ago.

www.jbpub.com/starlinks

StarLinks connects users of *In Quest of the Universe, 2E,* to an extensive astronomy web site developed by Jones and Bartlett Publishers. The site offers a variety of activities designed to enhance the learning process and to give students access to the most current astronomical information. You reach the StarLinks home page by entering the URL **http://www.jbpub.com/starlinks** into a Web browser such as Netscape Navigator or Microsoft Internet Explorer.

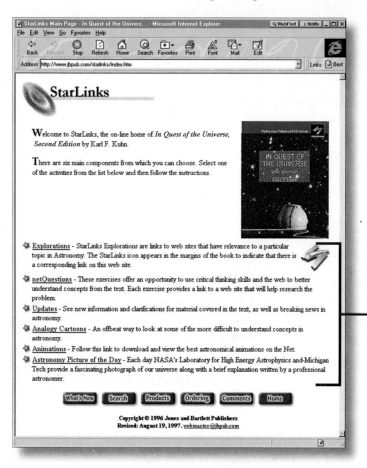

The StarLinks site offers unparalleled quality and reliability because:

- All of the Internet activities and resources are reviewed and handpicked by an astronomy instructor.
- The StarLinks page provides descriptions of each linked site, so you never surf alone.
- The StarLinks site is maintained in-house by Jones and Bartlett Publishers, so any broken links are quickly repaired or replaced.

You have a choice of six StarLinks activities and resources.

StarLinks Explorations direct students to useful astronomy sites on the net. The links enhance and constantly update the material presented in the text. Jones and Bartlett monitors the links daily to ensure there will always be a working and appropriate site at the other end of the link.

The StarLinks icon in the margin identifies key topics that are matched to WWW sites you can visit through StarLinks Explorations. These sites offer the latest information on a topic or expand on a related topic.

To find out more about StarLinks, please e-mail **StarLinks@jbpub.com,** *or call your Jones and Bartlett sales representative at 800 832-0034.*

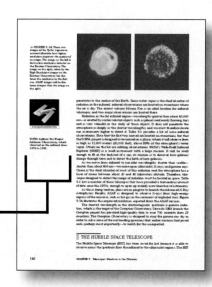

www.jbpub.com/starlinks

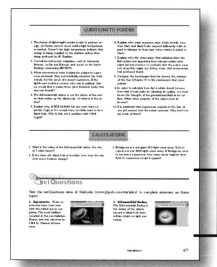

Each chapter ends with an introduction to the web-based exercises found on-line at StarLinks NetQuestions.

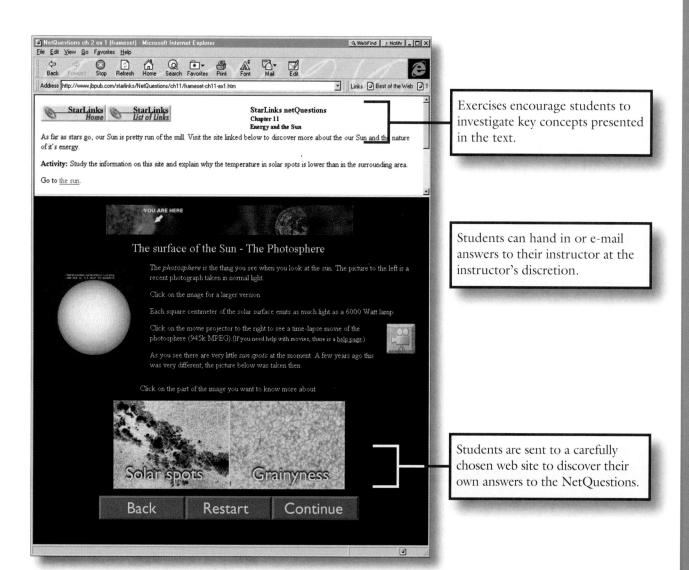

Exercises encourage students to investigate key concepts presented in the text.

Students can hand in or e-mail answers to their instructor at the instructor's discretion.

Students are sent to a carefully chosen web site to discover their own answers to the NetQuestions.

Contents in Brief

PROLOGUE
The Quest Ahead 1

CHAPTER 1
An Earth-Centered Universe 13

CHAPTER 2
A Sun-Centered Universe 39

CHAPTER 3
Galileo, Newton, and Einstein 63

CHAPTER 4
Light and the Electromagnetic Spectrum 95

CHAPTER 5
Telescopes: Windows to the Universe 121

CHAPTER 6
The Earth–Moon System 153

CHAPTER 7
A Planetary Overview 197

CHAPTER 8
The Terrestrial Planets 227

CHAPTER 9
The Jovian Planets 265

CHAPTER 10
Pluto and Solar System Debris 305

CHAPTER 11
The Sun 337

CHAPTER 12
Measuring the Properties of Stars 367

CHAPTER 13
Interstellar Matter and Star Formation 401

CHAPTER 14
The Lives and Deaths of Stars 423

CHAPTER 15
The Deaths of Massive Stars 449

CHAPTER 16
The Milky Way Galaxy 479

CHAPTER 17
The Diverse Galaxies 507

CHAPTER 18
Cosmology: The Nature of the Universe 543

Contents

PROLOGUE
THE QUEST AHEAD 1

The View from Earth 3
Questions, Answers, and Methods 4
From Earth to Galaxies 5
Units of Distance in Astronomy 6
■ CLOSE UP: Powers of Ten 8
The Scale of the Universe 8
Simplicity and the Unity of Nature 10
Astronomy Today 10

1 AN EARTH-CENTERED UNIVERSE 13

The Celestial Sphere 14
Constellations 15
Measuring the Positions of Celestial Objects 18
■ CLOSE UP: Celestial Coordinates 20
The Sun's Motion: How Long is a Year? 20
The Ecliptic 21
The Seasons 22
■ HISTORICAL NOTE: Leap Year and the Calendar 25
Scientific Models—A Geocentric Model 26
The Greek Celestial Model 27
Observation: The Planets 28
■ CLOSE UP: Why Is East on the Left in Sky Photographs? 30
Criteria for Scientific Models 31
Another Model 33
■ CLOSE UP: Astrology and Science 34
Model, Theory, and Hypothesis 35

2 A Sun-Centered System 39

The Marriage of Aristotle and Christianity 40
Nicolaus Copernicus 41
■ HISTORICAL NOTE: Copernicus and His Times 42
The Copernican System 42
Motions of the Planets 45
Comparing the Two Models 46
Parallax 47
Copernicus's Revolution 51
Tycho Brahe 51
Johannes Kepler 52
The Ellipse 53
Kepler's First Two Laws of Planetary Motion 54
Kepler's Third Law 55
■ HISTORICAL NOTE: Johannes Kepler 56
Kepler's Contribution 58
■ ACTIVITY: The Rotating Earth 61
■ ACTIVITY: The Radius of Mars's Orbit 61

3 GALILEO, NEWTON, AND EINSTEIN 63

Galileo Galilei and the Telescope 64
The Moon, the Sun, and the Stars 64
Satellites of Jupiter 65
The Phases of Venus 66
■ HISTORICAL NOTE: Galileo Galilei 68
Isaac Newton's Grand Synthesis 70
Newton's First Two Laws of Motion 70
An Important Digression—Mass and Weight 71
Back to Newton's Second Law 71
■ HISTORICAL NOTE: Isaac Newton 72
Newton's Third Law 73
Motion in a Circle 73
The Law of Universal Gravitation 75
Testing the Law of Universal Gravitation 75
■ CLOSE UP: Travel to the Moon 76
Newton's Laws and Kepler's Laws 77

The Center of Mass 79
The Tides 81
Rotation and Revolution of the Moon 83
Precession of the Earth 84
The Importance of Newton's Laws 85
Beyond Newton: How Science Progresses 86
General Relativity 86
Space Warp 87
■ CLOSE UP: The Special Theory of Relativity 88
Gravitation and Einstein 89
The Orbit of Mercury 89
■ HISTORICAL NOTE: Albert Einstein 90
The Correspondence Principle 91
■ ACTIVITY: Circular Motion 94

4 LIGHT AND THE ELECTRO-MAGNETIC SPECTRUM 95

Temperature Scales 96
The Wave Nature of Light 97
Wave Motion in General 98
Light as a Wave 99
Invisible Electromagnetic Radiation 100
The Colors of Planets and Stars 101
Color from Reflection—The Colors of Planets 101
Color as a Measure of Temperature 102
■ CLOSE UP: Blackbody Radiation 105
Spectra Examined Close Up 105
Kirchhoff's Laws 106
The Bohr Atom 106
■ HISTORICAL NOTE: Niels Bohr 108
Emission Spectra 109
■ CLOSE UP: Coexisting Competing Theories: The Nature of Light 110
Absorption Spectra—The Stars 110
■ CLOSE UP: The Balmer Series 112
The Doppler Effect 112
The Doppler Effect in Astronomy 114
The Doppler Effect as a Measurement Technique 115
Other Doppler Effect Measurements 117
Relative or Real Speed? 117
The Inverse Square Law 117

5 TELESCOPES: WINDOWS TO THE UNIVERSE 121

Refraction and Image Formation 122
The Refracting Telescope 123
Chromatic Aberration 124
The Powers of a Telescope 125
Angular Size and Magnifying Power 126
Light-Gathering Power 128
Resolving Power 129
The Reflecting Telescope 131
Large Optical Telescopes 132
■ CLOSE UP: Spinning a Giant Mirror 135
Telescope Accessories 135
Radio Telescopes 139
■ HISTORICAL NOTE: Radio Waves from Space 140
Interferometry 143
Detecting Other Electromagnetic Radiation 145
The Hubble Space Telescope 146
■ CLOSE UP: ET Life, Part I—SETI 148
■ CLOSE UP: ET Life, Part II—CETI 149

6 THE EARTH–MOON SYSTEM 153

Measuring the Size of the Earth and Moon 154
The Distance to the Moon 155
Summary: Two Measuring Techniques 159
The Moon's Changing Size 159
The Moon's Phases 160
Lunar Eclipses 162
Types of Lunar Eclipses 165
Solar Eclipses 166
The Partial Solar Eclipse 168
The Annular Eclipse 169
Earth 171
The Interior of the Earth 171
Earth's Magnetic Field 172
■ CLOSE UP: The Earth from Space 174
Plate Tectonics 176
Earth's Atmosphere 179
The Moon's Surface 181
■ CLOSE UP: The Far Side of the Moon 185
Theories of the Origin of the Moon 186
The Large Impact Theory 187
The History of the Moon 187
■ CLOSE UP: Measuring the Age of the Earth and Moon 189
■ ACTIVITY: Do-It-Yourself Phases 193
■ ACTIVITY: Observing the Moon's Phases 194
■ ACTIVITY: Observing a Solar Eclipse 195

7 A PLANETARY OVERVIEW 197

Distances in the Solar System 198
Measuring Distances in the Solar System 199
■ CLOSE UP: The Titius-Bode Law 200
Measuring the Mass of a Solar System Object 202
■ HISTORICAL NOTE: The Discovery of the Asteroids 203
Planetary Motions 204
Classifying the Planets 206
Size, Mass, and Density 206
Satellites and Rings 208
Rotations 208
Planetary Atmospheres and Escape Velocity 209
Gases and Escape Velocity 211
The Atmospheres of the Planets 212
The Formation of the Solar System 213
Evidential Clues from the Data 213
Evolutionary Theories 214
Catastrophic Theories 215
Present Evolutionary Theories 216
Explaining Other Clues 219
Planetary Systems Around Other Stars? 219
■ CLOSE UP: ET Life III—The Origin of Life 222

8 THE TERRESTRIAL PLANETS 227

Mercury 228
Mercury as Seen from Earth 228
Mercury via *Mariner*—Comparison with the Moon 229
Size, Mass, and Density 231
Mercury's Motions 233
Venus 236
Size, Mass, and Density 236
Venus's Motions 236
The Surface of Venus 237
■ CLOSE UP: Our Changing View of Venus 238
The Atmosphere of Venus 240

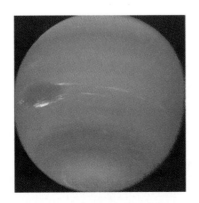

A Hypothesis Explaining Venus/Earth Differences 242
Mars 245
Mars as Seen from Earth 246
Size, Mass, and Density 247
Mars's Motions 247
Life on Mars 248
Invasion and Its Results 248
■ CLOSE UP: ET Live IV—*Viking*'s Search for Life 252
Atmospheric and Surface Conditions 255
The Moons of Mars 257
■ HISTORICAL NOTE: The Discovery of the Martian Moons 258
Why Explore? 260
■ ACTIVITY: Viewing Mercury, Venus, and Mars 263

9 THE JOVIAN PLANETS 265

Jupiter 266
Jupiter as Seen from Earth 266
Jupiter from Space 267
The Composition of Jupiter 269
■ CLOSE UP: ET Live V—Letters to Extraterrestrials 272
Energy from Jupiter 272
Jupiter's Moons 274
Jupiter's Ring 278
Saturn 280
Size, Mass, and Density 280
Saturn's Motions 280
Pioneer, Voyager, and *Cassini* 282
■ CLOSE UP: A Hypothesis to Explain Saturn's Excess Heat 283
Titan 284
Planetary Rings 285
The Origin of Rings 286
Uranus 288
■ HISTORICAL NOTE: William Herschel, Musician/Astronomer 290
Uranus's Orientation and Motion 292
Neptune 294
■ CLOSE UP: Shepherd Moons 296
Neptune's Moons and Rings 297
■ HISTORICAL NOTE: The Discovery of Neptune 300
■ ACTIVITY: Observing Jupiter and Saturn 304

10 PLUTO AND SOLAR SYSTEM DEBRIS 305

The Discovery of Pluto 306
Pluto as Seen from Earth 307
Pluto and Charon 308
A Former Moon of Neptune? 309
Solar System Debris 309

Asteroids 309
The Orbits of Asteroids 310
▪ *CLOSE UP: You Can Name an Asteroid* 312
The Origin of Asteroids 313
Comets 314
▪ *CLOSE UP: Chaos Theory* 315
Comet Orbits—Isaac Newton and Edmund Halley 315
The Nature of Comets 316
▪ *HISTORICAL NOTE: Astronomer Maria Mitchell, A Nineteenth-Century Feminist* 318
Comet Tails 319
The Oort Cloud and Kuiper Belt 321
▪ *HISTORICAL NOTE: Jan H. Oort, 1900–1992* 322
The Origin of Short-Period Comets 322
Meteors and Meteor Showers 323
Meteors 324
Meteoroids 325
Meteor Showers 326
Meteorites and Craters 329
▪ *CLOSE UP: Hit by a Meteorite?* 330
▪ *CLOSE UP: Meteors and Dinosaurs* 332
▪ *ACTIVITY: Observing Meteors* 335

11 THE SUN 337

Solar Data 338
Solar Energy 338
▪ *CLOSE UP: The Distance to the Sun* 339
The Source of the Sun's Energy 341
Solar Nuclear Reactions 341
▪ *CLOSE UP: Fission and Fusion Power on Earth* 342
The Sun's Interior 345
Pressure, Temperature, and Density 345
Hydrostatic Equilibrium 346
Energy Transport 347
The Neutrino Problem 349
The Solar Atmosphere 351
The Photosphere 351
▪ *CLOSE UP: Data Uncertainty in the Homestake Experiment* 352
▪ *CLOSE UP: Helioseismology* 355
The Chromosphere and Corona 356
The Solar Wind 358
Sunspots and the Solar Activity Cycle 359
A Model for the Sunspot Cycle 359
Solar Flares 361
▪ *ACTIVITY: Measuring the Diameter of the Sun* 364
▪ *ACTIVITY: Observing Sunspots* 364

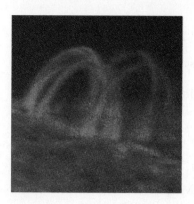

12 MEASURING THE PROPERTIES OF STARS 367

Stellar Luminosity 368
Apparent Magnitude 369
Distances to Stars—Parallax 372
▪ *CLOSE UP: Naming Stars* 373
▪ *CLOSE UP: A Long-Range Proposal* 374
Absolute Magnitude 374
Motions of Stars 375
Spectral Classes 377
▪ *CLOSE UP: Determining the Spectral Class of a Star* 378
The Hertzsprung-Russell Diagram 379
Spectroscopic Parallax 381
Luminosity Classes 382
Analysis of the Procedure 383
The Sizes of Stars 384
Multiple Star Systems 385
Visual Binaries 386
Spectroscopic Binaries 388
Eclipsing Binaries 390
Other Binary Classifications 390
Stellar Masses and Sizes from Binary Star Data 391
The Mass-Luminosity Relationship 394
Cepheid Variables as Distance Indicators 394
▪ *CLOSE UP: The Mathematics of the Mass-Luminosity Relationship* 395
▪ *HISTORICAL NOTE: Henrietta Leavitt* 397

13 INTERSTELLAR MATTER AND STAR FORMATION 401

The Interstellar Medium 402
Interstellar Dust 402
▪ *CLOSE UP: Holes in the Heavens?* 403
▪ *CLOSE UP: Blue Skies and Red Sunsets* 404
Interstellar Gas 405
Clouds and Nebulae 407

A Brief Woodland Visit 409
Star Birth 409
The Collapse of Interstellar Clouds 410
■ *CLOSE UP: A Celestial Godzilla?* 412
Protostars 412
Evolution toward the Main Sequence 413
Star Clusters 416
■ *CLOSE UP: ET Live VI—The Life Equation* 418
■ *ACTIVITY: Deep Sky Objects with a Small Telescope* 421

14 THE LIVES AND DEATHS OF STARS 423

Brown Dwarfs 424
Stellar Maturity 426
Stellar Nuclear Fusion 426
The Stellar Thermostat 427
Main Sequence Life 427
■ *CLOSE UP: Lifetimes on the Main Sequence* 428
Star Death 429
Flyweight Stars 430
Heavier than Flyweights—The Red Giant Stage 431
Electron Degeneracy 432
Lightweight Stars 433
The Helium Flash 433
Mass Loss from Red Giants 435
Planetary Nebulae 436
White Dwarfs 439
■ *HISTORICAL NOTE: Tycho Brahe's Nova* 442
Novae 442
The Chandrasekhar Limit 443
Type I Supernovae 444

15 THE DEATHS OF MASSIVE STARS 449

Middleweight and Heavyweight Stars 450
Type II Supernovae 450
Detecting Supernovae 454
■ *CLOSE UP: Supernovae from Lightweight Stars?* 455
SN1987A 457
Theory: The Neutron Star 458
Observation—The Discovery of Pulsars 458
■ *CLOSE UP: The Pulsar in SN1987A?* 459
Theory: The Lighthouse Model of Neutron Stars/Pulsars 460
Observation—The Crab Pulsar and Others 462
Middleweight Conclusion 463
General Relativity 464
A Binary Pulsar 466
■ *CLOSE UP: The Distance/Dispersion Relationship* 467

The Heavyweights 468
Black Holes 468
Properties of Black Holes 469
Detecting Black Holes 469
Our Relatives—The Stars 471
■ *CLOSE UP: Black Holes in Science, Science Fiction, and Nonsense* 472

16 THE MILKY WAY GALAXY 479

Our Galaxy 480
Globular Clusters 484
Components of the Galaxy 486
■ *HISTORICAL NOTE: The Shapley-Curtis Debate* 488
Galactic Motions 489
■ *CLOSE UP: ET Life VII—Where Are They?* 490
The Mass of the Galaxy 492
The Spiral Arms 492
■ *CLOSE UP: Calculating the Mass of the Inner Galaxy* 493
■ *CLOSE UP: The Milky Way: A Barred Spiral Galaxy* 496
Spiral Arm Theories 497
The Density Wave Theory 497
The Chain Reaction Theory 500
The Galactic Nucleus 500
The Evolution of the Galaxy 501
Age and Composition of the Galaxy 501
The Galaxy's History 502
■ *ACTIVITY: The Scale of the Galaxy* 506

17 THE DIVERSE GALAXIES 507

The Hubble Classification 509
Spiral Galaxies 509
Elliptical Galaxies 510
Irregular Galaxies 511
■ *HISTORICAL NOTE: Edwin Hubble* 512
Hubble's Tuning Fork Diagram 513
Measuring Galaxies 514
Distances Measured by Distance Indicators 514
The Hubble Law 517

- HISTORICAL NOTE: Milton Humason, Mule Driver/ Astronomer 518
- CLOSE UP: Observations, Assumptions, and Conclusions 521

The Hubble Law Used to Measure Distance 521
The Tully-Fisher Relation 522
- CLOSE UP: The Precision of Science 523

The Masses of Galaxies 523
Clusters of Galaxies—Missing Mass 524
The Origin of Galactic Types 526
The Cloud Density Theory 526
The Merger Theory 527
Look-Back Time 527
Active Galaxies 528
Quasars 529
Competing Theories for the Quasar Redshift 531
Seyfert Galaxies 532
Quasars and the Gravitational Lens 532
The Nature of Active Galaxies and Quasars 535

18 COSMOLOGY: THE NATURE OF THE UNIVERSE 543

The Search for Centers and Edges 544
Einstein's Universe 545
The Expanding Universe 547
What Is Expanding and What Is Not? 548
The Cosmological Redshift 548
- CLOSE UP: Wrong Explanation: The Doppler Effect 550

Olber's Paradox 550
Cosmological Assumptions 551
The Big Bang 552
Evidence: Background Radiation 553
- CLOSE UP: The Steady State Theory 555

The Age of the Universe 556
- CLOSE UP: Science, Cosmology, and Faith 557

The Future: Will Expansion Stop? 557
Evidence: Distant Galaxies 558
The Density of Matter in the Universe 560
The Inflationary Universe 561
The Flatness Problem 561
The Horizon Problem 561
- CLOSE UP: Stephen Hawking, the Ultimate Theoretician 563

The Grand Scale Structure of the Universe 565

APPENDIXES

A Units and Constants A-2
B Solar Data A-3
C Planetary Data A-4
D Planetary Satellites A-6
E The Brightest Stars A-8
F The Nearest Stars A-9
G The Constellations A-10
H Answers to Selected Questions and Try One Yourself Exercises A-12

Glossary G-1
Index I-1

PREFACE

Relatively few students who enroll in an introductory astronomy course are majoring in astronomy or another science. Instead, their astronomy course is one of very few science courses—and perhaps the only physical science course—they take during their college careers. It is important, therefore, that the class serve not only as an introduction to astronomy, but as an introduction to science in general.

In their astronomy course, students should not only come to understand the unique insight on the universe that astronomy provides, but they should also learn what science is, what scientists mean when they refer to a theory, how theories are tested and developed, and what motivates scientists. Because astronomy incorporates so much material from the other sciences, it is an ideal introduction to science.

This text employs the solar system-to-stars-to-galaxies approach. The advantages for the intended audience are many: students move from the more familiar to the less familiar; the historical development of science can be emphasized; and the heliocentric/geocentric conflict serves as an excellent introduction to the development of scientific theories.

In Quest of the Universe had its origin in the hardback text, *Astronomy: A Journey into Science. Quest* has continued that book's introduction-to-science theme. This second edition incorporates numerous changes, many as a result of feedback from users and reviewers, but it continues to be true to its roots—a book intended to introduce its audience to science.

UPDATED INFORMATION

This version of *Quest* has been updated to included numerous beautiful photographs from the Hubble Space Telescope, as well as some of the latest astronomical findings. In addition to numerous updates throughout the book, major changes include the following:

- Chapter 5, Telescopes, features an expanded discussion of the Hubble Space Telescope and other developing telescope sites.
- The Mars Pathfinder Mission is included in Chapter 8, The Terrestrial Planets.
- Results from the Jupiter Galileo Mission are described in Chapter 9, The Jovian Planets.
- New insights into the nature of comets have been provided by the Comet Hale-Bopp—Chapter 10.
- Chapter 13, Interstellar Matter and Star Formation, has been updated to describe a new understanding of the formation of protostars that was provided by HST images.

WEB ENHANCEMENT

This updated and enhanced web version provides students with web-integrated activities and direct links to World Wide Web resources. The starting point is *StarLinks*, Jones and Bartlett's own extensive astronomy home page. Students reach the *StarLinks* home page by entering the URL http://www.jbpub.com/starlinks into a World Wide Web browser such as Netscape Navigator or Microsoft Internet Explorer.

The *StarLinks* icon in the text's margins identifies important topics that are matched to WWW sites the students can visit through *StarLinks' Explorations* area. The author provides brief descriptions to place the links in context before the student connects to the site. Jones and Bartlett constantly monitors the links to ensure there will always be a working and appropriate site on-line.

At *StarLinks,* the web-integrated *netQuestions,* introduced at the end of each text chapter, provide the students with an opportunity to use the web and their own critical thinking skills to better understand concepts from the text. Each exercise sends the students to diverse web sites to help them in their research. Updates provide new information for material covered in the text, clarifications, and breaking news in astronomy. The *Analogy Cartoons* area offers an offbeat way to look at some of the more difficult to understand concepts in astronomy. *StarLinks* also includes links to the best of NASA's animations and movies, and the *Astronomy Picture of the Day.*

NEW TO THE SECOND EDITION

The most important change in this new edition of *Quest* is in organization. The first edition had 16 chapters; *Quest II* has 18. A chapter in stellar astronomy has been added and another in galactic astronomy, thus providing more balance in the coverage of the solar system, stellar systems, and galaxies. Cosmology now has a chapter devoted exclusively to it. The discussion of phases of the Moon has been moved from Chapter 1 (the geocentric system) to Chapter 6 (the Earth-Moon system). Many students are not adept at the spatial visualization that is required to fully understand lunar phases, and instructors who emphasize this topic asked me to place it later in the text. It fits logically with a study of eclipses.

Previous users of the text will notice the larger page size, which allows the photographs and illustrations to be enlarged. All of the art has been redrawn in full color, which improves the appearance of the text and serves a functional purpose as well. For example, the colors of stars are indicated on Hertzsprung-Russell diagrams, and in drawings of planetary orbits, each planet is assigned a different color to make the drawings clearer.

Naturally, the text has been updated to include the latest astronomical discoveries. Many photographs have been added making the book more visually appealing. New *Close Ups* and *Historical Notes* have been added to increase student interest.

Special efforts have been made to locate figures on the same page as their first reference in the text, and in no case does the reader have to turn more than one page to find a figure. This may seem like a minor feature, but it allows readers to consult a figure without having to search for it, thereby losing concentration. A related feature is a color cue located at the first reference to each figure. The cue enables the reader to return quickly to the text after consulting a figure. In addition, a person flipping through the text (reviewing for a test, perhaps) can quickly find the text reference for any photo or illustration.

Other new learning aids include the following:
- The number of questions at the end of each chapter has been increased significantly, and objective questions are now included.
- A study skills section has been added. Many astronomy students who are just beginning their college careers can profit greatly by improving their study habits. Instructors, please encourage your students to use this section, along with any other study hints you may give them.
- A data page has been added for each planet and for the Moon. These summaries will help students review.

FEATURES OF THE TEXT

As in the first edition of *Quest,* the logic of science is emphasized by focusing on the criteria for a good scientific model. Whenever possible, we examine the observations that support present theory. For example, there is a discussion of the methods of measuring masses and distances in the solar system; a chapter devoted to the processes involved in determining the properties of stars; and a careful discussion of the evidence for the nature of pulsars.

The heliocentric/geocentric question is just the first of many examples of conflicting scientific theories that are used to show how scientists evaluate hypotheses and develop theories. Other examples include solar system formation, star formation, the cause of galactic spiral arms, and cosmology. Astronomy is presented as a continuing, exciting activity of clashing ideas.

Close Ups, Historical Notes, and Activities

Many interesting aspects of astronomy are external to the main thrust of the text. Sometimes the material is too mathematical, and sometimes—although it is interesting—it is simply not of major importance. Chapters typically have two *Close Ups* that discuss complementary material. (A few examples: Chaos Theory, Radioactive Dating, Astrology and Science, and Lifetimes on the Main Sequence.)

Seven *Close Ups*—an increase from the previous edition—deal with extraterrestrial life. *ET Life Close Ups* are dispersed throughout the text, each appearing near material to which it relates. For example, the *Close Up* regarding the search for extraterrestrial intelligence is located near the discussion of radio telescopes. Instructors may choose to cover this material as a unit, treating it as another chapter. Or it may be omitted entirely.

Science is a human activity; students should become acquainted with the scientists behind the observations and theories. Accordingly, *Historical Notes* present personal accounts of selected astronomers

In an ideal world, all science courses would have laboratories. Because our world is real, optional activity boxes are included at the ends of most chapters. It is my experience that even if they are not assigned, a few students will take advantage of them and thereby better appreciate the active nature of science. A list of Activities—as well as *Close Ups* and *Historical Notes*—appears on the inside front cover.

Art Program

Today's student is visual-minded, and much attention has been given to the photo and illustration program. The larger pages of this edition allow us to include large, clear drawings and photographs. The use of color throughout not only makes the text more attractive, but helps to clarify difficult concepts.

End-of-Chapter Questions

This text contains many more questions at the end of each chapter than is common. Instructors can tell the students which ones they consider important, or they can assign some questions one term and others the next. There are three categories of questions: *Recall Questions, Questions to Ponder,* and *Calculations. Questions* in the first category, which has been expanded to include an average of 20 multiple-choice items per chapter, ask students to recall important facts, relationships, and theories.

Questions to Ponder are of three types: (1) questions that are more complex than the *Recall Questions* and ask students to relate a concept to others either in that chapter or in earlier chapters, (2) questions that ask students to evaluate their own positions on a question that has no right or wrong answer, and (3) research questions requiring students to consult other books.

A few quantitative questions—*Calculations*—give students a chance to practice numerical problems similar to the optional ones in the chapter or to apply quantitative ideas from previous chapters to material at hand. Answers to all even-numbered *Recall Questions* and *Calculations* are found in Appendix H.

Marginal Glossary and Marginal Notes

Students who will not become astronomers should spend their time studying the logic of science rather than the technical terminology of astronomy. Therefore, I have sought to minimize jargon. In deciding whether to introduce each new term, I asked two questions: (1) Is this word necessary for a clear discussion of the topic? and (2) Are the students likely to encounter the word in their reading in later life? If the answer to both questions is "No," I have not used the term in the text. When new terminology is introduced, it is defined in a "marginal glossary." This provides students with a quick, easy reference (and instructors with a quick outline of terms used). In addition, an alphabetical glossary is provided at the back of the book.

Notes to the student are located in the margins along with the marginal glossary. Some of these give additional facts about the material under discussion; some point the student to previous (or future) references in the text; and some provide study hints.

Optional Mathematical Examples

A number of optional mathematical examples are included. After each example is a parallel problem for the student, a "Try One Yourself." Since some instructors will want to omit them, the examples are set apart from the rest of the text so that some or all may readily be skipped without loss of continuity. In order to provide immediate feedback to students, the solution to each *Try One Yourself* is included in Appendix H.

INSTRUCTOR'S TOOL KIT

The Instructor's Tool Kit (ISBN 0-7637-0697-3) has been specifically developed to accompany the *Updated and Enhanced Web Version* of the *In Quest of the Universe 2nd Edition*. Featuring many valuable tools that will add value to lecture presentations, this easy-to-use kit contains everything in one package, a combined booklet and CD-ROM.

CD-ROM Tool Kit

Compatible with both Windows and Macintosh platforms, this CD-ROM contains easily retrievable files of animations and images, a comprehensive Microsoft Power Point Lecture Presentation, and an electronic test bank.

Media Archives allow instructors to access a variety of multimedia instructional aids. The following media elements can be accessed through separate file folders:

- 40 animations (in .avi and .mov format) based on line drawings from the text that will help illustrate challenging concepts for students.
- 15 of the best QuickTime movies from various NASA sites including ones on Galaxy Formation, Differential Rotation Near a Black Hole, the Rotating Nucleus of Comet Hale-Bopp, and many others.
- 20 analogy cartoons that provide an offbeat way to consider some of the more difficult concepts in astronomy.
- Over 200 electronic transparencies of images from the text.

The *Lecture Presentation* features a complete set of 500 carefully designed Power Point slides for each chapter in the text that may be utilized to increase lecture success. Embedded within individual slides are the 20 analogies described above, and numerous figures from the text. The Windows version of this PowerPoint presentation also features the 40 animated line drawings and the 15 NASA animations integrated for easy demonstration.

The *Electronic Test Bank* includes over 900 author-written multiple-choice questions and answers that have been substantially revised from the Second Edition.

Tool Kit Resource Guide

This booklet has been designed as a quick reference detailing contents of the CD-ROM. Included in it are lists of the instructional aids contained in the Files, an outline of the Lecture Presentation with numerous slides per page for cross-referencing, and the complete set of test questions and answers.

Additional Support

Supplementary Hubble Slides (ISBN 0-314-04966-5). This slide set includes 25 Hubble images not found in the text.

Slides to accompany *In Quest of the Universe* (ISBN 0-314-07667-0). This slide set features 200 images found in the text.

The Origin and Evolution of the Universe, Edited by Ben Zuckerman and Matthew A. Malkan (ISBN 0-7637-0030-4). This book presents the excitement of new discoveries in the larger context of cosmic evolution. Leading researchers, on the cutting edge of this field, explain these subjects in laymen's terms to the broader public. They offer the latest insights on the origin and evolution of the universe, chemical elements, galaxies, evolution of stars, planets and biological life. Contributers include Ned Wright, Alan Dressler, Fred Adams, Alexei Filippenko, Virginia Trimble, Chris McKay and Andrei Linde.

Study Guide to accompany *In Quest of the Universe* (ISBN 0-314-03053-0). Authored by Dr. Louis Winkler of the Pennsylvania State University.

Virtual Optical Bench CD-ROM, by Metec, Ltd., 0-7637-0339-7. *Virtual Optical Bench* (VOB) is an Optical CAD system, compatible for Windows, that simulates an optical bench. Far more than a slide show, VOB is simple enough for the beginning student to use while still containing many of the features and libraries that advanced students and professors require. With VOB one can perform simple experiments in geometrical optics, and construct entire systems such as spectrometers or fiberoptic systems with sources, lenses, fibers, and detectors.

ACKNOWLEDGEMENTS

Good teachers learn from their students. Attention to students' questions helps instructors to discover which concepts are most difficult and require special explanation. The same is true for textbook writers, and I thank my students for their questions.

An author's manuscript is just the beginning of a book. The labors and skills of many people are necessary to produce the final book. Jerry Westby provided major guidance for each of my three astronomy texts and has become a true friend. I am fortunate to have the team at Jones and Bartlett Publishers on my side for this enhanced Web version. The team includes Chris Hyde, Sponsoring Editor; Anne Spencer, Production Manager; Dean DeChambeau, Senior Development Editor; Kathryn Twombly, Project Manager; Rich Pirozzi, Marketing Manager; Mike Campbell, Director of Interactive Technology; Andrea Wasik, Web Designer; and Paul Lembo, Interactive Technology Project Editor. Thanks are also extended to Randy Miyake for his beautiful and instructive line art.

I am grateful to Dr. Louis Winkler of Pennsylvania State University for writing the Study Guide to accompany the text and for his suggestions. Thanks also to Dr. Thomas Jordan of Ball State University for his ideas concerning extraterrestrial life.

Thanks to my family—particularly my wife Sharon—for understanding my long hours in front of the MacIntosh and my late evening arrivals.

Reviewers play a major role in any successful book. A total of 49 reviewers read parts or all of the manuscripts for *Astronomy: A Journey into Science* and the first edition of *Quest*. In addition, my colleagues at Eastern Kentucky University provided valuable feedback as they used the book. Others wrote me with suggestions, corrections, and encouragement. Thanks to all. I am indebted to the reviewers listed below, who read parts or all of the manuscript for this edition.

William J. Boardman
Birmingham Southern College

Donald J. Bord
University of Michigan—Dearborn

Emerson Cannon
Salt Lake City Community College

Lon C. Hill
Broward Community College

Thomas M. Jordan
Ball State University

Robert C. Kennicutt, Jr.
University of Arizona

Robert J. Leacock
University of Florida

Paul-Emile Legault
Laurentian University

Scott Niven
Olympic College

Michael O'Shea
Kansas State University

Robert M. Rickett
Northeast Louisiana State University

Robert D. Schmidt
University of Nebraska—Omaha

Scott Temple
Cuesta College

Louis Winkler
Pennsylvania State University

Robert Zimmerman
University of Oregon

The final list shows the most recent group of astronomers/reviewers who helped with this revision. Their work is greatly appreciated.

Kurt S. J. Anderson
New Mexico State University

Paul J. Camp
Coastal Carolina University

J. Patrick Lestrada
Mississippi State University

Cynthia Peterson
University of Connecticut

John Silva
University of Massachusetts

John Oliver
University of Florida

I invite readers of *In Quest of the Universe* to write me with comments and suggestions. In particular, if you know of better illustrations or better examples of everyday phenomena that help students understand astronomical concepts, I would appreciate hearing from you. It is you who know best what should be changed in future editions. My address is:

Department of Physics and Astronomy
Eastern Kentucky University
Richmond, KY 40475

phykuhn@acs.eku.edu

Karl F. Kuhn

STUDY SKILLS
(Secrets for Students)

College is a busy and demanding time. Term papers, tests, reading assignments, and classes require commitment and perseverance. In addition, many of today's students are forced to work to pay for their education. This total load can become overwhelming and frustrating. Fortunately, there are ways to lighten the load, to manage your time efficiently, and to improve your knowledge and understanding (and thereby to increase your chances of getting good grades). First, there are two keys to success in any subject in college.

KEYS TO SUCCESS

- *Avoid procrastination.* (I wanted to write this section later, but I successfully avoided that temptation.) In your studies, procrastination is a dangerous habit. Putting off your work may provide temporary satisfaction because you thereby avoid something you do not want to do, but in the long run procrastination leads to stress, mistakes, and a subpar performance.

  Procrastination usually leads to long cramming sessions. It is much better to work for many short periods of time than to spend the same amount of time in one long session, which is usually what happens when you procrastinate. Avoiding procrastination is perhaps the number one key to success in college—and beyond.

- *Take charge of your education.* Some people watch the world go by. In order to succeed in education—and in life—you must take an active part in the process. You must be an actor and not a viewer, part of the cast and not of the audience. Your education is *yours,* and only you are responsible for it.

  Don't fall into the trap of excusing a poor performance by saying that it was someone else's fault or due to circumstances. *You* are the one who suffers if your education is less than it might be. For example, if you miss a class because your roommate turned off the alarm clock, it is easy to tell yourself that it isn't your fault, that your roommate is to blame. While this may be true, it is *you* and not your roommate who must suffer the consequences of the missed class. Therefore you must take the active role and repair the damage by going to see the professor or by getting notes from others in the class.

  When you find yourself beginning to make an excuse for a missed opportunity, stop immediately, and figure out how you can get the opportunity back. Education is not passive like watching TV; education is a *participatory* sport. Take charge of your education.

SPECIFIC SKILLS

This section offers some suggestions to help you manage your studies and improve your mastery not only of astronomy, but of other subjects you are studying. You are probably already doing some of the things I will suggest, but I hope to present a few new ideas that will help you. Read over the following suggestions. There are far too many to implement immediately, so pick a few that seem right for you under each category and put them into action. After they have become habits, re-read the section, pick a few more, and apply them. Using these ideas may pay huge dividends—not just in college, but throughout your life. Learning does not stop at graduation, and the ability to learn fast will serve you well throughout life.

Mastering new skills requires some work at first and may require that you break established habits. In the long run, though, the time you spend learning the new study skills presented here could save you far more time than you invest. More importantly, efficient study habits will result in higher grades because you will have increased your knowledge and understanding.

General Study Skills

- Develop the habit of studying on a daily basis. This is part of the "avoid procrastination" advice.
- Set aside specific times each day for study. Determine when you are most alert and use that time for study. Let your friends know that this is your study time, and you are not to be disturbed.
- Study for short periods and take frequent breaks, usually after an hour of study. Get up and move around. Do something completely different for a few minutes. This will help you stay alert and active.
- Have an area dedicated for study. It should include a well-lighted space with a desk and the study materials you need, such as a dictionary, thesaurus, paper, pens and pencils, a calculator, and a computer if you have one.
- Study each subject every day, or at least the day of the class, to avoid cramming for tests. Develop the habit of reviewing lecture material from a class the same day. Some courses require more work than others, so adjust your schedule accordingly.
- Look up new terms whose meaning is unclear to you in the glossaries of your textbooks or in the dictionary. Glossaries are preferred because they give the particular definition that is appropriate in the subject you are studying. In this book, new terms are defined in the margin when they are first introduced. A complete glossary is included at the end of the book.
- Don't assume that because you use a word every day you know its definition. Often a term is used in a much more restricted way in science, and its common definition is probably useless in a scientific context.
- If the instructor (or bookstore) has made a student study guide or computer study aid available, use it. At least, ask your instructor's advice about it.

Classes and Note Taking

- Before the lecture, read—or at least scan—the chapter the lecture will cover. This way you will be somewhat familiar with the concepts and can listen critically to what is being said rather than trying to write down everything.
- By learning the vocabulary of the discipline before the lecture, you can cut down on the amount you have to write—you won't have to write down a definition if you already know the word.
- Spend 5–10 minutes before each lecture reviewing the material you learned in the previous lecture. This will provide a context for the new material.

- Avoid missing classes. Try to sit near the front of the room, where there are fewer distractions.
- Develop a shorthand system of your own. Symbols such as = (equals), w/o (without), w (with), > (greater than), < (less than), ↑ (increases), and ↓ (decreases) can save you time. When you find that certain terms are repeated frequently, develop an abbreviation for them. Solar system might become ss.
- Omit vowels and abbreviate words to decrease writing time. (That is, omt vwls t ↓ wrtng tme.) This takes practice, but will pay off.
- Don't take down every word the professor says, but be sure your notes contain the main points, the supporting information, and important terms.
- Learn the mannerisms of the professor. He or she will usually have some way of indicating what ideas are important. For example, he or she may repeat them or write them on the board.
- Check any unclear points in your notes with a classmate, look them up in the textbook, or consult your professor.
- If the professor permits, use a tape recorder. Do not use this to replace note taking, however, for the act of writing notes keeps you focused on the lecture. Use the tape to fill any gaps you find in your notes.
- Review your notes soon after the lecture while it is still fresh in your mind. It is best to recopy your notes, inserting appropriate ideas and clarifications from the textbook. If time does not permit you to do this, leave room as you take notes so that you can add material later.
- Ask questions in class. If you are unclear about a point, other people probably are also, and they will be grateful for your question. If you are shy about asking questions in class, go up after lecture or visit your professor during office hours. Remember, *you* must take charge of *your* education.

Getting the Most from What You Read

- Before reading a chapter, look over the chapter outline or skim through the chapter. This will tell you what the material is about and provide a pattern for your thinking as you read.
- Take notes in the margin of the book or on a separate sheet of paper. Underline or highlight key points, but do not fall into the habit of extensive highlighting. In this text, new terms are in bold type and are defined in the margin. Thus, there is no need to highlight them. Instead, mark paragraphs or parts of paragraphs that contain explanations that are important and difficult, so that you can find them easily when you review.
- Pay attention to tables, charts, and figures. Astronomy is a visual science, and the illustrations will help you to understand and remember what is being discussed in the text. If your instructor does not make clear to you what numbers should be memorized from tables, be sure to ask. In some cases, he or she may want you to memorize values (such as diameters of planets), but in other cases may only require that you remember comparisons (larger or smaller). Ask, ask, ask what your professor expects.
- Examine carefully the logic involved in going from one step of an argument to another. Calculations of many quantities in astronomy depend on previous measurements, calculations, and assumptions. Be sure you understand these steps. If you don't, ask your professor for an explanation.
- Go over the end-of-chapter questions. Write out your answers as if you were taking a test. Only when you see your answer in writing will you know if you understand the material. If you cannot find the answer to a question, consult a classmate or ask your professor. The multiple-choice items are easy for you to score and will give you an idea how you would do with similar questions on a test.

Preparing for Tests

- Most important: Keep up with the professor day-to-day by reading any assignments and by reviewing your lecture notes. Don't depend upon cramming the night before the test. Instead, get plenty of rest that night.
- Find out how much of the test will come from lecture notes and how much from the text, and find out what type of test will be given.
- About a week before the test, start reviewing previous chapters and old lecture notes. Pay particular attention to how this material ties into later material. Try to see the big picture.
- If old exams are available, look at them to see what was emphasized. (If a different professor taught the course, you cannot depend on this helping you learn what was emphasized, but it should still help you learn the material.)
- Form a study group, but make sure it remains a *study* group and does not become a social group. Quiz each other and compare notes to be sure you have covered all of the material. Discuss what material you think is most important. A study group should be formed at least a week before the test. If you wait until the night before, you may panic when you find that someone has studied something you neglected. The panic will result in you doing worse than you would have.
- If review sessions are offered, be sure to go. If not, visit the professor or teaching assistant with any questions you may have. Again, do not wait until just before the test.
- Be sure you can define all terms and give examples of how they are used. Since new terms are in boldface type in this text, it is easy to page through and find them. If you do not understand a word's meaning, consult the margin glossary and study the sections to which the word pertains.
- If you are having more troubles than can be solved by study sessions and visits to the professor, find out if your school offers free or low-cost tutoring. If necessary, hire a private tutor. The best way to find one is to ask your professor or go to the office of the department that offers the course. Again, don't wait until you are helplessly lost. Tutors can't work miracles.
- If you have time, write your own tests, using questions of all types. If you are in a study group, swap tests.

Taking Tests

- Eat well and get plenty of exercise and sleep the day before each test.
- Remain calm during the test. Deep breathing may help.
- Look over the test briefly to see its format and help you budget your time. If it helps, quickly jot down any information you are afraid you might forget or that you see will be important in answering a question.
- Answer the questions that you know best first. Make sure, however, that you don't spend too much time on any one question or one that is worth only a few points.
- If the exam is a combination of multiple-choice and essay questions, answer the multiple-choice items first. These may remind you of facts you need to know for essay questions.
- Read each question carefully and answer only what is asked. Save time by not repeating the questions as your opening sentence to the answer. Get right to the point. For longer essay questions, jot down a quick outline to make sure you cover everything in a logical order.
- If you don't understand an item, ask the examiner. Don't be shy; take charge. Remember that even if the question was poorly written by the person who made up the test, it is your grade that will suffer if you misunderstand that question.

CONCLUDING COMMENTS

After reading these suggestions, you may feel overwhelmed. Not all of them may fit your situation, and even if they do, you can't expect to implement all of them immediately. Try one or two from each section. It is difficult to break old habits, but the good habits you develop through these suggestions may remain with you throughout life. It is worth the effort.

Go back over the list now, check off the things you already do, and mark the ones you will work on first. Don't procrastinate; take charge now!

PROGLOGUE

Supernove SN 1987A is in a nearby galaxy called the Large Magellanic Cloud. When the supernova was brightest, it was visible to the naked eye to people at southern latitudes. The bright area to the left and above the supernova is called the Tarantula nebula.

THE QUEST AHEAD

The View from Earth
Questions, Answers, and Methods
From Earth to Galaxies
Units of Distance in Astronomy

CLOSE UP: Powers of Ten
The Scale of the Universe
Simplicity and the Unity of Nature
Astronomy Today

•••••••• IT WAS WINDY and the roof was broken that February night in 1987. Ian had to open the roof by hand in order to use the telescope he had repaired. Shivering in the mountain air, he climbed a ladder and pushed the corrugated metal aside.

Ian Shelton was lonely on the peak of the mountain in Chile, but he had chosen this life. He had dropped out of two graduate schools and was working as resident astronomer at the Las Campanas Observatory. The pay was less than $15,000 a year, and his job description required that he live alone because there was only room for one person in the small quarters provided. Yes, he had visitors. Professional astronomers regularly traveled the three hours from Las Serena on the Pacific coast to use the 24-inch telescope that Ian kept in good repair. They only stayed a few weeks at the most, however, so they weren't the kind of friends most people have. His radio provided a link to civilization, but not a very good one, for its reception was very poor and long conversations on it were nearly impossible.

The 10-inch telescope he had repaired and learned to use was his almost constant companion. On his days off he preferred to stay on the mountain, using his *telescope at night and working on equipment in the observatory during the day. Except for the two years when he had returned to his native Canada because of an ailment the local people call "mountain fever," he had lived on the mountain for six years, and he was getting used to the loneliness.*

This night had started like most others. Winds up to 40 miles per hour pounded the cinderblock telescope shack, but once he had opened the roof, he was able to begin the work he liked: photographing the sky. This night he decided to photograph the Large Magellanic Cloud, which is not a cloud at all, but the nearest galaxy to the Milky Way. It and its companion, the Small Magellanic Cloud, are visible only from the Southern Hemisphere. To get a good bright photo, he decided to take a three-hour time exposure. (Long exposures are the norm for astronomers.) His telescope, however, was not like more modern telescopes that use motors and computers to guide them so that they automatically follow the stars. Instead, he had to look through the eyepiece constantly and guide the telescope manually. As he sat on the lonely mountaintop, with only the howling wind and his Pink Floyd tape as companions, Ian wished he had worn a heavy coat.

At three o'clock the sky suddenly went dark; the wind had blown the roof shut and jarred the telescope. What a night! Thinking that his photograph had been ruined, he considered going to bed, but then decided to develop the photographic plate and see what he had.

In the dim light of the darkroom, he examined the result. The Magellanic Cloud was apparent, but when things start going wrong, nothing stops the slide—there was a flaw on the plate. A bright blotch appeared near the edge of the galaxy's image. Of course, a bright star looks just like a blotch, but Ian knew the sky well enough to know that there is no bright star where the spot appeared. He had been using photographic plates of the same type for some time, however, and he knew that a flaw of this size would be extremely unusual. What was the alternative? A new star—a supernova? None bright enough to be seen with the naked eye had appeared in the last 383 years. If this was a supernova, it certainly was a bright one. It couldn't be.

The graduate school dropout argued with himself for 20 minutes before going outside to check the sky with his own eyes. There it was, the first naked-eye supernova since 1604, an object that will be studied by astronomers for years and one that has already caused us to change our theories about the death of stars. The unpromising night had yielded the astronomical event of the century, and Ian Shelton was the first to observe it.

A stronomy is the oldest of the sciences, and because it is still changing rapidly, it is also among the newest. It is a study rich in interesting personalities and exciting discoveries. Its long history and recent advances make it an excellent example of the progressive nature of science.

Science is fundamentally a quest for understanding, and in the case of astronomy, the subject of the quest is the entire universe. Astronomy's subject matter includes the full range of matter from smallest to largest. On the small end, it studies atoms and their internal workings, for astronomers need to know about the nuclear reactions that make *stars* shine. It studies Earth's neighbors, the planets, and in doing so helps us solve problems right here on Earth. For example, a study of weather patterns on Venus and Jupiter has advanced our understanding of weather on Earth. In some cases its results are more trivial, but nevertheless practical: materials developed for the Apollo missions to the Moon are now used in tennis rackets.

star. A self-luminous celestial object.

Planets shine by reflected sunlight, but stars have an internal source of energy for their light. Chapter 11 discusses this stellar energy source.

PROLOGUE   The Quest Ahead

Astronomy also studies the largest groupings of billions of stars, and it asks *big* questions, the answers to which tell us something about ourselves, both as individuals and as a species. Studying astronomy causes us to change our basic outlook as we better appreciate where humans fit into the magnificent universe. Our species and our planet seem to be dwarfed to insignificance by the immensity of the cosmos. In a sense, astronomy teaches humility.

At the same time, one marvels that humans have been able to learn so much about the universe in which we live, and that such tiny beings are able to comprehend the huge, complex universe. From this we may decide that size is not so important after all, at least not when compared to the intelligence and spirit of our species. Maybe we need not be so humble after all.

Science is a very human endeavor, and the science produced by a particular culture is in part a reflection of the characteristics of that culture. Science cannot be divorced from people. In astronomy the human aspects of science are very visible.

Before embarking on our quest to understand the universe, let us take a quick look at what lies ahead.

THE VIEW FROM EARTH

In our modern world, few of us can escape from city lights to sit under a clear sky and enjoy its splendors. Fortunately, our ancestors had the time and inclination to do so. The sky on a clear, dark night is a wondrous sight, and before electric lights and television, watching the sky must have been a very popular activity.

As your eyes adjust to the darkness, perhaps the most impressive spectacle you see is the myriad of twinkling stars in the sky, ranging from dim to bright and from lone stars to clusters of many stars (▶Figure P-1). The wondrous *Milky Way,* a hazy white area, may stretch across the sky (▶Figure P-2). If the Moon is visible, it presents a magnificent view, varying from night to night but always interesting.

Besides the hazy Milky Way, you can see a few smaller patches of dim light in the sky. We call them *nebulae* because of their indistinct, nebulous appearance. Photographs (▶Figure P-3) show that they have much more color than the naked eye can detect.

The patient observer is likely to see a meteor, a quick flash of light across the sky. Once called "falling stars," meteors are caused by rocky visitors from beyond the Earth that glow as they burn in our atmosphere. A few times in your life, you will be able to see a comet, which will probably appear as the largest object in the sky.

If you watch for a few hours, you will see that the stars are not stationary at all, but move across the sky, most of them rising in the east and setting in the west (▶Figure P-4). The entire sky seems to be on a bowl rotating around us. Why are we privileged to be at the center of this majestic spectacle?

As you watch the sky from night to night and begin to remember patterns of the positions of stars, you will see that a few "stars" do not stay in the same place relative to the others. These are the *planets*—wanderers on an apparently irregular path across the sky. You will make another observation as you watch the sky through the seasons: different stars are visible at different times of the year because the entire sky shifts to the west very slightly from night to night.

If you have the opportunity to use binoculars or a small telescope to view the heavens, a myriad of sights will be available. As we proceed with our quest for understanding in this text, we will point out many of the objects that are accessible to the backyard astronomer.

Milky Way. Historically, the diffuse band of light that stretches across the sky. Today the term refers to the **Milky Way Galaxy,** the group of a few hundred billion stars of which our Sun is one.

It is this group of stars that cause the diffuse band of light.

nebula (plural **nebulae**). An interstellar region of dust and/or gas.

planet. Any of the nine (so far known) large objects that revolve around the Sun.

▶ **FIGURE P-1.** The night sky illuminates a fir tree in this photo. Brighter stars appear larger in a photograph because they overexpose the film. Their apparent size is not related to their actual size.

▶ **FIGURE P-2.** This wide-angle photo shows the Milky Way in the constellations Sagittarius (left) and Scorpius (right).

Questions, Answers, and Methods

Ancient observers saw the sights we have just described. They wondered. We wonder at the same sights today. What are these objects, and where do we humans and our Earth fit into the realm of the universe?

Through the methods of astronomy, we are answering questions asked by

► **FIGURE P-3.** This is a photo (not a painting!) of a small area of the sky. The dark area to the right of center is called the Horsehead nebula because of its shape. It is an area of dust that obscures the bright nebula behind it.

those ancient observers as well as questions far beyond their imagining. As we find answers to our questions, we not only are filled with awe at the splendid universe, but we find more questions. The questions seem to be endless, providing excitement to those who are fortunate enough to be interested in the search for answers.

This text will try to describe not only the answers but also the methods used in the search. Studying the methods used by astronomers actually means studying scientific methods in general, for although each of the sciences uses different instruments, the basic methods of inquiry are similar for all the natural sciences. Thus we will explore the very nature of the scientific endeavor, showing how astronomers gather data, how those data are developed into theories, how various theories compete against one another, and how and why some theories are retained and others abandoned. Data and theories are the working materials of all the sciences, and examples of the interplay between data and theory will be an essential part of our study.

FROM EARTH TO GALAXIES

What we know in the 1990s about celestial objects would have astounded our forefathers. We know, of course, that the Moon is a solid planetlike object about one-fourth the diameter of the Earth. The planets range in size from tiny Pluto, with a diameter less than 20% of Earth's, to Jupiter, a giant more than 11 times wider than Earth.

The Moon is discussed in Chapter 6 and the planets in Chapters 7 through 10.

▶ **FIGURE P-4.** The constellation Orion is shown rising in the east. Because this is a time exposure, each star appears as a streak rather than a dot.

The Sun is described in Chapter 11.

The celestial object most important to us, the Sun, dwarfs even Jupiter. Its diameter is 10 times Jupiter's, 109 times Earth's. The nature of the Sun and the tremendous energy it produces have been mysteries through the centuries, but now we know that the Sun is a sphere made up almost entirely of hydrogen and helium and that its energy comes from nuclear fusion reactions similar to those in thermonuclear bombs here on Earth.

Chapters 12 through 15 discuss the stars.

What is truly astounding is that each of the stars that we see in the night sky is simply another sun, shining by the same processes that take place in our Sun. As you sit under the stars some night, try to think of each of those stars as a sun, and imagine how far away they must be to look as dim as they are. How many stars are there? Only about 5000 can be seen with the naked eye from Earth, but telescopes reveal hundreds of billions of stars clustered in a giant disk that we call the Galaxy (▶Figure P-5a). The Milky Way that stretches across our sky is caused by our view along the disk of the Galaxy (Figure P-5b).

Galaxies are the subject of Chapters 16 and 17.

Nebulae are involved in both the birth and the death of stars.

The nebulae are giant clouds of gas, illuminated by light from stars within them. Though we compare them to clouds, we know today that the gas within them is extremely sparse, more sparse than a laboratory vacuum here on Earth. The nebulae are visible to us only because they are very large; our line of sight passes through so much material that although any given part of it provides only slight illumination, the nebula as a whole is visible.

Other objects that were formerly called "nebulae" are separate galaxies from ours, groups of millions to hundreds of billions of stars (▶Figure P-6). They appear to the eye as tiny, hazy splotches only because they are so very far from us.

UNITS OF DISTANCE IN ASTRONOMY

In everyday life we use different units of distance for different types of measurements. We might use centimeters and meters (or inches and feet in the

PROLOGUE   The Quest Ahead

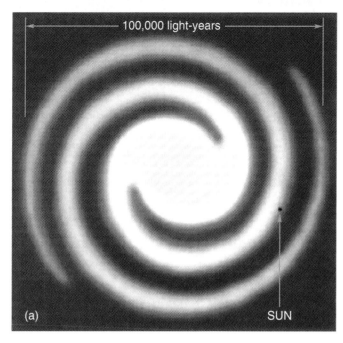
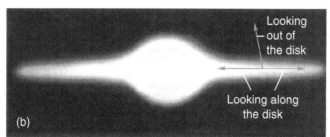

➤ **FIGURE P-5.** (a) Our Sun is one of a few hundred billion stars that form a spiral disk somewhat like this simplified drawing. (b) When we see the Milky Way in the sky, we are looking along the disk, and distant stars are not distinct but instead appear as a bright haze. When we look in other directions, we see right out of the Galaxy.

United States) to measure distances around the house. Although it would be possible to continue to use these units when we describe distances across the country or across the Earth, it is much more convenient to use kilometers (or miles) in these cases.

To measure distances in the solar system, kilometers and miles are too small to be convenient. In this case we use the **astronomical unit** (abbreviated *AU*), which is defined as the average distance between the Earth and the Sun. Thus we are able to say that Mars is 1.5 AU from the Sun, and that Venus gets as close as 0.3 AU to the Earth.

When we go from considering distances within the solar system to distances between stars, the astronomical unit is too small to be very useful. A unit that

astronomical unit. A distance equal to the average distance between the Earth and the Sun.

This is about 150 million kilometers or 93 million miles.

➤ **FIGURE P-6.** This distant spiral galaxy (named NGC 2997) is in the constellation Antlia. Although this galaxy is not visible to the naked eye, a few other galaxies are (including the Magellanic Clouds of the opening figure of this chapter). The bright stars in the photo are in our Galaxy and are much closer to us than NGC 2997, like flies on a car windshield.

Units of Distance in Astronomy

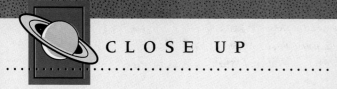

CLOSE UP

Powers of Ten

Numbers in science—and particularly in astronomy—are sometimes extremely small or extremely large. For example, the diameter of a typical atom is about 0.0000000002 meter, and the diameter of the Galaxy is about 1,000,000,000,000,000,000,000 meters. Both are rather awkward numbers to use. We can avoid such clumsy numbers by using a variety of units, such as the astronomical unit within the solar system and the light-year for distances between stars, but sometimes it is inconvenient to switch units simply to avoid large and small numbers, so scientists use *powers-of-ten* notation, also called *scientific* notation, or *exponential* notation. (The exponent is the power to which a number is raised.)

Scientific notation is simple because when the number 10 is raised to a positive power, the exponent is the amount of zeros in the number. For example:

$$10^1 = 10$$
$$10^2 = 100$$
$$10^3 = 1000$$

Written in meters, the diameter of the Galaxy contains 21 zeros, so it is written as 10^{21} meters.

If the number to be expressed in this notation is not a simple power of 10, it is written as follows:

$$2,100 = 2.1 \times 1000 = 2.1 \times 10^3$$
$$305,000 = 3.05 \times 100,000 = 3.05 \times 10^5$$

Notice that to change a number from scientific notation to regular notation, one simply moves the decimal point the number of digits indicated by the exponent, filling in zeros if necessary.

Rather than explain negative exponents, we will just give some examples and let you see the pattern:

$$10^0 = 1$$
$$10^{-1} = 0.1$$
$$10^{-2} = 0.01$$
$$10^{-3} = 0.001$$
$$2 \times 10^{-8} = 0.00000002$$
$$4.67 \times 10^{-5} = 0.0000467$$

In this case, the same rule is followed, moving the decimal point the number of places indicated by the exponent, but this time moving it to the left and supplying any needed zeros.

---

light-year. The distance light travels in a year.

This is about 6 million million miles.

is handy in this case is the **light-year,** defined as the distance light travels in one year. The speed of light is tremendous—300,000 kilometers/second or 186,000 miles/second—so a light-year is indeed a great distance. It is about 9.5 million million kilometers, or 9.5×10^{12} km. (See the accompanying Close Up for an explanation of this notation. You see here the reason it is needed.)

THE SCALE OF THE UNIVERSE

It is difficult for us to appreciate the sizes of objects in the universe and the distances between them (►Figure P-7). To try to get a sense of the objects that are the subject of astronomy, let us construct an imaginary scale model of part of the universe.

Suppose in our model we represent the Sun by a basketball, as in ►Figure P-8. On this scale, the Earth is the head of a pin, about 90 feet away and invisible in the photo. The Moon is a dot the size of a period on this page and is about 3 inches from Earth. Jupiter, the largest planet, is a large chewing gum ball about 150 yards away—the length of 1.5 football fields. Pluto, normally the farthest planet from the Sun, is a grain of sand about seven-tenths of a mile away. Thus the entire solar system occupies an area about 1.5 miles in diameter with a basketball-size Sun at the center.

Where is the nearest star (other than the Sun)? About 4500 miles away! If our model solar system is located in Atlanta, Georgia, the nearest star might be

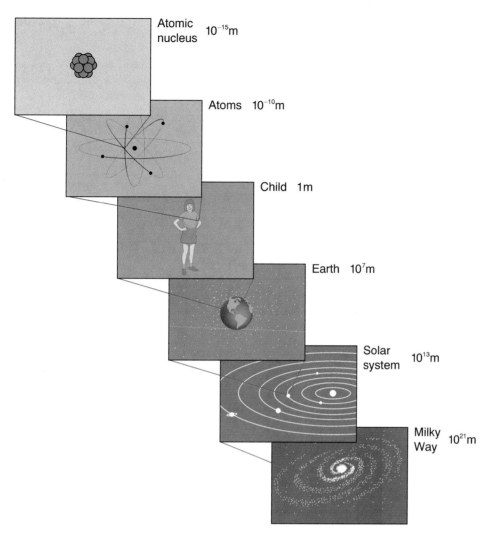

► **FIGURE P-7.** The use of powers-of-ten notation (see the Close Up) makes it much easier to describe the sizes of astronomical objects, but be careful that the notation does not obscure the tremendous differences in size.

► **FIGURE P-8.** If the Sun is a basketball, the Earth is the head of a pin 90 feet away.

in Honolulu, Hawaii (►Figure P-9). This is about the average distance between stars in our Galaxy, so we must imagine basketballs spread around randomly with approximately 4500 miles between adjacent ones. The diameter of the Galaxy on this scale would be about 100,000,000 miles. To imagine a scale model of the Galaxy, we would have to make the Sun much smaller; a 100,000,000-mile galaxy of basketballs is beyond our imagining.

By developing such models, we begin to appreciate the distances involved in the real universe. Such imaginary scale models will be constructed throughout the text. If you try to imagine the distances involved each time, the repetition will not be boring, but instead will be mind-expanding.

► **FIGURE P-9.** If our solar system with its basketball Sun is located in Atlanta, the nearest star is at the distance of Honolulu.

The Scale of the Universe

Simplicity and the Unity of Nature

It is our human nature to try to simplify things. In the sky we see a tremendous variety of objects and phenomena. Astronomy provides a method of seeing order in the apparent confusion. The more we learn about the objects that make up the universe, the more patterns we see, and the more order we find. As our knowledge of the universe expands, we become more and more aware of the unity of the cosmos.

Each of the natural sciences studies a different aspect of the universe, but because the unity exists, the various sciences overlap in many areas. Astronomy is particularly close to physics, with the two fields becoming indistinguishable at times. Likewise, astronomy and geology combine as we attempt to understand the formation, evolution, and surface features of the planets. Chemistry and biology become areas of the astronomer's concern as he or she studies the compositions of astronomical objects and the possibility of extraterrestrial life. And meteorology aids the astronomer is studying weather patterns on other planets.

astrophysics. Physics applied to extraterrestrial objects.

The application of physics in astronomy is now so common that "astrophysics" is often used synonymously with "astronomy."

ASTRONOMY TODAY

We live in exciting times. A few decades ago, humans walked and drove a vehicle on the Moon. As we move into the new millennium, a spacecraft (➤Figure P-10) is on its way to Saturn, where it will go into orbit and drop a probe into the atmosphere of Saturn's largest moon, Titan. Evidence is accumulating that life exists elsewhere, perhaps within our solar system. The Hubble Space Telescope is returning amazing images to us, images of things never seen before. Numerous new telescopes are under construction or being put into operation. Through the use of new instrumentation and new methods, we are discovering celestial objects that were undreamed of a few decades ago. The most distant galaxies are being observed and studied, and answers are being sought to questions as basic as the origin and the fate of the universe. Discoveries can be made only once, and the generations of people now alive are witnessing some of the most exciting discoveries ever.

➤ **FIGURE P-10.** In 2004 the *Casinni* spacecraft will release a probe toward Titan, the largest of Saturn's moons. As the nine-foot probe enters the atmosphere it will begin taking measurements. As it descends—first on a main parachute and then on a drogue chute for stability—various instruments will measure the temperature, pressure, density, and energy balance in the atmosphere.

Of what value is the science of astronomy? Will it help us advance toward worldwide prosperity? Probably not. Technological advances (sometimes beneficial and sometimes not) have followed almost every scientific advance. This, however, is not astronomy's purpose. Astronomy is a pure science rather than an applied science, and astronomers seek knowledge because knowledge is a reward in itself. Asking and answering questions is one of the things that makes humans different from the other animals with whom we share our Earth. We are curious because we are human; we study the heavens because we are curious.

CONCLUSION

We know today that the universe is more wondrous than the most imaginative dreamer of old could have envisioned. To fully appreciate the wonder, however, one must understand the questions asked, the methods used, and the results produced by modern astronomy.

In our quest to understand the universe, we will journey through the solar system, to the stars, and then to distant galaxies, and we will come to appreciate that although our Earth is but a tiny rock circling an ordinary star, we humans have significance in the universe. The significance lies not in our size, but—at least in part—in the fact that our tiny brains have the ability to comprehend the spacious universe.

RECALL QUESTIONS

1. How does a star differ from a planet?
   A. Stars are smaller than planets.
   B. Stars shine by their own light; planets don't.
   C. Stars move across the sky with greater speed than planets.
   D. Stars can be seen earlier in the evening than planets can.
   E. Stars appear dimmer in the sky than plants.

2. The Milky Way is
   A. the path planets take across the sky.
   B. a bright area in the northern part of the sky.
   C. a diffuse band of light caused primarily by nebulae.
   D. a diffuse band of light caused by the galaxy of which we are a part.
   E. a diffuse band of light caused by the remains of comets.

3. As we watch the sky during the night, most of the stars move across the sky
   A. from east to west.
   B. from west to east.
   C. from north to south.
   D. from south to north.

4. The smallest planet is _____ and the largest is _____ .
   A. Earth . . . Saturn
   B. Mercury . . . Earth
   C. Pluto . . . Earth
   D. Pluto . . . Jupiter
   E. Earth . . . Jupiter

5. The diameter of the Sun is about _____ the diameter of the largest planet.
   A. one-tenth
   B. one-half
   C. the same as
   D. 10 times
   E. 100 times

6. In astronomical units, how far is the Earth from the Sun?
   A. 0.5
   B. 1.0
   C. 1.5
   D. 3.0
   E. 93,000,000

7. A light-year is defined as
   A. 93,000,000 miles.
   B. the distance to the nearest star (other than the Sun).
   C. the time it takes light to travel from the Sun to the Earth.
   D. the time it takes light to travel to the nearest star (other than the Sun).
   E. the distance light travels in one year.

8. Which of the following lists the objects in correct size from smallest to largest?
   A. Earth, Sun, solar system, Galaxy
   B. Sun, Earth, solar system, Galaxy
   C. Earth, Sun, Galaxy, solar system
   D. Earth, Galaxy, Sun, solar system
   E. [None of the above.]

9. The fundamental purpose of astronomy is to
   A. improve the living standards of humans by applying astronomical knowledge.
   B. ensure the safety of the human species by understanding the heavens.

C. seek astronomical knowledge for its own sake.
D. [All of the above about equally.]
10. Name at least seven types of celestial objects that are visible to the naked eye.
11. What is a star?
12. What causes the Milky Way we see in the sky?
13. Describe a scale model that includes the Earth, Moon, Sun, and nearest star.
14. What is a galaxy?

QUESTIONS TO PONDER

1. Discuss the differences between the various natural sciences, using insights you have gained in studying other sciences.
2. Science never arrives at a final answer. As it answers one question, it just finds more. Why bother?
3. It might be argued that since astronomy does not produce anything useful for our lives, it does not deserve public funding. How do astronomers answer this? What do you think?
4. Look ahead to Figure 1.4, a drawing of Orion and Taurus. Locate the two stars at the hunter's shoulders and the three stars that make up his belt. Find these stars in the photograph in Figure P-4.

CHAPTER 1

This hand-colored engraving of Ursa Major appeared in *Uranometria* by Johann Bayer in 1603.

An Earth-Centered Universe

The Celestial Sphere
Constellations
Measuring the Positions of Celestial Objects
CLOSE UP: Celestial Coordinates
The Sun's Motion: How Long Is a Year?
The Ecliptic
The Seasons
HISTORICAL NOTE: Leap Year and the Calendar
Scientific Models—A Geocentric Model

The Greek Celestial Model
Observation: The Planets
CLOSE UP: Why Is East on the Left in Sky Photographs?
Model: Epicycles
Criteria for Scientific Models
Another Model
CLOSE UP: Astrology and Science
Model, Theory, and Hypothesis

•••••••• URSA MAJOR—*Latin for* Great Bear—*is a prominent constellation in the northern sky, although most people see and refer to only the Big Dipper portion of it. The group of stars that form Ursa Major was identified as a bear by ancient civilizations in North America, Europe, Asia, and Egypt. Historian Owen Gingerich has suggested that the labeling of these stars as a bear may have originated in Asia or Europe as far back as the ice age and then gradually spread to other cultures. The Greeks associated it with a bear before the time of Homer, and the constellation is mentioned in the* Odyssey. *According to the Greek story of the origin of the constellation, the bear had once been a nymph who attracted the attention of Zeus, the father of the other gods. This caused Hera (the wife and sister of Zeus) to be jealous and to change the nymph into a bear. Finally, to protect the bear from hunters, Zeus grabbed it by the tail and flung it into the sky. This is why Ursa Major has a longer tail than earthly bears do.*

One legend in native American culture held that hunters had chased the big bear onto a mountain from which it leaped into the sky. The bowl of our Big Dipper is now the bear, and the handle is made up of the hunters who followed.

In the last decade of the twentieth century, we have good explanations for what we see in the sky, and we understand the Earth's relationship to other astronomical objects. The search for this understanding began long ago, long before telescopes were invented. We will begin by examining what can be learned about the heavens without the use of modern instruments. Then we will consider a theory of the universe that is nearly 2000 years old.

Why does an astronomy text begin by looking back into history instead of plunging into today's astronomy? Furthermore, why start with an outmoded theory? There is good reason. In order to understand how astronomy—and science in general—works, we must look at its progression through the ages. We must see what led up to today's ideas.

Our starting place in the study of astronomy will be the same as that of people of long ago. We will consider the heavens as seen by the naked eye and will examine the observed motions of stars and planets. The first three chapters will consider two major theories that explain these motions, examining them for two different purposes: first, to answer the question of where the Earth fits into the scheme of things; second, to see how well these theories match the criteria for a good scientific theory. Finally, these chapters will show how and why one of the major theories won out over the other, and how the understanding of nature provided by the successful theory eventually allowed humans to travel beyond the Earth and leave footprints on the Moon.

THE CELESTIAL SPHERE

As we watch the sky during the night, we see some stars rising in the east (Figure P-4) and setting in the west (►Figure 1-1). Stars above the poles of the Earth move in concentric circles, centered on a spot in the sky above each pole (►Figure 1-2).

Another observation that can be made by even a casual observer is that the stars stay in the same patterns night after night and year after year. The Big Dipper seems to retain its shape through the ages as it moves around and around the North Star. It is easy to see why the ancients concluded that the stars act as if they are on a huge sphere that surrounds the Earth and rotates around it. ►Figure 1-3 illustrates this *celestial sphere.* Its axis of rotation passes through the sphere at the center of the circles of Figure 1-2. That photograph was taken from the Northern Hemisphere, and the point at the center of the circles, the *north celestial pole,* is exactly above the North Pole of the Earth. Above the Earth's South Pole, we find a corresponding point on the celestial sphere called the *south celestial pole.* The motions of the stars make it appear that the Earth is sitting still in the center of the celestial sphere as it rotates around us. To picture the motion of the celestial sphere, you might think of it as spinning on rods that extend straight out from the Earth's North and South Poles. These rods would connect to the celestial sphere at the north and south celestial poles.

celestial sphere. The sphere of heavenly objects that seems to center on the observer.

celestial pole. The point on the celestial sphere directly above a pole of the Earth.

▶ **FIGURE 1-1.** A time exposure shows the motion of stars as they set in the west. Figure P-4 is a similar photo of stars rising. You can make photographs like these with any camera capable of taking time exposures. Choose a dark, starry night and point the camera toward the part of the sky you want to photograph. Fix the camera so that it won't move, and open the shutter for an hour or two. Voilà!

▶ **FIGURE 1-2.** Notice that the stars in this time exposure of the northern sky seem to move around one point. Polaris, the North Star, formed the short, bright streak near the center of the circles. Color differences between the streaks were caused by variations in brightness, not by the stars' actual colors.

Constellations

It is natural, when we look at the sky, to look for order—for a pattern. All people feel this desire for order, and it is basic to science. Indeed, we can see (or imagine) patterns in the stars. The ancients saw similar patterns and identified them by associating them with beings in their particular mythology. ▶Figure 1-4a is a

The Celestial Sphere

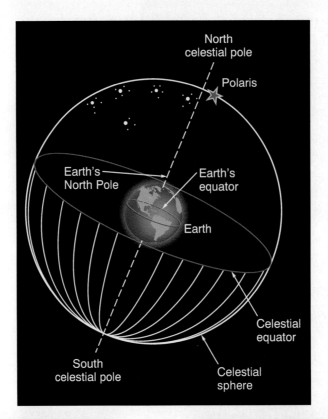

► **FIGURE 1-3.** Because of their daily motion, objects in the sky appear to be on a sphere surrounding the Earth. (The celestial sphere is really much larger compared to Earth than it appears in this diagram, of course.)

(a)

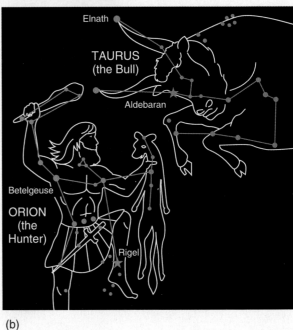
(b)

► **FIGURE 1-4.** (a) The stars of the constellation Orion are visible high in the winter sky. (b) Ancient Greeks pictured Orion as a hunter warding off Taurus the Bull.

constellation. An area of the sky containing a pattern of stars named for a particular object, animal, or person.

The word *constellation* comes from Latin, meaning "stars together."

photograph of the **constellation** Orion, and part (b) shows a drawing of the hunter Orion in Greek mythology. He is fighting off Taurus, the bull at the upper right. If you look at the evening sky in December, January, February, or March, you will see the stars of Orion. You may find it easy to imagine the three closely spaced stars as the belt of the hunter. And if you have a good dark sky, you can see the stars that form his sword's sheath hanging from the belt, as well as the stars forming his upraised arm.

CHAPTER 1 An Earth-Centered Universe

▶ **FIGURE 1-5.** The Big Dipper is part of the constellation Ursa Major. (Refer to the chapter-opening figure to see the full drawing of Ursa Major, the Big Bear.) The star at the end of the handle (Alkaid) is twice as far from us as the next star in the handle (Mizar, 80 light-years away). Notice also that a second star appears very close to Mizar; it is named Alcor.

The mythological creatures of most other constellations are much more difficult to imagine. Ursa Major, the Big Bear, is a constellation of the northern sky. (See the chapter-opening drawing.) Chances are, this group of stars will not remind you of a bear, but you will probably recognize the pattern of stars in the bear's rear and tail—the Big Dipper (▶Figure 1-5).

We know that as early as 2000 B.C. the Sumerians had defined a number of constellations, including a bull and a lion. As described in the story that opens this chapter, the constellation of the bear was common to many cultures. The 88 constellations that are used today were established by international agreement of astronomers. About half of these are constellations identified by the ancient Greeks, and the names we know them by are Latin translations of the original names. Today we realize that the constellations have no real identity. They are simply accidental patterns of stars, much the same as patterns you might have seen in the clouds when you were a young dreamer watching them pass overhead. The stars don't change their patterns as quickly as the clouds, however; so perhaps it was natural for ancient peoples to attribute real meaning to them. Besides, the stars were in the "heavens," and their association with the gods seemed natural.

Why do we say that the patterns are accidental? To begin with, the various stars are located at different distances from Earth. This means that if the Earth were in a greatly different position, we would see different patterns. For example, the stars of the Big Dipper are not all close to one another; one of them, named *Alkaid*, is more than twice as far from us as most of the other bright stars in the constellation.

In spite of their artificial nature, constellations are used by astronomers today to identify parts of the sky. We might say, for example, that Halley's comet was in the constellation Aquarius on Christmas Day, 1985. The ancient Greeks had no constellations in areas of the sky that did not have bright stars, nor in areas they could not see from their location in the Northern Hemisphere, so others have had to be added to their list. Today's 88 constellations fill the entire celestial sphere. In addition, the constellations of ancient cultures had poorly defined boundaries, so astronomers have had to establish definite ones. ▶Figure 1-6 shows the constellation Cygnus and its boundaries.

www.jbpub.com/starlinks

In addition, stars gradually move relative to one another, so constellations change shapes over thousands of years. (Figure 12-9 shows how the Big Dipper is changing.)

The Celestial Sphere

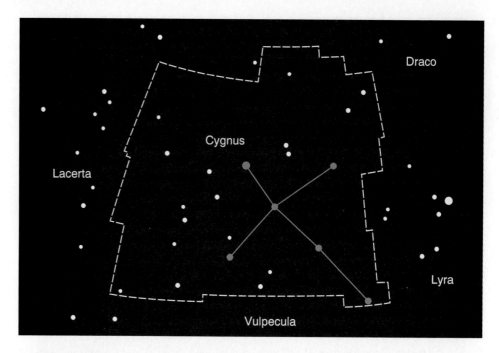

▶ **FIGURE 1-6.** The constellation Cygnus was seen as a swan, but we often call it the Northern Cross, as outlined here. The official boundaries of Cygnus are shown as white lines. (You can find Cygnus in the evening sky in late summer and early fall.)

Measuring the Positions of Celestial Objects

angular separation. Measured from the observer, the angle between lines toward two objects.

When people speak of objects in the sky, they sometimes talk about how far apart they are. What they are probably referring to is their **angular separation,** or the angle between the objects as seen from here on Earth ▶Figure 1-7 illustrates this angle. We say, for example, that the angular separation of the two stars at the ends of the arm of the Northern Cross (Figure 1-6) is about 16 degrees. As ▶Figure 1-8 indicates, this says nothing about the actual distance between the stars. It tells us their angular separation, but not their distance apart.

It is often necessary in astronomy to discuss angles much smaller than one degree. For this purpose, each degree is divided into 60 **minutes of arc** (or 60 **arcminutes,** or 60′), and each minute is divided into 60 **seconds of arc** (or 60 **arcseconds,** or 60″). Notice that although these units are very similar in definition to units of time, they are not units of time, but units of angle. As an example of

minute of arc. One-sixtieth of a degree of arc.

second of arc. One-sixtieth of a minute of arc.

▶ **FIGURE 1-7.** The two stars, when viewed from Earth, have an angular separation as shown.

▶ **FIGURE 1-8.** The angular separation of stars says nothing about their distances apart. All of the stars that lie along line X have the same angular separation from the stars that lie along line Y.

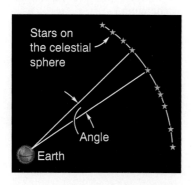

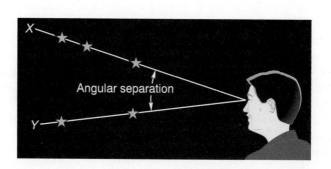

18 CHAPTER 1 An Earth-Centered Universe

the use of these smaller units, a good human eye can detect that two stars that appear close together are indeed two stars if they are separated by about 1 arcminute or more. Chapter 5 will explain how the use of a telescope allows us to detect such double stars if they are separated by as little as one arcsecond—3600 times smaller than one degree.

EXAMPLE

Mizar and Alcor are two stars in the Big Dipper that can be distinguished by the naked eye (see Figure 1-5). They are separated by 12 minutes of arc. Mizar, the brighter of the two, reveals itself in a telescope to be two stars, separated by 14 arcseconds. Express each of these angles in degrees.

Solution

Since 60 arcminutes equals one degree:

$$\frac{1°}{60'} = 1$$

Multiply the 12 arcminutes by this ratio, thus converting its value to degrees:

$$12' \times \frac{1°}{60'} = 0.20°$$

The second part of the problem is done in like manner, but another factor is needed to convert arcseconds to arcminutes:

$$14'' \times \frac{1'}{60''} \times \frac{1°}{60'} = 0.0039°$$

••••• **TRY ONE YOURSELF.** The top star in the head of Orion (Figure 1-4) is a telescopic double star with a separation of 4.5 seconds of arc. Express this angle in degrees.

The answer to each Try One Yourself is in Appendix H.

There is an easy way to estimate angles in the sky. Make a fist and hold it at arm's length. The angle you see between the opposite sides of your fist is about 10 degrees (see ➤Figure 1-9). For estimating smaller angles, the angle made by the end of your little finger held at arm's length is about one degree. You can use these rules to estimate the angular separations of stars.

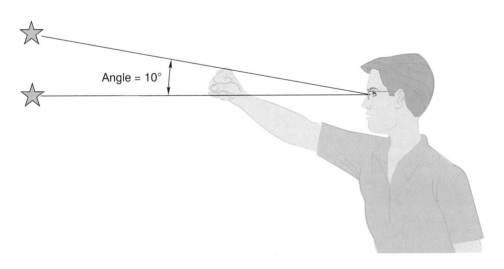

➤ FIGURE 1-9. Your fist held at arm's length yields an angle of about 10 degrees. Both the Sun and the Moon have a diameter of about one-half degree.

The Celestial Sphere

CLOSE UP

Celestial Coordinates

The celestial sphere is an imaginary sphere centered on the Earth. Astronomers specify locations of objects in the sky by a coordinate system on this sphere. The *equatorial coordinate system* describes the location of objects by the use of two coordinates, *declination* and the *right ascension*.

The declination of an object on the celestial sphere is its angle north or south of the celestial equator (see ➤Figure C1-1). Angles north of the equator are designated positive and those south are negative. Thus the scale ranges from +90 degrees at the North Pole to −90 degrees at the South Pole. For example, Sirius (the second brightest star in our sky after the Sun) has a declination of −16°43′.

The right ascension of an object states its angle around the sphere, measuring eastward from the vernal equinox (the location on the celestial equator where the Sun crosses it moving north). Instead of expressing the angle in degrees, however, it is stated in hours, minutes, and seconds (see ➤Figure C1-2). These units are similar to units of time, with 24 hours around the entire circle. Sirius has a right ascension of 6h 45m 6s, or—as it is usually written—$6^h45^m.1$, since 6 seconds is 0.1 minute.

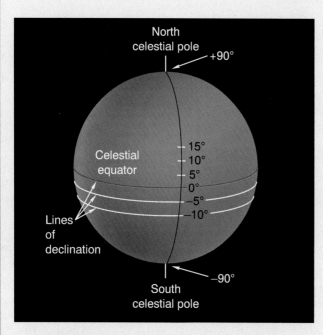

➤ **FIGURE C1-1.** Declination measures the angle of a star north or south of the celestial equator. Angles north are positive and angles south are negative.

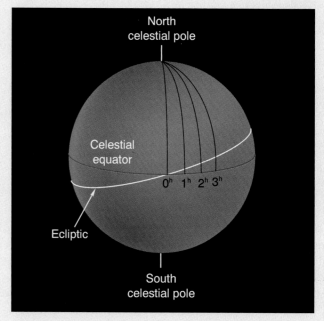

➤ **FIGURE C1-2.** Right ascension measures the angle around the celestial equator eastward from the vernal equinox, where the Sun crosses the equator moving northward. Angles are expressed in hours and parts of an hour, with 24 hours encompassing the entire circle.

THE SUN'S MOTION: HOW LONG IS A YEAR?

Like the stars, the Sun and Moon also seem to move around the Earth as the hours pass, rising in the east and setting in the west. Watching the Sun's motion over a few days might lead us to conclude that it is moving at the same rate as the stars, but if we carefully observe the stars in the sky immediately after sunset, we will see that they change as the weeks and months go by. It is as though the

Sun were out among the stars, but not staying in the same place on the sphere. Instead, it seems to move constantly eastward among the stars. To explain: If we have a map of the stars that surround the Earth, we can locate the position of the Sun on this map by observing the stars that appear just after the Sun sets and again just before the Sun rises. The Sun is between those two groups of stars. If we do this again two weeks later, we would find that the Sun has moved and is now farther toward the east on our map. So it appears that the Sun does not participate fully in the motion of the celestial sphere. It does appear to move around the Earth, but not quite as fast as the sphere of stars.

How long does the Sun take to get all the way around the sphere of stars? You could determine the approximate time simply by drawing the pattern of stars you see in the western sky after sunset tonight, and then waiting until you see that same pattern again. If you do that, you would probably find that the time is somewhere between 350 and 380 days. If you did this over a number of cycles, perhaps you could determine that the time is about 365 days. And, being an alert observer, you would undoubtedly notice that this cycle coincides with the cycle of the seasons here on Earth. Finally, you—or perhaps your descendants to whom you hand down your data—would decide that the time for the Sun's cycle through the stars exactly fits the cycle of the seasons, and that 365 days is the length of the year. (The time the Sun takes to return to the same place among the stars is actually not a whole number of days. It is about $365\frac{1}{4}$ days. Every four years we add one day to our year to make up for this difference. This is our leap year.)

See the Historical Note for more details on leap years.

Early in history people noticed that certain star patterns appear in the sky at the same time every year. Even before the development of the calendar, the arrival of these patterns was used as an indication of a coming change of seasons. For example, the arrival of the constellation Leo in the evening sky meant that spring was coming, and this warned people that even though the weather might not yet indicate it, warm days were on the way.

The Ecliptic

As the Sun moves among the stars, it traces out the same path year after year. ►Figure 1-10a is a map of the stars in a band above the Earth's equator ex-

▶ **FIGURE 1-10.** Part (a) is a map of the stars within 30 degrees of the equator. Picture this map wrapped around the Earth as shown in part (b).

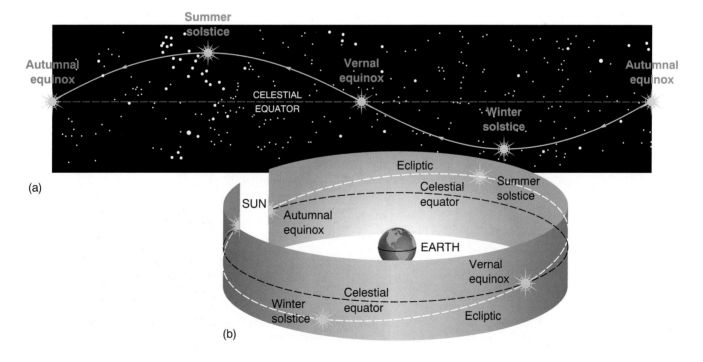

The Sun's Motion: How Long is a Year?

celestial equator. A line on the celestial sphere directly above the Earth's equator.

ecliptic. The apparent path of the Sun on the celestial sphere.

zodiac. The band that lies 9° on either side of the ecliptic on the celestial sphere.

tending 30 degrees on either side of the equator. The second part of the figure (b) shows the relationship of these stars to the Earth. In fact, the dashed line drawn straight across the map is directly above the Earth's equator and is called the *celestial equator.* The other line on the map shows the Sun's path among the stars and shows that it is sometimes north of the equator and sometimes south. The path that the Sun takes among the stars is called the *ecliptic,* and it is not the same as the celestial equator. (The name comes from the fact that an eclipse can occur only when the Moon is on or very near this line. This will be discussed further in Chapter 6.)

The constellations through which the Sun passes as it moves along the ecliptic are called constellations of the *zodiac.* ➤Figure 1-11 shows the 12 major constellations of the zodiac. (Actually, the Sun spends almost three weeks in a thirteenth constellation, Ophiuchus, which is a much longer time than it spends in Scorpius.)

Notice the months indicated on the ecliptic. These show the Sun's locations at various times. On March 21 the Sun is in the constellation Pisces and is crossing the equator on its way north. The changing position of the Sun on the celestial sphere is what causes the seasons, as will be explained below.

THE SEASONS

What causes the seasons? There are three easily observed differences between the behavior of the Sun in winter and summer:

- The Sun rises and sets farther north in the summer than in the winter. ➤Figure 1-12 shows the Sun's rising and setting points at various times of the year, as well as the Sun's path across the sky in each case. Although we may say that the Sun rises in the east, it actually rises exactly east only when it is on the celestial equator—at a particular time in March and again in September.
- The Sun is in the sky longer each day in summer than in winter. In December you may leave for an 8:00 A.M. class in near-darkness, and it may be dark again for your evening meal. Contrast this to June days, when it is more difficult to arise before the Sun, and darkness does not arrive until late in the evening. Figure 1-12 shows why this happens, for the Sun's path above the horizon is much longer in June than in December. This

There is no rule against looking forward to Chapter 2 and Figure 2-5 to see the explanation for the Sun's motions among the stars.

➤ **FIGURE 1-11.** The constellations of the zodiac lie along the ecliptic. Don't confuse the constellations of the zodiac with the astrological signs of the zodiac. Note that in addition to the 12 zodiacal constellations that most people are familiar with, there is a thirteenth—Ophiuchus—that the Sun passes through.

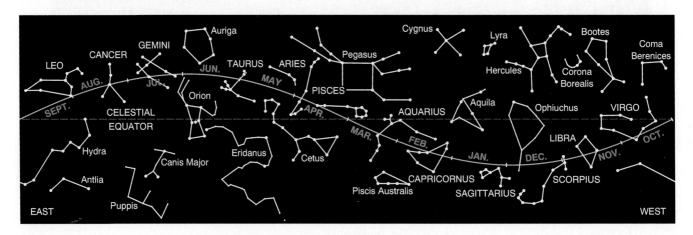

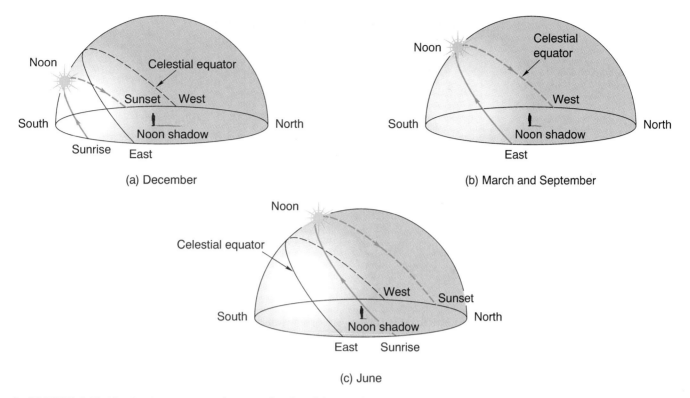

➤ FIGURE 1-12. The Sun's apparent path across the sky of the Northern Hemisphere in (a) December, (b) March or September, and (c) June. Notice that the Sun rises and sets at different places on each date and that its noontime altitude differs in each case.

effect is one reason for the seasons. Because the summer Sun is in the sky longer, more solar energy pours down on us each day, and less time is available at night for our surroundings to lose the heat they have gained.

■ The third difference in the Sun's observed behavior is the second reason for our seasonal differences. As Figure 1-12 illustrates, the Sun reaches a point higher in the sky in summer than in winter, and ➤Figure 1-13 emphasizes this by showing the Sun's path and its location at midday in late June and in late December. When the Sun is higher in the sky, its light hits the surface more perpendicularly. Consider shining a flashlight directly down onto a surface in one case and shining it at an angle to the surface in another (➤Figure 1-14). In the second case, the same amount of light is spread over more surface area, and thus each portion of the surface receives less light. The same is true for sunlight, so we receive more energy from the Sun in a given amount of time in June than in December.

If you live in the Southern Hemisphere, you know that this explanation is backward for your part of the Earth. In the Southern Hemisphere, the Sun gets higher in the sky in December than in June, and the seasons are reversed from those in the Northern Hemisphere. While people are enjoying summer in Canada, Australians are feeling the chill of winter.

To see how the path of the Sun along the ecliptic relates to the seasons, refer to the star map of Figure 1-11 and notice that the Sun is at its northernmost position on about June 21. If you picture the celestial sphere circling the Earth in late June, you can see that it will carry the Sun to a much higher position in the sky of North America. The Sun reaches its southernmost position on about December 22, and on that date it reaches the least *altitude* in the sky of the Northern Hemisphere (➤Figure 1-15).

The two dates discussed in the last paragraph are unique. During the spring the Sun gets higher and higher in the midday sky. Then about June 21—at the

A third factor that causes the seasons is that since the Sun is never high in the sky in winter, its light must pass through more atmosphere in winter than in summer. (See Figure 8-2 for this effect.)

A common misconception is that the Earth is closer to the Sun in summer. Not so! The Earth is closest to the Sun in January.

altitude. The height of a celestial object measured as an angle above the horizon.

The Seasons 23

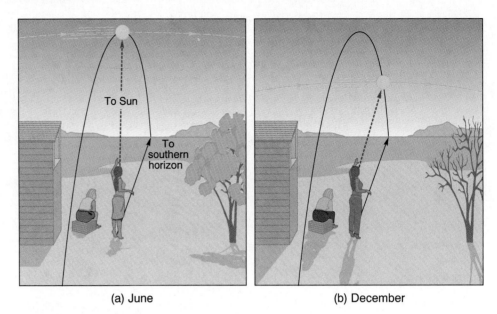

(a) June (b) December

➤ **FIGURE 1-13.** The Sun's apparent path across the sky in the Northern Hemisphere in (a) summer and (b) winter. In each case we have drawn a line from south to north straight over the person's head. The Sun moves across the sky along the yellow line and reaches a much higher position in the sky in summer than it does in winter.

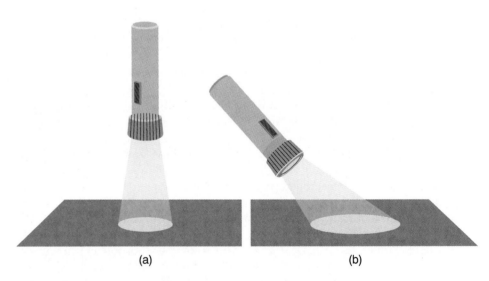

(a) (b)

➤ **FIGURE 1-14.** In (a) the light from the flashlight shines perpendicularly onto the surface, while in (b) it strikes the surface at an oblique angle. The fact that the same amount of light illuminates more surface in (b) means that each little part of the surface is less illuminated in that case. Relate this to the noontime Sun's position in the sky in summer and winter.

summer and winter solstice. The points on the celestial sphere where the Sun reaches its northernmost and southernmost positions, respectively.

summer solstice—it stops climbing, and after that it starts getting lower again. The reverse happens about December 22—at the **winter solstice.** (The word *solstice* is a conjunction of the Latin words "sol" meaning "Sun" and "sistere" meaning "to stand still.") It was named this because at the solstice the Sun *stops* and reverses its direction. For example, during the spring the Sun moves farther and farther northward from day to day. At summer solstice it stops moving north and begins moving south.)

The star map reveals two more unique events each year: Around March 21 the Sun crosses the celestial equator moving north, and about September 23 it crosses the equator moving south. On these dates every location on Earth ex-

CHAPTER 1 An Earth-Centered Universe

HISTORICAL NOTE

Leap Year and the Calendar

Our present calendar comes primarily from the Roman calendar. The calendar started its year in March; the Latin words for the numbers 7 through 10 are "septem," "octo," "novem," and "decem." By the time of Christ, January and February had been added, giving us the 12 months we have now. The months of the calendar were based on the Moon's period of revolution around the Earth and alternated in length between 29 and 30 days in order to match the average of $29\frac{1}{2}$ days between full Moons. Twelve of these lunar months totaled only 354 days, which was defined as the length of the standard year. To make up for the fact that this calendar quickly got out of synchronization with the seasons, an entire month was inserted about every three years—a sort of "leap month."

No single authority controlled the calendar, and by the time of Julius Caesar (100–44 B.C.), the date assigned to a specific day differed widely between different communities. In some countries, nearby communities did not even agree on what year it was. Caesar reformed the calendar in 46 B.C., making the months alternate between 31 and 30 days, except for February, which had 29. Thus the *Julian calendar* had 365 days. And (almost) like ours, it added one day at the end of February every four years to make it correspond more closely to the time the Sun takes to return to the vernal equinox.* This latter time determines the seasons and is called the *tropical year*.

The tropical year is 365.242199 days long. This means that the seasons on Earth repeat after that period of time, and the difference between the tropical year and the average 365.25 days of the Julian calendar caused the calendar to gradually get out of synchronization with the seasons. By the year 1582, the vernal equinox occurred on March 11 rather than March 25, as it had when Julius Caesar instituted the calendar. It was time for more reform. Pope Gregory XIII declared that 10 days would be dropped from the month of October, so that October 15 followed October 4 that year. That restored the vernal equinox to March 21, which corresponded to what the church wanted for establishing the date of Easter each year. To keep the calendar from having to be adjusted this way again, Pope Gregory instituted a new leap year rule: every year whose number was divisible by four would be a leap year unless that year was an even century year that was not divisible by 400 (such as 1800 or 1900). Thus 1700, 1800, and 1900 were not leap years, but the year 2000 will be a leap year.

Roman Catholic countries accepted the *Gregorian calendar*, but most other countries chose to stick with the old calendar. England changed in 1752, at which time it was necessary to omit 11 days. Russia did not adopt the Gregorian calendar until this century.

There has been one more change since Pope Gregory's time: the years 4000, 8000, and 12000, and so on will not be leap years, as they would have according to the original Gregorian calendar. The present calendar is accurate enough that it will not have to be revised for 20,000 years.

---

*The vernal equinox, defined later in this chapter, determines the moment when spring begins.

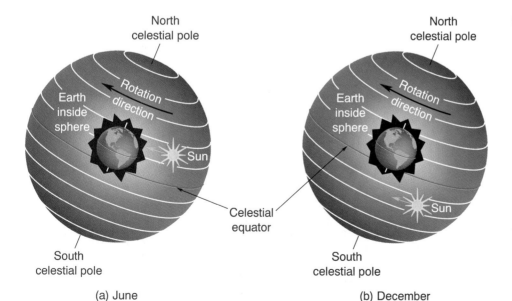

► **FIGURE 1-15.** The height the Sun reaches in the sky is explained by its position on the celestial sphere. In (a), the Sun is shown in its June position, and in (b) it is in its December position on the celestial sphere.

vernal and autumnal equinoxes. The points on the celestial sphere where the Sun crosses the celestial equator while moving north and south, respectively.

periences equal periods of day and night. The events are called the **vernal equinox** (or **spring equinox**) and the **autumnal equinox,** respectively. (*Equinox* means "equal night.")

Why did we estimate each of the four dates above rather than state them exactly? Because they may vary a day or so from year to year. The fact that there are not exactly 365 days per year causes our calendars to get out of synchronization with astronomical events until leap year allows us to catch up. In addition, your "today" may be the previous day or the next day somewhere else on the Earth.

SCIENTIFIC MODELS—A GEOCENTRIC MODEL

Thus far, this chapter has explained the celestial motions we see by describing *scientific models.* The idea of the stars residing on a giant celestial sphere is a model that explains the observation of the daily motion of the stars. Likewise, the changing position of the Sun in the sky explains the changing of the seasons.

A scientific model is not necessarily a physical model, in the sense of a model car or a model airplane. It may even be impossible to construct an actual physical object to represent a scientific model, because a scientific model is basically a mental picture that attempts to use analogy to explain a set of observations in nature. Thus we are able to say that the stars appear as if they are on a sphere rotating around the Earth. Later in the book we will encounter some scientific models that cannot be represented well by a physical construction.

scientific model. A theory that accounts for a set of observations in nature.

It is easy to combine our explanation for the seasons with our model of the celestial sphere. Imagine a track—perhaps like a railroad track—along the ecliptic. As the celestial sphere rotates around the Earth, the Sun stays on this track (▶ Figure 1-16), following the sphere's general motion, but gradually creeping back eastward along the track so that as the months pass, it changes its position among the stars. It moves along the track with such speed that in about 365 days it is back to where it started.

The Sun in our model moves from north to south of the equator and back again. This corresponds to the motion of the real Sun. We are not saying that there is actually—in real life—a track on which the Sun rides, but our model gives us this picture to help us feel comfortable with the motion of the Sun. That is one of the functions of a scientific model: to allow us to feel comfortable with observations. Our model does not actually explain the Sun's motion in the sense

▶ **FIGURE 1-16.** The Sun acts as if it follows a track around the celestial sphere. The sphere rotates toward the west as the Sun gradually moves along it toward the east.

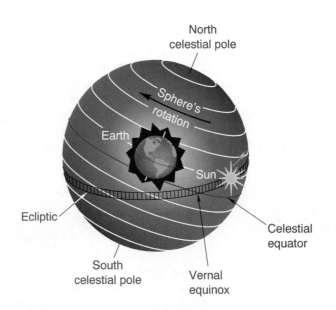

of telling us why it occurs, but it provides a mechanism that allows us to say, "OK, that makes sense now."

The model we have constructed is a *geocentric model;* an Earth-centered model. This is not today's model of the universe, however, as Chapter 2 will explain. In order to see how scientific models are developed, changed, and replaced, we will describe the geocentric model developed by the Greeks some 2000 years ago, and then we will examine reasons why the model has been replaced (but not entirely abandoned).

geocentric model. A model of the universe with the Earth at its center.

THE GREEK CELESTIAL MODEL

To trace the evolution of our present model of the system of the heavens, one must begin with the advanced Greek culture that lasted from about 600 B.C. until a few hundred years after Christ. Before looking directly at the Greek model, however, it is necessary to begin with a discussion of the world as seen by Aristotle, a Greek thinker who lived from about 384 to 322 B.C., but who still influences today's thinking.

Aristotle's system of the world made a great distinction between earthly things and the things of the heavens. He observed that matter on Earth and matter in the sky seem to behave fundamentally differently. For example, objects on Earth have an undeniable tendency to fall to the ground. They fall *down.* In fact, we might define the word "down" as the direction things fall. Things in the sky don't do that; instead they seem to move in circles around the Earth.

Aristotle spoke of the "natural" behavior of objects. He said that it is natural for earthly objects to move downward, that they seem *naturally* to seek the downward direction. On the other hand, it seems reasonable that heavenly objects move in circles. They never move downward. Something in the very nature of the objects seems to make them behave differently.

Aristotle saw another basic difference between the two types of objects: While earthly objects always seem to come to a stop, heavenly objects just keep going. Thus the natural motion of the two classes of objects appears to be different. This idea from Aristotle was a basic part of the world view of the Greeks, and it can be seen in their celestial models. They believed that there were two different sets of rules: one for earthly objects and one for celestial objects.

A celestial model developed by Greek thinkers before the time of Aristotle placed the stars on a sphere, similar to the model we developed. Instead of having the Sun move on a track on that sphere, however, the Sun was placed on another sphere that rotates around the Earth inside the sphere of the stars, as shown in ► Figure 1-17. To account for the fact that the Sun is sometimes seen north of the celestial equator and sometimes south, the Sun's sphere was made to turn on a different axis from the one on which the stellar (celestial) sphere rotates. The difference between the tilts of the two spheres is obvious in Figure 1-17.

There was good reason for the Greeks to use a sphere to carry the Sun, and the reason is linked to the Greeks' admiration of geometry. The love of geometry is traceable to the time of Pythagoras, who died around 500 B.C. He argued that the Earth must be a sphere because of the shape of the shadow it casts on the Moon during a lunar eclipse. The Greeks then carried the idea of spheres much further and developed a model of the Earth, Sun, Moon, and planets using many spheres rotating around one another. To the Greek thinkers, geometry was the queen of mathematics. They sought geometric explanations for all natural phenomena.

Pythagoras, of course, was the discoverer of the Pythagorean theorem.

The Greek model we will discuss most thoroughly is that of Claudius Ptolemy (► Figure 1-18), who lived around A.D. 150. In his book (which we call the *Almagest*), he presented a comprehensive model that lasted for more than 1300

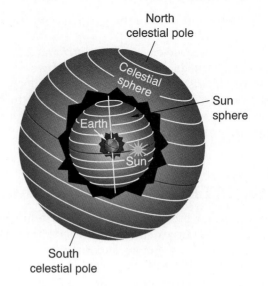

FIGURE 1-17. In order to account for the Sun's path around the Earth (and among the stars), the Greek model located the Sun on a sphere that moves around the stationary Earth inside the celestial sphere of stars. Notice that the axis of the Sun's sphere is tilted with respect to the axis of the celestial sphere.

Ptolemaic model. The theory of the heavens devised by Claudius Ptolemy.

Notice that we still find symbolism in the circle; we use the never-ending circle for our wedding rings.

FIGURE 1-18. Ptolemy.

The Greeks listed the Sun and Moon as planets. That made seven planets, and this is the origin of our week having seven days. In fact, some of the days are named after planets.

years after his death. Today we know this geocentric model as the ***Ptolemaic model.*** Ptolemy abandoned the spheres of earlier Greek models, for he saw no need to imagine actual physical objects carrying the Sun, Moon, and planets. He still spoke of the stars, however, as being on the celestial sphere.

In accordance with the thinking of Aristotle, people of Ptolemy's time thought that things in the sky (the "heavens") must be perfect. It seemed reasonable that the heavens would feature the circle, a symmetrical shape with no beginning and no end. Indeed, the stars seem to move in circles around the Earth. It seemed natural that they would lie in a spherical arrangement, and that the sphere would move around us at a constant speed. The motion and phases of the Moon are discussed in Chapter 6. As you probably know from observation, the Moon changes its position among the stars from day to day just as the Sun does (but at a different speed).

The Moon fit the Ptolemaic scheme perfectly. All that is necessary is that it moves around the Earth as the closest heavenly body. In fact, the Moon fit the overall Greek philosophical approach. Being the closest to the imperfect Earth, the Moon might be expected to have imperfections on it. Indeed, it has dark and light areas. It is somewhere between the imperfect Earth and the perfect heavens.

OBSERVATION: THE PLANETS

Thus far we have barely mentioned the other major class of objects in the sky: the planets. Without a telescope we can see five planets: Mercury, Venus, Mars, Saturn, and Jupiter. The Greeks, of course, knew of these planets. In fact, the word "planet" comes from a Greek word meaning "wanderer." And wander they do. Like the Sun and Moon, the planets move around among the stars on the celestial sphere. The Sun and Moon always move eastward among the stars, but the planets sometimes stop their eastward motion and move westward for a while. They lack the simple, uniform motion of the Sun and Moon.

➤Figure 1-19a is a photograph of the constellation Gemini on the night of August 30, 1992. Mars is the bright star at right center. Figure 1-19b is a photograph taken two weeks later, on the night of September 13. Notice that Mars has moved to the left. These photographs show the sky with east toward the left and south at the bottom. So Mars moved eastward (and slightly southward) with respect to the stars. A photograph taken a few weeks later would show that it was still moving eastward and southward. Things would be fine if Mars continued on in this uniform way. Uniform motions fit the idea of simple, circular

➤ **FIGURE 1-19.** (a) On August 30, 1992, Mars appeared as a bright star in the constellation Gemini, at right center in the photo. East is toward the left. (b) This photo of Gemini was taken on September 13, 1992. You can see that Mars moved toward the east (and a little southward) during the two weeks.

motions in the heavens. (The model can account for the southward motion of Mars if the plane of Mars's motions is placed at an angle with the celestial equator. Mars moves northward again later in the year. This is how the model explains the Sun's motion.)

The drawing in ➤Figure 1-20 shows the progress of Mars through the fall and winter sky of 1992–93 with its position marked for various dates. Notice that Mars continued its eastward motion until about December 1 and then started moving *backward*, heading west! It moved backward until about February 15 and then resumed its eastward motion again. It will continue moving eastward until December 1994, after which it will do another of its backward loops. We call this backward motion ***retrograde motion.*** Retrograde motion is characteristic of planets, including those discovered in modern times.

Although the planets move among the stars in a seemingly irregular manner, there are limits to where they move, for they never get more than a few degrees from the ecliptic. Mercury and Venus have an additional limitation on their mo-

retrograde motion. The east-to-west motion of a planet against the background of stars.

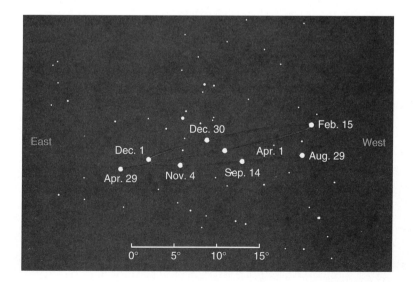

➤ **FIGURE 1-20.** This shows the motion of Mars through Gemini during its retrograde motion of 1992–93.

Observation: The Planets

CLOSE UP

Why Is East on the Left in Sky Photographs?

Maps of the Earth show north at the top, east on the right, and west on the left. Yet in Figure 1-19 east is toward the *left*. To understand why maps and photographs of the sky show east and west reversed from Earth maps, think of yourself lying face down on the ground with your head toward the north. Your right arm points toward the east. Suppose you now turn over to look up to the sky; your right arm is now toward the west! Thus the difference occurs because when we look at a map of the Earth we are looking down, but when we view a map of the sky we are looking up.

---

elongation. The angle in the sky from an object to the Sun.

tion. These two never appear very far from the position of the Sun in the sky. We only see them either in the western sky shortly after the Sun has set or in the eastern sky shortly before sunrise. Mercury appears so close to the Sun that it is hard to find, even if we know where to look. This is because even when Mercury is at its maximum *elongation,* we have to look for it in the semibright sky during dawn or twilight. You will never see Mercury or Venus high in the sky at night.

Any model of the planets must explain not only their observed retrograde motion, but why they stay near the ecliptic, and also the peculiar behavior of Mercury and Venus. How did the Greeks account for the planets' unique motions? Were the heavens perhaps imperfect, with the planets not moving uniformly in a smooth circle?

Model: Epicycles

Ptolemy was indeed able to use circles to make a model that fit the planets' motions. (Though this idea was not original with him, he did elaborate on it, and he is generally given credit for it.) It simply took more than one circle for each planet. ▶Figure 1-21 shows a rod extending out from Earth with a second rod attached to its end. The second rod can swing around its connecting point on the longer rod. At the other end of this second rod, we place Mars. Now we

▶ **FIGURE 1-21.** Mars seems to move in a circle on a second rod that rotates on the longer one.

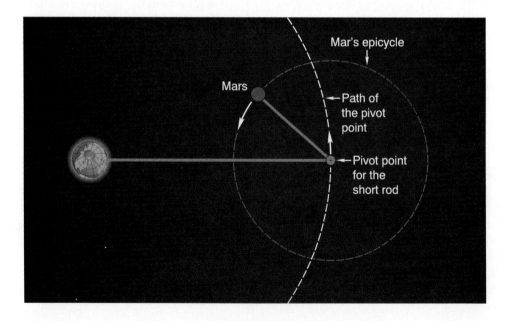

rotate the big rod uniformly around the Earth. At the same time, we rotate the small rod around its pivot point on the large one. By a correct choice of rotation speeds, an observed motion for Mars such as that shown in ➤Figure 1-22 can be produced.

A person on Earth viewing the motion of Mars among the stars sees the planet moving eastward most of the time. This corresponds to the long rod turning around counterclockwise in the drawing. Occasionally, however, the planet retrogrades. This occurs when the planet on the end of the little rod passes closest to Earth, so that it appears to back up for a short time.

The rods referred to here did not appear in the Ptolemaic model because the model did not use actual physical objects to guide the planets. Instead a point moving around the Earth served as the center of motion for the smaller circle of the planet's motion. This smaller circle is called an *epicycle*. In this way, Ptolemy was able to preserve the idea of perfect heavenly circles and perfect, uniform, heavenly speeds.

How, then, does the model explain why the planets never move far from the ecliptic? The answer is that the plane of their circular motion lies very close to the plane of the Sun's motion. The greater the angle between these two planes, the farther the planet will move from the ecliptic as it moves among the stars.

The different behavior of Mercury and Venus was explained by having the centers of their epicycles remain along a line between the Earth and the Sun (➤Figure 1-23). While the center of other planets' epicycles could be anywhere on their circle around the Earth without regard to where the Sun is, Mercury and Venus are special cases.

epicycle. The circular orbit of a planet in the Ptolemaic model, the center of which revolves around the Earth in another circle.

You might want to "cheat" and look ahead to the middle of Chapter 2 for today's explanation of the limits on where Mercury and Venus appear.

CRITERIA FOR SCIENTIFIC MODELS

The Greek model had many features that we look for today in a scientific model. We will discuss the Ptolemaic model in light of three criteria that are applied to scientific models today. Even though these are the standards of today's science and not of the thinking of Ptolemy's times, it is instructive to apply them to Ptolemy's model.

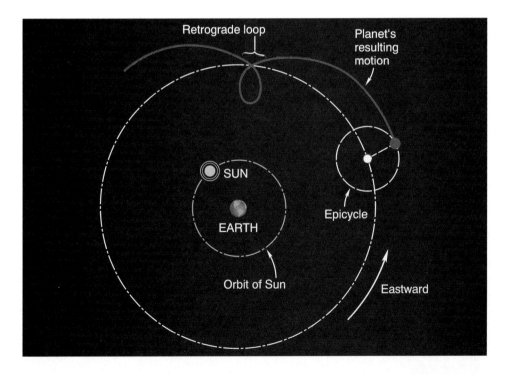

➤ FIGURE 1-22. Mars's motion on its epicycle results in a looping path, which—when seen from Earth—appears as retrograde motion.

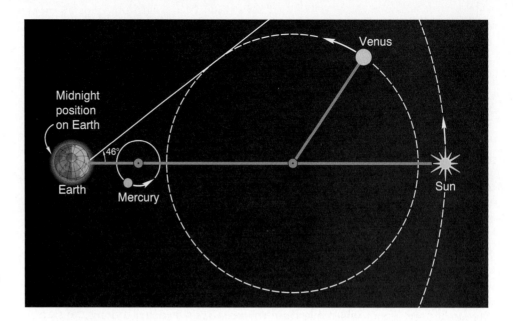

▸ **FIGURE 1-23.** In the Ptolemaic model, the centers of Mercury's and Venus's epicycles stay between the Earth and the Sun. This accounts for the fact that the planets never are seen far from the Sun. Venus reaches a maximum of 46 degrees elongation.

Actually, to make the model fit the data more accurately, Ptolemy either had to make the planets vary their speeds along their paths, or he had to move the Earth off-center among the planets' paths. He chose the latter method. After this and a few similar adjustments, the model fit pretty well.

The simplicity spoken of here is not accepted as a necessary attribute of beauty in art, so we are speaking of a different type of beauty.

■ The first criterion is that the model must fit the data. It must fit what is observed. The Ptolemaic model fits pretty well in that it gives an explanation for the motions of the heavenly objects. The stars' motion was explained by the rotation of the celestial sphere on which they reside. The Sun and the Moon were theorized to move in circles around the Earth. Explanation of the planets' motion required the use of epicycles and special treatment for Mercury and Venus, but when these features were included, the planets' motions were accounted for. So the Ptolemaic model passes the first test. It fits the available data.

■ The second criterion that today's scientists use in judging whether a model is a good one is as follows: The model must make predictions that allow it to be tested, and the model must be of such a nature that it would be possible to disprove it—that is, to show that it needs to be modified to fit new data or perhaps needs to be discarded entirely. Many testable predictions are built into the Ptolemaic model. First, the model made predictions for the locations of the planets at times in the future. One could use the model to predict that Jupiter would be at a particular place in the sky at a particular time the next year. Another example: the Ptolemaic model holds that the Earth is stationary. It would predict that as knowledge advances and new methods are found to measure the motion of the Earth—either rotational motion or motion through space—no motion would be found. If further experimentation did not confirm this, the model would have to be adjusted drastically or abandoned.

"Prediction" here does not necessarily refer to the future. It means that the theory itself indicates an observation that will either support it or disprove at least part of it. Sometimes the observation has already been made and needs only to be checked against the theory. For example, one might check the Ptolemaic model's prediction of where Jupiter was 10 years ago.

■ The third and final feature of a good model is that it should be aesthetically pleasing. This concept is difficult to define. It generally means that the model should be simple, neat, and beautiful. For example, Pythagoras is credited with the idea of placing the heavenly objects on a number of spheres rotating within one another. Later followers of Pythagoras wrote that the motion of these spheres resulted in "music of the spheres" that could be heard by those attuned to it. Such a claim would be quite unconventional—to say the least—in today's science. Nevertheless, the fact that people of that time spoke of such things does illustrate the necessity of a good scientific model having a pleasing quality or a beauty. Today's idea of beauty in a scientific model is more along the lines of symmetry and simplicity. A model should be as simple as possible; that is, it should contain the fewest arbitrary assumptions.

The principle that the best theory is the one that requires the fewest assumptions is often called *Occam's razor*. It is named for a fourteenth-century Franciscan monk-philosopher who stressed that in constructing an argument one should not go beyond what is logically required. Using Occam's razor one "cuts out" extraneous suppositions.

> **Occam's razor.** The principle that the best explanation is the one that requires the fewest unverifiable assumptions.

Let's test Ptolemy's model against this last criterion. In his model, the Sun and Moon travel around the Earth in different paths than the stars do, but they obey basically the same rules, moving from east to west around the Earth, and they all use circles. This idea that all objects obey the same rules would be described by scientists as a form of symmetry.

To explain the motions of planets, epicycles were required. In using epicycles, the model retained the use of circles, a shape that was dear to Greek thinking, but somehow one wonders if a better method than epicycles couldn't be found to represent what is observed. In addition, the fact that different rules are required for Mercury and Venus makes the Ptolemaic model less pleasing to the senses. The need for different rules for certain planets means that the model lacks an aspect of simplicity. How many new special rules will be needed? Each one makes the model less appealing in the scientific sense. So we see that the Ptolemaic model begins to look less attractive as a good scientific model.

> As pointed out in a previous margin note, Ptolemy's actual model was even more complicated and thus even less aesthetically pleasing.

Remember, though, that the primary rule for a good model is that it fits the observations. An aesthetically pleasing model that did not come close to the real world would be an almost useless model. The Ptolemaic model *did* fit the data, so we must judge it to be an acceptable model. It just lacks that certain neatness we would like.

We will return a number of times to a discussion of the features of a good scientific model.

ANOTHER MODEL

About 400 years before Ptolemy, the Greek philosopher Aristarchus had proposed a moving-Earth solution to explain the motions of the heavens. According to this model, the reason the sky seems to move westward is that the spherical Earth is spinning eastward. Why was this model not given much consideration? Ptolemy was aware of this model, so we'll let him answer the question:

> Now some people, although they have nothing to oppose to these [Ptolemy's] arguments, agree on something, as they think, more plausible. And it seems to them there is nothing against their supposing, for instance, the heavens immobile and the Earth as turning on the same axis [as the stars] from west to east very nearly one revolution per day. . . . But it has escaped their notice that, indeed, as far as the appearances of the stars are concerned, nothing would perhaps keep things from being in accordance with this simpler conjecture, but that in the light of what happens around us in the air such a notion would seem altogether absurd.

> Aristarchus's model also included the idea that the Earth orbits around the Sun instead of the Sun around the Earth. As we will see in the next chapter, an important observation contradicted this idea. (See "Parallax" in Chapter 2.)

Ptolemy was arguing that if the Earth turned on an axis, it would be moving through the air around it, and therefore a tremendous wind should be observed in the opposite direction. Notice, however, that Ptolemy refers to that model as being a "simpler conjecture." He saw that it was more aesthetically pleasing, but argued that it was not a good model because it presented obvious contradictions

Astrology and Science

Astrology is the belief that the relative positions of the Sun, planets, and Moon affect the destiny of humans. Often astrology is used to determine the nature of a person's personality or of certain circumstances in the future, and it is traditionally used for timing or resolving problems in marriage, business, medicine, or politics. To discover the astrological effect on an individual, an astrologer designs, or "casts," a horoscope.

Casting a horoscope involves determining the positions of the heavenly objects at the moment of the person's birth. To do so, the zodiac is divided into 12 equal-sized *signs*. Although these signs have the same names as the 12 principal constellations of the zodiac, they lack correspondence with the constellations in two ways. First, the constellations differ from one another in size, whereas all signs are the same size. The constellation Cancer, for example, is much smaller than Virgo or Pisces, but their signs are the same size. Second, the signs are at different locations in the sky from the constellations. At one time the positions corresponded fairly well, but because of *precession* (to be discussed in Chapter 3), the signs and constellations are now "out of whack" by about one position. Thus the sign of Aries is located mostly in the constellation Pisces. This means that a person born on March 22 is said by astrologers to be an Aries, while the Sun actually appeared to be in the constellation Pisces at the person's birth. (Some astrologers, called sidereal astrologers, use the constellations rather than the signs.)

Astrologers divide the sky above us into 12 *houses*, starting with the eastern horizon and proceeding around the sky. The houses do not rotate with the stars, but remain in the same position relative to Earth. Thus the first house stays above the eastern horizon. The houses are arbitrarily associated with various areas of our lives, such as children, health, and enemies. Likewise, the planets are associated with various good or evil influences. Mars, for example, is associated with war.

When the above information is determined for the moment of a person's birth, we have what is called the *natal chart* for that person. This chart could be made by anyone who knows the sky and has tables of solar, lunar, and planetary positions. It might be determined, for example, that the planet Mars was just above the eastern horizon at the time of your birth. So what? This is where interpretation by the astrologer comes in. After such interpretation, the astrologer is supposed to be able to tell you about your personality and life pattern.

The Criteria Applied to Astrology

Having briefly described the nature of astrology, we will now apply to it the criteria for a good scientific theory. The first criterion: Does astrology fit the observations? It claims to be able to account (at least in part) for our personalities and our characteristics. Can it? This question has been tested many times. Invariably, the tests show that astrologers are unable to identify the personality traits of people or to predict future events. (For example, read the report in the December 5, 1985 issue of *Nature* concerning the research done by Shawn Carlson at the University of California.) After numerous tests, we must conclude that astrology fails our first criterion.

The second criterion for a good scientific theory is that it makes verifiable predictions that could possibly prove the theory wrong. There is confusion as to whether astrology does this or not. In Carlson's study, astrologers claimed that if enough cases were considered, they could make verifiable predictions. Astrologers would say that it is not fair to test the accuracy of a horoscope of an individual person because a horoscope only predicts tendencies, and people do not necessarily follow their tendencies. In other words, you might be very different from what is predicted by your horoscope and still not contradict astrological predictions. If this is the case, there is no way that astrology can be proved incorrect. It therefore fails the second criterion.

The third criterion is one of aesthetics. Astrology does not pass this test because it is not a single, unified theory at all, but instead is a group of arbitrary rules as to the effect of heavenly objects on people. One could hardly call it aesthetically pleasing.

Since astrology is a faith system, it cannot be disproved to those who believe. People who believe in it do not care whether scientists can explain the belief or not.

Scientists do not accept astrology as a valid theory. Some scientists strongly object to the publication of astrological columns in newspapers, but others claim that astrology does no harm if people use it only as a pastime and do not let it actually determine what they do with their lives.

with the observation that the air stays where it is, along with everything loose on Earth. Ptolemy was unable to see that the Earth might be carrying the air and everything on it along as it rotates.

MODEL, THEORY, AND HYPOTHESIS

The three terms in the title of this section are words used not only by scientists, but by the general public. They mean something different in science from what they normally do in everyday language, however. We have been calling the work of Ptolemy a model and have discussed how this does not mean a physical model, but rather a developed set of ideas used to describe some aspect of nature. Instead of calling Ptolemy's creation a model, we could have called it a **theory**. The two words mean about the same thing in scientific usage, and we could have spoken of the "Ptolemaic theory."

Notice, then, that in science the word *theory* is used differently than in everyday language. In science a scheme is not usually called a theory until its ideas are shown to fit observed data successfully. In nonscientific use, the word *theory* is often used to refer to ideas that are much more fanciful and less secure. I might say, for example, that I have a theory about what caused yesterday's car wreck. In science we wouldn't call my car wreck idea a theory. We would call it a *hypothesis.*

A hypothesis is a guess—perhaps a very intelligent guess—at a theory or part of a theory. Generally, a theory starts as a hypothesis. The idea of epicycles to explain planetary motion had been used by Hipparchus, who lived about 300 years before Ptolemy. It was then a hypothesis. The hypothesis was developed and included with others to form Ptolemy's final model. According to today's scientific use of the words *theory* and *model,* Ptolemy's plan would not even have been given those names until it was shown to be able to explain the heavenly motions.

You might hear someone refer to Einstein's theory of relativity or to the theory of biological evolution and say, "Well, it is *only* a theory," meaning that you mustn't put too much faith in it. It is correct that these two concepts are only theories, but keep in mind what a scientist means in calling something a theory. Later chapters will show that Einstein's theory is well founded and has been experimentally verified many times. Although the theory of evolution will not be discussed here, virtually all biologists form the same conclusion about it.

theory. A hypothesis or set of hypotheses that have been well tested and verified.

hypothesis. A tentative explanation.

The distinction between hypothesis and theory is not totally clear-cut in science, though. Sometimes the word *theory* is used very loosely.

As Chapter 3 will explain, although it is called a law, Newton's law of gravity is no more a "law" and no less a "theory" than Einstein's theory.

CONCLUSION

There is a tendency to think that Greek science was bad science because it is not today's science. This is simply not true. It is true that the Ptolemaic model is more primitive than today's. The Greeks did not have the accurate observations and extensive data we have today. Their model, however, did fit the data they had. In fact, it fits the casual observations of most people today. Can you think of any direct evidence obtainable without a large telescope that contradicts Ptolemy's model? What observations can you personally make, without relying on reference materials, that will show the 2000-year-old Ptolemaic model to be a poor model?

RECALL QUESTIONS

1. During a single night, the Moon
   A. moves from west to east across the sky.
   B. moves from east to west across the sky.
   C. appears to hover in the sky above a given location on Earth.

2. Polaris is unique because it
   A. moves in a different direction than any other bright star.
   B. is the brightest star in the sky.
   C. twinkles more than any other bright star.
   D. is fairly bright and shows very little motion when viewed from Earth.
   E. [The premise is false. Polaris is not unique at all.]

3. A time exposure of the northern sky shows a circular pattern. The center of the circle is
   A. exactly at the north celestial pole.
   B. very close to the north celestial pole, but not exactly on it.
   C. the location of Polaris.
   D. the location of the nearest supernova.
   E. [Two of the above.]

4. In ancient times, people distinguished the planets from the stars because
   A. planets appear much brighter than any star.
   B. features on planets' surfaces could be seen whereas no star's features could be seen.
   C. planets move relative to the stars.
   D. planets differ in color from the stars.
   E. planets can be seen during the day.

5. Ptolemy's system of epicycles was used to explain the apparent
   A. daily motion of the stars.
   B. annual motion of the Sun around the sky.
   C. backward motion of the Moon through the sky.
   D. changing speeds and directions of the planets among the stars.
   E. motion of the Sun north and south of the celestial equator.

6. Which of the following can be seen from Indiana?
   A. Stars near the north celestial pole
   B. Stars near the ecliptic
   C. Stars near the celestial equator
   D. [All of the above can be seen from Indiana.]
   E. [None of the above can be seen from Indiana.]

7. Which of the planets known to the ancients can *never* be seen high overhead at night?
   A. Mercury and Venus only
   B. Mars, Jupiter, and Saturn only
   C. Saturn only
   D. Venus only
   E. [All of the planets *can* be seen high in the sky at some time of night.]

8. On the first day of spring, the Sun rises
   A. north of east.
   B. directly east.
   C. south of east.
   D. [Any of the above, depending upon your location on Earth.]

9. In January, the Sun rises
   A. south of east.
   B. directly east.
   C. north of east.
   D. [Any of the above, depending upon your location on Earth.]

10. What causes summer to be hotter than winter?
    A. The Earth is closer to the Sun in summer.
    B. The daylight period is longer in summer.
    C. The Sun gets higher in the sky in summer.
    D. [Both B and C above.]
    E. [All of the above.]

11. When we experience spring in the Northern Hemisphere, it is _____ in the Southern Hemisphere.
    A. also spring
    B. summer
    C. fall
    D. winter
    E. [No general statement can be made.]

12. The Sun's apparent path among the stars
    A. is south of the celestial equator.
    B. is right on the celestial equator.
    C. is north of the celestial equator.
    D. is south of the celestial equator part of the time and north of it part of the time.
    E. crosses the north celestial pole once each year.

13. In Australia, the longest daylight period occurs in late
    A. March.
    B. June.
    C. September.
    D. December.
    E. [No general statement can be made.]

14. The Sun crosses the celestial equator on the first day of
    A. winter.
    B. spring.
    C. summer.
    D. fall.
    E. [Two of the above.]

15. If the Earth rotated east-to-west instead of west-to-east,
    A. our seasons would occur in reverse order.
    B. we would not have seasons.
    C. daylight periods would be longer in winter and shorter in summer.
    D. winter in the Northern Hemisphere would occur in July.
    E. [None of the above.]

16. Which of the following planets appear(s) to move through the background of stars?
    A. Venus
    B. Mars
    C. Jupiter
    D. [Both A and B above.]
    E. [All of the above.]

17. The constellations of the zodiac
    A. lie (roughly) along the path of the Sun.
    B. are all about equal in size—30 degrees wide.
    C. are distributed fairly evenly over the northern portion of the celestial sphere.
    D. are distributed fairly evenly over the entire celestial sphere.
    E. [Both B and D above.]

18. The ecliptic and celestial equator intersect at two points called the
    A. equinoxes.   D. sidereal points.
    B. solstices.   E. poles.
    C. tropics.

19. Why have astronomers added modern constellations to the sky?
    A. To replace some that changed in shape through the years
    B. To replace some whose meaning has changed
    C. To combine previous small ones into large ones
    D. To name some where none had been designated
    E. [Three of the above answers are correct.]

20. In science, what is the difference between a *theory* and a *hypothesis*?
    A. A hypothesis is more fully developed than a theory.
    B. A theory is more fully developed than a hypothesis.
    C. A hypothesis is based on a model, while a theory isn't.
    D. There is essentially no difference. The words are interchangeable.

21. A minute of arc is
    A. a measure of how far the Sun moves during one minute of time.
    B. one-sixtieth of a degree.
    C. how far the Earth turns on its axis in one minute.
    D. 60 degrees.
    E. the angular diameter of the Sun.

22. There are _____ arcseconds in one degree of angle.
    A. 3.14159   D. 360
    B. 24   E. [None of the above.]
    C. 60

23. Thirty arcminutes is about _____ degrees.
    A. 0.008   D. 1800
    B. 0.5   E. [None of the above.]
    C. 180

24. In what century did Ptolemy live?

25. As you watch a planet during the night, in what direction does it appear to move across the sky?

26. How do stars near Polaris appear to move as we watch them through the night?

27. What is a constellation?

28. What observation convinced Pythagoras that the Earth was spherical?

29. What is the ecliptic?

30. What are the approximate dates of the summer and winter solstices and what is their significance?

31. What is the origin of the word *planet*?

32. In what direction across the background of stars, do the planets normally move?

33. Define *retrograde motion* of a planet.

34. Name the planets that are never seen far from the Sun in the sky.

35. What is an epicycle?

36. In what part of the sky must you look to find the planets Mercury and Venus?

37. Define *geocentric*.

38. List three criteria for a good scientific model.

39. Which of the criteria for a scientific model is the most important?

QUESTIONS TO PONDER

1. In your personal case, why are you probably less familiar with the night sky than your great-grandparents?

2. Explain how you can determine the length of the year by astronomical observation.

3. Explain why it is necessary to have a leap year every four years.

4. What is meant by saying that one criterion for a good scientific model is its aesthetic quality?

5. How did Ptolemy's model account for the Sun moving back and forth between the Northern and Southern Hemispheres?

6. How did the Ptolemaic model account for retrograde motion?

7. Explain why the Greeks used circles and spheres to account for heavenly motions.

8. How did the Ptolemaic model explain the different behavior of Mercury and Venus?

9. Discuss the relative importance of the three criteria for a good scientific model.

10. Describe one or two observations that would disprove the Ptolemaic model.

11. The star in Orion's left knee is much farther away than the other bright stars in the constellation (see Figure 1-4). If you look at Orion in the sky, and then move far into space in the direction to your left, how will that star shift relative to the others?

12. Which days of the week are named for which planets? You may have to consult another book for this because although you might guess (correctly) that Sunday was named for the Sun, some of the corresponding names do not sound alike in English.

13. Identify the position of the vernal equinox, autumnal equinox, and the two solstices on Figure 1-11.
14. Explain the concept of Occam's razor by giving an example of its use in distinguishing which is the better of two hypotheses. (You may make up two hypotheses about some everyday event if you wish.)
15. Examine Ptolemy's response to Aristarchus's model. Does he agree that the model works astronomically? Why does he oppose the model?

CALCULATIONS

1. Express an angle of 15 arcseconds in degrees.
2. Four degrees of angle is how many arcminutes?
3. Suppose one star is 3 arcminutes from another. What is this angle in degrees?

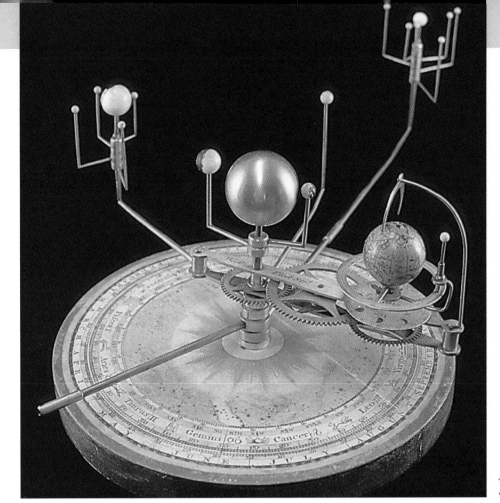

This early nineteenth-century device simulates celestial motions.

CHAPTER 2

A SUN-CENTERED SYSTEM

The Marriage of Aristotle and Christianity
Nicolaus Copernicus
HISTORICAL NOTE: *Copernicus and His Times*
The Copernican System
Motions of the Planets
Comparing the Two Models
Parallax
Copernicus's Revolution
Tycho Brahe

Johannes Kepler
The Ellipse
Kepler's First Two Laws of Planetary Motion
Kepler's Third Law
HISTORICAL NOTE: *Johannes Kepler*
Kepler's Contribution
ACTIVITY: *The Rotating Earth*
ACTIVITY: *The Radius of Mars's Orbit*

•······· *AT THE TIME OF MIKOLAJ KOPPERNIGK (1473–1543), Latin was the universal language in much the same way that English is today. It was the language that scholars learned in order to communicate with people from other countries. It was also common for people to "Latinize" their names. Thus Mikolaj Koppernigk became Nicolaus Copernicus. (Some people went further than this. One of Copernicus's teachers had been named Wodka, which means* vodka; *he Latinized his name to Abstemius, which means* abstainer.)

39

Copernicus entered the University of Bologna in Italy in 1496. It operated differently from today's universities in that the students were completely in charge. Students elected a rector who served a two-year term governing the university. At one time it was traditional for the students to celebrate the election of a new rector by tearing off his clothes and selling the pieces. Other traditions: In order to teach, professors had to swear to obey the rector. A professor could be fined for being late to class, for omitting anything from his lecture, or for lecturing too long.

This material is from Angus Armitage, The World of Copernicus *(The New American Library, 1947).*

The last chapter described the development of Ptolemy's geocentric system of the heavens. Recall that although this model is called the Ptolemaic model, it is not entirely due to Ptolemy. He worked in Alexandria, Egypt, and had access to the great Greek library there. His model was constructed by molding the thought and tradition of the past—much of it the work of Aristotle and Pythagoras—and adding his own insights to arrive at the final model. This is typical of scientific development. When we look with hindsight to the great advances in science, we see that they occur when the time is ripe for them, when previous thought in the area seems ready to be brought together with new insight to take a great step forward. It seems that all that is needed is the right person to take this step—the person who sees a little more clearly than others, is ready to take the leap of imagination into the future, and is able and willing to do the work and report the results.

Ptolemy's model, developed in about A.D. 125, served as the accepted model for more than 1300 years. This chapter will present a new model and examine the arguments for and against each of the two competing models. As will be seen, the new model finally won out because of adjustments that helped it pass the primary test of a scientific model: the model's ability to fit the observations accurately.

THE MARRIAGE OF ARISTOTLE AND CHRISTIANITY

During the thirteenth century, Saint Thomas Aquinas, one of the greatest theologians and philosophers of the Christian church, incorporated the works of Aristotle and Ptolemy into Christian thinking. Aquinas insisted that there must be no conflict between faith and reason, and he blended the natural philosophy of Aristotle with Christian revelation. For the Aristotelian and Ptolemaic ideas we have discussed, the blend was an easy one. The idea of an Earth-centered world fit comfortably with literal biblical interpretation, for it placed humans at the center of God's creation—the ultimate expression of the divine will.

As the last chapter pointed out, the idea of a central, unmoving Earth was natural for early humans. Through the work of Aquinas, this easily accepted idea was shown to fit perfectly with Christian beliefs. So Aristotle's science—and with it the Ptolemaic model—became even more entrenched in our Western culture. It was no longer just a natural, normal way of thinking about the world, but was part of Christian thinking and religious dogma.

Why have these early chapters presented material that seems to be more closely related to church history than to science? The reason is that science does not exist separately from culture and, in particular, from religious beliefs. Science and scientists are part of the society in which they live, and it is necessary to know something of the flavor of an age to appreciate the work of the thinkers of that age—including those thinkers who today might be labeled as scientists. Thus it is important to emphasize that after the thirteenth century, the teachings of Aristotle and the Ptolemaic model were an ingrained part of Western thinking.

A very important characteristic of the times was a great reliance on authority, particularly on authorities of the past. Today most of us have a much greater tendency to rely on our own thoughts, observations, experiences, and feelings than did people of the times we are discussing. Thus Aquinas was using the authority of the Bible, the authority of earlier churchmen, the authority of his superiors in the church, and the authority of Aristotle. Arguments were often settled by reference to authorities rather than by personal experience or independent experimentation.

Into this world came a man who was to cause a revolution in the way people think.

NICOLAUS COPERNICUS

At the time when Columbus was journeying to America, a brilliant Polish student was studying astronomy, mathematics, medicine, economics, law, and theology. As was common for scholars of that time, Nicolaus Copernicus (▶Figure 2-1) did not limit himself to studies in a particular discipline as we do today (perhaps necessarily). Later in life his primary responsibility was that of a churchman, canon of the cathedral of Frauenberg. He is known today, however, for initiating the revolution that finally resulted in replacing the geocentric system of Ptolemy with a *heliocentric*—Sun-centered—system.

For nearly 40 years, Copernicus worked on his heliocentric system. Even after all this work, he was slow to publish it. It was eventually published with the title, *On the Revolutions of the Heavenly Spheres*, but it is often called *De Revolutionibus*, a shortened form of the original Latin title.

Copernicus apparently decided to develop and publish his model for two primary reasons. First, as the centuries had passed, it was found that Ptolemy's predicted positions for celestial objects were not in agreement with the carefully observed positions. For example, Ptolemy's model could be used to predict the position of Saturn on some night in the future. The prediction of its position on a night a few years later was accurate enough, but when the prediction was made for a few centuries in the future, the predicted position might differ from the carefully observed position by as much as 2 degrees. This difference corresponds to about four times the diameter of the Moon (▶Figure 2-2). Ptolemy's model had been corrected a number of times through the ages to bring it up to date so that it would be fairly accurate for a while. These corrections were not refinements in the way it worked, but were "resettings" of each planet's position that were made to fit the latest data. A good model would not require such updating. Copernicus recognized this and sought a more accurate model.

A second reason Copernicus sought another model was that he did not believe that the Ptolemaic model was aesthetically pleasing enough. We will return to a discussion of the criteria for a good model after we have described the model developed by Copernicus.

▶ FIGURE 2-1. Nicolaus Copernicus (1473–1543).

heliocentric. Centered on the Sun.

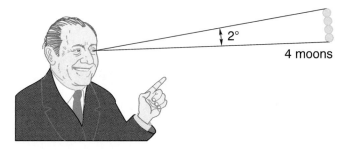

▶ FIGURE 2-2. When the difference between the predicted position of a planet and the observed position is 2 degrees, four Moons could fit between the two positions.

HISTORICAL NOTE

Copernicus and His Times

Mikolaj Koppernigk (or Nicolaus Copernicus by his Latinized name) was born in Torun, Poland, and was educated in Poland and Italy. During most of his life he worked for the church, serving as canon of the cathedral of Frauenberg, Poland. Copernicus lived during an exciting—and disturbing—time, as a list of some of his contemporaries suggests: Henry VIII, Martin Luther, Michelangelo, and Raphael.

As pointed out in the text, Copernicus was liberally educated in a number of subjects. His great interest in astronomy can be attributed to the close link between astronomy and religion at the time. There were two reasons for this link: First, it was important to the church that its feast days be celebrated at the right time, and this required a correct calendar—a matter for astronomers. (The dates of Easter and the feasts that follow it depend even today upon accurate determination of the time of the vernal equinox.) Second, at the time of Copernicus, astrology was considered a more important study than astronomy, but astrology cannot function without accurate data from astronomy. Thus astronomy was a necessary stepsister of what were considered more important subjects.

Copernicus lived long before the invention of the telescope. Measurements of the heavens had to be made with the naked eye, and yet accurate measurements were necessary. One important measurement was the exact time that a celestial object crossed the meridian. (The *meridian* is an imaginary line drawn on the sky from south to north passing directly over the observer's head.) Copernicus made this measurement by having his house constructed with a narrow slot in one of its walls. By placing himself at the correct location and looking through the slot, he could determine accurately when a given object was crossing (or *transiting*) the meridian.

It has long been known that planets change their brightnesses as they shift positions among the stars. The Ptolemaic system accounted for this with its epicycles, for they would cause a planet to change its distance from the Sun and from Earth. Copernicus noticed, however, that Mars seemed to change in brightness even more than would be predicted by that model. The epicycles of Ptolemy were simply not large enough to explain the great changes in Mars's brightness. Thus Ptolemy's model explained the basic phenomenon, but it did not explain it very precisely when accurate data were used. It was this observation that first prompted Copernicus to reconsider the heliocentric system of Aristarchus, which had lain hidden in obscurity for 2000 years.

Copernicus studied the ancient writings and became convinced that a Sun-centered system would not only provide better data for use by the church and by astrologers, but that it would also be a more aesthetically pleasing system. He linked astronomy very closely with his religious faith, arguing that placing the Sun at the center of everything is completely logical because the Sun is the source of light and life. The Creator, he said, would naturally place it at the center.

THE COPERNICAN SYSTEM

Try the first Activity at the end of this chapter to appreciate the change in outlook as one switches from a geocentric to a heliocentric system.

Copernicus's system revived many of the ideas of Aristarchus. (Recall the brief discussion near the end of the previous chapter.) Ptolemy had explained the daily motions of the heavenly objects as being due to the rotation of the celestial sphere from east to west around a stationary Earth. Like Aristarchus, Copernicus pointed out that a rotating Earth under a stationary sky produces the same observations as the Ptolemaic model (see ➤Figure 2-3). Ptolemy had answered that such a rotating Earth would produce a great wind. To solve this problem, Copernicus stated as an assumption that the air around the Earth simply follows the Earth around. How high above the Earth does this air extend? Copernicus does not answer the question, but notice that his theory requires that the air not extend all the way to the stars, or even to the Moon.

If the air extended to the Moon, it would tend to drag the Moon along with the Earth.

Copernicus's system is heliocentric, with the Sun at its center. The Earth assumes the role of just another one of the planets, all of them revolving around the Sun. The Earth becomes the third planet from the Sun. ➤Figure 2-4 shows the order of the planets that were known in the sixteenth century.

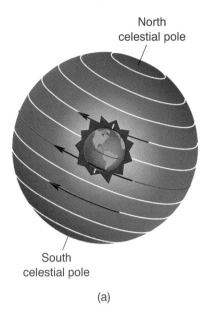
(a)

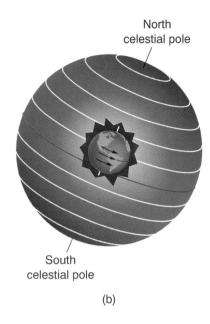
(b)

▶ FIGURE 2-3. (a) The celestial sphere rotating toward the west around a stationary Earth produces the same observed motion of stars as (b) the Earth rotating under a stationary celestial sphere.

If we consider the view of Figure 2-4 to be that of someone far out in space above the Northern Hemisphere of the Earth, the planets all move around in a counterclockwise direction as shown by the arrows in this figure. Another feature of Copernicus's model is that the planets closer to the Sun move faster than the planets farther out. In the figure, the lengths of the arrows represent the speeds of the planets, with the longer arrows indicating greater speeds.

▶Figure 2-5 shows the Earth in a circular orbit around the Sun. Its path does not show as a circle in the drawing because we are seeing it at an angle. If our point of view in the drawing were straight above the system, the Earth's orbit would appear as a circle, and the band of stars would likewise appear as a circle. Notice in the drawing that when viewed from Earth, the Sun can be thought of as located among certain stars. As the Earth continues around the

▶ FIGURE 2-4. The naked-eye planets, shown in order from the Sun, with their relative orbital speeds indicated by the length of the arrows.

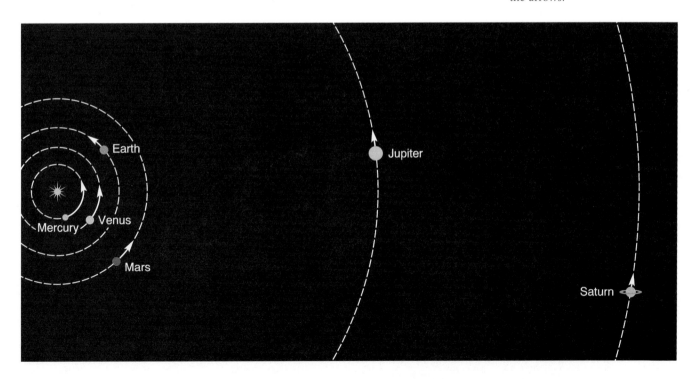

The Copernican System

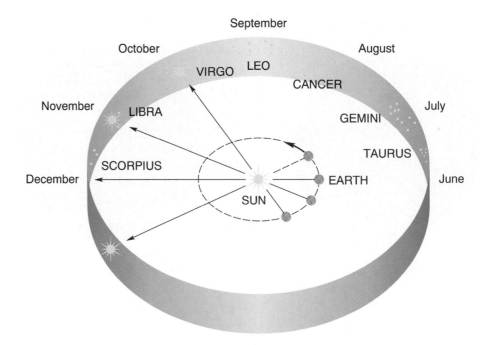

▶ FIGURE 2-5. As the Earth moves around the Sun during the year, the Sun appears to move among the background stars.

Sun, the position of the Sun among those stars appears to change. The Sun seems to be moving across the background of the stars. This, of course, is what is observed. Ptolemy explained it by having the Sun revolve around the Earth independent of the celestial sphere. Copernicus explained it by having the Earth move around a stationary Sun.

As pointed out in the last chapter, the Sun appears to move from among the stars south of the equator to those north of the equator and back again. To fit this observation, Copernicus's model had the plane of the Earth's equator tilted with respect to the plane of its orbit around the Sun (▶Figure 2-6). (Recall from Chapter 1 how and why this change in the Sun's apparent position causes the seasons.)

Look carefully at Figure 2-6 to see that the tilt of the Earth's axis would cause the ecliptic to be sometimes above and sometimes below the celestial equator.

▶ FIGURE 2-6. The Copernican model explains the Sun's apparent motion north and south of the equator by having the Earth's equator tilted with respect to the planet's orbit around the Sun.

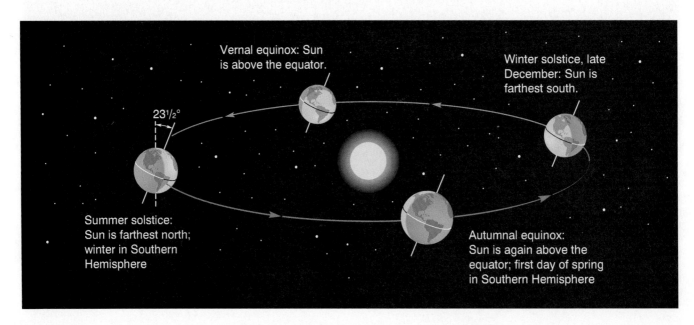

Motions of the Planets

As stated in the last chapter, someone watching the planets from Earth would see that these wanderers spend most of their time moving from west to east against the background of the stars. How does Copernicus's model handle this motion? ➤Figure 2-7 shows a number of positions of the Earth in its orbit, along with the corresponding positions of Mars. The lines from Earth to Mars illustrate the direction in which we on Earth see Mars. Look at positions 1, 2, and 3 for each planet. As time goes by, Mars appears to move among the background stars. Now picture a globe in the position of the Earth, and you will see that this motion of the planet (for the first three positions) is from west to east among the background stars. The model fits.

The retrograde motion of the planets presents a greater problem. On a schedule that is regular, but different for each planet, the planets stop their west-to-east motion and move for a time in the opposite direction. It was this retrograde motion that forced Ptolemy to use epicycles, and it is here that we see a major difference between the two models. Copernicus explained retrograde motion in an entirely different way.

Locations 4, 5, and 6 show Earth and Mars as they continue their motion. Look at what happens when Earth passes Mars. Mars no longer seems to be moving among the stars in the same direction it was before. It seems to have reversed its direction. This means that while the Earth is passing Mars, Mars appears to move from east to west. Retrograde motion!

To picture how this works, think of riding along a freeway in a fast lane and

➤ FIGURE 2-7. (a) While Earth moves from position 1 to position 3, Mars appears to be moving one direction among the stars. Then as Earth passes Mars, Mars appears to move backward. This is the heliocentric explanation for retrograde motion. (b) This is the view of Mars as seen from Earth for each position.

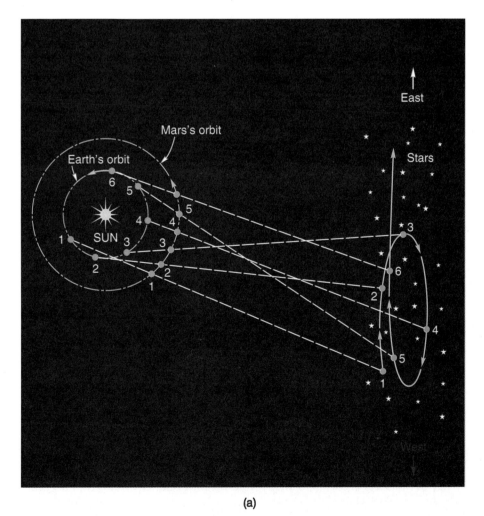

(a)

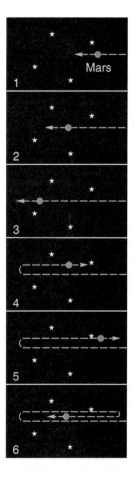

(b)

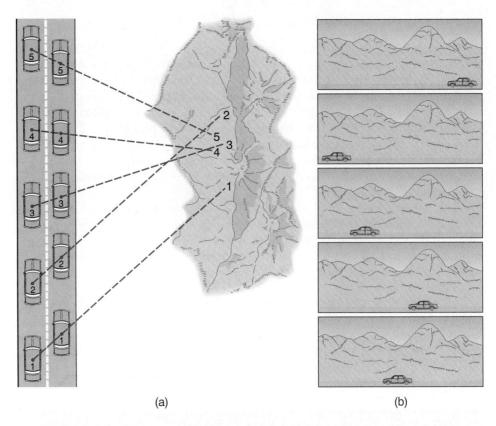

➤ **FIGURE 2-8.** As the car in the left lane speeds up and passes the other car, its driver sees the slower car appear to move backward against the distant mountains. Part (a) illustrates the lines of sight, and (b) shows what is seen from the car in the left lane.

quickly passing a slow-moving car. Now suppose you watch that car against a distant background—perhaps some trees across a field or even some distant mountains (➤Figure 2-8). As you are approaching the car, it will seem to move forward along the background. Then as you pass the car, you will see it appear to sweep backward against the background. In order to see this, you must be sure to concentrate on the background against which you see the car. That is what we are doing as we observe retrograde motion of a planet: we watch as it appears to move among faraway stars.

The Moon appears to complete about one revolution around the Earth each month. Copernicus accounted for this in his model by having the Moon on a sphere that revolves around the Earth in that time. For Copernicus, everything else circles the Sun but the Moon revolves around the Earth.

COMPARING THE TWO MODELS

1. *Accuracy in Fitting the Data*

The heliocentric model, as we have described it, is able to account qualitatively for the observed motions of the stars, the Moon, and the planets. If we are going to compare it to the Ptolemaic model, however, we must ask how accurately it fits the data. To be a good model, it must account for the most accurate positions recorded for the planets. Fitting a model to the observations is particularly difficult in the case of planetary retrograde motions. Did the positions of the planets according to the Copernican model correspond to accurate observations of the planets' positions? Well, not quite. There were differences between predicted positions and observed positions.

Copernicus sought to increase the accuracy of his model by inserting small epicycles for the celestial objects to move on. This was necessary because Copernicus assumed—like Ptolemy—that the planets move at a constant speed, and, in fact, the observed motions did not correspond exactly to planets moving at constant speed. The epicycles of Copernicus were smaller than those of Ptolemy, and the motion of a planet around them was much slower, so that the planets moved in a somewhat elongated circle.

Why did he not abandon the idea of circles altogether? The reason lies partly in the fact that the circle had such a long tradition. Beyond this, however, Copernicus felt that there was good reason for using the circle. He argued that the motions in the heavens are repetitive, and "a circle alone can thus restore the place of a body as it was."

So Copernicus was unable to abandon the epicycles of Ptolemy. And even with his epicycles, Copernicus's model has an error that typically comes to about 2 degrees over a few centuries of planetary motion. This is about the same error as in Ptolemy's model. How, then, do we choose between the two models? One way is to search for more data that might distinguish between them. If we could find an observation that could be explained by one model and not the other, we would have evidence upon which to build a case.

Parallax

Hold your thumb in front of your face while you close one eye and look at the wall across from you (➤Figure 2-9). Now, without moving your thumb, view the wall with the other eye. Wink with one eye and then the other. You will see that your thumb seems to move from one spot on the wall to the other. The explanation, of course, is that you are looking at it from a different location each time you change eyes. In general, if our observing location changes, nearby objects will appear to move with respect to distant objects. This observation is given the general name *parallax.*

parallax. The apparent shifting of nearby objects with respect to distant ones as the position of the observer changes.

Both the Copernican model and the Ptolemaic model held that the stars are on a sphere a great distance from the Earth. Neither model included the idea that some stars may be much farther from Earth than others. Parallax provided such evidence, as we will see.

➤Figure 2-10 shows the Earth at two positions on opposite sides of its orbit. Just as your thumb shifted its apparent position when you alternately winked

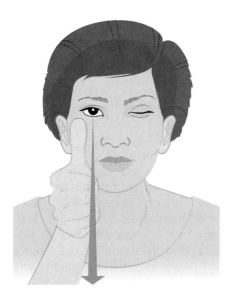

➤ **FIGURE 2-9.** To observe parallax, hold one thumb in front of you and look at it first with one eye and then with the other.

▶ **FIGURE 2-10.** As the Earth goes around the Sun, we see parallax of a nearby star as it shifts its position against background stars. The amount of parallax is very much exaggerated here. For the nearest star, the angle shown is only about 1.5 arcseconds (0.0004 degree).

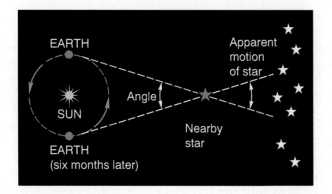

stellar parallax. The apparent annual shifting of nearby stars with respect to background stars. (Later this term will be used to refer to the angle of shift.)

your eyes, if stars differ in their distances from Earth, a nearby star would be expected to shift its position relative to very distant stars. Such **stellar parallax** was first observed in 1838. All of the stars are so far away, however, that the greatest annual shift observed for any star is only 1.5 arcseconds, and this is why stellar parallax was not observed earlier.

As we have said, parallax was not observed until it was well established that Ptolemy's model was unsatisfactory, but it serves as a good example of an observation that was contrary to that model. The Ptolemaic model held that the Earth does not move, but the observation of stellar parallax showed that it does. Although it would have surprised Copernicus to learn that the stars are at such different distances, this new idea could have easily been incorporated into his model, and in this case his model would have predicted stellar parallax.

No definitive observation could be made in the 1500s that would decide which of the two competing models fit the data more accurately. For that reason, using the evidence available in the 1500s, we still must call our comparison of the models a draw, based on the criterion of data fitting.

2. Predictive Power

As the example of parallax shows, predictions are inherent in models. They need not be stated by the model's designer.

The second prerequisite of a good scientific theory is that it must make verifiable predictions that might allow the theory to be disproved. The Copernican and Ptolemaic models predicted the existence and nonexistence of parallax, respectively, so both pass this test. But wouldn't all models make this type of prediction? Here is an example of a model that would not be considered a good scientific model because it does not make verifiable predictions. Suppose someone presents a model that says that the planets appear to move the way they do because of magical spirits that move them around at will. Using this model, whatever we later find out about a planet can be explained by saying that the magical spirits are doing it. There would be no way to prove this theory wrong. Whatever observation is made, one could attribute it to the whim of a magical spirit. This model cannot be accepted as a good scientific model because it does not contain within itself predictions that allow it to be disproved.

This seems to be an odd criterion. It says that a theory must contain the potential seeds of its own destruction. If a theory's prediction turns out to be correct, this serves as further evidence that the theory is good, but it does not "prove" the theory. On the other hand, a theory—or at least some aspects of a theory—can always be proven wrong by further evidence. We can disprove a theory, but we can never absolutely prove one.

astronomical unit (AU). A unit of distance equal to the average distance between the Earth and the Sun.

Before moving to the final criterion, let's look at another prediction made by the heliocentric system: the relative distances of planets from the Sun. Based on his theory, Copernicus was able to calculate such distances. We will not show here how this was done; an Activity at the end of this chapter shows the calculation for Mars. By such methods, Copernicus predicted that Mars is 1.5 times farther from the Sun than the Earth is. As the Prologue noted, to simplify discussion of distances in the solar system, the **astronomical unit** (abbreviated *AU*) is

48   CHAPTER 2   A Sun-Centered System

TABLE 2-1   Planetary Distances: Copernicus's Values versus Today's Values

Planet	Sun-to-Planet Distances	
	Copernicus's Values	Today's Values
Mercury	0.38 AU	0.387 AU
Venus	0.72	0.723
Earth	1.00	1.000
Mars	1.52	1.524
Jupiter	5.2	5.203
Saturn	9.2	9.539

defined as the distance from the Earth to the Sun. Since Mars is 1.5 times farther from the Sun than the Earth is, Mars is 1.5 AU from the Sun. Similar calculations for the other planets yielded for Copernicus the values shown in the center column of Table 2.1. The table also shows today's measurements.

Copernicus had no way to determine the distances in everyday units such as miles; he did not know the value of the astronomical unit. How do we know the actual distances today? One way to determine distance is to bounce a radar beam from a planet and measure the time it takes for the beam to return. Knowing the speed of the radar signal, we can calculate the distance to the planet and from this the value of the astronomical unit. Since the average distance from the Earth to the Sun is about 93,000,000 miles, that is the value of one astronomical unit.

The calculations of planetary distances based on the heliocentric model serve as an example of verifiable predictions made by the model. If later measurements had shown the distances to be wrong, the theory would have suffered a setback. The criterion that a theory must make verifiable (and therefore falsifiable) predictions is very important, and if this feature is lacking in a theory, the theory cannot be classified as a scientific theory.

There is a final important criterion that is used to compare scientific theories: aesthetics.

> We could have defined "the Martian Unit (MU)." This would have made Mars 1.0 MU from the Sun and Earth 0.66 MU from the Sun. We haven't done anything fishy in defining the AU. Its definition just makes the discussion simpler.

3. Aesthetics: Mercury and Venus

Copernicus worked for years to develop his heliocentric model. He, and other thinkers of his time and the years following his death, preferred it to the Ptolemaic model even though it had no particular advantage as far as the criteria we have discussed thus far. One of the reasons he preferred it was that he thought it was somewhat more accurate (which really is not the case). But his—and others'—real reason for preferring this model is due to our third criterion for a good model: neatness, simplicity, beauty, and aesthetic quality.

By the very nature of aesthetics, there can be disagreement over which of two things is more aesthetically pleasing. Let us, however, look at the two models in question and try to make a judgment. To make the comparison, it is instructive to see how the two models treated the motions of Mercury and Venus.

Recall that Mercury and Venus never get very far from the Sun in the sky. In order to explain this observation, the Ptolemaic theory treated these planets as special cases by requiring that the center of their epicycles remain along a line between the Earth and the Sun (Figure 1-23). The Copernican system explains the observation in a different manner. As ▶Figure 2-11 shows, neither planet can get very far from the Sun as seen from Earth simply because they orbit the Sun at a distance less than the Earth's distance. Notice that no matter where

> Venus is often called the "morning star" or "evening star" because it is so bright and obvious in the morning or evening sky. It easily outshines surrounding stars.

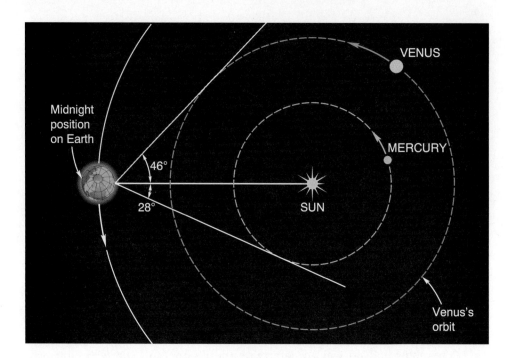

► **FIGURE 2-11.** This figure illustrates the heliocentric model's explanation for the fact that Mercury and Venus never get far from the Sun and can never be seen in the sky late at night. Compare it to Figure 1-23.

Venus is in its orbit, it can never appear farther than 46 degrees away from the Sun as viewed from Earth.

Now refer to the figure and locate the position on Earth where the time would be midnight. Notice that there would be no way for a person to see Mercury or Venus at midnight. Thus both models are able to explain why the observed motions of Mercury and Venus differ from those of the other planets, and both models are able to explain why Mercury and Venus are never seen high in the night sky. The Ptolemaic model, however, required a special rule to bring this about. In the Copernican model, this was a natural consequence of the fact that Mercury's and Venus's orbits are inside the Earth's orbit. They are not treated in any special manner. So the Copernican theory must be judged to be more aesthetically pleasing in this case.

A person on Mars would never see Mercury, Venus, or Earth around midnight.

Ptolemy used epicycles to explain retrograde motion. Each planet must be assigned an epicycle of a certain size, a speed for its motion around the epicycle, and another speed for the center of the epicycle moving around the Earth. The speeds that were assigned did not form a pattern, and, as we have seen, Mercury and Venus were special cases. Copernicus, on the other hand, saw a pattern in planetary speeds. According to his model, the farther from the Sun a planet is, the slower it moves in its orbit. This rule of motion was all that was needed to account for retrograde motion. Because of this simplicity, Copernicus and others felt that his was a better model for explaining the heavens.

This discussion of the two systems has almost neglected the fact that Copernicus also used epicycles in his model. They were much smaller epicycles than Ptolemy's, but he did find them necessary to make his model as accurate as he could. They were not needed to explain retrograde motion.

Recall from a margin note in the last chapter that Ptolemy's model actually contained further complications, such as the Earth being slightly off-center of the planets' motions. This contributed to Copernicus's conviction that his own system was simpler.

This leaves us with a dilemma: the basic idea of the heliocentric system would have to be judged as neater. But when one considers it in its full detail, it loses some of its beauty. So what is the final verdict? Probably in favor of the Copernican model, though the decision seems less than overwhelming.

Why were people of Copernicus's time concerned with how simple and orderly a theory is? This concern arose from a belief that a perfect God would not have created a disorderly, overly complicated universe. Thus the feeling that an aesthetically pleasing model is preferable to a less pleasing model was at least partly a religious feeling. Scientists today would express their desire for simple models in different terms, but the desire is still present. One of the reasons scientists search for a simple model is that our experience since the time of Copernicus tells us that when there are two competing models, the simpler one will be more likely to fit new data—data from observations not yet made. The belief that this pattern will continue with new models is a faith different from that of Copernicus and his supporters. It is a faith in the unity and beauty of nature, and a faith based on past experience.

COPERNICUS'S REVOLUTION

Copernicus died in 1543, just as *De Revolutionibus* was being released. To emphasize the importance of this work, it is instructive to consider the word "revolution." Today it has two primary meanings. One is the definition used in this chapter: the orbiting of one object around some point. The other everyday use of the word is in reference to an upheaval, as a social or political revolution. Before Copernicus's book was published, the word had only the first meaning. It was not used to describe upheavals. Recall the title of Copernicus's book: *On the Revolutions of the Heavenly Spheres*, or *De Revolutionibus*. This single book started such an upheaval in people's thinking that the word "revolution" took on a second meaning—the meaning that is perhaps more common today.

Copernicus began the revolution, but it was up to others to carry it forward to completion.

TYCHO BRAHE

Three years after Copernicus died, Tycho Brahe (pronounced "bra" or "bra-uh") was born in Denmark (➤Figure 2-12). When he was a teenager studying law, he developed an interest in astronomy, and in studying that subject he learned that both the Ptolemaic and Copernican models were based on recorded planetary data that were inaccurate. That is, he found discrepancies between different tables of observed planetary positions. He was convinced that before a decision could be made about which model was better, or before a completely new and better model could be devised, there was a need for more accurate observations of the positions of planets as they move across the background of stars.

At this time the telescope had not yet been invented, so all of Tycho's observations were made with the naked eye. In an observatory built for him by King Frederick II of Denmark, Tycho built the largest observing instruments yet constructed. The mural in ➤Figure 2-13 shows some of his equipment. The large quarter-circle in the foreground is drawn to the scale of the people in front. The person shown partially on the drawing at right center is looking past a pointer on the degree markings and through the slot in the wall at upper left. Because of the large size of the degree marks, Tycho was able to measure angles to an accuracy of better than 0.1 degree, far more accurate than any measurements up to his time and close to the limit the human eye can observe.

In addition to making such accurate measurements, Tycho was careful to determine just how great was the accuracy of each measurement. For example, he would not only state the position of a particular planet, but would also give a value indicating the amount of uncertainty in his measurement. When such data are later compared to a particular model to test the model for accuracy, it is possible to know how close the predictions of the model should be expected to come to the measurements reported. If Tycho stated that a particular measurement was accurate to 0.1 degree of angle, then all that could be expected of a model would be that its predictions should come within 0.1 degree of that measurement. On the other hand, if the model did not make the prediction

➤ FIGURE 2-12. Tycho Brahe (1546–1601).

People of Tycho's time are often known by their first names. For example, do you know the last name of Michelangelo?

One-tenth of a degree is the angle between opposite sides of a quarter when viewed from a distance of 50 feet.

► **FIGURE 2-13.** This mural depicts Tycho's quadrant, which was designed to measure the time and the elevation angle of an object crossing the meridian. The quadrant itself consists of the slot at upper left and the large quarter-circle. Tycho is shown as the observer at the edge of the mural at right center. In addition, he had himself painted on the quadrant (the large figure on the wall near the center) to remind his assistants that he was watching. Tycho's quadrant had a radius of over 2 meters.

Tycho was a partier! He died of a burst bladder that resulted from drinking too much.

within 0.1 degree of the measurement, it would have to be considered not accurate enough. The inclusion of a statement concerning the accuracy of measurements is now a common practice in science, but it wasn't in Tycho's time.

Night after night for 20 years, Tycho recorded data concerning the positions of planets, with particular emphasis on Mars. In addition to his accurate measurements, Tycho is known for something he did the year before he died: in 1600 he hired Johannes Kepler as an assistant.

JOHANNES KEPLER

Johannes Kepler (►Figure 2-14), who was born in what is now Germany, concentrated on the study of theology. During this study, he learned of the Copernican system and became an advocate for it.

In 1600 Kepler accepted a position as assistant to Tycho Brahe, who assigned him the job of working on models of planetary motion. After Tycho's death, Kepler took over most of Tycho's records and continued the search he began under Tycho. After working for four years and trying 70 different combinations of circles and epicycles, he finally was able to devise a combination for Mars that would predict its position—when compared to Tycho's observations—to within 0.13 degrees.

Kepler's accuracy of 0.13 degrees (8 arcminutes) is about one-fourth the diameter of the Moon. Recall that prior to this, a typical error between predicted positions and observed positions was about 2 degrees of angle. Kepler, however, was not satisfied. The error of 0.13 degrees still exceeded the likely error in Tycho's measurements (about 0.1 degree). Kepler knew enough about Tycho's methods to know that the data were accurate enough that an error of 8 arcminutes was too much, and he sought a model that would fit the data to the limits of its precision. This is a tribute both to Kepler's persistence and to his belief in the accuracy of Tycho's careful measurements.

Finally, Kepler decided to abandon the circle as the basic motion of the

►**FIGURE 2-14.** Johannes Kepler (1571–1630).

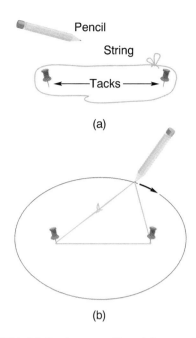

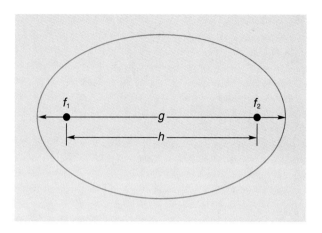

► FIGURE 2-15. To draw an ellipse (a) start with two tacks, a pencil, and a string. (b) Stretch the string taut with the pencil and swing it around.

► FIGURE 2-16. The points f_1 and f_2 are the foci of the ellipse. The eccentricity is calculated by dividing the distance between the foci (h) by the length of the major axis (g). Half of the major axis is called the semimajor axis and is what we have previously referred to as the average distance from the planet to the Sun.

planets and to try other shapes. He tried various ovals, with the speed of the planet changing in different ways as it went around the oval. After nine years of work, he found a shape that fit satisfactorily with the observed path of Mars. What's more, he found that the same basic shape worked not just for Mars, but for every planet for which he had data. The shape? An ellipse.

The Ellipse

An *ellipse*, at first glance, is nothing more than an oval. But it is more; not every oval is an ellipse. ► Figure 2-15 shows how to draw an ellipse. Stick two tacks into a board some distance apart and put a loop of string around them. Then use a pencil to stretch the string as shown. Finally, keeping the string taut, move the pencil around until you have completed the ellipse.

Each point where we placed a tack is called a *focus* of the ellipse. The plural of this word is "foci," so an ellipse has two foci.

Ellipses can be of various *eccentricities*. The eccentricity of an ellipse is defined as the ratio of the distance between the foci to the longest distance across the ellipse (see ► Figure 2-16). Essentially, the eccentricity tells us how "flat" the ellipse is. For example, if the tacks had been put closer together or a longer string had been used, the ellipse of Figure 2-15 would have had less eccentricity and might have looked like ► Figure 2-17a. And if the tacks had been farther apart or the string had been shorter, the ellipse would have had greater eccentricity and might have looked like Figure 2-17b. Although different ellipses have different eccentricities, it is important to see that definite rules govern the shape of an ellipse. It is not egg-shaped or two half-circles connected by straight lines.

An example of an elliptical shape you see every day is that of a circle seen at an angle. ► Figure 2-18 shows a round object seen in perspective. The shape the artist drew to represent the top of the can correctly is an ellipse. The eccentricity of such an apparent ellipse is changed by changing your angle of viewing the trash can. Viewing it from above, you see a circle, but by viewing it at an angle, you see its apparent shape become an eccentric ellipse.

ellipse. A geometrical shape every point of which is the same total distance from two fixed points (the foci).

focus of an ellipse. One of the two fixed points that define an ellipse. (See the definition of *ellipse*.)

eccentricity of an ellipse. The result obtained by dividing the distance between the foci by the longest distance across an ellipse (the *major axis*).

A circle has zero eccentricity.

Johannes Kepler

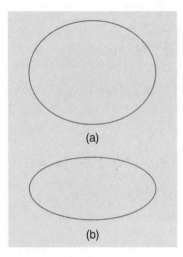

▶ FIGURE 2-17. (a) An ellipse with a small eccentricity. (b) A more eccentric ellipse.

▶ FIGURE 2-18. A circular shape, such as the top of a trash can, appears elliptical when viewed at an angle.

KEPLER'S FIRST TWO LAWS OF PLANETARY MOTION

Kepler published his first results in 1609 in his book *The New Astronomy*. Today we summarize his findings for planetary motion into three laws, the first of which is:

KEPLER'S FIRST LAW: Each planet's path around the Sun is an ellipse, with the Sun at one focus of the ellipse.

▶Figure 2-19 shows an elliptical path of a planet around the Sun. We have exaggerated the eccentricity of the ellipse, however, to make it more obvious. The elliptical paths of most planets are not very eccentric; they are nearly circles.

The second law tells us about the speed of a planet as it moves around its ellipse. Kepler found that a planet moves faster when it is closer to the Sun and slower when it is farther away. He was able to make a more definite statement than this, however, one that would allow a calculation of the speed:

KEPLER'S SECOND LAW: A planet moves along its elliptical path with a speed that changes in such a way that a line from the planet to the Sun sweeps out equal areas in equal intervals of time.

The second law takes some explanation. Suppose we consider that the Earth moves in an elliptical path such as that shown in ▶Figure 2-20. Further suppose

▶ FIGURE 2-19. Kepler's first law states that a planet moves in an elliptical orbit with the Sun at one focus of the ellipse. The orbits of actual planets are much more circular than the ellipse shown here.

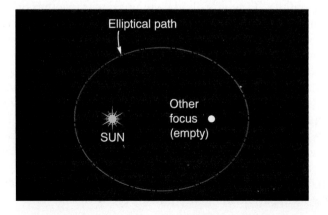

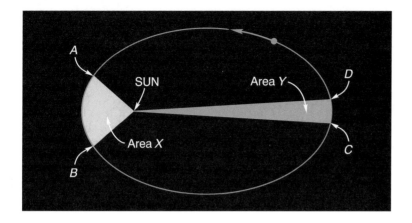

▶ FIGURE 2-20. Kepler's second law: A line from the Sun to a planet sweeps out equal areas in equal times. In the drawing, area X = area Y.

that point A in the figure represents the position of the Earth on January 1, and point B is its position on February 1. The yellow-shaded area represents the area "swept out" during those 31 days. Then suppose that on July 1 the planet is at point C. Kepler's second law tells us that in the next 31 days, the line from the Earth to the Sun must sweep out the same area as it did in January. Thus the Earth must move more slowly at its greater distance from the Sun, so that it gets only to point D on August 1. In fact, the law tells us that during any 31 days, the area swept out must be the same.

The time that is applied in Kepler's second law need not be 31 days, of course. The point of the law is that no matter what time interval is selected, if we compare the area swept out by the imaginary line during that amount of time at different places on the elliptical journey of the planet, we will get the same area everywhere. In Figure 2-20, we have again greatly exaggerated the eccentricity of the Earth's orbit. In actuality, the elliptical path of the Earth has a very small eccentricity, so much so that it is almost a circle. Copernicus's circle very nearly fits the path of the Earth.

To solve the problem of how to determine an unknown orbit while riding on a planet with an unknown orbit Kepler did the following: he selected observations of Mars on the same day in different years. This effectively froze the Earth in position. Then he did the same thing in different *Martian* years to map the orbit of Earth. Clever!

KEPLER'S THIRD LAW

The first two laws proposed by Kepler tell us about the paths of each individual planet. Although they propose the same basic shape for the path of each planet (the ellipse), they do not tell us anything about how the speed of one planet along its path compares to that of another. The third law does.

KEPLER'S THIRD LAW: The ratio of the cube of a planet's average distance from the Sun to the square of its orbital period is the same for each planet.

The third law is most easily understood when it is expressed in symbols. Let

a = average distance of a planet from the Sun

P = the planet's orbital period

and

C = a constant whose value depends on the physical units used for a and P

then

$$\frac{a^3}{P^2} = C$$

Kepler's third law states that a^3/P^2 has the same value for every planet. We'll do a sample calculation to show how the law can be used.

The distance in Kepler's third law is actually the *semimajor axis,* or half the length of the major axis of the orbit. This is very nearly the same as the planet's average distance from the Sun.

Kepler's Third Law 55

HISTORICAL NOTE

Johannes Kepler

Johannes Kepler was born in a small town in southern Germany in 1571. His father was poor and unreliable, his mother had a violent temper, and Johannes was a very sickly child. When he was four, he contracted smallpox and nearly died. The disease left him with poor eyesight, a condition that prevented him from enjoying the astronomical sights reported to him by Galileo. So rather than becoming an observer, he made his contributions in theoretical astronomy by applying his great mathematical abilities and his insights to the results of others' observations. The most accurate information was the nontelescopic data from Tycho Brahe, for whom Kepler worked during the last year of Tycho's life. (Tycho hired Kepler in the hope that the mathematician would be able to verify Tycho's Earth-centered model, but Kepler apparently took the position with the idea of gaining access to the accurate data he needed to test his own ideas.)

Today we know Kepler for the three laws to which we give his name, but these laws seem to have been developed almost accidentally as he searched for other, deeper patterns in the heavens. For example, his first great endeavor was to try to find out why the planets were spaced as they were. When he was about 24 years old and was working as a professor of mathematics, an idea occurred to him while he was lecturing. He was discussing the size of the largest circle that could be drawn inside a triangle and the smallest that could be drawn outside the triangle when he wondered if this could possibly be the basis for the spacings of the planets. After working with various shapes, he decided that it was not two-dimensional figures that were significant, but solids. It was known that only five symmetric solids were possible (the tetrahedron, cube, octahedron, dodecahedron, and icosahedron). That number is exactly what would be needed to fill the spaces between six planets. ▶Figure H2-1 shows a construction that illustrates the plan Kepler finally worked out. Kepler might have saved a lot of time if all of today's planets had been known then, for he would have had too many planets for the five solids.

Ptolemy and Copernicus had tried to work out a model for the heavenly motions, but Kepler went further. He also asked *why* the planets would have such motions. It was Kepler—not Isaac Newton, whom we will discuss in the next chapter—who first hypothesized that there was a force that kept the planets near the Sun: a force we today call gravity. In addition to this force, however, Kepler felt that there must be a force sweeping the planets around the Sun. Just like the gravitational force, this force should become weaker with distance. That is why, he argued, the more distant planets move around the Sun more slowly. It was his consideration of this sweeping force that led him to his third law. Although no such sweeping force exists, it helped Kepler find a rule that later led Newton to discover the forces that *do* exist.

Kepler remained unappreciated throughout his life. His difficulties were many: his first marriage was unhappy, his wife and one of his children died of smallpox, his mother was convicted of witchcraft, and he was forced to cast horoscopes—which he personally ridiculed—to support himself and his family. His consolation was that he believed that his scientific findings were important and would be recognized after he died.

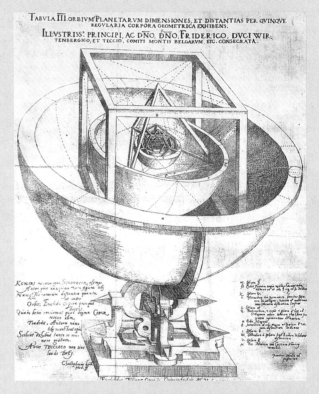

▶ **FIGURE H2-1.** Kepler worked out a plan using the five regular three-dimensional figures to fix the spheres of the six known planets at their correct distances.

EXAMPLE

Let us use the fact that the period of Mars's orbit is 1.88 years to calculate the average distance of Mars from the Sun. In using Kepler's third law, any units of measure can be used for the period and the distance. We will choose units that make the calculation simple, expressing time in units of the year so that P for the Earth is one year. We could state a in miles, but this would lead to very large numbers, so we will use the astronomical unit. Recall that one astronomical unit is defined as the average distance from the Earth to the Sun. Now let's apply the calculation to the Earth:

$$\frac{a^3}{P^2} = \frac{(1 \text{ AU})^3}{(1 \text{ year})^2} = 1 \text{ AU}^3/\text{year}^2$$

Kepler, of course, did not know the Earth-Sun distance. Today we know that an astronomical unit is about 93 million miles.

Notice the advantage of choosing the units we did: the ratio is equal to 1 for the Earth. That is, the numerical value for the square of the period is equal to that of the cube of the distance. $P^2 = a^3$ (in years and astronomical units).

Let's apply this to Mars. The third law says that a^3/P^2 is also equal to 1 for Mars when the above units are used. After rewriting the equation, we substitute the value of Mars's period and solve for its distance from the Sun.

$$\frac{a^3}{P^2} = 1 \text{ AU}^3/\text{year}^2$$

$$\frac{a^3}{(1.88 \text{ year})^2} = 1 \text{ AU}^3/\text{year}^2$$

$$a^3 = 3.53 \text{ AU}^3$$

$$a = 1.52 \text{ AU}$$

This agrees with the measured value of the average distance of Mars from the Sun.

TRY ONE YOURSELF. Calculate the radius of Venus's orbit using the fact that its period of revolution around the Sun is 0.615 year. (Venus's orbit is very close to circular, so its average distance from the Sun is essentially the radius of its orbit.)

Table 2-2 shows modern data for each of the planets known to Kepler. In each case, Kepler's third law tells us that to the limits of the accuracy of the data, the distance cubed is equal to the period squared.

Kepler had fairly accurate data for the orbital periods of the planets. The actual distances to planets were unknown to him, but remember that the Copernican system allows us to calculate the *relative* distances to the planets. For example, it was known that Mars is 1.52 times farther from the Sun than the Earth

TABLE 2-2 Testing Kepler's Third Law

Planet	Distance from Sun	Period of Revolution	Distance Cubed	Period Squared
Mercury	0.387 AU	0.241 yr	0.0580 AU^3	0.0581 yr^2
Venus	0.723	0.615	0.378	0.378
Earth	1.000	1.000	1.000	1.000
Mars	1.524	1.881	3.540	3.538
Jupiter	5.203	11.86	140.9	140.7
Saturn	9.539	29.46	868.0	867.9

is. This, however, is all that is needed to do calculations with Kepler's third law, because 1.52 is the radius of Mars's orbit in astronomical units. So Kepler didn't need to know the actual distances. To the limits of the accuracy of Brahe's data, Kepler's third law fit.

KEPLER'S CONTRIBUTION

Kepler's modification to the Copernican model brought it into conformity with the data; finally, the heliocentric theory worked better than the old geocentric theory. To achieve this fit to the data, however, it had been necessary to abandon the long-held idea of perfect circles in the heavens. Kepler's only reason for proposing ellipses for planetary motion was that they worked.

Our understanding of the solar system would be very unsatisfactory if it had remained where Kepler left it. At the time of Kepler, our scientific understanding of the world was just beginning to take major steps forward, and a man who was a contemporary of Kepler—Galileo Galilei—provided what was needed for the next leap in understanding. The discussion of Galileo's contributions will be left to the next chapter, which will show how they led in turn to the crowning achievements of Isaac Newton.

CONCLUSION

Two major models of the heavens have been presented: the geocentric model of Ptolemy and the heliocentric model as presented by Copernicus. Both models had simple explanations for the daily motion of the stars and for the annual motion of the Sun, but their explanations of planetary motions were more complicated. Copernicus's model had an aesthetic advantage in that it explained all planetary motions in the same way, whereas Ptolemy's model had special rules for Mercury and Venus. When the models were compared in the most important of scientific criteria—fitting the data—the heliocentric system of Copernicus was not significantly better than the old geocentric system, however. To accept Copernicus's model would mean taking an entirely different view of the world, for his model moved the Earth from its central position and reduced it to being just one of the planets. This fact, along with the model's lack of advantage in fitting the data, meant that it had little chance of acceptance.

Kepler's revision of Copernicus's theory solved its data-fitting problem by using ellipses rather than circles and by presenting simple rules for the speeds of planets. Using the criteria for scientific models, we judge Kepler's model a better one than Ptolemy's. Kepler's revision suffered one flaw, however: the laws he proposed did not "make sense" in that they did not correspond to anything else in nature. They made the model fit the data, and they were reasonably "neat and clean," but they seemed to have been pulled out of thin air. Before the heliocentric system could be fully accepted, it would have to be tied successfully to other phenomena in our experience.

RECALL QUESTIONS

1. It is _____ to observe Venus in the sky throughout the entire night from a point on the Earth's surface.
   A. sometimes possible
   B. always possible
   C. never possible

2. It is _____ to observe Jupiter in the sky throughout the entire night from a point on the Earth's surface.
   A. sometimes possible
   B. always possible
   C. never possible

3. If the plane of the Earth's equator were not tilted with respect to the ecliptic plane,

A. the daylight period on Earth would be the same year-round.
   B. there would be no seasonal variations on Earth.
   C. Earth's poles would not experience six-month-long nights.
   D. [All of the above.]

4. Which of the following is *not* a criterion for a good scientific theory?
   A. A theory should be aesthetically pleasing.
   B. A theory should be agreed upon by all knowledgeable scientists.
   C. A theory should fit present data.
   D. [None of the above; all are criteria for a good theory.]

5. The Copernican model explained retrograde motion as due to
   A. planets moving along epicycles.
   B. planets stopping their eastward motion, moving westward a while, and then resuming their eastward motion.
   C. different speeds of the Earth and another planet in their orbits.
   D. [Both A and B above.]
   E. [None of the above; the Copernican model was unable to explain retrograde motion.]

6. Why was Copernicus forced to use epicycles in his model?
   A. To account for retrograde motion
   B. To account for phases of the Moon
   C. To accurately predict the position of a planet
   D. [Both A and B above.]
   E. [All of the above.]

7. The observation that Mercury and Venus never get high in the night sky was explained by
   A. the geocentric model only.
   B. the heliocentric model only.
   C. both the geocentric and heliocentric models.

8. Which of the following could be calculated from the Copernican model but not from the Ptolemaic model?
   A. The length of the day on each planet
   B. The actual distance from the Sun to each planet
   C. The relative distance from the Sun to each planet
   D. The approximate position of a given planet in the sky a few years in the future.

9. If you lived on Jupiter, which of the following planets might you see high overhead during the night (assuming that your sky was clear enough that you could see the stars)?
   A. Mercury and Venus only
   B. Mercury, Venus, Earth, and Mars only
   C. All of the planets except Mercury and Venus
   D. All of the planets except Mercury, Venus, Earth, and Mars.

10. The Sun appears to move among the stars. Copernicus's model accounts for this as being due to
    A. the Earth's rotation on its axis.
    B. the Earth's revolution around the Sun.
    C. the actual motion of the Sun against distant stars.
    D. the Earth changing speed in its orbit.
    E. different planets moving at different speeds.

11. If the Earth's diameter were double what it is, stellar parallax would be
    A. easier to observe.
    B. more difficult to observe.
    C. the same, for stellar parallax does not depend upon the Earth's diameter.

12. Why did the existence of stellar parallax not convince people of Copernicus's time that his model was better than the geocentric model?
    A. Few people understood stellar parallax.
    B. Stellar parallax was understood, but ignored.
    C. Although most people understood the phenomenon, it was not clear that it provided evidence of the heliocentric theory.
    D. The phenomenon actually provides evidence for a geocentric theory.
    E. Stellar parallax had not yet been observed.

13. According to the heliocentric model, the reason the planets always appear to be near the ecliptic is that
    A. the ecliptic is only 23.5 degrees from the celestial equator.
    B. the planets revolve around the Sun in nearly the same plane.
    C. compared to the stars, the planets are near the Sun.
    D. the planets come much nearer to us than does the Sun.

14. The apparent motion of the planet Jupiter in the sky during a single night as seen by an Earthbound observer is
    A. from west to east, then backward a while, and then west to east again.
    B. a constant west-to-east motion.
    C. a constant east-to-west motion.
    D. [None of the above, for no motion is observed.]

15. According to Kepler's laws, a planet moves
    A. at a constant speed through its orbit.
    B. fastest when nearest the Sun.
    C. fastest when farthest from the Sun.
    D. [None of the above, for there is no simple rule.]

16. The primary reason that it is hotter in Australia in December than it is in July is that
    A. the Southern Hemisphere is tilted toward the Sun in December.
    B. the Earth is closer to the Sun in December than in July.
    C. the Earth is moving faster in its orbit in December than in July.
    D. the Earth is on the hotter side of the Sun in December.
    E. [Both B and C above.]

17. If the Earth were in an orbit closer to the Sun,
    A. the day would be longer.
    B. the day would be shorter.
    C. the year would be longer.
    D. the year would be shorter.
    E. [Two of the above are correct.]

18. Kepler's law of equal areas predicts that a planet moves fastest in its orbit when
    A. it is closest to the Sun.
    B. it is closest to the Earth.
    C. the Earth, Moon, and Sun are in a line.
    D. it is farthest from the Sun.
    E. [None of the above; the law makes no predictions as to speed.]

19. If there were another planet between Jupiter and Saturn,
    A. its "day" would be longer than Earth's.
    B. its "year" would be longer than Earth's.
    C. its *range* of temperature would be greater than Earth's.
    D. its speed in orbit would be greater than Earth's.

20. From the law of equal areas, one can predict that
    A. the Earth moves the same distance in January as in July (or in any other 31-day month).
    B. the Earth moves faster when closer to the Sun.
    C. Jupiter takes longer to circle the Sun than does Mars.
    D. [Both B and C above.]
    E. [All of the above.]

21. If a planet were found at a distance of 3 AU from the Sun, its sidereal period would be
    A. about 2.1 years.   D. 9 years.
    B. 3 years.   E. about 0.3 years
    C. about 5.2 years.

22. If a new planet were found with a period of revolution of 6 years, what would be its average distance from the Sun?
    A. About 2 AU   D. About 9 AU
    B. About 3.3 AU   E. 36 AU
    C. 6 AU

23. Identify or define: heliocentric, geocentric, astronomical unit, *De Revolutionibus*.

24. How die the Copernican model explain the daily motion of the heavens?

25. How did the Copernican model explain retrograde motion?

26. Why was Copernicus forced to use epicycles in his model?

27. Name a discovery made in the nineteenth century that the Copernican model fits (or can be made to fit), but the Ptolemaic model does not.

28. What rule did the Copernican model have concerning the speed of one planet compared to another?

29. Define parallax and describe a method of demonstrating it.

30. How did each model explain why Mercury and Venus are never seen high in the night sky?

31. What made Tycho Brahe decide that more accurate data were needed?

32. Explain how to draw an ellipse, including how to draw one with more or less eccentricity.

33. State and explain Kepler's second law, the law of equal areas.

34. List the planets known in Kepler's time in order of their speeds in orbit. List them in order of their distances from the Sun.

35. Copernicus had stated that planets farther from the Sun move slower than nearer ones. What did Kepler's third law add to this statement?

QUESTIONS TO PONDER

1. Were most of the ideas in Ptolemy's model developed by Ptolemy? Discuss.

2. Describe some features of Corpernicus's model that can be taken as an indication of his reluctance to break completely with the old mode of thinking.

3. Compare the two models as to their accuracy in fitting the data available in the sixteenth century.

4. Explain why the Copernican model is generally considered more aesthetically pleasing than the Ptolemaic model.

5. Discuss the reason why today's scientists seek aesthetically pleasing models.

6. Explain why Copernicus was dissatisfied with Ptolemy's model.

7. Explain what is meant when we say that the Sun is "in" a certain constellation, and how the Copernican model explains the Sun's change of position among the stars.

8. It might be possible to prove a theory false, but it is never possible to prove it true. Explain.

9. When you read how each of the models explained retrograde motion, you may have found it easier to understand the explanation of the Ptolemaic model than the explanation of the Copernican model. If so, how can Copernicus's explanation for retrograde motion be called the simpler one?

10. Ptolemy observed parallax of the Moon against the distant stars. Explain how this can be observed. (Hint: It is not due to the motion of the Moon or of the Earth around the Sun.)

11. Why is the Copernican explanation for the unusual motions of Venus and Mercury considered better than the Ptolemaic explanation?

12. Is a circle an ellipse? That is, would a planet moving in a circular path violate Kepler's laws?

13. The Earth is closest to the Sun in January. Why does this not cause us to experience our hottest weather in January?

CALCULATIONS

1. Calculate the distance to Jupiter in astronomical units. Data: Jupiter's sidereal period is 11.86 years, and quadrature occurs 87.5 days after opposition. (Hint: See the Activity "The Radius of Mars's Orbit.")

2. Mercury's period of revolution is 88 days, or 0.24 years. Use Kepler's third law to calculate its average distance from the Sun in astronomical units.

3. The planet Uranus was discovered in 1781. The semimajor axis of its orbit is 19.2 AU. Use this to calculate its period of revolution around the Sun.

4. A quarter at 50 feet spans an angle of about 0.1 degrees. About how many quarters could be placed side-by-side around a circle with a radius of 50 feet?

5. The eccentricity of Pluto's orbit is 0.25. Use a drawing of an ellipse to explain what this means quantitatively.

ACTIVITY

The Rotating Earth

On a clear dark night, find a good observing location and draw a quick sketch of some stars high over your head. Look for a pattern so that you can remember these stars later. If you can, mark your map so that it shows which stars are toward the north, east, south, and west.

Take about two hours off, and then go back to your observing location and find the stars you drew. How has their position changed? You probably know what this answer should be from having read the text. Your real job, however, is to stand under the stars and imagine that their motion is due to the rotation of the Earth. Try to "feel" the Earth turning under the stars. Which is easier to imagine, the stars on a sphere rotating around the Earth or the Earth spinning under the stars?

Finally, spend a half hour watching either a sunrise or a sunset. Or better yet, watch the Moon rise or set, particularly when it is full or nearly full. Now picture this phenomenon as explained by the heliocentric system. Try to think of the Earth turning instead of the Sun or Moon moving. Imagine yourself on a little ball that is turning so that the place where the Sun's light hits the ball changes.

ACTIVITY

The Radius of Mars's Orbit

Let us determine the radius of Mars's orbit from data available to Copernicus. To do so, you will make a scale drawing and will assume that the Earth and Mars move in circular orbits. Start by marking a location for the Sun on one side of a piece of paper. As in ➤Figure A2-1a, draw an arc of a circle about halfway across the paper to represent the orbit of the Earth. Then draw a line from the Sun horizontally across the paper. Suppose that when the Earth crosses your horizontal line, Mars is also crossing it, so the Sun and Mars are on opposite sides of the Earth. (Mars is then said to be in *opposition*.) Mars and the Earth are both moving in the direction shown in Figure A2-1a. Now suppose that Mars is observed night after night until it is at **quadrature**. Quadrature is defined as occurring when Mars is 90 degrees from the Sun in the sky, as shown in ➤Figure A2-1c. This is observed to occur 106 days after opposition. This observation, along with the sidereal period of Mars, is all that we need to determine the radius of its orbit. (The *sidereal period* of a planet is the time required for the planet to return to the same position among the stars as viewed from the Sun.) The sidereal period of Mars is 1.88 years.

➤ **FIGURE A2-1.** Steps involved in determining the radius of an outer planet's orbit. This figure is not drawn to scale, as yours must be.

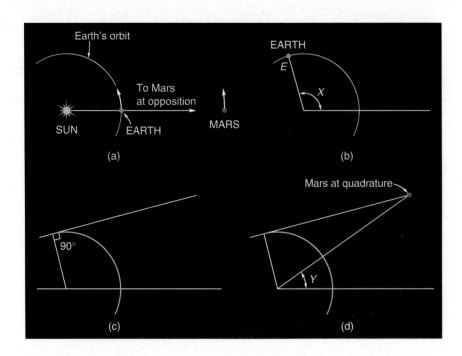

To draw your figure to scale, you must first calculate the angle through which the Earth moves in 106 days. This is obtained by dividing 106 days by 365 days and multiplying 360 degrees. Draw a line from the Sun at the angle you have calculated (angle X in ➤Figure A2-1b), and determine the position of Earth (point E) when Mars is at quadrature.

Since Mars is at quadrature when the Earth reaches point E, you should start at E and draw a line toward the right 90 degrees from the line to the Sun. This line points to the planet Mars. You don't know yet where Mars is located on this line. You can determine this by calculating the angle Mars moves through in 106 days. Use its period of 1.88 years to calculate this angle. (1.88 years is 686 days, so Mars moves 106/686 of a complete circle in 106 days.) Starting from the Sun, draw a line at the angle you have calculated (angle Y in ➤Figure A2-1d). Since Mars lies along this line, it must be located where the two lines cross.

Measure the distance from the Sun to Mars, divide this by the Earth's distance to the Sun and compare your result to that in Table 2-1. Are you close?

StarLinks netQuestions

Visit the netQuestions area of StarLinks (www.jbpub.com/starlinks) to complete exercises on these topics:

1. Orbital Theory Copernicus began the revolution in orbital theory, but it was Kepler's contribution that enabled the model to fit the data.

3. Kepler's Second Law Kepler's second law states that a planet moves along its elliptical path with a speed that changes in such a way that a line from the planet to the Sun sweeps out equal areas in equal intervals of time.

2. Kepler's Three Laws Kepler's three laws modified the Copernican model and paved the way for the achievements of Isaac Newton.

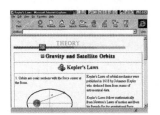

62   CHAPTER 2   A Sun-Centered System

CHAPTER 3

Neil A. Armstrong took this photo of Edwin E. Aldrin, Jr., on the Moon.

GALILEO, NEWTON, AND EINSTEIN

Galileo Galilei and the Telescope
The Moon, the Sun, and the Stars
Satellites of Jupiter
The Phases of Venus
HISTORICAL NOTE: Galileo Galilei
Isaac Newton's Grand Synthesis
Newton's First Two Laws of Motion
An Important Digression—Mass and Weight
Back to Newton's Second Law
HISTORICAL NOTE: Isaac Newton
Newton's Third Law
Motion in a Circle
The Law of Universal Gravitation
Testing the Law of Universal Gravitation
CLOSE UP: Travel to the Moon

Newton's Laws and Kepler's Laws
The Center of Mass
The Tides
Rotation and Revolution of the Moon
Precession of the Earth
The Importance of Newton's Laws
Beyond Newton: How Science Progresses
General Relativity
Space Warp
CLOSE UP: The Special Theory of Relativity
Gravitation and Einstein
The Orbit of Mercury
HISTORICAL NOTE: Albert Einstein
The Correspondence Principle
ACTIVITY: Circular Motion

•••••••• ON JULY 20, 1969, the first humans landed on another astronomical object. From the Sea of Tranquility they reported to mission headquarters in Houston, "Houston, Tranquility Base here. The Eagle has landed." Shortly thereafter, Neil A. Armstrong stepped onto the surface of the Moon saying, "That's one small step for a man, one giant leap for mankind." (Historians might note that although the words quoted here are what Armstrong meant to say, he actually left out the word "a.")

Armstrong and Edwin (Buzz) Aldrin, Jr., stayed on the Moon for 21.6 hours. When they left the lunar module, each wore a 90-kilogram spacesuit. On Earth this weighed 200 pounds, but on the Moon it weighed a mere 33 pounds. During their 2½ hours of walking on the Moon outside the lunar module, they deployed a number of instruments to measure various features of the Moon and unveiled a plaque that reads "Here Men From Planet Earth First Set Foot Upon the Moon. July 1969 A.D. We Came in Peace For All Mankind." As they left the Moon to rejoin Michael Collins in the command module and return to Earth, the plaque remained behind to express for ages to come a lofty ambition of the human race.

➤ **FIGURE 3-1.** Galileo Galilei (1564–1642).

Discoveries and Opinions of Galileo, by Stillman Drake, is a recommended source of information about Galileo.

Although Kepler's laws succeeded in adding accuracy to the heliocentric system, it was the work of Galileo and Isaac Newton that led to the final triumph of that system. This chapter will look first at how Galileo's use of the telescope established the validity of the heliocentric system and then at how Newton's new outlook on mechanical motion and his law of gravitation successfully explained the motions of celestial objects so that only a few unsolved problems remained. Those remaining problems were not solved until this century, when Einstein proposed an entirely new theory, the theory of general relativity.

GALILEO GALILEI AND THE TELESCOPE

Galileo Galilei (➤Figure 3-1) was born in Italy in 1564. He lived at the same time as Kepler, and the two men wrote to each other and exchanged ideas. Galileo's contribution to our understanding of the solar system consisted both of observations and theory. His observations were of particular value because he was the first person to use a telescope to study the sky. Galileo built his first telescope in 1609 shortly after hearing about telescopes being constructed in the Netherlands.

A number of Galileo's observations with his telescope relate to choosing between the geocentric and heliocentric theories:

    Mountains and valleys on the Moon (➤Figure 3-2)
    Sunspots—dark areas on the Sun that move across its surface
    More stars than can be observed with the naked eye
    Four moons of Jupiter
    The complete cycle of phases of the planet Venus (similar to the Moon's phases)

Each of these five observations will be discussed in turn to see how they affect the choice of models.

The Moon, the Sun, and the Stars

The first three observations in the list do not allow us to make a definite choice between the two theories, in that they provide no data that completely rule out one theory. All three, however, cast doubt on basic assumptions of the geocentric theory. The Ptolemaic idea of the perfection of the heavens was at the heart of the geocentric system, yet Galileo's telescope revealed Earth-like features on the Moon. The mountains there do not appear basically different from mountains on Earth. In the case of the Sun, the existence of dark spots did not fit at all with the idea of perfection in the celestial realm.

➤Figure 11-3 is a photograph of the Sun with visible sunspots.

➤ **FIGURE 3-2.** Even a small telescope reveals features such as mountains and craters on the Moon. This photo was taken in suburban Chicago with a 10-inch telescope.

Galileo's telescope also revealed that the sky contains many more stars than had previously been imagined. Recall that Thomas Aquinas had incorporated the Ptolemaic model into Christian theology. Part of this idea was the centrality of humans, not only in position, but in importance. Passages in the first book of the Bible can be interpreted to indicate that the stars were put in the sky for the exclusive purpose "to shed light on the Earth" (Gen. 1:17). Why, then, do stars exist that are so dim that they cannot even be seen by the unaided eye? What is their purpose? The existence of these stars seemed to undermine the Ptolemaic model, and with it a literal interpretation of the Bible. For this reason, many people in Galileo's time simply refused to look through a telescope at the stars.

Satellites of Jupiter

In January 1610, Galileo turned his attention to Jupiter. Near the planet, he saw three very faint stars. They were all in line, two east of the planet and one west. As he continued to observe them night after night, he noted that there were four stars, rather than three, and it became clear to him that they were not stars at all, but natural satellites—moons—of Jupiter. He concluded that they were objects that revolve around the planet just as the planets themselves revolve around the Sun in the heliocentric model. (Today we know these four satellites as the ***Galilean moons*** of Jupiter.) The fact that they never appear north or south of the planet indicated to Galileo that their orbital plane is aligned with the Earth. ➤Figure 3-3 shows that someone viewing Jupiter's satellites from Earth would see them moving back and forth from one side of the planet to the other.

The Ptolemaic model held that the Earth is the center of everything and that everything revolves around the Earth. Galileo's observation of these satellites indicated otherwise. In the heliocentric system, everything revolves around the Sun except for our Moon, which circles the Earth. It further holds that the Earth is just one of the planets. We see, therefore, that the heliocentric model is much more comfortable with this Galilean observation of a "solar system" in miniature.

In addition, the fact that Jupiter is able to move through space without leaving its satellites behind conflicts with the Aristotelian/Ptolemaic view, which held that if the Earth moved through space it would leave the Moon behind.

Galilean moons. The four natural satellites of Jupiter that were discovered by Galileo.

Notice that "moon" is not capitalized here. I am talking in general about satellites of planets. Earth's satellite is named "Moon," so I capitalize its name. Some books don't.

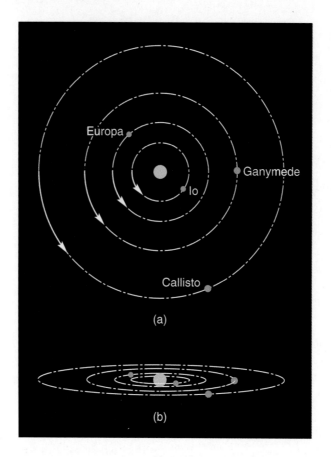

FIGURE 3-3. (a) The orbits of Jupiter's Galilean satellites as seen face on. (b) The same satellites and orbits seen almost edge on. From Earth, this is how this system appears.

The Phases of Venus

The observation that Galileo found most convincing in choosing between the contending models was his observation of a complete set of phases for the planet Venus. ➤Figure 3-4 shows four photographs of Venus at different times. Notice how it exhibits phases similar to the Moon. To the naked eye, Venus appears as a dot of light, but Galileo's telescope revealed its phases. As we will see, this observation was very important evidence for the heliocentric theory.

First, some definitions are needed for terms related to phases, either of a planet or of the Moon. When the entire disk of an object is seen lit, the object is said to be in *full* phase. Venus appears full in Figure 3-4a. (A full Moon is similar; we see the entire Moon lit.) When more than half but less than the

full (phase). The phase of a celestial object when the entire sunlit hemisphere is visible.

➤ FIGURE 3-4. When viewed through a telescope, Venus exhibits a full set of phases, including (a) full, (b) gibbous, and (c and d) crescent. Notice how the planet's apparent size changes as it goes through its phases.

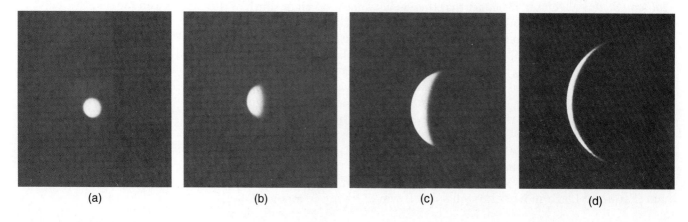

entire disk of an object is seen lit, it is said to be in a *gibbous* phase. Figure 3-4b shows a gibbous Venus. An object is said to be in a *crescent* phase when less than half of its lit disk is visible. Figures 3-4c and d both show Venus in crescent phases.

Recall that repeated observation of Venus leads us to conclude that Venus never gets far from the Sun in the sky. It always appears either in the east shortly before sunrise or in the west shortly after sunset, and it is never seen high overhead at night. The Ptolemaic model explained this by saying that the center of its epicycle always remains on a line that joins the Earth and the Sun (see ➤Figure 3-5a).

What phases of Venus would be seen according to the Ptolemaic model? Notice in Figure 3-5a that the sunlit side of Venus never faces the Earth. At times we should be able to see part of its illuminated surface, but never much of it. Venus should never get beyond a crescent phase. Galileo, however, saw Venus in a gibbous phase, an observation unexplainable by the Ptolemaic model. Let's look at the heliocentric model and see if it can account for the observation.

➤Figure 3-6 is a diagram of the orbit of Venus according to the heliocentric model. At point *X*, Venus would be seen in a crescent phase by an Earth-bound

gibbous (phase). The phase of a celestial object when between half and all of its sunlit hemisphere is visible.

crescent (phase). The phase of a celestial object when less than half of its sunlit hemisphere is visible.

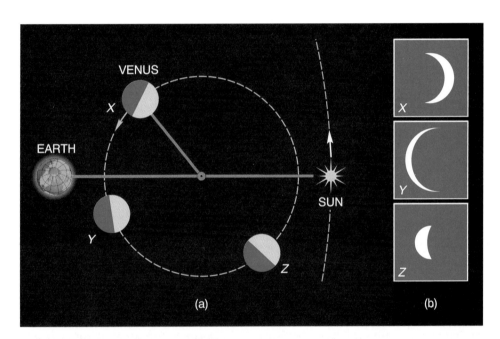

➤ **FIGURE 3-5.** (a) Venus's motion according to Ptolemy. (b) This is how Venus would appear from Earth when it is at each of the three points shown in part (a).

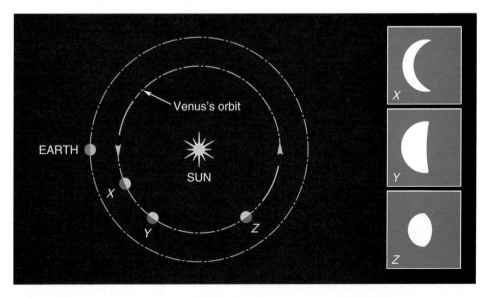

➤ **FIGURE 3-6.** Venus at various places in its orbit, according to the heliocentric model.

Galileo Galilei and the Telescope

HISTORICAL NOTE

Galileo Galilei

Galileo was born in Pisa, Italy, on February 15, 1564, the same year that Shakespeare was born (and Michelangelo died). He was the eldest of seven children born to Vincenzio Galilei, who was an accomplished musician and the author of works on musical theory. The family had been very influential and wealthy during the previous century, but Vincenzio was not a man of wealth. At age 12 Galileo went away to school, studying the usual course of Greek, Latin, and logic. When he was 17, he entered medical school but lost interest in it and took up mathematics. He was forced to leave school after four years and before graduating, however, because of a lack of money.

His creativity began to show when he was in medical school. While there he invented a pendulum device for measuring pulse rates. After leaving school, he continued to study mathematics, applying it to the physics of motion and of liquids. In 1589, at the age of 25, he was appointed professor of mathematics at Pisa and spent the next two decades as a college professor.

In his early years, Galileo had little interest in astronomy, but medical students whom he taught were required to learn some astronomy for use in medical astrology. He therefore became quite familiar with the Ptolemaic model. In 1597, however, he obtained a book published by Kepler concerning the Copernican theory. Galileo apparently preferred the Copernican theory from the time he learned of it but kept these ideas secret (except from Kepler and a few others) to avoid controversy. Not until his publication of *Letters on Sunspots* in 1613, well after many of his telescopic discoveries had been made, did he openly espouse Copernicus's ideas.

Galileo's announcement of support for the heliocentric system started an uproar that would cause him, and society in general, much grief. One factor contributing to the opposition engendered by the letters was that Galileo wrote them in Italian, his native language, while most scholarly writings of the time were written in the "international" language of Latin. Latin was the accepted language for scholarly papers because it was understood by well-educated people in all countries. The ordinary literate person, however, was less likely to be familiar enough with Latin to read books written in that language. Galileo chose to write in the language of the people of his country because he was convinced that people other than scholars could understand his arguments and his evidence that the Copernican system was correct.

Primarily because of this controversy, in 1616 the Roman Catholic church declared that the Copernican doctrine was "false and absurd" and issued a proclamation prohibiting Galileo from holding or defending it. Why did the church make such a strong statement? We must recall the times and the fact that the Ptolemaic theory and Aristotelian physics were a definite part of both religious doctrine and the general culture.

---

quarter (phase). The phase of a celestial object when half of its sunlit hemisphere is visible.

viewer. As it moves farther around and reaches point Y, it should appear in *quarter* phase. Then when it reaches point Z, it should appear in a gibbous shape, similar to the gibbous Moon. ➤Figure 3-7 illustrates the view of these phases from Earth. You can see that all phases are possible.

Thus the heliocentric system of Copernicus (and Kepler) is able to explain Galileo's observation of gibbous phases of Venus. Here, finally, we have an observation that gives us data for choosing one model over the other. A model must be able to explain the observations. Ptolemy's model cannot explain the gibbous phases that can be observed for Venus, and thus it must be rejected. The heliocentric model explains all the phases and therefore becomes the model of choice. One problem remains: there seems to be no logic supporting the laws of Kepler except that they work. We continue our quest for understanding the rules that govern the solar system by turning to the work of Isaac Newton, the man who "put it all together."

➤ **FIGURE 3-7.** You are an observer hovering in space just beyond the Earth. You see Venus at various locations in its orbit and can clearly observe its phases. Since you are north of the Earth, you see Earth in a thin crescent phase. This drawing corresponds to Copernicus's model and shows that the model predicts that all phases are possible and correspond to what is actually observed. Positions labeled *X, Y,* and *Z* correspond to positions in the previous figure. *(See Figure 3.7 opposite.)*

The Catholic church was not alone in voicing opposition to the Copernican theory. Martin Luther, who was a contemporary of Copernicus, had called Copernicus "the fool who would overturn the whole science of astronomy." John Calvin asked, "Who will venture to place the authority of Copernicus above that of the Holy Spirit?" To appreciate why religious leaders were so concerned about this issue, we must realize that they considered the salvation of the individual of paramount importance—more important than answering the question of what is the best world theory. They feared that the idea of a nongeocentric world might seem to undermine the supremacy of humans in God's plan, thereby confusing people and threatening their chance of salvation.

▷ **FIGURE H3-1.** This painting shows Galileo demonstrating his telescope in Venice.

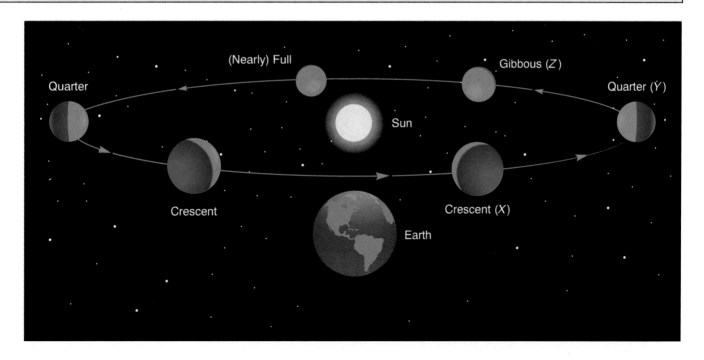

► FIGURE 3-8. Sir Isaac Newton (1642–1727) worked in many fields. In this painting he is shown experimenting with a prism to investigate the nature of light.

inertia. The property of an object whereby it tends to maintain whatever velocity it has.

The "net" force is the total force, taking into account all forces and their directions. Equal forces in opposite directions cancel one another.

accelerate. To change the speed or direction of motion of an object.

acceleration. A measure of how rapidly the speed or direction of motion of an object is changing.

ISAAC NEWTON'S GRAND SYNTHESIS

Galileo was a leader in breaking the bonds of Aristotelian and Ptolemaic thought. He and Kepler gave a new direction to the methods of stating natural laws, for their laws were stated in a manner that allowed measurement and testing by the methods of mathematics. In addition, Galileo referred constantly to experiments that would test his hypotheses. This was a new procedure in the study of nature. Prior to the times of Galileo, the primary method of discussing nature was to refer to the authorities, especially to Aristotle. Such a reliance on observation and experimentation rather than on authority is a cornerstone of science today. Its beginning is usually credited to Galileo.

The same year that Galileo died, Isaac Newton (►Figure 3-8) was born. Newton was the genius who took Galileo's findings and tied them together into one expansive theory of motion. Galileo had predicted this, writing that he had "opened up to this vast and most excellent science, of which my work is merely the beginning, ways and means by which other minds more acute than mine will explore its remote corners."

NEWTON'S FIRST TWO LAWS OF MOTION

Today we summarize Newton's conclusions concerning force and motion in three laws, known as "Newton's laws of motion." In science, the meaning of the word *force* is limited to mechanical pushes and pulls. (The "force of one's personality" does not qualify to be called a force.) The first of Newton's laws was built directly on conclusions Galileo had reached. Newton stated that an object has a tendency to continue in whatever motion it has. He said that a moving object stops only because something causes it to stop.

We have a name for the natural tendency of objects to keep moving: *inertia* is the tendency of an object to maintain its velocity. With this term in mind, we will state Newton's law formally:

NEWTON'S FIRST LAW (THE LAW OF INERTIA): Unless an object is acted upon by a net, outside force, the object will maintain a constant speed in a straight line.

We must point out here that the speed referred to includes the speed of zero. Thus if an object is at rest (at zero speed), it will continue to stay at rest unless a force causes it to move.

Newton's first law indicates that a force is needed to change the speed and/or the direction of an object's motion—to *accelerate* it. Newton's second law goes beyond this, telling us more about what a force is and how we can measure a force. It tells us how much force is necessary to produce a certain *acceleration* of an object.

Consider the brick shown in ►Figure 3-9 and imagine that the wheels allow the brick to move without friction. The brick is at rest in part (a). It has inertia, however, so it will stay at rest even though there is no friction. In (b), you give the brick a push. While you are pushing, the brick accelerates. What determines how great its acceleration is? The force you exert. If you exert a greater force, the brick's acceleration is greater (c). A tiny force will produce little acceleration. This idea gives us the first part of Newton's second law: The acceleration of an object is proportional to the force on it. "Proportional to" actually goes beyond what we deduced from our thought experiment. It tells us that twice the force will cause twice the acceleration.

Refer to ►Figure 3-10. Here we show one hand pushing on a frictionless brick as before and another hand pushing on a stack of two bricks. If the hands push with equal forces, which will cause the most acceleration? It should be

CHAPTER 3   Galileo, Newton, and Einstein

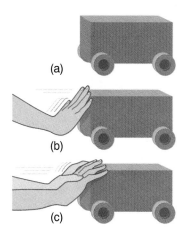

▶ FIGURE 3-9. (a) The wheeled brick will accelerate (b) if a force is exerted on it. (c) If twice as much force is exerted on it, it will accelerate at twice the rate.

▶ FIGURE 3-10. The same amount of force will give twice as much mass only half the acceleration.

intuitive that the greater acceleration will be produced by the hand pushing on the single brick. It is important to see that friction has nothing to do with this. It is tougher to accelerate two bricks than one simply because the two have more inertia. In fact, two bricks have twice as much inertia as one brick. This is an idea we didn't discuss above: some objects have more inertia than others. So before continuing with the second law, we must introduce the term *mass*.

An Important Digression—Mass and Weight

Mass is another of those terms, like speed, that is used in everyday language but is given a more specific definition in science. We have already laid the foundation for the scientific definition because an object's **mass** is simply the measure of the amount of inertia that object has. Instead of saying that one object has twice the inertia of another, we normally say that it has twice the mass.

It is important to say what mass is not. Mass is not volume. Suppose that the brick in the last figure is not a true brick, but is instead a piece of styrofoam in disguise. If the person does not know this, however, he will have quite a surprise when he pushes on the brick. He will find that it has much more acceleration than he expected. Why? Because a styrofoam brick has much less mass than a real brick even though they both have the same size—the same volume.

Mass also is *not* weight! In our examples of pushing the frictionless bricks, the weight of the brick was not a factor; weight is simply the force of gravity pulling downward on the bricks. Weight did not oppose the pushing hand. This is a subtle distinction, but a very important one.

The worldwide unit of mass measurement is the kilogram. At the International Bureau of Weights and Measures near Paris is a platinum cylinder that has, by definition, a mass of *exactly* one kilogram. To give you an idea of what one kilogram of mass is, a kilogram weighs about 2.2 pounds at the surface of the Earth. It is not correct to say that a kilogram is about *equal to* 2.2 pounds, because a pound is a unit that expresses weight and a kilogram is a unit that expresses mass. A kilogram weighs about 2.2 pounds on the surface of the Earth, but at some other location it might weigh a different amount.

mass. The quantity of inertia possessed by an object.

Rotational inertia—for a spinning object—involves more than just mass and will be discussed in a later chapter.

A kilogram on the Moon weighs only about one-third of a pound.

Back to Newton's Second Law

Figure 3-10 showed one hand pushing on one brick and another hand pushing on two. It was stated that we would expect less acceleration from the two bricks.

HISTORICAL NOTE

Isaac Newton

To give an idea of the importance of Isaac Newton (1642–1727) to the history of thought, we quote the English poet, Alexander Pope, who lived at the same time as Newton. He wrote:

Nature and nature's laws lay hid in night.
God said, Let Newton be! and all was light.

Newton was born prematurely on Christmas Day, 1642, in a small village in England. He was so small at birth that his mother said that he would have fit into a quart pot. His father, a fairly prosperous farmer, had died before Isaac was born, and his mother decided that he was too frail to become a farmer. He was not a particularly good student, however, "wasting" much of his time tinkering with mechanical things, including model windmills, sundials, and kites that carried lanterns and scared the local people at night. When her second husband died, Isaac's mother called him home from school to help on the farm, where he spent most of his teenage years.

At the age of 19 he was admitted to Cambridge University. There he studied mathematics and natural philosophy (known as science today). In 1665, after Newton had received his degree and was serving as a junior faculty member, England was swept by the bubonic plague. This incurable disease killed more than 10% of the population of London in only three months, and those who could afford to do so escaped it by moving away from population centers. Newton returned to his mother's home at Woolsthorpe. There he worked feverishly on science for the next two years. These must have been two of the most productive years in the history of science, for during this time Newton made discoveries in light and optics, in force and motion, in gravitation and planetary motion. He also devised a theory of color. To solve a problem in gravitation, he invented calculus. During this time he outlined what would become his major book, *Philosophiae Naturalis Principia Mathematica*, usually called *The Principia*. In later years, Newton wrote, "All this was in the two plague years of 1665 and 1666, for in those days I was in the prime of my age for invention, and minded Mathematics and Philosophy more than at any time since."

Newton's manner of attacking a problem was simply to concentrate his mind on it with such intensity that he finally solved it. His ability to concentrate must have been tremendous. As a result, he often appeared to be what we might call absentminded. It was not uncommon for him to work all day, forgetting to eat. The story is told that once when riding his horse, he got off the horse to unlatch a gate, led the horse through the gate, relatched the gate, and then led his horse home, forgetting to get back on. He was concentrating on some problem, and walking probably gave him more time to think than riding did, anyway.

As a young man, Newton was not interested in publishing his work. He seemed to want to discover the mysteries of nature simply for his own curiosity, and his friends often had to persuade him to publish his findings. Besides, he did not like having to defend his views against criticism, and he wanted to avoid getting involved in arguments over who was the first to make certain discoveries. Nevertheless he *did* get involved quite fiercely in such disputes, including one with Gottfried Leibniz over which of them first developed the calculus. This dispute lasted long after both had died.

It is interesting to note that Newton had a very practical side as well as being a theoretician. He served as Warden of the Mint and while in this office, he began the practice of making coins with small notches around their edges. (Check the quarters in your pocket.) This was done to discourage people from illegally shaving off and retaining the valuable metal of the coins.

In many cases, a person is not recognized as a genius until after his or her death. Such was not the case for Newton. He received honors and was given positions of authority. He was elected president of the Royal Society, an organization of scientists, in 1703 and every year thereafter until his death in 1727. In 1705 he was knighted, the first scientist to receive this honor. When he died, Sir Isaac Newton was buried in Westminster Abbey after a state funeral.

In fact, measuring the accelerations would show that if the forces were equal, the acceleration of the two bricks would be exactly half of that of the single brick, and the same force applied to three bricks would result in one-third the acceleration. Thus, acceleration is inversely proportional to the mass being accelerated. If the force is the same, more mass means less acceleration, in exact proportion.

We have thus far discussed the relationships between force and acceleration and between mass and acceleration. We can sum up these relationships in a mathematical statement:

$$\text{acceleration} = \frac{\text{net force}}{\text{mass}}$$

This tells us that the greater the net force, the greater the acceleration, but the greater the mass, the less the acceleration. It also makes it apparent that if the net force is zero, there is no acceleration, which agrees exactly with Newton's first law. The expression above is usually written:

$$\text{force} = \text{mass} \times \text{acceleration}$$

or in symbols:

$$F = ma$$

NEWTON'S THIRD LAW

Newton's third law is simple to state:

NEWTON'S THIRD LAW: When object X exerts a force on object Y, object Y exerts an equal and opposite force back on X.

This law seems very innocent but its implications are great. It is sometimes stated as: "For every action there is an equal and opposite reaction." In this form it is then applied to a wide variety of fields—politics, religion, sports. Newton meant the law only for physical forces, however, so our statement above makes its meaning much more explicit than the way it is commonly given.

You are probably sitting on a chair. In doing so, you exert a force downward on the chair. The third law tells us that the chair exerts an equal force upward on you. We often call the force you exert the action force and the force of the chair on you the reaction force, but these names are arbitrary. The chair's force could just as well be called the action force and your force on it the reaction force. The point is that one of these forces cannot exist without the other and neither "comes first."

Notice that the statement of Newton's third law indicates that two objects, (called X and Y here), are always involved in the application of forces. *Always.* A force cannot be exerted without an object to exert it and an object on which it is exerted. The word "object" here might refer to an individual atom, or it might refer to a collection of atoms, and the collection might be a gas or a liquid as well as a solid object. For example, if you hold your hand out of the window of a moving car to feel the force of air resistance, it is the air that exerts a force on your hand. The third law tells us that your hand therefore exerts a force on the air. This is not easy to see, but we know that your hand must deflect the air as it comes by and a force is necessary to deflect the air. Calling the air an object may seem odd, but air is simply a collection of objects—atoms.

MOTION IN A CIRCLE

Recall that an object is said to accelerate if either its speed or its direction of motion is changed. Thus far, our examples of acceleration have involved only changes in speed, not direction changes. Now we will look at the latter.

Recall also that Newton's second law says that a net force always produces an acceleration. Consider what happens if you whirl a rock on a string. You are exerting a force on the rock toward the center of its circular motion as you pull inward on the string (►Figure 3-11a). If you are careful, you might be able to

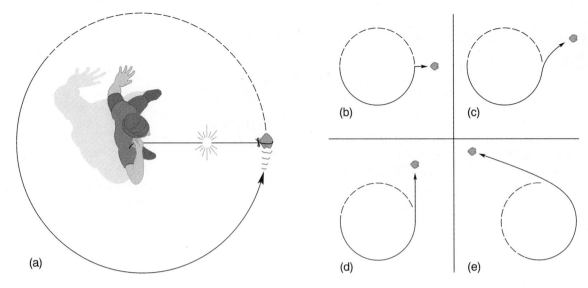

➤ **FIGURE 3-11.** (a) The string breaks as the rock is whirled in a circle. (b–e) Which way does the rock go after the string breaks?

You might try the Activity "Circular Motion" (at the end of the chapter) now.

centripetal force. The force directed toward the center of the curve along which the object is moving.

make the rock go around you with a constant speed. What about the acceleration that, according to Newton's law, must be produced by the force you are exerting? If the rock moves at a constant speed, the force is not causing a change in speed. Instead, it changes the *direction* of the rock's motion.

Motion of an object in a circle at constant speed is an example of acceleration by changing direction. It may seem odd to call this an acceleration, but notice that if acceleration is defined this way, Newton's law holds in all cases. When we define acceleration to include change in direction, the force you exert toward the center of the circle when you whirl a rock on a string does indeed cause an acceleration. The acceleration is in the direction of the force—toward the center of the circle.

Parts (b) through (e) of Figure 3-11 show a number of potential paths for the rock when the string breaks. Which of these is the way you expect the object to travel based on Newton's laws?

Once the string breaks, there is no horizontal force on the rock. If this is the case, there can be no horizontal acceleration of the rock. Thus, according to Newton's first law, it will continue in a straight line (Figure 3-11d). When the rock was moving in the circle, the only horizontal force on it was exerted by the string. (The rock's weight pulls down, but we are considering only horizontal forces and horizontal motions.) The horizontal acceleration produced was therefore due only to the string.

A force is necessary to make something move in a curve. We give this force a name, calling it the *centripetal* (meaning "center-seeking") *force.* This is not really a new kind of force, but rather it is the name we apply to a force if that force is causing something to move in a curve. For example, the centripetal force involved when you were whirling the rock was the force exerted by the string on the rock. The centripetal force on a car rounding a curve on a level roadway is the frictional force of the road acting on the tires in the direction toward the center of curvature. The roadway exerts a force on the car, keeping it in the curve. Do not think of centripetal force as another type of force similar to a push, a pull by a string, or a friction force. It might be any of those three or others; it is simply what we call whatever force is responsible for a motion along a curved path. Now we consider the force that causes the Moon to move in its (almost) circular path.

74 CHAPTER 3 Galileo, Newton, and Einstein

THE LAW OF UNIVERSAL GRAVITATION

The real importance of Newton's three laws of motion becomes evident when they are combined with another of his accomplishments: the law of universal gravitation, or "the law of gravity."

Newton's first law states that an object continues at the same speed in the same direction unless some unbalanced force acts on it. This law is applicable to objects on Earth, but what about objects in the sky? Aristotle believed that earthly laws do not hold in the heavens, but Newton sought to apply his laws of motion there too. It was during his great, productive years of 1665 and 1666 that, in Newton's words, he "began to think of gravity extending to the orb of the Moon." The Moon follows a nearly circular path around the Earth, so if the laws of motion apply to the Moon, a force must be exerted on it toward the center of its circle—a centripetal force.

Newton hypothesized that there is a force of attraction between the Earth and the Moon and that this serves as the centripetal force. What causes this force? Newton didn't know, but he hypothesized that it is the same force that causes an object here on Earth to fall to the ground. He proposed that there is an attractive force between every two objects in the universe and that the magnitude of this force depends on the masses of the two objects and the distances between their centers.

Let's stop a moment and emphasize what this means. Newton was saying that *every* object in the universe attracts every other one. This includes the book you are reading and the pen in your hand. According to Isaac Newton, they attract one another. Naturally, you can't feel the force pulling these two objects toward one another because they have such little mass and the strength of the force depends upon the amount of mass. When one of the attracting objects is the Earth, however, one of the masses is certainly large, and we experience the force. This is the force that causes what we call **weight**.

A more complete statement of what we now call Newton's law of universal gravitation (or the law of gravity) is as follows:

THE LAW OF UNIVERSAL GRAVITATION: Between every two objects there is an attractive force, the magnitude of which is directly proportional to the mass of each object and inversely proportional to the square of the distance between the centers of the objects.

It is this force, according to Newton, that not only makes objects fall to Earth, but keeps the Moon in orbit around the Earth and keeps the planets in orbit around the Sun. Newton proposed that this law, along with his laws of motion, could explain the motions of the planets as well as the falling of objects on Earth. If he could explain the planets' motions, this would clear up the mystery of why Kepler's laws worked and bring the motions of the heavenly planets within the realm of our scientific understanding.

Testing the Law of Universal Gravitation

How could Newton test his hypothesis of the existence of a force of attraction between masses? First, the part about the force being proportional to the masses of the objects is easy to test—in one case, anyway. A 10 kg object has twice the mass of a 5 kg object. Remember, we are speaking here of the objects' *masses*, not their weights. The law of gravity would predict that a 10 kg object has twice the weight of a 5 kg object, and indeed it does. Weight is proportional to mass, so the law seems to work when one of the objects is the Earth.

To test the force's dependence on distance, we must be able to compare forces on objects at different distances. The law states that the force is inversely proportional to the square of the distance between the objects' centers. For ex-

---

"Plato is my friend, Aristotle is my friend, but my better friend is truth."—Isaac Newton as an undergraduate.

More than 100 years after Newton published his law, Henry Cavendish was able to measure the gravitational force between two ordinary objects in a laboratory.

weight. The gravitational force between an object and the planetary body on which the object is located.

In equation form this is

$$F = \frac{Gm_1m_2}{d^2}$$

G here is a constant number that depends upon what units are being used for m_1, m_2, and d. (m_1 and m_2 are the masses of the objects, and d is the distance between their centers.)

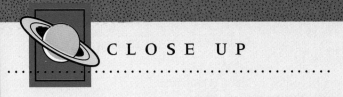

CLOSE UP

Travel to the Moon

> I believe that this nation should commit itself to achieving the goal before this decade is out, of landing a man on the Moon and returning him safely to Earth.
> —PRESIDENT JOHN F. KENNEDY, 1961

The motivation for President Kennedy's commitment to place a man on the Moon was not primarily scientific; it involved national and international politics. Nevertheless, science—particularly astronomy—certainly benefited from that program.

The gravitational aspects of a flight to the Moon are not vastly more complicated than those of Earth orbit. The mission begins with the spacecraft orbiting the Earth. As indicated in ▶ Figure C3-1, a rocket then launches the spacecraft out of Earth orbit to begin its trip to the Moon. As it gets farther from the Earth and nearer the Moon, the gravitational pull of the Moon on the spacecraft becomes significant. Recall that the law of gravity tells us that *every* object in the universe attracts every other. This means that the Moon exerts a gravitational force on a satellite even when it is in Earth orbit (and in fact on us here on the surface of the Earth). This force is very small compared to the gravitational attraction toward the Earth until the craft gets closer to the Moon. Along the path of the Moon-bound craft in Figure C3-1, we have drawn arrows indicating the relative amount of gravitational force exerted by the Earth and Moon.

There is one point where the gravitational pulls toward Earth and Moon exactly cancel one another. Up until this point, the force pulling the craft toward the Earth has been greater than that pulling it toward the Moon, and the spacecraft has been slowing down. After passing that point, the pull toward the Moon is greater, and the spacecraft speeds up. It is interesting to note that the astronauts feel no different when passing through that point. They simply feel weightless all the time. They are in free fall all the time, coasting right along with the craft, and as it changes speed, they change speed right along with it, completely unaware

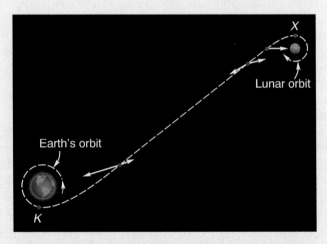

▶ **FIGURE C3-1.** A craft sent to orbit the Moon starts out in Earth orbit. A rocket is fired at point *K* to send it toward the Moon. At point *X*, a rocket is fired to put it in Moon orbit. The arrows along the path show the approximate relative gravitational pull to the Earth and Moon. Distances are vastly out of scale here.

We can now measure differences in weight at different locations on Earth.

ample, if the two objects are the Earth and your book and if you take your book to a location twice as far from the center of the Earth as it is now, the law says that the book will weigh only one-fourth (which is $1/2^2$) as much. At three times as far from the Earth, it will weigh one-ninth as much (see Figure 3-12).

Newton, of course, was unable to change an object's distance from Earth's center significantly. He could have taken an object up on a mountain and measured its weight, but his theory predicted that the weight change in this case would be so small that it would be unmeasurable with the methods he had available. Instead, he used an object already in the sky: the Moon. And rather than comparing forces directly, he compared accelerations produced by the forces.

An object at the surface of the Earth falls toward the ground with a certain acceleration. Newton knew that as an object gets higher above the Earth, this acceleration should change in the same way that the force does; that is, the acceleration should be less at greater distances from the Earth, and it should decrease as the square of the distance. The Moon's path is actually an ellipse, but it is close enough to circular that we can consider it a circle to do a rough test of the law of gravity. Newton knew that the distance from the center of the of any change. Finally, when the spacecraft has reached the appropriate point in its journey (at about X in Figure C3-1), a rocket is fired to slow it down so that it remains in orbit around the Moon. Without this firing, the craft would have so much speed that it would swing right past the Moon.

Now the astronauts are in orbit around the Moon. At this point the lunar module (LM—often pronounced "lem") disconnects from the command/service module that remains in orbit, fires its rockets to slow down, and descends to the Moon's surface. The lunar module does not have to be very large and does not need particularly powerful rockets. This is because the force of gravity on the surface of the Moon is only one-sixth of that on the Earth, so the fall toward the Moon is not as fast and the lift-off requires much less energy. In addition, the Moon does not have an atmosphere, so the frictional effects of an atmosphere need not be considered as they do for takeoff and landing on Earth.

To leave the Moon, part of the lunar module is left behind when a rocket launches the small remaining craft into lunar orbit to reconnect (to "dock") with the command module. Then rockets fire again to send the craft out of lunar orbit and back toward the Earth.

Notice that very little of the trip to and from the Moon is spent with the rockets firing. They are fired only to begin and end each portion of the trip and to make minor midcourse corrections. On one of the missions to the Moon, an astronaut spoke with his son by radio. His son asked who was driving. The reply was that Isaac Newton was doing most of the driving at the time.

The first manned *Apollo* flight took place in October 1968. In December of that year, three astronauts orbited the Moon, but did not land. Their mission provided the first whole-Earth photos ever made (▶Figure C3-2). In July 1969, astronauts Neil A. Armstrong and Edwin (Buzz) Aldrin, Jr., guided their lunar module "Eagle" onto the Moon at the Sea of Tranquillity and uttered the famous words, "Tranquillity Base here. The Eagle has landed." In all, 12 men visited the Moon between July 1969 and December 1972 when *Apollo 17* blasted off from the Moon. We no longer have the capability to journey to the Moon.

▶ **FIGURE C3-2.** Photos such as this, taken on one of the *Apollo* missions, changed our attitude about the Earth.

Earth to the Moon is about 60 times the distance from the center of the Earth to its surface. (Chapter 6 will explain how this was known.) According to his law, then, the centripetal acceleration of the Moon should be $(1/60^2)$, or $1/3600$ of the acceleration of gravity on Earth. But how could he measure the acceleration of the Moon? He did this by using the fact that the Moon moves in (nearly) a circle and therefore has a centripetal acceleration—it is accelerating toward the Earth.

By analyzing the Moon's path, Newton calculated the acceleration of the Moon toward the Earth and found that it is indeed $1/3600$ of the acceleration of an object near the Earth's surface. Thus he checked his hypothesis concerning gravitational force. This was one of the calculations that convinced him that his law was valid and that gravitation is the force that keeps the Moon in its orbit.

NEWTON'S LAWS AND KEPLER'S LAWS

The previous section has shown how Newton was able to use circular motion to apply his laws to the motion of the Moon. In fact, Newton was able to show that

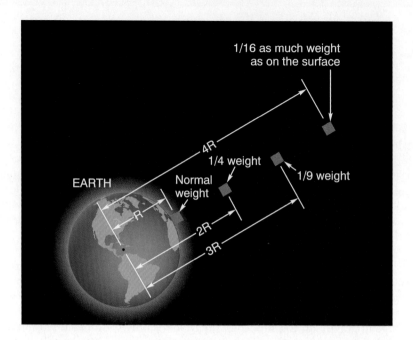

► FIGURE 3-12. When an object gets farther from Earth, its gravitational force toward Earth decreases as the square of the distance to Earth's center.

The derivation of Kepler's law from Newton's requires calculus.

based on his laws of motion and his law of gravity, objects in orbit around the Sun would be compelled to move in elliptical orbits. This, of course, is what Kepler had found to be their paths. The mathematics involved in Newton's calculations will not be shown here, but it is important to emphasize that if Kepler's first law, the law of ellipses, had not been known beforehand, it could have been derived from the laws formulated by Newton.

Kepler's second law states that as a planet orbits the Sun, a line from it to the Sun sweeps out equal areas in equal times. Again, we will not show the mathematics that prove that this law necessarily follows from the laws of Newton, but we will show, without math, that Newton's laws at least make Kepler's second law seem reasonable. ►Figure 3-13 shows the exaggerated elliptical path of a planet. Consider the planet when it is at position A, getting closer to the Sun as it moves along. The arrow pointing from the planet toward the Sun represents the gravitational force exerted on the planet by the Sun. Notice that when the planet is in this position, the direction of the force is not perpendicular to the motion of the planet; it is not simply a centripetal force. If it were a centripetal force it would be exerted perpendicular to the motion of the planet, and it would have only one effect: to cause the planet to curve in its path. Instead, it has two effects: it causes the planet to curve, and it also causes the

► FIGURE 3-13. At point A, the planet is moving closer to the Sun, and the gravitational force on it causes it to speed up as well as curve from a straight line. At point B, the force of gravity slows the planet as well as curving its path.

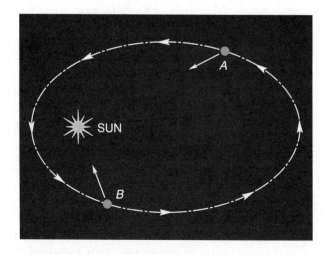

CHAPTER 3 Galileo, Newton, and Einstein

TABLE 3-1 Summary of Types of Motion

Motion	Type of Acceleration	Direction of Force and Acceleration
Linear	Change in speed	Forward (to speed up)
		Backward (to slow down)
Circular	Change in direction	Toward center of circle
Orbital	Change in speed and direction	Toward the object being orbited

planet to speed up. This is because it pulls *forward* as well as sideways on the planet.

At location *B* in the figure, the planet is moving away from the Sun. Here the arrow representing gravitational force shows us that the force pulls both backward and sideways on the planet, thus slowing it down as well as curving its path. So we see that as the planet moves toward the Sun, it speeds up, and as it moves away, it slows down. This is what Kepler's second law predicted.

We have not shown that Newton's laws would actually result in Kepler's law of equal areas, but only that both laws would result in the planet going fastest when it is closest to the Sun. In fact, it can be shown mathematically that application of Newton's laws of motion and his law of gravitation *necessarily* results in the law of equal areas. Here again, if Kepler's law had not already been known by Newton, he could have deduced it from his laws. Table 3-1 summarizes the types of motion as they were analyzed by Isaac Newton.

Finally, we come to Kepler's third law, the one that relates a planet's period of revolution to its distance from the Sun. Newton showed mathematically that if nature obeys an inverse square law for gravitation, planets must necessarily have the period-distance relationship of Kepler's third law. In addition, Newton added something to Kepler's third law, showing that the masses of the objects are important in the relationship, so Newton's work allows us to express Kepler's third law as

$\dfrac{a^3}{P^2} = KM$
  a = average radius of the orbit
  P = period of the orbit
  K = a constant whose value depends upon the units used in the equation
  M = the sum of the masses of the two objects

The mass of any planet is very small compared to the mass of the Sun, so the expression on the right side of the equation is essentially the same for all objects orbiting the Sun.

Science seeks to show that the various phenomena we observe are not independent of one another but are instead manifestations of a relatively few basic principles. We see here the success of Newton's work in relating the laws of Kepler to more basic ideas: those of gravitation and the motion and mass of objects. Once Newton's laws are known, all three of Kepler's necessarily follow. Before the discussion of Newton's laws, we must look at a change these laws brought in our understanding of the paths objects take as they orbit one another. A later section will discuss a particular success of the laws: the explanation of ocean tides.

THE CENTER OF MASS

Newton's third law tells us that when one object exerts a force on another, the second object exerts an equal force back on the first. Thus we should expect that the forces affect both objects. One object cannot just remain still while another

center of mass. The average location of the various masses in a system, weighted according to how far each is from that point.
This is the point that moves according to Newton's first law of motion when the entire system moves.

barycenter. The center of mass of two astronomical objects revolving around one another.

object orbits around it. Instead, the two objects turn about a point between them. This point is called the ***center of mass*** of the objects. (The center of mass is sometimes called the center of gravity; for practical purposes the two terms mean the same thing.) As a child, you probably played on a seesaw (or teeter-totter) with someone much heavier than yourself. Recall that the larger person had to sit closer to the pivot point in order to balance. In fact, both persons must sit so that their center of mass is at the pivot point of the seesaw if they are to balance on it.

➤Figure 3-14 shows a large ball and a small ball connected by a rod and held by a string tied to the rod. The two balls balance because the string is connected at the center of mass of the objects. If this contraption were thrown into the air with a spinning motion—like a baton—it would spin around that point, the center of mass. A planet and the Sun are somewhat like the two balls on the rod, but instead of being held near one another by a rod, they are held by the force of gravity.

Until now we have spoken of a planet orbiting the Sun as if the Sun stayed still and the planet moved around it. In fact, the two objects revolve around their common center of mass, or ***barycenter.*** The Sun is 330,000 times more massive than the Earth, however, so the barycenter of the Sun-Earth system is essentially in the center of the Sun. To examine a case where the center of mass has a more significant effect, we look at the Earth-Moon system.

To determine the location of the barycenter of the Earth-Moon system, we use the fact that the Earth's mass is 81 times that of the Moon. Think of a large person and a small person on a seesaw. Suppose one person weighs 50 pounds and the other 150 pounds. If one person is three times heavier than the other, the lighter person must sit three times farther from the pivot of the seesaw. Similarly, since the Moon's mass is $1/81$ of the Earth's, the Moon is 81 times farther from the center of mass of the Earth-Moon system than the Earth is. This means that the center of mass is about 5000 kilometers from Earth's center and 380,000 kilometers from the Moon. This point is inside the Earth ➤Figure 3-15 illustrates that it is about 1700 kilometers below the surface.

Historically, it was the relationship between the distances of Earth and Moon from the system's center of mass that allowed us to calculate the mass of the Moon. We have said that if we know that the Earth is 81 times more massive then the Moon, we can locate the center of mass. In actual practice, the logic

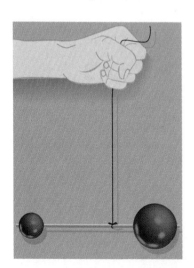

➤ **FIGURE 3-14.** The two balls at the ends of the rod balance at the center of mass of the device, where the string is connected.

➤ **FIGURE 3-15.** The center of mass of the Earth-Moon system is 4800 kilometers from the center of the Earth. If a model of the system were constructed to scale to sit on a weightless board, the construction would balance as shown.

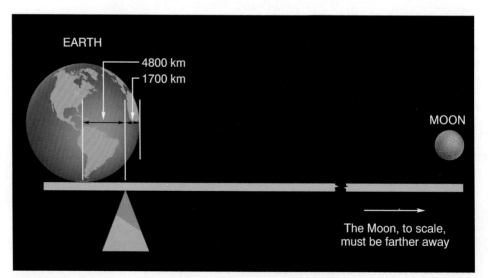

80   CHAPTER 3   Galileo, Newton, and Einstein

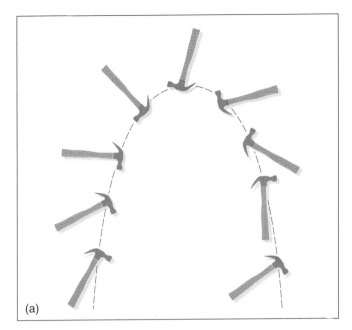 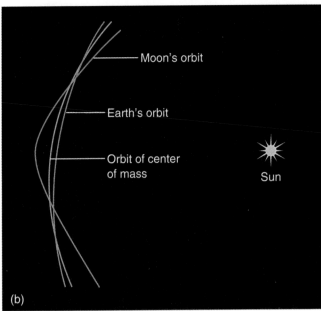

FIGURE 3-16. (a) The center of mass of a thrown hammer follows a smooth path as the hammer rotates around that point. (b) Likewise, it is the center of mass of the Earth-Moon system that orbits the Sun in an elliptical path. (By holding a straightedge up to the paths drawn, you can see the Earth and Moon are always curving toward the Sun during the orbit. This is because the Sun's gravitational force on either object is stronger than the gravitational force from the other object.)

proceeded the other way. The Moon's mass was calculated by determining the location of the center of mass. Today we know the mass of the Moon more accurately by observing its gravitational effect on space probes that have flown by it, but until the space age, the method described above was the most accurate. Chapter 12 will show that the center-of-mass method is what allows us to calculate the masses of stars.

Now consider the orbit of the Earth around the Sun. Kepler's laws would indicate that the Earth moves in an elliptical path. In fact, it is not the Earth, but the barycenter of the Earth-Moon system that follows the elliptical path. ➤Figure 3-16a illustrates the path of a hammer that has been thrown with a spinning motion. The center of mass of the hammer follows a regular path; similarly, the center of mass of a planet and its satellite is the point that follows Kepler's elliptical path (see Figure 3-16b). Kepler's law of ellipses, therefore, must be corrected as to the point that follows the ellipse.

Thus we see that the laws of Kepler, like most laws and theories in science, are approximations. The laws fit Kepler's data perfectly, but as more accurate and different observations became available, Kepler's laws had to be modified and improved. Kepler's laws as originally stated are accurate enough to justify their use in many instances, but they help remind us that scientific theories are always tentative and are subject to revision.

The location of the center of mass of the Earth-Moon system was determined by observing parallax of nearby planets due to the Earth's motion as the Moon went around. When observing Mars, the maximum parallax due to this motion is just 17 arcseconds.

THE TIDES

If you live near the seashore or visit it often, we don't need to describe the tides to you. But "landlubbers" don't often get a chance to experience the tides. ➤Figure 3-17 shows a seashore at low and high tide; the difference in depth of water is obvious.

Most locations on the Earth experience a high tide about every 12 hours and 25 minutes and a low tide midway between high tides. Thus on most days there are two high tides and two low tides. Newton realized that the force of gravity between the Earth and an object is not really a single force exerted by the entire

► FIGURE 3-17. Those of us who don't often get to the coast may not be familiar with the phenomenon of tides. Observe how much deeper the water is at high tide. The shadows are different because about six hours passed between the two photos.

Earth but the result of all the forces of gravitational attraction between the object and each part of the Earth—each little mass within the Earth. This idea applies to the force of gravity between the Earth and the Moon. The Moon exerts gravitational force not on the Earth as a whole but on each individual part of the Earth. We will consider only three little parts of the Earth. ►Figure 3-18 uses arrows to indicate the force of gravity between the Moon and a small mass at three different locations within the Earth. The mass on the side of the Earth nearest the Moon feels the greatest lunar gravitational force since it is closest to the Moon. The mass at the center of the Earth feels less force toward the Moon, and the mass at the far side of the Earth feels the least force.

Now each of these three parts of the Earth responds to the gravitational force toward the Moon, but the part closest to the Moon feels a greater force than the other two. There is more force toward the Moon on one kilogram of mass here than there is on one kilogram of mass at Earth's center, so the mass at the surface feels pulled away from the center. Water covers most of the Earth, and since it is liquid and free to flow, some water flows to the area under the Moon. As the water becomes deeper at that point, it causes a high tide.

The high tide on the other side of the Earth occurs because the center of the Earth feels a greater force toward the Moon than water on that side so the main body of the Earth is pulled away from that water, creating another high tide there.

►FIGURE 3-18. The gravitational force exerted by the Moon on a given amount of mass of Earth is greatest on the side nearest the Moon, less on the mass at the center of the Earth, and least on the mass on the side farthest from the Moon. Arrows here represent the forces.

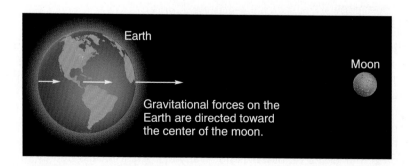

CHAPTER 3 Galileo, Newton, and Einstein

This differential gravitational pull on the various parts of the Earth results in two areas of the Earth experiencing high tides. The water that went to making those high tides has been pulled away from other parts of the Earth, so a low tide lies midway around the Earth between the areas of high tides.

One might think that since the Earth completes one rotation a day, there would be exactly two high tides each day. Instead, about 12 hours and 25 minutes pass between successive high tides, so that if a high tide occurs on my beach at 10:00 this morning, the high tide tomorrow morning will be at about 10:50. The reason for this is that the Moon is not stationary as the Earth rotates. Instead, the Moon moves through about $1/30$ of its cycle each day. Therefore the Earth must turn for the additional 50 minutes before a spot on the Earth returns to the same position with respect to the Moon. This, of course, is what causes the Moon to rise (or set) about 50 minutes later each day. ➤Figure 3-19 illustrates the Moon's motion as the Earth turns for 24 hours and 50 minutes.

As you might suspect, the actual tidal phenomenon is much more complicated than we have just described. The Earth is rotating and its landmasses disturb the flow of water. The shape of the shoreline, the depth of the water, and the location of the Moon all play a part in determining exactly when high and low tides occur at a particular location on Earth and just how high and how low those tides are.

In addition, the Sun causes tides on the Earth. Because the *difference* in the Sun's gravitational pull on opposite sides of the Earth is not very great, these tides are small and are not noticed independent of the Moon's tides. When they correspond to the Moon's tides (near the times of new and full Moon), however, we see extreme tides on Earth; these are known as **spring tides**. On the other hand, when the Sun's tides are 90° from the Moon's (near quarter Moon), they tend to cancel the Moon's, and this causes the change from low to high tide to be less than normal; such tides are called **neap tides**.

Rotation and Revolution of the Moon

The period of the Moon's rotation exactly matches its period of revolution. As a result, the Moon keeps the same face toward Earth at all times. The effect is caused by tidal forces. To understand it, recall that as the Earth rotates, it interacts with tides in the water. If there were no friction between the solid Earth and its oceans, one area of high tide would stay exactly under the Moon, and another would be on the opposite side of the Earth. The landmasses, however, exert forces on the water as the Earth rotates, causing the point of highest tide to be

A "differential" pull means that it is different at different points on the Earth.

spring tide. The greatest difference between high and low tide, occurring about twice a month when the lunar and solar tides correspond.

neap tide. The least difference between high and low tide, occurring when the solar tide partly cancels the lunar tide.

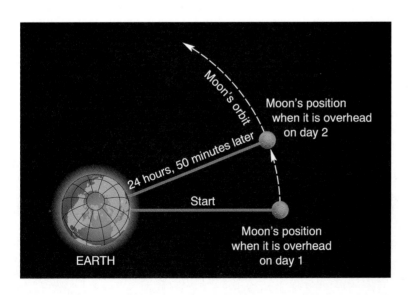

➤ FIGURE 3-19. The Earth is seen here from above the North Pole. Since the Moon moves in its orbit, the Earth must make more than one complete rotation for a point on its surface to return to the same position with respect to the Moon. About 24 hours and 50 minutes pass between the two positions shown.

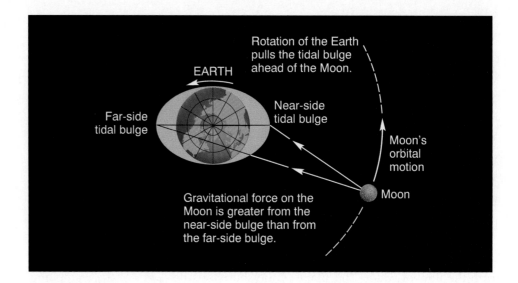

► FIGURE 3-20. The Earth's rotation tends to drag the tides along with it, so that a high tide is not directly under the Moon but instead is farther to the east.

Notice the next time that you are at the seashore that high tide occurs after the Moon is highest in the sky.

tidal friction. Friction forces that result from tides on a rotating object.

pushed from directly under the Moon (and directly opposite the Moon). ►Figure 3-20 illustrates this.

If the landmasses exert a force on the tidal bulges as the Earth turns, it follows from Newton's third law that the bulges exert an equal and opposite force on the Earth. This results in the Earth's rotation being slowed down because of the **tidal friction** exerted on it. The Earth is indeed turning more slowly today than it was years ago. Our days are very slightly longer than Shakespeare's were.

Tides also occur on the dry land. In this case, the rocks and dirt actually stretch to allow the surface to rise and fall. The maximum dry-land tide is about 9 inches; that is, the surface is about 9 inches farther from Earth's center at high tide than at low tide.

Just as the Moon causes tides on Earth, the Earth causes Moon tides. These are similar to the tides on the solid Earth. And just as tides on Earth are causing it to change its speed of rotation, the tides on the Moon have resulted in a change in its speed of rotation. At one time in the past, the Moon must have had a rotation period different from its revolution period. Through millions of years, the tides have slowed the Moon's rotation until it now keeps its same face toward the Earth.

Precession of the Earth

The Earth spins on its axis. Think of what happens when you spin a child's top on a smooth table. The axis of the top does not stay in the same orientation (unless you were able to begin the rotation around a perfectly vertical axis). Instead, the top wobbles around. What causes the wobble? The mathematics to describe this effect is far from simple, but it boils down to the fact that the top has a tendency to fall over, and its rotation prevents this from happening. Instead, the top wobbles around, keeping the same angle with the table's surface until friction slows it down. Anytime a spinning object feels a force trying to change the orientation of its axis, it will wobble. The wobble is called **precession.**

precession. The conical shifting of the axis of a rotating object.

It might seem that the Earth would not precess because there is nothing below it trying to pull it over. There is, however, a force on the Earth tending to change the orientation of its axis of rotation. Refer to ►Figure 3-21. The Earth drawn there is not spherical, but is "out of round," as is the real Earth, although the real Earth is much more spherical than the one in the figure. The Earth's diameter is about 26 miles greater across the equator than from pole to pole. This is caused simply by the fact that it is spinning. (Imagine what would happen if you made a ball of Jell-O® and got it spinning. The Earth is not Jell-O®, but it

CHAPTER 3 Galileo, Newton, and Einstein

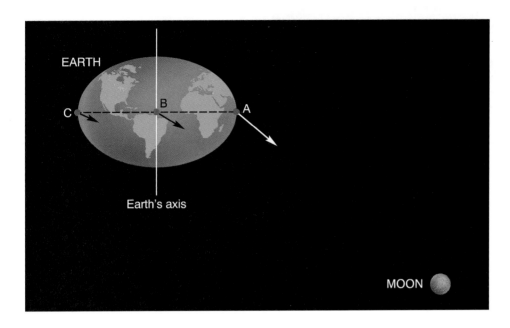

FIGURE 3-21. The Earth is not a perfect sphere, and as the Moon exerts a greater gravitational force on its nearer side, this tends to twist the Earth into a different orientation. This causes the Earth to precess.

is still flexible enough to show the effect.) The measure of how much a planet is "out of round" *(oblate)* is called its **oblateness.**

Finally, recall that the Moon exerts a gravitational force on each particle of the Earth. The arrows in Figure 3-21 illustrate this force at three places. Since point *A* is closer to the Moon than point *C*, the Moon exerts a greater force on a particle at *A*. This results in an overall force seeking to change the tilt of the spinning Earth's axis. Therefore we have what is needed to cause precession. The Earth's axis does indeed precess, although very slowly. While the child's top may complete a precession in about a second, the Earth requires about 26,000 years (see ➤Figure 3-22)!

What effect does this have on what we see in the sky? Right now, Polaris is the closest bright star to the Earth's north celestial pole. But it will not remain so forever. The pole will gradually change, and about 12,000 years from now the star Vega will be our "North Star."

A corresponding effect is that the position of the vernal equinox changes as the centuries pass. We discussed how this affects the practice of astrology in a Close Up in Chapter 1.

THE IMPORTANCE OF NEWTON'S LAWS

We have seen some of the value of Newton's laws. From them one can derive Kepler's laws. They allow us to explain tides and precession. In Chapter 9 a Historical Note tells the story of how they were used to predict the existence of the planet Neptune. From these examples you might conclude (correctly) that Newton's laws differ in their very nature from Kepler's laws. The latter explain a particular situation—the orbiting of the planet's around the Sun. Newton's laws, on the other hand, apply everywhere and to all objects. They are much more fundamental than Kepler's laws. Newton's laws not only explain motion, but they do so in a measurable, mathematical manner. This is important in science. We wish to measure things and to predict events quantitatively. Newton's laws do this.

Can we say that Newton's laws are *right*? Scientists usually avoid calling a theory right or wrong. But we can say that Newton's laws work and are therefore good laws. They fit the data, they make predictions that can be checked, and they fit with other laws to make an overall theory that is a simple, unified pack-

oblateness. A measure of the "flatness" of a planet, calculated by dividing the difference between the largest and smallest diameter by the largest diameter.

$$\text{oblateness} = \frac{(d_{\text{large}} - d_{\text{small}})}{d_{\text{large}}}$$

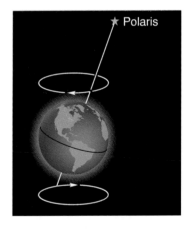

FIGURE 3-22. As the Earth spins, its axis precesses with a period of about 26,000 years, pointing to different "pole stars" over the centuries.

age (that we call "Newtonian mechanics" or "classical mechanics" to distinguish it from the mechanics of Einsteinian relativity).

Up until the time of Newton, Aristotle's idea of the separateness of the Earth and the heavens was the accepted world view. Isaac Newton was not the only one who thought differently, for from the time of Galileo the idea of the oneness of nature had begun to take shape. In fact, other thinkers of Newton's time were working on theories of gravitation. It was Newton, however, who put it all together. It is difficult to overemphasize the effect that Newton's work has had on our thinking.

For one thing, Newton's laws were the first ever that could be shown to hold for both the heavens and the Earth. The idea that there were two natures, one up there and one down here, had been a part of Western culture since before Greek times. Newton showed that the ancient idea was wrong; now it was possible to look upon the universe as ONE. In fact, the word universe begins with the prefix "uni," meaning "one, single" (*uni*t, *uni*fy, *uni*que, *uni*ted, and so on). Such a concept was foreign before the time of Newton. He showed us that we live in one cosmos, that nature is singular, that things up there might be expected to be like things down here. In a sense, Newton changed our world from a "duoverse" to a *uni*verse. Our modern science of astronomy could not exist without this basic understanding.

In addition, the work of Sir Isaac Newton gave us the idea that nature is explainable; that if we work hard enough at it, we have the ability to understand the many seemingly mysterious things that occur in nature. No longer do we have to fear the darkness, for the objects in the skies are part of our universe.

Change comes slowly. We can't pretend that Newton caused societal beliefs to change overnight, but he led us in a giant step forward. Many U.S. newspapers include a daily astrology column, but few newspapers have a daily article on astronomy. Even today, three hundred years after Newton, there are many who continue to look to the stars and planets for signs.

It has been claimed that in the United States more astrologers make a living casting horoscopes than astronomers do practicing science.

BEYOND NEWTON: HOW SCIENCE PROGRESSES

Mass was defined as the measure of the amount of inertia an object has. In stating the law of gravity, however, Newton proposed that mass is also the quantity that determines the strength of gravitational attraction. Why should the same quantity be the measure of two seemingly different physical properties? It is certainly not obvious that *inertia* (the resistance to a change in motion) should have anything to do with *gravitational attractiveness*. Yet the measures of these two properties are not just similar; they are identical at least to one part in 100 billion. Can this be simply a coincidence?

Scientists do not like such coincidences. If two things are so similar, they feel that there must be a reason. The attempt to explain this apparent coincidence is what led Albert Einstein (➤Figure 3-23) to develop his theory of general relativity. General relativity, developed more than two hundred years after Newton found the gravitational relationship, successfully relates the two concepts so that the puzzle of the coincidence no longer exists. In addition, it makes correct predictions in cases where Newton's law of gravity has been found to be in error. Thus, to continue the story, we move suddenly to the twentieth century.

➤ **FIGURE 3-23.** Albert Einstein (1879–1955) had a number of hobbies. One of his favorites was sailing.

GENERAL RELATIVITY

The general theory of relativity fundamentally changes the way we look at the phenomenon of gravity. The theory begins with a statement of the equivalence of gravity and acceleration.

CHAPTER 3   Galileo, Newton, and Einstein

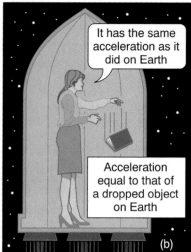

► **FIGURE 3-24.** Whether on Earth or in a spaceship far from Earth accelerating at the acceleration of gravity, the book will "fall" the same.

The woman in ►Figure 3-24a has dropped a book on the surface of the Earth, and in part (b) she drops the same book while in a spaceship that is far from Earth and accelerating in the direction shown. If her spaceship is accelerating at the same rate that falling objects accelerate here on Earth, she will observe the book falling toward her feet exactly the same as it did on Earth's surface. She won't be able to tell the difference between "falling" caused by the ship's acceleration and falling caused by gravity. And if she steps on a scale in such a spaceship, the scale will register the same weight as a scale on the surface of Earth. The *principle of equivalence* of Einstein's general theory of relativity tells us that there is no experiment whatsoever that she can do to distinguish between the two conditions. Being in a room that has an acceleration in the direction we call upward is indistinguishable from the force of gravity.

When we study black holes in Chapter 15, we will see that the general theory of relativity also predicts that light will curve in the presence of a massive object. That is, light will seem to be affected by gravity. Einstein predicted this in 1907, and observations made during a solar eclipse in 1919 confirmed that light passing near the Sun is indeed bent by the presence of the Sun. Since then this prediction has been confirmed in every experiment in which it has been tested.

The spaceship is far enough from Earth (and other large objects) that there is no observable gravitational force.

principle of equivalence. The statement that effects of the force of gravity are indistinguishable from those of acceleration.

Space Warp

Einstein's theory of general relativity explains the similarity of gravity and acceleration, as well as the bending of light near a massive object, by admitting to the curvature of space. Space warp is not easy to imagine because the world we experience is a three-dimensional one and we cannot imagine what dimension our world can curve into. We say that our world has three dimensions because we use three directions to specify the exact location of something. We can state these directions as north-south, east-west, and up-down. For example, I might state that to get from my office to my bed at home, I must go 4325 feet north, 5843 feet west, and 12 feet up to the second floor. Although other choices might be made for the three coordinates that specify the location of an object relative to me, I must always give three pieces of information in our three-dimensional world.

Imagine a land of two dimensions, "Flatland," on which two-dimensional creatures live. The surface of a desk might represent Flatland. Imagine that these creatures know only the two dimensions of their universe; they can perceive

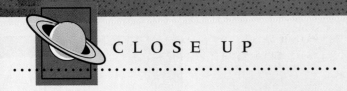

CLOSE UP

The Special Theory of Relativity

About 10 years before he developed the general theory of relativity, Einstein proposed the special theory of relativity. It is called *special* because it does not apply to accelerated motion, but only to the special case of uniform motion. The special theory of relativity is based on two postulates, the first of which states:

- All of the laws of science are the same for all non-accelerating observers, no matter what the speed of those observers.

Recall that people once thought that the laws of nature that govern objects here on Earth are different from those that rule the heavens. Then Newton showed that the same laws work for both, at least the mechanical laws of force and motion. Einstein's first postulate completes the progression; Einstein begins with the assumption that *all* laws, including those of electricity and light, are the same everywhere.

Einstein's second postulate concerns the speed of light:

- The speed of light is the same for all observers no matter what their motion relative to the source of the light.

This predicted behavior of light is different from the behavior of ordinary objects in our experience, for when we catch a baseball, we see it coming at us faster if we are moving toward the thrower than if we are standing still. If Einstein's postulate is true, this doesn't happen for light. ➤Figure C3-3 shows an imaginative case in which people on fast-moving spaceships are measuring the speed of light that comes from an Earthling's flashlight.

On the basis of his two postulates, Einstein reached a number of startling conclusions, including the statement that Newton's laws are not accurate at extremely great speeds. At such speeds, relativity predicts that three changes occur: (1) the observed mass of an object becomes greater than its mass when at rest, (2) the observed length of a fast-moving object becomes less than its length when measured at rest, and (3) the observed passage of time becomes slower for the fast-moving object.

So Einstein made two assumptions (postulates) and used these to make predictions. Fine, but are the predictions correct? Does the theory fit the data? These three predictions on the changes in mass, length, and time become significant only at speeds greater than those attained in everyday life, and they are therefore difficult to check experimentally. Nevertheless, experimental checks are possible in some cases. Every time a test has been conducted to check the special theory of relativity, the theory has passed the test. The theory *works!*

One more conclusion that is based on the postulates of special relativity should be mentioned: mass can change to energy and vice versa, and the conversion between the two is governed by the equation $E = mc^2$. (In this equation, E is energy, m is mass, and c is the speed of light.) The equation was dramatically verified in 1943 by the first explosion of a nuclear bomb, for the energy of these bombs comes from the conversion of mass to energy. A more peaceful example of mass-energy conversion is provided by nuclear power plants, which produce electrical energy based on Einstein's theory. Finally, we will see in Chapters 11 and 12 that nuclear energy is the source of the energy of the stars. The theory of special relativity is not a simple one to understand, but it is firmly established as a theory that astronomers must use if we are to advance in our quest to understand the universe.

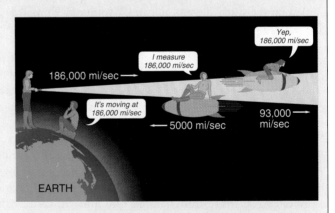

➤ **FIGURE C3-3.** The special theory of relativity predicts that if we measure the speed of light, we will get the same result regardless of the motion of the measurer. That speed is 186,000 miles per second, or 300,000 km/s.

---

north-south and east-west but have no conception of up-down. The location of everything in the universe of Flatland can be specified by saying how far it is along an east-west line and how far along a north-south line.

Imagine some tiny fleas on an expanding balloon. Instead of regular fleas, make the creatures Flatfleas and the balloon Flatland. Flatfleas, of course, have

► FIGURE 3-25. If fleas were truly two dimensional, they would have trouble imagining a curvature of their world.

no height and are only two-dimensional creatures. You might object that the surface of the balloon isn't flat, but if the balloon is large enough compared to the size of the Flatfleas, they would not easily perceive its curvature (►Figure 3-25). If one Flatflea realizes that his universe is curved, how can he explain this idea to his contemporaries? Saying that the universe is curved "downward" would have no meaning, for "up" and "down" are undefined in Flatland. We humans see that the balloon's surface is curved into the third dimension, but it would take a great stretch of the Flatfleas' imagination to think of a third dimension for it is not part of their everyday world.

By a similar analogy, we picture the curvature of *our* space. Suppose that the presence of a massive object causes space to be warped. We can picture space near the Sun as being warped analogous to the way the surface of a waterbed is warped by a bowling ball placed in its center. The waterbed's surface would be distorted so that a straight line following the surface would have to follow the distortion. Similarly, a beam of light passing near the bowling ball would follow the curvature of space caused by the massive object.

GRAVITATION AND EINSTEIN

Einstein proposed that instead of thinking of an attractive gravitational force between objects, we think of space being curved, so that as objects move, they follow that curvature. His theory of general relativity holds that a planet in orbit around the Sun is responding to the curvature of space that results from the Sun's mass. Space curvature thereby explains not only why objects fall and why planets orbit the Sun, but also why light bends near a massive object. (We will see later in the text that it also explains black holes.)

Here again we have two theories that explain a set of observations. How do we choose between them? First, we ask whether one better fits the data. We have seen that Einstein's theory explains the bending of light near a massive object and that Newton's does not. We now consider a final observation, one concerning Mercury's orbit.

The Orbit of Mercury

In 1859, 14 years after predicting the existence of Neptune, Urbain Le Verrier reported that Mercury's elliptical orbit *precesses;* that is, it does not keep the same orientation in space. ►Figure 3-26 illustrates a greatly exaggerated precession, showing the near point in the orbit (the *perihelion*) gradually sliding around the Sun. Mercury's orbit precesses very slowly—at the rate of 574 arcseconds per century—which is less than a degree per century. Calculations show that New-

precession (of an elliptical orbit). The change in orientation of the major axis of the eliptical path of an object.

perihelion. The point where an object in orbit about the Sun is closest to the Sun.

HISTORICAL NOTE

Albert Einstein

One of the greatest theoretical physicists of all time was considered a slow learner as a child. The young Albert Einstein found the formal, disciplinary schools of Germany at the end of the last century intimidating and boring, and he dropped out before completing high school. He studied at home, however, and learned to play the violin well enough that—although he played it only for his own enjoyment and relaxation—he became an accomplished violinist. His studies of geometry and science led him to conclude, at age 12, that the Bible is not literally true. That shock implanted in him a deep distrust of authority of any kind—a distrust that he carried with him throughout life.

On his second try, Einstein was granted admission to the Swiss Federal Institute of Technology in Zurich. At the university he often failed to attend class, preferring instead to study on his own, reading the classical works of theoretical physics. He was granted a Ph.D. in 1900, but it was two years before he found regular work—as a patent examiner in the Swiss Patent Office.

Einstein was similar to Isaac Newton in his lack of success in formal school and in the timing of his scientific work as well. Like Newton, Einstein's most revolutionary work was done during a very few years when he was in his early twenties. His work at the patent office was not particularly demanding, and in 1905, Einstein published four important papers in the prestigious German physics journal *Annalen der Physik*. Although the paper for which he is best known is the one concerning special relativity, he was awarded the Nobel Prize for another, concerning the photoelectric effect of light. Many physicists quickly recognized the importance of his work, but it was not until 1909 that Einstein was given a full-time academic position at the University of Zurich.

In 1903 Einstein married his college sweetheart, Mileva Maritsch, and they had two sons. When World War I began, Einstein had a position in Germany, but his wife and children were vacationing in Switzerland. They were unable to return to Berlin, and the forced separation resulted in a divorce in 1919. Later that year, Einstein married his cousin Elsa.

In 1916, Einstein published his general theory of relativity, but his relativity theories were slow to gain acceptance due to the lack of experimental verification. In 1919, however, the Royal Society of London announced that its scientific expedition to observe the solar eclipse of that year had verified Einstein's prediction of the amount starlight would bend near the Sun. The international acclaim that followed changed Einstein's life, for he was suddenly considered a genius. He began to travel more and more, giving lectures throughout the world.

Although Einstein continued his scientific work until he died in 1955, his fame also allowed him to exert influence in world affairs. Though a Jew and a critic of the political situation in Germany, he escaped the Nazis because he fortunately was visiting California when Hitler assumed power in 1933. Einstein renounced his German citizenship and never returned to his home country. The next year he became an American citizen.

It is ironic that Einstein's name is so closely linked to the atomic bomb. Although his theories predicted that mass could be converted to energy, Einstein was an avowed pacifist who worked untiringly to prevent war, which he saw as the ultimate scourge of humanity. He argued that the establishment of a world government was the only permanent solution.

Einstein disliked fame and the trappings that accompany it. In his travels to the Far East he refused to ride in rickshaws, feeling that to do so would be degrading to the person pulling him along. He preferred to be treated as a common person and disliked formal attire. He regularly gave important lectures in an open-collar shirt and was often seen near his Princeton home in rumpled clothes, carrying his violin.

ton's theory of gravity can account for the basic effect, because gravitational pulls by other planets, particularly Venus and Jupiter, would cause it. The total precession accounted for by these gravitational tugs amounts to 531 arcseconds per century, however, and this is 43 arcseconds short of the observed 574 arcseconds.

The unaccounted-for 43 arcseconds of precession per century presented a mystery for astronomers. One hypothesis held that there is another planet in the inner solar system that is responsible for the extra precession. This planet was given the name Vulcan, and extensive searches for it were carried out but it was never found.

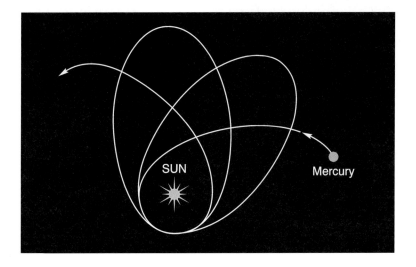

► **FIGURE 3-26.** This exaggerates the actual precession of the perihelion (the closest point to the Sun) of Mercury's orbit.

One of the first problems to which Einstein applied his new general relativity theory was the precession of Mercury's orbit. He found that the theory accounted precisely for the 574 arcseconds of precession. Einstein later wrote that "for a few days, I was beside myself with joyous excitement." Like Newton's theory, general relativity predicts the precession effect caused by the other planets, but unlike Newton's theory, it predicts additional precession due to properties of curved space. Thus the Sun itself caused the extra 43 arcseconds of precession. This was the first direct test of conflicting predictions made by the two theories, and the test showed Einstein's theory to be the better one.

The quotation is from Clifford M. Will, *Was Einstein Right?* (New York: Basic Books, 1986). Recommended.

The Correspondence Principle

There is a general principle in science concerning the replacement of old theories by new ones. The ***correspondence principle*** states that the predictions of the new theory must agree with those of the previous theory where the old one yields correct results. This is entirely reasonable; in the case we are discussing, it simply says that general relativity must not disagree with Newton's gravitational theory where the latter theory provided correct results. For example, Newton's theory predicts that the planets orbit the Sun in elliptical paths but that there are minor deviations from perfect ellipses (caused by gravitational attractions to other planets). We will see that irregularities in Uranus's orbit led to the discovery of Neptune. In fact, the discovery of Neptune after it was predicted by Urbain Le Verrier and John Adams (Historical Note, Chapter 9) provided a dramatic confirmation of Newton's theory of gravitation. Einstein's theory must also predict elliptical orbits, and it must not contradict the confirmed irregularities predicted by Newton's theory.

correspondence principle. The idea that predictions of a new theory must agree with the theory it replaces in cases where the previous theory has been found to be correct.

The general theory of relativity is in accord with the correspondence principle, for it does predict elliptical orbits with the variations that have been observed. The difference is that it views the orbits as being due to planets following their natural paths in a space that is warped by the presence of the Sun.

All tests of Einstein's relativity theory have confirmed it. The theory is a solid one and forms the basis of modern astronomy. Why then do we continue to talk about gravitation and Newton's ideas? The answer is the same as we saw in the case of the geocentric/heliocentric question: we use the more easily understood theory when it fits. The theory of relativity is not an easy one to understand, and since most of us do not need it to explain things in our everyday lives, few of us think in terms of curved space. Thus you'll see little mention of curved space in this text until the discussion of black holes, where Einstein's theory is essential.

Recall that we still speak of "sunrise" and "sunset" from the geocentric model.

CONCLUSION

In the previous two chapters, we introduced two primary models: the geocentric model of Ptolemy and the heliocentric model as presented by Copernicus and revised by Kepler. Galileo made observations with his telescopes that led to the conclusion that Kepler's model fit the data better and, in general, was the better theory. The theory suffered one flaw, however: Kepler's ellipses and his rules concerning the speeds of planets did not "make sense," in that they did not correspond to anything else in nature.

A more complete understanding of force and motion was necessary before Kepler's laws could be related to anything other than planetary motions. Newton's three laws of motion, along with the law of universal gravitation, gave humanity a different outlook on the world and provided scientists with a means of understanding the motions of objects, not only here on Earth but in the heavens as well.

Now we find that Newton's theories are not the final answer; they have been supplanted by Einstein's. It is interesting that during the times of Copernicus and then of Kepler and Galileo, people were confused and scandalized by the theories these men proposed. People's common sense told them that the Earth was the center of the universe, and they were convinced that anything else was nonsense. Nearly a century ago, Einstein proposed a new way of thinking about the universe. His theory of relativity seems to upset the order of the universe, even though scientists tell us that it adds order. It seems completely nonsensical and does not fit most people's experience at all. General relativity is firmly established in science, however, and perhaps someday it will become part of our everyday thinking.

RECALL QUESTIONS

1. Using his newly invented telescope, Galileo discovered all of the following except
   A. moons of Jupiter.
   B. phases of Venus.
   C. sunspots.
   D. stellar parallax.
   E. mountains on the Moon.

2. Which of the following planets can be seen (from Earth) in a crescent phase?
   A. Mercury
   B. Venus
   C. Mars
   D. [Two of the above.]
   E. [All of the above.]

3. Inertia is the tendency of an object to
   A. come to a stop as soon as possible.
   B. come to a stop at a rate that depends upon the nature of the object
   C. continue at the same velocity.
   D. fall to a location as close to the center of Earth as possible.
   E. have weight.

4. Suppose you are riding as a passenger in a car when the car stops suddenly. What pushes you forward?
   A. The force of inertia
   B. Your weight
   C. The car's forward motion
   D. Nothing pushes you forward.

5. If the same net force is applied to two different objects, one with a mass of 1 kg and the other with a mass of 2 kg,
   A. the 1 kg mass will have twice the acceleration of the other.
   B. the 1 kg mass will have more acceleration than the other, but not necessarily twice as much.
   C. both objects will have the same acceleration, for the force determines the acceleration.

6. If the radius of the Earth decreased with no change in its mass, your weight (as you stand on the surface) would
   A. increase. B. not change. C. decrease.

7. According to Newton, the natural motion of an object is
   A. a circle.   C. a straight line.
   B. an ellipse.   D. retrograde motion.

8. If you double the distance between two objects, how will the gravitational force exerted by one on the other be changed?
   A. It would be reduced to half as much.
   B. It would be reduced to one-third as much.
   C. It would be reduced to one-fourth as much.
   D. It would be reduced to one-ninth as much.

9. The gravitational attraction between an object and the Earth
   A. stops just beyond the Earth's atmosphere.
   B. extends to about halfway to the Moon.
   C. extends about five-sixths of the way to the Moon.
   D. extends to infinity.

10. The mass of the Moon is about $1/81$ that of the Earth. If you were on the Moon, though, you would weight $1/6$ of your Earth-weight. Why do you weigh more than $1/81$ of your Earth-weight?
    A. The Moon is made of different materials than Earth.

B. The Moon has a different density than Earth.
C. The Moon has no atmosphere, while Earth does.
D. The Moon is much smaller than Earth.
E. [Both A and B above.]

11. Which statement best describes the relationship between Newton's laws and Kepler's laws?
    A. Newton proved that Kepler was wrong.
    B. Newton's laws and Kepler's laws are now considered equally valuable.
    C. Newton's laws are more fundamental and more powerful than Kepler's laws.
    D. Neither Newton's laws nor Kepler's laws have any modern applications.

12. When you whirl a rock around on the end of a string, centripetal force
    A. pulls outward on the rock.
    B. pulls inward on the rock.
    C. pulls outward on your hand.
    D. [Both A and B above, and the two forces balance.]
    E. [All of the above, for all of the forces are equal.]

13. Kepler's third law states that the ratio of the cube of a planet's semimajor axis to the square of the planet's period of revolution is equal to a constant. Newton found that the value of that constant depends upon
    A. the sizes of the two objects.
    B. the masses of the two objects.
    C. the velocities of the two objects.

14. Newton checked his hypothesis concerning an inverse square law of gravitation by calculating
    A. the Moon's acceleration toward the Earth.
    B. the time required for the Moon to complete one orbit.
    C. the mass of the Earth.
    D. the mass of the Moon.
    E. [Both C and D above.]

15. The force of gravity is responsible for
    A. holding the planets near the Sun.
    B. the tides.
    C. holding the Moon near the Earth.
    D. [Two of the above.]
    E. [All of the above.]

16. If the Moon were covered with water, the number of tidal bulges on it would be
    A. one.
    B. two.
    C. [None.]

17. Tides on the Earth are primarily due to the mutual gravitational attraction between the Earth and
    A. the Moon.
    B. the Sun.
    C. Jupiter.
    D. [None of the above.]

18. How many high tides are observed most days at most seaports on Earth?
    A. One, caused by the Moon.
    B. Two, because the Sun and the Moon each cause one.
    C. One or two, depending upon the relative positions of the Earth, Moon, and Sun.
    D. Two, roughly 12 hours apart.

19. The gravitational force due to the Moon is exerted
    A. only upon the side of the Earth nearest the Moon.
    B. only upon the point on the Earth nearest the Moon.
    C. upon the center of the Earth only.
    D. upon the entire Earth.
    E. upon the water surfaces of the Earth but not upon the land surfaces.

20. The principle of equivalence of the theory of general relativity tells us that
    A. the force of gravity is equivalent to acceleration in the opposite direction.
    B. speeds measured in any location are equivalent.
    C. the speed of light is the same for all observers.
    D. Newton's laws are equivalent to Kepler's laws.
    E. Newton's laws are equivalent to Einstein's theories.

QUESTIONS TO PONDER

1. No celestial model we have discussed makes predictions concerning the smoothness of the Moon or the number of stars in the sky. Why did these two observations by Galileo have a bearing on the choice of models?

2. Figure 3-4 shows that Venus appears to be larger when it is in certain phases. Why does this occur?

3. Newton's laws tell us that no force is needed to keep something moving. Why, then, don't we turn off the engine of an airplane once the plane has reached the desired speed?

4. Use the idea of centripetal force to explain why acceleration was defined to include a change in direction as well as a change in speed. (Hint: Recall Newton's second law.)

5. Explain why the work of Newton had implications beyond science.

6. It is just coincidence that the Moon's periods of revolution and rotation are the same? If not, explain the cause.

7. Does the Earth orbit the Sun in an exact ellipse? Discuss.

8. The Sun also causes a tidal effect on the Earth. The force of gravity of the Sun on the Earth is greater than the force of gravity of the Moon on the Earth. Why, then, are Sun tides lower than Moon tides?

9. Over the years, under what conditions would the very highest tides occur? (An extremely high tide occurred in January 1987).

10. Venus is the greatest contributor to Mercury's orbital precession because the two are neighbors. The planet causing the next most effect is not Earth, however, but Jupiter. Hypothesize as to why this occurs.
11. Why do we teach Newton's law of gravitation even though general relativity is a more up-to-date explanation of the phenomena involved?
12. What causes weight?
13. Give an example of the correspondence principle in the case of the heliocentric theory replacing the geocentric theory.
14. Study Chapter 11 of Lewis Epstein's *Relativity Visualized* (San Francisco: Insight Press, 1983) and summarize his discussion of warped space.
15. The March 1993 issue of *Sky and Telescope* contains two articles concerning the Roman Catholic church's re-analysis of its treatment of Galileo. Use these articles to write a report on the issue.

CALCULATIONS

1. Suppose that you move three times as far from the center of the Earth as you are now. By what factor will your weight change?
2. On Earth's surface I am about 6500 kilometers from its center. If the force of gravity on me here is 150 pounds, how much will it be on me at a distance of 13,000 kilometers from the Earth's center?

ACTIVITY

Circular Motion

This should be done outside, away from anything breakable. Tie some object such as a shoe to the end of a fairly long (6 or 8 feet) string or rope. Now whirl the object in a horizontal circle around you. Feel the pull you must exert to keep the object in the circle. Whirl it faster. Do you have to increase the force you exert on the string?

Now let go of the string and carefully observe the path taken by the object. Forget the downward motion (caused by gravity) and concentrate on how the object travels horizontally. Figure 3-11 shows a number of potential paths for the object. Which of these did your object take?

StarLinks netQuestions

Visit the netQuestions area of StarLinks (www.jbpub.com/starlinks) to complete exercises on these topics:

1. The Moons of Jupiter
Galileo, the discoverer of four of Jupiter's moons, and the person for whom four of these moons are named, once said, "the Jupiter system represents the solar system in miniature."

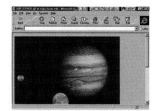

2. Tides The force of gravity between the Earth and the Moon has far reaching effects; primary among them are ocean tides.

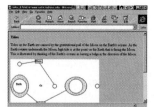

The sky's electromagnetic spectrum appears in stained glass at the Smithsonian Institute.

LIGHT AND THE ELECTRO-MAGNETIC SPECTRUM

Temperature Scales
The Wave Nature of Light
Wave Motion in General
Light as a Wave
Invisible Electromagnetic Radiation
The Colors of Planets and Stars
Color from Reflection—The Colors of Planets
Color as a Measure of Temperature
CLOSE UP: Blackbody Radiation
Spectra Examined Close Up
Kirchhoff's Laws
The Bohr Atom
HISTORICAL NOTE: Niels Bohr

Emission Spectra
CLOSE UP: Coexisting Competing Theories: The Nature of Light
Absorption Spectra—The Stars
CLOSE UP: The Balmer Series
The Doppler Effect
The Doppler Effect in Astronomy
The Doppler Effect as a Measurement Technique
Other Doppler Effect Measurements
Relative or Real Speed?
The Inverse Square Law

•••••••• THE GREATEST PHYSICIST from the time of Isaac Newton to Albert Einstein was probably James Clerk Maxwell. Maxwell was born into a wealthy family in Edinburgh, Scotland, in 1831. His genius was apparent early in his life, for at the age of 14, he published a paper in the proceedings of the Royal Society of Edinburgh. One of his first major achievements was an explanation for the rings of Saturn, in which he showed that they consist of small particles in orbit around the *planet. In 1885 Maxwell began a study of electricity and magnetism, and in the 1860s he discovered that it should be possible to produce a wave that combines electrical and magnetic effects, an* electromagnetic wave. *His analysis of this hypothetical wave showed that its speed would be 300,000 kilometers/second. Since this is the speed of light, Maxwell concluded that he had discovered the nature of light: light is an electromagnetic wave.*

The story is told that on the evening after he made his discovery, Maxwell went out walking with his wife-to-be, and she pointed out the beauty of the stars. He told her that she was with the only person in the world who understood what starlight was.

Nine years after Maxwell's death, radio waves were discovered and were shown to have properties similar to those of light. This verified Maxwell's prediction. The importance of Maxwell's work is indicated in the following quotation from the Nobel Prize winner Richard Feynman:

> From a long view of the history of mankind—seen from, say ten thousand years from now—there can be little doubt that the most significant event of the 19th century will be judged as Maxwell's discovery of the laws of electrodynamics. The American Civil War will pale into provincial insignificance in comparison with this important scientific event of the same decade.*

*The Feynman Lectures on Physics, *vol. 2, (Reading, Mass.: Addison-Wesley Publishing Co., 1964).

An important part of your study of astronomy is to learn about the fantastic objects that make up our universe, but perhaps more important is to see how astronomy functions by learning *how* we know what we know. How do we know that our Sun is a star like the thousands of others we see in the sky at night? How do we know what the stars are made of? How do we know their sizes? How can we measure a star's mass?

In fact, how can we learn anything about objects that we cannot hold, feel, weigh, and experiment with? By examining the radiation that they emit. This radiation, including not only visible light but many other types of radiation, carries to us a tremendous amount of information about the objects that it comes from. To understand how astronomers use radiation to answer questions about celestial objects, it is necessary to learn something about the radiation itself.

This chapter will examine the nature of light and show how we measure three major properties of stars: their temperatures, their compositions (that is, the chemical elements of which they are made), and their speeds relative to the Earth. This is quite a lot to learn about the stars, but it is just the beginning.

TEMPERATURE SCALES

One of the properties of a celestial object that astronomers measure is its temperature, so before going any farther, we must describe the temperature scales used in science. The scale in common use in the United States is the Fahrenheit scale, but most of us are at least somewhat familiar with the Celsius temperature scale, which defines the freezing point of water as 0°C and the boiling point of water as 100°C. The first two thermometers of ➤Figure 4-1 compare the Fahrenheit and Celsius scales from extremely low temperatures up to the boiling point of water. A third scale, the **Kelvin** scale, is the one most commonly used in science, however. This scale is based on the fact that there is a coldest temperature that can exist. This temperature is −273°C, and the Kelvin scale defines

Kelvin temperature scale. A temperature scale with its zero point at the coldest possible temperature ("absolute zero") and a degree that is the same size (same temperature difference) as the Celsius degree.

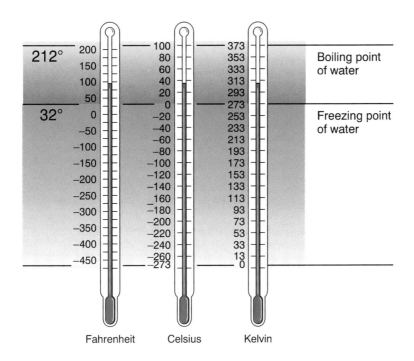

► FIGURE 4-1. A comparison of temperature scales.

this as its zero point. The intervals on the Kelvin scale are the same as on the Celsius scale, so in Kelvin temperature, the freezing point of water is 273 K and the boiling point is 373 K. (Notice that no degree symbol is included; the latter temperature is stated as "373 Kelvins.") Figure 4-1 includes the Kelvin temperature scale on the right.

THE WAVE NATURE OF LIGHT

►Figure 4-2 shows a beam of white light passing through a prism. The emerging light is separated into colors—into a *spectrum.* Isaac Newton showed that the prism does not add color to the light but rather that color is already contained in white light and that the prism merely separates the light into its colors. Analysis of the spectrum of light from stars is extremely important in astronomy.

spectrum. The order of colors or wavelengths produced when light is dispersed.

► FIGURE 4-2. A prism separates white light into its component colors.

The Wave Nature of Light

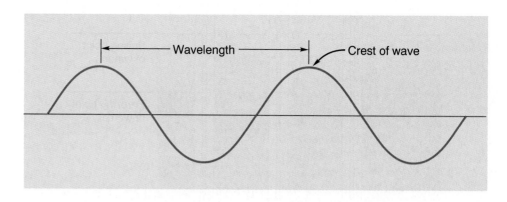

► FIGURE 4-3. Wavelength is the distance between successive crests of a wave.

What is it about light that allows it to be separated into colors and what is different about the various colors of light that result?

Wave Motion in General

Light acts like a wave. ►Figure 4-3 is a simplified drawing of a wave. As the figure indicates, the distance between successive peaks (crests) of the wave is called the *wavelength* of the wave. The wave in the drawing has a wavelength of 2¼ inches; waves you make by dipping your hand in a swimming pool might have a wavelength of about this value.

A wave does not sit still, however. Imagine a fisherman sitting on a pier watching waves pass underneath. As the waves move by, they cause the fisherman's cork to move up and down. This indicates that the water itself moves up and down rather than along the direction of the wave's motion. As the wave travels along the surface, the water's motion is primarily in the vertical direction and not along the direction of the wave.

Now suppose that the fisherman counts the number of times the cork moves up and down in one minute. He finds that the cork moves through a complete cycle 30 times each minute. We say that the *frequency* of the cork's motion is 30 cycles per minute and therefore that the frequency of the wave is 30 cycles/minute.

Suppose the fisherman measures the wavelength of the waves and finds it is 20 feet. He can use the fact that the wavelength is 20 feet and the frequency is 30 cycles/minute to calculate the speed with which the wave is moving. Each wave, from crest to crest, is 20 feet long, and 30 of these waves pass the fisherman each minute. This must mean that the waves move at a speed of 600 feet/minute. We multiply wavelength by frequency to obtain the speed of the wave. In equation form,

$$\text{wave speed} = \text{wavelength} \times \text{frequency}$$

or, using symbols,

$$v = \lambda \times f \qquad \begin{aligned} v &= \text{wave speed} \\ \lambda &= \text{wavelength} \\ f &= \text{frequency} \end{aligned}$$

wavelength. The distance from a point on a wave to the next corresponding point.

frequency. The number of repetitions per unit time.

λ is the lowercase Greek lambda.

This equation applies not only to water waves, but to all types of waves, including light and sound waves. Let's look at an example of its use in the case of sound.

EXAMPLE

Sound travels at a speed of about 335 meters/second in air. What is the wavelength of a sound that has a frequency of 262 cycles/second?

Solution Using the equation,

CHAPTER 4 Light and the Electro-Magnetic Spectrum

$$\text{wave speed} = \text{wavelength} \times \text{frequency}$$
$$335 \text{ m/s} = \lambda \times 262 \text{ cycles/s}$$

We now solve the equation for the wavelength and do the calculation:

$$\lambda = \frac{335 \text{ m/s}}{262 \text{ cycles/s}}$$
$$= 1.28 \text{ m}$$

So the length of the sound wave is about $1\tfrac{1}{4}$ meters.

••••• **TRY ONE YOURSELF.** What is the wavelength of a sound that has a frequency of 4000 cycles/second? (The speed of sound is the same for all frequencies, 335 m/s.)

hertz (abbreviated Hz). The unit of frequency equal to one cycle per second.

If you did the suggested exercise, you saw that a sound of higher frequency has a shorter wavelength, since the speeds are the same. The reason the example used sound rather than light is that sound waves have frequencies, velocities, and wavelengths within our everyday experience, whereas light waves don't. The next section returns to light and the spectrum produced when white light shines through the prism.

Light as a Wave

White light is made up of light of many different wavelengths. All wavelengths, however, travel at the same speed in a vacuum (and interstellar space is essentially a vacuum). That speed is tremendous, 3.00×10^8 meters/second, which is 300,000 kilometers/second or 186,000 miles/second. If a light beam could be made to travel around the Earth, it would circle the globe seven times in one second.

Light is extremely fast and its wavelengths are unimaginably short. We perceive light of different wavelengths as different colors. The wavelength of the reddest of red light is about 7×10^{-7} meters, or 0.0000007 meters. The wavelength decreases as one moves across the spectrum from red to violet, and the wavelength at the violet end of the spectrum is about 4×10^{-7} meters.

We have already seen that meters and kilometers are not convenient units with which to describe distances in the solar system, so we use astronomical units (AUs) there. In describing the wavelengths of light, a meter is much too long to be convenient, so scientists use another unit—the **nanometer.** One nanometer (abbreviated *nm*) is 10^{-9} meters. So the shortest violet wavelength and the longest red wavelength are 400 nm and 700 nm, respectively.

The same equation that relates the speed, wavelength, and frequency of other types of waves also applies to light: speed = wavelength × frequency. When the equation is used with the velocity of light to calculate frequencies of light waves, extremely high frequencies are obtained: 400 nm corresponds to 7.5×10^{14} cycles/second, and 700 nm corresponds to 4.3×10^{14} cycles/second. The frequency of visible light is tremendously large, larger than we can imagine.

The wavelength of a particular light determines its color, but since color is so subjective and people are unable to distinguish between two very similar colors, scientists describe light by referring to wavelength rather than color. They might mention the color in some cases, but this is to help us better picture the situation. Remember that light can be described more accurately than by simply calling it "red" or "green."

An important point: As we describe the light coming from astronomical objects, we will sometimes refer to its wavelength and sometimes to its frequency. Remember that if we know one, we can calculate the other.

A review of powers-of-ten notation is in a Close Up in the Prologue.

nanometer (abbreviated nm). A unit of length equal to 10^{-9} meters.

Angstrom (abbreviated Å). A unit of length equal to 10^{-10} m. There are 10 Angstroms in a nanometer.

It is probably not worth memorizing these frequencies of light waves, but it is handy to remember that the wavelength of visible light ranges from 400 nm (violet) to 700 nm (red).

The Wave Nature of Light

Invisible Electromagnetic Radiation

Can a light wave have a wavelength longer than 700 nm? Yes, although such a wave is not visible to us. The waves we see—visible light—are just a small part of a great range of waves that make up the ***electromagnetic spectrum.*** Waves somewhat longer than 700 nm (the approximate limit of red) are called *infrared* waves. ►Figure 4-4 shows the entire electromagnetic spectrum. Notice that the infrared region of the spectrum goes from 700 nm at the border of visible light up to about 10^{-4} m, which is a tenth of a millimeter or 100,000 nm. Electromagnetic waves longer than that are called *radio* waves.

Going the other way from visible light (toward shorter wavelengths), we first encounter *ultraviolet* waves, then *X-rays* and *gamma rays.*

It is important to emphasize that all of these types of waves (or rays, as certain portions of the spectrum are known) are essentially the same phenomenon. They differ in wavelength, and this causes some of their other properties to differ. For example, visible light is just that—visible. Ultraviolet is invisible to us but it kills living cells and causes our skin to tan or burn. On the other hand, we perceive infrared as heat radiation. And yes, the radio waves in the spectrum are the same radio waves we use to transmit messages on Earth. They are handy for carrying messages containing sound and pictures (in the case of television) for a number of reasons, including the fact that they pass through clouds and bend around obstacles.

Again, I emphasize that all of these various waves are electromagnetic waves, just as visible light is. We give the various regions different names because of their properties and the uses we have for them. Nature does not see the spectrum

electromagnetic spectrum. The entire array of electromagnetic waves.

Infrared, radio, ultraviolet, X-rays, and gamma rays are defined by their frequency and/or wavelength, as shown in Figure 4-4.

The waves are called "electromagnetic" because they consist of combined electric and magnetic waves that result when a charged particle accelerates.

►**FIGURE 4-4.** The electromagnetic spectrum is divided into a number of regions, depending upon the properties of the radiation. Region boundaries are not well defined. Notice the small portion of the spectrum occupied by visible light.

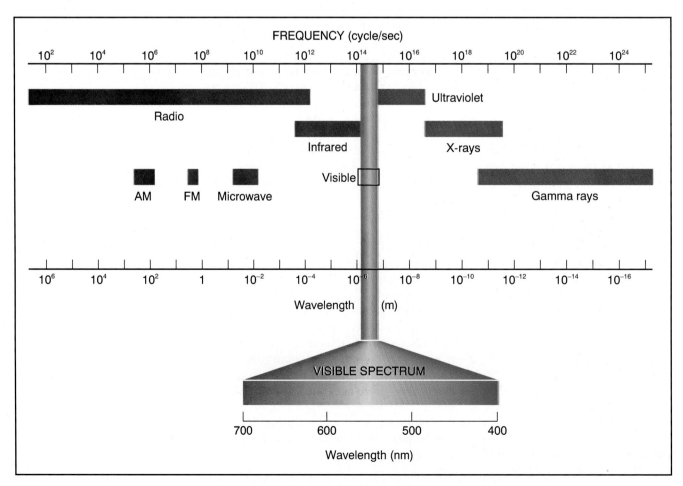

as seven or eight or ten different regions; all of these regions are simply electromagnetic waves.

The electromagnetic spectrum is important to astronomers because celestial objects emit waves of all the different regions of the spectrum. Notice that visible light is a very small fraction of the entire spectrum. We humans tend to regard it as the important part, but this only reveals our limited outlook. Astronomers learn a great deal from the *invisible* radiation emitted by objects in the heavens.

One of our problems with this invisible radiation is that most of it does not pass well through air so it does not reach the surface of the Earth. Our air is transparent to visible light and to part of the radio spectrum, but most of the rest of the electromagnetic spectrum is blocked to some degree. The chart in ➤Figure 4-5 shows the relative absorbency of the atmosphere to various regions of the spectrum. Where the purple region is highest, the least amount of radiation gets through. Notice that not much ultraviolet radiation penetrates to the surface. This is fortunate because, as you recall, this radiation damages living cells.

Astronomers, however, wish to detect and examine these nonpenetrating radiations from space. They accomplish this by using balloons to carry detectors high into the atmosphere or using artificial satellites to take them completely above the atmosphere. This will be covered in a later chapter.

Astronomers refer to *windows* in the atmosphere, saying that there is a visual window and a radio window. This means that our atmosphere allows radiation in these two regions of the spectrum to penetrate to the surface.

THE COLORS OF PLANETS AND STARS

This section returns to the visible spectrum to see how the spectrum from an object is analyzed in order to determine some properties of heavenly bodies. This analysis can be divided into two different categories. First, we look at the overall spectrum of light from the object. We typically call this the color of the object, but remember that when we look at the spectrum, we are really examining the actual wavelengths of light rather than just the color. The second way, to be discussed later in this chapter, involves examining individual parts of the spectrum.

Color from Reflection—The Colors of Planets

When you see a spectrum spread out before you—on a screen perhaps—you see a particular color coming from a given location on the spectrum. This is because the wavelength associated with that color is coming to your eye from that spot.

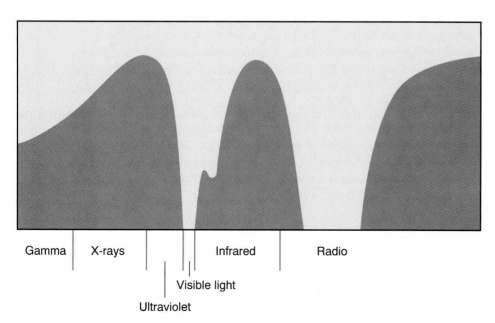

➤ FIGURE 4-5. The height of the purple region indicates the relative amount of radiation of a given wavelength blocked by the atmosphere from reaching Earth's surface. The atmosphere is transparent to two regions of the spectrum: visible light and part of the radio region.

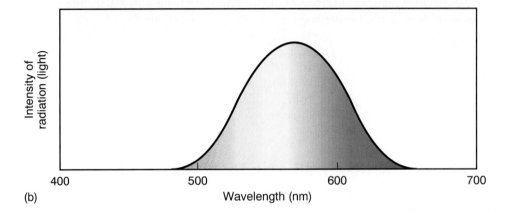

➤ **FIGURE 4-6.** (a) If light reflected from the lemons were sent through a prism to reveal its spectrum, we would see that the lemons reflect mostly yellow light. (b) This graph indicates the relative intensity of the various wavelengths of light from the lemons.

The color we see in most objects is not of a single wavelength, however. ➤Figure 4-6a might be the spectrum of light from some lemons. The spectrum contains many different wavelengths of light, but notice that there is no violet or blue light and that the center of the spectrum is indeed in the yellow. Part (b) of the figure shows a graph that indicates the intensity of light of each wavelength. Where the graph is higher, the light of the corresponding color is brighter.

The lemons have the spectrum illustrated in the figure because when white light strikes them, they absorb some of the wavelengths of the white light and reflect only those wavelengths we see in their spectrum (the wavelengths centered on the color yellow). This is what determines their color.

The planets have their colors because of a process like that described for the lemons. The rusty red color of Mars, for example, occurs because the material on its surface absorbs some of the wavelengths of sunlight and reflects a combination of wavelengths that looks rusty red to us.

➤Figure 4-7 shows the color spectrum and intensity/wavelength graph of the light from the red taillight of a car. The spectrum includes not only the many wavelengths of the red part of the spectrum but also some orange. In this case, the bulb inside the taillight emits white light, but part of that light is absorbed by the plastic cover. The light that gets through the cover has the spectrum in the figure.

The colors of the Sun and other stars are determined in part by a process somewhat like the taillight of the car, for light from a star is produced by emission within the star and some is absorbed as it passes through the outer layers of the star. We will look at this in more detail later.

Color as a Measure of Temperature

The light emitted by the Sun and other stars can be compared to light coming from a light bulb or from the element of an electric stove in an otherwise dark room. Consider what happens when you turn the burner of an electric stove to

➤ **FIGURE 4-7.** The spectrum of light from the taillight of this particular car. The light bulb emits white light, but the plastic cover over the bulb absorbs much of the light, letting pass some wavelengths in the red, orange, and yellow regions of the spectrum.

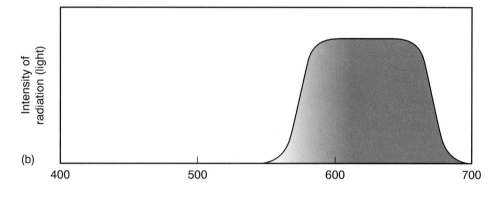

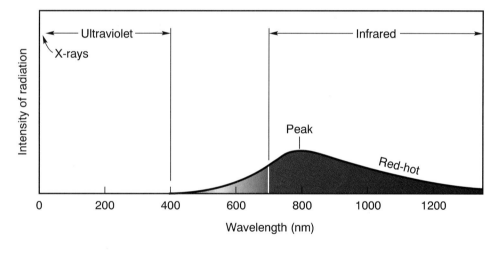

➤ **FIGURE 4-8.** The spectrum of a red-hot stove burner shows that it emits more infrared radiation than visible light.

a low setting. It glows a dull red. ➤Figure 4-8 is an intensity/wavelength graph of the stove burner glowing dull red. Notice that on this graph we have included not only the visible portion of the spectrum but quite a lot of the infrared. Recall that we experience infrared radiation as heat. The stove burner is emitting quite a bit of infrared radiation. (If you don't believe this, put your hand near it and feel the heat it radiates.) In fact, the graph indicates that more infrared radiation is being emitted than visible radiation, and the graph reaches its peak in the infrared portion of the spectrum. Some red light is emitted, but very little light from the center and violet end of the spectrum.

Now turn up the heat on the stove. The burner begins to take on an orange glow. The second curve from the bottom in ➤Figure 4-9 is an intensity/wavelength graph for this burner. Compare its curve to the bottom curve (the red-hot burner). First, the orange burner is emitting more radiation of all wavelengths. This should correspond to your experience, for you can feel that more

We want you to imagine a dark room because in a well-lit room, the lamp and stove element reflect light and you can see them even if they are turned off.

The Colors of Planets and Stars 103

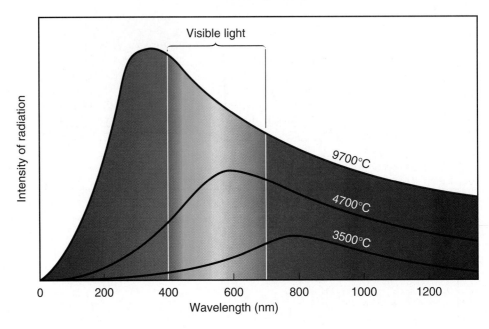

➢ **FIGURE 4-9.** The bottom curve represents a red-hot object and is the same as the curve in Figure 4-8. The next curve is "orange-hot." The third curve from the bottom is hotter still, and the top one is hottest, "white-hot."

➢ **FIGURE 4-10.** The graph is extended to temperatures found in stars. The 9700-degree object would be called "blue-hot," for it emits more radiation in the blue-violet region of the spectrum than in the other regions.

The Hubble Space Telescope imaged one of the coolest stars ever seen (upper right). It is a companion to the dwarf star at lower left. The surface temperature of the cool star may be as low as 2300 degrees Celsius. (The white bar is an effect produced by the camera.)

Astronomers usually call the intensity/wavelength graph of a star its *thermal spectrum*.

infrared is being emitted and you can see that more light is coming from the burner. Second, the peak of the graph has moved over toward the visible portion of the spectrum, toward the shorter wavelengths. (Remember, infrared radiation has longer wavelengths than visible light.)

This is about as far as you can go with an electric stove burner. If you have a lamp with a dimmer switch, you can take it to the next step and make it glow yellow-hot. The third curve from the bottom is the graph of such a lamp.

Finally, suppose you turn up the lamp all the way. It gets white-hot. It appears white because it is emitting light about equally in all visible regions of the spectrum. As the top curve of Figure 4-9 shows, its intensity/wavelength graph peaks near the center of the visible portion of the spectrum.

As this sequence indicates, the intensity/wavelength graph of an object emitting electromagnetic radiation can be used to detect its temperature. The graph in ➢Figure 4-10 shows three curves that represent widely different temperatures, including one hotter than any previously shown. The temperature corresponding to each curve is shown.

The peak of the intensity/wavelength curve for a light-emitting object always falls at a wavelength that depends upon the object's temperature. This is important to us in astronomy because by plotting this graph for a star we can know the temperature of its surface. The intensity/wavelength graph provides astronomers with an extremely valuable tool in their attempts to understand stars.

104   CHAPTER 4   Light and the Electro-Magnetic Spectrum

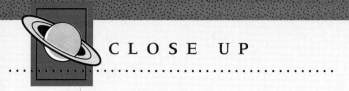

CLOSE UP

Blackbody Radiation

In the late nineteenth century, scientists made theoretical calculations of the radiation that would be emitted from an object that absorbed all wavelengths completely. Such an object was called a *blackbody,* and the radiation it emitted was called *blackbody radiation.* It was found that although stars are not perfect blackbodies, the theory applies very closely to them.

Two very important relationships derive from blackbody radiation theory, the Stefan-Boltzmann law and Wein's law.

The Stefan-Boltzmann Law

Figure 4-10 indicates that the hotter an object is, the more radiation it emits. In 1879 an Austrian physicist, Josef Stefan, discovered by experimentation the mathematical relationship between temperature and energy emitted. About five years later, another Austrian physicist, Ludwig Boltzmann, used blackbody theory to show why the relationship occurs. This rule, now called the *Stefan-Boltzmann law,* is as follows:

$$E = \sigma T^4$$

where E is the energy emitted each second per square meter of the surface, T is the temperature on the Kelvin scale, and σ (the Greek letter sigma) is a constant that relates the two quantities, called the *Stefan-Boltzmann constant.*

An example: Figure 4-10 shows a wavelength/intensity curve marked 4700°C and another marked 9700°C. These correspond to about 5000 K and 10,000 K, respectively. Since the temperature of the hotter star is twice that of the lower, the Stefan-Boltzmann law tells us that it emits 2^4, or 16, times more energy per square meter than does the cooler star.

Wien's Law

In 1893 a German physicist, Wilhelm Wien, discovered the mathematical relationship between the temperature of an object and the wavelength at which it emits maximum radiation. The location of the maximum emission is marked on Figure 4-8 and is easy to locate on any intensity/wavelength graph. *Wien's law* can be stated as

$$\lambda_{peak} = \frac{2,900,000}{T}$$

In the equation, the peak wavelength, λ_{peak}, is in nanometers when the temperature T is expressed in Kelvins. You can check that this equation applies to the lines of Figure 4-10.

The temperatures indicated in Figure 4-10 are typical of temperatures of stars, and the line that reaches its peak at the shortest wavelength on the graph represents the hottest star of the three. Notice that a star of this temperature emits more intense violet and blue light than light of the longer wavelengths. We say that it is a blue star. We often refer to stars by color; this is a quick way to indicate their temperature. A white star is hotter than a red star. In practice, of course, a red star does not appear red like a Christmas tree bulb, but it definitely has a red tint. If you have the opportunity to use a telescope to observe some pairs of closely spaced stars, you can see this color difference easily, but to the unpracticed naked eye, color differences between stars are not at all obvious. The important point is that by examining the thermal spectrum of a star, we can determine the temperature of the surface of the star without ever visiting it!

Beta Cygni (named Alberio) is the second brightest star in Cygnus and is a good example of a closely spaced pair of stars with obvious color differences.

SPECTRA EXAMINED CLOSE UP

The spectrum of visible light described above and shown in Figure 4-2 is called a *continuous spectrum.* Such a spectrum is produced when a solid object (in this case the filament of a lamp) is heated to a temperature great enough that the object emits visible light. Not all spectra are of this type, as was discovered nearly two hundred years ago.

continuous spectrum. A spectrum containing an entire range of wavelengths rather than separate, discrete wavelengths.

Kirchhoff's Laws

In 1814 Joseph von Fraunhofer, a German optician, used a prism to produce a solar spectrum. He noticed that the spectrum was not a continuous spectrum at all, but had a number of dark lines across it (►Figure 4-11). Fraunhofer had no explanation for these dark lines, but it was later discovered that the dark lines produced in the solar spectrum had resulted from the sunlight passing through cooler gases (in the Sun's and the Earth's atmospheres). Then in the mid-1800s some German chemists discovered that if gases are heated until they emit light, neither a continuous spectrum nor a spectrum with dark lines is produced; instead a spectrum made up of *bright* lines appears. Further, they discovered that each chemical element has its own distinctive pattern of lines (►Figure 4-12). This proved to be a very valuable way of identifying the makeup of an unknown substance and was soon developed into a standard technique that allows us to identify the composition of matter.

In the 1860s Gustav Kirchhoff formulated a set of rules, now called Kirchhoff's laws, which summarize how the three types of spectra are produced:

1. A hot, dense glowing object (a solid or a dense gas) emits a continuous spectrum.
2. A hot, low-density gas emits light of only certain wavelengths—a bright line spectrum.
3. When light having a continuous spectrum passes through a cool gas, dark lines appear in the continuous spectrum.

The dark lines that result when light passes through a cool gas (process 3) have the same wavelengths as the bright lines that are emitted if this same gas is heated (process 2).

Gustav Kirchhoff (1824–1887) was a German physicist and astronomer whose primary work was in the field of spectroscopy—the study of the spectrum.

THE BOHR ATOM

Kirchhoff's laws tell us how to produce the various types of spectra, but the science behind the laws—the connection between the laws and the nature of

► **FIGURE 4-11.** The Sun's spectrum may appear at first to be a continuous spectrum, but when it is magnified, we see dark lines across it where specific wavelengths do not reach us. Thus the solar spectrum is an absorption spectrum.

► **FIGURE 4-12.** This figure shows the visible emission spectrum of four different chemical elements. Each element emits specific wavelengths of light when it is heated. This applies not only to visible light (shown here) but also to infrared and ultraviolet.

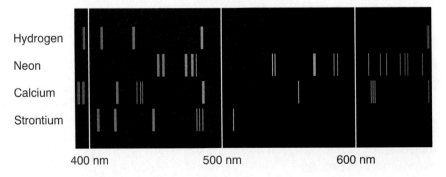

matter—remained a mystery in the nineteenth century. The connection was finally made in 1913, when a young Danish physicist, Niels Bohr, proposed a new model of the atom. His model describes the atom as having a **nucleus** with a positive electrical charge, circled by **electrons** with a negative electrical charge. Positive and negative electrical charges attract one another, and this electrical force holds the electrons in orbit around the nucleus.

The **Bohr atom**, as Niels Bohr's model is called, is based on three new ideas, or postulates:

1. *Electrons in orbit around a nucleus can have only certain specific energies.* To imagine different energies of electrons, imagine electrons orbiting at different distances. Since only certain energies are possible, we speak of "allowed" orbits for the electrons. The element hydrogen has only one electron but many possible energy levels and therefore many allowed orbits. The drawing in ➤Figure 4-13a depicts the hydrogen atom with its electron in the lowest orbit and also shows three other orbits this electron might have. The point of Bohr's first postulate is that the electron can have only these particular energies and therefore these particular orbits. This is far different from the solar system, where there are no limitations on possible positions of orbits. We will see later that asteroids and comets seem to have random, chaotic orbits. Bohr proposed the revolutionary idea that there is such a law that puts a limit on where electrons can orbit.

2. *An electron can move from one energy level to another, changing the energy of the atom.* In terms of electron orbits, when energy is added to an atom, the electron moves farther from the nucleus. On the other hand, the atom loses energy when an electron moves from an outer orbit to an inner orbit. This lost energy comes out of the atom in the form of electromagnetic radiation.

Our previous discussion of waves and light assumed that light acts as a simple long wave, similar to the waves you can make in a swimming pool. Theoretical work by Albert Einstein, Niels Bohr, and others showed that light is more complicated than this. The Bohr model holds that light is emitted not as continuous waves, but in tiny bursts of energy; each burst is emitted when an electron moves to an orbit closer to the nucleus. According to the Bohr model, light is emitted when an electron falls from an outer to an inner orbit. (We say "falls" because an electron is attracted toward the nucleus in a manner analogous to the way we are attracted to Earth and it seems natural to think of the electron's move inward as a

nucleus (of atom). The central, massive part of an atom.

electron. One of the negatively charged particles that orbit the nucleus of an atom.

Bohr atom. The model of the atom proposed by Niels Bohr; it contains electrons in orbit around a central nucleus and explains the emission of light.

You might think of the energy/distance relationship as follows: An electron "feels" a force of attraction toward the nucleus, and to pull it farther away requires energy.

➤ FIGURE 4-13. (a) The electron is in the lowest possible energy state—the lowest orbit. (b) When the atom gains energy (perhaps by collision with another atom), the electron jumps up to a higher orbit. (c) The electron then quickly falls down to its original orbit and emits energy in the form of a photon of electromagnetic radiation as it does so.

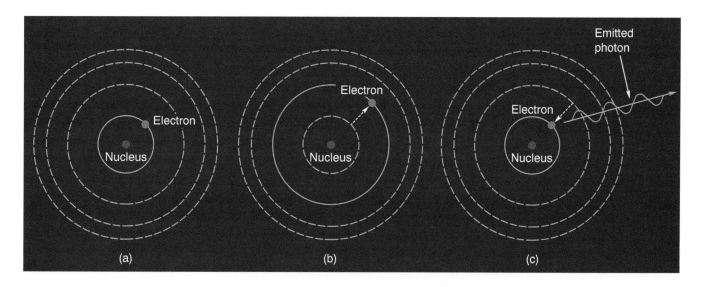

The Bohr Atom

HISTORICAL NOTE

Niels Bohr

Niels Bohr (Figure H4-1) was born in 1885 into a very cultured Danish home. His father was a physiologist and university professor, and both of his parents read at least four languages and were lovers of art and music. His father's interest in science led Niels to that subject, and he became known as a student who gave his utmost to every project—a reputation that continued throughout his life.

Niels was married in 1910. He and his wife Margarethe had six sons and a number of grandchildren. He was very family oriented and greatly enjoyed his children. Margarethe and Niels remained devoted to one another until Niels's death in 1962. One son, Aage, followed him in the study of physics, and the two worked together on a number of projects (and rode motorcycles together—one of Niels's hobbies).

In 1922 Niels Bohr was awarded the Nobel Prize in physics "for his services in the investigation of the structure of atoms and of the radiation emanating from them." In his acceptance speech he emphasized the limitations of his theory, and, indeed, he seems to have been more aware of its limitations than other scientists who worked with the theory.

One major project on which Niels and Aage Bohr worked was the development of the atomic bomb. Niels's mother was Jewish, and after Hitler's army overran Denmark, he and his family were forced to flee their native land to avoid arrest. Niels and Aage came to the United States and helped with the Manhattan Project (the code name for the bomb development effort).

Bohr's contribution to science goes far deeper than the development of the planetary model of the atom, as important as that is. His philosophical ideas on the nature of physical theory are perhaps his greatest contribution, but to discuss them here would require much more room than we have.

Niels Bohr was very interested in people and was greatly admired by all with whom he worked. He was a very soft-spoken, gentle man, and his work will influence scientific thinking for ages to come.

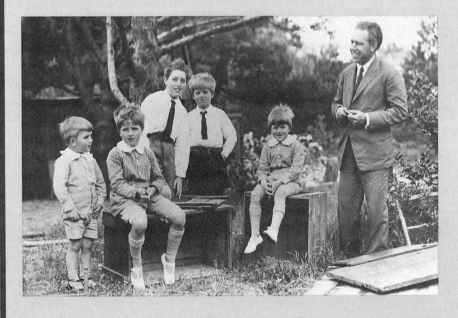

► FIGURE H4-1. Niels Bohr and his sons.

photon. The smallest possible amount of electromagnetic energy of a particular wavelength.

The applicable equation is $E = hf$, where E is the energy of the photon, h is a constant (called Planck's constant), and f is the frequency of the light.

fall.) These tiny bursts of electromagnetic energy are called **photons** of radiation. The energy of the photon depends upon the spacing between electron orbits.

3. *The energy of a photon determines the frequency (or wavelength) of light that is associated with the photon.* The greater the energy of the photon, the greater the frequency of light, and vice versa. Thus a photon of violet light has more energy than a photon of red light. (Recall that violet light has a greater frequency than red light.)

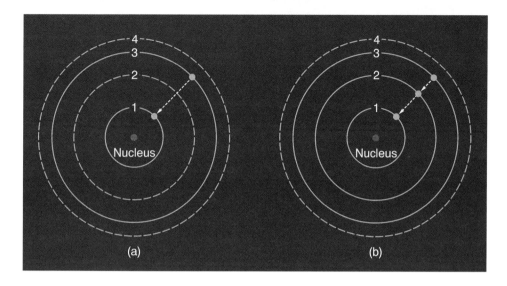

► FIGURE 4-14. (a) The electron may fall directly from orbit 3 to orbit 1, emitting a single photon. (b) The electron may fall from orbit 3 to orbit 1 in two steps, emitting two photons whose total energy is equal to the photon emitted in part (a).

Emission Spectra

The Bohr model of the atom can be used to explain why only certain wavelengths are seen in the spectrum of light emitted by a hot gas. In its normal, lowest energy state, the electron of a hydrogen atom is in its lowest possible orbit as indicated in Figure 4-13a. If this atom is given energy (perhaps by an electric current or by collisions with other atoms), the electron might jump to a more distant orbit. Figure 4-13b illustrates this. An atom will not stay in its energized state long. Quickly the electron falls down to a lower orbit, emitting a photon as it does, as shown in Figure 4-13c. The energy of this photon is exactly equal to the energy difference between the two orbits. Finally, since the energy of the photon determines the frequency of the radiation, the radiation coming from this atom must be of the corresponding frequency (and color).

We have described one atom emitting one photon. In an actual lamp that contains hot hydrogen gas, there are countless atoms being given energy and countless atoms emitting photons as they lose energy. Different atoms will have different amounts of energy, depending upon the energy they have absorbed (from a collision with another atom, for example). If a particular atom's energy corresponds to the electron being in the third orbit, the atom might release its energy in a single step as shown in ►Figure 4-14a, or it might release it a step at a time as shown in part (b) of the figure.

Suppose enough energy is available to cause electrons to move to the fourth orbit. ►Figure 4-15 shows all of the possible falls an electron might take to get back to the lowest orbit. Each of these falls corresponds to a certain specific energy and therefore to a certain specific frequency of emitted radiation. The electrons of some atoms will fall by some paths, and the electrons of other atoms will fall by other paths. As a result, radiation of a number of different frequencies will be emitted from the entire group of atoms. Not all frequencies will be emitted, however—just the ones that correspond to the electron jumps.

Hence, the spectrum from a heated gas is not a continuous spectrum. It contains only certain definite frequencies. We call such a spectrum an *emission spectrum*—the bright line spectrum mentioned earlier. An emission spectrum has a few individual bright lines instead of a continuous band of colors.

Refer back to Figure 4-12, which shows the emission spectra of four elements. Each spectrum is different because the allowed energy levels of the atoms are different for each chemical element. No two chemical elements have the same set of energy levels, and thus no two chemical elements have the same emission spectrum. This provides us with a valuable method of identifying ele-

The lowest energy state of an atom is usually called the *ground state*, and the energized states are called *excited states*.

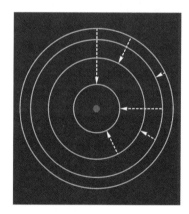

► FIGURE 4-15. These are the possible paths an electron may take to get from orbit 4 to orbit 1. Each fall corresponds to a different energy change and therefore to a photon of different frequency.

emission spectrum. A spectrum made up of discrete wavelengths rather than a continuous band of wavelengths.

The Bohr Atom

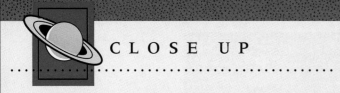

CLOSE UP

Coexisting Competing Theories: The Nature of Light

What is light? Isaac Newton's answer to this age-old question was that light is a series of particles that are emitted from a light source. The particles, he said, bounce from our surroundings (including distant surroundings such as planets) and into our eyes. However, in 1801—three quarters of a century after Newton's death—it was shown conclusively that light acts like a wave. As we have stated in the text, visible light is part of the electromagnetic spectrum, differing from other parts because it has a different wavelength.

Early in this century a competing picture of the nature of light was introduced, a picture that we have used in describing how the Bohr atom explains the emission of light by atoms. Bohr's theory tells us that atoms emit light in the form of photons, tiny bursts of energy rather than trains of waves. This photon theory is able to explain so much about the behavior of light that it is a definite part of today's theory of light and the electromagnetic spectrum. Notice that the photon theory gives us a picture of light that is closer to Newton's idea of particles than it is to the wave theory.

What about the wave theory, then? Has it been abandoned? The surprising thing is that the wave theory is still alive and flourishing, for some phenomena involving light can only be explained by thinking of light as a wave. We are forced to say that light has both a wave nature and a particle nature. Although this may seem unsatisfactory, we must face the fact that light is simply not like things we experience in everyday life where something must be either a wave or a particle. Light is not like this, and we cannot classify it simply as a wave or as a particle. Instead of saying what light *is*, we talk about how light *acts*. Here, then, we have an example in science where a new theory has not replaced an older one, but rather the two conflicting theories coexist to give us today's description of an important part of nature.

ments, since each has a unique spectral "fingerprint." This process has a number of important applications here on Earth, but since in this book we are more interested in the stars, we'll look at a stellar application.

Absorption Spectra—The Stars

Kirchhoff's laws tell us that dark line spectra result from light with a continuous spectrum passing through a gas. Let us examine this is in the case of the Sun.

photosphere. The region of the Sun from which visible radiation is emitted. We will discuss the Sun in Chapter 11.

The visible surface (the ***photosphere***) of the Sun emits a continuous spectrum. Even though the Sun is, in most ways, more like a gas than a solid, it produces a continuous spectrum rather than an emission spectrum. This is because as atoms are pushed together, their energy levels are broadened (as if slightly different orbits are allowed). As atoms become more and more tightly pressed together, their energy levels begin to overlap so that a full range of possible orbital energies is possible. Thus an entire range of photon energies is emitted by the atoms, and instead of separate, distinct spectral lines appearing in the spectrum, an entire range of frequencies appears.

Before the light from the Sun gets to us on Earth, it must pass through the relatively cooler atmosphere of the Sun as well as through the atmosphere of the Earth. The Sun does indeed have an atmosphere, and as we will see in a later chapter, its atmosphere is much deeper than Earth's. As the light passes through these gases, atoms of the gases absorb some of it. This absorption of energy raises the atom's energy level, but since only certain specific energy levels are possible, only certain amounts of energy can be absorbed by the atom. This results in the reverse of what we had before: instead of an atom emitting a photon as it releases energy, it absorbs a photon as it absorbs energy. And just as a hot gas emits photons of certain energies, the cooler gas of the atmosphere absorbs photons of the same energies.

If white light (a continuous spectrum) is passed through hydrogen gas that

► FIGURE 4-16. Part (a) shows the emission spectrum of an element, and (b) is the absorption spectrum of the same element. Notice that the absorbed wavelengths in (b) are the same as the emission lines in (a).

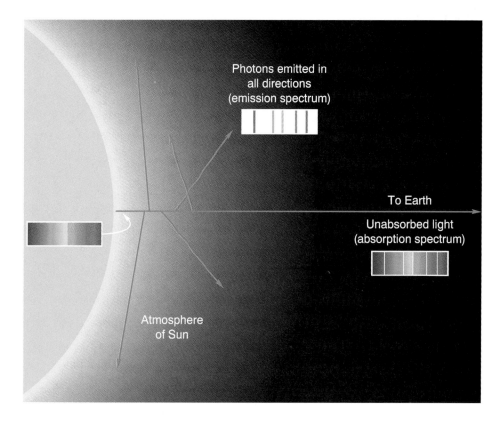

► FIGURE 4-17. Consider a ray of light leaving a point on the Sun and moving toward the Earth. As it passes through the Sun's atmosphere, some wavelengths are absorbed and then reemitted in random directions. This results in the solar spectrum being an absorption spectrum. Light from the Sun's atmosphere is an emission spectrum.

is too cool to emit light, the gas will absorb some frequencies of this light—the same frequencies it would emit as an emission spectrum. ►Figure 4-16a represents the emission spectrum of some element. Figure 4-16b shows the *absorption spectrum* that results when white light is passed through the cool gas of this same element. Notice that the dark lines of the absorption spectrum correspond exactly to the bright lines of the emission spectrum.

You might object that after the cool gas has absorbed radiation, it must reemit it. Shouldn't this then cancel out the absorption? No. As ►Figure 4-17 shows, the reemitted light is sent out in all directions. So certain frequencies of the light that was originally coming toward the Earth are scattered by the atmosphere of the Sun. This results in less light of those frequencies reaching us, and an absorption spectrum is observed.

The spectrum of the light that is reemitted is an emission spectrum. During a total eclipse of the Sun, light from the main body of the Sun is blocked out, and astronomers can see the light emitted by the Sun's atmosphere and examine its emission spectrum. It was by examining the emission spectrum from the gas near the Sun that astronomers discovered helium.

The Sun, and other stars as well, has a number of chemical elements in its atmosphere (that is, different kinds of atoms, each with its own pattern of electron orbits and spectral lines). As the white light passes through this gas, many frequencies are absorbed that correspond to the various chemical elements of the

absorption spectrum. A spectrum that is continuous except for certain discrete wavelengths.

This is the dark line spectrum of Kirchhoff's laws.

The elements in the Earth's atmosphere also absorb radiation, the frequencies of which depend upon what elements are in our atmosphere. But we know what those elements are and can take them into account.

The Bohr Atom 111

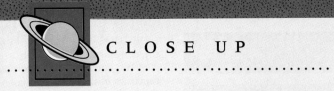

CLOSE UP

The Balmer Series

Notice in Figure 4-12 that there is a pattern to the spacing of bright lines in the hydrogen spectrum—they become progressively closer as they approach the blue end of the spectrum. The mathematical relationship that expresses this pattern was found in 1885 by Johann Jacob Balmer, a Swiss teacher. The lines visible in the figure are therefore called the **Balmer series** of spectral lines. The Balmer series served as the foundation for Bohr's work, and Bohr's model of the atom easily explains the pattern of the lines. In ►Figure C4-1, the spacing of the lines depicting the orbits of the hydrogen atom represents the relative energy of each of the levels. According to Bohr's model, there is less difference between the energy levels as one gets farther from the nucleus. Thus the levels are spaced closer together toward the top of the figure.

Suppose that an electron is in the lowest energy level, the ground state. This electron will jump to a higher state when a photon of the appropriate energy strikes it. Look at the left side of ►Figure C4-2. Five arrows point upward from the ground state, representing electron jumps to higher orbits. On each arrow is printed the wavelength of the photon corresponding to the jump indicated. Notice that each of these wavelengths is less than 400 nm, the shortest wavelength of visible light. They are in the ultraviolet region of the spectrum.

As a gas becomes hotter, more of its atoms have electrons in energy levels above the ground state. Suppose that hydrogen is at a temperature at which a significant number of its electrons are in energy level 2. The middle of Figure C4-2 shows the wavelength of photons that would cause electrons at this level to jump to a higher level. Notice that the wavelengths of the photons that cause jumps from the second level are within the visible range—from 400 to 700 nm—and since energy levels are more closely spaced toward the top of the figure, the wavelengths toward the blue end of the spectrum are closer together. These wavelengths correspond to the wavelengths of the Balmer series.

The first set of wavelengths described above, those in the ultraviolet region of the spectrum, form the *Lyman series*. As Figure C4-2 indicates, there is another series that falls in the infrared portion of the spectrum. It is called the **Paschen series**.

Thus far we have discussed the absorption of photons to produce an absorption spectrum. As we noted in the text, the emission spectrum of hydrogen is produced when electrons fall from higher energies to lower. Thus the spectrum shown in Figure 4-12 is an emission spectrum that resulted from electrons falling from higher energies to lower levels of the atom.

gas. By examining the complicated absorption spectrum that results, we are able to deduce what elements are in the star's atmosphere. Thus we answer a question that, just a century ago, was thought to be unanswerable. We now know what the surface layers of stars are made of! As you might appreciate from the complexity of the Sun's spectrum (look back at Figure 4-11), the analysis is fairly complicated, but it is now a common one in astronomy.

www.jbpub.com/starlinks

THE DOPPLER EFFECT

Have you ever stood near a road and listened to the sound of a car as it sped by you? Recall how the sound changed when the car passed. The change is especially noticeable for noisy, fast-moving race cars; perhaps you have heard it on televised auto races. This phenomenon has a parallel in astronomy—a very important one. To understand it, we will first consider water waves.

►Figure 4-18a is a photo of waves spreading from a disturbance on the surface of water. The waves move away from the source in a regular way and appear the same in all directions from the source. Figure 4-18b was made by moving a vibrating object toward the right as it makes waves on water. Look at the difference between the waves in front of and behind the moving source. Four important points can be made about this case, although only one of them is apparent in the photo.

112   CHAPTER 4   Light and the Electro-Magnetic Spectrum

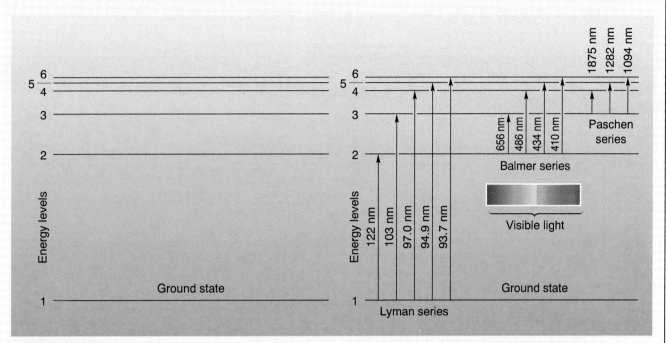

▶ FIGURE C4-1. The energy levels of the hydrogen atom. The levels are progressively closer in energy as they are farther from the nucleus.

▶ FIGURE C4-2. Electrons that move from the ground state to higher energy levels do so by absorbing photons of the wavelengths shown. Electron jumps for the first three series of the hydrogen atom's spectrum are represented here.

1. Even though the source of the waves is moving, the waves still travel at the same speed in all directions. The motion of a source does not push or pull the waves. It just disturbs the liquid, and the disturbance moves away at a speed that depends only on the characteristics of the liquid—water, in this case.
2. The wavelengths of the waves in front of the moving source are shorter than they would be if the source were stationary, and the wavelengths behind the moving source are longer. Recall that where the wavelength is shorter, the frequency is higher; and where the wavelength is longer, the frequency is lower. This means that if there were corks on the water, the corks in front would bounce up and down with a greater frequency than if the source were not moving, and corks behind the moving source would bounce with a lower frequency.

This is the same effect that causes the sound of a car to change as the car passes. The car emits sound just as the vibrating object produces water waves. When you are in front of the car, you are in the region of shorter wavelength and higher frequency. In the case of sound, high frequency means high pitch. After the car has passed, its sound waves are stretched in wavelength, causing you to hear a lower pitch.

This effect, which we see here both in water waves and in sound waves, is called the **Doppler effect,** named after Christian Doppler, the man who first ex-

Doppler effect. The observed change in wavelength from a source moving toward or away from the observer.

The Doppler Effect

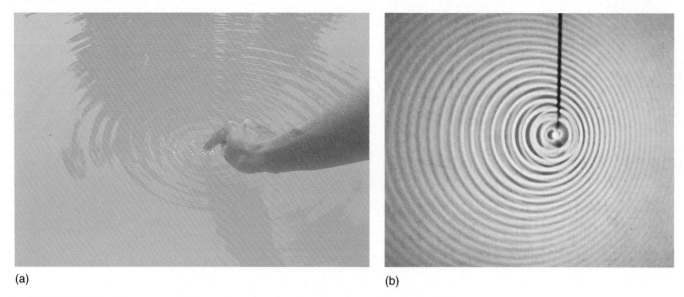

► **FIGURE 4-18.** (a) Waves spread evenly from their source. (b) If the source is moving to the right, the waves are compressed in front of the source and stretched behind it.

plained it. Before we apply it to light waves and astronomy, let's continue the list of important things to know about it.

3. The sound does indeed get louder as a car approaches (and the water waves get higher as the vibrator approaches a bobbing cork), but this is *not* what the Doppler effect is about. The Doppler effect refers only to the *change in frequency* (and therefore wavelength) of the wave.
4. The frequency does *not* get higher and higher as the source approaches at a uniform speed. Rather, the frequency is observed to be constantly higher than normal as the source approaches and constantly lower than normal as it recedes.

The Doppler Effect in Astronomy

We do not receive water waves and sound waves from the stars, but only electromagnetic waves. However, the Doppler effect also works for electromagnetic waves. This means that if an object is coming toward us, the light we receive from it will have a shorter than normal wavelength and one moving away will have a longer than normal wavelength. The reason you can't observe this for moving objects here on Earth is that the amount of the shortening and lengthening of the wave depends on the speed of the object *compared to the speed of the wave.*

For water waves, which move fairly slowly, you observe the Doppler effect even for a slow-moving wave source. In the case of sound, which has a speed much greater than water waves, you notice the Doppler effect only for fairly fast objects. A car going 70 miles per hour is moving at about 10% of the speed of sound.

In the case of light, you don't perceive the Doppler effect for a car traveling at 70 miles per hour because it is moving at only one ten-millionth the speed of light. To describe the Doppler effect for light, the example of a spaceship emitting green light is sometimes used. If the spaceship is moving away from us at a very great speed, the light we see from the lamp is stretched in wavelength so that it may appear red. The light is ***redshifted.*** If the spaceship is approaching, the light is ***blueshifted.***

redshift. A change in wavelength toward longer wavelengths.

blueshift. A change in wavelength toward shorter wavelengths.

114   CHAPTER 4   Light and the Electro-Magnetic Spectrum

► **FIGURE 4-19.** (a) Emission spectrum of hydrogen when the source is stationary with respect to the observer. (b) Absorption spectrum of hydrogen for a receding source. Note the shift of all lines toward redder colors.

The spaceship example does illustrate the Doppler effect, but it is very misleading in an important respect: Except for very distant galaxies, objects in the heavens do not move with speeds great enough to actually change their colors appreciably. The redshift or blueshift caused by the Doppler effect is very small in most cases. If the spectra of stars were continuous spectra, there would be few cases in which the Doppler effect could be used to detect motion. It is the spectral lines (usually the absorption lines) that make the Doppler effect such a powerful tool.

As an example of how we use absorption lines to detect the Doppler effect for stars, imagine that we record the spectrum of hydrogen gas in the laboratory. ►Figure 4-19a shows the spectrum. Figure 4-19b represents the spectrum of a star having only hydrogen in its atmosphere (which is unrealistic, but this is a simplified example). Notice that the absorption lines in the star's spectrum do not align exactly with the emission lines of the laboratory spectrum. They are shifted slightly toward the red. This indicates that the star is moving away from us.

There are three major differences between our example and a measurement of a real star: First, a real star's spectrum has many more spectral lines. Second, the Doppler shift is almost always much smaller than that indicated in the example. When we later show photos of actual spectra, they will usually be a magnified portion of a small part of the visible spectrum. Third, astronomers do not normally use color film in recording the spectrum. This may seem odd, but recall that color is not easy to describe accurately whereas wavelength is. The wavelengths of absorption lines can be measured very accurately and compared to the lines from a laboratory spectrum to determine the existence and the precise amount of the Doppler shift.

The Doppler Effect as a Measurement Technique

Thus far we have described the Doppler effect as a method for detecting whether a star is moving toward or away from us, but it is more powerful than this. From measurements of the *amount* of the shifting of the spectral lines, we can determine the **radial velocity** of the star relative to Earth. The radial velocity is the star's velocity toward or away from us, and must be distinguished from its *tangential velocity*, which is its velocity across our line of sight. ►Figure 4-20 distinguishes the two velocities. The Doppler effect provides a method to measure radial velocity, making it much easier to detect and measure than tangential velocity. To measure tangential velocity, we must look for motion of the star across our line of sight (like the yellow car in Figure 4-20), and this motion can be detected only for relatively nearby stars.

The following equation allows us to calculate radial velocity from Doppler shift data:

$$\frac{\Delta\lambda}{\lambda} = \frac{v}{c}$$

$\Delta\lambda$ = wavelength difference
λ = wavelength for stationary source
v = velocity of object
c = velocity of light

radial velocity. Velocity along the line of sight, toward or away from the observer.

tangential velocity. Velocity perpendicular to the line of sight.

This equation applies if the object is moving much slower than the speed of light, as is the case except for very distant galaxies. For these galaxies a slightly more complex equation is needed as Einstein's special relativity theory becomes important.

FIGURE 4-20. The red car has a radial (to-or-fro) velocity with respect to the observer, while the yellow car has a tangential (side-to-side) velocity. The Doppler effect cannot be used to detect tangential velocity.

Here λ is the wavelength of the spectral line from the stationary laboratory source, $\Delta\lambda$ is the difference in wavelength between the star's spectrum and the laboratory spectrum, v is the velocity of the object, and c is the speed of light. If we solve this equation for what we are usually calculating, the velocity of the object, we obtain

$$v = c(\Delta\lambda/\lambda)$$

To illustrate how easy this equation is to apply, let's do an example.

EXAMPLE

The wavelength of one of the most prominent spectral lines of hydrogen is 656.285 nm (this is in the red portion of the spectrum). In the spectrum of Regulus (the brightest star in the constellation Leo), the wavelength of this line is observed to appear longer by 0.0077 nm. Calculate the speed of Regulus relative to Earth and determine whether it is moving toward or away from us.

Solution

First, notice that the data indicates that the wavelength of the line in the spectrum of Regulus is *longer* by 0.0077 nm. This means that the star is moving *away* from us. To determine its speed, we will substitute the given values in the equation that relates Doppler shift and speed:

$$\begin{aligned} v &= c(\Delta\lambda/\lambda) \\ &= (3.0 \times 10^8 \text{ m/s})(0.0077 \text{ nm}/656.285 \text{ nm}) \\ &= 3500 \text{ m/s} \\ &= 3.5 \text{ km/s} \end{aligned}$$

The annual variation in the Doppler shift provides evidence for the Earth's motion around the Sun. The evidence was not available, of course, when the question was controversial.

So Regulus is moving away from Earth at a speed of about 3.5 kilometers/second. This is about 2 miles/second, or 8000 miles/hour, a typical speed for nearby stars. Since the Earth's speed in its orbit around the Sun is about 30 km/s, the Earth's motion would have to be taken into account in making the measurement. Our calculation assumed that the Earth was between the Sun and Regulus so that it had no motion toward or away from the star.

······TRY ONE YOURSELF. The nearest star to the Sun that is visible to the

CHAPTER 4 Light and the Electro-Magnetic Spectrum

naked eye is Alpha Centauri. With Earth's motion removed, the 656.285 nm line of hydrogen has a wavelength of 656.237 nm in Alpha Centauri's spectrum. Calculate the radial velocity of this star relative to the Sun and tell whether it is moving toward or away from the Sun.

Other Doppler Effect Measurements

The use of the Doppler effect is not limited to measuring the speeds of stars. Other applications include the following:

1. Measuring the rotation rate of the Sun. Galileo was the first to observe sunspots on the Sun, and he reported that they move across the Sun and thus provide a method to measure the Sun's rotation rate. The Doppler effect gives a second method. The measurement is done by examining the light from opposite sides of the Sun. Light from the side moving toward us is blueshifted, and light from the other side is redshifted.
2. In a similar manner, the rotation rates of planets (and the rings of Saturn) can be measured. The fact that the light in this case has been reflected by the planet (or rings) rather than emitted by it does not matter; the light is still shifted by the Doppler effect. In fact, the rotations of Mercury and Venus were first revealed by reflected radar waves.
3. Many stars are part of a two (or more) star system in which the stars orbit one another. If their orbits happen to be aligned so that each star moves alternately toward and away from our position in the universe, we can detect this motion by the red- and blueshift of each star's spectrum. When we discuss such *binary stars* in Chapter 12, we'll see that they are very important to us in our quest to learn more about stars.

Police radar catches speeders by bouncing waves from their cars and detecting the Doppler shift in the reflected waves.

binary star. A pair of stars gravitationally bound so that they orbit one another.

RELATIVE OR REAL SPEED?

The speed measured by the Doppler effect is the speed of the object *relative to the speed of the Earth.* As mentioned earlier, we must take the Earth's speed into account in any calculation made with the Doppler effect. You may argue that even if we do this, we are only measuring the speed of the object with respect to the Sun. What about the object's *real* speed? There is no such thing.

The reason for this last statement is that *all* speeds are relative to something. When you say that a car is moving at 50 miles/hour, you mean "relative to the surface of the Earth." When you walk up the aisle of a moving plane, you have one speed relative to the seated passengers, another relative to the Earth, another relative to the Moon, another to the Sun, and so on.

Just as all motion is relative, all nonmotion is too. When we say that something is at rest, we usually mean that it is at rest relative to the Earth. (Not always, however. You might tell your little brother to sit still in the back seat of your car even though the car is moving. In this case you are asking him to sit still relative to the car.) It is meaningless to say that something is absolutely at rest.

So there is no loss of meaning to our finding the velocity of an object relative to the Earth and then correcting for the Earth's motion around the Sun. All motion is relative.

This understanding of the relativity of motion is called *Galilean relativity* or *Newtonian relativity.* Einstein's theory of relativity goes much further than this.

THE INVERSE SQUARE LAW

Everyone has experienced the fact that the intensity, the brightness, of light decreases when we move farther from the source of the light. The decrease in

inverse square law. Any relationship in which some factor decreases as the square of the distance from its source.

The force of gravity follows such a law, as does the apparent brightness at some distance from a source of light.

intensity follows an *inverse square law* of radiation, a relationship that states that radiation spreading from a small source decreases in intensity as the inverse square of the distance from the source. That is, at twice the distance from the source, the intensity is one-fourth as much, and at three times the distance, the intensity is one-ninth as much. The requirement that the source be small is easily met in most astronomical cases. Even the Sun and stars qualify as small sources, for compared to our distance from them, their diameters are small.

CONCLUSION

As this chapter has shown, the spectra of stars are used in two very different manners. First, the spectra are treated as continuous spectra and their intensity/wavelength graph is examined. By measuring the wavelength of the maximum intensity of radiation, we can determine the temperature of a star's surface. In doing this analysis, we ignore the absorption lines. These lines, remember, are very narrow; their presence does not change the overall pattern of the intensity/wavelength graph.

Second, the absorption lines within the spectrum are examined. These lines allow us to determine the chemical composition of stars, and, when we use the Doppler effect, the shift in the lines allows us to calculate the radial speeds of stars relative to Earth as well as the speeds of rotation and revolution of celestial objects.

RECALL QUESTIONS

1. The frequency of visible light falls between that of
   A. infrared waves and radio waves.
   B. X-rays and cosmic rays.
   C. ultraviolet waves and X-rays
   D. short radio waves and long radio waves.
   E. ultraviolet waves and radio waves.

2. Infrared radiation differs from red light in
   A. intensity.
   B. wavelength.
   C. its speed in a vacuum.
   D. [All of the above.]
   E. [None of the above.]

3. The frequency at which a star emits the most light depends upon the star's
   A. distance from us.
   B. brightness.
   C. temperature.
   D. eccentricity.
   E. velocity toward or away from us.

4. In an infrared photo taken on a cool night, your skin will appear brighter than your clothes.
   A. Correct.
   B. Wrong, it depends upon the color of your clothes.
   C. Wrong, your clothes will appear brighter.
   D. Wrong, they will be equally bright.

5. Light waves of greater frequency have
   A. shorter wavelength.
   B. longer wavelength
   C. [Either of the above; there is no direct connection between frequency and wavelength.]

6. Which of the following does not have the same fundamental nature as visible light?
   A. X-rays
   B. Sound waves
   C. Ultraviolet radiation
   D. Infrared waves
   E. Radio waves

7. As an electron of an atom changes from one energy level to a higher energy level by absorbing a photon, the total energy of the atom
   A. increases.
   B. decreases.
   C. remains the same.

8. Which of the following is produced when white light is shined through a cool gas?
   A. An absorption spectrum
   B. A continuous spectrum
   C. An emission spectrum
   D. [All of the above.]
   E. [None of the above.]

9. The solar spectrum is which of the following?
   A. An absorption spectrum
   B. A continuous spectrum
   C. An emission spectrum
   D. [All of the above.]
   E. [None of the above.]

10. The spectrum of light from the Sun's atmosphere *seen during a total eclipse* is
    A. an emission spectrum.
    B. an absorption spectrum.
    C. a combination of emission spectrum and absorption spectrum.

D. a continuous spectrum.
E. a combination of all three types of spectra.

11. Analysis of a star's spectrum *cannot* determine
    A. the star's radial velocity.
    B. the star's tangential velocity.
    D. the chemical elements present in the star's atmosphere.
    D. [More than one of the above.]

12. According to the Doppler effect,
    A. sounds get louder as their source approaches and softer as it recedes.
    B. sounds get higher and higher in pitch as their source approaches and lower and lower as their source recedes.
    C. sounds are of constant higher pitch as their source approaches and of constant lower pitch as the source recedes.
    D. [Both A and B above.]
    E. [Both A and C above.]

13. The Doppler effect causes light from a source moving away to be
    A. shifted to shorter wavelengths.
    B. shifted to longer wavelengths.
    C. changed in velocity.
    D. [Both A and C above.]
    E. [Both B and C above.]

14. We can determine the elements in the atmosphere of a star by examining
    A. its color.
    B its absorption spectrum.
    C. the frequency at which it emits most energy.
    D. its temperature.
    E. its motion relative to us.

15. Which list shows the colors of stars from coolest to hottest?
    A. Red, white, blue
    B. White, blue, red
    C. Blue, white, red
    D. Red, blue, white
    E. Blue, red, white

16. The Doppler effect is used to
    A. measure the radial velocity of a star.
    B. detect and study binary stars.
    C. measure the rotation of the Sun.
    D. [Two of the above.]
    E. [All of the above.]

17. Sound waves cannot travel in a vacuum. How, then, do radio waves travel through interstellar space?
    A. They are extra-powerful sound waves.
    B. They are very high frequency sound waves.
    C. Radio waves are not sound waves at all.
    D. The question is a trick, for radio waves do *not* travel through interstellar space.
    E. Interstellar space is not a vacuum.

18. The energy of a photon is directly proportional to the light's
    A. wavelength.
    B. frequency.
    C. velocity.
    D. brightness.

19. In the Bohr model of the atom, light is emitted from an atom when
    A. an electron moves from an inner to an outer orbit.
    B. an atom gains energy.
    C. an electron moves from an outer to an inner orbit.
    D. one element reacts with another.
    E. [Both A and B above.]

20. The intensity/wavelength graph of a "blue-hot" object peaks in the
    A. infrared region.
    B. red region.
    C. yellow region.
    D. ultraviolet region.

21. The emission spectrum produced by the excited atoms of an element contains wavelengths that are
    A. the same for all elements.
    B. characteristic of the particular element.
    C. evenly distributed throughout the entire visible spectrum.
    D. different from the wavelengths in its absorption spectrum.
    E. [Both A and D above.]

22. Each element has its own characteristic spectrum because
    A. the speed of light differs for each element.
    B. some elements are at a higher temperature than others.
    C. atoms combine to form molecules, releasing different wavelengths depending on the elements involved.
    D. electron energy levels are different for different elements.
    E. hot solids, such as tungsten, emit a continuous spectrum.

23. The speed of sound is 335 m/s. What is the wavelength of a sound that has a frequency of 500 cycles/s?
    A. 0.67 m
    B. 1.49 m
    C. 165 m
    D. 835 m

24. Suppose the speed of a water wave is 12 inches per second and the wavelength is 4 inches. What is the frequency of the wave?
    A. 48 cycles/s
    B. 18 cycles/s
    C. 8 cycles/s
    D. 3 cycles/s
    E. ⅓ cycles/s

25. Define *wavelength* and *frequency* in the case of a wave.

26. Which has the higher frequency, light of 400 nm or light of 450 nm?

27. Approximately what are the least and greatest wavelengths of visible light?

28. Name six regions of the electromagnetic spectrum in order from longest to shortest wavelength. (Do not list the colors of visible light as separate regions of the spectrum.)

29. Which parts of the electromagnetic spectrum penetrate the atmosphere?

30. Define the following terms: electron, nucleus, photon.

31. State the three postulates upon which the Bohr model of the atom is based.

32. How is the energy of the photon related to the frequency of the light?

33. Why do the various elements each have different emission spectra?

34. Name the three different types of spectra and explain how each is produced.

QUESTIONS TO PONDER

1. In the case of a wave, the meanings of the terms *frequency* and *velocity* are often confused. Distinguish between the two, using water waves to give an example.

2. The period of a wave is defined as the amount of time required for the wave to complete one cycle. What, then, is the relationship between the period and the frequency of a wave?

3. It is especially easy to get a sunburn when skiing high in the mountains. How does the high altitude contribute to the danger of sunburn?

4. What does an object do to the white light that strikes it to give the object its color? How does this differ from the color of an object that emits its own light?

5. Draw an intensity/wavelength graph for a red star and another for a white star. Describe two ways in which they differ.

6. Graphs like that of Figure 4-10 are plotted for stars, but the radiation from stars is an absorption spectrum. Why doesn't the graph indicate less intensity (due to absorption) for some wavelengths?

CALCULATIONS

1. If the speed of a particular water wave is 12 meters/second and its wavelength is 3 meters, what is its frequency?

2. The speed of sound in air is about 335 m/s. What is the wavelength of a sound wave with a frequency of 256 cycles/second?

3. Express 500 nm in meters.

4. The 656.285 nm line of hydrogen is measured to be 656.305 nm in the spectrum of a certain star. Is this star approaching or receding from the Earth? Calculate its radial speed relative to the Earth.

5. If a certain star is moving away from Earth at 25 km/s, what will be the measured wavelength of a spectral line that has a wavelength of 500 nm for a stationary source?

The following questions use material from the Close Up, "Blackbody Radiation."

6. Suppose the Kelvin temperature of blackbody X is twice as great as that of blackbody y. Compare the energy released by equal areas of the two objects.

7. Use Wien's law to determine the peak wavelength emitted by an object with a temperature of 6000°C.

StarLinks netQuestions

Visit the netQuestions area of StarLinks (www.jbpub.com/starlinks) to complete exercises on these topics:

1. What is Spectroscopy? A white beam of light passes through a prism; a beam of multicolored light emerges on the opposite side. What is it about light that allows it to be separated into colors and why this so important for studying astronomy?

2. Radiation Laws In the late nineteenth century, scientists made theoretical calculations of the radiation that would be emitted from an object that absorbed all wavelengths completely. Such an object was called a blackbody, and the radiation it emitted was called blackbody radiation.

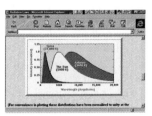

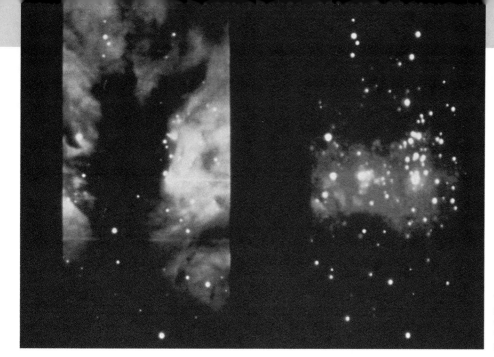

Two views of a region in the constellation Orion. The left one is what a visible-light telescope reveals. The other is an infrared image of the same portion of the sky.

Telescopes: Windows to the Universe

Refraction and Image Formation
The Refracting Telescope
Chromatic Aberration
The Powers of a Telescope
Angular Size and Magnifying Power
Light-Gathering Power
Resolving Power
The Reflecting Telescope
Large Optical Telescopes
CLOSE UP: Spinning a Giant Mirror

Telescope Accessories
Radio Telescopes
HISTORICAL NOTE: Radio Waves from Space
Interferometry
Detecting Other Electromagnetic Radiation
The Hubble Space Telescope
CLOSE UP: ET Life Part I—SETI
CLOSE UP: ET Life Part II—CETI

••••••• *LOOKS CAN BE DECEIVING. During the past few decades, astronomers have come face-to-face with this old saying. Up until the middle of this century, almost all of our knowledge of the sky came from light waves, the waves that we see. Therefore, photographs of the sky showed what our eyes see, just adding detail because telescopes can detect very dim light and magnify the image. Still, it seemed that "What we see is what we get."*

Then along came detectors of radiation in other portions of the spectrum. The chapter opening photo shows two views of a region called Orion B *displayed at the same scale and orientation. The image at left was taken with a "regular" visible light telescope; it shows what an extremely sensitive eye would see. The dark lane that runs vertically through the photo is caused by dust, which obscures stars within it. Infrared radiation penetrates dust, however, so the infrared image at the right looks very different from the visible light image. Which image shows reality? The answer is "Both, and more!" for the same picture taken in radio or ultraviolet radiation would look different still. We must conclude that our eyes don't tell us the whole story. In astronomy, what we see is just part of what we get.*

Galileo Galilei was the first to use a telescope to study the heavens systematically. Much was learned before Galileo's time by naked-eye observations, but Galileo's telescope changed astronomy—and our outlook on the universe—tremendously. As we make our telescopes larger and larger and take them into space, above the distortions of the Earth's atmosphere, we realize that we are just beginning to learn about the mysteries of our universe.

This century has brought a multitude of telescopic tools, and our view of the skies has expanded to parts of the spectrum well beyond visible light. The word *telescope* now includes instruments used to map the sky in all regions of the electromagnetic spectrum.

This chapter begins by describing some properties of light that are important to the understanding of visible-light telescopes. Then the use of such telescopes is discussed. Finally, the chapter concludes with a look at some of the nonvisible-light telescopes that are so indispensable to modern astronomy.

REFRACTION AND IMAGE FORMATION

The discussion of light thus far has concentrated on its wave properties. The effect of these properties will be examined later in this chapter, but for the moment we can forget about them and concentrate on a more obvious phenomenon: the path that light travels. That path is usually very simple, for light travels in a straight line as long as it remains in the same medium. It may, however, change direction upon entering a second medium. ►Figure 5-1a shows a ray of light passing through a wedge of glass. We see that the light travels in a straight line before and after passing through the glass and that it travels in a straight line inside the glass, but that the ray bends when it passes through each surface of the glass. Part (b) of that figure shows what happens when the two sides of the

The word medium *refers to the material that transmits the light.*

► **FIGURE 5-1.** When light passes through a wedge of glass, it is bent from its path. The smaller the angle of the wedge, the less the bending. (The "wedge" is really a prism, but we are concentrating here on the bending of the light rather than its separation into colors.) Notice that the bending occurs both when the light enters the glass and when it emerges.

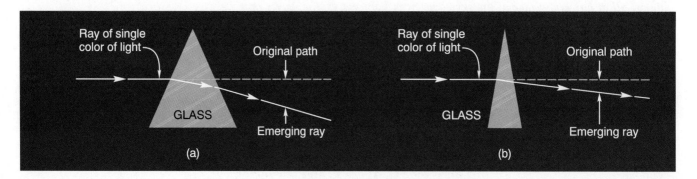

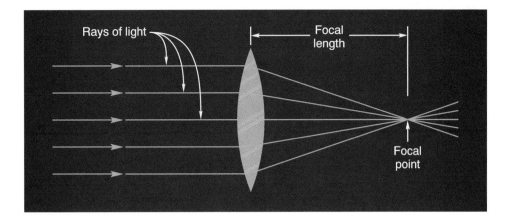

► **FIGURE 5-2.** A lens bends incoming rays of light toward a single point. When the incoming rays are parallel to the axis (as shown here), they cross at the lens's focal point.

wedge are more nearly parallel; the light bends less. The phenomenon of the bending of a wave as it passes from one medium into another is called *refraction*.

Two factors determine the amount of refraction that occurs when light crosses from one material into another. The first factor is the relative speeds of light in the two materials. In the last chapter we stated that the speed of light in a vacuum is 3.00×10^8 meters/second. In air, light's speed is just slightly slower. In glass, however, light travels at about 2×10^8 meters/second (more or less, depending upon the type of glass). Because of the change in speed, a ray of light may bend significantly when going from air into glass.

To understand the second factor, consider ► Figure 5-2, which shows a number of rays of light (that came from beyond the left side of the page) passing through a lens-shaped piece of glass. Notice that the rays that strike the surface of the glass at a glancing angle (farther from the perpendicular) bend more than those that hit it more "head-on." The central ray strikes the glass perpendicularly and does not bend at all. This is a general rule: the smaller the angle between the ray of light and the surface, the more the light bends upon passing through the surface.

Refer again to Figure 5-2. The surfaces of the lens have just the right curvature to cause all of the rays of light shown in the figure to pass through the same spot. To see the importance of this, imagine that each of the rays in the figure came from a distant star. If we put a piece of paper at the point where the light rays come together, all the light from that star that passes through our lens will come to a single point on the piece of paper. In fact, the rays of light coming from other stars will likewise come to a focus on the paper, forming an *image* of that area of the sky.

The *focal point* of a lens is that point where light from a very distant object comes to a focus. This is the point where the rays converge in Figure 5-2.

The *focal length* of the lens is the distance from the lens to the focal point. Depending upon the curvature of their surfaces, different lenses have different focal lengths.

THE REFRACTING TELESCOPE

Lenses, of course, are at the heart of the telescope—particularly the refracting telescope. (The reflecting telescope will be discussed later in this chapter.) The simplest use of a telescope (in principle, anyway) is as a lens for a camera. ► Figure 5-3a shows a camera mounted on a small telescope. What might not be obvious in the photograph is that the camera's normal lens has been removed. The camera is using the long–focal length lens of the telescope in place of its regular lens. Figure 5-3b shows the arrangement. The telescope lens simply re-

refraction. The bending of light as it crosses the boundary between two materials in which it travels at different speeds.

When light passes from the vacuum of space into Earth's atmosphere, it refracts, and this must be taken into account when measuring positions of celestial objects.

image. The visual counterpart of an object, formed by refraction or reflection of light from the object.

focal point (of a converging lens or mirror). The point at which light from a very distant object converges after being refracted or reflected.

focal length. The distance from the center of a lens or a mirror to its focal point.

 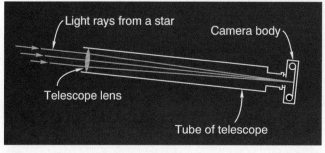

(a) (b)

▶ **FIGURE 5-3.** A simple way to use a telescope for photography is to let the telescope serve as the camera's lens. The image is focused directly on the film by the telescope's main lens.

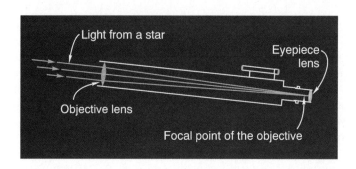

▶ **FIGURE 5-4.** When a telescope is used for direct viewing, the eyepiece is used as a magnifier to view the image formed by the objective lens. This image is formed at the focal point of the objective.

objective lens (or objective). The main light-gathering element—lens or mirror—of a telescope. It is also called the primary lens.

eyepiece. The magnifying lens (or combination of lenses) used to view the image formed by the objective of a telescope.

dispersion. The separation of light into its various wavelengths upon refraction.

chromatic aberration. The defect of optical systems that results in light of different colors being focused at different places.

places the regular camera lens and brings the image to a focus on the film. As we will see later, telescopes can be mounted so that they can track the stars across the sky and allow astronomers to take long time exposure photographs of the heavens. Long time exposures allow us to photograph much fainter objects than can be seen by simply looking through a telescope. Thus the use of a telescope with a camera is much more important to a professional astronomer than its use for direct viewing.

▶Figure 5-4 shows how a small telescope is used for direct observation. The primary lens, the lens through which the light passes first, is called the *objective lens,* or simply the *objective.* This lens brings the light to a focus at the focal point. This is the point where the film is located when the telescope is used with a camera. For direct viewing, a second lens, the *eyepiece,* is added just beyond the focal point. This lens simply acts as a magnifier to enlarge the image.

Chromatic Aberration

A prism separates light into its colors because different wavelengths of light are refracted different amounts. Except for the ray of light that goes straight through the center of a lens, rays go through a lens in much the same way as they go through a prism. And like the light going through the prism, the light passing through a lens separates into colors. This causes the lens to have a slightly different focal length for each wavelength of light. ▶Figure 5-5 exaggerates the effect but illustrates the idea.

Because the glass of a lens separates the colors, there is no single place where an image is exactly in focus. If the film of a camera is placed at the point where the red light focuses, the other colors will be out of focus, and the result will be an image with a fuzzy, bluish edge. This phenomenon, called **chromatic aberration,** occurs when a telescope is used for direct viewing as well as when it is used with a camera. The problem occurs not just in telescopes but in regular cameras as well. Fortunately, it can be corrected, at least in part.

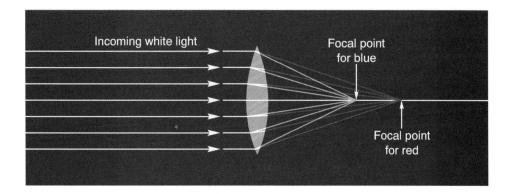

► **FIGURE 5-5.** Chromatic aberration. A lens exhibits a prism effect, separating white light into its colors. The lens therefore focuses each color at a different place. Only red and blue are shown here.

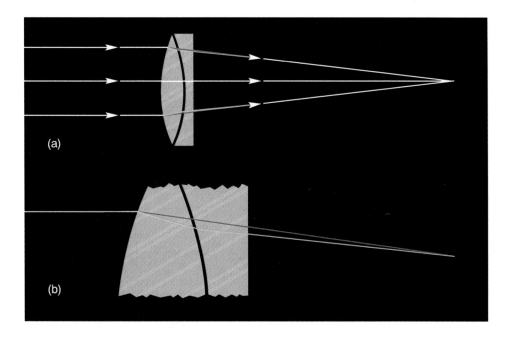

► **FIGURE 5-6.** (a) A lens can be corrected for chromatic aberration by the proper combination of lenses of different types of glass. Here the second lens brings the separated colors back together at the image. (b) This shows more detail of the upper portion of the lenses.

The amount of color separation that occurs when light passes through a lens depends not only on the curvature of the glass but also upon the type of glass. Some kinds of glass separate the colors more than other kinds. Telescope and camera manufacturers use this fact to correct for chromatic aberration. In all but the cheapest toy-store telescopes, the objectives of refracting telescopes are made of two lenses instead of one. As ►Figure 5-6a shows, the second lens has reverse curvature from the first. This curvature is not enough to undo all of the converging effect of the first, however, and the light is still brought to a focus. The second lens is made of a different type of glass than the first, and although it does not cancel out the bending of the light, it does cancel out most of the color separation as shown in Figure 5-6b. Such a combination of lenses is called an *achromatic lens.*

achromatic lens (or achromat). An optical element that has been corrected so that it is free of chromatic abberation.

THE POWERS OF A TELESCOPE

When most people think of a telescope's power, they think of magnification. Magnification, however, is only one of three major powers of a telescope and is the least important. Our discussion of the powers of a telescope begins with this least important one and then considers the other two powers: light-gathering power and resolving power. Again small telescopes will be described, but remem-

angular size (of an object). The angle between two lines drawn from the viewer to opposite sides of the object.

magnifying power (or magnification). The ratio of the angular size of an object when it is seen through the instrument to its angular size when seen with the naked eye.

Angular Size and Magnifying Power

The *angular size* of an object is the angle between two lines that start at the observer and go to opposite sides of the object. ➤Figure 5-7a shows someone looking at the Moon. The angle between the lines to the sides of the Moon is indicated. (Note that angular size is defined very similarly to angular separation in Chapter 1.) The angular size determines how big the image of the Moon is on the retina of your eye.

For a telescope (and binoculars and a number of other optical instruments), *magnifying power* or *magnification* is defined as the ratio of the angular size of an object when it is seen through the instrument to the object's angular size when seen with the naked eye. Figure 5-7b shows the angular size of the Moon as seen through a particular telescope. You might estimate from the angles in the figure that this telescope has a magnification of about six; that is, the telescope magnifies the object six times.

The magnification of a particular telescope depends upon the focal lengths of both the objective and the eyepiece. The magnification can be calculated using the following formula:

$$\text{magnifying power} = \frac{\text{focal length of objective}}{\text{focal length of eyepiece}}$$

or, in symbols,

➤ **FIGURE 5-7.** (a) The Moon seen by the naked eye. (b) In a telescope, the angular size of the Moon is apparently larger. The magnification of the telescope is the ratio of the two angular sizes.

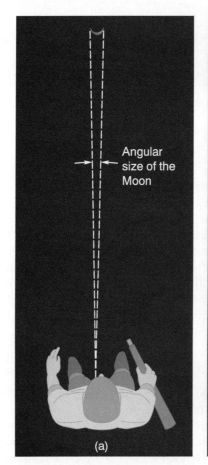
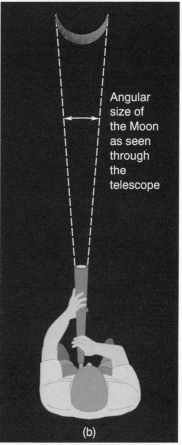

CHAPTER 5 Telescopes: Windows to the Universe

$$M = \frac{F_{\text{obj.}}}{F_{\text{eye.}}}$$

This means that the greatest magnification can be achieved by having a long–focal length objective and a short–focal length eyepiece.

EXAMPLE

The telescopes used in my introductory astronomy laboratory have objectives with focal lengths of 1250 millimeters. One eyepiece that is used has a focal length of 25 millimeters. What is the magnification produced by the telescope? What is the angular size of the Moon as seen through this telescope using the 25 mm eyepiece? (The naked-eye angular size of the Moon from Earth is about $1/2$ degree.)

Solution

To calculate the magnification, we use the equation that relates it to focal lengths:

$$M = \frac{F_{\text{obj.}}}{F_{\text{eye.}}}$$

$$= \frac{1250 \text{ mm}}{25 \text{ mm}}$$

$$= 50$$

Thus the magnifying power of the telescope is 50. We say that the magnification is 50 times, or 50×.

Now, since the Moon's angular size seen by the naked eye is $1/2$ degree and the telescope magnifies the Moon 50 times, the Moon's angular size in the telescope will be 50 times this, or 25 degrees.

......TRY ONE YOURSELF. What is the magnification produced by a telescope having an objective with a focal length of 1.5 meters when it is being used with an eyepiece having a focal length of 12 millimeters? (Hint: In doing the calculation, you must express the two lengths in the same units: either meters or millimeters.)

Notice that in both the example and the suggested problem, we hinted that one might use an eyepiece of a different focal length. Indeed, it is a minor matter to change the eyepiece of a telescope. ►Figure 5-8 shows someone putting an eyepiece into a small telescope. An eyepiece like this costs relatively little, so it is common to have a number of different eyepieces for a telescope in order to have different magnifications available. The obvious question is, "Why not always use the greatest magnification?" There are a number of answers to this question.

The first answer is that as the magnification increases, the ***field of view*** of the telescope decreases. Field of view refers to how much of the object is seen at one time. ►Figure 5-9 shows the decrease in field of view as the magnifying power increases. This is entirely reasonable, for if the object appears larger, not as much of it will be contained within the view of the telescope.

When viewing an object—the Moon, for example—you may wish to see the whole object rather than just a portion of it. If so, you will use an eyepiece with a long focal length, thus producing less magnification.

The other reasons why we do not always use the greatest magnification relate to the other powers of a telescope.

field of view. The actual angular width of the scene viewed by an optical instrument.

The Powers of a Telescope

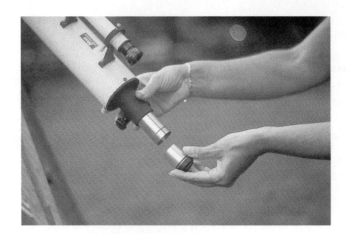

▶ **FIGURE 5-8.** Changing an eyepiece to change the magnification of the telescope is an easy matter.

(a) 50x   (b) 100x   (c) 250x

▶ **FIGURE 5-9.** Increasing the magnification decreases the field of view and makes the image darker. The entire telescopic view is shown in each case.

Light-Gathering Power

There is another difference in the three views in Figure 5-9. Notice that the more magnified the image, the darker it is. To see why this occurs, consider the part of the Moon we see in part (c). In part (a), the light from this portion of the Moon was concentrated on a small part of the image, but it is more spread out in the more magnified view in part (c). That is, the same light that covers only part of the image in part (a) has to illuminate the entire image in part (c). The image is therefore darker. There are two ways to make the image brighter and still retain this magnification. If one is taking a photograph, a longer time exposure can be used. When we do this, we allow the light's effect to accumulate on the film over a longer time, and the film becomes more exposed. The other way is to capture more light from the Moon in the first place. This can only be done by using a larger objective, which brings us to the second power of a telescope: the ***light-gathering power,*** which is often the most important power.

The light-gathering power of a telescope refers to the amount of light it collects from the object. When bright objects like the Moon and nearby planets are observed, the objects reflect so much light to Earth that lack of light is not a problem. Most objects that are observed with a telescope, however, are very faint, and to obtain an image, we need to capture as much light as possible from them. This is true whether we are using the telescope for photography or for direct viewing.

light-gathering power. A measure of the amount of light collected by an optical instrument.

Light-gathering power may seem an odd thing to call this power, for a telescope does not gather light in the sense of searching it out. It simply captures the light that hits the objective and brings that light to a focus.

The major way to gather more light is to use a telescope with a larger objective. The amount of light that strikes the objective simply depends upon the area of the objective. This is the primary reason it is desirable to have a telescope with a large-diameter objective and why the size of the objective is one of the features specified when discussing a telescope. For example, one might describe a particular telescope as a refractor with a 12-centimeter objective of focal length 140 centimeters. Notice, however, that although we state the *diameter* of the objective, it is the *area* that is important. Since the area of a circle is proportional to the square of the diameter, we must be careful in making comparisons of light-gathering power.

An example comparing the light received by two telescopes with objectives of different size will illustrate light-gathering power.

EXAMPLE

How does the light-gathering power of a telescope with a 6-inch objective lens compare to that of one with a 9-inch objective?

Solution

The area of a circle depends upon the square of its diameter, so the squares of the diameters of the telescopes must be compared in order to compare their light-gathering power. We'll set up a ratio of the squares of the two diameters:

$$\frac{9^2}{6^2} = \frac{81}{36}$$
$$= 2.25$$

Thus the light-gathering power of the larger telescope is more than twice that of the smaller.

••••• **TRY ONE YOURSELF.** If I trade my 3-inch telescope for a 5-inch model, by how many times do I increase my light-gathering power?

The desire to be able to see fainter and fainter objects in space has led us to make larger and larger telescopes. Before discussing large telescopes, however, another advantage of size, the third and last power, must be considered.

Resolving Power

One property of light that we assume in our everyday life is that it travels in straight lines (unless it reflects from a mirror or is refracted at a surface). If we could not assume this, we would be unsure whether an object we see is in front of us or behind us. Yet there are exceptions to this rule: light does not always travel in straight lines. The exceptions are usually unimportant but look at ➤Figure 5-10. This is a magnified photograph of the shadow of a regular household screw. The shadow was made by holding the screw a few meters from a screen and illuminating it with a small, bright light source. Notice that the edges of the shadow are not distinct and that in fact there are light and dark fringes near the edges.

In this case, the light passing near the edge of the object (the side of the screw) has "spread out" slightly. The effect is small and is seldom seen in everyday life because conditions must be right and you must look carefully. We will not discuss the reason that light acts this way except to point out that water waves behave similarly; they bend around corners. Such bending of waves as they pass by the edge of an obstacle is called ***diffraction.***

diffraction. The spreading of light upon passing the edge of an object.

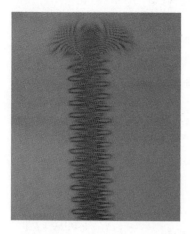

► **FIGURE 5-10.** If an extremely small light source is used, the shadow of an object will have light and dark fringes at its edge. This is due to diffraction of light. (A laser was used to make this photo, but laser light is not necessary to produce the effect.)

resolving power (or resolution). The smallest angular separation detectable with an instrument. Thus it is a measure of an instrument's ability to see detail.

The amount of diffraction that occurs when light passes through an opening depends upon two things: the wavelength of the light and the size of the opening. The longer the wavelength, the more diffraction; and the larger the opening, the less diffraction.

In a telescope, the objective itself forms the opening through which the light passes. The effect is small, but it is there: light that should bend regularly and accurately according to the laws of refraction fans out a slight amount. This results in the image not being exactly clear. The image of a star that should appear as a single point is blurred into a small spot. And when we increase the magnification, we simply make the blurred spot bigger and fainter.

►Figure 5-11 is a photograph of the Big Dipper and indicates what seems to be a single star in the Dipper's handle. If you have fairly good eyes and look at the Dipper in a clear dark sky, you can see that this star is not one, but two stars. We say that your eyes, and the clarity of the sky, allow you to *resolve* the pair of stars. Someone with poorer eyesight may be unable to resolve the pair. Now suppose you look at this pair of stars with a small telescope. ►Figure 5-12 shows what you see. The photograph shows three stars, two of them very close together. In fact, the two stars that are close together were seen as one star when viewed with the naked eye; it was the brighter of the naked-eye pair. The fainter of the naked-eye pair is the third star in the photograph, at the opposite side of the field of view. The telescope is able to resolve the group of stars into three.

The **resolving power** of an instrument is the smallest angular separation two stars can have and still be resolved as two by the instrument. Thus resolving power is described in terms of an angle.

What is it about a telescope that determines its resolving power? Naturally, the quality of the lenses is a major factor, but even with perfect optical components, the resolving power of a telescope is limited by diffraction. Since less diffraction occurs with a large objective, the maximum resolving power can be achieved by a telescope with a large-diameter objective.

The best human eye has a resolving power of about one minute of angle, or $1/60$ degree. A telescope with a 15-centimeter (about 6-inch) objective will have a maximum resolving power of about one second, or $1/3600$ degree. Based on size alone, the largest telescopes should have a resolving power far greater than this, but in fact the lack of clarity of the Earth's atmosphere becomes a major factor in limiting the resolution of large telescopes.

► **FIGURE 5-11.** Mizar and Alcor, stars in the handle of the Big Dipper, are seen as a single star by many people but can be resolved into two stars by those with better eyesight.

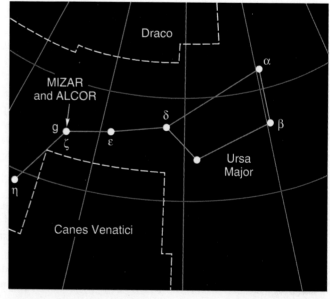

CHAPTER 5 Telescopes: Windows to the Universe

► FIGURE 5-12. Even a small telescope resolves Mizar into two stars (lower right). Alcor is at the upper left.

The lack of clarity of the atmosphere is caused by two factors: turbulence of the air and air pollution—the latter due to modern civilization or simply to dust. Recall that light is refracted as it passes from one material into another if there is a difference in its speed in the two materials. In fact, light travels at slightly different speeds in air at different temperatures. Our atmosphere always contains some amount of turbulence, and this causes air at various temperatures to move across the line of sight of a telescope. This results in the image moving slightly and places a limit on the resolution of even the largest telescope. Thus even the largest telescopes on Earth have a practical resolving power of about one-half arcsecond. As will be discussed later, the limit the atmosphere places on resolving power is the primary reason that the Hubble Space Telescope was built and put into orbit.

This atmospheric turbulence is what causes the twinkling of the stars when they are viewed with the naked eye.

THE REFLECTING TELESCOPE

As ►Figure 5-13 illustrates, an inwardly curved mirror, like a lens, will bring rays of light to a focus. This allows us to use it as the objective of a telescope. ►Figure 5-14 shows the arrangement devised by Isaac Newton. A small flat mirror is arranged in front of the objective mirror to deflect the light rays out to the eyepiece or camera body.

The largest refractor in existence is the 40-inch diameter telescope at Yerkes Observatory at Williams Bay, Wisconsin (►Figure 5-15). Reflecting telescopes

An inwardly curved mirror is said to be *concave*, so the objective mirror of a telescope is concave. A mirror with an outward curvature—such as the passenger-side mirror on many cars—is said to be *convex*.

► FIGURE 5-13. A curved mirror can bring incoming light rays to a focus. Again, the focal point is defined as the point where incoming rays that are parallel to the axis of the mirror converge.

► FIGURE 5-14. The Newtonian focal arrangement places a small flat mirror in the path of the reflected rays so that they are bounced off to the side and into the eyepiece (or camera or other instrument).

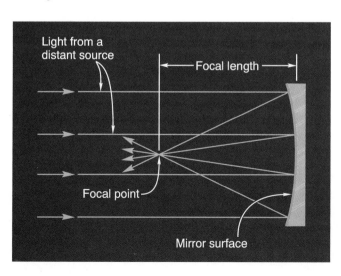
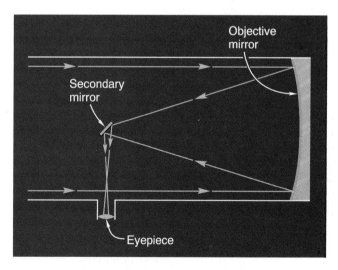

The Reflecting Telescope 131

► **FIGURE 5-15.** The 40-inch diameter Yerkes Observatory refractor. The telescope tube is nearly 20 feet long, indicating that the focal length of the objective is that long.

are made much larger than this, however. Before describing these telescopes, let's explain why reflectors can be made larger than refractors. The reasons include the following:

1. In order to be achromatic, a refractor requires two lenses. This means that four surfaces of glass have to be shaped correctly. A front-surface mirror, on the other hand, has only one critical surface. Since it is extremely important to obtain perfectly shaped surfaces, limiting the number of surfaces greatly simplifies the construction of the objective.

2. It is impossible to correct lenses completely for chromatic aberration. When light reflects from a mirror, however, all wavelengths reflect in exactly the same direction; thus reflectors automatically eliminate chromatic aberration problems.

3. Since a reflector's mirror is front-surfaced, the light does not pass through the glass of the mirror. Thus the glass does not need to be as perfect as that used for a refractor. It is difficult (and therefore expensive) to make large pieces of glass without tiny air bubbles or other imperfections.

For all of these reasons, a large reflector is much less expensive and more practical than a large refractor; as a result, all really large telescopes are reflectors.

LARGE OPTICAL TELESCOPES

Newtonian focus. The optical arrangement of a reflecting telescope in which a plane mirror is mounted along the axis of the telescope so that the mirror intercepts the light from the objective mirror and reflects it to the side.

Cassegrain focus. The optical arrangement of a reflecting telescope in which a mirror is mounted so that it intercepts the light from the objective mirror and reflects the light back through a hole in the center of the primary.

A reflecting telescope with an eyepiece arrangement like that in Figure 5-14 is called a *Newtonian telescope*, or is said to have a **Newtonian focus.** The Newtonian focus is a common one for small telescopes. ►Figure 5-16 shows the arrangement common in large telescopes, the **Cassegrain focus** (invented by G. Cassegrain, a French optician who lived at the time of Newton). Notice that the eyepiece or camera body is at the back of the telescope. In this arrangement, the secondary mirror is not a flat mirror but has an outward curvature. The effect of the curved secondary mirror is that an objective that actually has a short focal

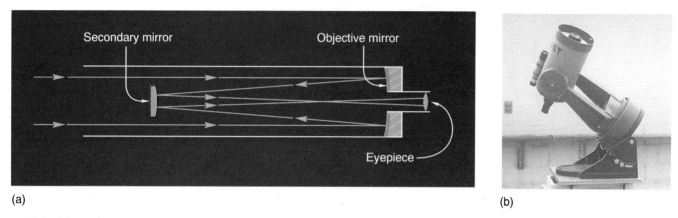

▶ **FIGURE 5-16.** (a) In the Cassegrain focal arrangement, the secondary mirror is curved outward (convex), and the light is reflected back through a hole in the objective or primary mirror. (b) This is a 5-inch diameter Cassegrain telescope.

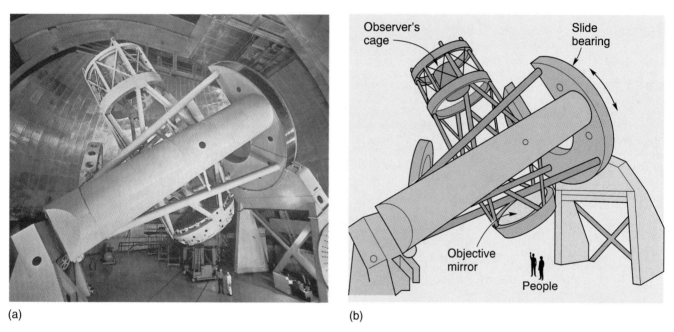

▶ **FIGURE 5-17.** The Hale telescope on Palomar Mountain. To appreciate the size of the telescope, look for the people at the bottom.

length can be given a longer effective focal length and thus can be contained in a short telescope.

In very large telescopes, observing is often done at the ***prime focus.*** This is the point where the light from the objective mirror comes to a focus, the focal point. ▶ Figure 5-17 shows the largest telescope in the United States, the 200-inch (5-meter) Hale telescope on Palomar Mountain. Look at the people in the photograph to appreciate the size of the instrument. The observer's cage is at the prime focus of the instrument, located at the top end of the telescope. The astronomer sits in the cage and is carried around with the telescope as it moves. The observer's cage does block some light to the objective mirror, but only a small fraction of it.

To achieve the best viewing conditions, it is advantageous to locate telescopes high in the mountains in dry, clear climates (▶Figures 5-18 and 5-19). This eliminates the blurring effects of the atmosphere as much as possible.

prime focus. The point in a telescope where the light from the objective is focused. This is the focal point of the objective.

Large Optical Telescopes

► **FIGURE 5-18.** Fifteen major telescopes of the European Southern Observatory cover the summit of this mountain in the Chilean Andes.

► **FIGURE 5-19.** The telescope domes of the Mauna Kea observatory are spread over the summit of Mauna Kea in Hawaii. The large dome in the foreground houses the 10-meter *Keck Telescope Facility*.

New telescope technology will soon produce a number of telescopes larger than the Hale telescope! The problem with making larger mirrors is that a large mirror tends to bend and sag when the telescope moves. The resulting change in the mirror's surface destroys image quality, of course. The accompanying Close Up describes how a laboratory at the University of Arizona is solving this problem.

A different new mirror design does not attempt to *prevent* bending. Rather, the mirrors of the four 8.2-meter telescopes of the European Southern Observatory's Very Large Telescope are designed to be flexible. The image they produce

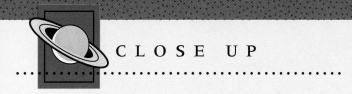

CLOSE UP

Spinning a Giant Mirror

The 5-meter mirror for the Hale Telescope was cast at the Corning Glass Works in 1934. It was nearly 60 years before a mirror this large was made again in the United States. In 1992, the Steward Observatory Mirror Laboratory transformed 10 tons of glass into a 6.5-meter mirror blank. The Steward Laboratory is making mirrors in a radically different way than ever before. In the past, mirrors were made by melting glass to form large pieces with flat surfaces. Then the working surface of the glass was carefully ground into a curve to form a concave mirror surface. For the 6.5-meter mirror, that would have meant that 12 tons of glass would have to be ground from the surface, and one year of extra work would be needed. Under the stands of the University of Arizona football stadium, technicians of the Steward Laboratory form the mirror surface by spinning the mirror as it cools.

Rotating the furnace—a technique called spin-casting—saves time and money in the production of large telescope mirrors. It shapes a natural curve in the surface of the molten glass in the same way that swirling a liquid in a glass causes a curved surface. The curvature created this way is close to the mirror's final parabolic shape, the rotation speed determining the amount of curvature. To create the 6.5-meter mirror, the furnace was spun at 7.4 revolutions per minute. This may seem slow, but remember that the mirror was 13 meters—30 feet—across.

The new mirrors are much thinner than the Hale telescope mirror, making them much lighter. The 6.5-meter mirror has 65% more light-gathering area than the Hale telescope mirror, but it is 40% lighter. The glass of the mirror surface averages only a little more than an inch thick; strength is provided by a honeycomb structure that extends back nearly 30 inches at the edge of the mirror and half that at the center.

Another advantage to spin-casting is that the mirrors can be made with a deeply curved, parabolic surface that gives them a focal length much shorter than conventional mirrors. The shorter focal length means that the mirrors require a much shorter, less expensive enclosure.

That first large Steward Laboratory mirror is being installed at the Multiple Mirror telescope on Mount Hopkins in southern Arizona. A second 6.5 meter mirror was completed in 1994 for the Magellan Project telescope on Las Companas, Chile. Early in 1997, the lab cast the first of two identical 8.4-meter mirrors destined for the Large Binocular Telescope Project on Mt. Graham, Arizona.

---

will be constantly monitored by a computer; to keep the image sharp under different conditions, the computer will control 180 motors that will push and pull on the back of each mirror to adjust its shape. This design is called *active optics*. The various instruments of the Very Large Telescope are scheduled to be put on-line from 1998 through 2001.

Meanwhile, a different type of active optics telescope began operating in 1993 on the 4200-meter (13,800-feet) summit of Mauna Kea in Hawaii (Figure 5-19). At this remote location, high above the clouds where few lights pollute the atmosphere, the sky is clear, calm, and dry 300 nights a year. ➤Figure 5-20 shows the multiple mirrors of the Keck telescope and ➤Figure 5-21 is an unusual photo of the telescope inside its dome. The telescope has a 10-meter aperture consisting of 36 segments, each weighing 880 pounds and each mounted separately and controlled by computer. Keck II is now under construction, and by the end of the century there will be two giant telescopes on the Hawaiian mountain.

active optics. A system that monitors and changes the shape of a telescope's objective to produce the best image.

Telescope Accessories

Telescopes have many other astronomical uses besides obtaining images of celestial objects. A few of these applications are described here. Each involves accessories that are used with the telescope.

➤ **FIGURE 5-20.** The 36 mirrors of the Keck telescope are reflecting the underside of the prime focus cage, where the secondary mirror is mounted.

➤ **FIGURE 5-21.** This "X-ray" of the dome of the Keck telescope was produced by taking a time exposure photo while the open dome rotated. The telescope is pointing straight up, and the objective mirrors are just below the "transparent" dome. The prime focus cage is visible at the top center of the frame.

► **FIGURE 5-22.** A CCD (charge-coupled device) is a rectangle of the semiconductor silicon. This one contains nearly 164,000 electric circuits that detect the intensity of light striking them. A computer analyzes the resulting data and produces an image.

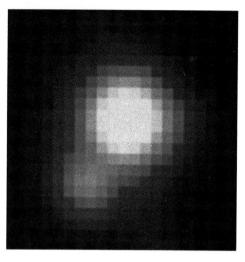

► **FIGURE 5-23.** This CCD image has been magnified so much that the individual pixels are obvious.

■ As an earlier section described, a camera can be attached to a telescope to take photos. The same basic principle is used in large professional telescopes, although the equipment and methods are more sophisticated. Photographic film is not a very efficient detector of light, however, for only about 5% of the light hitting the film causes the chemical reaction that results in an image. For this reason, astronomers often use various electronic light detectors, particularly the ***charge-coupled device (CCD)***. This device, about the size of a postage stamp (►Figure 5-22), is divided into small squares, each capable of detecting the intensity of the light that hits it. Just as a black-and-white newspaper photo (or the screen of an electronic game) is made up of many individual pixels, a CCD may have more than 4 million pixels. ►Figure 5-23 is a highly magnified CCD image in which individual pixels are visible.

The intensity of light collected in each pixel of a CCD is stored as a number in a computer. The data can then be used to show what the object would "look like" in a regular photograph, or they can be used to produce a false color image that reveals some other aspect of the object (►Figure 5-24). One must remember that a false color image does not tell us what the object actually looks like; instead it reveals some other property of the object. For example, a false color image may illustrate the intensity of radiation from the object by showing each brightness level as a different color. Many modern astronomical "photos" are actually CCD images, not standard photographs.

■ One important measurement that is made of celestial objects is the intensity of light that is received at various wavelengths, a procedure called ***photometry***. In the past this was done with a device similar to the light meter of

charge-coupled device (CCD). A small semiconductor that serves as a light detector by emitting electrons when it is struck by light. A computer uses the pattern of electron emission to form images.

photometry. The measurement of light intensity from a source, either the total intensity or the intensity at each of various wavelengths.

Large Optical Telescopes

➤ **FIGURE 5-24.** This false color image of the Moon not only emphasizes slight natural color differences, but compresses the spectrum from the ultraviolet to the near infrared so that it all shows as visible. The image was made from images from the spacecraft *Galileo*.

a camera, but today a CCD usually is used. The measurement can be made by placing filters in front of the light detector that allow only the wavelength of interest to pass through.

■ Chapter 4 pointed out that a tremendous amount of information about celestial objects is obtained by spectral analysis—the examination of light that has been separated into its various wavelengths. To obtain data for such an analysis, a *spectrometer* is connected to a telescope. This instrument uses a prism—or, more commonly, a *diffraction grating*—to separate light into its colors. A spectrometer produces either a photograph of the spectrum or numerical data about the intensity of light at various wavelengths.

An innovative new telescope design is used in the Hobby-Eberly Telescope (HET), recently put into operation at McDonald Observatory. Although it has the largest primary mirror in the world, using 91 hexagonal segments, it does not move the entire mirror to track objects across the sky. ➤Figure 5-25 illustrates how it works.

spectrometer. An instrument that separates electromagnetic radiation according to wavelength. (A spectrograph is a spectrometer that produces a photograph of the spectrum.)

diffraction grating. A device that uses the wave properties of electromagnetic radiation to separate the radiation into its various wavelengths.

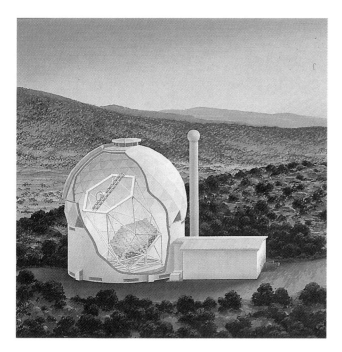

➤ **FIGURE 5-25.** The 11-meter-diameter objective of the Hobby-Eberly Telescope focuses light on a moving collector. (The collector is the device that spans across the hexagonal support at the upper left of the mirror.) Moving only the collector reduces by more than 10 times the amount of telescope mass that must be moved under precise control. The telescope's schedule is computer-controlled to allow many different types of observations to be made in a single night.

➤ **FIGURE 5-26.** The 140-foot (43-meter) radio telescope at the National Radio Astronomy Observatory near Green Bank, West Virginia.

RADIO TELESCOPES

Thus far the discussion has concentrated on optical telescopes—telescopes that gather visible light. Besides visible light, the type of radiation from space that best penetrates the atmosphere is radio waves. In 1931 a scientist working on radio transmission for Bell Laboratories noticed that static received by his antenna originated in the Milky Way. (See the accompanying Historical Note.) When better radio receivers were designed (largely during World War II), astronomers were able to pinpoint the sources of celestial radio waves, and the field of radio astronomy was born.

Two problems arise in examining radio waves from space. First, the intensity of radio waves from a star is much less than the intensity of light waves. Second, since the wavelengths of radio waves are a million times greater than the wavelengths of visible light, there is a corresponding decrease in the resolution of images made with radio waves. (Recall that diffraction is greater with longer wavelengths.)

Both of these problems are solved in the same way—by making radio telescopes extremely large. ➤Figure 5-26 shows the 140-foot radio telescope at the National Radio Astronomy Observatory at Green Bank, West Virginia. Notice that the telescope does not have a shiny reflecting surface. This is possible because of a feature of waves that goes hand-in-hand with the diffraction effect: Although longer wavelengths diffract more when going through an opening, they do not require as smooth a surface for reflection.

Radio telescopes are similar in principle to the satellite dishes we use to receive television signals from Earth satellites. In each case the reflector simply

www.jbpub.com/starlinks

Imperfections may not be more than one-tenth of the wavelength of the radiation being reflected. Thus for radio waves of wavelength 3 cm, 3 mm holes are okay.

HISTORICAL NOTE

Radio Waves from Space

In 1927 Karl Jansky received his B.A. in physics from the University of Wisconsin. After one year of graduate study, he was hired by Bell Laboratories to do research on radio communications. In 1927 the first telephone service across the Atlantic had opened, and the telephone signal was sent by radio. (The service cost $75 for three minutes!) Static was a common problem, however, and Jansky was assigned to design and build an antenna to find the source(s) of the static. He had no specific experience in radio or electrical engineering, but after much study and some dead ends, he built a 100-foot-long antenna with which he was able to detect weak radio signals and to determine the direction from which they came.

In January 1932 Jansky wrote in his monthly report that he detected "... a very steady continuous interference—the term 'static' doesn't quite fit it. It goes around the compass in 24 hours. During December this varying direction followed the sun." As the early months of 1932 passed, he found that the direction from which the signals came seemed to move around, getting further from the Sun. He anticipated that after the summer solstice, the apparent source would move closer to the Sun again, but instead it continued to move around the sky during the year. By August 1932 Jansky decided that the static was coming from a fixed place among the stars; it was some type of "star static."

Jansky's star static was the first detection of radio waves from space. The source of the waves was the center of the Milky Way Galaxy. (The *New Yorker* magazine said that "This is believed to be the longest distance anybody ever went to look for trouble.") Jansky had been slow to recognize the celestial nature of the source in part because he did not know astronomy well. Astronomers, in turn, were slow to recognize the significance of his work because they were unfamiliar with electronics and radio, and they could not imagine that celestial objects emit radio waves. Radio astronomy did not grow quickly after Jansky went on to other things, but today it is a major branch of astronomy and provides us with information about the heavens that could not be learned in other ways.

directs waves to a small detector located at the focal point of the "mirror." You can see the supports for the antenna in the photograph of the radio telescope.

The world's largest radio telescope, located in Arecibo, Puerto Rico, is not capable of independent movement. Shown in ➤ Figure 5-27, it is a telescope constructed by stretching wire mesh across a natural bowl between hills. This telescope is 300 meters (1000 feet) in diameter and scans the sky as it moves along with the Earth. Slight changes in the direction from which it detects radio signals can be achieved by moving its antenna, which hangs from the cables suspended above the bowl.

➤ FIGURE 5-27. (a) The radio telescope near Arecibo, Puerto Rico, is the world's largest. Its radio antenna is suspended above it from three supports. (b) This photo was taken under the Arecibo antenna.

(a)

(b)

The image formed by a radio telescope is not a normal photograph, of course. The radio telescope simply detects the intensity of radio signals from the area of the sky toward which it is pointed. One way to display and examine the data received is to plot a graph of the intensity of the radiation as the telescope moves across a small portion of the sky. ➤Figure 5-28a shows what such a plot might look like. A more complete image of the radio-emitting object can be obtained

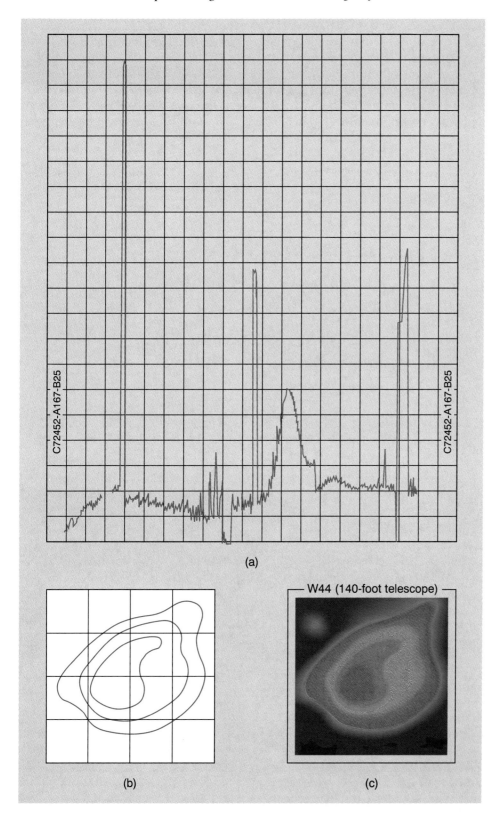

➤**FIGURE 5-28.** (a) A typical graph made by one scan of a radio telescope across a source. (b) A contour map can be made from a number of such scans. The strength of the radio signals is greatest in the center. (c) Color has been added to the contour to produce a false color image.

Radio Telescopes

by scanning the radio telescope back and forth across the celestial object and feeding the data into a computer that is programmed to represent the various intensities of the radio waves as different colors. Parts (b) and (c) of Figure 5-28 were produced in this manner and indicate the intensity of radio waves from a small portion of the sky.

At Green Bank, West Virginia, the first of a new generation of radio telescopes is being built. In a regular radio telescope, the detector and its supports not only block out a small portion of the waves, but they cause diffraction effects. The detector of the new Green Bank Telescope (GBT) will be located off to the side. This was done by constructing the reflecting surface in the shape of part of a much larger reflector, as shown in ➤Figure 5–29a. Part (b) of the figure is a drawing of the telescope at the Green Bank site, and ➤Figure 5-30 shows its size relative to the Statue of Liberty. The surface of the GBT will be 100 by 110 meters and will be composed of 2204 separate panels, each of which will be computer

Spark plugs in cars produce radio interference. Therefore vehicles with diesel engines are used near radio telescopes.

➤ FIGURE 5-29. (a) The reflecting surface of the Green Bank Telescope can be thought of as part of a much larger reflector. (b) The telescope will tower above other radio telescopes on the Green Bank site.

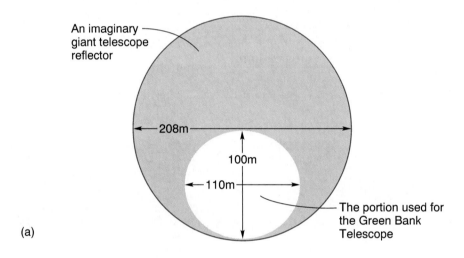

➤ FIGURE 5-30. The size of the Green Bank Telescope compared to the Statue of Liberty. The telescope is drawn here in an orientation to receive radio waves from straight above.

controlled. This adaptive capacity, the first use of active optics for radio telescopes, will permit the surface to be adjusted as the metal supports flex due to the telescope's motion. The resulting surface accuracy will allow the GBT to be useful at shorter wavelengths than would otherwise be possible. The telescope is scheduled for completion in 1999.

INTERFEROMETRY

Recall that the resolution of a telescope depends upon both the diameter of the telescope and the wavelength of the radiation. The greater the diameter of the telescope, the greater the resolution (assuming atmospheric clarity is not a factor), but the greater the wavelength, the poorer the resolution. Even though radio telescopes are very large, radio waves are so long that the best resolution from a single radio telescope is in the order of a number of arcminutes. This means that a radio source the size of a star would still appear in a radio telescope to be a blur as large as half the diameter of the Moon.

The solution to this problem lies in the fact that a giant radio telescope would retain the same resolution if only two portions of its outer surface were being used, as in ➤Figure 5-31a. Astronomers take advantage of this idea by combining two radio telescopes so that they act as one: in a sense, the two telescopes substitute for two portions of the outer part of a giant telescope, as in Figure 5-31b. In this way, they are able to obtain resolutions equal to that of a single large telescope.

Using two telescopes to act as one is not a simple matter, however. To understand the problem, refer to ➤Figure 5-32. Radio waves from a distant source are shown striking the dish of a radio telescope. Notice that they are reflected so that a single wave gets to the detector (located at the focal point of the dish) at the same time from all areas of the dish. This feature must be retained when two radio telescopes are used as one. Waves from each single dish must be combined in the correct relationship. We say that the waves from the two telescopes "interfere" with one another when they combine; therefore the technique of linking two (or more) telescopes so that they act as one is called *interferometry*. Due to today's extremely accurate atomic clocks, interferometry using widely separated radio telescope dishes has become possible.

The intensity of the signal received by only two spots would be much less than from the entire dish, of course.

interferometry. A procedure that allows a number of telescopes to be used as one by taking into account the time at which individual waves from an object strike each telescope.

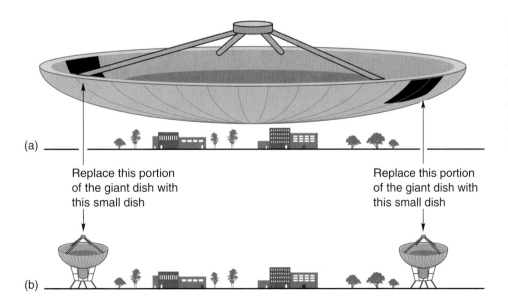

➤ FIGURE 5-31. Two small radio dishes (b) can be made to have the same resolution as a large radio telescope (a) that has a diameter equal to the distance between the two small ones. The strength of the signals detected will be much greater in the case of the single large telescope, of course.

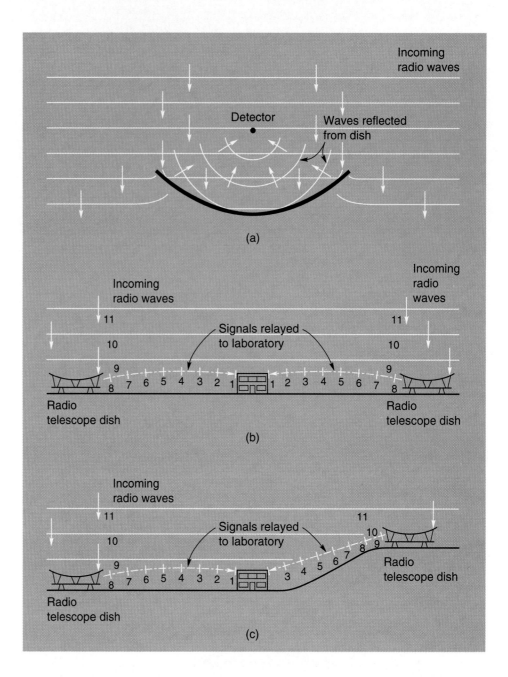

➤ **FIGURE 5-32.** (a) All portions of an incoming wave reach the detector of a radio telescope at the same time. (b) If two radio telescope dishes are to function as a single telescope, the waves must likewise reach the detector at the same time, or at least the time difference must be corrected for. (c) This is an obvious (and oversimplified) case of waves whose difference in reception times must be accounted for in the laboratory.

In the New Mexico desert is an array of radio telescopes used for interferometry. ➤Figure 5-33 is a photograph of part of this Very Large Array. The telescopes ride on a double pair of railroad tracks so that they can be moved and the arrangements changed.

The farther apart the telescopes, the better the resolution that can be obtained by interferometry. To achieve a longer baseline—distance between telescopes—a number of telescopes across the United States are being used as part of a single array, the Very Long Baseline Array (VLBA). The VLBA consists of 10 radio antennas, each 25 meters in diameter. The sites include Hawaii, New Hampshire, Washington, and the Virgin Islands. The control center is in Socorro, New Mexico. Very accurate atomic clocks are used to coordinate the signals received by such distant radio telescopes. The VLBA achieves resolutions of fractions of a milliarcsecond, 10,000 times better than earthbound optical telescopes. An angle of one milliarcsecond is less than the angular diameter of a dime at 1000 miles!

➢ FIGURE 5-33. The Very Large Array is spread over 15 miles of the New Mexico desert.

In 1997 the Japanese Space agency launched a radio satellite as part of the VLBI (Very Long Baseline Interferometry) program. The Japanese satellite is in an elliptical orbit that will provide a baseline three times longer than those achievable on Earth. This satellite, as well as others in the series, transmit their signals down to four Earth stations as they orbit.

Interferometry is also being employed in the newest optical telescopes. Recall that the Very Large Telescope of the European Southern Observatory is actually four separate telescopes rather than mirrors that focus light at the same point. To use the four telescopes as one, the signals from each must be combined using techniques similar to those originally developed for radio telescopes. Because the wavelength of visible light is much shorter than radio waves, matching waves from different sources is more critical for visible light waves. Therefore, optical interferometry is more difficult than radio interferometry.

The Center for High Angular Resolution Astronomy (CHARA) at Georgia State University is building an array of 5 telescopes of only 1-meter aperture each. They will be in a Y shaped array with arms 200 meters long. Because each telescope is relatively small, this system will cost much less than most new telescopes, but it is expected to produce a resolution of 0.2 milliarcseconds, better than any other optical telescope on Earth's surface.

DETECTING OTHER ELECTROMAGNETIC RADIATION

Visible light and radio waves pass through our atmosphere, and these two portions of the electromagnetic spectrum were the first that astronomers used in their quest to understand the heavens. But celestial objects emit radiation over the entire range from radio to gamma rays, and modern astronomy has tools that study each of the regions of the spectrum.

In the range of wavelengths from about 1200 nm to 40,000 nm (called the *near infrared*) are a number of narrow wavelength regions whose radiation

These wavelengths are called "near infrared" because they are the part of the infrared region that is near the visible portion of the spectrum.

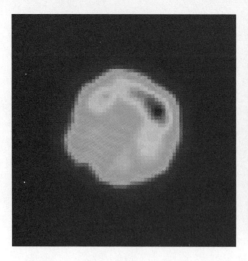
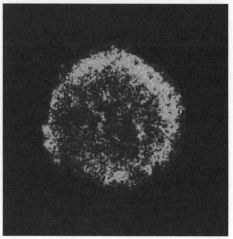

➤ **FIGURE 5-34.** These two images of the Tycho supernova remnant illustrate how higher resolution improves the quality of an image. The image on the left is from a low-resolution detector on the Einstein Observatory. The image on the right, taken by the High Resolution imager on the Einstein Observatory, has five times the resolution as the first one. AXAF images will be five times sharper than the image on the right.

SOFIA replaces the Kuiper Airborne Observatory, which observed in the infrared from 1974 to 1995.

penetrates to the surface of the Earth. Since water vapor is the chief absorber of radiation in the infrared, infrared observatories are located on mountains where the air is dry. The extinct volcano Mauna Kea is an ideal location for infrared telescopes, and two major observatories are located there.

Radiation in the far infrared region—wavelengths greater than about 40,000 nm—is emitted by cooler celestial objects such as planets and newly forming stars and is very valuable in the study of these objects. It does not penetrate the atmosphere as deeply as the shorter wavelengths, and we must therefore locate our instruments higher to detect it. Table 5-1 provides a list of some infrared observatories. Note that the first two named are located on mountains, but that the SOFIA project is designed to be carried on a plane, where it will observe from as high as 12,000 meters (41,000 feet), above 99% of the atmosphere's water vapor. Others on the list are orbiting observatories. NASA's Wide-Field Infrared Explorer (WIRE) is a small instrument with a large mission. It will be small enough to fit in the backseat of a car; its mission is to discover how galaxies change through time and to detect the birth of new galaxies.

As we move from infrared to consider wavelengths shorter than visible—shorter than about 400 nm—we come upon ultraviolet, X-rays, and gamma rays. Ozone is the chief absorber of most of this radiation and the atmosphere has a layer of ozone between about 20 and 40 kilometers altitude. Therefore, telescopes designed to detect this range of radiation must be located in space. Table 5-1 lists a number of these telescopes that have provided a tremendous amount of data since the 1970s, enough to open up entirely new branches of astronomy.

As this is being written, plans are in progress to launch the Advanced X-Ray Astrophysics Facility. AXAF is designed to observe X-rays from high-energy regions of the universe, such as hot gas in the remnants of exploded stars. Figure 5-34 illustrates the improved resolution expected from the AXAF mission.

The shortest wavelength in the electromagnetic spectrum is gamma radiation, which is the target of the Compton Observatory. Since its 1991 launch the Compton project has provided high-quality data to over 750 scientists from 23 countries. The Compton Observatory is designed to map the gamma ray sky in order to solve some of the outstanding questions that earlier missions had posed and—perhaps most importantly—to watch for the unexpected.

THE HUBBLE SPACE TELESCOPE

The Hubble Space Telescope (HST) has been saved for last because it is able to observe across the spectrum from the infrared to the ultraviolet regions. The HST

TABLE 5-1. Some Non-optical Observatories

Telescope	Sponsor	Aperture	Location	Dates
Infrared				
Wyoming Infrared Observatory	U. of Wyoming	2.3m	Wyoming mountains, 2940 meters elevation	1977-present
Infrared Telescope Facility	NASA	3 m	Mauna Kea, Hawaii 4200 meters elevation	1979-present
Kuiper Airborne Observatory	NASA	0.9 m (36 inches)	C-141 Cargo plane	1974-1995
Stratospheric Observatory for IR Astronomy (SOFIA)	NASA & German Space Agency	2.5 m	Boeing 747 airplane 12,000 m altitude	planned 2001 (expected 20-yr life)
Infrared Astronomical Satellite (IRAS)	U.S. (NASA), Netherlands, & United Kingdom	0.57 m	Earth Orbit	1983-1984
Infrared Space Observatory	European Space Agency		Earth Orbit	1996-present
Wide-Field Infrared Explorer	NASA	0.3 m	Earth Orbit	planned 1998
Ultraviolet				
International Ultraviolet Explorer	NASA, European Space Agency	0.45 m	Earth Orbit	1970-1996
Hopkins Ultraviolet Telescope	Johns Hopkins University		Flew aboard some Space Shuttle missions in the Astro Observatory	1990, 1995
Extreme Ultraviolet Explorer (EUVE)	NASA (UC Berkeley)	0.40 m	Earth Orbit	1992-present
Solar Extreme-ultraviolet Rocket and Spectrograph (SERTS)	NASA		Rocket launches from White Sands, NM	1995, 1996, continuing
X-ray				
Uhuru* or Small Astronomical Satellite (SAS)	NASA	0.048 square meters	Earth Orbit	1970-1973
High Energy Astrophysics Observatory (HEAO-1)	NASA		Earth Orbit	1977-1979
High Energy Astrophysics Observatory (Einstein)	NASA		Earth Orbit	1978-1981
European X-ray Observatory Satellite (EXOSAT)	European Space Agency		Earth Orbit	1983-1986
ROentgen SATellite (ROSAT)	Germany, U.S., United Kingdom	0.845 m	Earth Orbit	1990-present
Advanced X-Ray Astrophysics Facility (AXAF)	NASA		Earth Orbit, as high as 80,000 miles	planned 1998
Gamma Ray				
Compton Gamma-Ray Observatory	U.S., Germany, Netherlands, United Kingdom		Earth Orbit	1991-present

*Uhuru is Swahili for "freedom". The Uhuru satellite was launched from Kenya, Africa.

The Hubble Space Telescope

CLOSE UP

ET Life, Part I—SETI

Radio telescopes are designed and built to detect and measure radio waves coming from objects in space. If there are intelligent beings out there, these telescopes will also be useful both in detecting their presence and in communicating with them. In a Close Up in Chapter 13, we will describe various questions that must be answered before we can calculate the likelihood that extraterrestrial intelligence exists and how many life sites there are likely to be. Another way to answer the question of whether intelligent beings exist is to search for signals from such beings. SETI is the acronym for the Search for Extraterrestrial Intelligence.

Radio telescopes have been used on occasion to search the heavens for evidence of radio signals from intelligent beings. In 1960 American astronomer Frank Drake used a telescope of the National Radio Astronomy Observatory to search for signals from two nearby stars. The search, which Drake called Project Ozma, involved looking for unusual patterns in radio signals—patterns that were different from the signals emitted by inanimate objects such as stars and galaxies.

Since that time a number of astronomers have conducted searches. For example, in an experiment conducted in the mid-1970s, more than 600 nearby stars were watched for about 30 minutes each. In 1971 a group of astronomers and engineers made plans for an elaborate array of radio telescopes to be devoted to a search. Their proposal, called Project Cyclops, would have cost billions of dollars to put into operation.

Due to budget pressures, Congress canceled a search begun by NASA after $58 million had already been spent. This project was taken over by the SETI Institute, a private research group, and is now called Project Phoenix.

In order to search for signals from extraterrestrials astronomers have had to decide not only where to look in the sky, but on what frequencies to expect the signals. In a sense, we must know what "channel" the extraterrestrials are broadcasting on. Much study has gone into answering this question, and astronomers have selected certain ranges of frequencies as most likely. A primary consideration in making the decision is to select frequencies that are not emitted in great amounts by natural sources. Strong natural signals would drown out intelligent signals over long distances. Early SETI receivers had to "listen" to one frequency at a time, but the new receivers for MOP are capable of tuning to tens of millions of different frequencies simultaneously. This will make the search much faster than was previously dreamed possible.

Plans have been made in case we detect a signal that we verify as coming from an extraterrestrial being. There is even a protocol for involving international bodies in decisions about possible replies.

The "2" in the name of the Wide Field/Planetary Camera 2 is there because this instrument replaced an earlier version in the HST.

(➤Figure 5-35) has a modular design so that on subsequent shuttle missions astronauts can replace faulty or obsolete parts with new and/or improved instruments. This was fortunate, because soon after its 1990 launch, astronomers discovered that the objective mirror of the HST was slightly misshaped. "Major surgery" was performed in 1994 and the HST has performed flawlessly since.

The HST is roughly cylindrical in shape, 13.1 m end-to-end and 4.3 m in diameter at its widest point. Radiation enters the telescope through an opening below the open door at upper left in Figure 5-35. As of late 1997, the following instruments were aboard the spacecraft:

- The Wide Field/Planetary Camera 2 is actually four cameras. The "heart" of WF/PC2 consists of an L-shaped trio of wide-field sensors and a smaller, high resolution ("planetary") camera tucked in the square's remaining corner. This camera is the instrument that produces many of the beautiful astronomical images that you see in newspapers and popular literature, as well as in this book.
- The Space Telescope Imaging Spectrograph (STIS) can study celestial objects across a spectral range from the UV (115 nanometers) to the near-IR (1000 nanometers). The main advance in STIS is its capability to record the spectrum of many locations in a galaxy at the same time, rather than

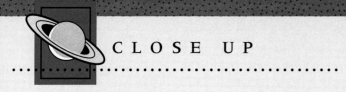

CLOSE UP

ET Life Part II—CETI

If our search for radio signals from a race of intelligent extraterrestrials is successful, it might succeed by detecting their stray, wasted radio signals. On the other hand, those beings may already be transmitting messages into space with the purpose of announcing their presence and telling others something about themselves. The same radio telescopes that are used to receive signals from space can be used to transmit radio signals. Considering the probable differences between the beings in different parts of the galaxy, what could one race of these beings communicate to another? The study of this question is called CETI, for Communication with Extraterrestrial Intelligence.

First, we must point out that if another intelligent race were found, the tremendous distances between stars would prohibit a dialogue. The nearest star is nearly five light-years away from us. It would require five years for our radio signals to reach a planet circling that star and another five years for the signal to return from beings on that planet. And the likelihood that life is so common that it exists near the nearest star is extremely remote. If exterrestrial life exists, the closest life sites are likely to be much farther away, making it impossible to get a reply to our signal during the span of one generation on earth.

At first glance, the language problems might seem insurmountable. But we do have something in common with every other race of beings that might exist: the physical universe and its laws, which are the same everywhere. Every intelligent race knows that hydrogen is the most common element and that an atom of hydrogen is made up of one proton and one electron. The prime numbers, those numbers that cannot be divided evenly by any other numbers but one and themselves, are the same in every language. Those studying the problem have concluded that an understandable message could indeed be sent if enough time were devoted to its transmission.

We Earthlings have already sent a very short message. In 1974 the reconditioned reflecting surface of the Arecibo telescope was rededicated, and at the ceremony, the telescope was used to transmit a message toward a cluster of 300,000 stars in the constellation Hercules. This transmission lasted only about 10 minutes, so the information that could be sent was very limited. The signal consisted of a series of Morse Code–type pluses containing 1679 data points. The number 1679 was chosen because, except for 1 and 1679, only one pair of numbers can be multiplied to obtain 1679: 23 and 73. If the data points are arranged into a rectangular array, there are only two ways to do it: either 23 across and 73 down, or vice versa. One way will produce no pattern, but the other will make a pattern that includes crude pictures and numbers to tell the being that intercepts the signal a little about those who sent the message.

If extraterrestrial beings near some star happen to detect our 10-minute message, will they be able to decipher it? Who knows? Although we cannot know how much of it they will be able to understand, we can be confident that they will know that it comes from an intelligent source rather than an inanimate object. If more time were available to transmit messages, a slower development of language would be used so that understanding would be much more likely, but with the time limitation that existed, scientists believed this message was the best that could be done.

When might we receive a reply? The cluster toward which the message was sent is 26,000 light-years away, so we need only wait 52,000 year for an answer!

---

observing one location at a time. As a result, STIS is much more efficient at obtaining scientific data than the earlier HST spectrographs.

- The Near Infrared Camera and Multi-Object Spectrometer (NICMOS) consists of three cameras that are designed for simultaneous operation. Since infrared radiation is heat radiation, NICMOS's surroundings have to be cooled to very low temperatures (as do all infrared telescopes). NICMOS keeps its detectors cold inside a thermally insulated container that contains frozen nitrogen ice. The coolant will last for years, much longer than any previous infrared space experiment.
- The Faint Object Camera, built by the European Space Agency, is extremely sensitive to dim light. Consider the faintest star that the naked eye can see. For this camera to observe a star one-millionth as bright as that, the light must be dimmed by a filter system to avoid saturating the camera's detectors.

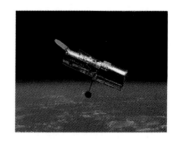

➤ **FIGURE 5-35.** The crew of the Shuttle mission STS-82 took this photo of the Hubble Space Telescope after they had released it.

Competition is keen for HST observing time. Only one of every ten proposals is accepted.

Although HST operates around the clock, not all of its time is spent observing. Each orbit lasts about 95 minutes, with time allocated for housekeeping functions and for observations. "Housekeeping" functions include turning the telescope to acquire a new target, or to avoid the Sun or Moon, switching communications antennas and data transmission modes, receiving command loads and sending data to Earth, and similar activities. Commands are sent to the HST several times a day to keep the telescope operating efficiently.

CONCLUSION

This chapter has shown how the phenomena of refraction and reflection of electromagnetic radiation allow us to gather radiation from dim stellar objects and to focus it to form an image. We saw that the powers of a telescope include not only magnification, but—more importantly—light-gathering power and resolving power. This analysis showed the importance of large telescopes and led to a discussion of reflecting telescopes, which can be made much larger than refractors.

Other-than-optical telescopes are becoming more and more important to our progress in understanding our universe by permitting us to observe objects that are invisible to the eye yet emit vast quantities of electromagnetic energy. New and different telescopes—including space telescopes—have provided us with information that was impossible to obtain by other means. As later chapters will show, entirely new celestial objects have been discovered in recent years by the new generation of telescopes. Undoubtedly, telescopes of the future will continue to bring us new and unexpected results and open whole new areas of exploration in astronomy.

Galileo's telescope began a revolution in astronomy nearly 400 years ago. Today the Hubble Space Telescope and other new telescopes are producing a comparable revolution. We do indeed live in exciting times.

RECALL QUESTIONS

1. The best site for an optical telescope is a place where the air is
   A. thin and dry.
   B. thin and moist.
   C. thick and dry.
   D. thick and moist.

2. Which of the following features determines the light-gathering power of a telescope?
   A. The diameter of the objective.
   B. The focal length of the objective.
   C. The focal length of the eyepiece.
   D. [Two of the above.]

3. Which of the following features determines the resolving power of a telescope?
   A. The diameter of the objective.
   B. The focal length of the objective.
   C. The focal length of the eyepiece.
   D. [Two of the above.]

4. Which of the following features determines the magnifying power of a telescope?
   A. The diameter of the objective.
   B. The focal length of the objective.
   C. The focal length of the eyepiece.
   D. [Two of the above.]

5. Which type of telescope is, in general, the largest?
   A. Infrared telescope.
   B. Visible-light telescope.
   C. Ultraviolet telescope.
   D. Radio telescope.
   E. [No general statement can be made.]

6. Which of the following telescopes has the greatest light-gathering power?

Telescope	Focal Length of Eyepiece	Focal Length of Objective	Diameter of Objective
A.	24 mm	150 cm	12 cm
B.	6 mm	100 cm	8 cm
C.	18 mm	125 cm	20 cm
D.	12 mm	90 cm	6 cm
E.	12 mm	100 cm	10 cm

7. Which of the telescopes in question 6 has the greatest magnification?

8. The objective of most radio telescopes is similar to the objective mirror of a reflecting optical telescope

A. in being concave in shape.
B. in its approximate diameter.
C. in being made of Pyrex glass.

9. The field of view of a telescope is
   A. the range of distance from the telescope over which it is in focus.
   B. the particular object being viewed by the telescope.
   C. the range of practical magnifying powers for the telescope.
   D. the range of wavelengths that can be detected by a particular telescope.
   E. the actual angular width of the scene viewed by the telescope.

10. Why are achromatic lenses used in optical telescopes?
    A. They reduce diffraction.
    B. They reduce color fringing.
    C. They produce greater magnification.
    D. They allow more light-gathering power.

11. Which of the following puts a limit to the useful magnification of a given telescope?
    A. Diffraction of light.
    B. Redshift of distant objects.
    C. The limit to how well lenses can be made.
    D. Reflection of light from parts of the telescope.

12. The resolving power of a telescope is a measure of its
    A. magnification under good conditions.
    B. overall quality.
    C. ability to distinguish details in an object.
    D. [All of the above.]

13. the main function of a telescope objective is to
    A. decrease chromatic aberration.
    B. collect light.
    C. disperse light.
    D. magnify images.
    E. [None of the above.]

14. Radio telescopes need not have finely polished surfaces because
    A. we are not interested in detail in the radio image.
    B. the speed of radio waves is less than that of light.
    C. the speed of radio waves is greater than that of light.
    D. radio telescopes can be used during the day.
    E. radio waves are much longer than light waves.

15. The primary purpose of a typical radio astronomer's work is to
    A. look for signals from other beings.
    B. send out radio waves to other beings.
    C. send out radio waves to be reflected back from stars and galaxies.
    D. receive radio waves sent out by radio sources.
    E. [All of the above.]

16. The eyepiece of a telescope is primarily used
    A. to collect as much light as possible.
    B. as a magnifier.
    C. as a prism to break light into its component colors.

17. A 40-inch telescope has _____ times the light-gathering power of a 10-inch telescope.
    A. 4
    B. 8
    C. 16
    D. 40
    E. [Either A, B, or C above, depending upon the eyepiece used.]

18. The Keck telescope is
    A. a large Earth-bound optical telescope.
    B. an orbiting telescope.
    C. a single radio telescope.
    D. an array of radio telescopes.

19. The Hubble telescope is
    A. a large Earth-bound optical telescope.
    B. an orbiting optical telescope.
    C. a single radio telescope.
    D. an array of radio telescopes.
    E. an orbiting infrared telescope.

20. The Arecibo telescope is
    A. a large Earth-bound optical telescope.
    B. an orbiting optical telescope.
    C. a single radio telescope.
    D. an array of radio telescopes.
    E. an orbiting infrared telescope.

QUESTIONS TO PONDER

1. Most backyard telescopes have a finderscope mounted on them—a small telescope that allows the observer to find a celestial object more easily. How would you expect the magnification and field of view of a finderscope to compare to that of the main telescope?

2. In the drawings showing light rays that come from very distant objects, the rays are represented as being parallel. If two rays come from a single point, how can they ever be parallel?

3. When you are searching for a stellar object in a backyard telescope, should you choose an eyepiece with a short or a long focal length? Why?

4. How is the field of view of a telescope changed when an eyepiece of longer focal length is used? How is the magnification changed?

5. Telescope A has a resolving power of 1.5 seconds, and telescope B has a resolving power of 2.0 seconds. Considering only their resolving power, which is the better telescope?

6. Suppose you and your neighbor each have a satellite dish to receive television signals from a satellite, but the signals from each dish are weak. Suppose further that you decide to combine the signals from the two dishes to produce a signal twice as strong. Why won't this work well?

7. Make a report on the progress of plans to use the Space Shuttle to install corrective optics (and make other repairs) to the Hubble Space Telescope. (Suggested sources: *Science News, Sky and Telescope,* and *Astronomy.*)

8. Write a report on plans for the Giant Metrewave Radio Telescope (GMRT) in India. (You might begin with *Sky and Telescope,* September 1992, p. 248.)

9. Some people argue that we should not beam radio messages into space because this will announce our location to any hostile beings who may detect the signal and they may then come and destroy or enslave us. What do you think?

CALCULATIONS

1. Suppose you have lenses of the following focal lengths: 30 cm, 10 cm, and 3 cm. If you wish to construct a telescope of maximum magnification, which two lenses would you use? Which would be the objective and which the eyepiece? What magnification would this telescope produce?

2. The pupil of your eye is the opening through which light enters. The maximum diameter of the pupil of a human eye is about 0.5 centimeter. How does the light-gathering power of two eyes compare to that of a telescope with a 10-centimeter objective?

3. The telescope at Mount Pastukhov in Russia has a 6-meter objective. Compare its light-gathering power to that of the 5-meter Hale telescope on Palomar Mountain.

4. In order to have four similar mirrors with the same light-gathering power as one 16-meter circular mirror, what must be the diameter of each of the four?

StarLinks netQuestions

Visit the netQuestions area of StarLinks (www.jbpub.com/starlinks) to complete exercises on these topics:

1. Our View of the Universe: Telescopes Telescopes are no longer limited to magnifying what we see; they include instruments that can map the sky in all regions of the electromagnetic spectrum.

2. Telescopes in Space? Instruments to measure and view the universe have changed considerably since Galileo first used a telescope to study the heavens systematically.

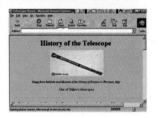

3. The Hubble Space Telescope (HST) The HST is returning amazing and often beautiful images of objects and events, some of which have never been seen before.

CHAPTER 6

The Moon at total eclipse. The red color is due to light that has passed through the Earth's atmosphere before striking and being reflected from the Moon.

THE EARTH-MOON SYSTEM

Measuring the Size of the Earth and Moon
The Distance to the Moon
Summary: Two Measuring Techniques
The Moon's Changing Size
The Moon's Phases
Lunar Eclipses
Types of Lunar Eclipses
Solar Eclipses
The Partial Solar Eclipse
The Annular Eclipse
Earth
The Interior of the Earth
Earth's Magnetic Field

CLOSE UP: The Earth from Space
Plate Tectonics
Earth's Atmosphere
The Moon's Surface
CLOSE UP: The Far Side of the Moon
Theories of the Origin of the Moon
The Large Impact Theory
The History of the Moon
CLOSE UP: Measuring the Age of the Earth and Moon
ACTIVITY: Do-It-Yourself Phases
ACTIVITY: Observing the Moon's Phases
ACTIVITY: Observing a Solar Eclipse

• • • • • • • • COLUMBUS'S FOURTH VOYAGE TO THE NEW WORLD was a particularly difficult one. He lost two of his four ships on the voyage, and the remaining two were infested with shipworms, forcing him to land in Jamaica. By February 1504 he had been trapped there for more than six months, and his attempts to barter with the natives for food were not going well. He learned from his almanac that an eclipse of the rising Moon was to occur in a few days, so he warned the *natives that God was angry at them for their laxity in supplying him with food and would punish them with famine and pestilence. Furthermore, God would demonstrate his intent by causing the Moon to become inflamed. Some natives were frightened, but others scoffed at Columbus's threats. When the Moon rose on the evening of February 29, 1504, the eclipse had already begun, so the Moon appeared to have a piece missing. As the eclipse grew, the Moon became darker and took on a red glow. (The photo shows a more recent total lunar eclipse.) On seeing this, the natives came running to Columbus with supplies and begged him to ask God to forgive them. Columbus knew that the eclipse would remain total for about an hour, so he told them that he would speak with God. When the eclipse reached maximum, he returned to the natives and told them that God had agreed to forgive them and would soon remove the inflammation from the Moon. From that night, Columbus was supplied with whatever he needed.*

Lunar eclipses also served a much more practical purpose for sailors of that time. Although a ship's latitude could be determined easily by measuring the altitude of Polaris, determination of longitude depended upon knowing the difference in local time between the ship's location and some location at a known longitude. Columbus attempted to use the eclipse to make this determination, for his almanac listed the local time of the eclipse at locations in Europe. Unfortunately, he made an error in his calculations, and the error confirmed his belief that he had sailed to Asia. He went to his death not knowing that he had traveled to a world that was unknown in the old country.

Astronomers have learned a great deal about the Earth and Moon since the advent of the space program a generation ago, but you might be surprised to learn that simple naked-eye observations had allowed people to calculate the size of the Earth and both the size of the Moon and the distance to it long before we even journeyed around the Earth.

This chapter begins by describing how the size of the Earth was measured more than two thousand years ago and then how the Moon's size and distance were measured. The phases of the Moon are discussed as well as solar and lunar eclipses. Finally, the last sections consider the Earth and Moon as planetary objects and examine some of their gross features. From this, some conclusions can be drawn about the origin of the Moon.

MEASURING THE SIZE OF THE EARTH AND MOON

Syene was located near the Nile River in Egypt and is now named Aswan.

zenith. The point in the sky located directly overhead.

stadium. An ancient Greek unit of length, perhaps equal to 0.15 to 0.2 kilometers. Various stadia were in use.

Eratosthenes (276–195 B.C.) devised a clever way to measure the size of the Earth. It was known that at noon on the first day of summer, the Sun shone straight down a well in the town of Syene. Thus the Sun was directly overhead, or at the **zenith** at that time. The city of Alexandria lies about 500 miles north of Syene, and at noon on that day, the Sun was located at an angle of 7 degrees away from the zenith (see ➤ Figure 6-1).

It was known that the Sun is very far away, so Eratosthenes could consider that sunlight striking both towns was coming from the same direction. As Figure 6-1 illustrates, this showed that Alexandria was 7 degrees around the Earth from Syene. Since 7 degrees is about 1/50 of a full circle (7/360), the circumference of the Earth should be 50 times the distance from Syene to Alexandria. The Greek unit of distance at the time was the **stadium** (plural: stadia), and measurements along the surface indicated that Alexandria and Syene were 5000 stadia apart.

CHAPTER 6 The Earth-Moon System

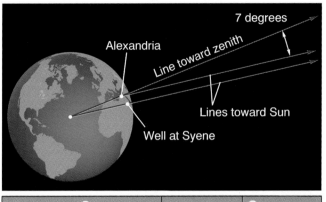

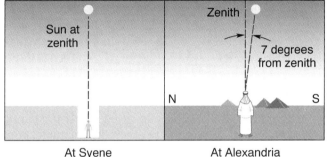

▶ FIGURE 6-1. It was observed that when the Sun was overhead at Syene, it was 7 degrees from overhead at Alexandria. This permitted the circumference of the Earth to be calculated.

Thus it was calculated that the distance around the Earth was 50 × 5000 stadia, or 250,000 stadia.

It would be nice to know how accurate Eratosthenes' measurement was. The problem is that we don't know the precise length of the stadium he used. Units were not well defined at the time. Besides, the distance between the towns was determined by measuring how long it took runners to go from one city to the other. Today we don't consider this a particularly good way to measure distances. The important point, however, is that the method is correct, and the Greeks were able to arrive at a realistic value for the circumference of the Earth. Today we know this distance to be about 40,000 kilometers. This means that the diameter is about 13,000 kilometers—8000 miles. This value is worth memorizing. A more accurate value can be found in Appendix C.

Eratosthenes apparently used a stadium equal to about 0.2 kilometers. If so, he measured the Earth's circumference to be about 50,000 kilometers, which is 25% too high.

The Distance to the Moon

The size of the Moon cannot be measured directly, of course. To calculate it, we need to know the distance to the Moon. ▶Figure 6-2 illustrates the idea.

By permission of Johnny Hart and Creators Syndicate, Inc.

▶ FIGURE 6-2. Both characters are right. To determine size, one must know the distance, and the word "looney" comes from "lunar."

Measuring the Size of the Earth and Moon

Recall the discussion of parallax in Chapter 2. Parallax is the phenomenon we observe when we hold up a thumb and look at it first with one eye and then with the other. The thumb's position changes with reference to the background when we do this. The Moon exhibits parallax when seen from different positions on the Earth. For example, if a person in Chicago and another person in Paris, France, happen to be looking at the Moon at the same time, they will observe it in slightly different positions against the background of stars. ➤Figure 6-3 shows this effect but exaggerates it greatly. In reality, the Moon is far enough away that its parallactic shift among the stars is very small. But the shift can be observed, and the observation allows us to measure the distance from the Earth to the Moon.

Using parallax, Ptolemy determined that the distance from the Earth to the Moon is 27.3 Earth diameters—very close to today's value of 30.16 for the average distance to the Moon. Taking Earth's diameter to be 13,000 kilometers, this puts the Moon about 390,000 kilometers from Earth. Now that the distance to the Moon is established, the question of how we know its size can be considered.

When we look up at the Moon from Earth, we have no way of judging its actual size. If you ask children how big the Moon looks, one might say that it is about the size of her play ball. Another might claim that it is gumball size. Just how do we judge the size of something we see? For example, suppose you see two objects like those in ➤Figure 6-4a. Could you tell me how large they are?

> 390,000 kilometers equal 240,000 miles.

➤ **FIGURE 6-3.** When viewed from two different places on Earth (*A* and *B* in the figure), the Moon seems to be at two different places among the stars. The effect is greatly exaggerated in the drawings.

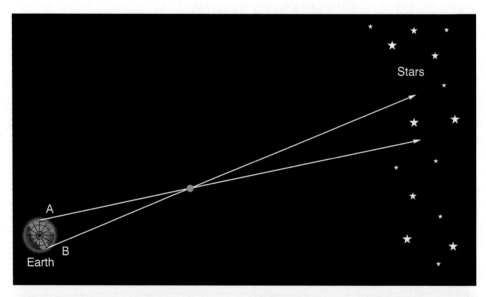

As seen from point A   As seen from point B

▶ FIGURE 6-4. (a) It is difficult to judge the real size of the two balls here. They are actually the same size. (b) Here we can use the surroundings to judge the size of the balls.

One looks larger than the other, but in fact they are the same size and one looks smaller simply because it is farther from the camera. When we view an object at a distance, we estimate its size in a number of ways. One way is by comparing it to objects of known size that are near it. That makes the judgment of the size of the two balls in Figure 6-4b easier.

When we look at the sky, there are no familiar objects near the things we see. In this case, we can estimate the size of an object only if we know how far away it is. By combining knowledge of its angular size and its distance, we make a judgment as to the object's size.

Recall that the angular size of an object is the angle between two lines that start at the observer and go to opposite sides of the object. ▶Figure 6-5 indicates the angular size of the Moon. Refer again to Figure 6-4a. As you look at the two balls, the angular size of the ball on the left is less than that of the ball on the right. So, not knowing its distance, you would say that it looks smaller.

The angular size of the Moon seen from Earth is very close to $1/2$ degree. Since there are 360 degrees in a complete circle, this means that about 720 Moons could be fitted in a circle around the Earth (▶Figure 6-6).

▶ FIGURE 6-6. A total of about 720 Moons could be drawn around the Earth if they were drawn at the correct distance. They wouldn't fit here, because these are drawn much too close to Earth.

▶ FIGURE 6-5. The Moon's angular diameter is about $1/2$ degree.

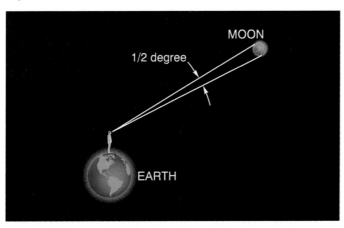

Measuring the Size of the Earth and Moon   157

Now the distance to the Moon and the angular size of the Moon (as seen from Earth) can be used to calculate the diameter of the Moon. This can be done with a simple equation relating the angular size, the distance, and the width of the object. (The term *width* is used here instead of diameter so that the equation will be general for all objects. In the case of the Moon, this will be the diameter.) The equation:

$$\text{width} = \frac{(\text{angular size in degrees}) \times (\text{distance})}{57.3}$$

This "small-angle formula" is very convenient and will be used again later in the book.

In symbols:

$$W = \frac{\theta d}{57.3} \quad \begin{aligned} W &= \text{width} \\ \theta &= \text{angular size in degrees} \\ d &= \text{distance away} \end{aligned}$$

This equation cannot be used accurately for large angles (more than about 5 degrees), but in most cases in astronomy, angular sizes are much less than this. Before using the equation for the Moon, let's use it in a terrestrial example.

EXAMPLE

Suppose you look at a picture on the wall across a large room. You happen to have a device with you that allows you to measure the angular width of the picture, and you determine it to be 1.5 degrees. Suppose further that you know that the picture is 5 meters away from you. How wide is the picture? Substituting into the equation:

Solution

$$\begin{aligned} W &= \frac{\theta d}{57.3} \\ &= \frac{1.5 \times 5 \text{ m}}{57.3} \\ &= .13 \text{m, or 13 cm} \end{aligned}$$

TRY ONE YOURSELF. Use the equation to calculate the diameter of the Moon, using the fact that the Moon's angular size is 0.52 degree and its distance is 384,000 kilometers (more accurate values than given above).

▶ **FIGURE 6-7.** The larger ball has a diameter four times the smaller. Would you say that its size is four times larger? Its volume is actually 64 times greater.

The small-angle formula, when used with the data in the "Try One Yourself," yields 3480 kilometers (2160 miles) for the Moon's diameter. This is very close to one-fourth of the diameter of the Earth. ▶Figure 6-7 shows two balls, one four times the diameter of the other. Form a mental image of these two balls. You would probably not say that the large one is four times the size of the small one. The word *size* is an indefinite term, so one can't tell whether it refers to diameter, area, or volume. *Size* will not be used in a quantitative sense here; that is, an object that has four times the diameter of another will not be described by saying that its *size* is four times greater.

To appreciate the scale of the Moon-Earth system, imagine the Earth to be a large grapefruit (about 5 inches in diameter). On this scale, the Moon would be a Ping-Pong ball 12 feet away. Between the Earth and Moon, there would be nothing but empty space.

SUMMARY: TWO MEASURING TECHNIQUES

As this study of astronomy proceeds, it will describe a number of relationships that allow us to measure features of objects in the heavens. Two have been introduced here. First, the way parallax was used to measure the distance to the Moon was described. This method is often called triangulation because it involves using a triangle to find a distance. As a later chapter will explain, this method is also used to measure distances to nearby stars. Parallax is an important phenomenon in astronomy. Notice that using it involves a relationship among three quantities: the size of the baseline, the angle of parallax, and the distance to the object. Knowing any two of the three, one can calculate the third.

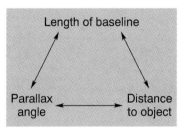

A second relationship involves angular size, actual size, and distance. Again, if you know any two of the quantities, you can calculate the third. This is an important relationship, and later chapters will explain its use in measuring the sizes of planets.

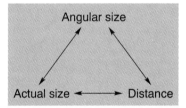

The Moon's Changing Size

▶Figure 6-8 shows two photographs of the full Moon taken at two different times by the same camera. The photographs make it obvious that the apparent size of the Moon changes. But apparent size depends on distance and actual size.

▶ FIGURE 6-8. These are two photos of the Moon put side by side; one was taken when the Moon is closest to Earth, the other when it is farthest.

perigee. The point in the orbit of an Earth satellite where it is closest to Earth.

apogee. The point in the orbit of an Earth satellite where it is farthest from Earth.

rotation. The spinning of an object about an axis that passes through it.

revolution. The orbiting of one object around another.

phases (of the Moon). The changing appearance of the Moon during its cycle, caused by the relative positions of the Earth, Moon, and Sun.

The words "month" and "Moon" have the same root, as you might guess by their similarity.

Terms related to the phases of the Moon are defined based on their elongation as shown in Table 6-1.

The actual size of the Moon doesn't change, of course, so our distance from the Moon must change. This occurs because the Moon's orbit is an ellipse, just like the orbits of all orbiting objects. And the Moon's orbit is fairly eccentric; that is, it is not very circular. The larger apparent size occurs when the Moon is at its *perigee,* or closest to the Earth, a distance of 363,300 kilometers. The smaller apparent size occurs at the Moon's maximum distance from Earth, 405,500 kilometers—when the Moon is at its *apogee.*

THE MOON'S PHASES

The Moon orbits the Earth in such a way that its same face points toward Earth at all times. "The Man in the Moon" always faces us. At first thought, you might be tempted to say that the Moon does not *rotate,* but this is not so. Consider ➤Figure 6-9. The Moon is shown with a spot on its surface. In (a), you see that if the Moon did not rotate, this spot would always face the same way in space. In (b), that spot continues to face the Earth as the Moon goes around its orbit. But this means that the Moon must rotate once for every *revolution* around the Earth. The fact that the rotation period and revolution period of the Moon are exactly equal seems remarkable and cannot be attributed to mere coincidence. In fact, this is another phenomenon that can be explained by the law of universal gravitation, as will be described later in this chapter.

The photographs in ➤Figure 6-10 show the *phases* of the Moon at various times during a period of about a month. The cause for the Moon's phases is fairly straightforward and has been known since antiquity. To explain the Moon's phases, we need only consider three objects: the Earth, the Sun, and the Moon.

The Moon circles the Earth, completing one orbit in slightly less than a month. This causes its position in the sky relative to the Sun to change with time. ➤Figure 6-11 shows various positions of the Moon in its orbit around the Earth. The Sun is out of the picture, far to the left. The drawings of the Earth and the Moon are dark on the side away from the Sun, because sunlight does not reach that side of them. Consider the Moon in position A. Most of the side that faces the Earth is dark, and only a small portion of that side is lit by the Sun. Figure 6-10a shows how the Moon appears from Earth when it is in position A. We call such a Moon a *waxing crescent* Moon.

When the Moon is at position B on its monthly trip around the Earth, we call it a *first-quarter* Moon, for if we start the cycle when the Moon is at position H, it is now one-quarter of the way around.

Position C will appear from Earth like the photograph of Figure 6-10c and

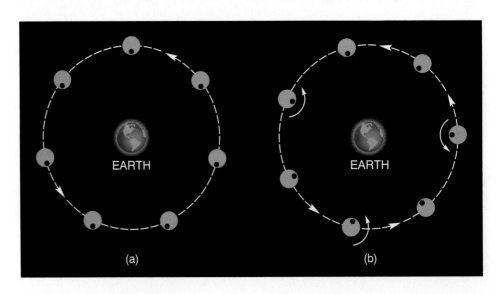

➤ FIGURE 6-9. The Moon rotates as it revolves. If it did not rotate, a dot on its surface would always point in the same direction in space as in (a). Instead, the Moon always keeps the same face pointed toward Earth (b).

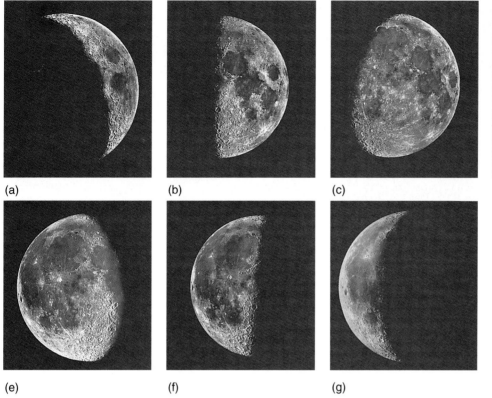

(a) (b) (c) (d)

(e) (f) (g)

▶ **FIGURE 6-10.** These photos of the Moon in phases correspond to the positions shown in Figure 6-11.

is called a *waxing gibbous* Moon. The word "waxing" is derived from an old German term that means "growing." Between points H and D, the Moon is waxing; the visible portion is growing from night to night from a thin crescent near H toward the **full Moon** (when the Moon is at position D). When we see the Moon in a gibbous phase, most of its sunlit side is facing the Earth.

Observe the photograph of the Moon when it is in position E. It is again in a gibbous phase, but here the gibbous phase is called **waning gibbous** because from night to night the lit portion that we observe is decreasing (waning) in size.

At position F the Moon is again in a quarter phase, the **third quarter,** or **last quarter.** Then around position G we have the **waning crescent** phase, and finally the Moon is back to where we start the cycle, at position H. Because we (arbitrarily) begin the cycle here, we call this a **new Moon.** In this position, the Moon is not visible in our sky because no sunlight strikes the side facing Earth. Only at this phase can an eclipse of the Sun occur. (In the next section we will see why such an eclipse does not occur every time there is a new Moon.)

The phases of the Moon are the subject of two Activities at the end of this chapter.

TABLE 6-1 Terms Relating to Moon Phases

Phase	Elongation (in degrees)
Waxing	0–180° east
Waning	0–180° west
Crescent	0–90° east or west
Gibbous	90°–180° east or west
New	0
First quarter	90° east
Full	180°
Third quarter	90° west

elongation. The angle of the Moon (or a planet) from the Sun in the sky.

The angle is illustrated in Figure 6-11.

The Moon's Phases

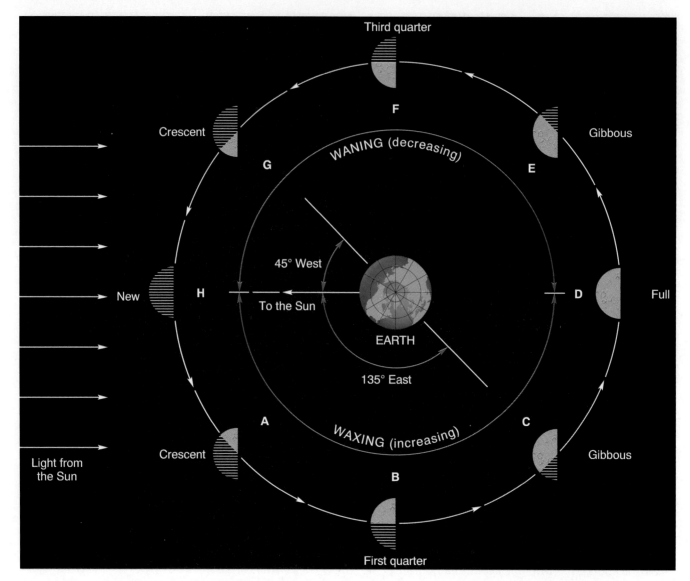

▶ **FIGURE 6-11.** The Moon in various phases. No light reaches the half of the Moon shown as black. The shaded portion, though lit, cannot be seen from Earth. The elongation of the Moon, stated as an angle either east or west of the Sun, is indicated for the waxing gibbous phase and for the waning crescent phase. We are viewing the system from above the north, so east is counterclockwise.

sidereal period. The amount of time required for one revolution (or rotation) of a celestial object with respect to the distant stars.

synodic period. The time interval between successive similar alignments of a celestial object with respect to the Sun.

lunar month. The Moon's synodic period, or the time between successive similar phases.

The lunar months is $29^d 12^h 44^m 2^s.8$.

Because the Earth moves in its orbit while the Moon revolves around it, we must distinguish between two revolution periods of the Moon. Refer to ▶Figure 6-12. The drawing shows the Earth in two positions, labeled A and B. At position A, the Moon is full. About $27^1/_3$ days later, when the Earth reaches position B, the Moon has completed one *sidereal* revolution, one revolution with respect to the distant stars. Notice, though, that the Moon is not full at this time. Slightly more than two additional days will be required for the Moon to reach the full phase again. At this point it will have completed one *synodic* revolution, and a *lunar month* will have passed. The lunar month is about $29^1/_2$ days.

LUNAR ECLIPSES

During its orbit of the Earth, the Moon sometimes enters the Earth's shadow; when this occurs, sunlight is blocked from reaching the Moon. This phenomenon

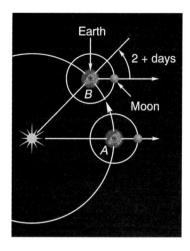

► **FIGURE 6-12.** This drawing shows the difference between the Moon's sidereal period and its synodic period. When the Earth is at point *A*, the Moon is full. At point *B*, the Moon has completed one sidereal period, but about two more days are required for it to reach full phase again, at which time one synodic period will have been completed.

is known as a ***lunar eclipse.*** You might wonder why this doesn't happen at the time of every full Moon. There are several reasons. First, as we have seen, the Moon and Earth are very small compared to their distance apart: the Moon is 30 Earth diameters away. Thus it is unlikely that they will align so accurately that one eclipses the other. Think of trying to align a grapefruit, a Ping-Pong ball 12 feet away, and a distant object.

Another reason a lunar eclipse does not occur during each lunar orbit is that the Earth's shadow is smaller than the Earth. This is because the source of light—the Sun—is so much larger than the Earth. Look at ►Figure 6-13. Although distances are not nearly to scale, the principle illustrated is correct. Notice that there is a cone of darkness behind the Earth. When the Moon is in this area (at point *A*, for example), it is in the full shadow of the Earth. This full shadow of

lunar eclipse. An eclipse in which the Moon passes into the shadow of the Earth.

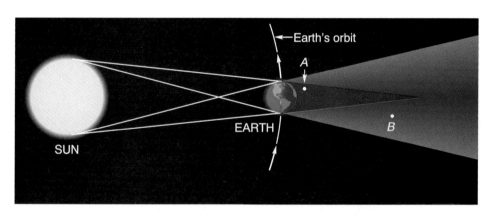

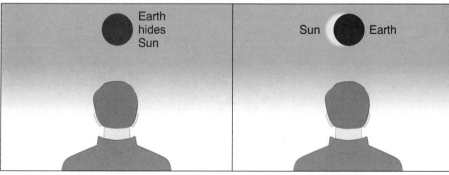

This observer is looking at the Earth and Sun from point *A*, in the umbra of the Earth's shadow.

This observer is looking at the Earth and Sun from point *B*, in the penumbra of the Earth's shadow.

► **FIGURE 6-13.** Point *A* is in the umbra of the Earth's shadow. Point *B* is in the penumbra, where light from part of the Sun hits it. Distances are not to scale in the drawing.

Lunar Eclipses

umbra. The portion of a shadow that receives no direct light from the light source.

penumbra. The portion of a shadow that receives direct light from only part of the light source.

the Earth—called the **umbra**—tapers down to a point. At the distance of the Moon, the width of the umbra is only three-fourths of the diameter of the Earth. So the Moon is less likely to pass through the shadow than if the shadow were the size of the Earth.

If the Moon is at point *B* of Figure 6-13, on the other hand, it is only in partial shadow, for light from the lower part of the Sun is hitting it. When the Moon is here, in the **penumbra**, it will not receive the full light from the Sun and will appear dim to Moon-watchers on Earth. The penumbral shadow increases in size at greater distances from Earth, but it is not equally dark across its width. Right next to the umbra, the shadow is very dark, but it gets brighter and brighter out toward its edge. When the Moon passes through the outer penumbra, we don't even notice the darkening.

There is a more important factor in explaining why a lunar eclipse does not occur at each full Moon, however. The plane of revolution of the Moon is tilted compared to the plane of the Earth's revolution around the Sun. Consider ➤Figure 6-14a, which shows both the Earth's and the Moon's orbit. The tilt of the Moon's orbit with respect to the Earth's is actually only 5 degrees, but we have exaggerated it in the drawing. Notice that when the Earth is at positions *B* and *D*, its shadow cannot hit the Moon. In fact, only when the Earth is near points *A* and *C* can the Moon pass through its shadow. These points represent the two

➤ **FIGURE 6-14.** (a) When the Earth is very near either *A* or *C*, lunar eclipses can occur, but when it is at other points in its orbit, the Moon does not pass through its shadow. (b) This imaginative drawing shows a flying fish (the Moon) "orbiting" a duck (the Earth) while the duck swims around a beachball (the Sun). The Moon-fish at the left never comes between the Sun and the Earth, so no eclipse occurs, but the orbit of the Moon-fish at the right is such that an eclipse will occur.

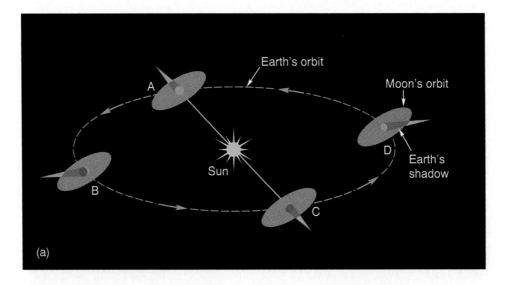

eclipse seasons that occur each year. Thus in most cases of a full Moon, the Moon will not be in the Earth's shadow but will be either north or south of it. The plane of the Moon's orbit changes relatively little as the year progresses, so eclipses can only occur about twice a year.

Types of Lunar Eclipses

The Earth's umbra gets smaller as the distance from Earth increases. At the Moon's average distance, the umbra has a diameter of about 9200 kilometers. Since the diameter of the Moon is less than 3500 kilometers, the Moon can easily be covered by the umbra. But the Moon might not pass right through the umbra. ➤Figure 6-15 shows three possible paths of the Moon through the shadow of the Earth. If the Moon moves along path *A*, it will only pass through the penumbra, producing a *penumbral lunar eclipse.* In this case, it will darken slightly as it does so, but such an effect will not be obvious from Earth and will only be noticeable if the Moon passes into the darkest part of the penumbra, near the umbra.

If the Moon follows path *B*, it will slowly darken as it moves toward the umbra. ➤Figure 6-16 is a triple exposure of the Moon during an eclipse. The Moon is moving along a path such as path *B*, and the exposure at the right side of the photo shows the Moon while only part of it is in the Earth's umbral shadow. As the Moon continues to move into the umbra, the shadow slowly moves across its surface until the Moon appears as it does in the center exposure, where we see a *total lunar eclipse.* Depending upon the Moon's distance from Earth, it may take an hour from the time of first contact with the umbra until the eclipse reaches totality. The Moon can stay in the shadow for up to about 1½ hours. On the left side of Figure 6-16, we see the Moon leaving the Earth's umbra again.

If the Moon follows path *C* of Figure 6-15, it will never be entirely covered by the umbra, and we will see only a *partial lunar eclipse.* The dark shadow will creep across the Moon, covering (in the case shown) only the top part of the Moon.

An eclipse of the Moon, especially a total eclipse, is a beautiful sight. The totally eclipsed Moon is not completely dark, however. Even when the Moon is

eclipse season. A time of the year during which a solar or lunar eclipse is possible. These occur when the Earth is at points A and C in Figure 6–14(a).

Eclipse seasons are slightly less than six months apart because the orientation of the Moon's orbit changes slightly as time passes.

penumbral lunar eclipse. An eclipse of the Moon in which the Moon passes through the Earth's penumbra but not through its umbra.

total lunar eclipse. An eclipse of the Moon in which the Moon is completely in the umbra of the Earth's shadow.

partial lunar eclipse. An eclipse of the Moon in which only part of the Moon passes through the umbra of the Earth's shadow.

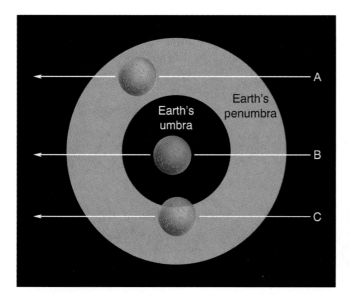

➤ FIGURE 6-15. Three of the possible paths of the Moon through the Earth's shadow. Path *A* produces only a penumbral eclipse. Path *B* produces a total eclipse and *C* produces a partial eclipse.

➤ FIGURE 6-16. This is a triple-exposure photo of the Moon taken before, during, and after the total eclipse of August 16, 1989. (The photo was taken with a 4-inch refractor with 30-second exposures.)

▶ **FIGURE 6-17.** Though the Moon is in total eclipse, some light is refracted toward the Moon by the Earth's atmosphere. As the light passes through the atmosphere, however, the blue end of the spectrum is scattered away more than the red is, so most of the light that makes it to the Moon is red. (This selective scattering is the same reason that sunsets are red.)

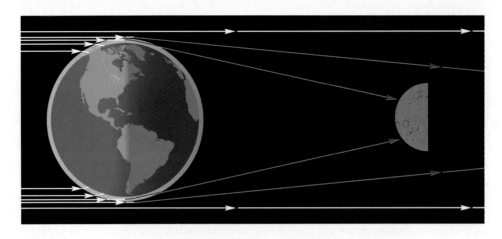

TABLE 6-2   Dates of Lunar Eclipses, 1994–2000

Date	Type of Eclipse	Visible from Part of North America?
May 25, 1994	Partial	Yes
April 15, 1995	Partial	No
April 4, 1996	Total	Yes
Sept. 27, 1996	Total	Yes
Mar. 24, 1997	Partial	Yes
Sept. 16, 1997	Total	No
July 28, 1999	Partial	Yes
Jan. 21, 2000	Total	Yes
July 16, 2000	Total	No

completely in the umbra, some sunlight strikes the Moon. This light has been bent by the Earth's atmosphere as shown in ▶Figure 6-17. Light that passes through many miles of atmosphere is red in color. For this reason, the sunlight reaching the totally eclipsed Moon is red, and the Moon appears a dark red color.

Table 6-2 shows the dates of coming lunar eclipses. You cannot be sure that you will be able to see any particular lunar eclipse, however, for two reasons. First, in order to be able to see it, you must be on the dark side of the Earth when the eclipse occurs. So, on the average, only half the people on Earth have a chance to see a given lunar eclipse. For this reason, the last column of the table indicates whether each eclipse will be visible from at least part of North America. Second, there is the weather factor. A cloudy night can ruin a long-planned eclipse-viewing party.

SOLAR ECLIPSES

We have seen that a lunar eclipse occurs when the shadow of the Earth falls on the full Moon. An eclipse of the Sun—a *solar eclipse*—occurs when the Moon, in its new phase, passes directly between the Sun and the Earth so that the Moon's shadow falls on the Earth. There is a major difference between these events, however. The Earth's size is such that its umbral shadow reaches back into space nearly a million miles, and at the distance of the Moon, it is easily large enough to cover the entire Moon. The umbral shadow of the Moon, however, reaches only about one-fourth that far, or about 377,000 kilometers. But

solar eclipse (or eclipse of the Sun). An eclipse in which light from the Sun is blocked by the Moon.

377,000 km = 234,000 mi and
384,000 km = 239,000 mi.

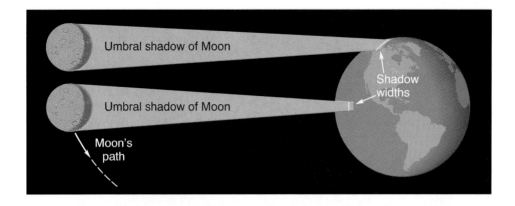

► **FIGURE 6-18.** If the Moon's umbra strikes the Earth at an angle, a wider area on Earth will experience a total eclipse.

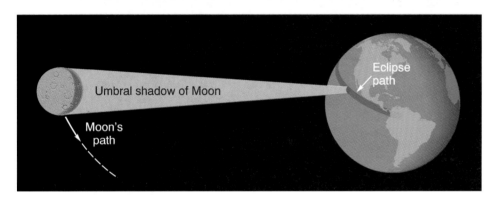

► **FIGURE 6-19.** The Moon's motion causes the path of a total solar eclipse to sweep across the Earth. The eclipse shown moves primarily across land and therefore would be seen by many people.

recall that the average distance from the Earth to the Moon is 384,000 kilometers. Thus if the Moon stayed at this distance, its umbra would never reach the Earth. A dark shadow of the Moon would never fall on the Earth.

The Moon, though, follows an eccentric orbit, coming as close as 363,300 kilometers to Earth. So it does get close enough that its umbra can reach the Earth. When this occurs, we can experience one of the most spectacular of natural phenomena, a *total solar eclipse.* ►Figure 6-18 shows two cases of the Sun, Moon, and Earth being aligned when the Moon is close enough for its umbra to reach the Earth. Even when the Moon is at its closest, the width of its umbral shadow at the Earth's distance is only about 130 kilometers. If the shadow hits the Earth as in the lower case in Figure 6-18, its width will be between 130 kilometers and about 270 kilometers. If the shadow happens to strike the Earth as in the upper case, the shadow on the surface may measure wider than 270 kilometers, but it seldom exceeds 400 kilometers.

This explains why relatively few people ever experience a total solar eclipse. Only people in that small area of Earth covered by the umbra see the Sun entirely hidden by the Moon. As the Moon moves along, its shadow swings in an arc across the surface of the Earth (see ►Figure 6-19). Thus there is a strip across the Earth's surface in which the total eclipse may be seen. This strip may be many thousands of kilometers long, but its width can reach a maximum of only about 400 kilometers. You have to be within this strip at the exact moment of totality, in clear weather, to see the total eclipse of the Sun.

At totality, the sky is dark enough that planets and the brightest stars can be seen in the sky. The appearance of the Sun is shown in ►Figure 6-20. What you see around the dark disk where the Moon blocks out the Sun is the glowing outer atmosphere of the Sun, called the *corona.* This is a layer of gas that extends for millions of miles above what normally appears to be the surface of the Sun. The gas glows because of its high temperature, but the glow is so much dimmer than the light we receive from the main body of the Sun that it is observed only

total solar eclipse. An eclipse in which light from the normally visible portion of the Sun (the photosphere) is completely blocked by the Moon.

130 km = 80 mi, 270 km = 170 mi, and 400 km = 250 mi.

corona. The outer atmosphere of the Sun (to be discussed in Chapter 11).

Solar Eclipses

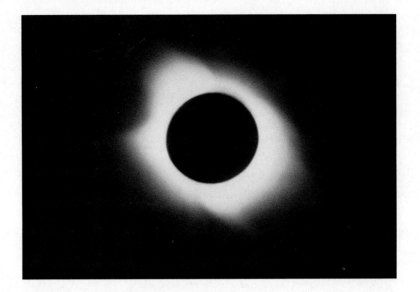

► **FIGURE 6-20.** During a total solar eclipse, the glowing light of the Sun's atmosphere—the corona—is visible. This photo of the June 30, 1991 eclipse was taken from the window of a DC-10 41,000 feet above the ground.

TABLE 6-3 Dates of Solar Eclipses, 1994–2000

Date	Location	Type of Eclipse
May 10, 1994	Pacific, U.S.	Annular
Nov. 3, 1994	South America	Total
April 29, 1995	South America	Annular
Oct. 24, 1995	India, S.E. Asia	Total
March 9, 1997	Siberia	Total
Feb. 26, 1998	South America	Total
Aug. 22, 1998	Indian Ocean	Annular
Feb. 16, 1999	Australia	Annular
Aug. 11, 1999	Europe to India	Total

during an eclipse. The opportunity to observe the corona is one of the scientific values of an eclipse, although today we are able to block out the Sun by artificial means in order to observe its outer layers. We will discuss the Sun further in Chapter 11.

A total solar eclipse is truly an awesome experience. As Table 6-3 shows, it will be some time before one is visible in the United States. If you get a chance to travel to the path of totality of a solar eclipse, don't pass it up. It is one of nature's grandest spectacles.

The Partial Solar Eclipse

►Figure 6-21 shows not only the Moon's umbra but also its penumbra. People on Earth who are in the umbra see a total eclipse, but anyone within the penumbra will see a ***partial solar eclipse.*** The penumbra covers a much greater portion of the Earth's surface, stretching about 3000 kilometers (2000 miles) from the central path of totality, so most of us have the opportunity to see a number of partial solar eclipses during our lifetimes.

In a partial eclipse, the dark disk of the Moon moves across the Sun, but its path is not perfectly aligned with the Sun and it does not move across the center of the Sun. The closer you are to the path of totality, the more of the Sun will be blocked out by the Moon.

partial solar eclipse. An eclipse in which only part of the Sun's disk is covered by the Moon.

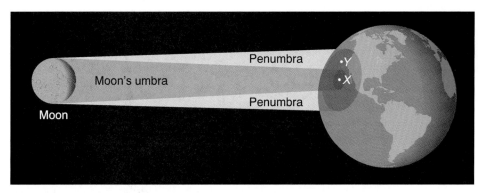

(a)

(b)

▶ FIGURE 6-21. (a) A person at point X sees a total solar eclipse while a person at Y sees a partial eclipse, with only the southern part of the Sun blocked by the Moon. (b) This series of photos shows the progression of a partial eclipse as would be seen by a person standing at point Y.

The Annular Eclipse

Remember that a total solar eclipse can occur only when two conditions are present: the Moon is directly between the Sun and the Earth, and the Moon is close enough to the Earth that its umbral shadow reaches the Earth. When the Moon is at its average distance from the Earth, it is a little too far away to cause a total eclipse on Earth. Therefore somewhat fewer than half the solar eclipses that occur are total. ▶Figure 6-22 illustrates what happens when the Moon is too far from Earth to allow a total eclipse. A person at point P on the Earth will see a partial eclipse where the disk of the Moon crosses that of the Sun.

Think of what an observer at point A on Earth would see in this case. The Moon is so far away that its disk is not large enough to cover the Sun even when it is directly centered on the Sun. The photograph in ▶Figure 6-23 shows what the observer would see at eclipse maximum. The Latin word *annulus* means *ring*, and from the figure you can see why such an eclipse is called an **annular eclipse.** Note the spelling; this is not an annual eclipse. Slightly over half of the solar eclipses are annular. Table 6-3 indicates which of the coming eclipses will be total and which annular. ▶Figure 6-24 shows the paths of total solar eclipses through 2017. Notice that the next total solar eclipse to cross the continental United States will occur on August 21, 2017 (and then the next will be on April 8, 2024).

annular eclipse. An eclipse in which the Moon is too far from Earth for its disk to cover that of the Sun completely, so the outer edge of the Sun is seen as a ring.

There are about twice as many total or annular solar eclipses as total lunar eclipses, but you are much less likely to see a solar eclipse because of the narrow path of the shadow.

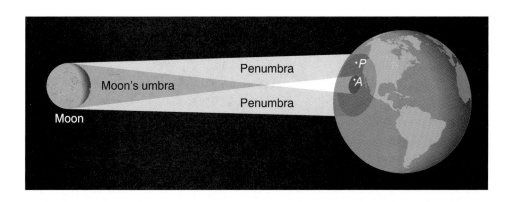

▶ FIGURE 6-22. When the Moon is far away during a solar eclipse, the eclipse will be annular. The person at point A sees the annular eclipse, while the person at P sees a partial eclipse.

Solar Eclipses

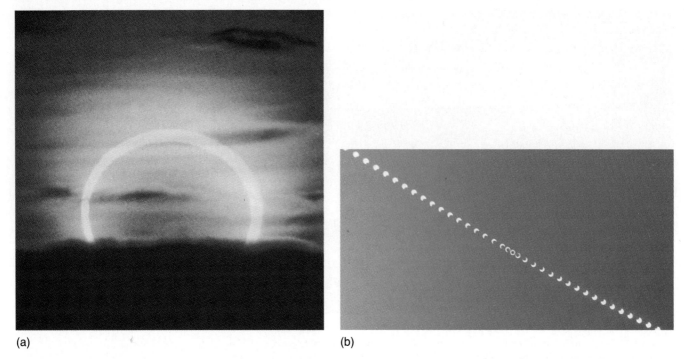

(a) (b)

▶ **FIGURE 6-23.** (a) During an annular eclipse, we can see the entire ring—annulus—of the Sun around the Moon. This photo shows the annular eclipse of May 30, 1984. The irregularity of the ring is due to mountains and valleys on the Moon's surface. (b) This series of photos shows the progression of the Moon's motion across the Sun during an annular eclipse.

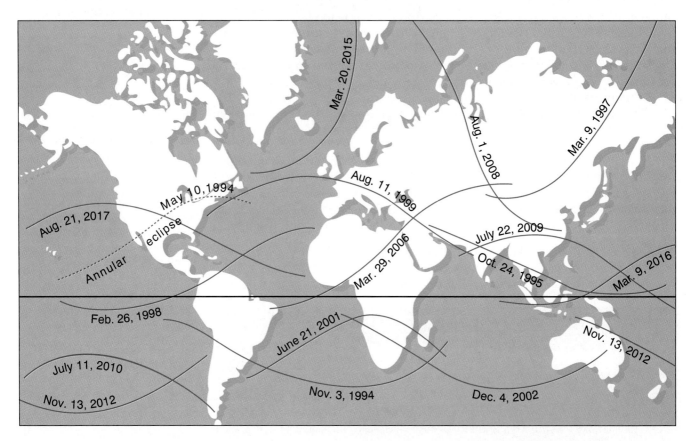

▶ **FIGURE 6-24.** The map shows the path of the Moon's shadow during total solar eclipses through 2017. The annular eclipse of 1994 is shown as a dotted line.

170   CHAPTER 6   The Earth-Moon System

EARTH

The next few chapters will focus on the details of the objects that make up the solar system. Before moving outward from the Earth and Moon, however, a close examination of these two objects is in order to provide a basis for comparison.

The Interior of the Earth

How can we know what is inside the Earth? One of the first properties of the interior that would aid our quest is the *density* of the Earth. The density of an object is defined as the ratio of the object's mass to its volume. To do the calculation, we need to know the Earth's mass and volume. As we have seen, the diameter of the Earth has been known for centuries, and, of course, the space program has greatly increased the accuracy of our measurements of the Earth's size and shape.

The mass of the Earth was first determined by applying Kepler's third law (as revised by Newton) to the period and radius of the Moon's orbit. We saw in Chapter 3 that the period of revolution of an orbiting object depends only upon the size of the orbit and total mass of the objects. Today we can calculate the gravitational pull on orbiting satellites and use this to further refine our values for the Earth's mass. Values of the mass and diameter of the Earth are given on page 180.

Using the mass and volume of the Earth, we calculate its density to be 5.52 grams per cubic centimeter (gm/cm^3). For comparison, the density of water is 1 gm/cm^3, aluminum is 2.7 gm/cm^3, and iron is 7.8 gm/cm^3.

A comparison closer to the situation of the Earth, however, is a snowball with a rock inside. Like the Earth, the snowball is made up of different materials at different levels of its interior. We learn about the makeup of the Earth's interior primarily by detecting two types of waves that result from earthquakes. These waves travel through the Earth, and by analyzing their times of travel from distant earthquakes, geologists can deduce some properties of the materials that lie deep within the Earth.

The interior of the Earth is made up of three layers, as shown in ➤ Figure 6-25. The *crust*, which is the outer layer, extends to a depth of less than 100 kilometers and is made up of the common rocks with which we are familiar here on the surface. The density of the crust is about 2.5 to 3 grams per cubic centimeter (gm/cm^3).

The *mantle* extends nearly halfway to the center of the Earth, about 2900 kilometers below the surface. This layer, although it is solid, is able to flow very slowly when steady pressures are exerted on it, but it will crack and move suddenly under extreme sudden pressures. The mantle is more dense (3 to 9 gm/cm^3) than the crust, and therefore the crust is able to float on it.

The *core* of the Earth seems to be divided into two parts, a liquid outer core and a solid inner core. The core is even more dense than the mantle, ranging from 9 gm/cm^3 to 13 gm/cm^3. Because of its high density, the core of the Earth is thought to be made up primarily of iron and nickel, the most common heavy elements. The average density of the Earth is 5.52 gm/cm^3, or more than five times the density of water.

The pattern of increasing density of materials within the Earth tells us something of the Earth's past, for such *differentiation* could only have come about when the Earth was in a molten state, when the heavier elements would have sunk through the less dense layers. The molten state must have resulted from heating during the formation of the planet and from energy released by radioactive elements early in the Earth's history.

---

density. The ratio of mass to volume.

For example, if you take a 1 cm cube of metal (1 cm^3 of metal) and determine that its mass is 5 g, then its density is 5 g/cm^3.

The applicable equation is

$$\frac{a^3}{P^2} = KM,$$

where a is the average radius, P is the period, and M is the sum of the masses of the objects—the Moon and Earth in this case. See Chapter 3 for how we know what fraction of this mass is the Earth's.

crust (of the Earth). The thin, outermost layer of the Earth.

mantle (of the Earth). The thick, solid layer between the crust and the core of the Earth.

core (of the Earth). The central part of the Earth, probably consisting of a solid inner core surrounded by a liquid outer core.

differentiation. The sinking of denser materials toward the center of planets or other objects.

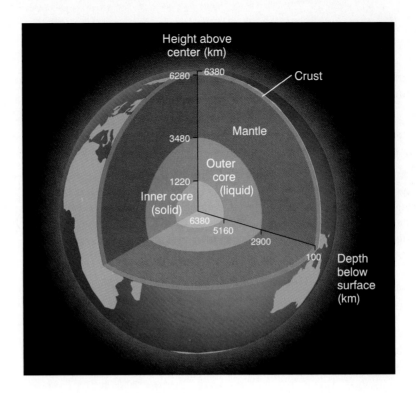

> **FIGURE 6-25.** The interior of the Earth, showing its primary layers.

Earth's Magnetic Field

If a piece of paper is laid over a magnet and filings of iron are sprinkled on the paper, a pattern such as that shown in ➤Figure 6-26b results. The area around a magnet where a magnetic force is felt is called a ***magnetic field***. In a magnetic field, a small magnet aligns with the field.

When a magnet is suspended so that it is free to rotate near the Earth, it always aligns so that opposite ends point in specific directions, indicating that the magnet is aligning with a magnetic field that is related to the Earth. By plotting the direction of the magnetic field at various places on and off the Earth's surface, we can determine that the Earth's magnetic field has a shape similar to that of a bar magnet, as shown in ➤Figure 6-27. Notice that the poles of the Earth's

magnetic field. A region of space where magnetic forces can be detected.

A magnetic compass is simply a magnet that is free to rotate so that it can align with a magnetic field.

> **FIGURE 6-26.** (a) An edge-on view of iron filings being sprinkled onto a piece of paper that covers a bar magnet. (b) The resulting magnetic field pattern made by the iron filings. The outline of the magnet is obvious, but notice also the pattern the filings form beyond the magnet.

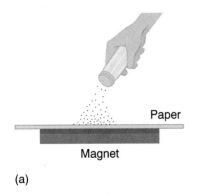

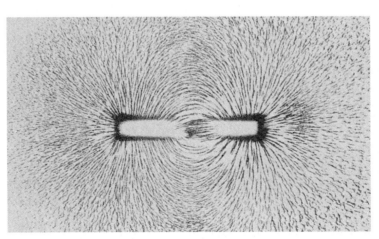

172   CHAPTER 6   The Earth-Moon System

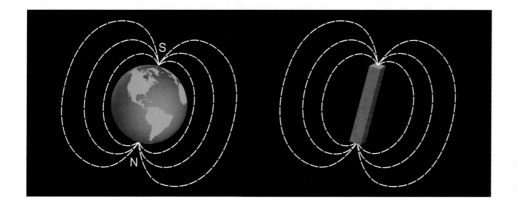

▶ FIGURE 6-27. The magnetic field of the Earth is similar to that of a bar magnet.

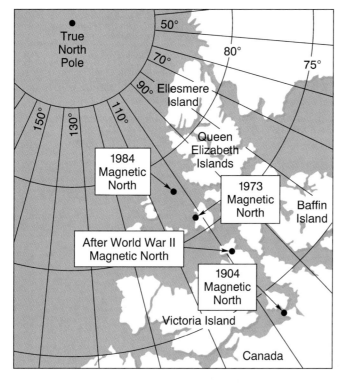

▶ FIGURE 6-28. The Earth's magnetic poles are not located at its poles of rotation. The location of the north magnetic pole is shown here; note also that its location changes with time.

magnetic field—those points toward which the magnetic field converges—are located near (but not exactly at) the poles of the rotation axis of the Earth.

One might picture the Earth as containing within it a bar magnet that is not quite aligned with its rotation axis. We know that the magnetic field of the Earth is not caused by a bar magnet inside, but the origin of the field is not completely understood. Magnetic fields (even those near a magnet) are caused by moving electric charges, and we are confident that the cause of the Earth's field depends upon the existence of Earth's molten iron core. Apparently, circulation within this core causes electric currents that result in the magnetic field. We know that the locations of the Earth's magnetic poles wandered in the past (▶Figure 6-28), and the magnetic field has even undergone complete reversals. At least in part, these changes are due to movements of the Earth's crust with respect to the interior. It has been hypothesized that a major reason for the reversals was a change in the Earth's rotation rate that resulted from bombardment with meteoriods and from volcanic eruptions. Many questions remain before hypotheses concerning pole wandering can be confirmed or denied.

One effect of the Earth's magnetic field became the first scientific discovery of the space age. Early spacecraft discovered the existence of electrically charged

You can produce a magnetic field by causing a current (moving electric charge) in a coiled wire.

The model that explains the Earth's (and other planets' magnetic fields) as due to currents within an iron core is called the *dynamo effect*.

CLOSE UP

The Earth from Space

We Earthlings have learned much about neighboring planets by sending spacecraft by them, putting craft into orbit around them, and sending landers to their surfaces. Let us consider what could be learned about Earth if we lived on another planet and used similar technology to study the Earth.

The atmosphere of the Earth and its cloud layer hinder detailed viewing of the surface from afar. Inhabitants of nearby planets, however, could detect differences between our continents and oceans, and they could see the Earth's white polar caps (as we see those on Mars). In order to determine what makes up the continents, oceans, and polar caps, our neighbors could analyze the spectra of light reflected from these areas. They would be able to determine that most of our planet is covered by water, and that our polar caps are frozen water. In addition, they would see our changing cloud cover and determine that it, too, is made up of water.

Could they detect signs of life? Probably not. Although we think of ourselves as important, we have not changed our planet in a way that would be obvious to an observer on another planet or moon. The only sign of life they are likely to detect from afar is our electromagnetic noise, i.e., radio and television broadcasts.

As an alien spacecraft approached Earth and went into orbit around it, it would obtain views such as those in ▶Figure C6-1. Now its occupants might be able to see signs of life. Perhaps their first visible evi-

▶ **FIGURE C6-1.** These photos of Earth show how beautiful our planet is. In the photo at the left, we can clearly see the mountains of Egypt's Sinai Peninsula. Compare this to a map of the region. At the right we see a storm system over the ocean. The windowsill of the spacecraft is at the bottom and the Earth's horizon is at the top.

---

particles swarming in doughnut-shaped regions high above the Earth's surface (▶Figure 6-29). The reason for these "Van Allen belts" is well understood. Just as an electric current (moving electric charges) causes a magnetic field, a magnetic field exerts a force on charged particles. When charged particles enter a magnetic field, they are forced to move in a spiral around the lines of the field as shown in ▶Figure 6-30. In the case of the Van Allen belts, the charged particles (protons and electrons) come primarily from the Sun and have been captured by the magnetic field of the Earth.

We see the effect of the particles trapped within the Earth's magnetic field in the **auroras** that are often visible in the skies near the North and South Poles of the Earth (▶Figure 6-31). These beautiful displays of light are the result of disturbances of the Earth's magnetic field that cause some of the particles to

aurora. Light radiated in the upper atmosphere due to impact from charged particles.

dence of life would come when they viewed the dark side of our planet, for urban areas would be visible at night because of the wasted light that escapes from them into space (➤Figure C6-2).

If landers were sent to Earth to look for life, their ease in finding it would obviously depend upon where they landed. But if they were equipped like our *Viking* missions that landed on Mars to search for life, they would not only be capable of photographing our large animals, but they would be able to detect organic molecules in the soil (or ice) no matter where they landed.

➤ **FIGURE C6-2.** This is a composite of a number of photos showing the United States from space at night.

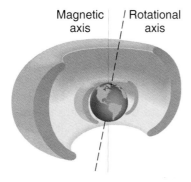

➤ **FIGURE 6-29.** The Van Allen belts are regions where the magnetic field of the Earth traps charged particles. They surround the Earth except near the poles.

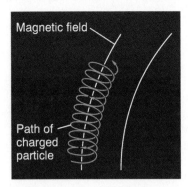

▶ **FIGURE 6-30.** Charged particles are forced to move in spirals around the lines of a magnetic field. This causes them to become trapped in the Earth's field.

▶ **FIGURE 6-31.** An aurora is caused by charged particles trapped in the Earth's magnetic field striking atoms in the upper atmosphere.

follow the magnetic field lines down to the atmosphere, where they collide with atoms of the air and cause it to glow. We will discuss the reason for such disturbances when we study the Sun in Chapter 11.

PLATE TECTONICS

You may have noticed while looking at a map of the Earth that there seems to be a rough fit between the eastern edge of the American continents and the western edge of Europe and Africa. Early in this century it was proposed that the continents were once in contact. No acceptable mechanism for continents moving relative to one another *(continental drift)* was obvious, however, and the idea was put aside.

Later, geologists discovered that there is a line near the center of the Atlantic Ocean where lava flows upward, forming an extensive range of underwater mountains (▶Figure 6-32). They discovered further that as the lava solidified, magnetic material in it oriented in specific directions depending upon the direction of the Earth's magnetic field at the time. Recall that the magnetic field of the Earth has changed through the ages. From the magnetic properties of the material, they were able to determine roughly when it solidified. Then by examining the seafloor at various distances from the center of the *rift zone* in the mid-Atlantic, they learned that the seafloor has gradually spread from the rift. Finally, laser light was bounced from satellites to detect any relative motion between continents, and it was discovered that Europe and North America are moving apart at the rate of 2 to 4 centimeters per year. From this evidence, our present theory of *plate tectonics* developed.

The Earth has about a dozen tectonic plates, sections of the Earth's crust (and perhaps upper mantle) that extend about 50 to 100 kilometers deep. ▶Figure 6-33 shows the major plates.

continental drift. The gradual motion of the continents relative to one another.

Alfred Wegener, a German meteorologist, is credited with first developing the idea of continental drift.

rift zone. A place where tectonic plates are being pushed apart, normally by molten material being forced up out of the mantle.

plate tectonics. The motion of sections of the Earth's crust across the underlying mantle.

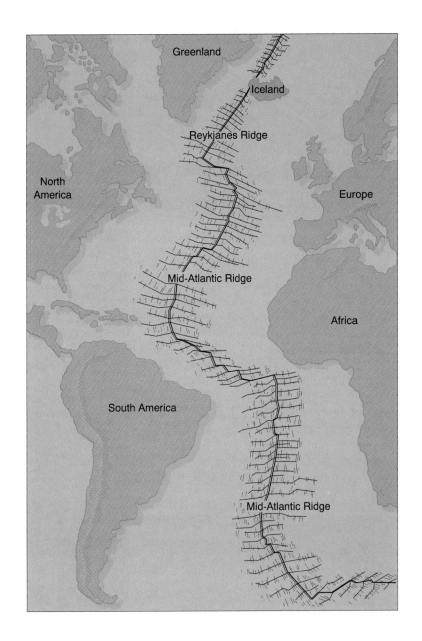

➤ **FIGURE 6-32.** If the Atlantic Ocean were drained, we could see the Mid-Atlantic Range of mountains and the rift down its center. As lava is forced out of the rift, the plates are pushed aside, causing continental drift.

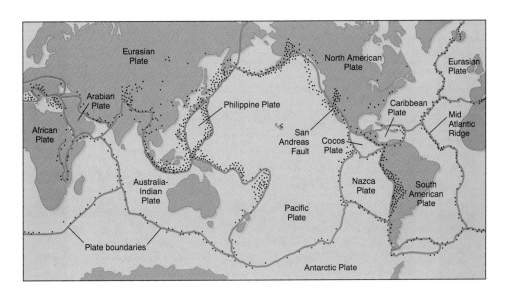

➤ **FIGURE 6-33.** The major tectonic plates of the Earth. The dots represent earthquake locations. Note that earthquakes tend to occur near plate boundaries.

Plate Tectonics 177

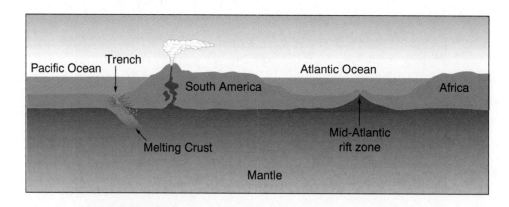

► **FIGURE 6-34.** As the South American Plate is pushed westward (to the left here) from the mid-Atlantic rift, it is pushed against the Nazca Plate under the Pacific Ocean. This forces that plate down into the mantle, resulting in the volcanoes of the Andes Mountains.

If the plates are spreading from places where lava flows from beneath the crust, they must be jamming together at some other places. ►Figure 6-34 shows how one plate might be pushed below another at such a location. South America is on a plate that is moving westward from the rift zone in the mid-Atlantic. At the western boundary of South America, this plate is forced against the Nazca Plate (Figure 6-33), which is moving eastward. Consequently, the Nazca Plate gets pushed under the South American continent. Two major effects have resulted from these movements: (1) A deep ocean trench has opened off the western coast of South America. (2) As the material of the Nazca Plate is pushed downward, it melts and low-density rock is forced upward, erupting from the Earth's surface as volcanic lava. Over millions of years this volcanic action has raised the Andes mountain range that runs the length of the western side of the continent. The volcanoes of the west coast of North America and those of Japan were formed in the same manner.

In some places, plates are slipping by one another. This is the primary motion that occurs between the Pacific Plate and the North American Plate. The slippage

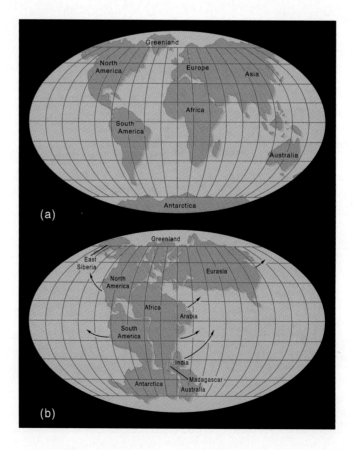

► **FIGURE 6-35.** (a) The Earth today and (b) about 200 million years ago. The motion of the continents in (b) is indicated by arrows.

is not smooth and continuous, however. Instead, forces build up over a number of years until motion occurs very suddenly, causing an earthquake.

By projecting the motion of the plates backward in time, we can conclude that some 200 million years ago, a map of the Earth's continents looked somewhat like that in ►Figure 6-35b. Today's continents started separating shortly after that and continue to move apart today. The great mountains, the canyons, and indeed the continents that may appear to us as permanent monuments are instead transitional stages of an ever-changing planet.

EARTH'S ATMOSPHERE

Compared to the size of the Earth, the atmosphere reaches a very small distance above the surface. The atmosphere gets thinner and thinner farther from Earth so it is impossible to put a definite boundary on it, but at a distance of 100 or 150 kilometers above the surface, the atmosphere is essentially nonexistent.

Our atmosphere consists of approximately 80% nitrogen and 20% oxygen. Other constituents such as water vapor, carbon dioxide, and ozone make up a very small percentage of the atmosphere even though they are very important to life on Earth. We will discuss the atmospheric importance of carbon dioxide when we discuss the planet Venus in Chapter 8.

Most of the mass of the atmosphere—about 75%—lies within 11 kilometers (7 miles) of the surface. This portion of the atmosphere, called the *troposphere,* is where all of our weather occurs. The troposphere receives most of its heat from infrared radiation emitted by the ground, so the troposphere is cooler as one gets higher. ►Figure 6-36 illustrates the temperature of the atmosphere at different heights.

Centered at about 50 kilometers above the surface is the ozone layer. Ozone is a molecule containing three oxygen atoms, and it is an efficient absorber of ultraviolet radiation (UV) from the Sun. This absorption is the reason that temperature reaches a peak at the ozone layer.

troposphere. The lowest level of the Earth's (and some other planets') atmosphere.

Commercial planes fly at the top of the troposphere in order to avoid the turbulence of weather in that layer.

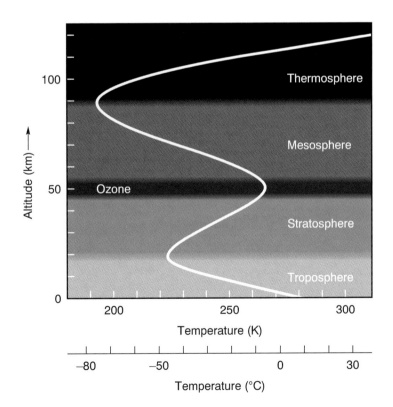

► FIGURE 6-36. The temperature of the atmosphere varies with altitude because of the way solar energy is absorbed by the different layers.

Earth's Atmosphere

⊕ EARTH
DATA PAGE

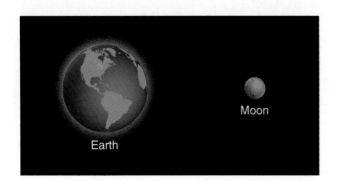

Earth	Value
Equatorial diameter	12,756 km
Oblateness*	0.0034
Mass	5.97×10^{24} kg
Density	5.52 gm/cm^3
Escape velocity	11.18 km/s
Sidereal rotation period	23.934 hours
Solar day	24.0 hours
Albedo	0.31
Tilt of equator to orbital plane	23.44°
Orbit	
Semimajor axis	1.496×10^8 km
Eccentricity	0.0167
Sidereal period	365.26 days
Moons	1

*Recall that oblateness tells us how "out of round" an object is. The Earth's polar diameter is 12,714 km.

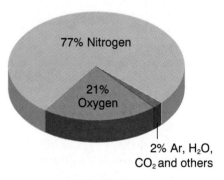

Atmospheric Gases

77% Nitrogen
21% Oxygen
2% Ar, H$_2$O, CO$_2$ and others

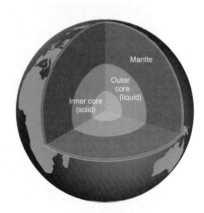

Interior

QUICK FACTS

*Third planet from Sun. Very circular orbit. Except for Pluto, has largest satellite compared to its size. Only planet with liquid water on surface. Active interior.

*This Quick Facts Box may be trivial for Earth, but for other planets (Chapters 8 and 9) it should help you review.

The ozone layer is extremely important to life on Earth, for most life has developed while being sheltered from all but a little UV. Ultraviolet radiation breaks apart molecules that make up living tissue, as you have experienced if you have ever been sunburned, a condition resulting from a fairly small amount of UV. As we are frequently warned, too much exposure to UV can cause skin cancer. Modern civilization releases into the atmosphere chemicals that are reducing the amount of ozone, primarily chlorofluorocarbons from aerosol cans and chlorine from various sources. The United Nations is seeking international agreements to limit the amount of damaging chemicals that are released, but even if we suddenly and drastically reduce the release of these gases, the problem will not be solved, for chlorofluorocarbons remain in the atmosphere for over a hundred years, continuing to damage the ozone layer.

THE MOON'S SURFACE

The surface of the Moon can be divided into the *maria* (singular *mare*) and the mountainous, cratered regions. From Earth, the maria appear darker than the other regions. The far side of the Moon, first seen when a Soviet spacecraft photographed it in 1959, is covered almost completely with craters.

Until the middle of this century, it was assumed that lunar craters were formed the same way almost all earthly craters are, by volcanic action, instead of by impacts of *meteorites* from space. Astronomers had two primary reasons for thinking that the craters were volcanic: (1) Since almost all craters on Earth are formed that way, it was reasonable to assume that lunar craters were similar. (2) Lunar craters are very circular, but if you form a crater by throwing a rock into sand, the crater will only be circular if the rock is thrown straight down. Otherwise the crater will be elongated. Since meteorites would be likely to hit the Moon's surface at various angles, one would expect many impact craters to be somewhat elliptical in shape if they were formed by meteorite impact.

One observation, on the other hand, provided evidence against a volcanic origin for the craters: The floors of lunar craters are lower than the surrounding surface (as we can tell by measuring the shadows cast by the crater walls). Volcanic craters typically occur at the top of volcanic mountains, and their floors are higher in elevation than the surrounding territory.

We now realize that almost all (and perhaps all) lunar craters are the result of impacts. The arguments against an impact origin are answered as follows:

1. Earth has few impact craters because its atmosphere prevents any but the largest meteorites from reaching its surface. Small meteorites burn up in the air or are slowed enough by the air that they do not produce craters. In addition, craters produced far back in Earth's history have been eroded so that they are no longer noticeable. Lunar craters, on the other hand, do not suffer erosion on the airless Moon. Now that we have observed the Earth from space, we have found many more impact craters on Earth.
2. The argument that we would expect some impact craters to be elongated does not apply to lunar craters because the latter are not formed simply from material being "splashed" away by the impact. Instead, an approaching meteorite is pulled toward the Moon by the force of gravity and strikes with such great speed that it penetrates below the surface, compressing and heating the Moon's material until an explosion occurs. Such an explosion is similar to a nuclear bomb ignited below the surface of the ground; it results in a circular crater in spite of the angle of the meteorite's fall.

Much of the material thrown upward by the explosion falls back into the crater and forms its floor. Other material is thrown away from the crater to form

mare (plural **maria**). Any of the lowlands of the Moon or Mars that resemble a sea when viewed from Earth.

meteorite. An interplanetary chunk of matter that has struck a planet or moon.

Until about 1940, Meteor Crater in Arizona (Chapter 10) was thought to have a volcanic origin.

MOON
DATA PAGE

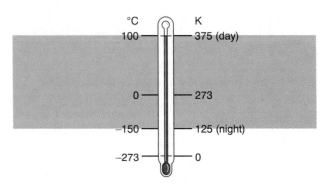

Surface Temperature Extremes

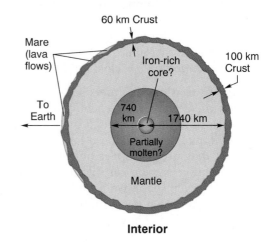

Interior

Moon	Value	Compared to Earth
Diameter	3476 km	0.27
Oblateness	0	
Mass	7.35×10^{22} kg	0.0123
Density	3.34 gm/cm^3	0.61
Surface gravity		0.165
Escape velocity	2.4 km/s	0.21
Sidereal rotation period	27.322 days	
Synodic period (phases)	29.531 days	
Surface temperature	$-170°C$ to $130°C$	
Albedo	0.07	0.2
Tilt of equator to orbital plane	6.68°	
Orbit		
Average distance from Earth	384,400 km (center-to-center)	
Closest distance	363,300 km	
Farthest distance	405,500 km	

QUICK FACTS

One-fourth Earth's diameter. Surface craters were caused by meteorite impact. Maria were caused by lava flow. No atmosphere. Weak magnetic field. Likely to have formed as a result of a collision between Earth and a large object early in Earth's history.

(a) (b)

▶ FIGURE 6-37. (a) Light-colored rays can be seen radiating from the prominent crater (named Tycho) near the bottom of this photo of the full Moon. They were formed by material ejected from the crater. (b) A close up of Tycho shows that it has a prominent central peak, the result of its surface rebounding after the impact of the meteorite that caused it.

the *rays* that are seen radiating from some craters (▶Figure 6-37a). The peaks found in the center of many craters (Figure 6-37b) were caused by a rebound of the surface after the explosion. Both the low floor levels and the central peaks are in accord with the idea that the craters are impact craters.

Volcanic action did occur in the Moon's past, however, and it resulted in the Moon's maria. They are caused by the flow of dark lava onto lowland areas of the Moon. Most of the maria are roughly circular, leading us to believe that they were originally the floors of very large craters. We know that the dark lava that fills them was not produced by the impacts that formed the craters, however, for within the maria are old, smaller craters that have themselves been mostly covered by lava. Thus the lava flowed from beneath the surface after the small craters had been formed inside the giant ones.

The top few centimeters of the Moon's surface are made up of loose powdery lava, small rocks, and mostly spherical pieces of glass, the result of bombardment by countless meteorites through the ages. ▶Figure 6-38 is a photograph of a footprint made in the dust of the Sea of Tranquility during the first visit to the Moon by a human, on July 20, 1969.

The crust of the Moon (▶Figure 6-39) ranges in depth from about 60 to 100 kilometers and is thinner on the side facing the Earth than on the other side. When the maria were formed, molten lava released from within the Moon flowed toward the side with the thinner crust. It is no accident that the side of the Moon with more maria faces the Earth, for lava has a greater density than the rocks of the highland areas. Recall that tidal forces are what caused the Moon to slow its rotation. These same forces acted on the Moon's uneven distribution of mass to cause the more dense side to face the Earth.

If you have studied geology, you know that earthly mountains were formed by the motion of plates within the Earth and by volcanic action. The lunar mountains were formed differently. They are simply the results of millions of ancient craters, one on top of another. Mountain ranges that border maria are the walls of giant craters whose floors are now covered with lava.

lunar ray. A bright streak on the Moon caused by material ejected from a crater.

Many of the glassy spheres resulted from the melting and solidification of rock upon ejection from a crater.

The Moon's Surface 183

▶ **Figure 6-38.** This bootprint on the Moon's surface was made by an *Apollo 11* astronaut, one of the first two humans to visit our nearest neighbor in space. The astronaut's prints will remain on the Moon for millions of years, for the only erosion that occurs there is due to tiny meteorites and is extremely slow.

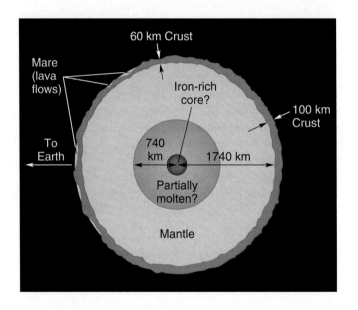

▶ **Figure 6-39.** Studies of the Moon's interior have been made by striking the surface to produce "moonquakes" and then investigating the vibrations that result at other points on the surface. Cracks under the mare allowed lava to reach the floor of the ancient craters.

The sensors on the Moon to measure the intensity of quakes are called *seismographs*.

The density of the Moon is 3.36 grams per cubic centimeter. This means that if it has an iron core, that core must be small. In addition, the magnetic field of the Moon is less than one ten-thousandth of the Earth's, so this also indicates that the Moon cannot have a large molten iron core. It is thought that any iron core must be less than 700 kilometers in diameter. Questions remain, however: First, we are not certain that planetary magnetic fields are due to molten iron cores, and second, some rocks brought back by *Apollo* astronauts were magnetized more than we would have expected in such a weak magnetic field. Perhaps the Moon's field was stronger in the past, or perhaps these rocks were magnetized by the Earth's or the Sun's magnetic fields.

The *Apollo* missions left sensors on the Moon to measure moonquakes. Some of these quakes were artificially produced by striking the Moon at various places, but we have detected about 3000 natural moonquakes a year. These were very weak, much weaker than our earthquakes, but they tell us that the interior of the Moon is essentially dead, with no major changes taking place.

CLOSE UP

The Far Side of the Moon

▶ Figure C6-3 is an image that includes most of the far side of the Moon, the side that is never seen from Earth. Although the image is in false color, you can see immediately that this side is different from the side that faces Earth. The most obvious difference is that the far side has fewer maria. (The large green area is not really a mare. If you look closely, you can see that it is filled with craters. The colors are explained below.) To understand the reason for the lack of maria, refer to Figure 6-39, which shows a cross section of the Moon. The crust is thicker on the side opposite Earth. Maria are thought to have formed between 4 billion and 2.5 billion years ago as the result of volcanism. Lava is more likely to be forced to the surface where the crust is thinner, and since the side of the Moon that faces Earth has a thinner crust, that is where most of the volcanism occurred.

The image of Figure C6-3 is a composite of images made by *Mariner 10* in 1973 (as it passed the Moon on the way to Venus and Mercury) and the Jupiter-bound *Galileo* spacecraft in December 1990. The image was made from separate photos that were taken with different filters on the cameras. This technique allows astronomers to determine the mineral content of the surface. For example, green and yellow indicate a large amount of iron and magnesium. Blue indicates a high abundance of titanium oxide, a mineral that is associated with maria. Knowing the mineral composition of each location helps astronomers in their quest to understand the Moon's past.

▶ **FIGURE C6-3.** The far side of the Moon is shown here in false color. The prominent bull's-eye crater toward the left is called Mare Orientale. Different colors represent different mineral abundances.

The Moon's Surface

THEORIES OF THE ORIGIN OF THE MOON

Evidence indicates that the Moon formed about 4.6 billion years ago. (The next section and its accompanying Close Up explain some of the evidence.) Until recently, there have been three theories of the origin of the Moon, called the *double planet*, the *fission*, and the *capture* theories. We will briefly describe each of these theories and look at the evidence to see which fits the data best. Then, we will introduce a newer theory that seems to work better than any of the other three.

■ The *double planet theory,* which was suggested in the early 1800s, is the oldest. It holds that as the Earth formed from a spinning disk of material, not all of that material coalesced to form the Earth. A small part of it was left orbiting the Earth and formed into the Moon. In Chapter 7 we will discuss theories of the origin of the solar system and will see that this idea is entirely consistent with those theories.

double planet theory. A theory that holds that the Moon was formed at the same time as the Earth.

A simple comparison of densities seems to rule out the double planet theory, for if the Moon formed along with the Earth, the two bodies should have about the same density. The Earth's density, however, is 5.5 grams per cubic centimeter, much greater than the Moon's 3.3 grams per cubic centimeter.

■ In 1878, the astronomer Sir George Howard Darwin, son of the biologist Charles Darwin, proposed that the Moon was once part of the Earth and broke (or *fissioned*) from it due to forces caused by a fast rotation and solar tides. This **fission hypothesis** proposed the large basin of the Pacific Ocean as the place from which the Moon was ejected.

fission theory. A theory that holds that the Moon formed when material was spun off from the Earth.

The difference in density between the Earth and the Moon might at first glance seem to rule out the fission theory along with the double planet theory, but the crust of the Earth does have a density close to that of the Moon. If the Moon formed from material from the Earth's crust, we would expect its density to be just as we find it.

There is a problem with the fission theory, however. Astronomers have difficulty explaining how an object as massive as the Moon might have been pulled out of—or thrown off—the Earth. No satisfactory mechanism for this event has been proposed. In addition, the Moon does not orbit in the plane of the Earth's equator as it would if it were ejected from a spinning Earth.

■ Early in this century, another theory was proposed. It holds that the Moon was originally a separate astronomical object that happened to come near the Earth and was captured by the Earth's gravitational field so that it settled into orbit as the Moon. This is the **capture theory.**

capture theory. A theory that holds that the Moon was originally solar system debris that was captured by Earth.

There are also problems with the capture theory. If one astronomical object comes close to another, each of their paths will be changed by the gravitational force between them (▶Figure 6-40), but one will not capture the other unless there is contact between the two or unless a third object is involved, so that the interaction of the three objects results in one of them being slowed down to an orbital speed. Such a near-collision between three objects seems highly unlikely.

Although we have known the density of the Moon for a long time, its chemical composition was not well known until the *Apollo* astronauts brought back soil and rock samples. This new data posed new problems for the three theories. In many ways, the chemical composition of the Moon is similar to that of the Earth's crust, for both have about the same proportions of some of the major elements: silicon, magnesium, iron, and manganese. The Moon, however, has far smaller proportions of easily vaporized *(volatile)* substances than does the Earth. The Moon has much higher proportions of nonvolatiles (such as alumi-

volatile. Capable of being vaporized at a relatively low temperature.

CHAPTER 6 The Earth-Moon System

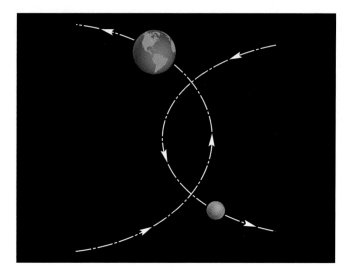

► **FIGURE 6-40.** If the Earth and another object (in this figure, a smaller one) had a near-collision, each object's path would be altered, but they would not begin to orbit one another.

num and titanium), which require a very high temperature to vaporize, than does the Earth's crust. The differences in chemical composition, along with the Moon's lack of an iron core, seem to rule out both the fission theory and the double planet theory.

Thus none of the theories for the origin of the Moon seemed to fit all of the data. The question remained, "How did the Moon get there?"

The Large Impact Theory

In the 1970s A. G. W. Cameron and William Ward of Harvard proposed a new theory. They proposed that early in the Earth's history, it was struck at a glancing angle by a large object, that the impact resulted in a fusion of the two objects, and that material was thrown off of the two to form the present Moon. Computer simulation of such a collision shows that if the impacting object has a mass nearly as great as Mars, heat resulting from the collision would vaporize material and eject enough of it into orbit to account for the mass of the Moon, once the material coalesced in its orbit around the Earth. The *large impact theory,* as it is called, is able to explain both the similarities and the differences in the compositions of the Earth and the Moon.

Since the mid-1980s, a consensus has been building among astronomers that the large impact theory fits the data better than the other three theories. Recent theoretical work on the formation of the planets indicates that without large impacts, the Earth would rotate every 200 hours instead of every 24 hours. A glancing impact by a large object explains its present rotation rate. Like all new theories, the large impact theory will be tested against both existing data and new data as the years pass. Although it will probably have to undergo modification, it appears that astronomers may finally have found the answer to the age-old question of the origin of the Moon.

large impact theory. A theory that holds that the Moon formed as the result of an impact between a large object and the Earth.

THE HISTORY OF THE MOON

The history of the Moon can be pieced together in a number of ways. For example, when we observe that one crater overlaps another (►Figure 6-41), we know that the overlapping crater was formed after the other. Likewise we know that the crater Tycho was formed relatively recently because its rays overlap the craters around it. Lunar rays darken with time, providing more information on

The color change is caused by sunlight and bombardment by tiny meteorites.

► **FIGURE 6-41.** In this photo, a small crater overlaps a larger one (named Gassendi), so that we know that it was formed after Gassendi was. Note that Gassendi has a central peak.

radioactive dating. Any of a number of procedures that examine the radioactivity of a substance to determine its age.

The volcanic action described here consisted of lava seeping through cracks in the surface, so volcanic mountains did not form.

In Chapter 7 we will discuss the formation of the solar system and the sweeping up of its original matter into planets and moons.

micrometeorite. A tiny meteorite.

the order of lunar events. Such examination of the surface only provides information about the order of events, however; it does not allow us to determine how long ago the events happened. Reliable assessments of time scales had to wait for the *Apollo* missions and the 840 pounds of lunar material they brought back to Earth. The accompanying Close Up describes a procedure that uses radioactivity of rocks to determine their age. Such **radioactive dating** techniques have been indispensable in forming a model of the Moon's history.

The Moon formed about 4.6 billion years ago. (The oldest rocks we found there are 4.42 billion years old.) We know that its surface was molten a few million years after the Moon was formed and conclude that the surface was probably molten from its formation, in agreement with the large impact theory. As the Moon cooled and solidified, cratering marked its surface. Most craters were formed between 4.2 and 3.9 billion years ago. Giant impacts that occurred near the end of the cratering period produced the areas where we see maria today. After most of the cratering ended, the interior of the Moon became hot enough (probably due to radioactivity) that molten lava flowed from beneath the surface and gathered in the floors of the giant craters. Rocks gathered from the maria are between 3.1 and 3.8 billion years old, leading us to believe that this volcanic stage ended about 3.1 billion years ago.

Cratering continues today, but at a rate much reduced from the Moon's early history. Three rocks have been found on Earth that are thought to have been ejected from the Moon by impacts within the past few million years. Our region of the solar system has been swept clear of most large chunks of matter, however, so meteorites large enough to produce noticeable craters on the Moon are now infrequent. We will see in Chapter 10 that the Earth is constantly being struck by debris from space, but much of it burns up in the atmosphere and never reaches the surface. The Moon has no atmosphere, however, so meteorites large enough to produce small craters must still strike it occasionally, although no new crater has ever been observed. In addition, **micrometeorites** strike the Moon's surface, further pulverizing its soil. Except for these—and some recent visits by humans and their machines—the surface of the Moon changes very little. This is fortunate for astronomers, for the surface becomes a book in which we can read the Moon's distant history.

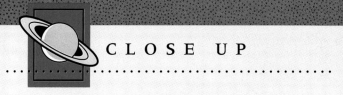

CLOSE UP

Measuring the Age of the Earth and Moon

From rocks brought back from the Moon by *Apollo* astronauts, we know that the Moon is at least 4.5 billion years old. We have found rocks on Earth that are 3.9 billion years old. How do we determine the age of these rocks—how do we "date" them? The method used is called *radioactive dating*, and to understand it, we must first look at the makeup of the nuclei of atoms.

There are two primary particles within the nucleus: protons and neutrons. The number of protons in a nucleus determines what element the atom is. For example, any atom with only one proton in its nucleus is necessarily hydrogen; two protons, helium; and so forth, up to uranium, which has 92 protons, more than any other naturally occurring element. Although different atoms of the same element all have the same number of protons, the number of neutrons in their nuclei may differ. Such different forms of an element are called *isotopes* of the element. To specify which isotope we are talking about, we state the name of the element and the total number of protons and neutrons the nucleus contains. Thus uranium-238 has a total of 238 protons and neutrons in its nucleus (and it has 92 protons because every uranium nucleus has 92 protons).

The nuclei of some isotopes are unstable, in that their nuclei change spontaneously. When this happens, some isotopes emit a gamma ray and others emit a nuclear particle. The gamma ray or particle emitted is the radiation that comes from a radioactive material. We say that the isotope *decays*, but the term *decay* here does not refer to deterioration such as wood undergoes when it decays. Rather it simply means that the isotope spontaneously emits radiation and in doing so changes to another isotope. For example, when rubidium-87 undergoes radioactive decay, it changes into strontium-87. Each isotope has its own characteristic rate of decay, which is called the *half-life* of the isotope—the amount of time needed for *half* of the isotope to emit radiation and change to another isotope. Half-lives range from tiny fractions of a second to billions of years. Isotopes with long half-lives are the ones that are useful in dating geological samples.

The basic idea of radioactive dating is quite simple. Consider uranium-238, which has a half-life of 4.5 billion years. If we begin with a pure sample of uranium-283, we know that in 4.5 billion years half of it will have decayed to another isotope (and then by a series of quicker decays to lead-206). Thus by comparing the percentage of uranium-238 and lead-206 in a sample that was once pure uranium-238, we can tell how much time has passed since the sample was pure. The problem, of course, is that we must know that the sample was pure at the beginning.

There are a number of radioactive dating techniques, all of which depend upon an assumption about the original condition of the matter. One technique uses the fact that when molten material solidifies, the crystals that are formed have certain specific chemical elements in them. If uranium is present when crystallization takes place, certain types of newly formed crystals will contain uranium but no lead. If we examine rocks containing such crystals and find lead-206, we can be confident that this resulted from radioactive decay of uranium-238. The relative percentages of the two isotopes then allow us to calculate the time elapsed since the crystals formed.

Fortunately, the dating of a sample of rock does not depend upon just one isotope. For example, the radioactive isotope rubidium-87 is found in crystals along with its decay product strontium-87. From the relative percentages of these two isotopes, a second value can be found for the age of the sample, thus providing a way to check the value found by uranium dating. Other techniques depend upon other isotopes, for example, potassium-40 decaying into argon-40 with a half-life of 1.3 billion years.

Note that radioactive dating does not actually tell us the age of the Moon or the Earth, but only the minimum age, for the Moon and Earth existed before the rocks solidified. Knowledge about the solidification of matter and theories of planetary formation are used to determine how much time elapsed between the formation of the planetary body and the solidification of its surface rocks.

CONCLUSION

To determine the size of the Moon, we must first know its distance, and to determine this distance, we must know the size of the Earth. By measuring the position of the Sun as seen from different locations on Earth, Eratosthenes was able to measure the Earth's size in the third century B.C. A few centuries later, Ptolemy successfully used the method of parallax to measure the distance to the Moon, and he was therefore able to calculate its size.

The familiar phases of the Moon are caused by the relative positions of the Sun, Earth, and Moon and have been understood for more than two thousand years. Likewise, as the Moon orbits the Earth, alignments of the two bodies with the Sun produce both solar and lunar eclipses, exciting events that were mysterious to primitive people but are easily explained by simple geometry.

The Moon and Earth are astronomical studies themselves. Astronomers use knowledge gained by studying the Earth as a planet to help them understand other objects in the solar system. With the aid of what we learned from the *Apollo* missions, we are able to explain the features on the surface of the Moon and are becoming more and more confident that it originated in a tremendous impact between some object and the Earth. We are also piecing together an outline of the Moon's history from the information we have obtained from telescopic observations and from visits during the 1960s and 1970s.

RECALL QUESTIONS

1. Knowledge of which of the following will allow us to calculate the diameter of the Moon?
   A. The Moon's distance and speed
   B. The Moon's angular size and speed
   C. The Moon's angular size and distance
   D. All three—distance, speed, and angular size—must be known.
   E. [None of the above.]

2. Suppose some object is known to be 6 kilometers away and is observed to have an angular size of 0.25 degrees. What is its actual size?
   A. 26 kilometers
   B. 57.3 kilometers
   C. 60 kilometers
   D. 1.5 kilometers
   E. [None of the above.]

3. Eratosthenes calculated the size of the Earth from
   A. angles to the Sun from locations a measured distance apart.
   B. angles to the Moon from locations a measured distance apart.
   C. its angular size and distance from the Sun.
   D. its orbital speed and distance from the Sun.
   E. its calculated rotational speed.

4. The crescent shape in which the Moon often appears is caused by the
   A. direction from which sunlight strikes the Moon.
   B. shadow of the Earth on the Moon.
   C. spinning of the Moon on its axis.
   D. particular side of the Moon that happens to be facing the Earth.

5. If the Moon was new last Saturday, what phase will it be this Saturday?
   A. Waning crescent
   B. Waxing gibbous
   C. At or very near first quarrter
   D. At or very near full
   E. [Any of the above, depending upon other factors.]

6. If you observe the Moon rising in the east as the Sun is setting in the west, then you know that the phase of the Moon must be
   A. new.
   B. first quarter.
   C. full.
   D. third quarter.
   E. [Any of the above, depending upon other factors.]

7. In which case(s) can the Earth, Sun, and Moon be in a straight line?
   A. Full Moon only
   B. First quarter and third quarter
   C. New Moon only
   D. Both full Moon and new Moon
   E. This can happen at any phase of the Moon.

8. Paris is about one-fourth of the way around Earth from Chicago. On a night when people in Chicago see a first-quarter Moon, people in Paris see
   A. a new Moon.
   B. a first-quarter Moon.
   C. a full Moon.
   D. a third-quarter Moon.
   E. [Any of the above, depending upon the time of night.]

9. If the Moon revolved around the Earth in the opposite direction, but with the same period,

A. it would not exhibit phases.
   B. it would exhibit crescent phases but not gibbous phases.
   C. it would exhibit gibbous phases but not crescent phases.
   D. we would see it rise in the west and set in the east.
   E. [None of the above.]

10. Suppose that astronauts land somewhere on the Moon when it is new for Earth-bound observers. Which of the following would be true?
    A. Earth would appear full to the astronauts (assuming that they could see it).
    B. Around the landing site there might be bright sunlight.
    C. The landing site might be dark.
    D. [All of the above.]
    E. [None of the above.]

11. If the Moon always orbited directly above the equator, we would
    A. have eclipses every month.
    B. never have eclipses.
    C. have eclipses only at solstice.
    D. have eclipses only at equinox.

12. A lunar eclipse occurs
    A. only around sunset.
    B. only near midnight.
    C. only near sunrise.
    D. at any time of night.
    E. at any time of day or night.

13. The *Moon* is _____ during an annular eclipse as/than during a total solar eclipse.
    A. farther from Earth
    B. closer to Earth
    C. the same distance from Earth
    D. [No general statement can be made.]

14. The *Sun* is _____ during an annular eclipse as/than during a total solar eclipse.
    A. farther from Earth
    B. closer to Earth
    C. the same distance from Earth
    D. [No general statement can be made.]

15. Earth doesn't experience an eclipse of the Sun every month because
    A. sometimes the Moon is too far away.
    B. the Moon always keeps its same side toward the Earth.
    C. the Moon's orbit is not in the same plane as the Earth's orbit.
    D. you have to be in the right place to see a solar eclipse.

16. You are likely to see more of which type of eclipse during your lifetime?
    A. Solar eclipses
    B. Lunar eclipses
    C. [No general statement can be made.]

17. At the same time an astronaut on the Moon sees a solar eclipse, observers on Earth can
    A. see a lunar eclipse.
    B. also see a solar eclipse.
    C. see either a solar or a lunar eclipse, depending on Earth's orientation.

18. The density of an object is defined as
    A. its thickness.
    B. how much solid material the object contains.
    C. its mass.
    D. its volume.
    E. the ratio of its mass to its volume.

19. The Earth's atmosphere is made up of about
    A. 80% oxygen and 20% nitrogen.
    B. 50% oxygen and 50% nitrogen.
    C. 20% oxygen and 80% nitrogen.
    D. equal amounts of oxygen, nitrogen, and carbon dioxide.
    E. equal amounts of oxygen, hydrogen, and carbon dioxide.

20. Auroras results from
    A. the Earth's magnetic field and its rotation.
    B. the Earth's magnetic field and its revolution around the Sun.
    C. the Earth's magnetic field and the solar wind.
    D. the solar wind and the Sun's rotation.
    E. the motion of the Moon around the Earth.

21. Maria are
    A. lunar mountains.
    B. lunar highlands.
    C. flat plains.
    D. near the lunar poles, and nowhere else.
    E. [More than one of the above.]

22. Which of the following theories of the Moon's origin seems to fit the data best?
    A. The capture theory
    B. The fission theory
    C. The double planet theory
    D. The large impact theory
    E. [Either A or C above.]

23. The fact that the density of the Moon is somewhat different from that of the Earth is an argument against which theory of the origin of the Moon?
    A. The capture theory
    B. The fission theory
    C. The double planet theory
    D. The large impact theory
    E. [Both B and C above.]

24. Define angular size and zenith.

25. When we see an unfamiliar object at a distance, how can we judge its size?

26. What is the angular size of the Moon? How does this compare to the angular size of the Sun?

27. Explain how Eratosthenes measured the diameter of the Earth.

28. Name eight different phases of the Moon in the order in which they occur.

29. Define umbra and penumbra.

30. Total (and annular) solar eclipses occur more frequently than do total lunar eclipses. Why, then, will

you probably observe many more lunar eclipses than solar eclipses during your lifetime?

31. Describe the appearance of a totally eclipsed Moon. Why is it not completely hidden from view?
32. At what phase of the Moon does a lunar eclipse occur? A solar eclipse?
33. Why might a partial solar eclipse go unnoticed by most people?
34. Why are some solar eclipses annular rather than total?
35. Show on a sketch the relative positions and sizes of the Earth's core, mantle, and crust.
36. What is a magnetic field, and how can one be detected?
37. What are the Van Allen belts and what causes them?
38. List some evidence for the theory of plate tectonics.
39. Name and describe four theories for the origin of the Moon. Which best fits present data?
40. What caused the craters and rays on the Moon?
41. Explain how we can determine the relative order in which events occurred in the formation of the Moon's surface.

QUESTIONS TO PONDER

1. The angular size of your finger held at arm's length is about 2 degrees. On some clear night use this fact to determine the angular distance between the two pointer stars of the Big Dipper. (See the star charts at the back of the book to help you find the dipper.)
2. If you observe the Moon with first one eye and then the other, do you detect a parallax shift against the stars? Why or why not?
3. Why must a baseline larger than the Earth's diameter be used to measure the distance to stars by the method of triangulation?
4. In what direction do we see the Moon move across the sky during the night? In what direction does it move relative to the stars? Explain the discrepancy.
5. At about what time does the full Moon rise? Set?
6. If you see the Moon high overhead shortly after sunset, about what phase is it in?
7. Will the Moon appear crescent or gibbous when it is at postion *X* in Figure 6-42? At position *Y*? At about what time will the Moon rise if it is at position *Y*?
8. When people in Chicago see a total lunar eclipse, what type of eclipse do people one-fourth of the way around the Earth see? (Assume that the Moon is visible in both skies.)
9. Refer to Figures 6-15 and 6-16. Identify the position the Moon must have in Figure 6-15 in order for it to appear as it does in the right-hand exposure of Figure 6-16.
10. Why does the duration of a lunar eclipse depend upon the Moon's distance from Earth?
11. Discuss the danger involved in viewing a solar eclipse and describe four ways to view such an eclipse safely. (Hint: See the Activity, "Observing a Solar Eclipse.")
12. Write a description of the reports that might be sent back by extraterrestrials as they view Earth through the window of their approaching spaceship.
13. We say that the Moon's maria were formed by volcanic action, but that the craters of the Moon are not volcanic in origin. Explain this apparent contradiction.

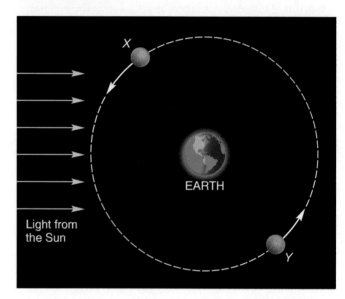

▶ FIGURE 6-42. (Question 7) Is the Moon crescent or gibbous at point *X*? At *Y*? What time does it rise when at *Y*?

14. ▶Figure 6-43 is a multiple exposure of the Moon and Venus as they set one night over Tulsa. Explain why the Moon's distance from Venus changes as time passes.
15. ▶Figure 6-44 was taken by the *Galileo* spacecraft as it passed the Earth-Moon system in December 1992. Judging from the relative sizes of the objects and their locations in the photo, answer the following questions:
    A. Which is closer to *Galileo's* camera, the Earth or the Moon?
    B. As seen from the spacecraft, is the Moon in the crescent, quarter, gibbous, or full phase?
    C. As seen from the Earth, is the Moon in the crescent, quarter, gibbous, or full phase?
    D. As seen from the Moon, is the Earth in the crescent, quarter, gibbous, or full phase?

▷ **FIGURE 6-43.** For Question 14.

▷ **FIGURE 6-44.** For Question 15.

CALCULATIONS

1. Compare the distances reached by the umbral shadows of the Moon and the Earth. Compare the diameters of the two bodies. Discuss any relationship you see.

2. Assume that under the conditions described for Eratosthenes' measurement of the Earth's size, the Sun was *exactly* 7 degrees from the zenith at Alexandria. Now at a time when the Moon was directly above the well in Syene, would it have been exactly 7 degrees from the zenith in Alexandria? Why or why not? If not, would its angular distance from the zenith have been greater or less than 7 degrees?

3. If Eratosthenes had observed the Sun to be 14 degrees from the zenith at Alexandria (instead of 7 degrees), would his value for the Earth's circumference have been greater, smaller, or the same?

ACTIVITY

Do-It-Yourself Phases

This is an important activity and is worth the trouble if you want to understand Moon phases. We will simulate the Sun/Earth/Moon system as it is shown in Figure 6-11. For the Moon you will need something like an orange, a grapefruit, or a softball. The softball would be best because it is rounder than the other choices. For the Sun, you can use a bright light across an otherwise dark room. (You could go outside and use the real Sun, but you must be careful not to look directly at it. This method would work best when the Sun is low in the sky.) Your head will be the Earth, and one of your eyes will be you.

Hold the ball out at arm's length so that it is nearly

Activity 193

▶ **FIGURE A6-1.** The person is holding a ball that represents the Moon at first quarter. When doing this, you should put the light bulb farther away than indicated here, perhaps 10 to 15 feet away.

between you and your Sun. Now observe it as you move it around to the left until it is at 90 degrees to the Sun, the first-quarter position, as shown in ▶Figure A6-1. Did you see its growing crescent as you were moving it? Continue to move it around your head and observe it as it changes phase. (When it gets directly behind you from the Sun, you will eclipse it.)

Now fix the Moon at the first-quarter position by laying it down on something in that position. Turn your head (and body) slowly around and around toward the left to simulate the Earth rotating. When the Sun is directly in front of you, it is noontime. As you lose sight of the Sun, it is sunset, about 6:00 P.M. It is midnight when the back of your head is toward the Sun. When the Moon is at first quarter and it is sunset on Earth, where do you see the Moon?

Put the moon in various other positions and observe it as you rotate your head to simulate different times of day. Answer the following questions:

1. If the Moon is at third quarter, at about what time will it rise? At about what time will it set?

2. At about what time will a full Moon appear highest overhead?

3. If you see the Moon in the sky in midafternoon, about what phase will it be?

ACTIVITY

Observing the Moon's Phases

The Moon and its phases are a very easy-to-observe astronomical phenomenon. Actually observing the changing phases of the Moon will make them much more real to you than simply reading about them.

This exercise will take a number of nights. Find a calendar or daily newspaper that lists Moon phases, or ask your instructor for this information, so that you can begin your observations three or four nights after the new Moon. Observations should start at sunset or shortly thereafter. At least four observations should be made, continuing for at least three more clear nights during the two weeks following your first night.

It is important that the observations be made from exactly the same place and at exactly the same time of night. Stand at the same place, not just in the same parking lot, for example. You might even go so far as to mark your location with chalk.

1. On each night of your observations, after you have arrived at your observing location, use a full sheet of paper to make a sketch of the position of the Moon relative to buildings, trees, and the like on the horizon. This sketch should show how things look to you and should not be a map. Label buildings, such as "Student Union." Also include prominent stars you see near the Moon. Draw the Moon in the shape it appears, and with the correct apparent size relative to objects on the horizon. Finally, write the date and time on your paper.

2. On one of the nights, repeat the observation after waiting about an hour. Use the same sketch and show the new position of the Moon.

3. When you have finished your four (or more) observations, make a general statement about how the Moon changed position and phase from one night to another. Look for a pattern in the Moon's behavior and explain that pattern based on the Moon's motion around the Earth.

4. You might try continuing your observation for an entire phase cycle (about a month), and note any problems that arise or any modifications you have to make to your observing scheme.

ACTIVITY

Observing a Solar Eclipse

There is a misconception that the Sun emits especially harmful rays of some kind during a solar eclipse. This is a particularly geocentric idea, for it would mean that somehow the Sun knows when Earth's Moon is about to block sunlight from the Earth. Naturally, the radiation emitted by the Sun during an eclipse is no different from that emitted at any other time.

Like most misconceptions, however, there is an element of truth in this idea. In fact, the Sun continuously emits radiation that is harmful to our eyes: *infrared* radiation.* If you were to look at the Sun anytime, this radiation would harm your eyes, but normally you are not able to look at the Sun. If you attempt it, your eye will quickly close because of the intense light. When the Sun is nearly totally eclipsed, however, its light is dim enough that you are able to look at it. So during an eclipse it would be possible for a person to stare directly at the Sun for some time, all the while unknowingly absorbing the harmful infrared rays in his or her eyes.

There are a number of safe methods of observing the Sun during an eclipse. First, we might use a telescope with a solar filter attached. This is a filter that blocks out some 99.99% of the Sun's light, allowing just enough through for us to see the Sun.

A more convenient way to use a telescope to observe an eclipse is illustrated in ►Figure A6-2, which shows the author using a telescope to project the partially eclipsed Sun onto a screen. In this case, a reflecting device was mounted on the telescope to cause the image to appear off to the side. This photo was taken when the Moon had progressed much of the way across the solar disk. It was near noon at the time but the sky had still not darkened noticeably. Even a little of the Sun's light is enough to give us a bright day on Earth.

A third method of safely observing a solar eclipse—one requiring little equipment—is by pinhole projection. To use this method, all you need is a piece of cardboard with a hole in it and a piece of paper to use as a screen. ►Figure A6-3 illustrates the method. Try holes of different sizes, from one millimeter to one made with a paper punch. (A hole made by a pin will probably be too small.) To see the image better, you might use large pieces of cardboard to shield your screen from reflected light or work in a dark room that has an opening facing the Sun. Block all of the opening except for your pinhole.

A fourth method of observing an eclipse is to obtain a safe filter through which you can view the Sun directly. Extreme care must be exercised here, however. If your filter does not block the Sun's infrared radiation, it may damage your eyes. (Smoked glass is definitely not recommended.) To avoid the possibility of injury, it is recommended that you be very sure of your filter or that you use one of the other methods.

---

*The Sun also emits ultraviolet radiation that can harm the eyes. The primary danger during solar eclipses is infrared radiation, however.

►FIGURE A6-2. The telescope has a mirror (called a *star diagonal*) attached to it so that it projects the image of the partially eclipsed Sun onto the screen at the side.

►FIGURE A6-3. A pinhole projector can be used to view a solar eclipse, but you must shield your screen better from scattered light than is done here.

StarLinks netQuestions

Visit the netQuestions area of StarLinks (www.jbpub.com/starlinks) to complete exercises on these topics:

1. The Phases of the Moon The moon's changing appearance (phases) is caused by the relative positions of the Earth, Moon, and the Sun.

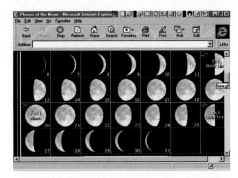

2. Lunar Eclipses Have you ever wondered how the Moon becomes shaded during an eclipse? What colors will show?

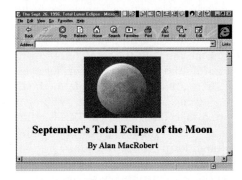

3. Europe and North America Are Moving Apart! Europe and North America are moving apart at a rate of about 2-4 centimeters per year. The theory of plate tectonics helps explain this movement.

4. The Total Solar Eclipse of 1999 A total solar eclipse occurs when light from the normally visible portion of the Sun is completely blocked by the moon, and is one of the most spectacular natural phenomena that we can experience. Get ready for the one coming in 1999.

CHAPTER 7

A PLANETARY OVERVIEW

Distances in the Solar System
Measuring Distances in the Solar System
CLOSE UP: The Titius-Bode Law
HISTORICAL NOTE: The Discovery of the Asteroids
Measuring the Mass of a Solar System Object
Planetary Motions
Classifying the Planets
Size, Mass, and Density
Satellites and Rings
Rotations
Planetary Atmospheres and Escape Velocity

Gases and Escape Velocity
The Atmospheres of the Planets
The Formation of the Solar System
Evidential Clues from the Data
Evolutionary Theories
Catastrophic Theories
Present Evolutionary Theories
Explaining Other Clues
Planetary Systems Around Other Stars?
CLOSE UP: ET Life III—The Origin of Life

.... LIZ, YOUR AUTHOR'S GRANDMOTHER, lived quite a life! She was born in 1878, and as a young adult she delighted at her first sight of an automobile, a "horseless carriage." In her thirties she saw what had been the dream of people for centuries—a flying machine. Later in her life she experienced a less promising "advance," the atomic bomb.

Liz saw her first television set in the late 1940s, and although she never became a television addict, she appreciated its value, for she remembered marveling at radio when she was a little girl.

In 1957, as Liz's 79th birthday approached, the Soviet Union launched Sputnik, the first Earth satellite. Liz asked her grandson about satellites, and together

197

they stared at the sky in wonder thinking about the object above them. Two years later the Soviets sent a spacecraft around the Moon, and she saw the first photographs ever taken of the back side of that distant body.

In 1961, the young President Kennedy announced a plan to land a person on the Moon! At the time, the United States had not even placed a person in orbit, but during the 1968 Christmas season Liz shared via television the excitement of the Apollo 8 *crew as they circled the Moon and then returned to Earth. (The chapter-opening photograph was taken during this voyage. The Earth is above the Moon's horizon.)*

On July 20, 1969, Neil Armstrong stepped onto the Moon from the Apollo 11 *Lunar Module. It was too late at night for Liz to be able to watch the events on television, but she learned the next morning that Armstrong had said, ''One small step for man, one giant leap for mankind.''*

Liz died in 1970, having seen humans advance from riding behind horses to flying to the Moon. Will any other generation in history experience as much change as Liz's did?

Previous chapters have pointed out that there are patterns among the planets of our solar system. For example, Kepler's third law tells us of the relationship between a planet's distance from the Sun and its period of revolution. Before turning to the individual planets, an examination of other patterns of similarities and differences among the planets will be helpful.

DISTANCES IN THE SOLAR SYSTEM

Figure 7-1 shows the planets as disks, but they are actually spheres. In trying to imagine their comparative sizes, think of them that way.

To say that the Sun is the largest object in the solar system is a gross understatement. We can almost say that the Sun *is* the solar system, so great are its size and mass compared to the other objects. ➤Figure 7-1 shows the planets drawn to scale. At the bottom of the drawing, you see the partial disk of the Sun. The Sun is so large that if it had been drawn as a complete circle fitting the page, many of the planets would have been nearly too small to see. The Sun's diameter is about 1,390,000 kilometers, while the Earth's diameter is about 13,000 kilometers. Thus the diameter of the Sun is more than 100 times that of Earth. To better picture this, think of the Sun as an object the size of a basketball, a sphere about one foot in diameter. On this scale the Earth would be a BB, about the size of the head of a shirtpin, an eighth of an inch in diameter (➤Figure 7-2).

Jupiter, the largest planet, is much larger than the Earth. Its diameter, in fact, is about 11 times that of the Earth. On our scale, with the Sun as a basketball, Jupiter would have a diameter of about 1¼ inches, which is about the size of a Ping-Pong ball. It's still not much compared to the Sun.

As we discuss the sizes of solar system objects and the distances between them, try to form a mental picture of the relative distances rather than just memorizing the values.

Pluto is the smallest planet, with a diameter about one-fifth that of Earth. In our scale model it would be a grain of sand, less than ¹⁄₃₂ inch across! Appendix C lists the actual sizes of the planets along with their sizes compared to the Sun and the Earth.

Now let us consider the distances between the solar system objects. Table 7-1 shows the average distance of each of the planets from the Sun in astronomical units and according to our model. To continue the model in which the Sun is a basketball, we might put the basketball at one end of a tennis court. A BB at the opposite end of the tennis court would be the Earth. Two football fields away is the Ping-Pong ball Jupiter. Pluto would be a grain of sand a mile away! Between these objects we put nothing—or at least almost nothing. There are

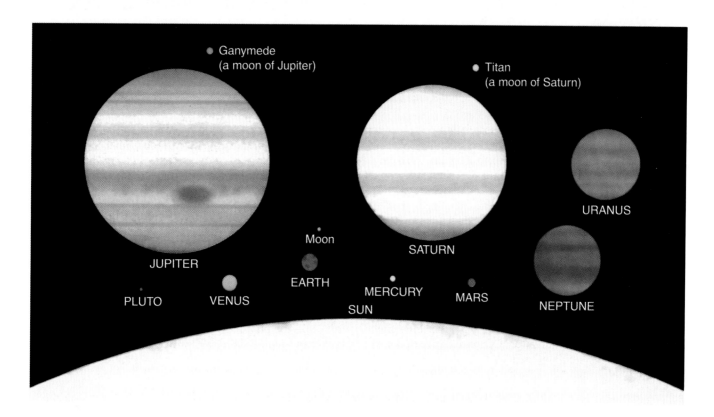

► **FIGURE 7-1.** This shows the Sun, planets, and a few of the large moons drawn to scale. The rings of Saturn are not included because they are not visible when Saturn is in this orientation.

TABLE 7-1 Average Distances of the Planets from the Sun

Object	Distance from the Sun (AU)	On Our Scale Diameter	On Our Scale Distance from Sun
Sun	—	1 foot	—
Mercury	.39	0.04 inch	45 feet
Venus	.72	0.1 inch	80 feet
Earth	1.0	0.1 inch	110 feet
Mars	1.52	0.05 inch	170 feet
Jupiter	5.20	1 inch	200 yards
Saturn	9.54	0.9 inch	350 yards
Uranus	19.19	0.4 inch	0.4 mile
Neptune	30.06	0.4 inch	0.6 mile
Pluto	39.44	0.02 inch	0.8 mile

► **FIGURE 7-2.** If the Sun were the size of a basketball, the Earth would be the size of the head of a shirtpin.

only the other planets, all smaller than Jupiter, and some smaller objects.

Nearly 3000 objects that are too small to include in our scale model have been discovered in the solar system. The largest of these **asteroids** has a diameter of about 1000 kilometers, or 600 miles. Perhaps another 100,000 much smaller asteroids orbit the Sun, most of them in the asteroid belt between Mars and Jupiter.

asteroid. Any of the thousands of minor planets that orbit the Sun.

Measuring Distances in the Solar System

A previous chapter showed that Copernicus used geometry to calculate the relative distances to the planets. That is, he was able to calculate that Mars is 1.5

Distances in the Solar System   199

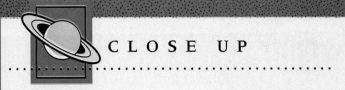

CLOSE UP

The Titius-Bode Law

As pointed out in Chapter 2, the relative distances to the planets were known in Copernicus's time. We saw that Kepler used this data to formulate his third law. From the time of Copernicus, people have wondered why the planets are at the distances they are. Is there any pattern to the distances?

In 1766 a German astronomer named Johann Titius found a mathematical relationship for the distances from the Sun to the various planets. The rule was publicized by Johann Bode, the director of the Berlin Observatory, in 1772, and is known today as the Titius-Bode law or simply Bode's law. Table C7-1 illustrates how the law works. Column 1 shows a series of numbers starting with zero, jumping to three, and then doubling in value thereafter. Column 2 was obtained by adding 4 to each of those values. Finally, to get column 3, we divide each of the column 2 values by 10. Now compare these figures to the measured distances of each of the planets from the Sun.

The table shows that the Titius-Bode law fits fairly well, except that there is a gap: No planet is found at 2.8 AU from the Sun. The law seems to indicate that there should be a planet between Mars and Jupiter and, further, that the planet should be 2.8 AU from the Sun. In addition, the law predicts that if other planets are found beyond Saturn, the next one will be about 19.6 AU from the Sun.

Bode published the law in 1772. Nine years later—in 1781—the planet Uranus was discovered by William Herschel in England. Its distance from the Sun was

TABLE C7-1 Planetary Distances According to the Titius-Bode Law Compared to Today's Values

A Series of Numbers	Add 4	Divide by 10	Today's Measured Distance (AU)	Planet
0	4	.4	.39	Mercury
3	7	.7	.72	Venus
6	10	1.0	1.0	Earth
12	16	1.6	1.52	Mars
24	28	2.8		
48	52	5.2	5.20	Jupiter
96	100	10.0	9.54	Saturn

The radar signal is typically a burst of 400 kilowatts of power, but the returning signal is only 10^{-21} watt.

AU from the Sun although he could not determine the value of an astronomical unit. Today we can measure the distances to planets using radar. We bounce radar signals from a planet and measure the time required for the signal to reach the planet and bounce back. Then, knowing that radar travels at the speed of light, 3×10^5 kilometers/second, we can calculate the distance to the planet.

EXAMPLE

Suppose that a radar signal is bounced from Mars. The signal returns to Earth 22 minutes after being transmitted. How far away is Mars?

Solution First, we realize that 22 minutes is the time the signal takes to reach Mars and return to Earth. So a one-way trip requires 11 minutes. Now let's change 11 minutes to seconds (since our radar speed is given in kilometers/second).

19.2 AU. The Titius-Bode law predicted 19.6 AU. A close fit!

With this confirmation of the validity of the Titius-Bode law, a group of German astronomers (who called themselves the Celestial Police) divided the zodiac into a number of regions, planning to assign a specific region to each of a number of astronomers who would systematically search for the missing planet at 2.8 AU. The searchers did not find it, however. Instead, a monk who was working on a different project discovered the largest of the asteroids at the distance predicted for Bode's missing planet. (See the Historical Note in this chapter.) Although it was first thought that this was the missing planet, the discovery within a few years of other objects at about the same distance made it obvious that things were not this simple. The Titius-Bode law could not account for the large number of planets between Mars and Jupiter.

Two more planets have been discovered since the discovery of the asteroids. How well do they fit the Titius-Bode law? Table C7-2 shows the complete comparison. Notice that Neptune does not fit the prediction at all but that Pluto comes fairly close. So here we see that the law no longer fits all of the data.

Today the Titius-Bode law is considered little more than a curiosity rather than a scientific law.* There are a number of reasons for this: The law is not accurate even for the planets it fits, it does not fit Neptune at all, and it is not internally consistent. (Note that the number in column 1 of Table C7-1 is not doubled in one case.) Surely, though, it is not just coincidence that the law fits even in its limited way. This chapter

TABLE C7-2 How Celestial Objects Fit the Titius-Bode Law

Bode's Law Prediction	Today's Measured Distance (AU)	Object
.4	.39	Mercury
.7	.72	Venus
1.0	1.0	Earth
1.6	1.52	Mars
2.8	2.8	Ceres
5.2	5.20	Jupiter
10.0	9.54	Saturn
19.6	19.19	Uranus
	30.06	Neptune
38.8	39.44	Pluto

shows that theories proposed for the formation of the solar system account for the fact that the more distant a planet is from the Sun, the farther it is from other planets. Still, the fact that the Titius-Bode law fits as well as it does must be considered a coincidence. Astronomers may be able to judge its significance better when, at some future date, they are able to observe planetary spacing around other stars.

---

*Relationships such as the Titius-Bode law are said to be *empirical*. This means that they are found to work, but they are not related to any theoretical framework; we don't know why they work. The Titius-Bode law isn't a particularly good empirical law, however, for the reasons noted above.

$$11 \text{ min} \times \frac{60 \text{ s}}{1 \text{ min}} = 660 \text{ s}$$

Now,

$$\text{velocity} = \frac{\text{distance}}{\text{time}}$$

So

$$\begin{aligned}\text{distance} &= \text{velocity} \times \text{time} \\ &= (3.0 \times 10^5 \text{ km/s}) \times (660 \text{ s}) \\ &= 2.0 \times 10^8 \text{ km}\end{aligned}$$

To check that this is a reasonable answer, recall that one astronomical unit is 1.5 × 10⁸ kilometers. So our calculated distance is 1⅓ AU. Since the orbit of Mars

Since Mars is about 1.5 AU from the Sun, the distance from Earth to Mars varies from about 0.5 AU to 2.5 AU.

Distances in the Solar System

is 1.5 AU from the Sun, we see that at some point in its orbit, it is possible for Mars to be at our calculated distance from Earth.

> ••••• **TRY ONE YOURSELF.** When Venus is at its closest to Earth, it requires about 4.8 minutes for a radar signal to travel to Venus and back. What is the distance to Venus? Convert the answer to astronomical units and check it with the correct distance given in Table 7-1.

When the nearest planet, Venus, is closest to Earth, a radar signal still requires nearly 5 minutes to get there and back. The great distances in the solar system become more obvious when we consider that if such a signal could be emitted in New York City and reflected from something in Washington, D.C., only 0.002 seconds would be required for the round trip.

MEASURING THE MASS OF A SOLAR SYSTEM OBJECT

How do we know the masses of the Sun and planets? To answer this we must return to Kepler's third law, which relates each planet's distance from the Sun to its period of revolution. We saw in Chapter 3 that Newton's formulation of Kepler's third law was more complete than the original statement by Kepler. The equation for Newton's version of the third law is

$$\frac{a^3}{P^2} = K(m_1 + m_2)$$

a = semimajor axis of the orbit
P = period of planet's orbit
K = constant, value depends on units of other quantities
m_1 = mass of one orbiting object
m_2 = mass of the other object

Since the mass of even the largest planet, Jupiter, is less than 0.001 times the mass of the Sun, the sum of the two masses on the right is essentially equal to the mass of the Sun and is therefore the same for each planet. Thus for objects in orbit around the Sun we can write the equation as

$$\frac{a^3}{P^2} \cong KM$$

a = average distance to the Sun
P = period of the planet's orbit
K = constant, depending on units
M = mass of the Sun
($\cong$ means "approximately equal to")

Recall that the semimajor axis of a planet's elliptical orbit is essentially its average distance from the Sun.

This is an example of the correspondence principle, which states that a new theory must make the same predictions as the old one in applications where the old one worked. (See Chapter 3.)

Notice that Newton's statement does not actually conflict with Kepler's until great accuracy is demanded. Since the Sun's mass is constant, the value on the right side of the equation is very nearly the same for each of the planets, just as Kepler said. Newton's statement reduces to Kepler's when data are used that are no more precise than Kepler had. Newton's statement of the law, however, allows us to calculate something else—the mass of the Sun. All we need to know in order to do this is the semimajor axis of one planet's elliptical orbit and that planet's period of revolution around the Sun.

There is even more value to the equation. Kepler's third law, as completed by Newton, applies to any system of orbiting objects. Recall that Galileo had compared Jupiter's system of moons to the solar system. Here the equation lets

HISTORICAL NOTE

The Discovery of the Asteroids

Johannes Kepler once proposed that there might be an undiscovered planet between Mars and Jupiter, because the large distance between their orbits does not follow the pattern of other orbits. The Titius-Bode law also seemed to predict such a planet, and this led Francis von Zach, a German baron, to plan a systematic search for the planet. Giuseppe Piazzi, a Sicilian astronomer and monk, was one of the astronomers who had been chosen to search in one of the sectors into which von Zach had divided the sky. Before he was notified where he was to search, however, Piazzi discovered what he first thought was an uncharted star in Taurus. The object he discovered (on January 1, 1801) was very dim, far too dim to see with the naked eye. He named it Ceres after the goddess of the harvest and of Sicily. Continuing to observe it, he saw that it moved among the stars, and by January 24 he decided that he had discovered a comet. He wrote two other astronomers (including Bode) of his discovery, but on February 11 he became sick and was unable to continue his observations. By the time Bode and the other astronomer received their letters (in late March), the object was too near the Sun to be observed.

Bode was convinced that the hypothesized new planet had been discovered, but he also realized that it would not be visible again until fall and by that time it would have moved so much that astronomers would have a difficult time finding it again. This was because relatively few observations had been made of the object's position, not enough for the mathematicians of the time to calculate its orbit. Fortunately, a young mathematician named Carl Frederick Gauss, one of the greatest mathematicians ever, had recently worked out a new method of calculating orbits. He worked on Bode's project for months and was able to predict some December positions for the object. On December 31, the last day of the year in which Piazzi had seen it on the first day, von Zach—the man who had sought to organize the original search—rediscovered the object.

The elation over finding the predicted planet did not last long, however, for another "planet" was found in nearly the same orbit about a year later. Its discoverer, Heinrich Olbers, was looking for Ceres when he discovered another object that moved. He sent the results of a few nights' observations to Gauss, and the mathematician calculated its orbit. The object was given the name Pallas, and a new classification of celestial objects had been found: the new objects were called *asteroids*.

By 1890 about three hundred asteroids had been found using the tedious method of searching the skies and comparing the observations to star charts, looking for uncharted objects. In 1891 a new method was introduced: a time exposure photograph of a small portion of the sky was taken, and the photograph was searched for any tiny streaks. The streaks (Figure H7-1) would be caused by objects that did not move along with the stars. These objects were then watched very closely and their orbits determined. Using such methods, well over 3000 asteroids are now known and named, and it is predicted that some 100,000 asteroids are visible in our largest telescopes.

► **FIGURE H7-1.** The two streaks on the time exposure photo are caused by the motion of two asteroids as the camera follows the stars' apparent motions across the sky.

TABLE 7-2 Percentages of the Total Mass of the Solar System	
Object	Percentage of Solar System's Mass
Sun	99.85
Jupiter	.095
Other planets	.039
Satellites of planets	.00005
Comets	.01 (?)
Asteroids, etc.	.0000005 (?)

us calculate the mass of Jupiter, which is the central object in this case. As we will see, every planet except Mercury and Venus has at least one natural satellite. Thus, to calculate the mass of one of those planets, we need only know the distance and period of revolution of at least one of its satellites.

What about Mercury and Venus, which have no moons? Their masses have been calculated on a few occasions by observing their effects on the orbits of passing asteroids and comets. No asteroid or comet has passed close enough to provide highly accurate data, however, and thus the accuracy of the calculations was limited until space probes flew by these planets. If a space probe is put into orbit around a planet, the equation above applies to it and allows us to calculate the mass of the planet. In practice, the space probe does not actually have to be put into orbit. By analyzing how the gravitational force of the planet changes the direction and speed of a probe during a flyby, the planet's mass can be calculated, although by a more complicated method than the equation we have used.

To calculate the mass of a planet from its effect on an asteroid or comet, astronomers must use more complicated methods than Kepler's third law.

When we consider the masses of the objects that make up the solar system, we must be impressed by the fact that the Sun makes up almost the entire system. Table 7-2 shows the masses by percentages of the total; the Sun's mass is almost 99.9 percent of the total. Jupiter makes up most of the rest, having more than twice as much mass as the remainder of the planets combined.

PLANETARY MOTIONS

➤Figure 7-3 illustrates the orbits of the planets drawn to scale. They are all ellipses, as Kepler had written. Most are very nearly circular, but Pluto's orbit is eccentric enough that it overlaps the orbit of Neptune. In 1979 Pluto moved to the location in its orbit where it is inside Neptune's orbit. Until 1999 it will remain closer to the Sun than Neptune is. So until 1999, you should answer the appropriate trivia question by listing Neptune as the planet most distant from the Sun.

Figure 7-3 indicates that each of the planets revolves around the Sun in the same direction. When viewed from far above the Earth's North Pole, this direction is counterclockwise. We saw in previous chapters that when viewed from this perspective, the Earth rotates on its axis in this same counterclockwise direction and that the Moon also orbits the Earth in a counterclockwise direction. We might ask if this pattern holds elsewhere in the solar system. Yes, in most cases. All of the planets except Venus, Uranus, and Pluto rotate in a counterclockwise direction as seen from above the Earth's North Pole.

Most of the other planets have natural satellites revolving around them, just

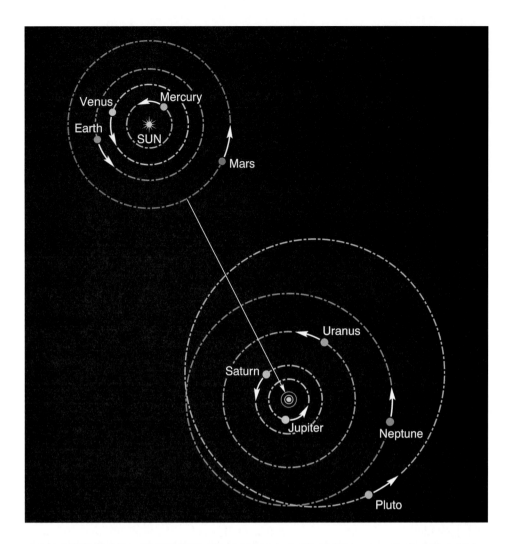

▶ FIGURE 7-3. The orbits of the planets are ellipses, according to Kepler's first law, but most are very nearly circular. The obvious exception is Pluto, which for about 20 years of its 248-year period is closer to the Sun than Neptune is.

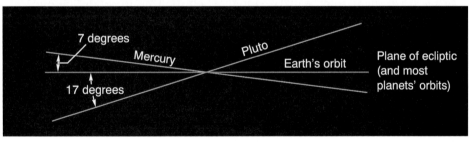

▶ FIGURE 7-4. The orbits of most of the planets are in the same plane as the Earth's (the ecliptic), but Mercury's plane is inclined at 7 degrees and Pluto's at 17 degrees.

as the Earth does. The direction of revolution of most of these satellites is also counterclockwise, although there are a number of exceptions.

The elliptical paths of all of the planets are very nearly in the same plane. This means that we can draw them on a piece of paper without having to foreshorten their paths. ▶Figure 7-4 illustrates the angles between the planes of the various planets' orbits and the plane of the Earth's orbit (this angle is known as the planet's *inclination*). Notice that Pluto's orbit is the most "out of line." We will see that Pluto is unusual in a number of other ways.

The fact that the planets have their orbits in basically the same plane, that they all orbit in the same direction, that most of them rotate in that same direction, and that most of their satellites revolve in that direction cannot be simple coincidence. We will recall these similarities when we discuss theories concerning

inclination (of a planet's orbit). The angle between the plane of a planet's orbit and the ecliptic plane.

Planetary Motions

TABLE 7-3 Eccentricities of the Orbits of the Planets and Ceres

Planet	Eccentricity of Orbit
Mercury	0.206
Venus	0.007
Earth	0.017
Mars	0.093
(Ceres)	0.077
Jupiter	0.048
Saturn	0.056
Uranus	0.047
Neptune	0.009
Pluto	0.250

the formation of the solar system; we must be sure that the theories explain these properties.

Chapter 2 showed that the eccentricity of an elliptical orbit is a measure of the "out of roundness" of the orbit. The eccentricities of the planets' (and Ceres's) orbits are given in Table 7-3. Notice how much Pluto and Mercury differ from the other planets in eccentricity.

CLASSIFYING THE PLANETS

When the properties of individual planets are examined in the following chapters, it will be clear that they divide easily into two groups. It is convenient to classify the four innermost planets—Mercury, Venus, Earth, and Mars—in one group, which we call the terrestrial planets because of their similarity to Earth ("terra"). The next four planets—Jupiter, Saturn, Uranus, and Neptune—are called the Jovian planets because of their similarity to Jupiter. As already noted, Pluto is unusual in a number of ways. Indeed, it does not fit well into either of the two categories of planets, although it is sometimes classified as a terrestrial planet. A closer examination of some of the properties of the various planets will reveal even more cases in which Pluto fits into neither category.

Jupiter was named after the most powerful of the Roman gods. The name "Jove" is derived from the Latin genitive form of Jupiter: Jovis.

Size, Mass, and Density

➤Figure 7-5 shows the diameters of the planets, in kilometers and compared to Earth. Notice that although the four terrestrial planets differ quite a bit from one another, they are all much smaller than the Jovian planets. Although Pluto is out beyond the Jovian planets, it has a size more like the terrestrials. In fact, it is the smallest of the planets.

The masses of the planets present even bigger differences between the terrestrial and Jovian planets. Look at the planetary masses shown in ➤Figure 7-6.

Many people have a tendency to skip over tables and graphs. You are not expected to memorize the values given, but a few minutes looking at patterns and thinking of their meaning will yield much knowledge about the solar system. Study the values of the masses of the planets in terms of Earth's mass. Notice the tremendous difference between the two classifications of planets. Earth is the most massive of the terrestrial planets, but the least massive Jovian planet has more than 14 times the mass of Earth. Again Jupiter stands out as the giant.

CHAPTER 7 A Planetary Overview

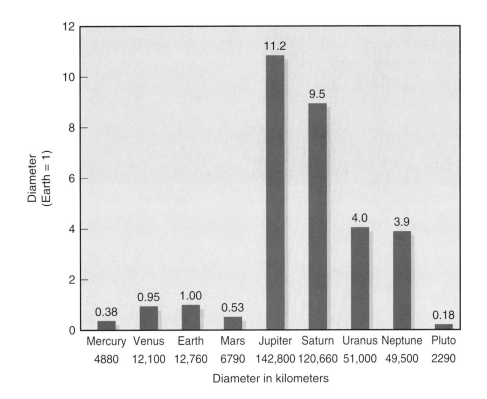

► FIGURE 7-5. A plot of planetary diameters (Earth's diameter equals one here) makes obvious the distinction between the terrestrial and Jovian planets. The diameter of each planet is shown in kilometers below the planet's name.

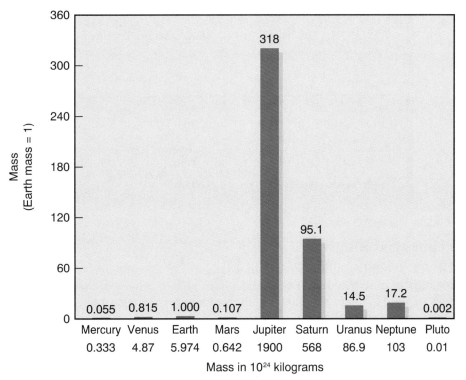

► FIGURE 7-6. When we plot the masses of the planets, we see that most of their total mass is in Jupiter. On this graph we have made Earth's mass equal to one. The masses are given in 10^{24} kilograms below each planet's name. To find the mass in kilograms, multiply the value given by 10^{24}.

A density graph (►Figure 7-7) shows another difference between the terrestrial planets and the Jovian planets. The terrestrials are more dense. This is because they are primarily solid, rocky objects, while the Jovians are composed primarily of liquid. At one time Jovian planets were commonly called "gas planets," but now we know that they actually contain much more liquid than gas.

Classifying the Planets

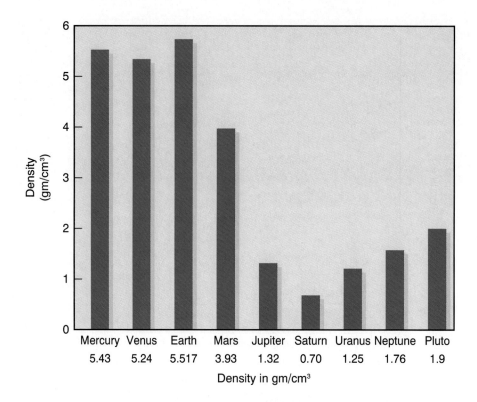

► FIGURE 7-7. The average density of the terrestrial planets is significantly greater than that of the Jovian planets. Density values are given in grams per cubic centimeter below the names of the planets.

TABLE 7-4 The Number of Known Planetary Satellites

Planet	Known Satellites	Planet	Known Satellites
Mercury	0	Jupiter	16 + r*
Venus	0	Saturn	18 + r
Earth	1	Uranus	15 + r
Mars	2	Neptune	8 + r
		Pluto	1

*An "r" indicates that the planet has a ring system.

Satellites and Rings

Table 7-4 shows the number of natural satellites of each planet. Notice that although there is no obvious pattern, the Jovian planets have more satellites. More details of the planetary satellites are found in Appendix D and in discussions of the planets in future chapters.

The table also indicates that the Jovian planets all have rings. A planetary ring is simply planet-orbiting debris ranging in size from a fraction of a centimeter to several meters. Motions of particles within the rings are extremely complex due largely to gravitational interactions with nearby planetary satellites. In a sense, each of the particles of a ring may be considered a satellite of the planet, but if we do so, counting satellites becomes meaningless. So we will continue to speak only of the larger "moons" as being planetary satellites.

Rotations

In discussing the rotation periods of the planets we must distinguish between a solar day and a sidereal day. The **solar day** is defined as the time between suc-

solar day. The amount of time that elapses between successive passages of the Sun across the meridian.

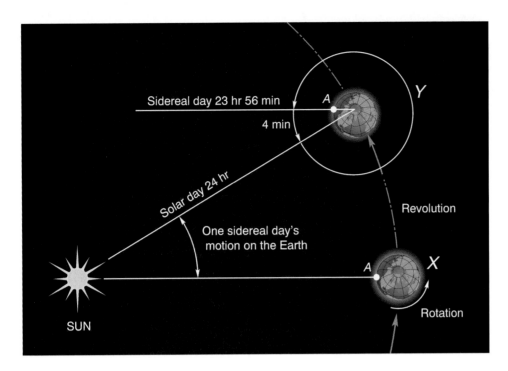

► FIGURE 7-8. When the Earth has rotated once with respect to the stars, it has not completed a rotation with respect to the Sun. Thus the sidereal day is shorter than the solar day. (The drawing greatly exaggerates the distance the Earth moves in one day.)

cessive passages of the Sun across the *meridian,* so that the length of a solar day on Earth is 24 hours. This is not the same amount of time that the Earth takes to complete one rotation, however. With respect to the stars, the Earth rotates once in 23 hours, 56 minutes.

Refer to ►Figure 7-8 to see why there is a difference between the solar day and the *sidereal day,* which is the amount of time required for an object to complete one rotation with respect to the stars. In that figure we have designated a particular location on the Earth as A. When the Earth is at location X in the figure, point A is directly under the Sun. When the Earth has turned so that point A is again under the Sun, one solar day will have passed. The Earth moves around the Sun as it rotates, however, moving about $1/365$ of the way around the Sun in one day. In the figure we have exaggerated how far it moves so that you can see the effect. Notice that when the Earth has rotated once with respect to the stars, moving from X to Y, point A is not directly under the Sun. The Earth's sidereal period is 23 hours and 56 minutes, but the Earth has to rotate for another 4 minutes to complete a solar day.

►Figure 7-9 shows the sidereal periods of rotation of the planets. Although the rotation periods of the terrestrials differ considerably from one another, notice that all of the Jovian planets rotate faster than the fastest of the terrestrials.

meridian. An imaginary line that runs from north to south, passing through the observer's zenith.

sidereal day. The amount of time that passes between successive passages of a given star across the meridian.

If you divide the number of minutes in a day by 365, you get 3.95 minutes. Thus Earth's sidereal and solar days differ by that amount.

PLANETARY ATMOSPHERES AND ESCAPE VELOCITY

Long before people visited the Moon, we knew that it contained no air and no water. This had been predicted by applying Newton's law of gravity, and the same law can also be applied to make predictions concerning planetary atmospheres. To see the connection between the law of gravity and an object's lack of atmosphere, we will first discuss jumping off the Earth. This discussion leads to an idea that will help us understand not only why some planets have no atmosphere, but—in Chapter 15—what a black hole is.

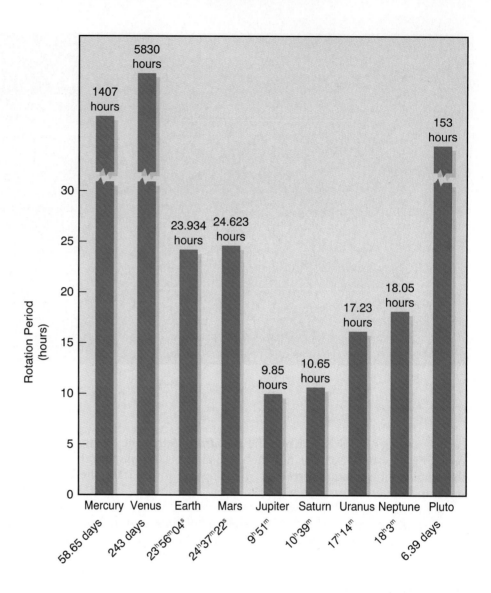

▶ **FIGURE 7-9.** The rotation periods of Mercury, Venus, and Pluto are so much greater than those of the other planets that their bars go far off the graph.

escape velocity. The minimum velocity an object must have in order to escape the gravitational attraction of an object such as a planet.

We will start by imagining an Earth with no air. On such an Earth, if we throw something upward, it will not be slowed by air friction. It will still feel the effect of gravity, however, so it will slow down, stop, and then fall back to Earth. So we throw it harder. It rises farther, and as it gets higher, the force of gravity on it is less. Thus its rate of slowing—its deceleration—is less at greater heights. Could we throw the object fast enough so that gravity never stops it and brings it back down? The answer is yes. We can calculate from the laws of motion and gravitation that the speed needed is about 11 kilometers per second. An object fired upward from Earth at this speed or greater will continue to rise, slowing down all the time, but never stopping. We call this speed the *escape velocity* from Earth.

When we consider that the speed of sound is only about 0.3 kilometer per second, we see that the escape velocity from Earth is a tremendous speed. The reason that we imagined an Earth without air friction is that, in practice, if we fired an object from the surface at 11 kilometers/second, it would be slowed—and probably destroyed—by air friction. In the space program we have fired objects into space with a velocity exceeding escape velocity; every space probe that has been sent to other planets has completely escaped the Earth. The probes were not destroyed by air friction because rockets carried them above the atmosphere before increasing their speed to escape velocity.

The escape velocity of an astronomical object depends on the gravitational

210   CHAPTER 7   A Planetary Overview

force at the object's surface (or from whatever height we are launching the projectile). Recall that the gravitational force at the surface of the Moon is only one-sixth of that at the surface of Earth. A 120-pound astronaut weighs only about 20 pounds on the Moon. The escape velocity from the Moon is therefore less than that from Earth. It is only about 2.5 kilometers per second—eight times the speed of sound in air at the Earth's surface.

To see what escape velocity has to do with the question of the atmosphere of an astronomical object, the nature of a gas must briefly be discussed.

Gases and Escape Velocity

The states of matter in our normal experience are three: solid, liquid, and gas. Some understanding of the gaseous phase is necessary to understand planetary atmospheres. To envision a gas, picture a swarm of bees buzzing around a hive. The bees represent individual molecules of a gas. This analogy is faulty in a few ways, however. First, compared to their size, the molecules of a gas are much, much farther apart than are the bees. Second, molecules move in straight lines until they collide. Then they bounce off and move straight again. Finally, remember that molecules are moving through completely empty space. There is nothing between them. Nothing.

There are three additional things we must keep in mind about the molecules of a gas:

1. As gas molecules bounce around, different molecules have different speeds at any given time. Some will be moving fast and some slow. In this sense, they are like the bees.
2. The average speed of the molecules depends on the temperature of the gas. Gases at higher temperature have faster-moving molecules.
3. At the same temperature, less massive molecules have greater speed. For example, since a molecule of oxygen has less mass than a molecule of carbon dioxide, if we have a gas that is a mixture of oxygen and carbon dioxide, the oxygen will, on the average, be moving faster.

Now let's consider the atmosphere of the Earth. Our atmosphere is held near the Earth by gravitational forces. Consider a molecule in the upper part of our atmosphere. At great heights above Earth, the atmosphere has a low density; we say that it is "thin." That means that the molecules are much farther apart than down here at the surface. Suppose that at some instant a particular molecule up there happens to be moving away from Earth. There are very few other molecules around, so a collision is unlikely and our molecule acts just like any other object moving away from Earth. The force of gravity slows it down. Whether the molecule will return to Earth or escape depends upon how the speed of the molecule compares to the Earth's escape velocity. If the molecule's speed is greater than escape speed, the molecule is gone, never to return to Earth.

Obviously, since an atmosphere exists on Earth, the velocities reached by molecules of the air do not exceed escape velocity. Recall, however, that molecules of lower mass have greater speeds. Hydrogen molecules have less mass than those of any other element. It is therefore no coincidence that there is little hydrogen in the Earth's atmosphere: the temperature of the upper atmosphere is great enough for hydrogen molecules to escape. Any hydrogen that is released into the Earth's atmosphere eventually will be lost. The chemical element hydrogen does not exist alone in our atmosphere, but only as part of more massive molecules. A molecule of water vapor, for example, contains hydrogen.

As noted earlier, the escape velocity from the surface of the Moon is about 2.5 kilometers per second. At the temperatures reached on the sunlit side of the Moon, all but the most massive gases attain speeds greater than this, and therefore the Moon has essentially no atmosphere. The *Apollo* astronauts were not surprised to find no air to breathe when they landed on the Moon.

> Phobos, a natural satellite of Mars, is tiny by astronomical standards and has an escape velocity of only about 50 kilometers per hour. You can't jump this fast, but if you have a good arm, you could throw a ball from Phobos fast enough that it would never return.

> In fact, temperature is defined by the average energy of molecules.

The Atmospheres of the Planets

The average speed of a particular type of molecule depends upon the temperature of the gas, but at any given time some molecules will be traveling faster than average. This means that although the average speed of a particular gas may be less than the escape velocity from a planet, the gas may still gradually escape because the speed of a small fraction of its molecules exceeds the escape velocity. Because of this, we must use a multiple of the average speed in considering whether or not a gas will escape from a planet. Theory shows that rather than the average speed, the value we should use in determining whether the planet will retain the gas for billions of years is 10 times the average speed of the molecules of that gas.

▶Figure 7-10 is a graph of the average speeds of various molecules versus their temperatures. The dashed lines represent 10 times the average molecular speed. All the planets, as well as some planetary satellites, are plotted on the graph at their respective temperatures and escape velocities. A planet can retain a gas if the planet lies *above* the dashed line for that gas. Find the Moon on the graph and notice that every gas has an average molecular speed that will allow it to escape. The planet Mercury is similar. On the other hand, only hydrogen and helium escape from Earth and Venus; the four Jovian planets retain all of their gases.

Table 7-5 summarizes the differences we have discussed between the terrestrial and Jovian planets.

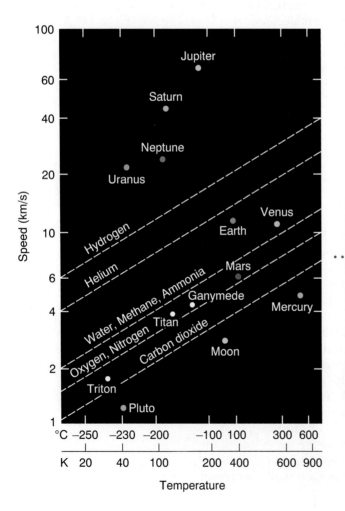

▶ **FIGURE 7-10.** This graph shows how the speeds of various gases depend upon their temperatures. The dashed lines represent 10 times the gases' average speeds. All the planets and some planetary satellites are indicated at their corresponding temperatures and escape velocities.

TABLE 7-5 Characteristics of the Jovian and Terrestrial Planets

Terrestrials	Jovians
Near the Sun	Far from the Sun
Small	Large
Mostly solid	Mostly liquid and gas
Low mass	Great mass
Slow rotation	Fast rotation
No rings	Rings
High density	Low density
Thin atmosphere	Dense atmosphere
Few moons	Many moons

THE FORMATION OF THE SOLAR SYSTEM

More than five billion years ago the atoms and molecules that now make up the planets—and our own bodies—were dispersed in a gigantic cloud of dust and gas. In a later chapter we will study how such an interstellar cloud condenses to form a star, but with what we have learned about the patterns within the solar system, we can discuss the formation of the system. A study of the beginnings of the solar system is interesting as an example of the way science in general (and astronomy in particular) progresses, because the theory is still in its early development and many gaps remain. The search for answers here resembles a mystery story where there are many clues; new ones appear all the time and some of the clues seem to contradict others.

There are two main categories of competing theories to explain the origin of the solar system: evolutionary theories and catastrophic theories. This section examines the evidence for each and shows why one is gaining favor among astronomers. First, however, it will be helpful to review the clues, many of which were discussed earlier.

Notice that we are concerned here only with the formation of the solar system, and not the formation of the Galaxy or the origin of the universe, both of which occurred much earlier. These questions are considered in later chapters.

Evidential Clues from the Data

The members of the solar system exhibit a number of patterns that must be explained by any successful theory of the system's origin. In addition, a theory should be able to account for exceptions to the patterns. Here is a list of significant data that must be explained.

1. All of the planets revolve around the Sun in the same direction (which is the direction the Sun rotates), and all planetary orbits are nearly circular except that of Pluto (and, to a lesser degree, Mercury).
2. All of the planets lie in nearly the same plane of revolution.
3. Most of the planets rotate in the same direction as they orbit the Sun, the exceptions being Venus, Uranus, and Pluto.
4. The majority of planetary satellites revolve around their parent planet in the same direction as the planets revolve around the Sun (and as the planets rotate). In addition, most satellites' orbits are in the equatorial plane of their planet.
5. There is a pattern in the spacing of the planets as one moves out from the Sun, with each planet being about twice as far from the Sun as the previous planet.
6. The chemical compositions of the planets have similarities, but a pattern of differences also exists, in that the outer planets contain more *volatile elements* and are less dense than the inner planets.
7. All of the planets and moons that have a solid surface show evidence of craters, similar to those on our Moon.
8. All of the Jovian planets have ring systems.
9. Asteroids, comets, and meteoroids populate the system along with the planets, and each category of objects has its own pattern of motion and location in the system.
10. The planets have more total angular momentum (to be described later in this chapter) than does the Sun, even though the Sun has most of the mass.
11. Recent evidence indicates that planetary systems in various stages of development may exist around other stars.

The pattern of the spacing of planets was discussed in a Close Up, The Titus-Bode Law, near the beginning of this chapter.

volatile element. A chemical element that exists in a gaseous state at a relatively low temperature.

nonvolatile element. An element that is gaseous only at a high temperature and condenses to liquid or solid when the temperature decreases.

As astronomers try to solve the mystery of the origin of the Sun and its companions, they must be sure their theories fit these clues. In addition, any successful theory of the origin of the solar system should coincide with the theory of the formation of stars.

Evolutionary Theories

There is no single evolutionary theory for the solar system's origin, but there are a number of theories that have in common the idea that the solar system came about as part of a natural sequence of events. These theories have their beginning with one proposed by René Descartes in 1644. He proposed that the solar system formed out of a gigantic whirlpool, or vortex, in some type of universal fluid and that the planets formed out of small eddies in the fluid. This theory was rather elementary and contained no specifics as to the nature of the universal fluid. It did, however, explain the observation that the planets all revolve in the same plane—that plane being the plane of Descartes's vortex.

After Isaac Newton showed that Descartes's theory would not obey the rules of Newtonian mechanics, Immanuel Kant (in 1755) used Newtonian mechanics to show that a rotating gas cloud would form into a disk as it contracts under gravitational forces (➤Figure 7-11). Thus the philosophical "universal fluid" of Descartes became a real gas working under natural laws to explain the disk aspect of the solar system (our clue #2). In 1796, Pierre Simon de Laplace, a French mathematician, introduced the idea that such a rotating disk would break up into rings similar to the rings of Saturn and that perhaps these rings could form into the individual planets while the Sun was being formed from material in the center.

Application of Newtonian mechanics to such a contracting gas cloud, however, caused another problem. To understand this problem, consider what happens when a spinning ice skater pulls in his arms. His rotation speed increases greatly (➤Figure 7-12). This increase in speed is predicted by Newton's laws and is a result of the law of *conservation of angular momentum.* We will not define angular momentum mathematically, but will simply state that the angular momentum of a rotating (or revolving) object is greater if the object is rotating (or revolving) faster or if the object is farther from the axis of its rotation/revolution. As the skater pulls in his arms, he decreases their distance from the axis of rotation; in the process, he decreases their angular momentum. To make up for this, his entire body increases its rotation speed, keeping the total angular momentum approximately constant ("conserved"). (The total angular momentum

angular momentum. The tendency of a rotating or revolving object to continue its motion.

conservation of angular momentum. A law that states that the angular momentum of a system will not change unless an outside force is exerted on the system.

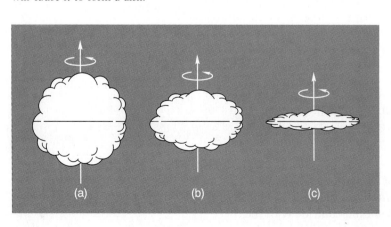

➤ FIGURE 7-12. If an ice skater begins a spin with his arms extended, he will spin faster and faster as he draws his arms in. This effect is explained by the law of conservation of angular momentum.

➤ FIGURE 7-11. As a cloud contracts, its rotational motion will cause it to form a disk.

would remain perfectly constant if it were not for air resistance and small friction forces with the ice.)

The law of conservation of angular momentum must also apply to a contracting, rotating cloud of gas. Like the ice skater, the cloud speeds up its rotation as its parts come closer to the center of rotation. When calculations are made for a cloud contracting to form the Sun and planets, we find that the Sun should rotate much faster than it does; it should spin around in a few hours. Chapter 12 will provide evidence that the Sun's period of rotation is close to a month. This means that the angular momentum possessed by the Sun is much less than the theory would predict. In fact, the total angular momentum of the planets (because of their greater distance from center) is observed to be much greater than the angular momentum of the Sun. This should not occur, according to Newton's laws.

The contradiction of these well-established laws caused the evolutionary theories to lose favor early in this century. The opposing theory was a catastrophic theory.

Catastrophic Theories

Contrary to what the name may imply, a ***catastrophic theory*** does not refer to a disaster, but rather to an unusual event—in this case the formation of the solar system by an unusual incident. In 1745 Georges Louis de Buffon proposed such an event: the passage of a comet close to the Sun. Buffon proposed that the comet pulled material out of the Sun to form the planets. In Buffon's time, comets were thought to be quite massive, but in this century we learned that a comet's mass is not great enough to cause this breakup of the Sun. However, his basic idea—that a massive object exerted gravitational forces on the Sun, pulling material out and causing it to sweep around the Sun until it eventually coalesced to form the planets—still seemed a reasonable hypothesis. Such an event as the passage of a massive object so near the Sun would be very unusual but not impossible.

More recently, it was suggested that the Sun was once part of a triple-star system, with the three stars revolving around one another. As we will see, such star systems are common, so this in itself is not a farfetched idea. This particular catastrophic theory holds that the configuration was unstable and that one of the stars came close enough to cause a tidal disruption of the Sun that produced the planets. The close approach of this star also caused the Sun to be flung away from the other two stars.

Starting around the 1930s, astronomers began to find major problems with catastrophic theories. First, calculations showed that material pulled from the Sun would be so hot that it would dissipate rather than condense to form planets. A second problem involved deuterium, a form of hydrogen. Even the outer portions of the Sun are too hot for deuterium to be stable, so not much deuterium exists in the Sun. Much more deuterium is found on the planets than in the Sun, indicating that the material of the planets could not have been part of the Sun.

Finally, as will be explained later, evidence is accumulating that other nearby stars have planetary systems around them. A catastrophic theory would predict that such systems are rare since they are produced by unusual events. If we find planetary systems elsewhere, there is probably some common process that forms them.

At the same time as these problems were becoming apparent, a solution appeared for the angular momentum problem of the evolutionary theories; as a result, catastrophic theories have been nearly abandoned in favor of modern evolutionary theories.

catastrophic theory. A theory of the formation of the solar system that involves an unusual incident such as the collision of the Sun with another star.

PRESENT EVOLUTIONARY THEORIES

In the 1940s the German physicist Carl von Weizsäcker showed that a gas rotating in a disk around the Sun would rotate differentially (the inner portion moving faster than the outer) and that this would result in the formation of eddies as shown in ➤Figure 7-13. As the figure shows, the eddies would be larger at greater distances from the Sun. According to his view, these eddies are the beginnings of planet formation, and the eddies therefore explain the pattern of distances between the planets.

A real breakthrough occurred when it was realized that a mechanism exists to account for why the Sun does not rotate faster than it does. Before turning to this mechanism, let's consider the beginning of the scenario that today's theory envisions.

Chapter 13 will explain that new stars form from enormous interstellar clouds of gas and dust. ➤Figure 7-14 shows the Rosette nebula, a stellar nursery where newly formed stars can be seen.

Recall that when an interstellar dust cloud collapses, any slight rotation that it had at the beginning will result in a greatly increased speed of the central portion (explained by the conservation of angular momentum). The material in the center will become a star—the Sun in the case of our solar system. In the meantime, the matter surrounding the newly forming Sun is condensing into a disk.

As the gases in the disk cool, they begin to condense to liquids and solids, just as water vapor condenses on the cool side of your iced drink glass. Non-volatile elements such as iron and silicon condense first to form small chunks of matter, or dust grains. Each of these grains has its own elliptical orbit about the center, and as time passes, more matter condenses onto its surface. The orbits of

➤ FIGURE 7-13. Carl von Weizsäcker showed that eddies would form in a rotating gas cloud and that the eddies nearer the center would be smaller.

FIGURE 7-14. The bright bluish stars visible in the Rosette nebula are apparently hot, young ones forming from dense dust clouds. The nebula is about 3000 light-years away.

these tiny objects are elliptical, so that they intersect one another. The resulting collisions between particles have two effects: (1) particles involved in gentle collisions (as if they were rubbing shoulders with one another as they orbit) occasionally stick together and form larger particles, and (2) particles are forced into orbits that are more nearly circular.

As the matter sticks together, small chunks grow into larger chunks. Their increased mass causes nearby particles and molecules of gas to feel a greater gravitational force toward them. Since this force is still very small, the coalescing is a very slow process, but, gradually, larger particles—now called *planetesimals*—sweep up smaller ones. Some planetesimals, resembling miniature solar systems, have dust and gas orbiting them—material that eventually condenses to become the moons we know today.

As the force of gravity shrinks a celestial object, gravitational energy causes the object to heat up. A simple case of gravitational energy being converted to thermal energy occurs whenever you drop something. The object hits the floor and heats up slightly. (The heating is very slight, and to experience it, you should probably cheat and *throw* the object down to the floor a number of times.) Perhaps you are able to visualize a release of heat when an object falls from the heavens onto a planet, but the same effect occurs when gravitational forces cause the collapse of a cloud of gas and dust. The material heats up as it falls.

This heating effect occurs with our solar-system-in-formation. The material that falls inward to form the Sun gets hot, and the high temperatures near the new Sun will not permit condensation of the more volatile elements. This means that the planets that form in the inner solar system are made primarily of nonvolatile, dense material.

Farther out, matter orbiting the new Sun (the *protosun*) is moving at a more leisurely pace, and the swirling eddies around protoplanets are more prominent. The situation at this time is illustrated in ➤ Figure 7-15.

A particularly large outer planet, Jupiter, gravitationally stirs the nearby planetesimals of the inner system so that the weak gravitational forces between them cannot pull them together. Today's asteroids are the remaining planetesimals.

While planet formation is taking place, the Sun continues to heat up. It heats the gas in the inner solar system and causes electrons there to be separated from their atoms, forming charged atoms *(ions)* and electrons. A magnetic field does

planetesimal. One of the small objects that formed from the original material of the solar system and from which a planet developed.

Chapter 9 will present evidence that Jupiter has not yet lost all of the excess thermal energy that resulted from its formation.

ion. An electrically charged atom or molecule.

Present Evolutionary Theories

▶ **FIGURE 7-15.** At the stage of development shown here, planetesimals have formed in the inner solar system, and large eddies of gas and dust remain at greater distances from the protosun.

not exert a force on an uncharged object, but if a magnetic field line sweeps by a charged object, a force will be exerted on that object. This is what must have slowed the Sun's rotation; the magnetic field of the rapidly rotating Sun exerted a force on the ions in the inner solar system, tending to sweep them around with it. Newton's third law tells us, however, that if the Sun's magnetic field exerts a force to increase the rotational speed of these particles, they must exert a force back on the Sun to decrease its rotational speed. So it is the magnetic field of the Sun, discovered rather recently, that provides the explanation for why the Sun rotates so slowly—a fact that was once a stumbling block for evolutionary theories.

The solar system of our story is getting close to what is seen today. Gas and dust are still more plentiful between the planets, however, than in today's solar system, and the inner solar system contains much more hydrogen and other volatile gases than exist there today. To help answer the question of how these gases were moved to the outer solar system and how in general the system was "cleaned up," we can again look into space at interstellar dust clouds.

In these clouds we see stars at various stages of formation. There is evidence that many newly forming stars go through a period of instability during which their **stellar wind** increases in intensity. The stellar wind, called the solar wind in the case of our star, consists of an outflow of nuclear particles from the star. It continues throughout a star's lifetime, as Chapter 14 will explain. If the instabilities we observe in other stars occurred during the formation of the Sun, the

stellar wind. The flow of nuclear particles from a star.

pulses of solar wind would sweep the volatile gases from the inner solar system. Even without this increased activity, it is expected that the solar wind would gradually move this material outward, but if the Sun did go through this period, there is certainly no difficulty explaining why hydrogen and helium exist on the outer planets but not the inner. Once in the outer system, this material would gradually be swept up by the giant planets there.

Explaining Other Clues

As millions of years passed, remaining planetesimals fell to the planets and moons, resulting in the craters we see on these objects today.

Comets are thought to be material that coalesced in the outer solar system, the remnants of small eddies. These objects would feel the gravitational forces of Jupiter and Saturn, and many would fall into those planets. Small objects that formed beyond the giant planets' orbits, however, would be accelerated by Jupiter and Saturn as those planets passed nearby and would be pushed outward. As Chapter 10 will explain, there is reason to think that great numbers of comets exist in a region far beyond the most distant planet.

For information on the distant shell of comets, refer to the Oort cloud in Chapter 10.

Notice that the theories explain that nonvolatile elements would condense in the inner solar system but that volatiles would be swept outward by the solar wind. This accounts for the differences in the planets' chemical composition. In fact, astronomers find that when compression forces are taken into account in calculating density, planets closest to the Sun contain the most dense and least volatile material, as would be expected from the theories.

Further confirmation of evolutionary theories is found in Jupiter's Galilean satellites. As Chapter 9 points out, these satellites also decrease in density and increase in volatile elements as we move outward from Jupiter. The formation of Jupiter and its moons must have resembled the formation of the solar system; thus we see the same density pattern in Jupiter's system.

The next two chapters will discuss each planet in turn and will point out some exceptions to the patterns described here. Some of the exceptions are easy to explain using evolutionary theories. Others cannot be explained by these theories and require a hypothesis of collisions—"catastrophes"—within the early solar system.

Catastrophes may well have played a part in the formation of the solar system, but the evidence indicates that it was a fairly minor part, involving a relatively few objects, and that the overall formation of the system in which we live was evolutionary in nature. Nonetheless, the origin of the solar system is poorly understood. Pieces continue to fall into place but we still have much to learn.

PLANETARY SYSTEMS AROUND OTHER STARS?

Is the existence of our planetary system unusual, or is it common for stars to have planets? Four categories of evidence can help answer this question. We'll examine each in turn.

- *Binary stars.* As Chapter 3 showed, gravitationally connected objects revolve around their common center of mass. The case we discussed there was the Earth-Moon system. Among the stars, we observe many cases of **binary star systems** in which two stars revolve around one another in this manner. In some cases, only one star of a binary system is visible, but we can deduce the existence of the dimmer star from the motion of the visible one. If we see a star that appears to wiggle in its position or that exhibits an elliptical motion, we can conclude that it is in orbit with another object. This provides us with a possible method of detecting the presence of a large planet

binary star system. A pair of stars that are gravitationally linked so that they orbit one another.

Recall that radial velocities are measured by the Doppler effect and that radial motion can therefore be measured with greater sensitivity than can tangential motion.

in orbit around another star. If we hope to find a planet by such means, we must look at nearby stars.

A star named Barnard's star is the second closest star to the Sun. In the first half of this century, a back-and-forth motion was reported for Barnard's star. The motion is very slight, however, and detecting it involved comparing photographs taken over long periods of time under changing conditions. Most astronomers considered the data very suspect, for the photographs were taken under different conditions with instruments that had been changed over the course of the observations. Recent measurements of changes in the *radial* velocity of Barnard's star, however, agree with the original conclusions and indicate that the star does indeed have large planets revolving around it. (Large planets, of course, would cause the star to move more than small ones would.) Astronomers have found other wobbling stars, but in each case the wobble is very small and is difficult to verify.

- *Infrared companion.* The star T Tauri has a companion that emits significant radiation only in the infrared region. The companion has too little mass for it to become a star itself, yet its infrared radiation indicates that it has a high temperature. A possible explanation for this high temperature is that it is a giant planet in the process of formation, with dust and gas still falling into it. If so, this scenario lends support for evolutionary theories of planet formation—at least of large Jupiter-size planets.

- *Dust disks.* Disks of dust about the size of the solar system have been detected around a number of stars. The most obvious case is that of β (the Greek letter beta) Pictoris. ➤Figure 7-16 shows a false color photo of an infrared image of β Pictoris. The bright central star was blocked out so that the radiation from the dimmer disk that surrounds it could be observed. ➤Figure 7-17 shows another photo of β Pictoris, this one a composite of 10 red-light CCD images. Studies of the bright band that extends from lower left to upper right show that it contains fine dust with an icy consistency, which is consistent with our theory of the development of the solar system. Other evidence comes from the Hubble Space Telescope, which has found numerous cases of dust disks around new stars, making it appear that at least half of the Sun-like stars in the Galaxy have planetary systems.

➤ **FIGURE 7-16.** This is an infrared photo of β Pictoris. The star itself has been blocked out, and the disk of particles around it is visible.

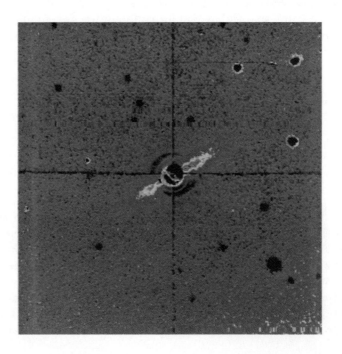

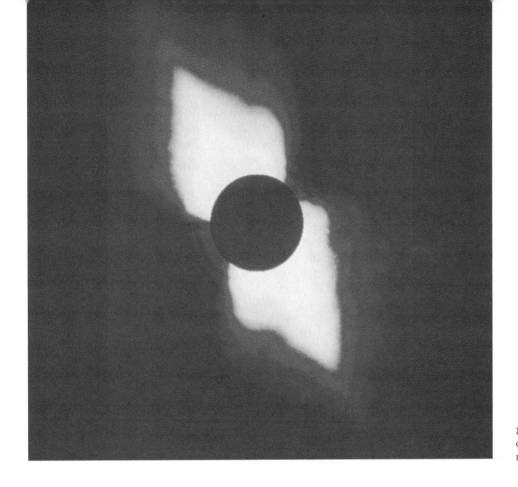

► **FIGURE 7-17.** This composite of CCD images of β Pictoris was made in red light.

■ *Pulsar companion.* In January 1992 a team of astronomers reported that they had found variations in the rate of the signals from a ***pulsar.*** Pulsars emit bursts of radio waves that normally are very constant in their frequency of pulsation. The variations in the frequency of this pulsar can be explained if it has a companion with 10 times the mass of Earth orbiting with a period of six months about the same distance from the pulsar as Venus is from the Sun. If this finding is confirmed, other pulsars will be examined for similar evidence, and we will have more pieces to add to the puzzle of whether planetary systems are common.

pulsar. A celestial object of small angular size that emits pulses of radio waves with a regular period between about 0.03 and 5 seconds.

Pulsars will be described in Chapter 15.

If we find that planetary systems are common around other stars, we will be even more confident that our solar system formed by an evolutionary rather than a catastrophic process. The reasoning is as follows. If catastrophic processes are necessary for the formation of planetary systems, relatively few such systems would be expected since the special circumstances needed are, by their nature, rare. On the other hand, if stars form by the process described by evolutionary theories, we would expect planetary systems to be common. Notice that although the formation of planetary systems by catastrophic events is unlikely, this in itself is not an argument against such a scenario in the case of the solar system. If there is any possibility at all that a catastrophic event can cause a planetary system, it could well have happened here. If we find that evolutionary development of planetary systems is common, though, the discovery will lend support for hypotheses that postulate that this is what occurred in our own system.

www.jbpub.com/starlinks

Planetary Systems Around Other Stars?

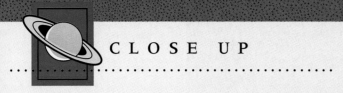

CLOSE UP

ET Life III—The Origin of Life

The question of the origin of life on Earth concerns not only biology but also astronomy, for its answer will help us determine the probability of the existence of extraterrestrial life. The theory of evolution, under development since the last half of the nineteenth century, explains how higher forms of life evolve from more primitive forms, but this still did not answer the question of the beginnings of life. Then some 50 years ago, J. B. S. Haldane, a Scottish biochemist, and A. P. Oparin, a Russian biochemist, proposed that soon after the Earth's formation, the necessary chemical elements were present for complex molecules to form—molecules that are needed for life. The seas of the early Earth were thought to be made up primarily of water, methane, ammonia, and hydrogen, and the two men hypothesized that these molecules would spontaneously collect into more complex, organic (carbon-based) molecules.

The Haldane/Oparin hypothesis remained an interesting conjecture until 1950, when Harold Urey (the Nobel Prize winner who discovered deuterium) suggested that his graduate student, Stanley Miller, test the hypothesis by simulating the conditions of the early Earth. Miller put a mixture of water, hydrogen, methane, and ammonia into a sealed container. He heated the liquid—the young Earth was hot—and used an electrical source to produce sparks in the gas above the liquid. The sparks simulated lightning, which must have been common in the Earth's young atmosphere.

After Miller's apparatus had heated and sparked for a week, the mixture had turned dark brown. Analysis showed that it now contained large amounts of four different amino acids (complex organic molecules that form the basis for proteins). He also found fatty acids and urea (a molecule that is necessary in many life processes).

The Miller-Urey experiment showed that given the right chemicals and a source of energy, chemical reactions will occur that produce the building blocks of life. Since then, researchers have learned that the early Earth's atmosphere was made up primarily of carbon dioxide, nitrogen, and water vapor rather than the four compounds used in Miller's experiment. When these compounds are used in the sealed apparatus, even more complex organic molecules are produced. Furthermore, if instead of using electrical sparks for energy, one uses ultraviolet light (which strikes Earth from the Sun), the same results are achieved.

It must be emphasized that the Miller experiment did not produce life, only organic molecules. The experiment confirmed the Haldane/Oparin hypothesis, however, and showed that such molecules will form easily in a short time. The results of the experiment would indicate that the early Earth must have contained seas made up of an organic goo, perhaps like chicken bouillon.

More recently, traces of amino acids have been found in some meteorites. In addition, there is spectroscopic evidence that organic molecules exist in interstellar clouds. Thus the molecular building blocks of life seem to be common in the universe. How does life form from these blocks? We don't know, but it seems that nature's deck may be stacked in the direction of life.

CONCLUSION

This chapter has shown that our solar system contains a myriad of objects, vastly different from one another. Similarities are also present, however, and enable the planets to be divided into two categories: terrestrial and Jovian. Also described were some of the characteristics that prohibit us from listing the (normally) most distant planet, Pluto, in either category.

One of the most important topics discussed during our overview of the solar system was how we measure some of the properties of the objects circling the Sun, including distances between them, their sizes, and their masses.

The patterns we observe in the solar system, along with what we know about the formation of stars, allow us to put together a reasonably detailed story of the origin of the solar system. Although many questions remain, observations of newly forming stars are confirming our theories of the development of our system.

The next three chapters will examine each of the planets in more detail and will look at the lesser objects within the solar system—comets, meteoroids, and asteroids.

RECALL QUESTIONS

1. Which planet is most massive?
   A. Mercury
   B. Mars
   C. Earth
   D. Jupiter
   E. Saturn

2. The only planet whose orbit is more eccentric than Mercury's is
   A. Saturn.
   B. Earth.
   C. Pluto.
   D. Venus.
   E. Neptune.

3. Whether a planet or moon has an atmosphere depends upon the planet's (or moon's)
   A. orbital speed.
   B. temperature.
   C. escape velocity.
   D. [Both A and C above.]
   E. [Both B and C above.]

4. Which planet has its plane of rotation tilted most with respect to its plane of revolution?
   A. Uranus
   B. Earth
   C. Venus
   D. Mars
   E. Mercury

5. Venus might be called Earth's sister planet because it is similar to the Earth in
   A. size.
   B. mass.
   C. rotation period.
   D. [Both A and B above.]
   E. [Both A and C above.]

6. Saturn is one of the _____ planets.
   A. Jovian
   B. inner
   C. inferior
   D. minor

7. Which of the following statements is true of all of the planets?
   A. They rotate on their axes and revolve around the Sun.
   B. They rotate in the same direction.
   C. They have at least one moon.
   D. Their axes point toward Polaris.
   E. [More than one of the above is true of all of the planets.]

8. Saturn's density is
   A. less than that of Jupiter.
   B. more than that of Jupiter.
   C. similar to the Earth's.
   D. greater than that of Earth.
   E. [Two of the above.]

9. Which of the following lists the four planets from smallest to largest?
   A. Mars, Mercury, Earth, Uranus
   B. Mercury, Uranus, Mars, Earth
   C. Uranus, Mercury, Mars, Earth
   D. Mars, Mercury, Earth, Uranus
   E. [None of the above.]

10. Compared to Jovian planets, terrestrial planets have a
    A. more rocky composition.
    B. lower density.
    C. more rapid rotation.
    D. larger size.
    E. [More than one of the above.]

11. Which of the following is true of Jovian planets?
    A. They have low average densities compared to terrestrial planets.
    B. Their orbits are closer to the Sun than the asteroids' orbits.
    C. They have craters in old surfaces.
    D. They have smaller diameters than terrestrial planets do.
    E. They have fewer satellites than terrestrial planets do.

12. Most asteroids orbit the Sun
    A. between Earth and Mars.
    B. between Mars and Jupiter.
    C. between Jupiter and Saturn.
    D. beyond the orbit of Saturn.
    E. [None of the above. No general statement can be made.]

13. Distances to the planets are measured today by the use of
    A. geometry.
    B. calculus.
    C. spacecraft flybys.
    D. analysis of the motion of their moons.
    E. radar.

14. The mass of Jupiter was first calculated
    A. using its distance from the Sun and its revolution period.
    B. using its angular size and distance from the Earth.
    C. using data from spacecraft flybys.
    D. by analysis of the motion of its moons.
    E. [Two of the above.]

15. Which is a longer time on Earth?
    A. A sidereal day
    B. A solar day
    C. [Either of the above, depending upon the time of year.]
    D. [Neither of the above, for they are the same.]

16. At greater distance from the surface of a planet, the escape velocity from that planet
    A. becomes less.
    B. remains the same.
    C. becomes greater.
    D. [Neither of the above, for the behavior of different planets is different in this regard.]

17. At the same temperature, the average speed of hydrogen molecules is _____ that of oxygen molecules.
    A. less than
    B. the same as
    C. greater than
    D. [No general statement can be made.]

18. The escape velocity from the top of Earth's atmosphere is _____ the escape velocity from the surface of the Moon.
    A. less than

    B. the same as
    C. greater than
    D. [No general statement can be made, for the escape velocity depends upon temperature.]
19. Which planet (of those listed) gets closest in distance to the Earth?
    A. Jupiter   D. Saturn
    B. Mercury   E. Mars
    C. Venus
20. If planetary systems are caused as proposed by the catastrophic theories, there should be
    A. many planetary systems besides ours.
    B. few planetary systems besides ours.
    C. [Neither of the above; the theories would make no predictions in this regard.]
21. Evolutionary theories now account for the slow rotation rate of the Sun by pointing to
    A. the slowing effect on the Sun of the solar wind.
    B. friction within the gases involved, which would prevent the Sun from rotating fast.
    C. the effect of the inner planets on the Sun.
    D. the effect of the large planets—particularly Jupiter—on the Sun.
    E. the conservation of angular momentum, which predicts a slowly rotating Sun when it formed.
22. Which of the following observations *cannot* be accounted for by evolutionary theories of solar system formation?
    A. All of the planets revolve around the Sun in the same direction that it rotates.
    B. All of the planets revolve in nearly the same place.
    C. Planets farther from the Sun are farther apart.
    D. The outer planets contain more volatile elements than the inner planets do.
    E. [All of the above are accounted for by evolutionary theories.]
23. After the evolutionary theory of the formation of the solar system was proposed, it was almost dismissed because it seemingly could not explain
    A. planetary masses.
    B. planetary distances from the Sun.
    C. the existence of comets.
    D. why some planets—particularly Jupiter—have a strong magnetic field.
    E. the observed rotation rate of the Sun.
24. According to the evolutionary theories of solar system formation, the outer planets contain much more hydrogen and helium than the inner planets because these elements
    A. never fell in near the Sun.
    B. condensed quickly to liquids and solids and remained far from the Sun.
    C. were blown away from the inner solar system by the solar wind.
    D. [Both A and B above.]
    E. [All of the above.]
25. Astronomers are now reasonably confident that the planets of the solar system
    A. formed when a comet pulled material from the Sun.
    B. formed when another star passed very close to the Sun.
    C. evolved from a rotating disk when the Sun was forming.
    D. [None of the above. There is currently no satisfactory explanation for the origin of the planets.]
26. In a previous chapter we saw that Kepler's third law relates a planet's period to its distance from the Sun. This law was expanded by Isaac Newton to include what other quantity?
27. How does the Sun's mass compare to the total mass of all other objects in the solar system?
28. What is the largest planet, and how does it compare in size and mass to the Earth?
29. What was the first celestial object discovered after Bode's law was proposed? Did it fit predictions made by the law? (Hint: See the Close Up.)
30. How do we know the masses of the planets?
31. Distinguish between a sidereal day and a solar day. Which is longer on Earth?
32. The planets' directions of rotation and revolution have certain features in common. Describe these features. Which planets have an unusual direction of rotation?
33. Name the terrestrial planets and the Jovian planets. Why are the latter called *Jovian?*
34. In what ways are the terrestrial planets similar to one another but different from the Jovians?
35. How does the eccentricity of the Earth's orbit compare to that of other planets?
36. How does the Earth compare to the other planets in density?
37. What is meant when we say that the escape velocity from the Earth is 11 kilometers per second?
38. What two factors determine the speed of the molecules of a gas?
39. Explain why hydrogen escapes from the Earth's atmosphere but carbon dioxide does not.
40. How do evolutionary theories explain the pattern of chemical abundance among the planets?
41. If definite evidence were found that planetary systems exist around other stars, would this lend support to either the catastrophic or evolutionary theories? If so, which?
42. How do evolutionary theories account for asteroids? The Oort cloud?
43. Describe the evidence that planetary systems exist around other stars.

QUESTIONS TO PONDER

1. State Bode's law and discuss how well it fits the three criteria of a good scientific theory. (Hint: See the Close Up.)
2. Kepler's third law allows us to calculate the total mass of a planet and one of its satellites. How can we tell what fraction of the total mass is the planet and what fraction is the moon?
3. Is the escape velocity from Earth the same whether we are considering a point just above the atmosphere or a point higher up? Explain.
4. The compositions of the Sun and of the Earth are similar in some ways and different in some ways. How are these similarities and differences explained by the evolutionary theories of the solar system's formation?
5. How do evolutionary theories explain observations 1, 2, and 4 (as given in the section "Evidential Clues from the Data")? How do catastrophic theories explain these observations?
6. Explain the problem the law of conservation of angular momentum presents for evolutionary theories and describe how present theory accounts for the observations.
7. Describe similarities in properties between Jupiter's system and the entire solar system. How do the evolutionary theories explain these similarities?
8. Consult other books and report on the catastrophic theory of M. Woolfson, proposed in 1960.
9. If we wish to find planets near other stars, why not just use telescopes to look for these planets directly?
10. Write a report on the discussion about different origins of life in the June 1992 issue of *Sky and Telescope*.

CALCULATIONS

1. The eccentricity of Pluto's orbit is 0.25. Use a drawing of an ellipse to explain what this means quantitatively.
2. If another planet were found beyond Pluto, how far would Bode's law predict it to be from the Sun (assuming it is the next planet in his scheme)?
3. At a time when Jupiter is 9.0×10^8 km from Earth, its angular size is 33 seconds of arc. Use this data to calculate the diameter of Jupiter, and then check your answer in Appendix C.
4. Suppose that a radar signal is bounced from Jupiter. It returns to Earth 100 minutes after being sent. How far away is Jupiter at the time of the measurement?

StarLinks netQuestions

Visit the netQuestions area of StarLinks (www.jbpub.com/starlinks) to complete exercises on these topics:

1. Solar System The solar system is made up of many different and fascinating objects.

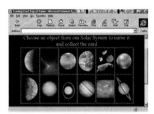

2. Jupiter Jupiter, named for a Roman god, is the second largest planet in the Solar System (the sun being the largest).

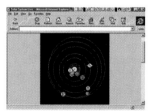

THE TERRESTRIAL PLANETS

Mercury
Mercury as Seen from Earth
Mercury via *Mariner*—Comparison with the Moon
Size, Mass, and Density
Mercury's Motions
Venus
Size, Mass, and Density
Venus's Motions
The Surface of Venus
CLOSE UP: Our Changing View of Venus
The Atmosphere of Venus
A Hypothesis Explaining Venus/Earth Differences

Mars
Mars as Seen from Earth
Size, Mass, and Density
Mars's Motions
Life on Mars
Invasion and Its Results
CLOSE UP: ET Life IV—Viking's *Search for Life*
Atmospheric and Surface Conditions
The Moons of Mars
HISTORICAL NOTE: The Discovery of the Martian Moons
Why Explore?
ACTIVITY: Viewing Mercury, Venus and Mars

MARS HAS TWO MOONS, Phobos and Deimos. They would be interesting places to visit, for both are very small and irregularly shaped. Their mass is so little, you would weigh almost nothing standing on their surfaces. Deimos, the smaller moon, varies in radius from 5 kilometers to 8 kilometers, and a person who weighs 120 pounds on Earth would weigh only about 1 ounce at the lowest spot on Deimos and 0.4 ounces at the highest point.

On Deimos, you could jump up and not come down for many minutes. The escape velocity from the highest point on Deimos is about 13 miles per hour. You *could easily throw a ball completely off this moon, never to return. Throwing the ball over the horizon at a lower speed would result in the ball going into orbit, and since the distance around Deimos is about 50 kilometers (30 miles), such a ball would take about three hours to complete its orbit. A baseball game on Deimos would be quite strange.*

Our trip to the planets will begin with the planet closest to the Sun and will then proceed outward. As each planet is discussed in turn, you should compare that planet with Earth and with planets previously discussed. Concentrate on remembering patterns and comparisons between planets rather than memorizing numerical facts.

MERCURY

In legend, Mercury was the god of roads and travel and the messenger of the Roman gods. He wore winged sandals and delivered his messages with godlike speed. It is for this speedster that the fastest planet of the solar system is named.

Mercury as Seen from Earth

Although Mercury was one of the planets known to the ancients, it is the naked-eye planet least seen by people on Earth. Even Copernicus supposedly lamented near the end of his life that he had not seen it. Mercury is so hard to see because it is so close to the Sun that it can be seen with the naked eye only either shortly after sunset in the western sky or shortly before sunrise in the eastern sky (see ➤Figure 8-1).

When Mercury is viewed by telescope, we are able to see that it exhibits phases like Venus and the Moon. Surface detail is difficult to discern, however, primarily because when Mercury is near the horizon, its light is passing through so much of the Earth's atmosphere. ➤Figure 8-2 illustrates this situation. The best telescopic views of Mercury actually are made when it is high overhead during the day, but even then, surface features are not well defined, and the telescope shows only that Mercury's surface contains bright and dark areas. Early

➤ FIGURE 8-1. (a) Mercury can only be seen from Earth either shortly before sunrise (shown) or shortly after sunset. Mercury's maximum elongation (its maximum angle from the Sun) is about 28 degrees. (b) Possible positions of Mercury and Venus in the evening relative to the horizon. Mercury never appears in a really dark sky.

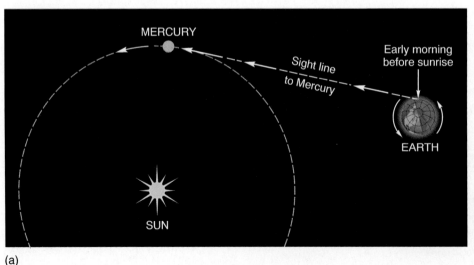
(a)

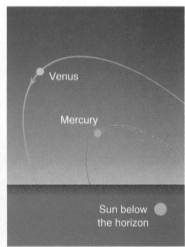
(b)

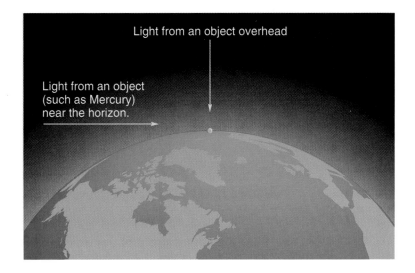

► FIGURE 8-2. The thickness of Earth's atmosphere is exaggerated here, but notice that light from an object near the horizon must pass through more atmosphere than light from an overhead object. The dot on the earth's surface is the location of a person looking at the two objects.

in this century some reports indicated a cratered surface, but this idea was not well accepted. Details of Mercury's surface features were not seen clearly until the planet was visited by space probes.

Mercury via *Mariner*—Comparison with the Moon

In November 1973, *Mariner 10* left Earth on a mission to Mercury. After passing near Venus, it swept by Mercury in March 1974, taking numerous photographs. Its orbit was then adjusted so that after passing Mercury, it orbited the Sun with a period just twice that of Mercury's, coming back near Mercury once during each of its orbits. *Mariner's* systems lasted long enough to enable it to take photographs on two passes of Mercury after the first, providing us with more than 4000 photographs of Mercury.

►Figure 8-3 is a mosaic of *Mariner's* views of Mercury. Notice its similarity to our Moon; both are covered with impact craters produced by debris from space. We find, however, that the walls of the craters of Mercury are less steep than those of the Moon so that Mercury's craters are less prominent than the Moon's (►Figure 8-4). This would be expected, because Mercury's surface gravity is about twice that of the Moon and loose material will not stack as steeply under the greater gravitational force.

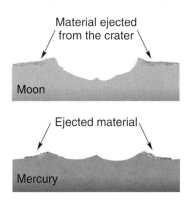

Like the craters on the Moon, most (or all) of Mercury's craters were formed by impacts with infalling objects—meteorites.

► FIGURE 8-3. A mosaic of photos of Mercury taken by *Mariner 10*.

► FIGURE 8-4. Craters on the Moon have steeper walls than those on Mercury, and because of the greater surface gravity on Mercury, material ejected from the crater does not travel as far.

Mercury

In comparing photographs of Mercury and the Moon, notice that ray patterns of material ejected from craters are less extensive on Mercury than on the Moon. When craters were formed by meteorite impacts, the greater gravitational force on the surface of Mercury kept the material from being thrown as far from the craters.

Notice also that Mercury lacks the large maria we see on the Moon. Instead, Mercury's craters are fairly evenly spread across the surface, separated by smooth plains. This difference between the two surfaces is related to differences in the rate at which the objects cooled after formation. As we discussed in Chapter 7, the terrestrial planets and the Moon began as balls of molten rock. As they cooled, a solid crust formed on their surfaces. Debris falling from space struck the crust with enough energy to penetrate it and let lava flow up through the break. The results of these lava flows are obvious on the Moon, where large maria were formed from the lava that welled from beneath the surface.

A larger object cools more slowly than a small one, however, so Mercury's crust formed more slowly than the Moon's. For a long time after the crust started forming, meteoritic debris was still able to penetrate it, allowing lava to flow out and obliterate older craters. This resulted in the plains we see between the craters.

Another difference between Mercury's surface and the Moon's is the great number of long cliffs (called *scarps*) that are found on Mercury (➤Figure 8-5). These are found at many locations around the planet and suggest that after its crust hardened during its formation, Mercury shrank a little, causing the cliffs to be formed.

Mariner photographs revealed a large impact crater named the Caloris Basin on Mercury's surface (➤Figure 8-6). It consists of a number of concentric rings somewhat like those of Mare Orientale on the Moon (➤Figure 8-7), although the Mercurian feature is larger. Both "bull's eyes" were caused by large objects striking the surfaces; the impacts resulted in shock waves that caused the rings of cliffs. On the opposite side of Mercury from Caloris Basin, we find a jumbled, wavy area strange enough that it has been dubbed "weird terrain." This was probably the outcome of the shock wave that was sent around the planet as a result of the Caloris Basin impact. When the waves met on the other side of the planet, they caused a permanent disruption of the surface there. We will see another consequence of the Caloris Basin impact when we discuss Mercury's motions.

scarps. Cliffs in a line. They are found on Mercury, Earth, Mars, and the Moon.

➤ **FIGURE 8-5.** Near the upper right center of this photo, one can see a line of cliffs called scarps. These are much more extensive on Mercury than on the Moon.

► **FIGURE 8-6.** The center of the Caloris Basin lies just out of the photo toward the left. Notice the ring pattern that was formed by the impact of a large object (now buried under the surface). The basin is about 1400 kilometers (850 miles) in diameter.

► **FIGURE 8-7.** Mare Orientale, on the Moon, is similar to the Caloris Basin, but is only about 1000 kilometers (620 miles) in diameter. Mare Orientale is clearly visible in Figure C6-3 on page 185.

Data from *Mariner* confirmed that Mercury has negligible atmosphere, as was expected considering the escape velocity from Mercury's surface and the high daytime temperatures caused by its proximity to the Sun. Refer to Figure 7-10 and notice that although the escape velocity from the surface of Mercury is greater than from the Moon, Mercury's temperature is higher, and therefore we would expect none of the gases shown to be found on Mercury.

Size, Mass, and Density

Mercury is the second smallest planet, exceeding only Pluto in size. In fact, there are two natural satellites in the solar system that are larger than tiny Mercury. (►Figure 8-8 compares Mercury with some other solar system objects.) With a diameter of about 4880 kilometers, Mercury's total surface area is only slightly greater than that of the Atlantic Ocean.

4880 km = 3030 miles.

The mass of Mercury, accurately determined from its gravitational pull on the *Mariner 10* probe, is only 0.055 of the mass of the Earth. Its average density is 5.43 times the density of water, slightly less than the Earth's 5.52.

The density of an astronomical object is determined not only by the material of which it is composed but by how much that material is compressed by the gravitational field of the object. For example, an object on the surface of Mercury feels less gravitational pull downward than it would on Earth. Thus, if Mercury were made up of the same elements as the Earth, matter below the surface would be less compressed by material above, and Mercury would have a density considerably less than the Earth's. The fact that Mercury's average density is only slightly less than Earth's means that Mercury has a higher concentration of the

▶ **FIGURE 8-8.** The size of Mercury is compared to the sizes of some other solar system objects. Pluto is smaller than Mercury, but one of Jupiter's and one of Saturn's moons are larger. (In the case of moons, the parent planet is shown in parentheses.)

heavy elements than the Earth does. Evidence from *Mariner* indicates, however, that the rocks on Mercury's surface are similar to Earth rocks so this higher-density material must be below the surface. The answer must be that Mercury has a very large iron core, one that accounts for 65 to 70% of the planet's mass.
▶ Figure 8-9 shows Mercury's theorized interior and compares it to Earth's. Astronomers speculate that early in Mercury's history, a catastrophic collision with a large asteroid blasted away much of Mercury's rocky mantle, leaving behind a planet with a greater-than-normal percentage of iron.

Before the *Mariner 10* mission, Mercury was thought to have no magnetic field. To see why, recall that the Earth's field is caused by the rotation of the planet and its metallic core. Mercury's rotation rate was thought to be too slow to produce a magnetic field.

Mariner 10 did detect a magnetic field on Mercury, although it is not very strong—about 1% as strong as the Earth's field. The field appears to be shaped like the Earth's, with magnetic poles nearly aligned with the planet's spin axis. The presence of the magnetic field has been a puzzle since it was detected in 1974. First, it led astronomers to hypothesize that Mercury's metallic core is molten, because the **dynamo effect** (Chapter 6)—the predominant explanation for planets' magnetic fields—requires molten magnetic material within the planet. More recent measurements indicate that Mercury does not release more heat than it receives from the Sun. Therefore the planet's interior must no longer retain heat from its formation. Thus its core must not be molten. Many astronomers still think that some form of the dynamo effect is responsible for the magnetic field, but the question remains open.

dynamo effect. The generation of magnetic fields due to circulating electric charges, such as in an electric generator.

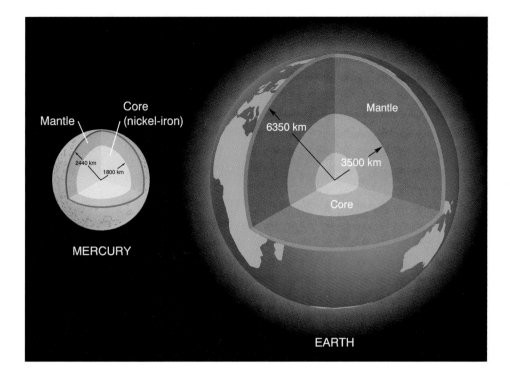

► FIGURE 8-9. Mercury's core occupies a greater portion of the planet's volume than does Earth's core.

Mercury's Motions

Being the closest planet to the Sun, Mercury circles the Sun in less time (88 days) than any other planet and moves faster in its orbit than any other (average speed: 48 kilometers/second). Again, except for Pluto, Mercury's orbit is the most eccentric of the planets. Its distance from the Sun varies from 47 million kilometers to 71 million kilometers.

Because of Mercury's proximity to the Sun and its elongated orbit, it exhibits an interesting rotation rate. Recall that because of tidal effects, the Moon keeps its same face toward the Earth. Astronomers once thought that Mercury likewise points the same face toward the Sun at all times. Radar observations, however, indicate that this is not the case. They show that Mercury rotates on its axis once every 58.65 Earth days, which is precisely two-thirds of its orbital period of 87.97 days. This means that the planet rotates exactly 1½ times for every time it goes around the Sun.

Why would this pattern occur? The answer is that Mercury is not perfectly balanced; one side is more massive than the other. Because of this, the Sun exerts a torque (a turning force) on the planet, especially when it is closest to the Sun (at *perihelion*). After countless revolutions this has resulted in the rotational period of the planet being coupled with its revolution period. ►Figure 8-10 shows this situation; note the light spot on Mercury's surface, which has been added to illustrate that opposite sides of the planet face the Sun at each perihelion passage.

The point marked on Mercury's surface in Figure 8-10 could represent Caloris Basin, for that Mercurian feature falls directly under the Sun at every other perihelion position. This is unlikely to be a coincidence. It is hypothesized that the object that hit Mercury's surface and created the basin must have been made of dense material and that the object remains under the planet's surface, causing Mercury to be lopsided and therefore to have its periods of rotation and revolution coupled as they are.

Mercury's solar day is far different from its sidereal day. Refer again to Figure 8-10 to see why. Notice that after two-thirds of an orbit, Mercury has com-

48 km/s is equivalent to 110,000 mi/hr.

perihelion. The point in its orbit when a planet (or other object) is closest to the Sun.

aphelion. The point in its orbit when a planet (or other object) is farthest from the Sun.

The Latin word *calor* means *heat.* A unit for measuring heat is the *calorie.*

Mercury 233

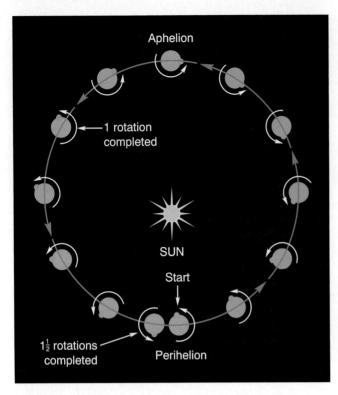

➤ **FIGURE 8-10.** Mercury's rotation period is such that a point on its surface that is under the Sun at one perihelion position is opposite the Sun on the next.

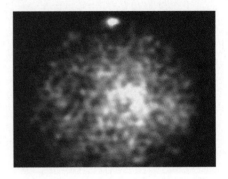

➤ **FIGURE 8-11.** Radar reflections from Mercury show strong reflectivity from the planet's north pole. The reflections (using 3.5-centimeter wavelength radar) can be explained if ice exists at the pole.

pleted one rotation with respect to the distant stars. This does not mean that it has completed one solar day, however. Mercury does not complete one rotation with respect to the Sun until 176 Earth days have passed. Notice that a person standing at the light spot when the planet is at the position marked "start" would see the Sun at its highest point in the Mercurian sky. As time passed, the Sun would get lower, finally setting when the planet has completed half a revolution. When Mercury reaches "start" again, it is midnight for that person. It takes 88 days for Mercury to complete its orbit and that constitutes only half of a day for the person on the planet. A solar day on Mercury is therefore 176 Earth days (or two Mercurian years)!

Imagine the Earth with no atmosphere to shield us from the Sun and with a daylight period lasting 88 of our days. This is the situation on Mercury. Temperatures there reach as high as about 450°C. The element lead melts at about 330°C, so lead would be molten at midday on Mercury.

450°C is about 800°F.

On the other hand, the nighttime side of Mercury also has no atmosphere to hold in the heat during its long nights. It therefore gets as cold as −150°C (which is −250°F). Mercury's temperature variations are much greater than any other planet's.

"Almost exactly" in this case means within a few arcminutes.

Mercury's poles are thought to remain in constant cold, because the planet's equator aligns almost exactly with its orbit. This means that Mercury has no seasons as we define seasons on Earth. Therefore even if the polar surface were perfectly smooth, only a very small amount of sunlight would strike it. But the surface isn't smooth, and the bottoms of some craters never receive sunlight. Indications are that the surface inside these polar craters hovers forever at a temperature around −150°C, about the same temperature reached by the nighttime side of the planet.

albedo. The fraction of incident sunlight that an object reflects.

Recent radar results from Mercury show a very high **albedo** for the polar regions (➤Figure 8-11). The most plausible explanation for the reflections is that

☿ MERCURY
DATA PAGE

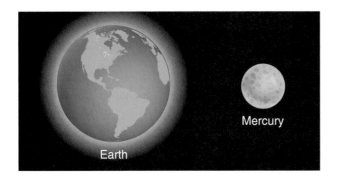

Mercury	Value	Compared to Earth
Equatorial diameter	4878 km	0.39
Oblateness	0	
Mass	3.302×10^{23} kg	0.055
Density	5.43 gm/cm^3	0.98
Surface gravity		0.38
Escape velocity	4.25 km/s	0.39
Sidereal rotation period	58.65 days	58.65
Solar day	176 days	176
Surface temperature	$-150°C$ to $+450°C$	
Albedo	0.11	.4
Tilt of equator to orbital plane	2° (?)	
Orbit		
Semimajor axis	5.79×10^7 km	0.387
Eccentricity	0.2056	12.3
Inclination to ecliptic	7.0°	
Sidereal period	87.97 days	0.24
Moons	0	

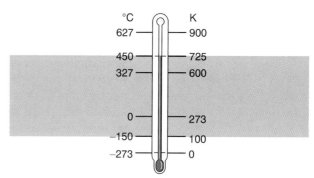

Surface Temperature Extremes

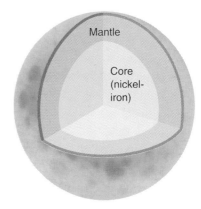

Interior

QUICK FACTS

Of the planets, only Pluto is smaller (Mercury is one-third Earth's diameter). Closest to Sun. Most eccentric orbit (except Pluto). Greatest difference between maximum and minimum temperatures. Difficult to see from Earth. Almost no atmosphere. Surface cratered somewhat like Moon. Large iron core. Period of revolution 1½ times period of rotation. Magnetic field not fully explained.

Mercury has ice at its poles. Even cold ice should evaporate over the ages, however, and the question of how ice could remain so long remains a mystery at the time of this writing.

VENUS

Venus is never seen farther than 47 degrees from the Sun, so—like Mercury—it is only visible either in the evening sky after sunset or in the morning sky before sunrise. Except for the Sun and Moon, however, Venus can be the brightest object in our sky. Seeing such a bright object above the horizon has fooled many people into thinking they were seeing something else. During World War II, pilots of fighter planes sometimes shot at Venus thinking that it was an enemy plane. Today reports of UFOs often increase when Venus is at its brightest in the sky.

Venus is such a beautiful sight that it is little wonder that the ancients named it after the goddess of beauty and love. We'll see that as a possible home for transplanted Earthlings, however, it is not such a beautiful planet after all.

Size, Mass, and Density

Venus has historically been known as the Earth's sister. It is similar to Earth in many ways: its diameter is 95% of Earth's, its mass is 82% of Earth's, and of all the planets, its orbit is located closest to us.

Through a telescope, Venus looks as smooth as a big white cue ball. No markings are visible. Images like the one in Figure 8-12 greatly exaggerate slight color variations.

From the values for its diameter and mass, we can calculate that Venus has a density 5.24 times that of water—not much different from Earth's 5.52. It would have a slightly higher density if its material were compressed as much as Earth material. Soviet spacecraft that landed on Venus have shown that its surface rocks have about the same composition as Earth rocks, and we therefore conclude that to have a density as high as it does, Venus must have a very dense interior and is probably differentiated, with a metallic core.

No magnetic field has been detected on Venus, and based on the sensitivity of the instruments that have searched for it, if one does exist, its strength must be less than 1/10,000 of Earth's. Because of Venus's slow rotation, we would not expect it to have a strong magnetic field, although according to present theories of the cause of planetary magnetic fields, the field should be strong enough for us to detect. Recall that Earth's field is known to have reversed its direction in the geologic past. Perhaps the magnetic field of Venus is now in the process of reversing, which would explain why it is so weak.

Venus's Motions

Venus orbits the Sun in an almost circular orbit—more nearly circular than any other planet. Its period is 225 days, resulting in a very nearly constant orbital speed of about 35 kilometers/second.

35 km/s is 78,000 mi/hr.

The surface of Venus is covered by dense clouds so it is not visible from Earth. Radar, however, allows astronomers to learn about nearby planets by bouncing radio waves from them. Radar can penetrate Venus's cloud cover, and since 1961 we have been bouncing radar signals from its surface. This is how we first learned about its rotation rate and its surface features.

The use of radar to study Venus told us that its rotation is very unusual. The planet rotates *backward* compared to most other rotations in the solar system. That is, if we view the solar system from above the Earth's North Pole, we see that all planets circle the Sun counterclockwise and that all the planets except Venus, Uranus, and Pluto rotate in this same direction. Venus's rotation is very slow, with a sidereal rotation period of 243 days. Its period of revolution around

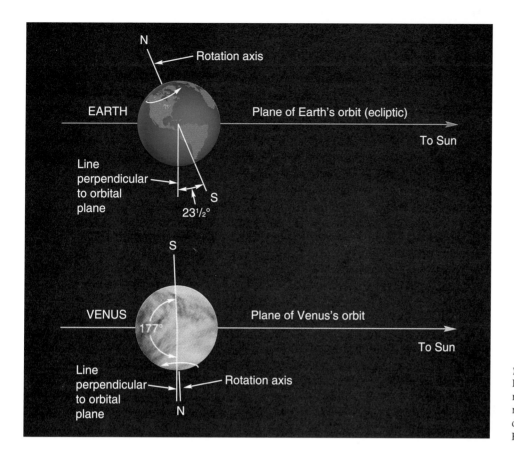

FIGURE 8-12. An arrow on Earth's surface indicates its rotation. Notice that Venus's rotation arrow points in the same direction around its pole as does Earth's arrow.

the Sun, however, is only 225 days. These rotation and revolution rates result in the solar day on Venus being about 117 Earth days.

Tables showing the tilt of Venus's equator relative to its orbital plane list an angle of 177 degrees. Since a tilt of 180 degrees would turn a planet completely over, wouldn't a tilt of 177 degrees be the same as a 3-degree tilt? Not really. To show the difference, we first need to define what we mean by the north pole of a planet. After all, only the Earth has its axis oriented so that Polaris, the North Star, is nearly above its North Pole. To decide which pole of a planet is to be called north, imagine we are standing on the planet facing one of its poles. If when we do this the rotation of the planet carries us around to our right, we are facing the north pole. The north pole of Venus, as defined in this manner, lies on the other side of the plane of the ecliptic from the Earth's North Pole. ➤Figure 8-12 illustrates this difference. Thus we usually list the tilt of Venus's axis as 177 degrees rather than 3 degrees. The fact that the angle is greater than 90 degrees tells us that Venus has a backward rotation.

Another way to define the north pole of a planet: Grab the planet with your right hand in a manner such that your fingers point in the direction of its rotation. Your thumb will then point to the north pole.

The Surface of Venus

Since 1962 a total of 19 spacecraft have visited Venus. From 1962 to 1975, three *Mariner* spacecraft from the United States flew by the planet, obtaining data as they did so. The Soviet Union landed 11 spacecraft on Venus, and some of them produced close-up photos of the surface (➤Figure 8-13). The photos indicate that the surface is rock-strewn, at least at the landing sites. They showed that rocks in some locations are more weathered than in others, indicating that they have been on the surface for a longer period of time. The sharp-edged rocks were a surprise to scientists studying Venus until they realized that the wind is fairly calm at the surface. If there had been great winds at the surface, the dense

Venus

CLOSE UP

Our Changing View of Venus

Galileo was the first to observe the phases of Venus. He argued that the fact that it exhibits a gibbous phase proves that it orbits the Sun and not the Earth. Although its phases are not visible to the naked eye, they have a definite effect on how bright Venus appears to us. When Venus is between the Earth and the Sun, it is closer to us than any other planet gets. It is then in a new phase, however, and cannot be seen from Earth. Just before and just after the new phase, its crescent is so small that the planet appears dim to the naked eye. On the other hand, when it is nearly full and we see almost its entire disk, it is on the other side of the Sun from us and is so far away that it appears dim again.

Venus appears brightest at an intermediate position, when it shows less than a full face but is fairly close to Earth (▶Figure C8-1). This position occurs when Venus is 39 degrees from the Sun, about 36 days before and after its new phase. Table A8-2 in the Activity at the end of the chapter shows when Venus will be at its brightest over the next few years.

In a telescope, the angular diameter of Venus varies from 10 seconds of arc when it is most distant to 64 seconds of arc when it is closest. The photographs of the planet in Figure 3-4 show the great change in its angular diameter as seen from Earth

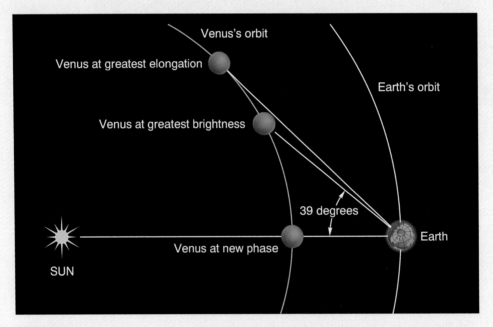

▶ **FIGURE C8-1.** Venus is most brilliant when its elongation is 39 degrees, which occurs about 36 days before or after (shown here) its new phase. (Recall that a planet's elongation is the angle between it and the Sun.)

▶ **FIGURE 8-13.** This photo of the surface of Venus was taken by *Venera 13*. The object at the bottom is part of the lander.

CHAPTER 8 The Terrestrial Planets

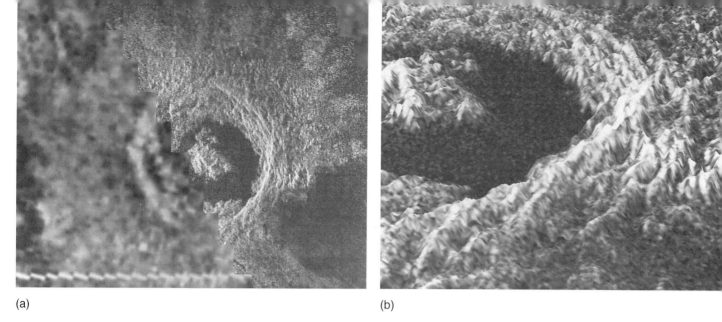

> **FIGURE 8-14.** (a) The Venusian crater Golubkina is shown as a mosaic of two images. The image on the left side of the photo was produced by the Soviet *Venera 15/16* craft that discovered the crater, and the one on the right was produced from *Magellan* data. The crater, named after Anna Golubkina (1864–1927), a Soviet sculptor, is about 34 kilometers across. Its features, including the central peak, terraced inner walls, and surrounding ejected material, are characteristic of impact craters on the Earth, Moon, and Mars. (b) This three-dimensional image was made from *Magellan* data and enhances the height of structural features of Golubkina Crater.

atmosphere (see the next section) would have caused considerable dust movement and erosion of the rocks.

The surface of Venus is not visible from above its atmosphere, but before we ever visited Venus, we had produced Venusian maps using radar from Earth. Radar penetrates the atmosphere, and from the characteristics of the returned signals, we are able to make maps of the surface. *Pioneer Venus 1* was put in orbit around Venus in 1978 and continued sending radar data until 1992. This allowed us to make higher resolution maps of the surface. Just as radar from airport control towers is used to determine the distances to airplanes, radar from Venus orbiters can determine the distances from the orbiter to points on the surface of the planet. From 1990 to 1994 the *Magellan* orbiter returned images of much higher resolution (Figure 8-14), so that objects smaller than 100 meters could be detected. *Magellan* mapped 98% of the surface of Venus.

About two-thirds of the surface of Venus is covered by rolling hills with craters here and there. Highlands occupy less than 10% of the surface and lower-lying areas make up the rest. Remember that maps of the surface (➢Figure 8-15) are made from radar data, so their color is not the color that would appear in a photo. Instead, colors are used to indicate altitude. In Figure 8-15, brown/red indicates the highest altitude and blue/violet the lowest.

Do not be fooled by the blue in the map of Venus. All of Venus, including the rolling plains that make up most of its surface, is dryer than the driest desert on Earth. Earth has rolling water; Venus, rolling desert.

Venus has about a thousand craters (➢Figure 8-16) that are larger than a few kilometers in diameter. This is many more than are found on Earth, but far fewer than are found on the Moon. We know that the terrestrial planets experienced more frequent cratering early in the history of the solar system than they do now, for there was much more interplanetary debris. The cratering rate decreased as this debris was swept from space. This allows us to determine the age of a planet's surface, for volcanic action and motion of tectonic plates remove old craters. Venus has no craters older than about 800 million years, and we conclude that the average age of the planet's surface is no more than about 500 million years, about twice as old as Earth's, but much younger than the Moon's.

Adjectives for Venus include "Venusian," "Venerian," and "Cytherean." "Venereal" is not appropriate.

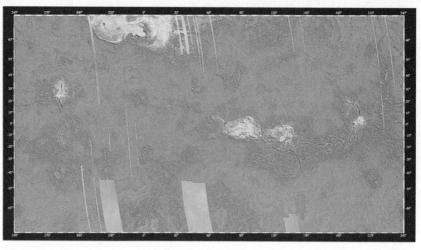

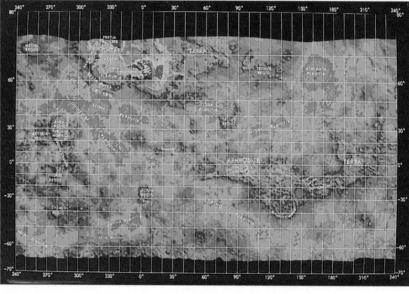

➢ **FIGURE 8-15.** (a) A color-coded map of Venus's surface shows the highest elevations in brown/red and the lowest in blue/violet. (b) All but one Venusian feature are named for women, as is the planet. Aphrodite Terra, the second largest "continent," is named for the Greek equivalent of Venus.

Probes landed by the USSR contained instruments to measure the composition of the Venusian soil and rocks.

On Venus's surface we find evidence of past volcanic and tectonic activity, including mountains, volcanoes, and large lava flows (➢Figure 8-17). As a surface becomes smoother due to weathering, it becomes a poorer reflector of radar. This allows us to compare the age of lava flows over the planet. At the time of this writing, no evidence has been found of present volcanic activity.

Notice how the shapes of some rocks in Figure 8-13 indicate that they fit together like pieces of a puzzle. Apparently, these are pieces of lava that fractured as it cooled and solidified. Most of the surface of Venus is covered with lava rock. Venus has no large tectonic plates. If it did, we would see mountain ranges as we find on Earth. Instead, Venus's wrinkled surface appears to be the result of a thin crust that moves and flexes over most of its area.

The Atmosphere of Venus

The length of a day on Venus and the lack of water on its surface are not the only features that make the planet a poor candidate for the title of Earth's twin sister. The Soviet *Venera* landers confirmed what we had already begun to learn about the unusual atmosphere of Venus.

The Earth's atmosphere is composed of nearly 80% nitrogen and 20% oxygen, with small amounts of water, carbon dioxide, and ozone. Venus, on the

➤ **FIGURE 8-16.** The thick atmosphere of Venus shields the surface from small interplanetary debris, but impacts of larger objects result in craters such as these. The foreground crater (named Howe) is 37 kilometers wide. Most of the floors of the craters are flat because lava has flooded them.

➤ **FIGURE 8-17.** The lava near Maat Mons is the youngest crustal material yet found on Venus. The mountain is 8 kilometers high, exaggerated 10 times in this image made from *Magellan* radar data.

other hand, has an atmosphere made up of about 96% carbon dioxide, 3.5% nitrogen, and small amounts of water, sulfuric acid, and hydrochloric acid—the same acids that are in car batteries! In fact, the clouds we see on Venus are made up in large part of sulfuric acid droplets. Venus is indeed inhospitable.

The upper atmosphere of Venus is very windy. The winds there reach speeds up to 350 kilometers/hour. The wind blows in the direction of the planet's rotation, but remember that the planet is rotating very slowly. As one moves lower in the atmosphere, the wind velocity decreases until it is nearly zero at the surface. We do not fully understand the causes of Venus's wind patterns. Undoubt-

350 km/hr is 225 mi/hr.

Venus

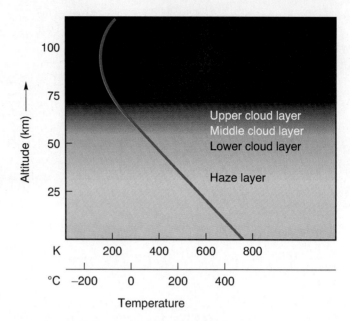

► FIGURE 8-18. At the top of Venus's atmosphere, the temperature is about 170 K (−100°C). It increases smoothly to about 730 K (460°C) at the surface. The cloud layer causes the surface to appear orange.

The gas of Venus's atmosphere is so thick that if we could survive its hazards, we could strap on wings and fly.

edly, the slow motion of the Sun across Venus's surface (because of its slow rotation rate) is the basic cause, but the details are not known. Further study of the weather on Venus would not only help us to know that planet better but would increase our understanding of weather patterns on Earth.

As the space probes descended through the atmosphere of Venus, it was not only the acidic atmosphere and the great winds that presented a hazard. The space probes were also subjected to tremendous pressures, for the atmospheric pressure on the surface of Venus is about 90 times that on the surface of Earth. As if the acid atmosphere and high pressure were not enough, Venus is also inhospitable because of its high temperatures: about 460°C (850°F) near the surface.

The clouds of Venus form a layer between altitudes of about 50 and 70 kilometers (►Figure 8-18). A layer of haze extends from the cloud layer down to about 30 kilometers. From there to the ground, the Venusian atmosphere is surprisingly clear. The light that filters through the clouds, however, has a distinct yellow/orange hue, caused primarily by the sulfur dust in the clouds. Refer back to Figure 8-13 and note the color of the rocks and the base of the spacecraft. This color is not the natural color of the objects, but is the same result you would get if you turned out all the white lights in a room and lit only a yellow light—everything would have a yellow tint. If the surface of Venus were lit with white light, it would appear gray.

A Hypothesis Explaining Venus/Earth Differences

As noted earlier, both Earth and Venus have small amounts of water vapor in their atmospheres. Venus's atmospheric water is about all the water that exists on the planet, however. Why is this planet, so close to the size of the Earth, so different from Earth?

Terrestrial planets get much of their atmospheres from gases released from their interiors, primarily through volcanic action. Volcanoes on Earth release large amounts of both water vapor and carbon dioxide, and it is likely that this is also the case for volcanoes on Venus. Both planets are theorized to have once had water on their surfaces and substantial amounts of carbon dioxide in their atmospheres. Most of the carbon dioxide on Earth is now dissolved in the oceans and in rocks such as limestone, which is found in the oceans. Venus, however, would have had a higher surface temperature due to its position nearer the Sun, and this is thought to have prevented water from condensing to liquid form. Thus the carbon dioxide remained in the atmosphere. The clouds of water vapor,

If oceans once formed on Venus, they quickly evaporated in the heat.

along with large amounts of carbon dioxide, resulted in an effect we experience only to a mild degree here on Earth: the greenhouse effect.

Did you ever notice that our coldest nights occur when the sky is clear? When we have cloudy skies at night, the difference between day and night temperatures is minimized. The reason for this is that in order for the Earth's surface to cool down at night, it must radiate into space a portion of the heat it gained during the day. The primary heat loss occurs in the form of infrared radiation. Infrared radiation, however, does not readily pass through water, so the clouds reflect the infrared waves back to Earth. Thus clouds form a sort of blanket for the Earth.

Early in Earth's and Venus's history, both planets experienced this blanketing effect, but because of Venus's higher temperature, it had more water vapor than Earth did. In addition, Venus had large amounts of carbon dioxide in its atmosphere. This gas is also opaque to infrared radiation. Visible light, on the other hand, is not blocked by carbon dioxide. Thus, although most of the visible light from the Sun was reflected by the clouds of Venus, some of it passed through the atmosphere and was absorbed by the surface, causing the surface to heat up. Meanwhile, high in Venus's atmosphere, ultraviolet radiation from the Sun was breaking water molecules into their constituents: hydrogen and oxygen. The hydrogen escaped from the planet, and the oxygen combined with other elements in the atmosphere. This left the atmosphere with large amounts of carbon dioxide, which continued to trap infrared radiation.

The hotter Venus's surface got, the more infrared radiation it emitted. Because of the carbon dioxide in the atmosphere, though, much of this radiation could not escape. Thus the planet continued to heat up. The high surface temperature on Venus baked further carbon dioxide out of the surface rocks. So a chain reaction resulted: the more carbon dioxide released, the hotter the surface became and the more carbon dioxide was released.

The planet did not continue heating forever, of course. The atmosphere was not *completely* opaque to infrared. A small fraction of the infrared radiation that hit it was reemitted into space and that energy was lost to the planet. The amount of solar energy being absorbed by Venus was constant, while the amount being released depended upon the temperature of the planet. Thus Venus continued to heat up until the amount of radiant energy leaving it was equal to the amount striking it. At this point an equilibrium condition was reached. ▶Figure 8-19 illustrates the situation.

This explanation for why Venus developed an atmosphere so different from Earth's is not fully accepted, as indicated by the fact that I call it a hypothesis. The effect illustrated in Figure 8-19—the greenhouse effect—is a well-established theory.

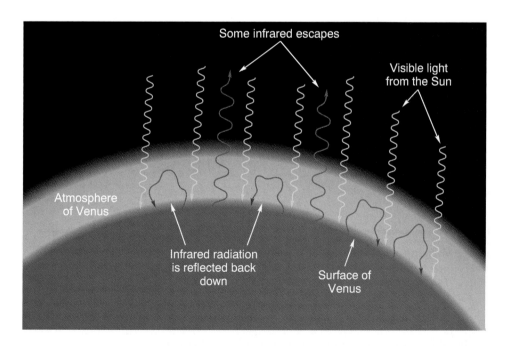

▶ **FIGURE 8-19.** Some visible light penetrates the atmosphere of Venus and reaches the surface, heating it. Most of the infrared radiation emitted by the surface, however, is reflected back by the atmosphere, causing Venus to be very hot.

♀ VENUS
DATA PAGE

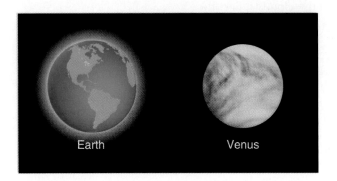

Venus	Value	Compared to Earth
Equatorial diameter	12,104 km	0.95
Oblateness	0	
Mass	4.869×10^{24} kg	0.82
Density	5.24 gm/cm^3	0.95
Surface gravity		0.90
Escape velocity	10.36 km/s	0.93
Sidereal rotation period	243 days	243
Solar day	117 days	117
Surface temperature	+460°C	
Albedo	0.75	2.4
Tilt of equator to orbital plane	177.3°	7.6
Orbit		
Semimajor axis	1.082×10^8 km	0.723
Eccentricity	0.0068	0.41
Inclination to ecliptic	3.39°	
Sidereal period	224.7 days	0.62
Moons	0	

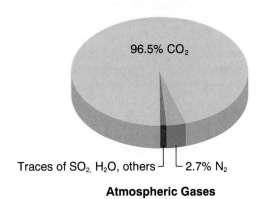

96.5% CO_2
Traces of SO_2, H_2O, others — 2.7% N_2

Atmospheric Gases

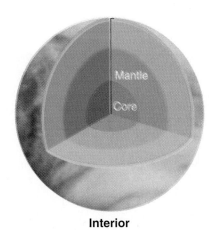

Interior

QUICK FACTS

Morning and evening star; Earth's twin in size, mass, and density. Surface is hidden by cloud layer, mapped by radar. Slow retrograde rotation. Dry, hot surface. No large tectonic plates. Surface primarily solidified lava. Surface older than Earth's, younger than Moon's. Dense CO_2 atmosphere. Greenhouse effect causes high atmospheric temperature.

Venus continues to have high temperatures due to trapped infrared radiation. Such heating is called the ***greenhouse effect*** because it is somewhat similar to what happens in greenhouses on Earth. Sunlight penetrates the glass of a greenhouse's roof and walls and is absorbed by the plants and soil within. These surfaces then emit infrared radiation. Glass, like water vapor and carbon dioxide, is a poor transmitter of infrared, so the air in a greenhouse is kept warmer than the outside air. This is not the entire story in the case of a greenhouse, however. A very important factor in the heating of greenhouses is that the walls prohibit the warm air inside from mixing with the cooler outside air. Since this trapping of the air is not a factor on a planet, the heating of Venus and of a greenhouse are not truly similar.

The greenhouse effect is also at work in Earth's atmosphere. If we calculate what Earth's temperature would be due to the amount of solar radiation that strikes it and the energy it would emit into space due to its temperature, we obtain a temperature 35 Celsius degrees colder than it actually is. The greenhouse effect, primarily caused by water and carbon dioxide, makes up the difference. As we burn more and more fuel, we add more carbon dioxide to the atmosphere. The carbon dioxide (and other greenhouse-causing chemicals) in Earth's atmosphere is increasing at the rate of about 0.5% per year, and many scientists fear the effect this may have in the future. Computer calculations of the effect of an increase in carbon dioxide depend greatly on a number of assumptions, but various studies indicate that if we double the amount of carbon dioxide in the atmosphere, we will cause an increase in Earth's temperature of somewhere between 2.8 and 5.2 Celsius degrees.

The great disparity between conditions on Venus and on Earth appears to be due to a fairly slight difference in temperature far back in their histories. This difference was magnified by the greenhouse effect so that now Venus's atmosphere might be compared to Hell. We know that an increase of just a few degrees in Earth's temperature would cause major disruptions in our way of living because of changes in growing conditions, but the greater fear is that a change of just a few degrees may cause a runaway greenhouse effect, destroying Earth as we know it. More study is needed before we know at what point a runaway greenhouse effect would begin, but can we afford to take chances?

Climatologists report that, worldwide, the five warmest years in the last one hundred occurred during the 1980s. Working with scientists from other fields, including astronomy, they are trying to determine whether this represents a chance fluctuation or whether it is due to the greenhouse effect. If the greenhouse effect is responsible, significant changes in our lifestyles may be necessary to reverse the process.

As we leave Venus and move outward from the Sun, we next encounter Earth. This planet is so important that it was included in a previous chapter and will therefore be skipped here. That makes Mars next.

greenhouse effect. The effect by which infrared radiation is trapped within a planet's atmosphere through the action of particles, such as carbon dioxide molecules, within that atmosphere.

The Stefan-Boltzmann law (Close Up, Chapter 4) allows us to calculate the energy radiated from Earth due to its temperature.

Plant growth has the effect of decreasing atmospheric CO_2. We are destroying tropical forests at the rate of 1 acre per second!

So much solar radiation is reflected from the clouds of Venus that if Venus had no greenhouse effect, its atmosphere would be cooler than Earth's.

MARS

The ancient Greeks called the fourth planet from the Sun Ares, the name of their mythical god of war. Our names for the planets, however, come from their Roman names, and indeed, in Roman mythology the god of war is Mars. Why is this planet associated with warfare? Perhaps because of the association of its red color with blood. Mars is indeed red. Earlier in this text red stars were mentioned, but the red stars that are visible to the naked eye are not very red at all compared to the planet of war.

Mars as Seen from Earth

Mars is the only other planet with surface features that can be seen from Earth. The amount of surface detail that is visible, however, depends on a number of factors. First, the surface of Mars is often obscured by dust storms on that planet, which will be discussed later. In addition, surface visibility varies greatly depending on Mars's distance from Earth.

As you might expect, Mars is best seen when it is directly opposite the Sun in the sky. When a planet is in such a position, 180 degrees from the Sun, it is closest to Earth and is said to be in *opposition.* This happens about every 2.2 years for Mars. But since Mars's orbit is so eccentric, the distance from Earth to Mars at opposition might be as little as 55 million kilometers or as great as 100 million kilometers.

➤Figure 8-20 shows three views of Mars, the one on the right taken by the Hubble Space Telescope. The red color of the light areas is obvious. Before the advent of color photos such as these, a number of observers claimed that the darker areas appear green. These markings were observed as early as 1660, and the rotation rate of Mars was determined from their motion. In addition, you can see a white cap at the pole of the planet. Observing this cap as Mars orbits the Sun, we can watch it diminish in size as that pole faces the Sun and then grow again when that pole faces away from the Sun. This effect is similar to the Arctic and Antarctic areas on Earth. The tilt of Mars's axis is very similar to Earth's, and we are observing these poles as they experience the Martian summer and winter.

Other seasonal changes can be observed on Mars. Large parts of the planet change periodically from a dark to a light color and back, depending on the position of the planet in its orbit. Since the darker areas appear green (to some observers, at least), this color change led to speculation that there is vegetation on the planet and that it changes color in response to seasonal growth. If there is vegetable life, could there be animal life on Mars? What about intelligent life, then? Before exploring the possibility of life on Mars, we will consider the planet itself.

> We had to say "other" because Earth's surface can be seen from Earth.

> **opposition.** The configuration of a planet when it is opposite the Sun in our sky. That is, the objects are aligned as follows: Sun-Earth-planet.

> **syzygy.** A straight line arrangement of three celestial objects.

> The reason for this reported green color is probably because green is the complementary color to red. Stare at something red for a little while and then close your eyes. You'll see green.

➤ **FIGURE 8-20.** (a) This 1971 photo of Mars shows its rusty color. The small polar cap indicates that the photo was taken during the Martian summer in that hemisphere. (b) This 1988 image of the same side was made with CCD technology, which can produce an image of higher resolution. (c) This 1997 image taken by the Hubble Space Telescope shows the northern hemisphere of Mars and clearly shows how much better resolution the HST provides. The colors, however, are falsely exaggerated.

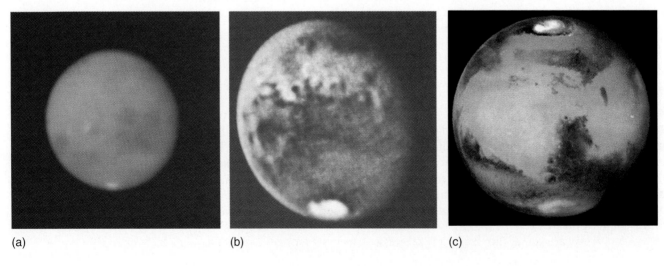

► **FIGURE 8-21.** This mosaic of photos shows the objects with their correct relative sizes.

Size, Mass, and Density

Mars is a small planet, closer in size to our Moon than to the Earth. ►Figure 8-21 compares Mars to Mercury, Venus, Earth, and the Moon. Mars falls between Mercury and Venus in size, having a diameter about half of Earth's. This makes the surface area of Mars about one-fourth that of Earth. The volume of Mars, then, is about one-eighth of Earth's—$(1/2)^3$. Now recall that both Mercury and Venus have densities comparable to Earth's. If Mars follows this pattern, its mass should be about one-eighth that of Earth, which would make the ratio of its mass to its volume about like Earth's. The mass of Mars, however, is only one-tenth of Earth's, and its density is only 3.95 times that of water, or less than three-fourths of Earth's density.

Area depends on the square of the diameter and $(1/2)^2$ is $1/4$.

Mars's Motions

Mars orbits the Sun at an average distance of 1.524 AU (about 228 million kilometers) from the Sun. Its orbit, however, is fairly eccentric, and its distance from the Sun varies from about 210 million kilometers to 250 million kilometers. From Kepler's third law we can calculate that Mars requires 1.88 years to complete its orbit around the Sun.

The eccentricity of Mars's orbit is 0.093, while Earth's is 0.017.

We saw that a solar day on Mercury was 176 Earth days, and on Venus 117 Earth days. In this sense, it is Mars that should be called Earth's sibling for it has a sidereal rotation period of 24 hours, 37 minutes and its day is 24 hours, 40 minutes.

You should be able to figure out why the two periods are nearly the same for Mars.

The equator of Mars is tilted 25.2 degrees with respect to its orbital plane. This is very close to Earth's 23.4 degrees. Since the tilt of a planet's axis causes opposite seasons in the planet's two hemispheres, we expect such seasons on Mars and,

▶ **FIGURE 8-22.** Mars is closer to the Sun during summer in its southern hemisphere. In this drawing both the tilt of Mars's axis and the change in its distance from the Sun are exaggerated.

indeed, this is the case. Another feature of Mars's motion affects its seasons, however. The eccentricity of Mars's orbit causes it to be much closer to the Sun at some times of its year than at others. It turns out that Mars is 19% closer to the Sun during the northern hemisphere's winter than during its summer (see ▶Figure 8-22). Being closer to the Sun in winter and farther in summer means that seasonal temperature variations are moderated in the northern hemisphere.

In the southern hemisphere, the reverse is true. Mars is closer to the Sun during summertime, and farther from the Sun in winter. Thus the southern hemisphere experiences greater seasonal temperature shifts than the northern hemisphere.

The same effect occurs for Earth, which is closest to the Sun in January (wintertime in the Northern Hemisphere). The Earth's orbit is so nearly circular, though, that we are only about 3.3% closer to the Sun at one time than the other (rather than 19%).

Life on Mars

Speculation about life on other planets—particularly Mars—is nothing new. The famous mathematician Carl Frederick Gauss was so convinced that intelligent life existed on Mars that in 1802 he proposed that a huge sign be marked in the snows of Siberia to signal the Martian beings.

During the favorable (close) opposition of Mars in 1863, Father Angelo Secchi, an Italian astronomer, observed Mars and drew a colored map of the planet's surface. The map showed some lines that Father Secchi called *canali*, an Italian word best translated as *channels*. Then in 1877 the director of the Rome Observatory, Father Giovanni Schiaparelli, observed Mars and drew a more elaborate map showing many *canali*. In translation to English, however, Schiaparelli's *canali* became *canals*, rather than channels, giving the implication of artificial waterways. This struck a responsive chord with the public, and Schiaparelli's findings were taken as confirmation of an intelligent race of beings on Mars.

Lowell's observatory is on Mars Hill, now within the city of Flagstaff. The observatory is still engaged in active research, financed largely by the Lowell family. It is open at times to the public.

In December 1907, the *Wall Street Journal* said that one of the most important events of the year was the proof that there was intelligent life on Mars.

A headline in the November 9, 1913, *New York Times* read "THEORY THAT MARTIANS EXIST STRONGLY CORROBORATED."

In 1894 an American astronomer, Percival Lowell (1855–1916), built an observatory on a hill near Flagstaff, Arizona, to concentrate on the study of Mars and its intelligent life. Lowell's drawings of Mars include as many as 500 canals (▶Figure 8-23). Recall that white polar caps are easily visible on Mars and that they change in size according to the season. Lowell believed that the canals were built to transport water from the polar caps to the drier parts of the planet.

Other astronomers failed to see canals, however, even though some of them were using telescopes much larger than Lowell's and Schiaparelli's. In 1894, as Lowell was completing his observatory, the astronomer Edward Barnard said, "to save my soul I can't believe in the canals as Schiaparelli draws them." Barnard was probably the keenest telescopic observer of his day, and he reported to friends that he saw craters on Mars. For fear of ridicule, he did not publish his drawings. Lowell's assistant, Vesto Slipher, also believed that he saw Martian craters, but he also did not publish his views.

In 1898, H. G. Wells wrote *The War of the Worlds*, describing a fictional invasion of Earth by Martians. When Orson Welles dramatized this novel in a very realistic radio broadcast in 1938, he made it appear that a radio music show was being interrupted frequently to report on the invasion as it occurred. Many people did not hear (or forgot about) the announcement at the beginning of the show that it was a dramatization, and there was widespread panic, especially in New Jersey, where the invaders were supposed to be landing.

Invasion and Its Results

A series of Martian invasions did start in the late 1960s and is still in progress. The invasion, however, is proceeding in the opposite direction than H. G. Wells

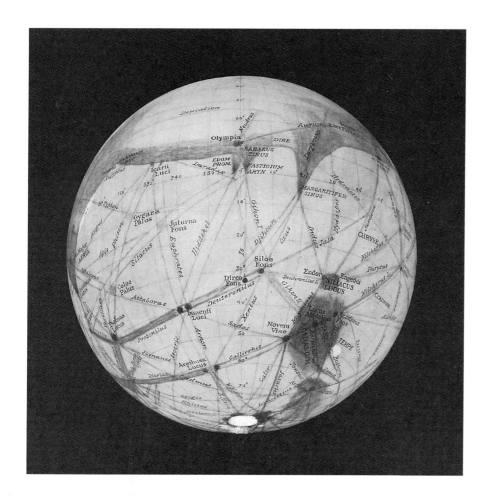

► FIGURE 8-23. This map of Mars was drawn by Percival Lowell in 1907.

envisioned, and it is a peaceful endeavor, although competition between major powers has certainly been a factor. In 1965 *Mariner 4* passed by Mars, sending back 21 pictures of the surface. These pictures ended speculation about canals, for none were observed. The planet was seen to be covered by deserts and craters. Other questions were raised, however, for the pictures confirmed that major dust storms are common on Mars, yet the surface contains numerous craters—craters that should have been worn down by the constant pounding of wind and dust. We know now that erosion does not occur very quickly because the atmosphere of Mars is extremely thin, so that the pressure at the surface is about 1/200 of the air pressure at Earth's surface. Dust stirred up by this thin atmosphere must be extremely fine and not at all like a sandstorm in one of our deserts.

Mariner 4 was followed in 1969 by *Mariners 6* and *7* and in 1971 by *Mariner 9*. The latter spacecraft went into orbit around Mars. It arrived while a great dust storm was occurring, and astronomers feared that its batteries would be depleted before the dust settled and the surface could be seen. Fortunately, the storm ended, and much was learned about the surface, particularly about the spectacular volcanoes and canyons.

The largest of the volcanoes is Olympus Mons. ►Figure 8-24(a) shows the gigantic cliffs around its base, and ►Figure 8-25 compares it to two of the largest mountains on Earth. Its height of 15 miles is twice that of our largest mountain, and its base would cover much of the area of Washington and Oregon. In ►Figure 8-26 Olympus Mons and its three companion volcanoes are superimposed on those states. These volcanoes would be very prominent in the western United States!

Venus also has at least one mountain larger than any earthly mountain. Why do these planets have such tall volcanoes? On all three planets, volcanoes are

The base of Olympus Mons has a diameter of 400 miles!

Mars   249

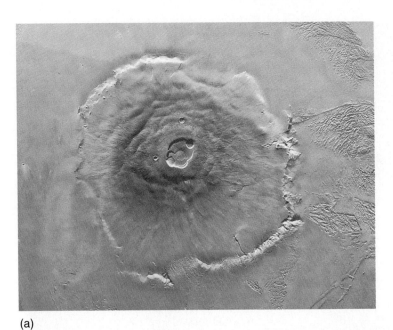

(a)

(b)

➤ **FIGURE 8-24.** (a) A computer-enhanced photo of Olympus Mons. (b) The peak of Olympus Mons extends above the white clouds of frozen carbon dioxide.

➤ **FIGURE 8-25.** Olympus Mons compared to Mount Everest and the volcanic island of Hawaii.

formed above hot spots that lie deep within the planets. From time to time, material wells up from these hot spots, spilling lava out and building up the volcanoes. The reason that Mars and Venus have larger mountains than Earth has to do with the motions of their crusts. On Earth, the crust is divided into plates that move across the underlying material. This means that the crust moves across a given hot spot and the volcanic action slowly shifts along the surface. This is why we often find earthly volcanoes in a straight line, with only the volcano on the end being active. The crusts of Venus and Mars do not contain large tectonic plates, so the same part of the crust stays above a particular hot spot. This causes the volcano to grow higher and higher.

Recall that Venus also lacks large tectonic plates. The reason in that case is that Venus's crust is too thin to be strong enough to remain in large pieces. Mars lacks tectonic plates for just the opposite reason. Mars is a small planet and

➤ **FIGURE 8-26.** A composite photo of Olympus Mons and three companion volcanoes has been overlaid on the western United States to show comparative sizes.

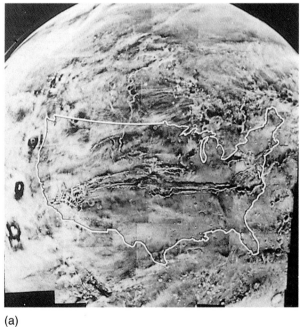

(a)

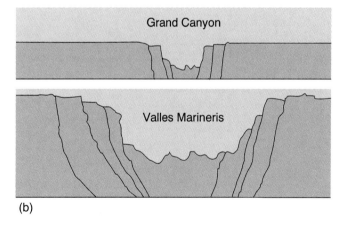

(b)

➤ **FIGURE 8-27.** (a) Valles Marineris would stretch nearly across the United States. (b) In width and depth, it dwarfs the Grand Canyon.

therefore its interior cooled quickly. Its crust is now so thick that it is too strong to be broken into individual pieces.

Stretching away from the area of Olympus Mons is a canyon that more than matches the volcano. If placed on Earth, *Valles Marineris* (named for the *Mariner* spacecraft) would stretch all the way across the United States, as shown in ➤Figure 8-27a. Part (b) of the figure shows its size relative to the Grand Canyon.

On the seventh anniversary of humans' first landing on the Moon, *Viking 1* landed on a cratered plain of Mars and took the first close-up photographs of the

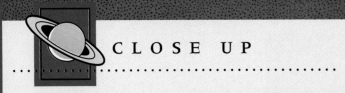

CLOSE UP

ET Life IV—Viking's Search for Life

In the late summer of 1975, *Viking 1* and *Viking 2* were launched from Cape Kennedy (now Cape Canaveral) toward Mars. After traveling to Mars, *Viking 1* separated into two parts, one part staying in orbit and another descending to a soft landing on the planet. *Viking 1* landed on Mars on July 22, 1976, seven years after our first step on the Moon. Then in August, *Viking 2* landed nearly on the other side of the planet.

One of the primary purposes of the *Viking* mission was to look for signs of life on our neighboring planet. Various tests were conducted toward this end. The first was made by a television camera that scanned the area for any signs of plant or animal life or even for footprints. Scientists did not really expect that such large life-forms existed on Mars, so the negative results were not particularly disappointing. The other tests, however, were more sophisticated and were designed to detect less obvious life-forms.

To perform these experiments, each lander contained an arm with a scoop at its end (▶ Figure C8-2) so that it could retrieve some soil from the surface. The first of these experiments was called the "labeled release experiment." It was performed by taking about a teaspoon of soil and dampening it with a rich nutrient that should have been absorbed by any organism in the soil. Some of the carbon atoms in this nutrient were carbon 14, a radioactive form of carbon. It was thought that after an organism absorbed nutrients, it should release carbon to the air, including some of the radioactive carbon. Such radioactivity can be easily detected.

To try to ensure that a positive result from this experiment would be due to biological processes rather than chemical processes, other soil samples were heated to 300°F before being fed the nutrient. This was done to sterilize the soil and kill any living matter that might be in it. In this way, the investigators could see whether the unheated soil acted differently from soil that had been heated to kill life-forms. (In a procedure like this, the sample that was sterilized is called the *control*. Such a procedure allows a comparison to be made and is common in all branches of science.)

When the labeled release experiment was tried on unsterilized soil by both *Viking 1* and *2*, the radioactivity of the air increased quickly. In the case of the control, the change in radioactivity was much less. Thus the experiment seemed to indicate that life was present. But there was a problem: the radioactivity showed up *too* quickly, faster than should occur if it had been caused by an organism absorbing and releasing the carbon. And when more nutrient was added, there was no further increase in the radioactivity of the air. This argued *against* the presence of life in the soil.

Although the initial results of the experiment caused some excitement among the researchers monitoring the results from Earth, they finally concluded that the release of carbon was caused by simple chemical reactions with elements in the soil and did not involve any biological processes. The surface of Mars is thought to have plentiful oxygen, enough to rust iron. A hydrogen peroxide molecule is a water mole-

---

We'll see the reason for the red color later.

surface. ▶ Figure 8-28 shows the landscape near *Viking 1* in approximately true color. The red color we see from Earth is real!

The *Viking* orbiting modules confirmed something that had been seen by *Mariner 9*: it appears that running water was once common on Mars. ▶ Figure 8-29 shows dry riverbeds that look very similar to the arroyos found in our desert southwest. Arroyos are formed on Earth by infrequent flows of large amounts of water from rainstorms. At present, there can be no rainfall on Mars, however, because atmospheric pressure on Mars is much too low for liquid water to exist. The Martian arroyos are so similar to those found on Earth that we conclude that at one time there must have been liquid water flowing on Mars. Where would the water have gone? The answer lies, at least partly, in those white polar caps.

The polar caps of Mars consist of two parts: a water ice base that is covered during the winter by frozen carbon dioxide. In summer the carbon dioxide evaporates, leaving behind the ice. The temperature on Mars does not rise high enough to melt that ice. If it did, there would still be only a small fraction of the water found on Earth.

cule with an extra oxygen atom. It seems likely that hydrogen peroxide is present in the soil and that this chemical reacted with the carbon to produce carbon dioxide, which was then detected by the radiation monitors. Once the reaction used up the hydrogen peroxide from the Martian soil, the reaction stopped and no more carbon was released. The reason the control sample did not show the activity was that the heat broke down the hydrogen peroxide before it was exposed to the nutrient containing carbon.

Another experiment aboard the *Viking* was a test for respiration. Here, the soil sample was put in a container with inert gases (gases that don't react chemically), and a nutrient was added. Finally, the gas was examined for changes in its chemical composition. If the living organism released any gases, they would be detected. When the experiment was performed, no more new gases appeared than would be expected from normal chemical reactions.

A common procedure in laboratories on Earth is to use a *mass spectrometer* to search for very small quantities of given chemicals in a sample. The *Viking* landers contained mass spectrometers capable of finding organic molecules even if they made up only a few billionths of the sample. All life on Earth contains these very large molecules that use carbon as their foundation. No such molecules were found.

We have learned a great deal about the surface of Mars since the original *Vikings* landed there. This knowledge could be used to improve the experiments on a future mission and avoid the confusion created by the hydrogen peroxide reactions. Is there life on Mars? Before we landed, most scientists doubted it. The experiment was judged worth trying, however, for if we had found life, the knowledge we could have gained by examining a life-form different from our own would have been tremendous.

The *Vikings* did not find life, but on the other hand they did not show absolutely that life does not exist on Mars. We still do not know the answer to the question, but we must admit that the chances appear slimmer now than they were before the *Viking* experiments.

➤ FIGURE C8-2. *Viking 1*'s sampler scoop is at the end of the arm extending from right.

➤ FIGURE 8-28. The horizon is about 2 miles away in this photo from *Viking 1*. Color photos were made by combining three photos, each made with a different color filter on the camera.

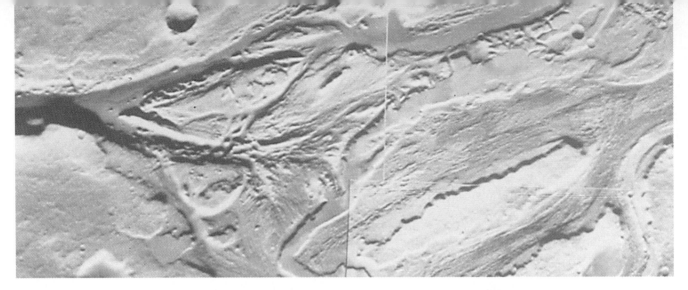

➤ FIGURE 8-29. What appear to be dry riverbeds on Mars lead us to believe that water once flowed on the surface.

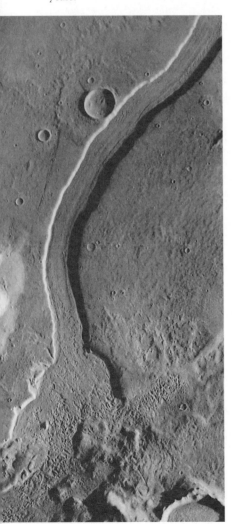

➤ FIGURE 8-30. The Martian channel shown here is 800 feet deep. It is thought to have been formed by the motion of a glacier sometime in the last 300 million years.

Other water seems to be hidden in permafrost below the surface of Mars. We have no way at present of knowing how much water is there and whether this water and the water in the polar caps are enough to have provided the moisture for a warmer, more hospitable Mars at some time in its history. An alternative hypothesis is that the riverbeds were formed in brief, cataclysmic periods when ice was melted by meteorite impacts or volcanic heating. Look at ➤Figure 8-30. The channel stretching down the photo is 800 feet deep. Some astronomers hypothesize that this channel—and others like it—were formed by the motions of glaciers during an ice age on Mars.

A total of 12 meteorites found on Earth have been identified by astronomers as having come from Mars (based on their composition and on gases trapped inside). The meteorites are causing controversy among scientists. Some who study the meteorites claim to find evidence that some of them were under water while they were on Mars. Further, they claim that the meteorites contain the remains of single-cell organisms, showing that life once existed on Mars. Others dispute this, however, and explain the observed properties of the meteorites in other ways. ➤Figure 8-31 shows the latest meteorite to cause excitement about the possibility of life on Mars.

➤ FIGURE 8-31. (a) Some scientists claim that this meteorite-from-Mars shows evidence of ancient life inside. It is hypothesized to have stayed more or less undisturbed on Mars until a huge asteroid or comet smacked into that planet 15 million years ago. The rock wandered in space until about 13,000 years ago, when it fell on the Antarctic ice sheet, where it was found in 1984. (b) The same meteorite after being cut. The entire meteorite has a mass of 1.94 kilograms, and the cut section measures 10 cm by 6 cm.

(a)

(b)

Atmospheric and Surface Conditions

Near the Martian equator, noontime temperatures reach as high as 20°C, a comfortable temperature for humans. At night, however, the temperature at the same location might drop to −140°C. The thin atmosphere is the reason for this extreme difference in temperature. As we saw in the discussion of Venus and Earth, a planet's atmosphere shields it from the Sun during the day and serves as a blanket at night by reflecting back some infrared radiation toward the surface. The amount of shielding and blanketing is determined by the amount and type of atmosphere. The atmosphere of Mars is 95% carbon dioxide, and one might suppose that Mars would have a greenhouse effect as Venus does. The low atmospheric pressure at the surface of Mars, however, means that there is simply too little atmosphere of any type to significantly moderate the temperature on Mars.

20°C equals 68°F and −140°C equals −220°F.

The escape velocity from Mars is 5 kilometers/second, less than half of Earth's escape velocity. Even though Mars is colder than Earth, almost all of the water vapor, methane, and ammonia have escaped the atmosphere along with the less massive gases.

Refer again to Figure 7-10.

Another difference between Mars and Earth is that Earth has a layer of ozone that prevents most of the Sun's ultraviolet radiation from reaching the surface. On Mars, there is little ozone so the Sun's intense ultraviolet radiation passes through the atmosphere and breaks up water vapor into hydrogen and oxygen. The hydrogen escapes, and the oxygen enters into chemical reactions with other elements.

One of the elements with which the oxygen combines is iron. The surface of Mars is rich with iron, and when oxygen and iron react, we get a compound that is usually a nuisance to us on Earth—rust. It is this rust that gives Mars its characteristic red color. The surface is rusty! The views of Mars in ➤Figure 8-32 were taken by the Hubble Space Telescope when Mars was in opposition in 1997, at a distance of 60 million miles. The photos show different views of Mars and make the red, rusty color obvious.

➤ **FIGURE 8-32.** The three views of Mars (taken by the Hubble Space Telescope) show its northern polar cap during the transition from spring to summer in the Martian northern hemisphere.

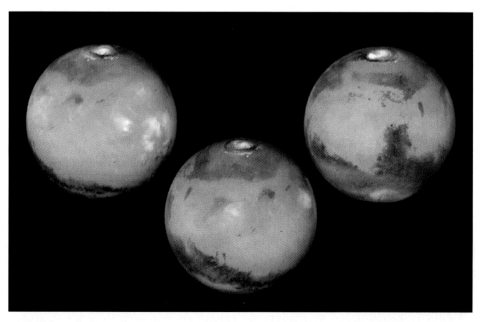

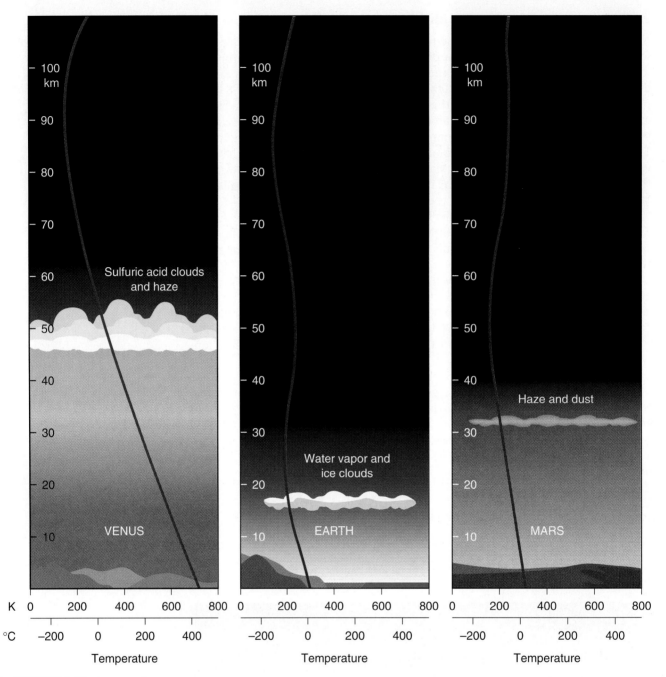

► **FIGURE 8-33.** A comparison of the atmospheres of three of the terrestrial planets. Mercury is omitted because it has almost no atmosphere.

From the *Viking* landers, we also learned what causes the seasonal color variations. The fine grains that make up the dust storms are of a lighter color than the underlying surface. In the springtime, this dust is stripped away by the wind, exposing the darker surface.

So there is no vegetation on Mars. In fact, the *Viking* landers found no signs of any vegetation at all. Nor did they find any animal life or even life of microscopic size. The *Viking* landers scooped up some of the Martian soil to analyze its chemical composition and to look for signs of microbial life. They found no organic chemicals whatsoever in the soil, meaning that they found no evidence for life ever existing on Mars. It is important to understand that this is not the same as saying they found evidence that there was never life on Mars. No current evidence for life was found. Perhaps our experiments were too limited in what they were searching for, or perhaps we simply need to look in other locations.

We are forced to conclude, however, that there is very little chance that life now exists on Mars.

On July 4, 1997, the *Pathfinder* spacecraft landed on Mars. It immediately sent back photos that showed the rock-strewn area where it landed. The photos further confirmed that Mars once had a great amount of water on its surface, and NASA scientists think that the region where *Pathfinder* landed may have once been covered with water a few meters deep.

The next day, *Pathfinder*'s roving robot, named Sojourner, began a slow journey to investigate the composition of some of the many rocks in the area (➤Figure 8-34). At the time of writing the major surprise has been that Mars is more similar to Earth than had been thought. The first rock investigated turned out to be rich in silica, the quartz material found in sand. Such a rock may have been brought to the surface by volcanic action or by meteorite impact.

The Global Surveyor was scheduled to begin orbiting Mars in September of 1997. *Pathfinder* and Surveyor will probably continue to change the way that we think of our neighbor, the red planet.

The Moons of Mars

Mars has two natural satellites named Phobos and Deimos. Both are small and irregularly shaped. Phobos, for example, is 28 kilometers across its longest dimension, but it is 23 kilometers and 20 kilometers across its other dimensions. Thus it is shaped like a potato. Deimos is even smaller and is similarly shaped.

This is 17 × 14 × 12 miles.

The surfaces of both moons of Mars are very dark, similar to those of many asteroids. Their densities are 1.9 and 2.1 times that of water (for Phobos and Deimos, respectively), also similar to the densities of the rocky asteroids. Such similarities lead us to believe that these satellites are captured asteroids rather than objects that were formed in orbit around Mars during the formation of the solar system. A capture like this could have taken place as the asteroid passed very close to Mars and was either slowed by friction with the Martian atmosphere, by collision with a smaller asteroid, or by gravitational pull from another asteroid. Such an event may not seem likely, but given the number of asteroids that have passed by Mars over billions of years, it is certainly a possibility.

The Martian moons are not only small, but they orbit very close to the planet and have short periods of revolution, about 0.3 days for Phobos and 1.3 days for Deimos. Both revolve in the same direction as most other solar system objects, counterclockwise as seen from north of Earth.

The counterclockwise motion of our Moon results in it moving slowly eastward among the stars as viewed from Earth. As we watch the Moon during a single night, however, this eastward motion is overwhelmed by its apparent westward motion that results from the rotation of the Earth. The eastward mo-

➤ FIGURE 8-34. This photo was taken soon after the Sojourner rover moved onto the Martian surface at the Carl Sagan Memorial Site. The rock that Sojourner is examining was dubbed Yogi.

HISTORICAL NOTE

The Discovery of the Martian Moons

Although Johannes Kepler made important contributions to science, he was very nonscientific in some ways. For example, he practiced numerology, the study of occult meanings of numbers. One of the patterns he saw in the heavens concerned the numbers of satellites of the various planets. Mercury and Venus have none, Earth one, and Jupiter four (known at that time). Between Earth and Jupiter was Mars, and Kepler proposed that Mars must have two moons if it is to fit the pattern. Scientists don't use logic like this today, but that was Kepler's way. From the time of Kepler, popular thought held that Mars must have two moons circling it. The idea appears a number of times in literature, most notably in Jonathan Swift's *Gulliver's Travels*, where Swift fills in such details as the moons' size (small) and orbital periods (10 hours and 21.5 hours).

There was no evidence for the existence of any satellites of Mars until 1877 when Asaph Hall, the astronomer of the newly constructed U.S. Naval Observatory near Washington, D.C., decided to search very near the planet for moons. To do so, he used a disk that blocked the planet's brightness from his view. There is a controversy about how quickly he found the moons, but Carl Sagan reports that Hall was unsuccessful for the first few nights and was about to quit when his wife encouraged him to continue searching. He did find the two moons and named them Phobos (Fear) and Deimos (Terror), the names of the horses that pulled the chariot of the Greek god of war.

In his book, *The Cosmic Connection*, published before *Viking* sent back photographs of the moons, Sagan states that a feature on one of the moons should be named after Mrs. Hall. Look at ➤Figure H8-1, a photograph of Phobos. The large crater is about 10 kilometers across and is named Stickney, Angelina Hall's maiden name, and the name proposed by Sagan. Another large crater on Phobos is named for her husband.

The actual orbital periods of the two moons are 7.7 hours and 30.3 hours. Although Swift had predicted 10 hours and 21.5 hours, he hit close enough to the actual times that some people believe that he had special knowledge of some sort. Actually, it was reasonable for him to assume that the moons, if they existed, must be close to the planet and must be small. Otherwise they would already have been discovered. And if they are close to the planet, Kepler's laws tell us that their periods of revolution must be short. There is no reason to hypothesize strange explanations for Swift's guesses. He just knew about Kepler's work.

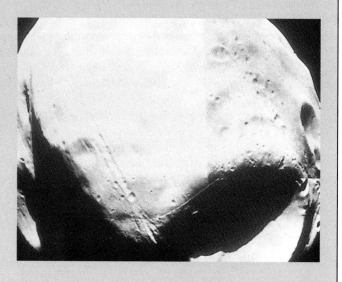

➤ **FIGURE H8-1.** The large, prominent crater is named Stickney, the maiden name of Mrs. Hall, the wife of the discoverer of Mars's moons.

We will explain why the equatorial diameter of a planet might be greater than its polar diameter when we discuss Jupiter, a noticeably out-of-round planet.

tion is far from obvious to the casual observer during one night. Things would be quite different for a Martian. Recall that Mars rotates with a period of about 24½ hours in a counterclockwise direction. Phobos takes only about 8 hours to circle Mars, however. As a result, an observer on Mars would see stars move across the sky just as we do on Earth, but that person would see Phobos rise in the west and move across the sky toward the east! Deimos would hover overhead for long periods of time, moving only very slowly toward the west.

On the Mars data page you will see that Mars has an oblateness of 0.009, which means that its polar diameter is 0.009 smaller than its equatorial diameter. Mars is the most oblate of the terrestrial planets, so these planets are all very nearly perfectly spherical. We have seen that Phobos and Deimos are far from spherical. Before explaining why they are not spheres, we should first explain why one might expect celestial objects to be spherical in the first place.

♂ MARS
DATA PAGE

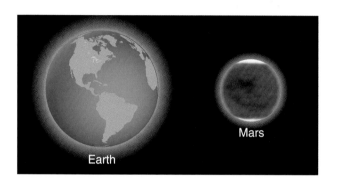

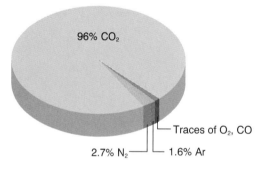

Atmospheric Gases

Mars	Value	Compared to Earth
Equatorial diameter	6786 km	0.53
Oblateness	0.009	
Mass	6.419×10^{23} kg	0.11
Density	3.95 gm/cm^3	0.71
Surface gravity		0.379
Escape velocity	5.02 km/s	0.45
Sidereal rotation period	24.623 hours	1.03
Solar day	24.67 hours	1.03
Surface temperature	$-140°$C to $+20°$C	
Albedo	0.15	0.5
Tilt of equator to orbital plane	25.20°	1.08
Orbit		
Semimajor axis	2.28×10^8 km	1.524
Eccentricity	0.0934	5.6
Inclination to ecliptic	1.85°	
Sidereal period	1.881 years	1.881
Moons	2	

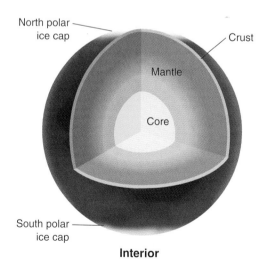

Interior

QUICK FACTS

The red planet. Half Earth's diameter. Least dense of terrestrial planets. Rotation rate and equatorial tilt very similar to Earth's. Seasons. Water–carbon dioxide polar caps. Has largest mountain (Olympus Mons) and canyon (Valles Marineris) known. Low-pressure CO_2 atmosphere. Evidence of warmer, wetter past.

A celestial object tends to be spherical because the force of gravity pulls its parts toward its center. As a result, the object takes the shape that will give it the smallest possible surface area. This shape is a sphere. Consider what would happen if the Earth were a cube. In this case, a rock at the corner of the cube could get closer to Earth's center by moving to the center of one face of the cube. In that event, the Earth would not remain a cube. Earthly mountains, which are formed by geological phenomena within the Earth (such as volcanic action), are forever being disturbed by wind and rain so that their parts can be pulled closer to the center, making the planet a more perfect sphere.

The same tendency exists in the case of Phobos and Deimos, but the gravitational force is too small to form these small objects into a spherical shape. On the surface of Mars's two moons, the gravitational force toward the center is much too small to pull the satellite into a sphere. The strength of the rock of which the moon is made resists the weak gravitational force that tends to reshape it.

> Most asteroids are likewise nonspherical.

Phobos and Deimos are more interesting to us than their size might seem to indicate. In fact, we may visit them someday. When people go to Mars, some plans call for using Phobos as a landing area before proceeding to the surface of the planet. The moon's weak gravitational field will make landing and taking off easy. At the beginning of the chapter, we mentioned a baseball game on Deimos. Perhaps someone will play on Phobos during your lifetime.

WHY EXPLORE?

The Martian invasion continues. In spite of the failure of the *Mars Observer* in 1993, ambitious plans are being made for the further exploration of Mars. In 1989, President Bush established a goal of a human expedition to Mars. This venture will probably be international in scope, and the earliest it could occur would be the first decade of the twenty-first century. It may be preceded by an unmanned mission that returns samples from the planet.

We have learned much about our planetary neighbors by observing them from Earth, but there is a limit to what we can learn without sending spacecraft to visit these worlds. Some people question the value of using our resources for these endeavors. There are two answers to their questions.

First, seeking knowledge about our universe is one of the things that separates humans from any other known creature. It is our nature to explore, just as it is our nature to enjoy music and art. If we never achieve one practical benefit from planetary exploration, such exploration would be valuable simply because seeking new knowledge is one of our highest aspirations. The amount of money that should be spent on such endeavors might be a subject for discussion, but when we compare the total amount of money spent on the basic sciences to what is spent on warfare, we see that the money spent on science is trivial.

Second, planetary exploration offers many practical benefits. Few would argue that knowledge of Earth is without practical benefit, but many people do not realize that a study of other planets is a valuable source of knowledge about our Earth. If we did not study other planets, we would be severely limited in what we would know about our own planet. For example, by learning more about the greenhouse effect on Venus and the lack of such an effect on Mars, we will find out how serious the threat of a runaway greenhouse effect is here on Earth. We cannot afford to experiment with our planet, but an examination of other planets provides us with just such an experiment. By studying weather systems of other planets, we learn about our own. Was there once life on Mars? If so, why does it not seem to be there now? Could the same thing happen on Earth? Surely such knowledge is of practical benefit.

CONCLUSION

Throughout history, people have wondered about the planets. With the advent of telescopes, particularly the larger ones, we began to learn something about their surfaces. It was the planetary visits of the space age, however, that made us think of these planets as *places*, rather than as celestial objects. We now have detailed maps of their surfaces, and landers have invaded Venus and Mars to begin exploration.

As we learn more about the other terrestrial planets, we are finding that in many ways they are similar, for they share a common history of formation, but we also find major differences. Some of the differences exist because the planets differ in size and mass. These factors determine whether a planet will retain an atmosphere, and the atmosphere of a planet determines such properties as the planet's range of temperature. Other differences occur simply because of the planets' different distances from the Sun.

In the next chapter, we move out to the Jovian planets. Each of these giants forms another piece of the puzzle of how the solar system—including the inhabitants of its third planet—came to be.

RECALL QUESTIONS

1. Mercury's atmosphere is
   A. mostly hydrogen.
   B. mostly hydrogen and helium.
   C. mostly nitrogen and oxygen.
   D. virtually nonexistent.

2. Mercury's rotation and revolution are linked so that Caloris Basin is
   A. directly under the Sun at every perihelion passage.
   B. directly under the Sun at alternate perihelion passages.
   C. never directly under the Sun.

3. Mercury's surface is difficult to map from Earth because
   A. Mercury is such a small planet.
   B. Mercury rotates so slowly.
   C. the surface is hidden below a thick cloud cover.
   D. Mercury is always close to the sun.
   E. [All of the above.]

4. Mercury's diameter is about
   A. one-fifth of Earth's.
   B. one-third of Earth's.
   C. the same as Earth's.
   D. twice that of Earth's.
   E. more than twice that of Earth's.

5. The property of Mercury that makes its temperature variations greater than those of any other planet is primarily
   A. its lack of an atmosphere.
   B. its proximity to the Sun.
   C. its small size.
   D. the carbon dioxide in its atmosphere.
   E. [The statement is false; Mercury is so close to the Sun that it is always hot.]

6. *Mariner* spacecraft found that Mercury has _____, and this leads us to conclude that the planet _____.
   A. no magnetic field . . . rotates very slowly
   B. no magnetic field . . . has little or no iron in it
   C. a magnetic field . . . rotates faster than we had thought
   D. a magnetic field . . . has a metallic core
   E. [Both A and B above.]

7. The greenhouse effect heats a planet because
   A. more sunlight strikes the planet's surface than normal.
   B. the surface of the planet is darker than normal.
   C. infrared radiation is trapped by the planet's atmosphere.
   D. cloud cover prevents the atmosphere from escaping.
   E. cloud cover prevents visible light from striking the surface.

8. In which of the following ways is Venus most like the Earth?
   A. Its period of rotation
   B. Its average surface temperature
   C. Its average density
   D. The length of its day
   E. The composition of its clouds

9. When we see Venus in the sky, the light we receive is sunlight reflecting from

A. its oceans.
   B. its solid surface.
   C. the top of its cloud layer.
   D. [Both A and B above.]
   E. [All of the above.]

10. How is it possible that Venus's surface may be hotter than Mercury's?
    A. It is closer to the Sun.
    B. Venus's larger area absorbs more heat.
    C. Venus rotates in a retrograde direction.
    D. Venus's lack of atmosphere allows sunlight to hit the surface without reflection.
    E. Venus has quite a lot of carbon dioxide in its atmosphere.

11. The radar mapper that began orbiting Venus in 1990 is named
    A. *Galileo.*
    B. *Galileo II.*
    C. *Venusian.*
    D. *Voyager.*
    E. *Magellan.*

12. Mars is least similar to Earth in
    A. the tilt of its equator to its orbital plane.
    B. its period of rotation.
    C. that Mars's surface is not hard.
    D. its atmosphere.

13. The two moons of Mars are named
    A. Ceres and Pallas.
    B. Io and Europa.
    C. Galileo and Copernicus.
    D. Helios and Juno.
    E. Phobos and Deimos.

14. The seasonal color changes on Mars probably result from
    A. vegetation.
    B. rain.
    C. dust movement.
    D. dry ice.
    E. [Both A and B above.]

15. The Martian polar caps
    A. are all water ice.
    B. are all frozen carbon dioxide ("dry ice").
    C. are a combination of carbon dioxide and water ice.
    D. completely disappear during the Martian summer.
    E. [Two of the above.]

16. Which of the following statements about the Martian satellites is true?
    A. Both are large satellites (nearly the size of Earth's Moon).
    B. Both are small satellites compared to Earth's.
    C. One is very large and the other is small.
    D. Mars has only one satellite.

17. The main source of erosion on Mars today is
    A. flowing liquids in the channels.
    B. ice floes that cover the entire planet in winter.
    C. microscopic organisms in a layer just below the surface.
    D. giant dust storms.
    E. [The statement is false; there is no erosion on Mars.]

18. The channels of Mars (observed in the nineteenth century) were found to be
    A. irrigation ditches dug by now-extinct Martians.
    B. faults in the Martian crust.
    C. straight mountain ranges.
    D. optical illusions.

19. Mars's moons are not spherical like Earth's Moon because
    A. of cratering.
    B. their gravitational force is not strong enough.
    C. they are much older than our Moon.
    D. they are much younger than our Moon.
    E. [The statement is false; they are spherical.]

20. Olympus Mons has a base _____ miles in diameter and a height of _____ miles.
    A. 5000 . . . 40
    B. 400 . . . 15
    C. 20 . . . 2
    D. 10 . . . 1

21. Why do Venus and Mars have larger volcanoes than Earth?
    A. Differences in atmospheres result in volcanoes being larger.
    B. Movement of the Earth's crust has prevented our volcanoes from growing as large as those on the other planets.
    C. Erosion on Earth is more prominent than on either of the other planets.
    D. [The statement is false; Earth's volcanoes are larger.]

22. Water is thought to be present on Mars
    A. in its polar caps.
    B. in permafrost below its surface.
    C. in liquid form on its surface.
    D. [Both A and B above.]
    E. [All of the above.]

23. If we list the terrestrial planets in order of increasing atmospheric pressure at the surface, the list should read
    A. Mercury, Venus, Earth, Mars.
    B. Mercury, Earth, Venus, Mars.
    C. Mercury, Earth, Mars, Venus.
    D. Mercury, Mars, Earth, Venus.
    E. Mars, Mercury, Earth, Venus.

24. The escape velocity from the surface is least for which of the following planets?
    A. Mercury
    B. Venus
    C. Earth
    D. Mars

25. Which is the smallest of the terrestrial planets?

26. Which planet is most similar to the Earth in size? In rotation period?

27. Why is the surface of Mercury difficult to see from Earth?

28. What is unusual about the revolution and rotation rates of Mercury and why has this occurred?

29. How was the most accurate value for Mercury's mass obtained?

30. Explain why Mercury's surface experiences such great extremes of temperature.

31. In what ways is Venus the Earth's sister planet? In what ways is it different?

32. What is the greenhouse effect? Why is the effect not an exact analogy to a greenhouse on Earth?

33. Describe the surfaces of Mercury, Venus, and Mars.

34. How does the tilt of Mars's axis compare to Earth's?

35. What is meant by a planet being in opposition? In inferior conjunction?

36. Of what are Mars's polar caps composed?

QUESTIONS TO PONDER

1. What would be the effect on a planet's temperature if its atmosphere reflected most of the visible light striking it but let infrared radiation pass through?
2. Explain why we can say that *Viking* did not prove that life does not exist on Mars.
3. A planet that rotates in the same direction in which it orbits has a longer solar day than sidereal day. Give one example of this and explain why it occurs.
4. The difference between the length of the solar day and the sidereal day is much greater for Earth than for Mars. Why?
5. Mars has no ozone layer protecting its surface from ultraviolet radiation. Our industrial society may be destroying the ozone layer here on Earth. Report on the latest research concerning depletion of the Earth's ozone layer.
6. Report on the latest results from the *Mars Observer*.
7. It has been suggested that we may be able to "terraform" Venus and/or Mars so that they are habitable by humans. Assuming that we have (or will have) the technology to do so, what do you think of the idea?

CALCULATIONS

1. If a planet has a diameter one-third of Earth's, how does its surface area compare to Earth's? Its volume?
2. Jupiter's diameter is 11 times Earth's diameter. Compare the volume of Jupiter to the volume of Earth.
3. Show that a density of 1 gram/cm^3 corresponds to 1000 kg/m^3.

ACTIVITY

Viewing Mercury, Venus, and Mars

Table A8-1 shows the dates between 1998 and 2000 when Mercury will be situated for best viewing. Between these dates Mercury rises at least 1½ hours before sunrise or sets at least 1½ hours after sunset. The longer the rising or setting time is separated from sunrise or sunset, the easier it is to find Mercury in the sky. Mercury is highest in the sky about midway between each pair of dates, and the longer the time between the two dates, the better the viewing will be at the midpoint. For example, in 1998, the December period provides a better viewing opportunity than in March. Mercury will be highest around December 18.

Within about 45 minutes of sunrise and sunset, the sky is so bright that Mercury is particularly difficult to see with the naked eye. Thus, if you are viewing in the morning, you should begin your search at least an hour before sunrise. In the evenings, there is no need to begin searching until about 45 minutes after sunset. Mercury is never easy to find for the first time, and you must have a very clear sky and a low horizon. Look low in the sky in the area lit by the Sun, searching for what appears to be a dim star.

Table A8-2 shows favorable dates for viewing Venus in the morning and evening sky. Between each pair of dates,

TABLE A8-1 Favorable Dates for Viewing Mercury

Year	Evening Viewings	Morning Viewings
1998	Mar. 16–Mar. 23	Jan. 1–Jan. 11
		Dec. 12–Dec. 25
1999	Feb. 28–Mar. 3	Nov. 26–Dec. 10
2000	Feb. 10–Feb. 19	July 24–Aug. 4

Venus rises at least one hour before the Sun or sets at least one hour after. The dates of Venus's greatest brilliance are shown, but between the dates indicated, Venus should be easy to find any time the sky is clear. Venus is bright enough that you can find it as close as 20 minutes to sunrise or sunset.

Table A8-3 shows some rising and setting times for Mars. Here is an example of how to use the table: Suppose you wish to know where to look for Mars in November of 1998. Since it was rising at the beginning of morning twilight the previous July and will rise at

TABLE A8-2 Favorable Dates for Viewing Venus

Year	Range of Dates	Morning or Evening
1998	Jan. 23–Sept. 15	M
1999	Jan. 1–July 30	E
1999–2000	Aug. 29–Mar. 7	M
2000–2001	Aug. 15–Mar. 23	E
2001	Apr. 5–Nov. 20	M
2002	Mar. 10–Sep. 25	E
2002–2003	Nov. 10–May 25	M

TABLE A8-3 Selected Rising and Setting Times for Mars

Date	Description
Late Feb. 1998	Setting at end of evening twilight
Late July 1998	Rising at beginning of morning twilight
Mid-Feb. 1999	Rising at midnight
Late April 1999	Rising at sunset, setting at sunrise
Late July 1999	Setting at midnight

*When midnight is indicated, the reference is to standard time (rather than daylight saving time).

midnight the next February, you can figure that it will rise about 2:00 or 3:00 A.M. in November (because November is about midway between the two). Thus you must look for it in the early morning in the eastern sky. Note that Mars is not visible from the dates when it is described as "Setting at end of evening twilight" until the dates described as "Rising at the beginning of morning twilight."

As you look at Mars, observe its color compared to nearby stars. It should be obvious why Mars is called the red planet.

If a telescope is available to you, use it to observe the planets. You should be able to see the phases of Venus (and perhaps Mercury). The color of Mars appears even more obvious in a telescope, and when Mars is at opposition and closest to Earth (when it is rising at sunset), some observers report that they can see the white polar caps even in a small telescope.

Helpful advice is available for the amateur astronomer in a number of monthly magazines, particularly *Astronomy* and *Sky and Telescope*. These include diagrams of planetary locations and hints on viewing. The two listed here are highly recommended. In addition, helpful planetary information, including positions during each month of the year, can be found in reputable almanacs such as *The World Almanac and Book of Facts*.

StarLinks netQuestions

Visit the netQuestions area of StarLinks (www.jbpub.com/starlinks) to complete exercises on these topics:

1. All About Mercury
Mercury, the fastest planet in the solar system, was named after the Roman god of roads and travel, who delivered his messages with godlike speed.

2. Explore With the Mars Pathfinder The Mars Pathfinder sent back fascinating images that confirm Mars once had a great amount of water on its surface. The roving robot, Sojourner, wheeled around the surface and found rocks with compositions very similar to Earth rocks.

3. The Night Sky As you've learned, the position of the planets in the night sky is constantly changing.

The painting depicts *Cassini* during the Saturn Orbit Insertion burn just after the main engine has begun firing. The spacecraft is moving out of the plane of the page and to the right and is firing to reduce the spacecraft velocity with respect to Saturn, so it can be captured by Saturn's gravity.

THE JOVIAN PLANETS

Jupiter
Jupiter as Seen from Earth
Jupiter from Space
The Composition of Jupiter
CLOSE UP: ET Life V—Letters to Extraterrestrials
Energy from Jupiter
Jupiter's Moons
Jupiter's Ring
Saturn
Size, Mass, and Density
Saturn's Motions
CLOSE UP: A Hypothesis to Explain Saturn's Excess Heat
Pioneer, Voyager, and Cassini

Titan
Planetary Rings
The Origin of Rings
Uranus
HISTORICAL NOTE: William Herschel, Musician/Astronomer
Uranus's Orientation and Motion
Neptune
CLOSE UP: Shepherd Moons
Neptune's Moons and Rings
HISTORICAL NOTE: The Discovery of Neptune
ACTIVITY: Observing Jupiter and Saturn

••••••• IN OCTOBER 1997, *the spacecraft* Cassini *was launched toward Saturn. Its trajectory will get the spacecraft to Saturn in 2004.* Cassini *(mass = 5650 kilograms—the mass of a small school bus) was launched toward Venus rather than Saturn in order to take advantage of a "gravitational assist" from Venus. Gravitational assists use the gravitational field of a planet to increase the speed of a passing spacecraft.* Cassini *will get two assists from Venus, one from Earth, and one from Jupiter before it finally heads toward Saturn. Without such assists,* Cassini *would have taken decades to get to Saturn.*

Nineteen days before arrival at Saturn, Cassini *will pass close to Phoebe, Saturn's most distant satellite. Of all the moons in the solar system, Phoebe is one of the most curious. Its inclined, retrograde orbit suggests that the moon may be a captured object, perhaps an old comet or asteroid that wandered close to Saturn in ages long past.*

Gravitational assists speed up a spacecraft but slow down the planet. When the *Voyager* spacecraft got an assist from Jupiter, *Voyager* gained 36,000 miles per hour at a cost of slowing down Jupiter by 1 foot every trillion years!

When Cassini *reaches Saturn, it will approach to an altitude only one-sixth the diameter of Saturn itself. The spacecraft will cross Saturn's rings (in a "gap," of course) and then ignite its rockets in order to slow down enough to be captured by Saturn's gravity in a 5-month-long orbit. During the rocket burn (shown in the chapter-opening figure),* Cassini *will decrease its speed by more than 600 meters per second (1300 miles per hour).*

The tour of Saturn will continue for four years, as Cassini *makes about 60 orbits with various orientations, from as close as three Saturn radii to more than seven. It will collect large amounts of data about the planet, its rings, and its moons, and it will make more than 30 flybys of Saturn's moon Titan—the second largest moon in the solar system. In 2004* Cassini *will drop a probe (supplied by the European Space Agency) to Titan; the probe will explore Titan's clouds, atmosphere, and surface.*

In this Chapter, attention turns to the planets of the outer solar system—the Jovian planets. (Pluto will be reserved for the next chapter.) Although Mercury, Venus, and Mars are certainly strange worlds for us Earthlings, this chapter will show that the outer planets are even stranger. Continuing our journey out from the Sun, we will take these planets in order. There is still a lot that astronomers do not know about our neighboring worlds, and you will see that as we get farther and farther from our home base, less and less is known about the objects there.

JUPITER

The size of a planet is not apparent to the naked eye, but long before the invention of the telescope, Copernicus had deduced that Jupiter was larger than Venus even though Venus at its brightest is brighter than Jupiter. From the two planets' relative brightnesses and distances, he concluded that since Jupiter is so much farther away, it would have to be much larger to appear so bright if it shines only by reflected light. Galileo observed the planets' angular sizes with a telescope and was able to determine that Jupiter was indeed larger, for he could use their angular sizes and relative distances to calculate their relative sizes. Today we know that Jupiter is the largest object in the solar system besides the Sun. It is appropriate, then, that it is named after the king of the Roman gods.

Copernicus simply assumed that every planet reflects the same percentage of the light that hits it.

Jupiter as Seen from Earth

Jupiter's mass is calculated by observing the radii of the orbits of its moons and their periods of revolution, as explained in Chapter 7. And Jupiter is massive! It has more than twice the combined mass of all the other planets, their moons, and the asteroids. When we compare Jupiter to the Earth, we find that Jupiter is 318 times more massive than our little planet.

In size, Jupiter is even more remarkable. Look back at Figure 7-1 to see how it compares in size to the other planets. Jupiter's diameter is about 11 times that of the Earth, making its volume about 1300 times that of the Earth.

While the Earth has a density about twice as great as common rock, Jupiter's density is 1.3 times that of water, or one-fourth of Earth's density. This means that Jupiter must be composed of a higher percentage of light elements than is the Earth.

You might have a tendency to picture Jupiter as a lumbering, slow-moving giant. Indeed, it is 5.2 AU from the Sun and takes nearly 12 years to circle the Sun. Its rotation is not slow, however. This gigantic planet spins on its axis once every 9 hours and 50 minutes. This is the first Jovian planet we have studied,

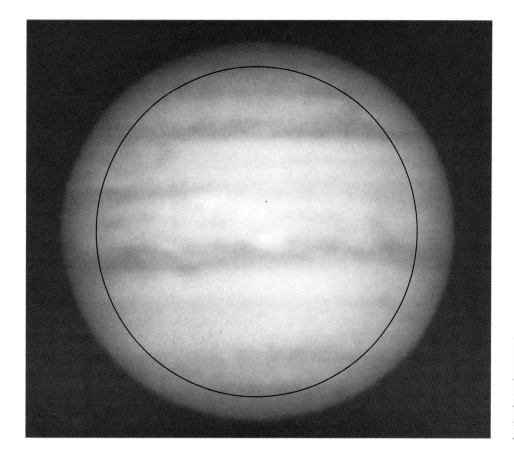

► FIGURE 9-1. The bands around Jupiter are easily visible in this photo taken from Earth. The red spot is called—imaginatively—"the red spot." The circle has been drawn to show how "out of round" Jupiter is.

and we will find that fast rotation is common for Jovians. Their rotational velocities are much greater than the terrestrial planets'.

Things are not quite this simple, however. Through even a small telescope we can see that Jupiter has dark and light bands parallel to its equator. The photograph in ►Figure 9-1, which was taken with a large Earth-based telescope, shows obvious parallel bands around the planet. Observations show that the bands near the equator rotate slightly faster than those nearer the poles. The band at Jupiter's equator has a rotation period of 9 hours, 50 minutes. Bands closest to the poles complete one rotation in 9 hours, 56 minutes. This "spreading" of the rotation, or **differential rotation,** indicates that the visible surface of Jupiter is not solid.

Another thing about Jupiter that can be observed fairly easily is that it is "out of round." A circle has been drawn on the photo of Figure 9-1 to make it obvious that Jupiter's equatorial diameter is definitely greater than its polar diameter. In fact, it is 6% greater. Recall that both Earth and Mars are oblate, but not nearly as much as Jupiter. The cause for the great oblateness of Jupiter is its great rotation rate. (Picture a spinning ball of Jello; its spin would cause it to flatten out.)

differential rotation. Rotation of an object in which different parts have different periods of rotation.

Jupiter from Space

In December 1973 *Pioneer 10* flew to within 130,000 kilometers of the surface of Jupiter. Then in December 1974 its twin, *Pioneer 11*, came within 50,000 kilometers of the surface. These two craft sent back great amounts of data from which a computer produced black and white photographs of the surface. In addition, the craft contained instruments to detect charged particles, radiation from the planet, and the magnetic field of Jupiter. After passing Jupiter, *Pioneer 11* continued on to Saturn, also giving us our first close view of that planet. Both

Because color filters were used when the black and white photos were taken, color images could be reconstructed.

> FIGURE 9-2. This photo of Jupiter was taken by *Voyager 1* at a distance of 40 million kilometers from the planet.

The spot is actually more gray than red and is difficult to see in a small telescope.

of these spacecraft are now on their way out of the solar system, to glide endlessly through space.

Knowledge gained from the *Pioneer* missions helped in the design of *Voyager I* and *II*, which visited Jupiter in 1979. They sent back 33,000 images including that of ➤Figure 9-2. Since 1995 the most recent visitor to Jupiter, the spacecraft *Galileo*, has been orbiting the planet and passing close to its moons. Upon reaching Jupiter, *Galileo* dropped a probe into the planet's atmosphere, and some of the information that follows has been learned from *Galileo* and its probe.

As we pointed out in the last chapter, if we knew more about weather on Venus, we would know more about weather on our own planet. Jupiter may provide an even better study of weather systems than does Venus, for weather patterns in the upper atmosphere of Jupiter are far removed from any solid surface Jupiter may have and therefore are almost unaffected by complications produced by surface irregularities. Jupiter's weather system should therefore be simpler in nature and allow us to learn what happens when surface features are not a major factor. (The practice of examining a simple system to form a working hypothesis about a complicated one is common in science.)

The fact that surface features have little effect on Jupiter's upper atmosphere is probably what allows Jupiter's weather patterns to last for such long periods. The prime example of this is the giant red spot on Jupiter (see ➤Figure 9-3). This spot was seen as early as the mid-1600s, so it has lasted for more than 300 years. The spot is about 40,000 kilometers in length and nearly 15,000 kilometers across. When we realize that the Earth is about 13,000 kilometers in diameter, we can appreciate the immensity of this feature. From time to time over the centuries, the red spot has faded in intensity, but it has never disappeared completely.

Data from Jupiter indicate that the red spot is a high-pressure storm system similar to the much smaller Earth systems reported by your local weather forecaster. High-pressure systems in Earth's Northern Hemisphere rotate in a clockwise direction, and in the Southern Hemisphere they rotate in the opposite direction. The red spot is in Jupiter's southern hemisphere, and indeed it rotates counterclockwise—with a period of about six days. Figure 9-3 is a close-up view of part of the spot; we can see the swirling currents at its edges. In addition, the figure shows the size of Earth relative to the red spot.

➢ **FIGURE 9-3.** Jupiter's red spot (at upper right) is seen with the great turbulence around it. Earth is superimposed at the same scale to help us appreciate the size of the spot.

The Composition of Jupiter

The terrestrial planets are composed mostly of hard, rocky material. Jupiter is different. The compositions of Jupiter and the other Jovian planets is more similar to the Sun than to the terrestrial planets. The *Galileo* probe measured Jupiter's composition to be about 86% hydrogen and 14% helium, with small amounts of methane, ammonia, and water vapor. Astronomers think that this data is representative of the nebula from which the solar system formed. This implies that helium has not rained down or settled toward the center of the planet as much as it seems to have done on Saturn, where helium is less abundant. The theory of planetary evolution must now take into account the fact that there has been little change in helium abundance in the Jovian atmosphere since the birth of the solar system.

Galileo detected small amounts of heavy elements—carbon, nitrogen, and sulfur—suggesting that meteorites and other small bodies have contributed to the planet's composition. Few complex organic compounds were evident so the likelihood of finding life as we know it here on Earth is extremely remote.

The colors seen in Jupiter's upper atmosphere apparently are the result of chemical reactions induced by sunlight and/or by lightning in its atmosphere. The light areas around the planet are high-pressure areas where gas is rising from inside Jupiter, and the dark bands are low-pressure regions where the gas descends. The rapid rotation of Jupiter moves the regions around the planet so that it has a banded appearance.

As might be expected from the fact that the planet has differential rotation, measurements made by *Voyager* show that neighboring bands of Jupiter's atmosphere are moving at different speeds around the planet. As a result of these

The *Galileo* probe did not detect the thick dense clouds that were expected. It may be that the site where the probe descended is not typical.

Jupiter

FIGURE 9-4. This photo of the bands of Jupiter shows the swirling that results from the different speeds of adjacent bands.

different speeds, there is considerable swirling at the boundaries between bands. The stormy nature of these swirls is obvious in the view of the bands in ▶Figures 9-4 and 9-5.

The gaseous atmosphere of Jupiter is fairly thin, only a few thousand miles in depth. Naturally, the lower you go in the atmosphere, the greater the pressure is (because the gas at lower elevation supports the gas above). On Jupiter, the pressure soon becomes so great that the hydrogen no longer acts as a gas but starts acting like a liquid. The molecules are forced so close together that the hydrogen acquires a liquid nature; it becomes liquid hydrogen. On the Earth we have a gaseous atmosphere above a liquid ocean. On Jupiter, however, there is no distinct boundary between the liquid and the gas. At the top of the atmosphere, we would classify the hydrogen as a gas. Much lower, we would call it a liquid. In between, the gas becomes more and more dense as we move deeper and deeper into the atmosphere. If we could travel inward from the gaseous region, we would find ourselves in thicker and thicker gas until we finally decided that we were no longer in a gas but were in a liquid.

At perhaps 15,000 kilometers below the clouds, there is another change in Jupiter's composition, caused by the increasingly greater pressures and temperatures. At these depths, electrons move easily from one atom to another, making the hydrogen a good electrical conductor. Because it conducts electricity like a metal, we call it *liquid metallic hydrogen.* Hydrogen in this form cannot be produced on Earth because we cannot create such a combination of high temperature and high pressure, but it is predicted by atomic theories. As ▶Figure 9-6 indicates, most of the planet is made up of this state of matter.

Figure 9-6 shows a core of heavy elements at the center of Jupiter. We have no direct evidence that such a core exists, but it is hypothesized on the basis that there must have been heavy elements in the original material of which Jupiter is made. These heavy elements exist on the moons of Jupiter, and there is good reason to think that they also exist on the planet. If so, they would have sunk to the center of Jupiter. The size of this portion of the planet is unknown, but it is thought to have a diameter perhaps as small as the Earth's or as much as a few times the Earth's. In any case, it makes up a very small portion, perhaps 1%, of the entire planet.

When the *Pioneer* spacecraft passed Jupiter, we learned of that planet's very strong magnetic field. *Voyager* provided more data, and then in February 1992, the *Ulysses* spacecraft passed by the planet and found that Jupiter's magnetic field is even larger and stronger than had previously been thought. Jupiter's magnetic field is nearly 20,000 times stronger than Earth's!

The *Ulysses* spacecraft was built by the European Space Agency. It will orbit the Sun and gather data regarding that star.

▷ FIGURE 9-5 As are many photos we see of Jupiter, this one is color-enhanced to emphasize subtle differences in the real colors.

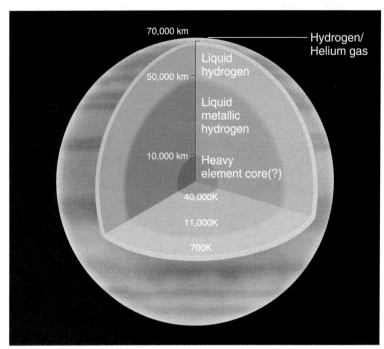

▷ FIGURE 9-6. The interior of Jupiter consists mostly of liquid metallic hydrogen, possibly with a core made up of heavy elements.

Recall from Chapters 6 and 8 that the dynamo theory of planetary magnetic fields holds that in order for a planet to have a strong magnetic field, two conditions are necessary: rotation and electrically conductive material within the planet. The terrestrial planets appear to have iron (or nickel-iron) cores that are good electrical conductors. Jupiter does not depend upon its core to produce its magnetic field, for the liquid metallic hydrogen that makes up much of the planet is responsible—along with Jupiter's fast rotation—for its strong magnetic field.

Like the Earth's magnetic field, Jupiter's field deflects the solar wind around the planet as well as trapping some of the charged particles of the wind in belts around it. Jupiter's **magnetosphere** extends to perhaps 15 million kilometers from the planet, enveloping most of Jupiter's satellites. (The field has presented problems to the electronic circuits of spacecraft that passed through it.) ▷Figure 9-7 illustrates the radiation belts of Jupiter.

magnetosphere. The volume of space in which the motion of charged particles is controlled by the magnetic field of the planet rather than by the solar wind.

▷ FIGURE 9-7. Jupiter's radiation belts are flattened by the planet's rapid rotation. Compare them to the Van Allen radiation belts in Chapter 6.

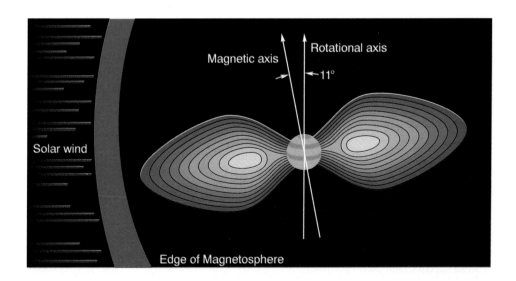

Jupiter

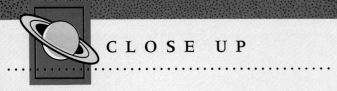

CLOSE UP

ET Life V—Letters to Extraterrestrials

Pioneer Plaques

In late 1971, Dr. Carl Sagan learned that the trajectories of the *Pioneer 10* and *Pioneer 11* spacecraft would take them out of the solar system and that it might be possible to include on them a message to extraterrestrials. Sagan called NASA authorities and within three weeks got approval to put a plaque on both *Pioneer 10*, which was scheduled to be launched the following March, and *Pioneer 11*, which was to be launched a year later. Carl and his wife Linda, along with Dr. Frank Drake, designed the message, which was etched on a 6 × 9-inch gold-anodized aluminum plaque. ➤Figure C9-1 shows their design.

Some of the message on the plaque is easily recognizable (to humans, at least). The umbrella-shaped object behind the man and woman is an outline of the *Pioneer* spacecraft drawn to scale with the people. Thus the finder can determine the size of a human. (The man on this scale is 5 feet 9½ inches tall.)

The two circles at the upper left represent hydrogen atoms emitting radiation of a particular wavelength. Binary numbers on the plaque (not visible in this small rendition) use this wavelength to show the size of the woman and of the solar system at the bottom, where the *Pioneer* spacecraft is shown leaving the third planet.

The spidery-looking feature at the left shows the directions from the solar system to pulsars, pulsating sources of waves we detect in deep space. The binary numbers along the lines represent the period of the pulses expressed in terms of the period of the wave from hydrogen. By analyzing the directions and pe-

➤ **FIGURE C9-1.** This plaque aboard *Pioneer 10* and *11* carried a message to extraterrestrials—and to Earthlings.

Energy from Jupiter

A number of puzzles concerning Jupiter remain besides the composition of its interior. One puzzle results from the fact that Jupiter emits more energy than it absorbs from the Sun. This was known before the *Pioneer* and *Voyager* missions, and those missions only confirmed the observations. Let's explain why this is a problem.

We would expect the radiation coming from a planet to be simply the sum of the solar radiation that is reflected from the planet and the infrared radiation that is emitted as the planet reemits the absorbed solar radiation. We don't expect a planet to have an energy source of its own. It is easy to calculate the amount of solar energy that strikes any planet. We use our knowledge of how much solar

When we speak of solar radiation, we include—along with visible light—ultraviolet, infrared, and all other types of radiation.

272 CHAPTER 9 The Jovian Planets

riods of the pulsars, the finders could tell where our solar system is located in the Milky Way. In addition, since pulsars slow down with time, they could tell when the craft was launched.

Could extraterrestrials interpret the message? Curiously enough, since hydrogen atoms and pulsars are common to all creatures of the universe, it is thought that they would be more likely to interpret the pulsar sketch than the human figures, which we recognize immediately.

Voyager Records

In 1977, *Voyager 1* and *2* were launched into space to rendevous with Jupiter and Saturn. Plans were later adjusted to allow *Voyager 2* to continue to Uranus and Neptune. As this is being written in the summer of 1997, *Voyager 1* has moved past 67 AU from the Sun. Sometime later in 1997 it will overtake the *Pioneer 10* spacecraft as mankind's farthest object from the Sun. Since Pluto does not get farther than 50 AU from the Sun, *Voyager 1* is well outside our system by that measure.

With more time available to plan a message, a group of people including the three who prepared the *Pioneer* plaque designed a much more complete message to go aboard *Voyager 1* and *2*. The messages on *Voyager* are contained on two copper records, designed to be played at $16\frac{2}{3}$ revolutions per minute. Instructions for playing the records are included in pictures on the cover. On the records are 90 minutes of the world's greatest music, 118 pictures, and greetings in nearly 60 languages (including one nonhuman message from a humpback whale).

The music on the *Voyager* includes parts of compositions from Bach *(Brandenburg Concerto No. 2* and *The Well-Tempered Clavier)*, Beethoven (Symphony No. 5, String Quartet No. 13 in B-flat), and Mozart *(The Magic Flute).* In addition it includes Chuck Berry's "Johnny B. Goode," Australian Aborigine songs, a Peruvian wedding song, and numerous other selections from the cultures of our world.

The records contain pictures as well as sound. The pictures include many photographs of scenes of nature here on Earth, of the Earth from space, of the Earth as changed by humans, of people in various activities, and of the biology of humans.

Will the Message Be Found?

The messages were not put aboard the *Pioneer* and *Voyager* craft with high hopes of them ever being found. They were sent in much the same manner as a child puts a message in a bottle and throws it into the sea. In fact, the space messages have even less chance than the child's bottle of being found. Even if the Galaxy abounds in intelligent life, the emptiness of space and the slow speed of the spacecraft make the chances slim indeed. The soonest any of the craft will pass within two light-years of a star is 40,000 years!

The messages are really a symbol of hope, a shout that "we are here." As stated in *Murmurs of Earth*—which tells the story of the *Voyager* records—after the Earth has been reduced to a charred cinder by an expanded and brighter Sun, the messages will continue to travel through space, "preserving a murmur of an ancient civilization that once flourished—perhaps before moving on to greater deeds and other worlds—on the distant planet Earth."*

---

*From Carl Sagan and F. Drake, A. Druyan, T. Ferris, J. Lomberg, and L. Sagan, *Murmurs of Earth* (New York: Ballantine Books, 1978).

---

energy strikes a square mile of Earth's upper atmosphere, along with the inverse square law, to calculate the energy that strikes a square mile of that planet's surface (or atmosphere). Then we multiply this value by the cross-sectional area of the planet to get the total solar energy that strikes that planet.

When the energy that strikes a planet is compared to the energy reflected and emitted from the planet, we expect to find that the energy coming from the planet is the same as the energy absorbed. If it isn't, the planet must either be heating up as it gains energy, or cooling as it loses energy.

Jupiter, though, does not behave as expected; it emits about twice as much energy as it absorbs. There are three possible ways for it to do this. One possibility is that it may have an internal energy source such as chemical reactions or a

The inverse square law of radiation was discussed in Chapter 4.

Recall that in discussing Venus, we said that the greenhouse effect results in the planet being so hot that it emits as much energy as it absorbs.

source of radioactivity within it. There is no reason to believe that much energy is produced in Jupiter by either of these methods, but it was once hypothesized that Jupiter acts like a miniature star, with significant nuclear fusion reactions going on within it. (Nuclear fusion reactions provide the energy of the stars.) Further calculations showed that Jupiter is not massive enough and does not have sufficient internal pressures or temperatures to support fusion reactions, however.

A second possibility: recall from Chapter 7 that as an object shrinks due to gravitational force, it heats up. Calculations show that Jupiter may still be shrinking and producing heat. If this is the case, however, the amount of shrinking would still not be enough to explain all of the extra energy from Jupiter.

It is now thought that the excess energy from Jupiter is energy left over from its formation. By now, the smaller planets have lost their excess heat, but Jupiter's immense size has served to insulate its interior so that it cools slowly, and the cooling continues even today. Computer models show a temperature of 40,000°C for Jupiter's center.

Jupiter would have to be nearly 100 times more massive to support nuclear fusion.

Do not get the idea that Jupiter gets significantly cooler from year to year. The planet is so large that the extra energy emitted from it corresponds to an extremely small temperature change.

Jupiter's Moons

Jupiter's family of 16 (or more) moons can be divided into three groups. ➤Figure 9-8 shows that one group—the eight outermost moons—orbits in a different direction than the four inner moons (and most objects in the solar system), and that their orbits are fairly eccentric. Astronomers hypothesize that these moons are captured asteroids. That is, they once orbited the Sun, but came close enough to the Jovian system that they were captured. In its early history, Jupiter probably had a more extensive atmosphere, which might have been sufficient to slow

Most of our knowledge of Jupiter's system comes from the two Voyager missions, and most of the photographs in this section of the text were taken by those spacecraft.

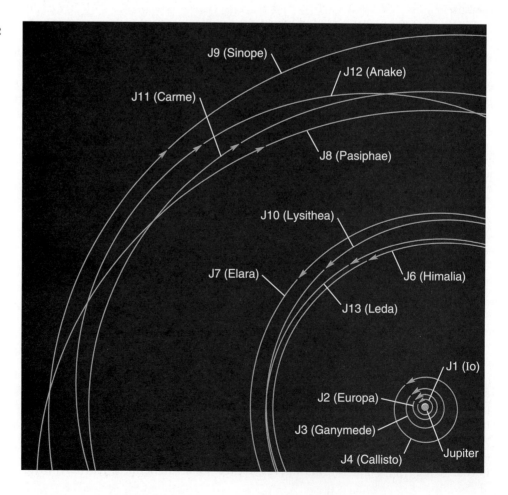

➤ **FIGURE 9-8.** The orbits of 12 of Jupiter's satellites. (Satellites are numbered in the order of their discovery.) Data on planetary satellites are given in Appendix D.

▶ **FIGURE 9-9.** Compare the sizes of the Galilean moons to other objects we have studied.

down the asteroids so they could be captured. The moons have black surfaces, as do many asteroids, providing further support for the hypothesis. It is suggested that these outer moons are divided into two groups of four each because only two asteroids were captured, each of which broke into pieces when it hit Jupiter's atmosphere.

Four moons do not show in the figure, for they are very close to Jupiter, inside the smallest orbit shown. These satellites are generally referred to as fragmented moonlets.

The four satellites shown orbiting in (nearly) circular orbits are the Galilean moons. The smallest of these is 5000 times more massive than the largest of the rest. ▶Figure 9-9 shows the relative sizes of the four Galilean moons, along with the four terrestrial planets, Earth's Moon, and Saturn's largest moon.

The Galilean moon nearest Jupiter is Io. ▶Figure 9-10 shows its mottled surface; Io is a strange sight. Its yellow-orange color is caused by the element sulfur, which covers its surface. The biggest surprise to astronomers came when Linda Moribito of NASA, while examining *Voyager* photographs, discovered an active volcano on Io. Further examination showed that Io has many active volcanoes or geysers. The two *Voyager* craft photographed eight or nine active volcanoes (see ▶Figure 9-11). The circles and dark spots you see in the photograph of Io are not impact craters; they are the results of volcanoes. In fact, the volcanoes explain the lack of impact craters, for the surface of Io is constantly being changed by the release of hot sulfur from below the surface.

We must ask what is the source of the energy for these volcanoes. In the case of geysers and volcanoes on Earth, the energy source is primarily internal heat retained from the Earth's formation and energy released by radioactive materials within the Earth. But a small object cannot retain heat nearly as long as

Recall that Galileo discovered the four moons named for him.

The volcanoes on Io might be more appropriately called geysers, for we usually associate volcanoes on Earth with mountains and the eruptions on Io do not come from mountain tops.

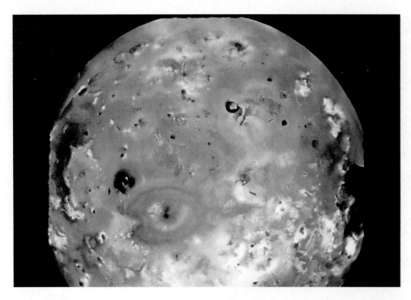

▶FIGURE 9-10. Io, the innermost of the Galilean satellites, has a number of active volcanoes on its sulfur surface.

▶FIGURE 9-11. The upper part of the photo shows one of Io's volcanoes erupting.

An interesting example of the confirmation of a scientific prediction occurred here. Just three days prior to the discovery of Io's volcano, S. Peale published a paper predicting that tidal heating of Io would result in volcanism.

This is an example of how we can deduce characteristics of a surface from its different reflectivities in different spectral regions.

a large one, and Io would have lost any heat that resulted from its formation. Nor can radioactivity account for the tremendous volcanic activity; rather, the heat within Io is produced by tidal forces. These tides result from the eccentricity of Io's orbit, so that the moon is sometimes slightly closer to Jupiter than at other times. Even a small difference in distance from massive Jupiter results in powerful, varying tidal pulls on Io. The resulting "kneading" of Io adds heat to it until the molten interior bursts through cracks in the crust—hence, volcanoes.

▶Figure 9-12 shows two images of Io taken by the Hubble Space Telescope. The one on the left was taken in visible light, and the one on the right shows the same surface of Io in ultraviolet radiation. Some areas are bright in visible light but dark in ultraviolet. Sulfur dioxide frost absorbs ultraviolet, but reflects visible light quite well, and this is thought to cause the differences between the two images.

Io is surrounded by a halo of sodium atoms, some of which are swept away from the moon by Jupiter's magnetic field, resulting in a faint ring of sodium near Io's orbit.

The mass of Io has been calculated based on its gravitational attraction for passing spacecraft, and its density is calculated to be about 3.5 times that of water. This leads us to conclude that the satellite is composed mostly of rock, with a

▶FIGURE 9-12. The image of Io at the left was made in visible light, while the one at the right is by ultraviolet light. Both show the same surface of Io. Areas that are bright in visible light but dark in ultraviolet are probably covered with sulfur dioxide frost.

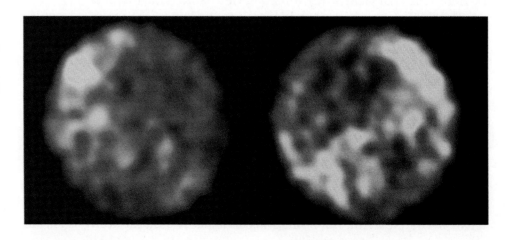

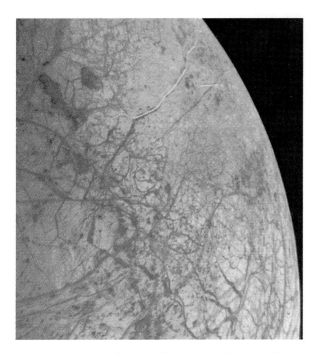

▶ FIGURE 9-13. The lack of craters on Europa indicates that the surface is active.

▶ FIGURE 9-14. Ganymede is larger than Mercury.

relatively shallow layer of sulfur on its surface. No water is found on Io; Io is a strange, volcanic desert.

Europa, 1.5 times as far from Jupiter as Io, presents a far different picture. Its surface is not sulfur, but something more familiar to us—ice. ▶Figure 9-13 shows its cracked-billiard-ball appearance. Europa's density is slightly less than Io's, but again we must conclude that it is mostly rock covered by an ocean of frozen water.

Note the lack of craters on Europa. Only three were found in *Voyager* photographs. Again, this does not indicate that meteoroids have not struck Europa but rather that its surface is active so that craters do not remain visible for long. Although it is farther from Jupiter than Io is, Europa also experiences tidal heating. There appears to be at least one volcano on Europa, although what looks like a volcano may simply be water spraying out through a crack in the moon's crust, forced out by pressure resulting from tidal flexing. The cracks in Europa's surface are hypothesized to be another result of this flexing.

Ganymede is the largest moon in the solar system; it is larger than the planet Mercury. On its surface (▶Figure 9-14) we see ice, but notice how different Ganymede is from Europa. There are craters, indicating that its surface is not as active. Its ice is also darker than Europa's. This darker color is the dust from meteorite falls that spreads across the surface of the satellite. On Europa, the surface is constantly being refreshed with water from below, causing the meteorite dust to be spread through a much deeper layer of the satellite. The less active surface of Ganymede, on the other hand, leaves the dust on the surface. Notice, though, that there are light-colored streaks on Ganymede. These are where cracks have formed and icy slush from below has welled up to fill the cracks.

The outermost of the Galilean moons is Callisto (▶Figure 9-15). Now we see a more familiar cratered surface. At its greater distance from Jupiter, Callisto experiences little tidal heating and has a very inactive surface. The newer craters are the whitest, showing where a meteorite impact has brought clean ice to the surface. The large white crater near the top of the photograph is named Valhalla and is the largest impact crater known in the solar system.

➤ **FIGURE 9-15.** Callisto, the Galilean satellite that is farthest from Jupiter, has the least active surface.

➤ **FIGURE 9-16.** The ring system of Jupiter was imaged by the Galileo spacecraft on November 9, 1996. The ring clearly shows a structure that had only been hinted at in the Voyager images.

As one moves from the innermost Galilean moon to the outermost, patterns of change are obvious. The farther the moon is from Jupiter,

1. the less active the surface,
2. the lower the average density of the moon, and
3. the greater the proportion of water.

Numbers 2 and 3 listed here are related, for since water has a lower density than most other substances on the moons, the more water on a moon, the lower its density.

Recall that when we studied the origin of the solar system in Chapter 7, we saw similar patterns of change among the planets as we moved from Mercury to the outer planets. This confirms Galileo's intuition that the Jupiter system represents the solar system in miniature.

Jupiter's Ring

A surprising discovery by *Voyager 1* was that Jupiter has a ring. Figure 9-16 shows the ring as photographed by *Galileo*. The photograph was made when the camera was in Jupiter's shadow. The light that reached the camera from the ring was therefore scattered forward by the material of the ring; since large chunks of matter cannot scatter light in this manner, the ring is known to be made of very tiny particles (as small as particles of cigarette smoke). The ring is fairly close to Jupiter, extending to about 0.8 planetary radii from the planet's surface.

Calculations indicate that the particles of Jupiter's ring could not have been there since the formation of the solar system because radiation pressure from the Sun, as well as forces from Jupiter's strong magnetic field, gradually send some particles down into the planet and others away into space. Therefore the ring is thought to be continually replenished, probably from small moonlets within it or near it. Two of these moonlets, Adrastea and Metis, which are not seen in Figure 9-16, orbit through the outer portion of the ring. We will discuss planetary rings further when we discuss Saturn and its very prominent system of rings.

2 JUPITER
DATA PAGE

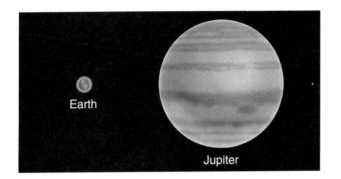

Jupiter	Value	Compared to Earth
Equatorial diameter*	142,980 km	11.2
Oblateness	0.064	
Mass	1.899×10^{27} kg	318
Density	1.326 gm/cm^3	0.24
Surface gravity*		2.36
Escape velocity	59.5 km/s	5.3
Sidereal rotation period	9.925 hours	0.41
Albedo	0.34	1.1
Tilt of equator to orbital plane	3.12°	0.13
Orbit		
Semimajor axis	7.78×10^8 km	5.203
Eccentricity	0.0485	2.9
Inclination to ecliptic	1.3°	
Sidereal period	11.86 years	11.86
Moons	16	

*Measured where the pressure is one atmosphere—Earth's pressure at sea level

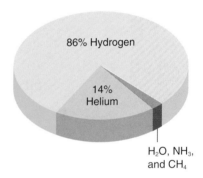

Atmospheric Gases

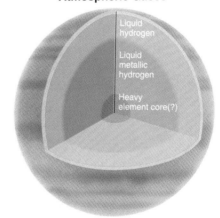

Interior

QUICK FACTS

The largest planet. Contains more than twice the mass of all other planets combined. Fast rotation. Atmosphere shows bands and the red spot and is mostly hydrogen and helium. Giant magnetic field, due to liquid metallic hydrogen that makes up much of the planet. Emits more energy than it receives. Has a large family of moons and a faint ring.

Huygens (1629–1695) was a Dutch physicist and astronomer who made major advances in the fields of optics and light.

Jupiter's average density is nearly twice that of Saturn.

SATURN

Probably the most impressive object visible in a small telescope is the planet Saturn. Galileo called Saturn "the planet with ears" and was unable to explain what appeared to be bumps on opposite sides of it. Some 50 years later, Christian Huygens recognized that the "ears" were due to rings around the planet.

Size, Mass, and Density

Saturn has an equatorial diameter of 120,000 kilometers, not much smaller than Jupiter's, but more than nine times that of the Earth (➤Figure 9-17). Except for its obvious rings, Saturn is in many ways similar in appearance to Jupiter. If the planets are indeed similar, we would suspect that Saturn has a density like Jupiter's, but in fact Saturn's density is less, only 0.7 as much as water. As we will see later, the low density of Saturn is probably due to a less dense core and a lower percentage of liquid metallic hydrogen.

The composition of Saturn is about the same as Jupiter's: $4/5$ hydrogen and $1/5$ helium, with only about 1% of heavier materials. This is about the same as the composition of the Sun. The similarity is not a coincidence, for all of these objects formed from the same material—the interstellar cloud that collapsed to form the solar system. Recall from Chapter 7 that the reason the terrestrial planets have less hydrogen and helium is that these volatile gases were swept away from the hot, inner portions of the system.

Saturn's Motions

Saturn orbits the Sun at 9.5 AU; its distance from the Earth varies from about 8.5 AU to 10.5 AU. Its appearance from the Earth varies greatly, but not because of changes in distance. ➤Figure 9-18 shows a number of views of Saturn taken at various times during half of its 29.5-year period of revolution. To see the reason for the change in appearance of its rings, think of Saturn revolving at its great distance from the Earth and Sun as shown in ➤Figure 9-19. The rings of

➤ FIGURE 9-17. Here photos of Saturn, Earth, and Jupiter are reproduced to scale. Jupiter's diameter is 11.2 times Earth's, while Saturn's diameter is 9.5 times Earth's.

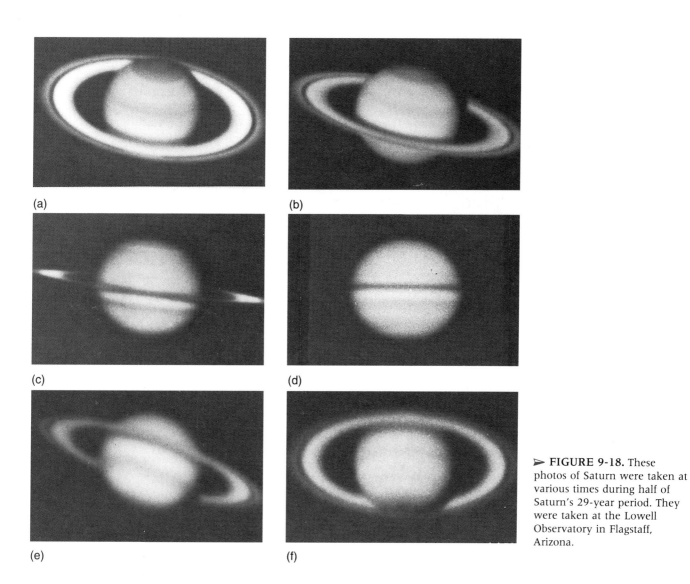

▶ FIGURE 9-18. These photos of Saturn were taken at various times during half of Saturn's 29-year period. They were taken at the Lowell Observatory in Flagstaff, Arizona.

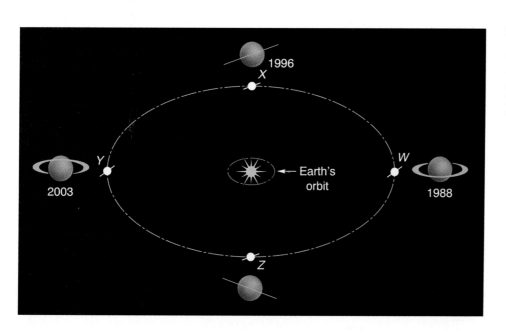

▶ FIGURE 9-19. Saturn is shown at four points in its orbit. The outer drawings show its appearance from Earth when it is at each of those locations. At points X and Z the rings are edge-on to Earth.

Saturn

Saturn are in the plane of the planet's equator, and the planet is tilted 27 degrees with respect to its orbital plane. Saturn keeps this same tilt while moving around the Sun (and the Earth), so when it is viewed from the Earth, we see the rings at different orientations at different times.

When we consider the rotation of Saturn, we find two more similarities to Jupiter. The various belts of Saturn's atmosphere rotate at different rates, just as in the case of Jupiter, and its equatorial rotation rate of 10 hours and 39 minutes is close to Jupiter's 9 hours and 50 minutes.

Recall that Jupiter's oblateness was explained by its high rotation rate. Saturn has a similar rotation rate but is even more oblate. This is because the gravitational field at its surface is weaker than at Jupiter's; since there is less gravitational force tending to keep the planet in a spherical shape, Saturn is more "out of round" than any other planet.

> Saturn's oblateness is 0.102 so that its equatorial diameter is 10% greater than its polar diameter.

Pioneer, Voyager, and Cassini

As you might suspect, we have learned much about Saturn from space probes. *Pioneer 11* passed Saturn in September 1979, *Voyager 1* about a year later, and *Voyager 2* a year after that. (*Pioneer 10* swung around Jupiter in such a manner that its path did not take it to Saturn.) Knowledge gained from each of the first two probes was used to guide experimenters in decisions concerning the following ones.

The atmosphere of Saturn is similar to that of Jupiter except that Saturn has a slightly lower percentage of helium. Since Saturn is smaller, one must descend deeper into its interior to reach a pressure great enough for hydrogen to form a liquid metallic state. Whereas liquid metallic hydrogen extends out about two-thirds of the way to Jupiter's cloud tops, it extends only about halfway to Saturn's clouds. As a result, Saturn has a weaker magnetic field than Jupiter does. ➤Figure 9-20 shows a cross section of Saturn.

Except for its rings, Saturn appears much more bland than Jupiter for two reasons. First, because of its greater distance from the Sun, Saturn is much colder than Jupiter, and the extreme cold inhibits the chemical reactions that give Jupiter's atmosphere its varied colors. Secondly, a layer of methane haze above

> Saturn's magnetic field is only 5% as strong as Jupiter's.

> Saturn's cloud tops have a temperature of −180°C.

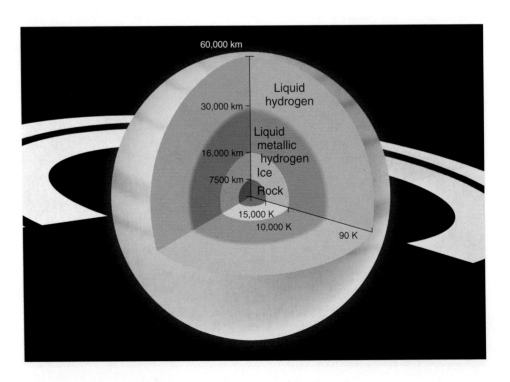

➤ FIGURE 9-20. Saturn's interior resembles Jupiter's but Saturn contains a smaller percentage of liquid metallic hydrogen.

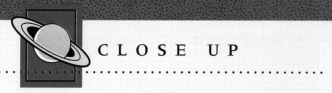

CLOSE UP

A Hypothesis to Explain Saturn's Excess Heat

All of the planets were hot after their formation. Jupiter is still radiating away its excess heat. Smaller planets cool more quickly, however, and we calculate that Saturn would have lost most of the energy from its formation. Thus this energy cannot be used to explain why Saturn radiates twice as much energy as it receives from the Sun.

Saturn also has another unusual feature that astronomers must explain: analysis by the *Voyager* spacecraft showed that Saturn has less helium in its upper atmosphere than Jupiter has. By mass, Saturn's upper atmosphere contains 11% helium, much less than the nearly 20% on Jupiter. Astronomers had expected Saturn to have about the same composition as the original material of the solar system—the composition of Jupiter and the Sun.

An interesting hypothesis explains both of these anomalies. Some astronomers propose that the cooling of Saturn's atmosphere causes helium to condense to liquid form and rain downward. This would explain the low percentage of helium in the upper atmosphere. In addition, as the helium droplets fall, they lose gravitational energy, and this energy is converted to thermal energy. This energy, the hypothesis holds, results in Saturn emitting more energy than it absorbs.

Saturn's cloud tops blurs out color differences. *Voyager* photos show that Saturn has atmospheric features similar to Jupiter, however, and that its winds reach speeds three to four times faster than Jupiter's. ➤Figure 9-21 is a color-enhanced photo taken by the Hubble Space Telescope. It shows a major storm (white) that occurred on Saturn in 1990.

Like Jupiter, Saturn radiates more heat than it absorbs. We once thought that the source of the heat is the same as Jupiter's—leftover heat from the planet's formation—but calculations show that because of its smaller size, Saturn would have cooled quickly enough that it would not still be radiating excess heat. The reason for Saturn's excess heat is not fully understood, but the accompanying Close Up describes one hypothesis.

Present plans call for another visit to Saturn, this one by the spacecraft *Cassini*, which will be launched in 1997 and will arrive at Saturn in 2004. It will go into orbit around the planet and drop a probe into the atmosphere of Saturn's largest moon, Titan.

➤ FIGURE 9-21. The Hubble Space Telescope took this photo of Saturn in 1990. The white storm system near Saturn's equator appeared that year and lasted a number of months.

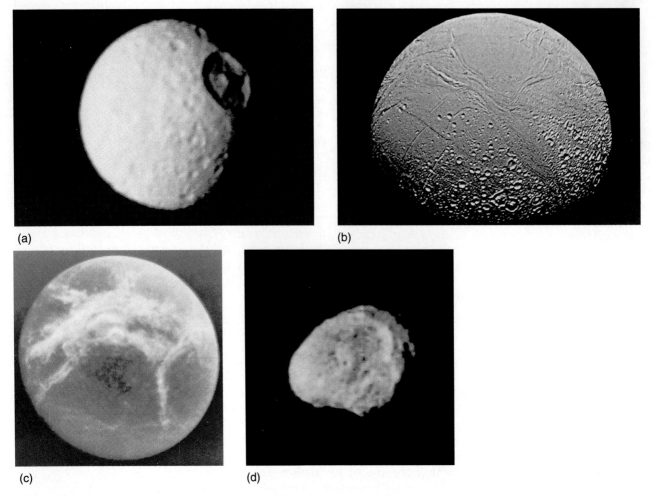

► **FIGURE 9-22.** A number of Saturn's moons: (a) Mimas, 392 km in diameter. Note the large impact crater. (b) Enceladus (500 km) shows some areas with fewer craters, suggesting that volcanic action took place fairly recently there. (c) Dione (1120 km) is in synchronous rotation so that it keeps its same face toward Saturn. This is the side that trails, and it is less cratered than the opposite side. (d) Little Hyperion (350 × 200 km) is irregular in shape and tumbles as it orbits Saturn. All of the moons shown here have surfaces of water ice.

Titan

Confirmation of Saturn's 18th moon was announced July 1990.

Saturn has at least 18 moons, most of which consist of dirty ice such as we see on some of Jupiter's moons. Although each of Saturn's moons (►Figures 9-22 and 9-23) is unique, we will discuss only Titan, the largest. Refer to Appendix D for details concerning the moons of Saturn as well as other moons of the solar system.

Titan (►Figure 9-23) may turn out to be the most interesting moon in the solar system, but our knowledge of its surface is limited by the haze that covers it. The atmosphere of Titan is mostly nitrogen, with perhaps 1% methane (and a trace of argon). When sunlight strikes methane, it can cause the formation of large organic molecules. If this is happening on Titan, these molecules must drift down to the surface. Radar bounced from its surface varies in intensity as time passes, indicating that the surface is not uniform and that therefore the moon is probably not completely covered by an ocean of organic soup. Perhaps it has lakes or large seas of organic material. This raises the question of whether life might have formed from these organic molecules.

Titan's diameter is 5150 km (3200 miles), making it the second largest moon in the solar system, after Ganymede.

Titan is only slightly larger than Mercury, which has no appreciable atmosphere. Titan, however, has an atmosphere more dense than Earth's. How can this be? Recall that an object can retain an atmosphere only if the escape velocity

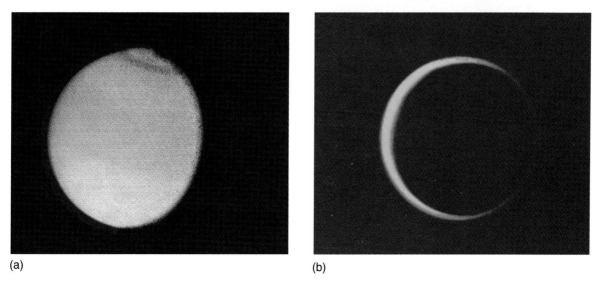

(a)   (b)

▶ **FIGURE 9-23.** Titan. (a) The difference in color between the northern and southern hemispheres results from differences in cloud thicknesses between the two atmospheres. (b) This is a photo of the night side of the moon and shows that Titan has an extensive atmosphere.

from its surface is greater than the speed of gaseous molecules there. The escape velocity from Titan is somewhat less than from Mercury (Figure 7-10). The difference in temperature between the two objects allows Titan to retain an atmosphere, however. On Mercury's Sun side, we find temperatures up to 450°C, but the temperature of Titan's surface is only about −220°C. Thus the molecular speed of the gas molecules on Titan is less than the escape velocity from the moon, and it has retained its atmosphere.

Planetary Rings

Notice in Figure 9-18 that the rings of Saturn are not visible when they are edge-on to us. This is because they are very thin; most of them are less than 100 meters across. The rings, however, are not solid sheets but are made up of small particles of water ice or perhaps rocky particles coated with ice. The great amount of empty space between the particles in the rings means that if they were compressed, their thickness would be reduced to less than a meter.

Each of the particles that make up the rings of Saturn revolves around the planet according to Kepler's laws. Thus, particles nearer the planet move faster than those farther out. Each of the particles is, in a sense, a separate satellite of Saturn.

Three distinct rings are visible from Earth, and long before the advent of space flight they were named (from outer to inner) rings A, B, and C. Photographs from space indicate that there are many more rings than this, as is evident in ▶Figure 9-24. A number of reasons have been given for the existence of the spaces between the rings, and it is now evident that different spaces have different causes. Some explanations are quite complicated, but others are much simpler.

One of the spaces that is easily explained is the one between rings A and B, known as *Cassini's division* after the French astronomer G. D. Cassini. To explain this space between rings, we must compare the motion of a particle in the Cassini division to the motion of Mimas, one of Saturn's moons. A particle in Cassini's division would orbit the planet with a period of 11 hours, 17½ minutes. This is just half of the orbital period of Mimas. Thus any particle found in the division will be at the same place in its orbit each time Mimas passes near it. Each time

In proportion to their diameter, Saturn's rings are much flatter than a compact disk.

▶ FIGURE 9-24. *Voyager* photos revealed that the rings of Saturn are very complex.

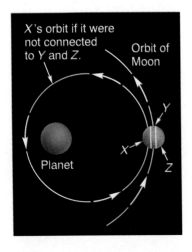

▶ FIGURE 9-25. The moon is divided into three parts for analysis. Part X is closer to the planet than the rest of the moon and would have to move faster if it were a separate satellite. If there were no attractive force toward the rest of the planet, at X's present speed, it would fall inward and begin an eccentric orbit.

this happens, Mimas exerts a slight gravitational tug on the particle, deflecting it slightly from its path. If Mimas did this at random times during the orbiting of the particle, the gravitational effect would be random, and the particle's path would not be severely affected. Because of the synchronous relationship between the periods, however, the tug is repeated regularly, and the overall effect is to pull the particle out of its orbit. This is what is happening to particles that drift into Cassini's division, and it is the reason the division exists.

If such synchronous gravitational tugs were the only mechanisms at work determining the structure of Saturn's rings, Cassini's division might be completely clear of particles, but in fact it is not. Many other forces are at work, including gravitational forces from other moons and particles as well as electromagnetic forces from Saturn's magnetic field.

Other features of the rings are explained by the existence of small moons orbiting near the rings. These satellites, called *shepherd moons,* cause some of the rings to keep their shape. Shepherd moons are discussed in a Close Up in this chapter. Still other features of the rings seem to be similar to spiral wave patterns seen in galaxies. Various hypotheses have been used to explain the various features of the rings, but no comprehensive theory has yet been proposed.

The Origin of Rings

The origin of the rings of Saturn is not well understood. The most likely scenario is that an icy moon (or moons) once orbited near Saturn and was shattered in a collision with a passing asteroid or comet. The particles of the shattered moon eventually dispersed in a ring around the planet. This latter event may seem unlikely, but in fact we know that small objects in orbit around a planet will arrange themselves in a flat ring. Whether or not this is the true origin of the rings, we do understand why the rings have not formed into moons. To see why, we must review the effect of tides.

Recall how the gravitational force between the Earth and the Moon causes tides on both the Earth and the Moon. It is not simply the existence of a gravitational force that causes the tides, however, but the fact that there is a *difference* in the amount of gravitational force exerted on masses that are at different places on the Earth (or the Moon). We considered tides in Chapter 6, but we will now look at them in another way, focusing here on Kepler's laws.

According to Kepler's third law—expanded to apply to planets and moons—if two objects are orbiting a planet, the object closer to the planet must move faster in its orbit if it is to stay in that orbit. ▶Figure 9-25 illustrates a moon in

♄ SATURN
DATA PAGE

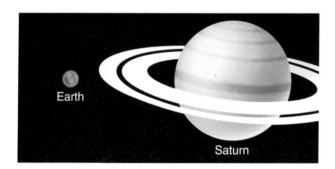

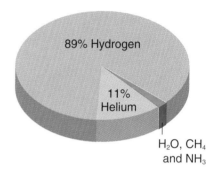

Atmospheric Gases

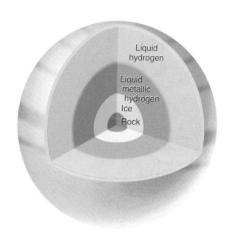

Interior

Saturn	Value	Compared to Earth
Equatorial diameter*	120,540 km	9.5
Oblateness	0.102	
Mass	5.684×10^{26} kg	95.1
Density	0.69 gm/cm^3	0.12
Surface gravity*		0.916
Escape velocity	35.6 km/s	3.2
Sidereal rotation period	10.65 hours	0.43
Albedo	0.34	1.1
Tilt of equator to orbital plane	26.73°	1.1
Orbit		
Semimajor axis	1.427×10^9 km	9.54
Eccentricity	0.0556	3.3
Inclination to ecliptic	2.49°	
Sidereal period	29.46 years	29.46
Moons	18	

*Measured where the pressure is one atmosphere—Earth's pressure at sea level

QUICK FACTS

The ringed planet. Diameter is 85% of Jupiter's, but density is only half. Primarily hydrogen and helium, like Jupiter. Magnetic field 1/20 as strong as Jupiter's. Emits more energy than it receives—not fully explained. Large family of moons. Rings have detailed features and, seen from Earth, change greatly during its year. Spaceship *Cassini* will start sending back data from Saturn in 2004.

a circular orbit around a planet, but in the figure we have divided the moon into three parts, each at a different distance from the planet. Kepler's law tells us that if the three parts of our moon were not connected, part X would have to move faster. If it were to move at the same speed as part Y, it would fall inward toward the planet. Likewise, part Z should travel more slowly than Y, and if it were to travel at the same speed as Y, it would fly out of Y's orbit. This, in fact, is just another way to look at the cause of tides, for each part of the moon tends to do just what we have described.

Why doesn't a moon fly apart because of this tidal effect? The answer, of course, is that the moon has its own forces holding it together. A rock on its surface is pulled by the force of gravity toward the center of the moon. Artificial satellites, on the other hand, are held together by the strength of the materials of which they are made.

It is important to see that it is the *relative difference* in distance from the planet that causes the effect. If a moon is far from its planet, the outside edge may be only a fraction of a percent farther from the planet than the center of the moon is. If the same moon is located close to its planet, its outer edge will be a greater fraction of the distance from the planet than its center is. This causes the effect of tides to be greater if a moon is nearer a planet. There is a critical distance, called the **Roche limit,** inside which the tidal force on a moon will be greater than the moon's own gravitational force. Inside this distance a large moon will be unable to hold itself together.

Roche limit. The minimum radius at which a satellite (held together by gravitational forces) may orbit without being broken apart by tidal forces.

Artificial satellites are held together by the strength of their materials, so the Roche limit does not apply to them.

Now we come to Saturn. Until 1859 it was thought that Saturn's rings might be solid sheets of material, but calculations of Saturn's Roche limit showed that they are inside that limit and therefore the gravitational force between particles within the rings is less than the tidal force that tends to pull them apart. That they are made up of separate particles was experimentally confirmed in 1895 by means of the Doppler effect, which showed that different parts of the rings have different speeds. Because the rings are inside the Roche limit, no moons have formed from the particles.

The particles of the rings vary in size from small grains to irregularly shaped pieces more than a meter across. It is interesting to note that if all of the material of the rings of Saturn were formed into a single moon, the moon would be about the mass of Janus (one of the smallest of Saturn's moons) and only $1/_{20,000}$ the mass of the Earth's Moon.

URANUS

Uranus (YOOR-uh-nuss) was unknown to the ancients even though they undoubtedly saw it, for it is barely visible to the naked eye under perfect viewing conditions. It was plotted on star charts made by telescope as early as 1690. In anything but large, high-resolution telescopes, however, Uranus appears only as a speck of light rather than a disk. This, coupled with the fact that it moves very slowly, caused it to go unnoticed as a planet until 1781 when the English astronomer William Herschel (1738–1822) noticed that this particular "star" seemed to have a size. He thought at first that it was a comet, but after calculations showed that the planet's orbit was nearly circular, Herschel realized that he had discovered a new planet.

▶ **FIGURE 9-26.** Uranus as seen from Earth, with three of its moons.

Herschel proposed that the new planet be named "Georgium Sidus" (the Georgian star) after King George III, but the name didn't take. Johann Bode (of Bode's law fame) suggested "Uranus," after the mythical Greek god of the sky.

The size of Uranus is difficult to determine from the Earth because its disk is so indistinct in a telescope (▶Figure 9-26). Recall that the size of an astronomical object can be determined from its distance and angular size. If the image is indistinct, the angular size cannot be determined accurately, of course. The first reliable value for Uranus's diameter came from a telescope aboard a high-altitude balloon operated by Princeton University. The balloon lifted this telescope to an altitude of 15 miles, above most of the distortion caused by the Earth's atmosphere.

⊕ URANUS
DATA PAGE

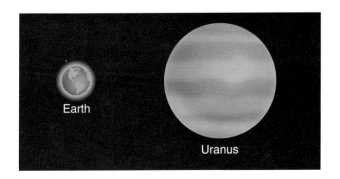

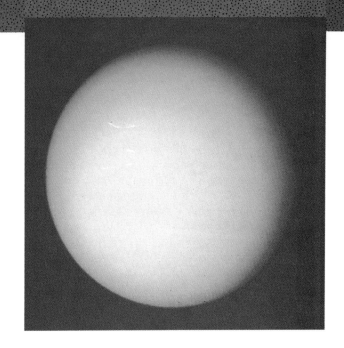

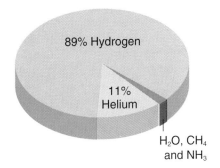

Atmospheric Gases

Uranus	Value	Compared to Earth
Equatorial diameter	51,000 km	4.0
Oblateness	0.024	
Mass	8.698×10^{25} kg	14.5
Density	1.2 gm/cm^3	0.23
Surface gravity		0.93
Escape velocity	21.1 km/s	1.9
Sidereal rotation period	17.2 hours	
Albedo	0.34	1.1
Tilt of equator to orbital plane	97.86°	
Orbit		
Semimajor axis	2.870×10^9 km	19.19
Eccentricity	0.0472	2.8
Inclination to ecliptic	0.77°	
Sidereal period	84.01 years	84.01
Moons	15	

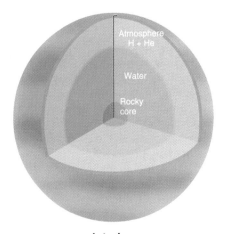

Interior

QUICK FACTS

Four times Earth's diameter, but only one-third of Jupiter's. Composition like other Jovian planets. Almost featureless atmosphere. Far greater tilt of axis than any other planet. Retrograde rotation. Uniform temperature. Many small moons, no large one.

HISTORICAL NOTE

William Herschel, Musician/Astronomer

William Herschel, who has been called the greatest observational astronomer ever and the Father of Stellar Astronomy, was a musician. At least that was his training and how he earned his living for a great part of his life. His father Isaac was an oboist in the band of the Hanoverian Foot Guards. (Hanover was a province in the former state of Prussia in Germany.) Armies at that time were accompanied by a band, whose members did not participate in the actual fighting but provided fanfares and spirit. Friederich Wilhelm Herschel, who adopted the name William when he moved to England, was one of Isaac's five children. All of the children were raised as musicians, and William learned to play the violin as soon as he could hold the small one his father obtained for him. When he was 14, he joined his father and older brother in the military band.

The semimilitary life was not for him, however, so he and his brother requested a discharge, and just after William turned 19, they moved to England together. They were almost penniless and without a good command of the English language, but their musical abilities allowed them to get work copying music and tutoring. William served for a brief time as instructor to another military band but spent most of his young adult life teaching, performing, and composing music. Herschel's notes refer to a number of symphonies he composed, but little of his music remains today. It attracted attention in England at the time, however, and Herschel was able to join in the fashionable society of London and Edinburgh.

One of the interests that William inherited from his father was the study of astronomy. He began to build his own telescopes, and after teaching as many as eight music lessons a day, he would spend the evenings observing the heavens. By 1778 he had built an excellent reflecting telescope about six inches in diameter. This was no easy task, for construction of the mirror involved metal casting.

Herschel's life changed after March 13, 1781. On that night he saw a hazy patch in the sky and upon observing it over a few nights, he saw that it moved. What he thought was a comet turned out to be the first planet to be discovered in recorded history.

William Herschel was not immediately accepted by the scientific community. His telescopes were of such quality that he was able to see things in the heavens that other astronomers couldn't, and many did not believe his claims about the details of what he was seeing. There seems to have been some professional jealousy toward this musician who, in his writings on astronomy, did not use the jargon accepted by the profession. Herschel's skills as a telescope maker and as an observer won out, however, and his discovery of Uranus could not be denied. He was soon recognized and celebrated by scientific societies. Within a year after he discovered Uranus, he received a pension from King George III so that he could devote full time to astronomy, his former hobby.

While he was still devoting most of his time to music, Herschel had brought his sister Caroline to live with him. (She served as a vocalist at some of his performances.) Now that he was an astronomer, she became a very important assistant and an accomplished astronomer herself. Together they made the first thorough study of stars and nebulae (faint, hazy objects not understood at the time). One important discovery they made was the existence of double stars—stars in orbit about one another. After Herschel's death, the orbits of these stars were analyzed in enough detail to determine that the stars obey Newton's laws of motion and gravitation, an indication that these laws are valid not only here in our solar system but even among the distant stars.

Herschel continued serious observing until he was nearly 70. After his death at the age of 84, Caroline compiled a large catalog of his observations and received many honors for her own astronomical knowledge.

William Herschel did not marry until the age of 49. Four years later he had a son, John, who himself became a noted astronomer.

---

occultation. The passing of one astronomical object in front of another.

51,000 km = 32,000 mi. This is four times Earth's diameter.

An improved determination of the diameter of Uranus was made in 1977 when Uranus was observed passing in front of a star. Such an event, when a moon or planet eclipses a star, is called an **occultation** and provides us with a valuable measuring method. Since the speed of Uranus in its orbit was known, all that had to be measured during the occultation in order to calculate Uranus's diameter was the time during which the star was occulted.

The diameter of Uranus is 51,000 kilometers, about half the diameter of Saturn and one-third that of Jupiter. Long before *Voyager 2* passed Uranus in

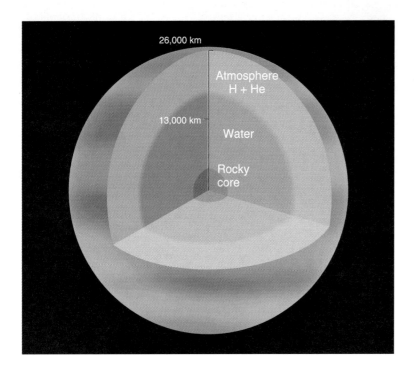

► FIGURE 9-27. A model of Uranus's interior.

1986, the mass of the planet had been calculated by applying Kepler's third law to its five known moons. When its density is then calculated, we find that Uranus has a density of 1.2 times that of water, greater than Saturn but slightly less than Jupiter. Taking into account the surface gravity on Uranus (which is less than on Saturn or Jupiter), we can calculate what Uranus's density would be if the planet were made up entirely of hydrogen and helium. Astronomers once thought that a dense rocky core about the size of the Earth was necessary to produce Uranus's measured density, but recently researchers have subjected material that matches the composition of Uranus to a pressure two million times that of Earth's atmosphere and found that it becomes dense enough to account for the planet's density. Thus it may be that Uranus has a very small rocky core or no core at all. ►Figure 9-27 shows a theoretical model of Uranus's interior with a small rocky core.

The atmosphere of Uranus is similar to those of Jupiter and Saturn, consisting primarily of hydrogen and helium, with some methane. ►Figure 9-28 is a *Voyager* photo of Uranus. The reason for the planet's blue coloring is that the methane absorbs red light and reflects the rest of the spectrum. The atmospheres of Jupiter and Saturn also contain methane, but they have high cloud layers that reflect sunlight before it reaches the higher concentrations of methane. Uranus does not have these cloud layers.

The purpose of the Uranus observations during the occultation of 1977 was to determine the planet's size more accurately. A surprise was in store for the observers, however. A short while before Uranus was expected to occult the star, the light from the star blinked five times. Then after the occultation, similarly timed blinks were observed but in the opposite order. From this, astronomers concluded that Uranus has at least five rings that are invisible from the Earth but are dense enough to obscure the light from a star behind them. By comparing the results obtained by a number of observatories, details of the rings' orientations and sizes could be determined. Use of the same method during an occultation in 1978 revealed a total of nine rings. ►Figure 9-29, taken by *Voyager*, indicates the complexity of the ring system.

The rings of Uranus cannot be seen from the Earth because they reflect only about 5% of the sunlight that hits them. The rings must be made of material as

Uranus

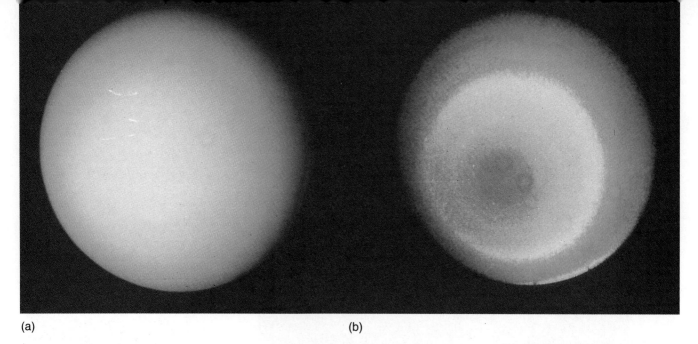

➤ **FIGURE 9-28.** (a) Images of Uranus taken by *Voyager 2* under ordinary light show a featureless planet. (b) When color is enhanced by computer processing techniques, Uranus is seen to have zonal flow patterns in its atmosphere. One of Uranus's poles is at the center of the red spot.

➤ **FIGURE 9-29.** *Voyager's* photograph of the rings revealed a great number of ringlets. This is a backlit view. That is, the camera is on the opposite side of the rings from the Sun, so that the light we see has been scattered in passing through the rings.

dark as soot. By comparison, Saturn's rings reflect 80% of the light that hits them.

Uranus's Orientation and Motion

Recall that Venus is unusual in that it rotates in a backward direction: clockwise as observed from the north. Except for Venus, every other planet we have studied rotates normally: counterclockwise as observed from the north. In addition, all planets we have studied, including Venus, have their equatorial plane tilted less than 30 degrees to their plane of revolution around the Sun. Here is where Uranus is unique. Its equatorial plane is tilted nearly 90 degrees to its plane of

The tilt of Uranus's orbit is listed as 98°, indicating its retrograde rotation. See ➤Figure 9-31.

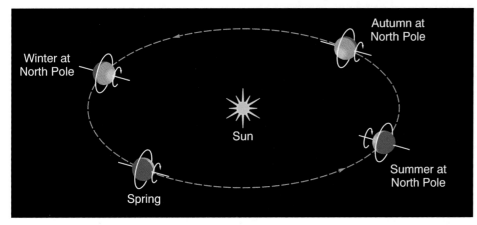

► **FIGURE 9-30.** Uranus's axis is tilted so that its poles point nearly directly at the Sun at times.

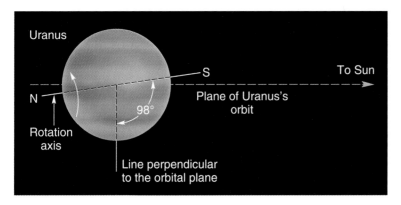

► **FIGURE 9-31.** Uranus's axis is tilted 98°, which tells us that the planet has retrograde rotation.

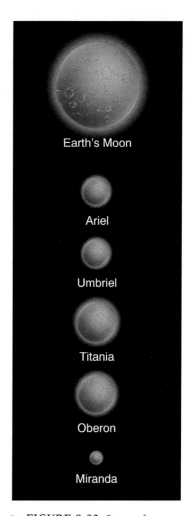

► **FIGURE 9-32.** Some of Uranus's moons compared in size to Earth's.

revolution. This means that during its 84-year orbit, its north pole at one time points almost directly to the Sun and at another time faces nearly away from the Sun. ►Figure 9-30 illustrates this. Remember that the tilt of a planet's axis is what causes the planet to have seasons. One might think that such a large tilt of Uranus's axis would have a great effect on its weather. Data from *Voyager* reveal that Uranus has a fairly uniform temperature over its entire surface (about −200°C), however, indicating that the atmosphere is continually stirred up, with winds moving from one hemisphere to another.

Features of Uranus's surface are not visible from the Earth, so until *Voyager* approached Uranus, there was uncertainty about its period of rotation. *Voyager* photographs told us that, as on Jupiter and Saturn, there are various bands of clouds on Uranus and that these bands rotate with different periods—from 16 hours at the equator to nearly 28 hours at the poles. Uranus has about the same chemical makeup as Jupiter, about three-fourths hydrogen with most of the remainder helium.

Uranus's magnetic field is comparable to Saturn's. Probably it originates in electric currents within the planet's layer of water. The magnetic field is unusual in that its axis is tilted 55 degrees with respect to the planet's rotation axis. No other planet has such a large angle between the two axes, although Neptune's— at 50 degrees—is close.

Five moons of Uranus (►Figure 9-32) were known before *Voyager*, and now we know of 10 more. All are low-density, icy worlds. The innermost, Miranda, is perhaps the strangest looking object in the solar system. ►Figure 9-33 shows

On Earth, there is little air exchange between the Northern and Southern Hemispheres.

The corresponding angle for Earth is 12°.

Uranus 293

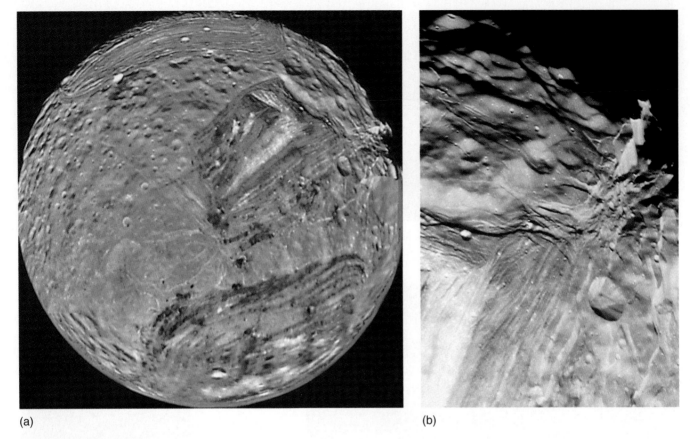

(a) (b)

▶ **FIGURE 9-33.** Miranda. It is hypothesized that an impact once tore Miranda apart. (a) The V-shaped feature is called the Chevron. (b) Detail of the upper right portion of part (a) shows a cliff that is 5 kilometers high.

The *Voyager* spacecraft were originally designed to operate for only five years. *Voyager 2*, launched in 1979, is now expected to communicate with us until 2010. Way to go, NASA!

its varied terrain. You may wonder who has been farming it. Because of its racetrack-like grooves, one feature has been named Circus Maximus after the arena of ancient Rome. One possible explanation for the strange features of Miranda is that it experienced a tremendous collision with another object and was broken into pieces that later fell back together.

Two of Uranus's moons are shepherd moons (see the Close Up), keeping one of the rings in formation. Material in the rings is very sparse, however, so much so that there is more material in Cassini's division of Saturn's system than in all of the rings of Uranus! Nine rings had been observed (by occultations) from the Earth; only one more was found by *Voyager*. The 10 rings, however, contain hundreds of smaller ringlets, as seen in Figure 9-29.

NEPTUNE

Uranus and Neptune are similar in many ways. The first similarity is size; they are very nearly the same. Uranus appears as only a small, indistinct disk in a telescope, and when we consider that Neptune is half again as far away from Earth, it is no surprise that Neptune appears even more indistinct from the Earth, even in the largest telescope. Most of what we know about Neptune was learned when *Voyager 2* passed it in August 1989 and sent back more than 9000 images as well as other data.

The composition of Neptune matches that of Uranus, and its blue color (▶Figure 9-34) has the same cause—methane in its upper atmosphere. Light penetrates deeper into Neptune's atmosphere, however, resulting in a deeper

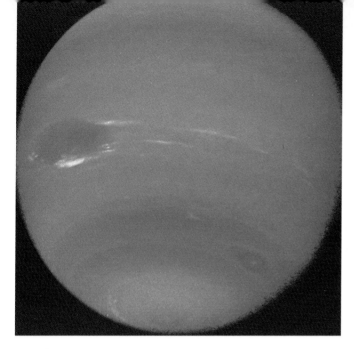

► **FIGURE 9-34.** Neptune was photographed by *Voyager 2* in August 1989. The photographs revealed the previously undiscovered cloud features, including the Great Dark Spot at left center.

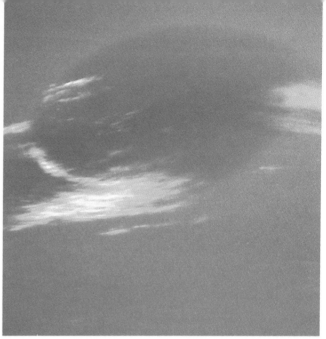

► **FIGURE 9-35.** The Great Dark Spot of Neptune. The narrow clouds are made up of crystals of methane.

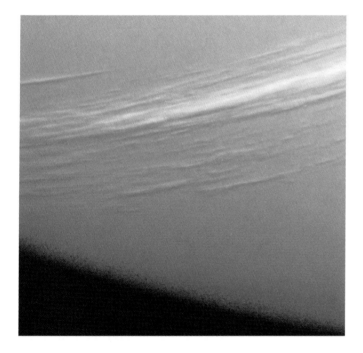

► **FIGURE 9-36.** The methane clouds are at a higher altitude than the surrounding atmosphere, as indicated by the shadows cast on the other side from the Sun.

blue than Uranus. The other major difference in appearance is that while Uranus is almost featureless, Neptune exhibits weather patterns. It has parallel bands around it, and its *dark spot* (►Figure 9-35) is similar in appearance to Jupiter's red spot. Like the red spot, it is in the planet's southern hemisphere and rotates counterclockwise. The dark spot is about the same size relative to Neptune as the Great Red Spot is relative to Jupiter.

Neptune radiates more internal heat than Uranus, although the cause of the heat is not understood. This heat, however, is what drives the weather on Neptune, stirring the liquids and gases of the planet so that winds reach speeds of 700 miles per hour.

The wispy white clouds seen on Neptune are thought to be crystals of methane. ►Figure 9-36 shows that the clouds are higher than the surrounding at-

Neptune's dark spot is approximately Earth-size.

Neptune receives only $1/900$ as much solar energy per square meter as Earth does.

CLOSE UP

Shepherd Moons

The 1977 occultation that resulted in the discovery of Uranus's rings showed that they are very narrow and that their edges are well defined. Soon after this discovery, two astronomers—Peter Goldreich and Scott Tremaine—proposed a mechanism to explain why a ring might stay in a narrow band rather than spreading out over space.

They hypothesized that there are a pair of moons, one orbiting just inside and another just outside the narrow ring (Figure C9-2). Kepler's third law tells us that the moon closer to the planet moves faster than the particles of the ring. As that moon catches and passes the particles on the inside portion of the ring, gravitational pull from it increases the particles' energy just a little, causing them to move outward in their orbit. (This energy comes from the moon, and therefore the moon loses energy and moves inward a bit. The moon is so much more massive than the particles, however, that the change in its path is almost insignificant.)

In an opposite manner, as particles on the outside edge of the ring pass the more distant moon, they lose energy and move inward. The effect on each particle is very small, much less than is indicated by the figure. Nevertheless, Goldreich and Tremaine showed that the effect is significant enough that as the moons continue to orbit near the rings, they force the particles together; in other words, the moons act as shepherds for the flock of particles in the ring.

The scientific community awaited *Voyager*'s passing by the rings of Uranus to see if it found such moons. The answer came earlier than expected. *Voyager 2* at that time had not yet passed Saturn. When it did, it photographed Saturn's rings, and on each side of Saturn's narrow F ring was a moon (Figure C9-3). Calculations showed that the two moons near the F ring would indeed perform the function of shepherds for that ring.

In January 1986, *Voyager 2* passed Uranus and returned a remarkable photograph (Figure C9-4) of the epsilon ring of Uranus along with two shepherds. Other rings around Uranus exist without the aid of shepherd moons, so although Goldreich and Tremaine were right in their prediction of the action of shepherd moons, this mechanism is only a part of what is happening in planetary ring systems.

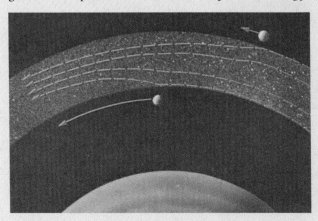

FIGURE C9-2. As the inner moon passes the inner portion of the ring, it forces ring particles outward. The outer moon forces particles inward. The effect is greatly exaggerated here.

---

mosphere, for they cast shadows. One of the clouds (Figure 9-35) seems to be permanently associated with the Great Dark Spot and is probably caused by methane rising with other gases in the spot until it freezes, becoming visible as a white cloud.

The rotation rate of Neptune was uncertain until *Voyager* measured it. Neptune exhibits the most extreme differential rotation of all the planets: near the equator it rotates in 18 hours, but near the poles its rotation period is only about 12 hours. These great differences are confined to the upper few percent of the atmosphere, however, Neptune's magnetic field rotates with a period of 16 hours, 3 minutes. Since the magnetic field is thought to result from electric currents in liquid layers deeper within the planet, this is taken as Neptune's basic rotation rate.

The temperature of Neptune's surface is remarkably uniform at the poles and the equator, −216°C. This is another similarity to Uranus and was a surprise to astronomers, for Neptune's axis is tilted less than 30 degrees to its orbit rather

Some features on Neptune have periods as long as 18 hours.

Inside the Roche Limit?

The shepherd moons of Saturn's F ring and those orbiting Uranus are within the Roche limit of each planet. How can they exist there without being pulled apart by gravitational forces? There are a number of answers to this question. First, it is an oversimplification to refer to a single Roche limit. For a smaller moon, the limit is closer to the planet than it is for a larger moon. The shepherds are small moons. In addition, the position of the limit depends upon the density of the moon. But most important, the Roche limit concerns only moons that are held together by gravitational forces. A single large rock can exist inside the Roche limit, for it is held together by cohesive forces within the material. Each shepherd moon may well be a single large rock rather than being made of smaller particles held together by gravity.

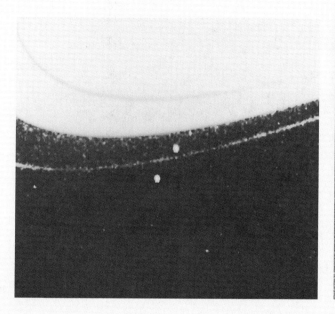

➤ FIGURE C9-3. *Voyager* photographed these shepherd moons near the F ring of Saturn.

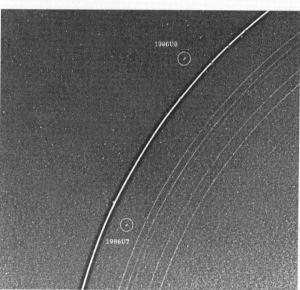

➤ FIGURE C9-4. Moons 1987U7 and 1987U8 were photographed shepherding a ring of Uranus.

than the 98-degrees tilt of Uranus. The reasons for many of the similarities and differences between these two planets remain unknown.

Neptune is presumed to have an interior much like that of Uranus, but its central rocky core is probably larger, thus resulting in its greater average density (see table on Neptune data page). ➤ Figure 9-37 illustrates one model of Neptune's interior.

Neptune's Moons and Rings

Before *Voyager's* visit, Neptune was known to have two moons, both unusual. Triton, its largest moon (and the seventh largest in the solar system), revolves around the planet in a clockwise direction as seen from the north. Although a number of smaller moons of Jovian planets move in retrograde orbits, Triton is the only major moon to do so. The other moon known prior to *Voyager*, Nereid, orbits in the "right" direction, but it has the most eccentric orbit of any known

Jupiter's Galilean moons, our Moon, and Saturn's Titan are larger then Triton. All seven of these are larger than Pluto.

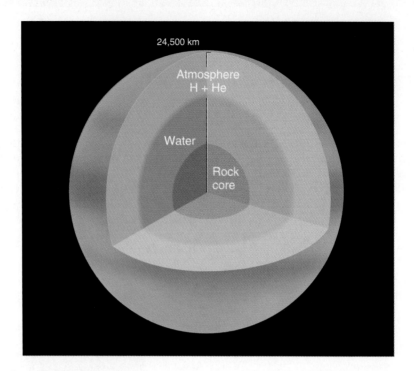

► **FIGURE 9-37.** Neptune's interior is probably similar to that of Uranus (Figure 9-27), although Neptune may have a larger core.

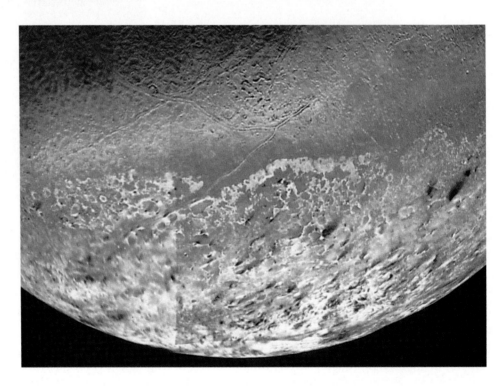

► **FIGURE 9-38.** The southern part of Triton has been exposed to sunlight for some 30 years, and erupting volcanoes (not obvious here) were photographed by *Voyager*.

moon in the solar system, eccentric enough that its distance from the planet is five times greater at some times than at others. Its 360-day orbital period is the longest of any moon of its size (340-kilometer diameter) in the solar system.

Voyager showed us that the surface of Triton is as unusual as its orbit. ► Figure 9-38 shows the region surrounding its south pole, which is near the bottom of the photograph in an area of very irregular terrain. Other parts of the moon are much smoother, as seen in the top part of the photograph. The surface consists primarily of water ice, with some nitrogen and methane frost. Three observations lead us to believe that the surface is very young: (1) It is very light in color.

NEPTUNE
DATA PAGE

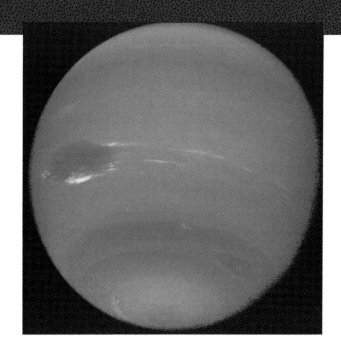

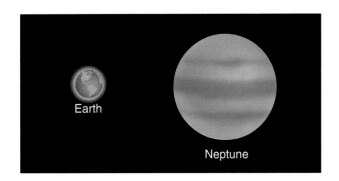

Neptune	Value	Compared to Earth
Equatorial diameter	49,500 km	3.9
Oblateness	0.027	
Mass	1.03×10^{26} kg	17.2
Density	1.76 gm/cm^3	.31
Surface gravity		1.22
Escape velocity	24.6 km/s	2.2
Sidereal rotation period	16.05 hours	
Albedo	0.29	0.94
Tilt of equator to orbital plane	28.8°	1.3
Orbit		
Semimajor axis	4.497×10^9 km	30.06
Eccentricity	0.0086	0.5
Inclination to ecliptic	1.77°	
Sidereal period	164.79 years	164.79
Moons	8	

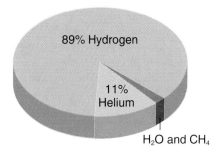

Atmospheric Gases

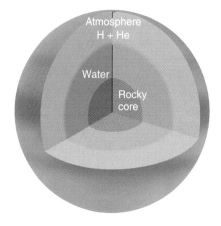

Interior

QUICK FACTS

Blue surface due to methane. Jupiter-like atmospheric features, including Great Dark Spot. Composition is thought to be similar to other Jovian planets. Unexplained internal heat source. Extremely differential rotation. Largest moon, Triton, has retrograde revolution. The other major moon, Nereid, has the most eccentric orbit of any moon in the solar system. Lumpy rings.

HISTORICAL NOTE

The Discovery of Neptune

The discovery of Neptune is especially interesting because it reveals some very human aspects of science.

After Uranus was discovered in 1781, astronomers examined old star charts and found that Uranus had been plotted on charts as far back as 1690. Those who had plotted it had mistaken it for just another star, but the positions they marked allowed later astronomers to calculate the orbit of the planet. As time went on, however, it was clear that Uranus was not following the orbit predicted by Newton's laws, even with the gravitational effects of Jupiter and Saturn taken into account. Most astronomers thought that an unknown planet was disturbing Uranus's orbit, but others thought that perhaps the law of gravity acts differently at great distances from the Sun.

John C. Adams, a 26-year-old British mathematician, took up the problem of calculating the position of the hypothesized planet. In late September 1845, after two years of calculations, he took his results to Sir George Airy, the British Astronomer Royal, telling him that the planet would be found near a certain position in the constellation Aquarius. Adams first had problems contacting Airy. He must have felt like a student trying to find his adviser, for the first time he tried to deliver his results in person Airy was out of the country. Another time, Airy was out of his office. A third time, he was at dinner and his butler would not let him be disturbed. Adams left messages but they never reached Airy.

Airy was one of the astronomers who doubted the existence of the new planet, thinking that Newton's laws were inexact at the distance of Uranus. When Airy finally got one of Adams's messages, he therefore refused to use valuable time on the large telescope for what he thought would be a fruitless search. Adams's youth and inexperience also seem to have contributed to Airy's skepticism.

During the same period, the French mathematician Urbain Le Verrier was making similar calculations of Uranus's orbit. In November 1845 Le Verrier published the first part of his calculations in a scientific journal. Then in June 1846 he published his final results. Although neither knew of the other's work, Adams's and Le Verrier's predicted positions for the planet were within one degree of one another.

Similar to Adams's experience with his own countrymen, Le Verrier was unable to get French astronomers to search for the planet. Finally, on September 23, 1846—one year after Adams had first requested a search—Le Verrier asked Johann Galle in Berlin to look for the planet. Galle quickly convinced the director of the observatory where he worked of the worthiness of the project, and the new planet was discovered on the very first night of searching.

Credit for scientific discovery goes to the first person who documents his or her results. (Today such documentation normally takes the form of publication in a scientific journal.) There is little doubt, however, that if Airy had used Adams's calculations to find Neptune, the prediction would be credited to Adams. For years, controversy swirled about the priority of prediction. Recall from your study of history that England and France are historical rivals. So in this case, not only was the reputation of the two mathematicians involved but also national prestige. (Similar disagreements have arisen in this century concerning scientific discoveries made in the United States and the former Soviet Union.) Today the two men are given joint credit, and we find the story of the discovery interesting enough to retell it in nearly every astronomy text.

This makes Triton the third object, after Earth and Io, that we know to be volcanically active.

When methane is exposed to sunlight for a long time, darker organic compounds are formed. (2) There are very few craters, indicating that they have been obliterated by surface activity. (3) Active volcanoes have been seen on the moon.

A current hypothesis explains many of the unusual properties of Triton and Nereid. The mass of Triton was determined by its gravitational influence on *Voyager*, and we find that its density (1.76 gm/cm^3) is about the same as the density of Pluto. It could be that both Triton and Nereid are objects that were captured by Neptune after the initial formation of the solar system. If so, Triton probably once had an eccentric orbit like Nereid. If its orbit came very close to Neptune, tidal forces would eventually cause it to settle into a circular orbit such as the one in which we find it. In addition, the tidal forces would heat the interior of Triton by continually distorting its shape. This heat would result in the volcanic action that has covered old craters. Much more remains to be learned about Neptune's unusual moons before we can be confident of this hypothesis.

Data on Neptune's moons can be found in Appendix D.

► **FIGURE 9-39.** Neptune's rings are seen clearly in this photo. Note their "lumpiness."

Voyager discovered six more moons orbiting Neptune, one of them larger than Nereid. It had not been seen from Earth because it is very close to the planet. The other five moons are very small.

Stellar occultations observed in 1984 had revealed that Neptune has rings, but the rings were thought to be incomplete, not circling the entire planet. *Voyager* photographs revealed that they are indeed complete rings (►Figure 9-39), but that they are "lumpy," perhaps as a result of undiscovered moons orbiting with them. Thus we find that each of the Jovian planets has a ring system.

CONCLUSION

The Jovian planets certainly are different from the worlds of the inner solar system. Compared to the terrestrial planets, the Jovians are larger, more massive, more fluid (rather than solid), less dense, and rotate faster. Their many moons are vastly different from one another; the differences are caused at least partly by their distances from the Sun and from their planet.

As we learn more about the Jovian planets, we are even more amazed at these beautiful giants. We see great similarities between them, but we also see great differences. Jupiter, Saturn, Uranus, and Neptune are similar enough that we classify them into a single group, but each is an individual world with its own history.

One more planet is yet to be discussed—Pluto. This planet is such an enigma that it has been placed in the next chapter, along with various small objects that are part of our solar system.

There are many mysteries remaining in the solar system. These mysteries are what make its study so interesting. If we were to run out of mysteries, science would cease to exist, and we would have lost one of the pursuits that makes life so exciting.

RECALL QUESTIONS

1. The most abundant constituent of Jupiter's atmosphere is
   A. helium.
   B. carbon dioxide.
   C. ammonia.
   D. hydrogen.
   E. sulfuric acid.

2. The great red spot is
   A. a continent.
   B. a storm.
   C. an optical illusion.
   D. a shadow of one of Jupiter's moons on its surface.
   E. a mountain protruding above Jupiter's atmosphere.

3. The chemical composition of Jupiter is most similar to which of the following?
   A. The Earth
   B. The Sun
   C. Mars
   D. The Moon
   E. Venus

4. A planet is said to have differential rotation if
   A. it changes speed when orbiting the Sun.
   B. it changes its rate of spin from time to time.
   C. different parts of it have different rotation periods.
   D. its rotation rate is far different from the planets closest to it.
   E. most of its moons revolve around it in a different direction than it rotates.

5. Jupiter's weather patterns should be simpler than Earth's because
   A. Jupiter is so large.
   B. Jupiter is so massive.
   C. Jupiter is made up primarily of hydrogen.
   D. Jupiter has a small ring.
   E. Jupiter's core is a small fraction of the total planet.

6. The bands around Jupiter are due to the fact that the planet
   A. rotates differentially.
   B. has a fairly low rotation rate.
   C. has so many moons.
   D. is made up primarily of hydrogen.
   E. has no solid core.

7. Most of the extra energy that Jupiter emits is thought to be energy
   A. left over from its formation.
   B. that results from the fact that it is still shrinking.
   C. that results from radioactive materials within it.
   D. that results from the greenhouse effect.
   E. [All of the above are considered about equally responsible.]

8. The energy for the volcanic activity on Io results from
   A. radioactive substances within it.
   B. chemical reactions.
   C. tidal action.
   D. energy radiated from Jupiter and the Sun.
   E. [Both A and B above.]

9. Saturn's rings are
   A. composed of ice particles.
   B. in the plane of the planet's equator.
   C. within the planet's Roche limit.
   D. [All of the above.]

10. Titan is able to retain an atmosphere even though it is just slightly larger than Mercury because it is
    A. close to Saturn.
    B. very dense.
    C. far from Saturn.
    D. far from the Sun.
    E. [Three of the above.]

11. Saturn has rings instead of additional moons because of
    A. magnetic fields.
    B. tidal forces.
    C. electrical forces.
    D. radiation.
    E. the proximity of Jupiter.

12. Saturn is unique in that it
    A. is the only planet with rings.
    B. is the brightest planet in our sky.
    C. is the only Jovian planet with more than four moons.
    D. has the least average density of the planets.
    E. [Three of the above.]

13. Liquid metallic hydrogen
    A. can exist only at great pressures.
    B. exists both in Jupiter and Saturn.
    C. is a conductor of electricity.
    D. [Two of the above.]
    E. [All of the above.]

14. Uranus is unique in that
    A. it has the least average density of the planets.
    B. its poles point nearly toward the Sun at some times.
    C. it is the most distant of the Jovian planets.
    D. its north pole is south of its orbital plane.
    E. it has a more eccentric orbit than any other planet.

15. The fact that Uranus has a fairly uniform temperature over its surface indicates that
    A. it has a great angle between its equatorial plane and its orbital plane.
    B. it has no solid core.
    C. it has a large solid core.
    D. its winds move from one hemisphere to the other.
    E. [Both A and B above.]

16. The oblateness of Jupiter and Saturn is caused primarily by
    A. expansion forces from within.
    B. rotation.
    C. revolution around the Sun.
    D. tidal forces from their moons.
    E. storms in their atmospheres.

17. Observation of the occultation of a star by a planet allows us to determine

A. a planet's diameter.
   B. whether a planet has rings.
   C. the chemical composition of a planet.
   D. [Both A and B above.]
   E. [Both B and C above.]
18. The existence of Neptune was proposed
    A. on the basis of photographs.
    B. because of its effect on Pluto's orbit.
    C. because of its effect on Uranus's orbit.
    D. on the basis of Bode's law.
    E. [None of the above; it was discovered by accident.]
19. Which of the following is a spacecraft that will analyze Jupiter's atmosphere?
    A. *Viking*   D. *Mariner*
    B. *Galileo* E. *Mir*
    C. *Skylab*
20. We say that Jupiter has a rotation rate of 9 hours, 50 minutes. What is it on the planet that has this rotation rate?
21. What is the *Galileo* probe?
22. Starting at the cloud tops and descending downward, describe Jupiter's interior. What leads us to think that Jupiter has a core made up of heavy elements?
23. What causes the swirling between the bands we see in Jupiter's atmosphere?
24. Explain why, as the years go by, Saturn appears so different when viewed from the Earth.
25. Describe the rings of Saturn and explain the Roche limit.
26. Describe two ways the diameter of a planet can be measured from the Earth.
27. Why were the rings of Uranus not observed directly from the Earth? How were they observed?
28. What is unusual about the orientation of Uranus?
29. How do the densities of the Jovian planets compare to the densities of the terrestrial planets?
30. What causes the heating of the inner Galilean moons that results in volcanism? On which moons is this most prevalent?
31. In what ways are Triton and Nereid unusual?

QUESTIONS TO PONDER

1. Discuss the problem of Jupiter's energy balance and list possible explanations.
2. If there have been developments regarding the *Galileo* probe since this writing, look them up and describe them.
3. How does knowing the composition of the surface of a planet allow us to determine its size?
4. Does one pole of every planet point to Polaris? What is meant by the direction *north*? What defines which pole of a planet is its north pole?
5. A series of concentric rings can be seen on Callisto around a crater. Name two other occurrences of this phenomenon in the solar system.
6. How can it be that Titan has an atmosphere, while Mercury (only slightly smaller) has almost none?
7. Use the data in Appendixes C and D to plot the average distances of the moons of Neptune from the planet. Show the radius of Neptune on the scale.
8. In your experience, which cools faster, a cup of coffee or a large pot of coffee? Why does this happen? Relate this to the differences in cooling rates between large and small planets.
9. At the time of this writing, there were plans for a craft to be launched toward Neptune in July 2002. Write a report on any updates to these plans. (The magazines *Sky and Telescope* and *Astronomy* are suggested sources.)
10. At the time of this writing it appeared that scientists would not be able to unfurl the high-gain antenna on the *Galileo* spacecraft and that this would limit the number of photos returned to about 3,000 rather than the 50,000 planned. Report on the status of the *Galileo* mission.

CALCULATIONS

1. If an object has a diameter equal to four times the Earth's diameter, how would its volume compare to the Earth's?
2. If an object has a mass equal to three Earth masses and a volume five times that of the Earth, how would its density compare to the Earth's?
3. Suppose a planet has an angular diameter of 11 arc-seconds when it is 7.5 AU from the Earth. Use the small-angle formula to calculate its diameter.

ACTIVITY

Observing Jupiter and Saturn

Jupiter and Saturn are beautiful objects to observe if you have access to a telescope. Even with binoculars you may be able to see the four Galilean moons of Jupiter and the rings of Saturn. Make a careful sketch of the location of Jupiter's moons and then view them a few hours later or the next night. You will be able to see that they have moved. In addition, see if you can tell that Jupiter and Saturn are not perfectly round. Look for dark bands across the disk of Jupiter.

Table A9-2 shows selected rising times for Jupiter. For dates between those shown, you can find Jupiter as shown in the following example: Suppose you wish to find Jupiter in late September, 2000. The table shows Jupiter rising at midnight in mid August and at sunset in late October. Since late September is about midway between the two dates, Jupiter should rise around 9 or 10 P.M. in late September. When 'noon' is given for the rising time of Jupiter, you can figure that it sets around midnight. Table A9-1 shows some rising and setting times for Saturn.

In each table, midnight refers to standard time rather than daylight saving time.

Although these tables should be of help to you in finding the planets, a better plan would be to pick up an issue of either *Astronomy* or *Sky and Telescope*, two monthly magazines that contain maps and hints for viewing the planets and other celestial objects.

TABLE A9-1 Rising and Setting Times for Saturn, 2000

Date*	Rising or Setting Time
Mid Feb.	Sets at midnight
Late Apr.	Sets in evening twilight
Mid June	Rises in morning twilight
Early Aug.	Rises at midnight
Late Oct.	Rises in evening twilight
Early Dec.	Sets in morning twilight

*The chart can be used for three to four years before or after 2000 by subtracting a week for each year before 2000 and adding a week for each year after. For example, in late July of 1998 Saturn will rise at midnight.

TABLE A9-2 Selected Rising Times of Jupiter

Date		Rising Time
1998:	Late Feb.	Sunrise
	Late June	Midnight
	Mid Sep.	Sunset
	Early Dec.	Noon
1999:	Early Apr.	Sunrise
	Mid July	Midnight
	Late Oct.	Sunset
2000:	Mid Jan.	Noon
	Early May	Sunrise
	Mid Aug.	Midnight
	Late Nov.	Sunset
2001:	Early Mar.	Noon
	Mid June	Sunrise
	Mid Sep.	Midnight
2002:	Early Jan.	Sunset
	Late Apr.	Noon
	Mid July	Sunrise
	Late Oct.	Midnight

StarLinks netQuestions

Visit the netQuestions area of StarLinks (www.jbpub.com/starlinks) to complete exercises on these topics:

1. Jupiter from Voyager I Voyager I was launched in 1977, and along with the Voyager II, relayed to Earth over 30,000 photographs of Jupiter, and much more information about Jupiter and its moons.

2. Neptune and the Great Dark Spot Voyager 2 passed Neptune in August of 1989, and sent to Earth more than 9,000 images and a multitude of data, giving us most of what we know about this planet.

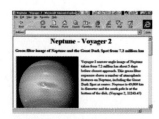

The Hale-Bopp comet as it appeared from Scotland on April 5, 1997. (Tim Schroder took this picture at Gordonstoun, Morayshire, using a 25-second exposure.)

PLUTO AND SOLAR SYSTEM DEBRIS

The Discovery of Pluto
Pluto as Seen from Earth
Pluto and Charon
A Former Moon of Neptune?
Solar System Debris
Asteroids
The Orbits of Asteroids
CLOSE UP: You Can Name an Asteroid
The Origin of Asteroids
Comets
CLOSE UP: Chaos Theory
Comet Orbits—Isaac Newton and Edmund Halley
The Nature of Comets

HISTORICAL NOTE: Astronomer Maria Mitchell, a Nineteenth-Century Feminist
Comet Tails
The Oort Cloud and Kuiper Belt
HISTORICAL NOTE: Jan H. Oort, 1900–1992
The Origin of Short-Period Comets
Meteors and Meteor Showers
Meteors
Meteoroids
Meteor Showers
Meteorites and Craters
CLOSE UP: Hit by a Meteorite?
CLOSE UP: Meteors and Dinosaurs
ACTIVITY: Observing Meteors

.... IN THEIR NOVEL, *Lucifer's Hammer, Larry Niven and Jerry Pournelle write of two characters who do not know one another, but who co-discover a comet. One, Tim Hammer, is rich and owns the most expensive astronomical equipment. The other, Gavin Brown, works with a homemade telescope.*

In real life, on July 22, 1995, two astronomers co-discovered Comet 1995o1, better known as Comet Hale-Bopp. Alan Hale is a professional astronomer who observes known comets about once a week, recording and reporting their changing brightness to the Central Bureau for Astronomical Telegrams in Cambridge, Massachusetts. At the time the comet was discovered, Tom Bopp worked for a construction materials company, and although he is a serious amateur astronomer, he did not own a telescope when the discovery was made. He was at a star party with friends and was using a friend's telescope.

Each astronomer, the professional and the amateur, found a faint, fuzzy object in the sky, an object that was new to him and that did not appear on any sky chart he had. After observing it for a few hours and finding that it was moving among the stars, each man contacted the Central Bureau to report his findings. In fiction, the comet was named Hammer-Brown; in real life it is Hale-Bopp. Strange, but true.

A chapter has been devoted to the four terrestrial planets and another to the four Jovian planets. That leaves only Pluto. This planet was not included among the others because it is so different from them that it fits in neither of the above categories. We will discuss it here before moving on to the debris of the solar system: the asteroids, comets, and meteoroids.

THE DISCOVERY OF PLUTO

The fact that astronomers were able to predict the existence and location of Neptune based on irregularities in the orbit of Uranus led them to try the method once more. (See the Historical Note, "The Discovery of Neptune," in Chapter 9.) Analysis of the orbital data of Uranus led to the conclusion that although the gravitational pull from Neptune accounted for about 98% of the variation from Uranus's expected orbit, there still were unexplained irregularities. A number of astronomers used the data to predict a ninth planet. The one whose work led to success was Percival Lowell, a successful businessman-turned-astronomer. Lowell had built an observatory near Flagstaff, Arizona, in 1894, and in 1905 he made his prediction of the existence of the new planet. He used the Lowell Observatory telescope to search for the disk of "Planet X" until he died in 1916, but he had no success.

This is the same Percival Lowell who thought he had found evidence for life on Mars.

In the 1920s a new photographic telescope, donated by Percival Lowell's brother, was installed at Lowell Observatory. On April 1, 1929, Clyde W. Tombaugh started a new search for the predicted planet. Tombaugh, however, used a different method of searching. Instead of looking for a small, faint disk, he concentrated on looking for the motion a planet must exhibit.

Tombaugh searched for the moving planet with the aid of an instrument called a *blink comparator*. This instrument is essentially a microscope containing a mirror that can be flipped quickly, allowing the observer to look alternately at two different photographs of the sky. The astronomer takes photographs of the same area of the sky a few days apart. The two photographs are then arranged in the comparator so that the stars in each photograph appear at the same place as he or she shifts from viewing one to viewing the other. If a moving object such as a planet is in the photographs, it will appear to jump from one spot to another as the astronomer changes views.

Searching for an object in this manner is very tedious work because the comparator must be scanned slowly over one pair of photographs after another. Tombaugh's search was especially difficult because the predicted position of Planet X was in Gemini, which is near the Milky Way and therefore in a region with a great number of faint stars. As a result, each photograph contained some 300,000 star images. Nevertheless, Tombaugh was successful.

On February 18, 1930, more than 10 months after he began, Tombaugh detected a difference indicated by the arrows in the two photographs in ➤Figure 10-1. Consider that in searching for the planet, Tombaugh had to examine not

➤ **FIGURE 10-1.** These are the two photos on which Tombaugh discovered Pluto. Notice how difficult it is to detect the change in Pluto's position, which is indicated here by the arrows.

just this pair of photographs but numerous pairs. (Of course, he had the blink comparator.)

Tombaugh announced the discovery of the new planet on March 13—the 75th anniversary of Lowell's birth and the 149th anniversary of the discovery of Uranus. It was named Pluto after the mythical Greek god of the underworld.

Pluto was discovered nearly 6 degrees from where Lowell had predicted it, and it soon became clear that its size was so small that its mass must also be small—too small to cause the irregularities that had been seen in Uranus's orbit. More recent analysis of the orbital data used by Lowell and his contemporaries indicates that the irregularities they perceived were not caused by another planet at all but were simply variations due to the limited accuracy of the data. This is an example of the importance of knowing the amount of uncertainty involved in a measurement.

Thus we must conclude that Pluto was discovered by accident. If Lowell's calculations and predictions had not been made, however, the search would not have been carried on so diligently, and the planet probably would not have been discovered until much later.

After the discovery of Pluto was confirmed by other astronomers, Tombaugh went to college, enrolling as a freshman at the University of Kansas.

Neptune's discovery was not an accident, of course, but a triumph of the ability of Newton's theories to make predictions.

Pluto as Seen from Earth

Pluto was as close to Earth in 1989 as it has been for 248 years, but its image in a telescope was very small and fuzzy even then. It has a very eccentric orbit, more eccentric than any other planet, so that although its average distance from the Sun is about 40 AU, its distance ranges from 30 AU to about 50 AU. Recall from Chapter 7 that Pluto's orbit has another unusual aspect: it is tilted 17 degrees to the ecliptic, whereas no other planet's orbit is tilted more than 7 degrees.

Stellar occultations revealed that Pluto has an atmosphere, and spectroscopic examination of the light from Pluto indicates the presence of methane. It is probable, however, that this atmosphere is not present during Pluto's complete orbit, for when Pluto is at aphelion it receives so little energy from the Sun that its surface is probably below the temperature at which methane freezes. Thus its atmosphere is probably a temporary phenomenon that occurs only when it comes within a certain distance of the Sun.

In 1956 astronomers observed that Pluto's brightness changes slightly every 6.4 days, leading to the conclusion that it has a dark area on its surface and that its period of rotation is 6.4 days. Observations of an occultation in 1965 had placed an upper limit on the planet's size, but not until the discovery of Pluto's moon did we have an opportunity to learn much about Pluto's mass and size.

Recall that from 1979 to 1999 Pluto is inside Neptune's orbit. Refer back to Figure 7-3.

Recall that *aphelion* refers to the point in a planet's orbit when the planet is farthest from the Sun.

The Discovery of Pluto

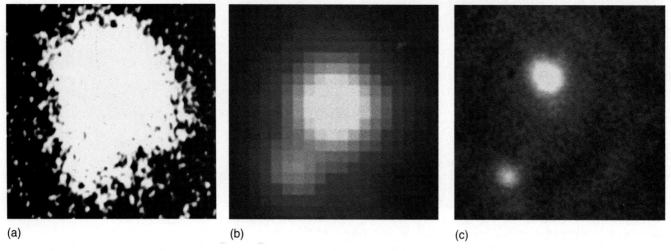

► **FIGURE 10-2.** (a) James Christy concluded from a photo such as this that Pluto has a moon, although Pluto and Charon cannot be seen as separate objects. Charon causes the "bump" at the lower left. (b) In this Earth-based CCD photo, two objects can definitely be seen. (c) This image was made by the Hubble Space Telescope.

Pluto and Charon

In 1978, James W. Christy of the U.S. Naval Observatory was analyzing data and noticed that Pluto seemed to have a bump on one side. ►Figure 10-2a shows this bump and indicates the limited resolution of photographs of this distant, tiny planet. Continued observation showed the bump moved from one side to another, and it was concluded that Pluto has a moon. The moon was named Charon (pronounced KEHR-on), after the mythical boatman who ferried souls to the underworld to be judged by Pluto, and also after Christy's wife, Charlene. Figure 10-2b is a more recent ground-based photo of Pluto and Charon. In this photo, taken with a CCD camera, two separate objects are obvious.

Charon's orbit is tilted at 61 degrees to Pluto's orbit around the Sun. It orbits Pluto every 6.4 days, the same as Pluto's period of revolution. ►Figure 10-3a shows the orbits of Charon and Pluto in 1978. As Pluto continued to move to the left in its orbit, our view of the system shifted so that between 1985 and 1990, Charon occulted Pluto during each revolution (Figure 10-3b). The duration of each occultation allowed astronomers to calculate the sizes of the two objects with greater accuracy. Finally, in 1991 the Hubble Space Telescope provided the

The two photos of Pluto on page 311 were taken by the HST and are the first photos to reveal surface features on the planet.

Our Moon was the first case where we saw synchronous rotation and revolution.

This series of occultations occurs only twice during each of Pluto's orbits of the Sun—once every 124 years.

► **FIGURE 10-3.** The plane of Charon's orbit is tilted at a great angle with respect to Pluto's path around the Sun. (a) This represents the view from Earth when Charon was discovered. (b) A decade later Pluto had moved such that Charon passed in front of and behind Pluto.

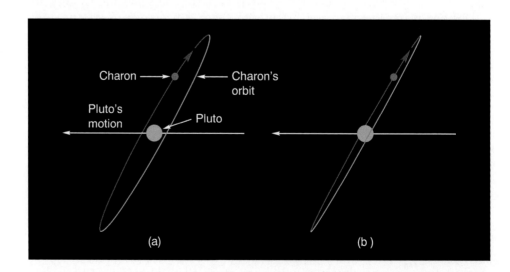

CHAPTER 10 Pluto and Solar System Debris

best photos yet of the system (Figure 10-2c). Still, a lack of knowledge of the nature of Pluto's atmosphere limits the accuracy of measurements of the planet's diameter. Its diameter may be as little as 2362 kilometers (if the atmosphere is hazy) or as great as 2412 kilometers (if the atmosphere is clear and we are seeing the surface).

Pluto's mass is about 12 times that of Charon, but only a fifth of our Moon's mass. Its density (between 1.8 and 2.1 grams/cm^3) leads us to conclude that its interior must consist of ice and rock in about equal amounts. Pluto and Charon are drawn to scale according to our best data in Figure 10-3.

Charon's diameter is about 1200 kilometers.

A Former Moon of Neptune?

For a number of reasons, including its very eccentric orbit and the fact that it is so much smaller than any Jovian planet, some astronomers have proposed that Pluto was once a moon of Neptune. Jupiter, Saturn, Neptune, and the Earth all have moons that are larger than Pluto. Perhaps a collision or near passage of some unknown celestial object caused Pluto to be ejected from the Neptunian system. Such a collision would be highly unlikely but not impossible.

The discovery of Charon made it seem less likely that Pluto was once a moon of Neptune, for how could Pluto itself have gotten a moon? The density of Charon (1.2 to 1.3 grams/cm^3), however, is much less than that of Pluto, indicating that the two did not form out of the same material, but that Charon was indeed captured by Pluto after the planet already existed. So the origin of Pluto is still open to question.

A more recent proposal is that Pluto and Charon should be classed as asteroids rather than as a planet and moon. A number of asteroids are known to exist as gravitationally bound pairs. The suggestion indicates how arbitrary our classification of celestial objects is. Nature is unified; we divide objects within it into discrete groups only to help our understanding.

Although the orbits of Neptune and Pluto intersect on a two-dimensional drawing, the two objects don't actually cross paths.

Recall the "double planet" and "capture" theories for the origin of our Moon, discussed in Chapter 6.

Present NASA plans call for launching a craft toward Pluto in 2001, to arrive in 2015.

SOLAR SYSTEM DEBRIS

One dictionary defines debris as "an accumulation of fragments of rock"—not far from what we will be describing. It could be said that the solar system is made up of one large object (the Sun), a number of medium-sized objects (the planets), and debris. The medium-sized objects (as well as some smaller moons) were discussed in the last two chapters, and the large object will be discussed in a later chapter. Now we will look at the debris. The material of which the debris is composed is not basically different from that which makes up the planets. Rather it is material that did not become part of the Sun or the planets when the solar system was being formed. The debris comes in a number of forms, including asteroids, meteoroids, comets, and dust.

ASTEROIDS

Since the first night of the nineteenth century, when Father Giuseppe Piazzi discovered Ceres, we have accurately determined the orbits of some 5000 asteroids. Approximate orbits are known for thousands more. The photograph in ▶ Figure 10-4 shows the motion of an asteroid across the background of stars; it is from photographs like this that most asteroids are discovered and their orbits analyzed. Astronomers estimate that about 100,000 different asteroids appear on today's photographs of the heavens. These objects traditionally have been called *minor planets*, indicating their stature as objects that orbit the Sun.

▶ FIGURE 10-4. The telescope was following the stars during this time exposure. The two streaks (identified with arrows) are asteroids.

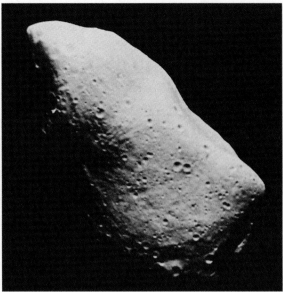

▶ FIGURE 10-5. The asteroid Gaspra was photographed by *Galileo* on the spacecraft's trip to Jupiter. Gaspra is about 19 kilometers (12 miles) end-to-end.

Only when the orbit of an asteroid is determined well enough that its location can be predicted on succeeding oppositions do we give it a name. Since calculations of asteroid orbits are no trivial task and usually are of no great value to astronomical knowledge, most asteroids go nameless.

Ceres is by far the largest of the asteroids, being greater than 1000 kilometers (600 miles) in diameter. In fact, Ceres makes up about 30% of the entire mass of all asteroids. Two others (Pallas and Vesta) have diameters greater than 300 kilometers, about 30 more are between 200 and 300 kilometers in diameter, and about 100 are larger than 100 kilometers. All the rest are smaller.

The masses of the three largest asteroids are calculated by observing the perturbations they cause on smaller asteroids passing nearby. Masses of other asteroids are estimated based on their sizes and expected densities.

For some time astronomers observing from Earth have seen that the brightness of most asteroids changes with periods that are typically a few hours. They conclude that asteroids are not spherical, but are irregularly shaped, which causes them to reflect different amounts of light toward Earth as they rotate. In October 1990 the *Galileo* spacecraft passed 16,000 kilometers from the asteroid Gaspra and took the photo of ▶Figure 10-5. In all, 16 images were made, and they show that this $19 \times 12 \times 11$ kilometer asteroid rotates with a period of about four hours. It is apparently covered with about a meter of dusty, rocky soil.

Gaspra is one of a type of asteroid that is bright and has a reddish tint. These asteroids are more reflective than our Moon. Others are dark like a lump of coal.

The Orbits of Asteroids

The direction of motion of all of the asteroids is, like that of the planets, counterclockwise as viewed from north of the solar system. Most of their planes of revolution are near the plane of the ecliptic, although we know of one with a plane of revolution 64 degrees from Earth's.

Most of the asteroids orbit the Sun at distances from 2.2 to 3.3 AU (in what is called the ***asteroid belt***). This corresponds, by Kepler's third law, to periods of from 3.3 to 6 years. Most of the asteroids in the asteroid belt have fairly circular orbits, although not as circular as the orbits of the major planets.

Ceres was named for the patron goddess of Sicily by its discoverer, Giuseppe Piazzi. German astronomers wanted to call it Juno or Hera, while some French astronomers wanted to call it Piazzi. Piazzi the man won out over Piazzi the name.

The bright reddish asteroids are called S-type. Bright nonreddish ones are M-type, and dark ones are C-type.

asteroid belt. The region between Mars and Jupiter where most asteroids orbit.

♇ PLUTO
DATA PAGE

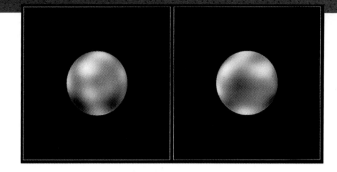

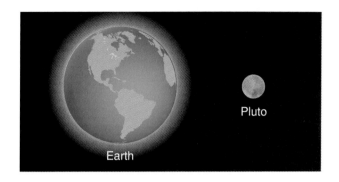

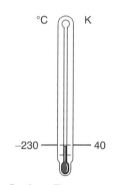

Surface Temperature

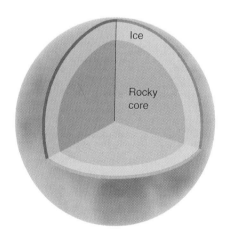

Interior

Pluto	Value	Compared to Earth
Diameter	2360–2410 km	0.2
Mass	1.3×10^{22} kg	0.002
Density	1.8–2.1 gm/cm^3	0.35
Surface gravity		0.05
Escape velocity	1.2 km/s	0.1
Sidereal rotation period	6.39 days	6.4
Surface temperature	$-230°$C	
Albedo	0.5	1.3
Tilt of equator to orbital plane	119°	
Orbit		
Semimajor axis	5.900×10^9 km	39.44
Eccentricity	0.250	15
Inclination to ecliptic	17.2°	
Sidereal period	248.5 years	248.5
Moons	1	

QUICK FACTS

The most distant planet. Most eccentric orbit, loops inside Neptune's orbit. Orbit has by far the greatest angle to the ecliptic of any planet. Receives $1/1600$ the intensity of sunlight that Earth receives. Occultations by its moon, Charon, provided a method of determining its size and mass. Charon apparently was captured.

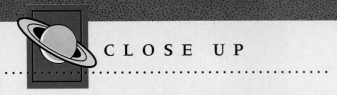

CLOSE UP

You Can Name an Asteroid

Name an asteroid? That sounds like a magazine advertisement. But the fact is that only about 3000 of the perhaps 100,000 asteroids that appear on photographs (and thus have a diameter of about 1 kilometer or greater) have been named. Why? In order to name an asteroid, you must know its orbit accurately. You determine its orbit by taking photographs of it and perhaps searching photographs taken by other astronomers who were studying other objects. After you have determined the orbit, you must wait until the asteroid has completed at least one more cycle around the Sun in order to check your predicted orbit.

If you have indeed calculated a reliable orbit, you will be given the privilege of naming the asteroid. The official name of the asteroid will include not only the name you give, but a number indicating the order of discovery. For example, 1 Ceres and 2 Pallas. These first asteroids were named after gods of Roman and Greek mythology. Names from mythology quickly ran out, and today almost any name is acceptable. Asteroid number 1001 is Gaussia, named after the mathematician who discovered the method of calculating orbits that made possible the determination of the orbit of Ceres. As you can verify in the internationally accepted catalog of asteroids, the Russian *Ephemerides of Minor Planets*, asteroid number 1814 is named for Bach and 1815 for Beethoven. Others are named Debussy, McCartney, and Clapton. There is no Elvis yet. Some of the more recently named asteroids honor the seven astronauts lost in the *Challenger* explosion in January 1986.

If the 100,000 asteroids that appear in photographs were in exactly the same plane and were spread evenly across the asteroid belt, neighboring asteroids would be separated by 2 million kilometers or more than a million miles. In reality, the asteroids are irregularly spaced in the asteroid belt and are not all in the same plane. So 2 million kilometers is a low estimate for the average distance between the 100,000 asteroids large enough to appear in photographs. There are many more asteroids smaller than this, but even so, the asteroids present no major hazard to spacecraft. Compared to the popular picture of a crowded asteroid belt, the belt is almost empty. To illustrate this, imagine a scale model in which the largest asteroid—Ceres—is the size of a grain of sand. In this case, a normal-size asteroid would be one meter away and would be too small to be seen (➤Figure 10-6).

➤ **FIGURE 10-6.** If Ceres, which has a diameter of about 1000 kilometers, were the size of a grain of sand, its nearest neighbor would typically be a meter away, but it would be too small to see, as are the individual grains of sand that Ty is pouring for Mandy and Jim.

312   CHAPTER 10   Pluto and Solar System Debris

► **FIGURE 10-7.** Kirkwood gaps appear where asteroids would have periods of one-third and one-half of Jupiter's.

Not all of the asteroids orbit the Sun between Mars and Jupiter. About 50 with diameters larger than 1 kilometer are known to have orbits eccentric enough that they cross the Earth's orbit; these are known as ***Apollo asteroids.*** Some of these have passed fairly close to the Earth in recent history. On March 22, 1989, an asteroid passed within about a million kilometers of the Earth. This is slightly more than twice the distance of the Moon from Earth. The asteroid was estimated to be about 1 kilometer in diameter. When we discuss meteors, we will describe what happens when a large asteroid hits the Earth.

Apollo asteroids. Asteroids that cross the Earth's orbit.

As the orbits of more and more asteroids were calculated during the nineteenth century, it became clear that they are not evenly distributed across the asteroid belt. At certain distances from the Sun, there are gaps in the belt. ►Figure 10-7 illustrates the situation. One very prominent gap occurs at 2.50 AU and another at 3.28 AU. Daniel Kirkwood, an American astronomer, first explained these in 1866. He noticed that 3.28 AU corresponds to a period of revolution of 5.93 years, which is just half of Jupiter's period of 11.86 years. An asteroid at 2.50 AU would orbit the Sun in 3.95 years, which is one-third of Jupiter's period. We saw a similar situation in the last chapter in the explanation for Cassini's division in Saturn's rings. In both cases, synchronous tugs from a large object on smaller particles gradually move those particles out of their orbits and result in gaps.

Jupiter has cleared out other regions of the asteroid belt where it exerts a regular gravitational pull on objects that may once have been there. Besides the gaps at locations corresponding to one-third and one-half of Jupiter's period, there are also major gaps corresponding to two-fifths and three-fifths of the period of the giant planet.

The Origin of Asteroids

Astronomers once thought that asteroids are the remains of the explosion of a planet. This theory has been abandoned today for two reasons. First, there is no known mechanism by which a planet could explode. Second, if all of the asteroids were combined into one object, that object would be only about 1500 kilometers in diameter—much less than the 3500-kilometer diameter of our Moon. Such a small object does not fit the pattern of planetary sizes.

It is considered much more probable that the asteroids are simply primordial material that never formed into a planet. There is a good reason that the material in the region of the asteroid belt did not form into a planet; the reason is Jupiter. The gravitational pull from that planet causes a continual stirring effect on the objects in the asteroid belt. This prevents the small gravitational forces that exist between the asteroids from pulling them together to form a larger object.

Even today, there is evidence that Jupiter causes collisions between asteroids that result in their fragmentation. In fact, some of the asteroids that have orbits outside the main belt are thought to have resulted from such collisions. Analyses of the orbits of such asteroids have identified groups ("families") whose orbits, when traced backward, indicate that they were once together. They apparently broke apart as the result of a collision with another asteroid caused by the gravitational force of Jupiter.

Asteroids will be discussed again later in the chapter when we take up the question of the origin of meteoroids. Now we turn to another category of solar system debris—comets.

COMETS

One of the most spectacular astronomical sights available to the naked eye is a comet. ➤Figure 10-8 is a photograph of Comet West, which made its appearance in 1976. Persons who have not seen a comet commonly assume from photographs such as Figure 10-8 that the comet streaks across the sky in the direction opposite to its tail. Perhaps from seeing streaks behind fast-moving characters in

➤ **FIGURE 10-8.** Comet West, discovered by Richard West of European Southern Observatory in 1976.

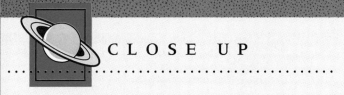

CLOSE UP

Chaos Theory

Many phenomena in nature are very complicated and involve numerous objects and complex motions. Think of the path taken by smoke as it climbs from the end of a cigarette. The motions of the myriad particles in the ring system of a planet or the motions of the asteroids in the asteroid belt are other examples. Sometimes a single large object that is acted on by a great number of other objects exhibits what seems to be unusual motions. An example is Hyperion, a moon of Saturn that tumbles in erratic motions as it orbits the planet. Scientists historically have assumed that if they could analyze the multitude of forces on these objects, their seemingly random motion could be understood using Newton's laws, but it appeared that the complete understanding of some complicated systems was beyond our capability.

A new method of analysis, called *chaos theory*, is finding success in investigating some of these systems. The theory's name comes from its ability to explain seemingly chaotic situations. It uses the power of computers to find patterns within a system's complexity. Chaos theory is still in its infancy, but it is becoming a major subject of research.

cartoons, we assume that the comet in the figure is moving downward at great speed. In fact, if you see a comet in the sky, you observe no rapid motion at all. Unless you observe carefully, the comet will appear to stay in the same place among the stars, having only the motion across our sky caused by the rotation of the Earth. (It does move among the stars, of course, and its motion can easily be seen over a few days.) Don't confuse a comet with a meteor, which *does* streak across the sky as will be discussed later.

Comet Orbits—Isaac Newton and Edmund Halley

Isaac Newton proposed that comets orbit the Sun according to his laws of universal gravitation and motion just as the planets do. He concluded that since comets were visible from Earth for only short periods of time (typically a few months), their orbits were very eccentric, that is, very elongated.

Edmund Halley was a friend of Newton. In fact, he talked Newton into publishing the *Principia*, personally financed the book, and was the only person for whom Newton expressed appreciation in the book. Halley used Newton's methods, his own observations, and descriptions of previous comet sightings to calculate the orbits of a number of comets. He noticed that the orbits of the comets of 1531, 1607, and 1682 were very similar and suspected that they might be the same comet, but he was at first confused by the fact that the time that elapsed between one appearance of the comet and the next was not always the same. When he realized that the comet's path would be changed slightly by the gravitational pull of a planet—particularly Jupiter or Saturn—when the comet passed nearby, he hypothesized that these three comets were in fact three appearances of the same comet. He predicted that the comet would reappear in 1758.

On Christmas night of 1758, the comet was sighted. In honor of its predictor, it was named Halley's comet. The brilliant Edmund Halley, however, had died 16 years earlier at the age of 85 and was unable to see his hypothesis verified. By investigating reports of comets in literature, we can now trace Halley's comet (or *Comet Halley*) back as far as the year 239 B.C.

Today we discover about a dozen comets each year. Of the nearly 1000 whose orbits have been calculated, about 100 have a period of revolution around the Sun of less than 200 years. Most have extremely long periods of up to a million years.

www.jbpub.com/starlinks

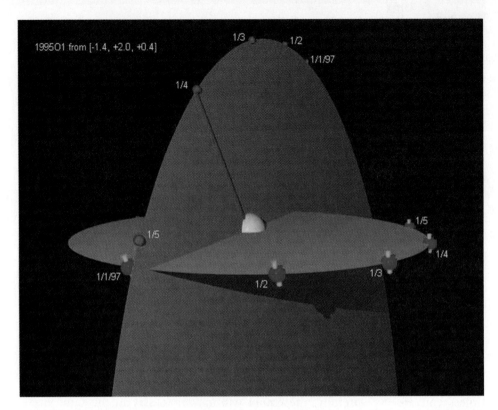

➢ **FIGURE 10-9.** The orbits of Comet 1995o1, commonly known as Comet Hale-Bopp, and the Earth. The dates listed use the European system of day/month/year.

The planes of revolution of comets are not limited to the plane of the planets but are randomly oriented, so comets sweep past the Sun from all directions. ➢Figure 10-9 is a diagram of the orbit of Comet Hale-Bopp, which reached perihelion in 1997. Many of us will remember it as the naked-eye comet of a lifetime. The gray surface in the figure is the Earth's orbit, and the blue spheres show the location of the Earth from January 1 to May 1, 1997. (The dates in the figure use the European notation, so that 1/3/97 is March 1.) The red dot represents the comet and shows its location on the same dates. You can see the importance of the relative positions of the Earth and a comet in determining the comet's visibility. ➢Figure 10-10 shows the beauty of Comet Hale-Bopp.

The Nature of Comets

coma (KOH-mah). The part of a comet's head made up of diffuse gas and dust.

nucleus (of comet). The solid chunk of a comet, located in the head.

tail (of comet). The gas and/or dust swept away from a comet's head.

We normally think of a comet as having three parts. What is commonly called the head of a comet consists of the ***coma*** and the ***nucleus.*** Sweeping away from the coma is the comet's ***tail,*** which varies greatly in size and appearance from comet to comet. The coma of a comet may be as large as a million kilometers in diameter—almost as large as the Sun. Some comet tails have been as long as an astronomical unit.

In 1950, Harvard astronomer Fred L. Whipple proposed what remains today as the basic model of a comet. He proposed that the nucleus resembles a dirty snowball made up of water, ice, frozen carbon dioxide, a few other frozen substances, and small solid grains—the "dirt." Observations of comets since that time have confirmed his model, but have caused it to change to include a crusty layer on the surface of the nucleus with the ices inside. When the comet is far

➢ **FIGURE 10-10.** In this March 29, 1997 photo of Comet Hale-Bopp the blue ion tale is obvious. It is straight and streams directly away from the Sun. The white dust tail is curved.

from the Sun, the nucleus (which is only a few miles in diameter) is all there is. As the comet approaches the Sun, it becomes warmer and the ices inside melt and vaporize. These materials then break through the crust forming a geyser. These gases, along with dust particles that have been torn away, hover near the nucleus, held there by the small gravitational field of the nucleus. This is what makes up the coma.

➢Figure 10-11 is a close up of the coma of Comet Hale-Bopp. This comet has a fairly large nucleus—25 miles across—and the nucleus spins around about once every 12 hours. As it spins, parts of it shoot away in a few geysers. As the ejected material spirals away from the nucleus, it causes the appearance of rings around the nucleus. These are the "layers" seen in the photo. They were visible even in small telescopes in March and April 1997. ➢Figure 10-12 shows a cross-sectional view of the nucleus of a comet.

The coma and tail of a comet are mostly empty space. When Comet Halley came through the inner solar system in the late 1980s, the European space community launched the spacecraft *Giotto* to intercept it. *Giotto* passed within 600 kilometers (350 miles) of the nucleus, going right through the coma. *Giotto* found that the coma is billions of times less dense than the atmosphere of Earth at sea level. The entire mass of a typical comet is less than one-billionth the mass of the Earth, and more than 99.99% of this mass is in the nucleus. The reason that we can see the coma at all is that its molecules and dust particles reflect sunlight to us. In addition, ultraviolet light from the Sun causes the molecules to fluoresce in much the same manner that a "black light" poster glows in ultraviolet light. There are very few molecules and dust particles in any particular volume of the coma, but the coma is extremely large—hundreds of thousands of kilometers in diameter—so there is a lot of material along any particular line of sight.

Comets Halley and Hale-Bopp are now moving through the outer portions of the solar system. Comet Haley will go almost to Pluto's orbit and will return in 2061. Hale-Bopp is in a much larger orbit; it will return about 3000 years from now.

The spacecraft *Giotto* was named after the artist who included Comet Halley as the star of Bethlehem in his fresco of the nativity scene. He had just seen a return of the comet a few years earlier, in 1301.

Comets 317

HISTORICAL NOTE

Astronomer Maria Mitchell, A Nineteenth-Century Feminist

Maria (pronounced ma-RYE-a) Mitchell (1818–89) grew up on Nantucket Island, Massachusetts, and learned astronomy both from her father and from her readings while she worked as a librarian. In 1847, while observing the sky from her rooftop, she discovered a comet. The discovery resulted in her becoming the first woman elected to the American Academy of Arts and Sciences (although the all-male membership refused to name her a "fellow," calling her instead an "honorary member").

The king of Denmark awarded a gold medal to each discoverer of a comet, and receiving the medal brought Mitchell some fame. Her astronomical work remained on the amateur level, however, until her mother died, when finally at the age of 40 she sought a professional career. Vassar College had just been founded, and Maria Mitchell was hired to teach.

Mitchell was an early believer in women's rights and saw astronomy and science as an avenue for liberation and a way for women to break from tradition. Her outspoken nature was revealed on one occasion when her wish to observe the Moon as it occulted a star conflicted with something her college president wanted her to do. She wrote the following to him:

> My good natured President, I want to hear you preach tomorrow, and I also want to see the moon pass over Aldebran. Can't you let me do both? Will you stop at eleventhly or twelfthly? Or, why need you show us *all* sides of a subject? The moon never turns to us other than the one side we see, and did you ever know a finer moon? If I could stop the moon and do no more harm than Joshua did, I wouldn't ask such a favor of you, knowing, as I do, what a difficult thing it is for you to pause, when you are once started, and knowing also, that I never want you to do so—except this once. Yours with all regret, even if it doesn't appear—M. Mitchell. (Mitchell's underscores)

This story is from E. A. Daniels, "Maria Mitchell, the Star of Vassar College," *Star Date*, November/December 1989. To subscribe to this delightful magazine, write Star Date, RLM 15.308, University of Texas at Austin, Austin, Texas 78712.

➢ FIGURE 10-11. The layers below the nucleus of Comet Hale-Bopp are caused by the slow spinning of the nucleus as material jets from its surface. The red color was added to enhance the image.

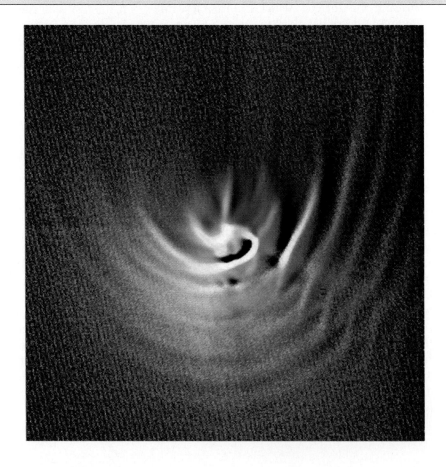

CHAPTER 10 Pluto and Solar System Debris

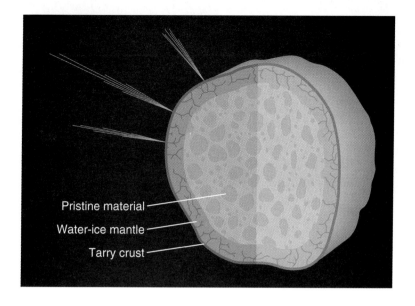

▶ **FIGURE 10-12.** The nucleus of Comet Halley (and presumably other comets) is made up of a mixture of ice and dust surrounded by an ice mantle and covered with a dark crust. Material from inside is ejected through the crust as the comet is heated by the Sun.

Comet Tails

As indicated earlier, the tail of a comet does not necessarily follow the head through space. Rather, a comet's tail always points away from the Sun. Thus after a comet has passed the Sun, its tail actually *leads* it through space (▶Figure 10-13). This takes some explaining.

▶ **FIGURE 10-13.** In general, a comet's tails point away from the Sun. Notice that the tails are largest shortly after the comet has passed the Sun.

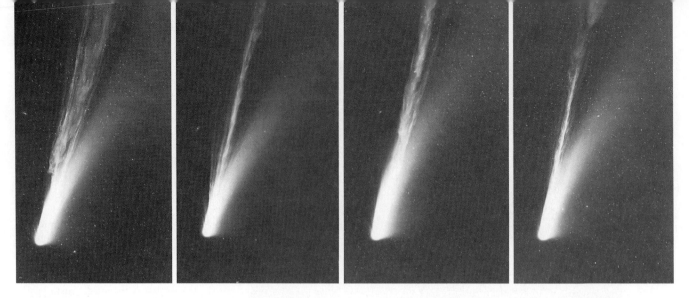

► **FIGURE 10-14.** The two tails of Comet Mrkos are obvious in these photos, taken on different dates during its appearance in 1957.

ion. A charged atom or molecule resulting from the atom's loss or gain of an electron.

First, comets have two tails although one or both may be very small and they may change greatly as time passes. ►Figure 10-14 shows four photographs of a 1957 comet. Notice that one tail is very straight while the other curves away from the comet. The straight tail consists of charged molecules *(ions)* being swept away from the comet by the solar wind—nuclear particles that are always being emitted by the Sun. These molecules move away at a great speed and form a straight tail.

The curved tail is caused by the grains of dust in the coma being pushed away by the weak pressure of solar radiation. This radiation pressure is not detectable by us in everyday life, but light does indeed exert a tiny force on objects it strikes. In the weak gravitational field of the comet's nucleus, the force is great enough to move dust particles away from the head. They move much more slowly than the molecules that form the ion tail, however, and ►Figure 10-15 shows why this results in the dust tail being curved. Dust particles that left the comet when it was at location X have reached X', those that left at Y are at Y', and so on.

► **FIGURE 10-15.** Dust that was blown off the comet when it was at point X is now at X'. Dust blown off at Y is at Y'. This causes the dust tail to be curved.

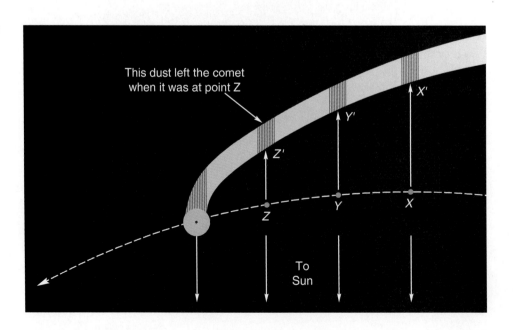

320 CHAPTER 10 Pluto and Solar System Debris

Sometimes only one tail of a comet is visible because one is behind the other as seen from Earth. In addition, some comets are not as dusty as others and do not contain a prominent dust tail.

In general, the tails of comets point away from the Sun regardless of the motion of the comet nucleus because the agents responsible for them (the solar wind and light) are emitted radially from the Sun. Comet tails are typically 10^7 to 10^8 kilometers long and may be as long as 1 AU.

If a comet loses material as its gas and dust are pushed off to form its tails, comets must have limited lifetimes. Comet Halley spews about 25 to 30 tons of its mass through its jets each second when it is close to the Sun. This sounds like a great deal, but it means that the nucleus loses less than 1% of its mass on each pass of the Sun. Comet Halley is now on its way toward the outer portion of its elliptical orbit, and since its jets are now inactive, it is no longer losing mass. (It will continue out past Neptune's orbit before returning to our region of the solar system in 2061, as shown in Figure 10-9.)

Comets do gradually lose their ices, however. It is thought that some comets finally evaporate away all of their nuclei so that they just fizzle out. Other comets, after they lose all of their volatile materials, become simple chunks of rock and probably would be classified as asteroids. A third way that comets die is to fall into the Sun. A comet that comes close to the Sun is slowed by the Sun's atmosphere, and after a number of passes, it will be slowed enough that it falls into the Sun.

THE OORT CLOUD AND KUIPER BELT

Thus far we have been considering only short-period comets—those with periods of less than a few hundred years. Most comets have much longer periods and approach us from far beyond the orbits of the planets. About a dozen comets are discovered each year; almost all are long-period comets.

In 1950 the Dutch astronomer Jan Oort proposed that great numbers of comets, all of which orbit the Sun, exist in a region of space that lies from 10,000 AU to 100,000 AU from the Sun (➤Figure 10-16). This shell of comets

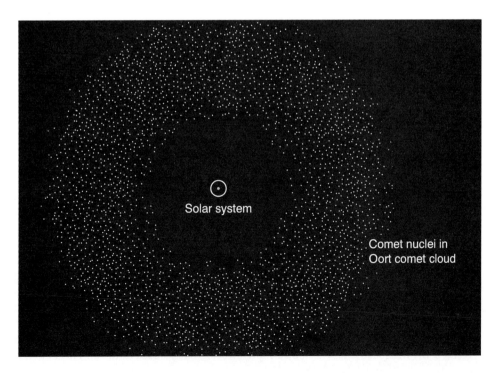

➤ FIGURE 10-16. A comet cloud—called the Oort cloud—is hypothesized to exist in a spherical shell between 10,000 and 100,000 AU from the planetary part of the solar system.

HISTORICAL NOTE

Jan H. Oort, 1900–1992

On November 5, 1992, Jan Oort died at the age of 92 in his home country, the Netherlands. Although he is best known for the cloud of comets that bears his name, he considered his work with distant comets a sidelight to his other astronomical work. He first drew international attention in 1927, when he and Bertil Lindblad of Sweden discovered that the Milky Way Galaxy rotates (Chapter 16).

Oort's accomplishments are numerous. He was among the first to realize the value of radio astronomy and was the driving force behind the Westerbork Radio Telescope. His work resulted in establishing the link between the Crab nebula that we observe and the supernova observed by the Chinese in 1054 (Chapter 15). Oort served as director of the Leiden Observatory from 1945 until his retirement in 1970. His work continued, however, and he published an article in *Mercury* magazine as recently as the March/April 1992 issue.

Oort cloud. The theorized spherical shell lying between 10,000 and 100,000 AU from the Sun containing billions of comet nuclei.

surrounding the solar system has come to be known as the **Oort cloud**. Naturally, the Oort cloud is too far away for its comets—which are simply nuclei at that distance—to be visible from Earth.

Many of the comets we observe are in elliptical orbits with extremely long periods. They obey Kepler's second law, however, which means that they move at extremely low speeds when they are far from the Sun. If this is true, they spend the overwhelming majority of the time well beyond planetary distances. For example, a comet with a period of 500,000 years would spend about 499,998 of those years beyond the orbits of the planets. Thus many of the comets in the Oort cloud are there simply in accordance with Kepler's laws.

In addition, there must be many comets in the Oort cloud whose orbits never approach the inner solar system. This would not necessarily mean circular orbits, for a comet could vary from 10,000 AU to 100,000 AU from the Sun and still remain in the Oort cloud. From time to time one of the comets of this category must pass near another comet, causing it to change its path. This may result in a more circular path, or it may result in the comet being moved into an orbit that takes it toward the inner solar system. In addition, although the outer Oort cloud stretches only about one-third of the way to what is now the nearest star, every few million years a star passes closer to the Sun than this, and gravitational forces from the star cause changes in the orbits of comets. Astronomers believe some comets are deflected into the inner solar system by this method. These comets from the Oort cloud become long-period comets.

Kuiper belt. A band of comets hypothesized to exist closer to the solar system than the Oort cloud.

The Oort cloud does not explain all comets, however, and in 1951 Gerard Kuiper proposed that a second, smaller band of comets must exist inside the Oort cloud. The first object in this **Kuiper belt** was discovered in 1992 (▶Figure 10-17).

The Origin of Short-Period Comets

The age of the solar system is about 5 billion years. Why have those 100 comets with periods of less than 200 years not evaporated away their ices so that they are no longer observed? The answer must be that these short-period comets are relative newcomers to the inner solar system. We must ask where they come from, and an obvious hypothesis is that long-period comets sometimes become short-period comets. How might this occur?

If the path of a comet depended solely on its gravitational attraction toward the Sun, there would be no way in which its orbit could be changed to make it a short-period comet. Other objects (particularly the massive Jupiter) affect a comet's orbit, however. The combined effects of the gravitational forces from

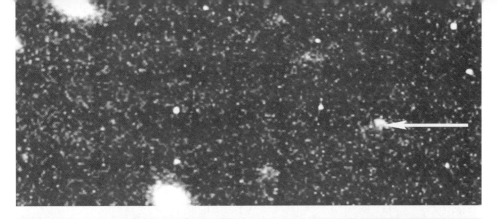
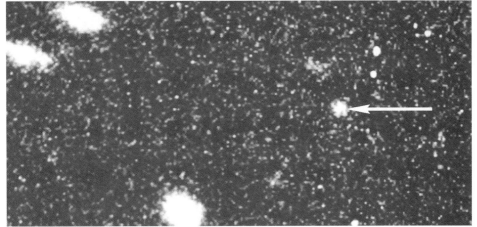

▶ **FIGURE 10-17.** The arrow points to an object that moved during a two-hour period on August 30, 1992. The object was given the name 1992 QB$_1$ and is thought to be the first object observed in the Kuiper belt.

Jupiter and the Sun can cause one of these comets to change its orbit so that it becomes a short-period comet. ▶Figure 10-18 illustrates one way this could occur. All of the short-period comets were captured in the inner solar system by such a mechanism. On the other hand, the Sun and a planet can have the effect of changing the orbit of a comet so that it leaves the solar system and the Oort cloud entirely.

Astronomers have no idea how many comets are in the Oort cloud and the Kuiper belt. Estimates range from a million to a trillion. We must remember that the clouds are simply hypotheses to explain three observations. First, new comets are continually joining the inner solar system. Second, long-period comets pass the Sun from all directions. Their observed motions lead us to conclude that they have their origin in a regularly distributed cloud around the Sun. Finally, the orbits of the vast majority of comets indicate that they are gravitationally bound to the Sun.

The large range of estimates illustrates how little we know of the Oort cloud.

The word *cloud* in reference to comets in orbit around the Sun can be misleading. When you think of a cloud on Earth, you may think of a volume fairly crowded with water droplets. But the Oort cloud and Kuiper belt are far from crowded. We pointed out that the asteroid belt is mostly empty space, but it is crowded compared to the Oort cloud. If the Oort cloud contains a trillion comets—a high estimate—there would still be an average distance of 16 AU (1.5 billion miles) between comets. A future interstellar traveler stranded in the middle of the Oort cloud with a telescope probably would not be able to find a comet nucleus.

Most comet orbits are either elliptical or parabolic, not hyperbolic. The latter would indicate that the comet is not bound to the Sun.

METEORS AND METEOR SHOWERS

Almost everyone has seen the flash of light in the sky that is sometimes called a *falling star* or *shooting star*. This phenomenon is better termed a **meteor** because obviously it is not a star. (Recall the size of our Sun—a typical star—compared

meteor. The phenomenon of a streak in the sky caused by the burning of a rock or dust particle as it falls.

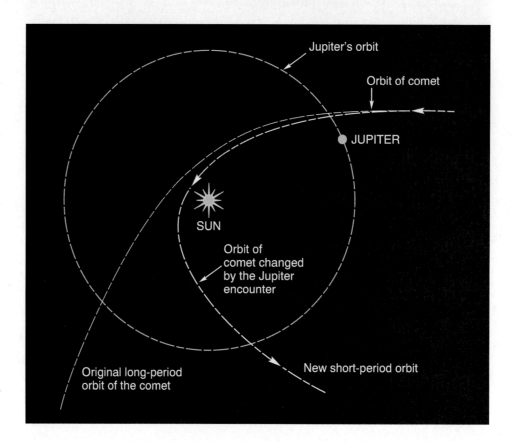

▶ FIGURE 10-18. Gravitational attraction to a planet—Jupiter, here—can cause a comet to alter its orbit, perhaps changing a long-period comet to a short-period one.

to the Earth.) The streak of light in the photograph in ▶Figure 10-19 is a meteor. The nature of these sudden flashes of light across the sky must have been of great concern to people since the beginning of time. The idea that they are caused by rocks falling from the heavens can be found in ancient writings of the Romans, the Greeks, the Chinese, and in the Old Testament, but some people found the idea hard to accept.

The first confirmation by modern science that rocks do indeed fall from the heavens occurred on April 26, 1803. Citizens of the small town of L'Aigle, France, saw an exceedingly bright meteor that exploded and formed a shower of 2000 to 3000 fragments that fell to Earth. Reportedly, some fragments were still warm when found. The French Academy of Science sent the respected physicist J. B. Biot to investigate the incident, and his report confirmed the ancient writings.

Before describing the stones and the flashes of light in any more detail, some definitions are in order. A meteor is the phenomenon of the flash of light; it is not the object itself. The object out in space that causes the meteor in our atmosphere is called a *meteoroid* and is in orbit around the Sun. Finally, if the object survives its fall through the atmosphere and lands on Earth, it is called a *meteorite*.

meteoroid. An interplanetary chunk of matter smaller than an asteroid.

meteorite. An interplanetary chunk of matter after it has hit a planet or moon.

Meteors

Most meteors are very dim. You might see one out of the corner of your eye and wonder if you saw something or not. Others, however, known as *fireballs*, are very bright and might cause a long streak across the sky. In fact, the brightest are brighter than a full Moon. The light is the result of the meteoroid burning itself up as it passes through our atmosphere. As the stone enters the atmosphere, it experiences friction caused by the molecules of air. This heats up the object as well as the air it passes through. At first glance, this may seem odd to you if you have held your hand out of a car window and experienced the *cooling* effect of

fireball. An extremely bright meteor. More than 10,000 fireballs can be seen each day over the Earth.

➤ FIGURE 10-19. The streak of light is a meteor.

the air striking your hand. The difference is in the speeds of the objects—your hand and the meteoroid. A meteoroid's speed is typically 50 kilometers/second (100,000 miles/hour). Thus the air molecules are striking the object at tremendous speeds. This causes the surface of the object to heat up until it vaporizes, streaming its atoms in its wake. These atoms, along with the similarly heated air, glow like the gas in a fluorescent lamp and present us with the phenomenon known as a meteor.

It is difficult for an individual observer to estimate the height or speed of a meteor, but if two observers at different locations each record the same meteor on a photograph, they can use triangulation to calculate these quantities. By means of such measurements, it has been determined that the typical meteor begins to glow at a height of about 130 kilometers (80 miles) and burns out at about 80 kilometers (50 miles). The speed of a meteoroid as it enters the Earth's atmosphere might be anywhere from about 10 kilometers/second to about 70 kilometers/second.

These speeds correspond to 20,000 mi/hr and 150,000 mi/hr.

Meteoroids

Most meteoroids vaporize completely in the atmosphere and never reach the Earth's surface. The energy source that produces the light we see is simply the motion of the meteoroid. By calculating how much light is emitted from the meteoroid as it burns up and by estimating how much of its original energy of motion changes to light (rather than heat), we can calculate the mass of the original meteoroid. Most meteors are produced by meteoroids with masses ranging from a few milligrams (a grain of sand) to a few grams (a marble-size rock).

Under ideal viewing conditions, a person looking for meteors can see an average of five to eight meteors per hour. Since a meteor can be seen only if it is within 150 to 200 kilometers of the viewer, we can calculate that over the entire Earth there must be about 25,000,000 meteors a day bright enough to be seen by the naked eye. The number visible in telescopes would be hundreds of

The "energy of motion" referred to here is called the *kinetic energy* of the meteoroid.

It is estimated that only 1 in 1 million meteoroids that hit the atmosphere survives to reach the surface.

billions. Although the average meteor may have a mass of only a fraction of a gram, it is estimated that 1000 tons of meteoritic material hit the Earth each day.

If a meteor trail is recorded from more than one location, its path can be determined accurately enough to calculate the path of the original meteoroid before it was slowed by the atmosphere. In this way we find that, as expected, most meteoroids are simply tiny particles orbiting the Sun. Meteoroids differ from asteroids in that most asteroids orbit the Sun close to the plane of the ecliptic, while small meteoroids do not suffer this limitation; their orbit around the Sun may be in any orientation. A majority have very eccentric orbits rather than the nearly circular orbits of most asteroids.

This does not rule out the asteroid belt as the origin of these meteoroids, however. It is thought that many small meteoroids are debris from collisions between asteroids. Such collisions would break the asteroids into a number of smaller pieces, including pieces as small as sand grains. These tiny pieces would come out of the collision in all directions and move in elliptical orbits at all orientations with the ecliptic.

Many meteors, however, are from a source other than the asteroids—they are due to material evaporated from a comet's nucleus and then blown off the comet by the action of the Sun. This leads us to a discussion of meteor showers.

➤ FIGURE 10-20. This photo shows the Leonid meteor shower of November 1966. The streaks radiating from one area of the sky are all part of the shower.

meteor shower. The phenomenon of a large group of meteors seeming to come from a particular area of the celestial sphere.

Meteor Showers

On some nights, we see many more meteors than normal, and if we observe closely, we see that there is a pattern to the directions of the meteors. ➤Figure 10-20 is a time exposure showing the phenomenon. The streaks are meteors. Notice that they seem to point to (or rather, *from*) one point in the sky. This phenomenon is called a ***meteor shower*** and is caused by the Earth passing through a swarm of small meteoroids. ➤Figure 10-21 illustrates this. A cluster of tiny particles is shown in the Earth's path. These tiny meteoroids cause the shower.

To see why the meteor shower seems to originate in a certain constellation, think of the Earth moving through space. When it encounters the swarm of

➤ FIGURE 10-21. The Earth here is moving toward the constellation Leo and is about to encounter a swarm of meteoritic particles.

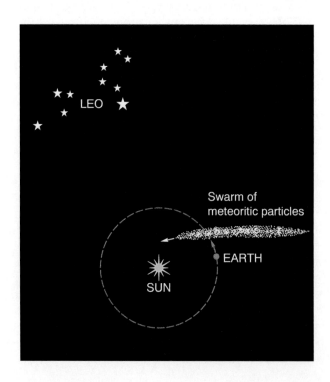

326   CHAPTER 10 Pluto and Solar System Debris

meteoroids, the Earth is moving in a particular direction toward some constellation in the sky. The meteor trail seems to originate in that constellation or close to it. The shower may not appear to come exactly from the constellation toward which the Earth is moving because the swarm itself has a speed and therefore its direction of hitting the Earth is determined by a combination of the Earth's and the swarm's direction of motion.

Meteor showers are named after the constellation from which they seem to come. Notice that the meteors in Figure 10-20 radiate from a single point off the upper right of the photograph (this point is known as the *radiant*). They appear to come from within the constellation Leo, and the shower is therefore known as the Leonid meteor shower. The meteor shower about to occur in Figure 10-21 will likewise be a Leonid shower.

The particles that cause a meteor shower strike the Earth along nearly parallel paths, yet they appear to diverge from a point. The reason for this apparent contradiction is the same as the reason that parallel strips of a long, straight highway seem to diverge as they near a viewer standing in its center (▶Figure 10-22). The particles forming the shower are coming into the Earth's atmosphere along parallel paths, but as they near us, they seem to be spreading out.

For centuries we have known that some meteor showers repeat regularly each year, but the origin of the showers was not known. Then in 1866 it was shown that the particles that cause the Perseid meteor shower (which occurs around August 12 and appears to originate in the constellation Perseus) have almost exactly the same orbital path that an 1862 comet had. It was concluded that the meteoroids of the shower were simply particles that had long ago come from the comet and formed its tail. Those particles were in orbit about the Sun when their paths intersected the Earth's. Today we are able to associate most of the major annual meteor showers with some comet. Table 10–1 lists the major showers and their associated comets.

radiant (of a meteor shower). The point in the sky from which the meteors of a shower appear to radiate.

▶ FIGURE 10-22. (a) Meteors in a meteor shower seem to come from a single point in the sky. (b) The divergence of the meteors is the same phenomenon seen by a person standing on a long, straight highway. The road seems to diverge as it comes toward the observer.

Meteors and Meteor Showers

TABLE 10-1 Some Major Meteor Showers

Shower	Date of Maximum*	Associated Comet	Expected Hourly Rate
Quadrantids	Jan. 3	??	40
Lyrids	April 21	Comet 1861 I	15
Eta Aquarids	May 4	Comet Halley	20
Delta Aquarids	July 30	??	20
Perseids	Aug. 12	Comet 1862 III	50
Draconids	Oct. 9	Comet Giacobini-Zinner	15
Orionids	Oct. 21	Comet Halley	20
Leonids	Nov. 17	Comet 1866 I	15
Geminids	Dec. 13	Asteroid Phaethon**	50

*The dates given are for the approximate date of maximum and may vary slightly from year to year.
**This asteroid is in the same orbit as the meteoroids and is thought to be the remains of the comet's nucleus.

The table also indicates the average number of meteors expected to be seen during the maximum of the shower. The intensity of some showers changes greatly from year to year, however. ➤Figure 10-23 shows why this occurs. In part (a), the particles that had been blown from the comet have become spread fairly evenly around the comet's orbit, so each time the Earth passes through this orbit, about the same number of particles strike our planet.

In part (b) of the figure, however, the particles are clumped in one region of the comet's orbit. If the Earth and this swarm happen to meet, we see much greater meteor activity. The Leonid showers are irregular in intensity from year to year. In 1833, 1866, and again in 1899, the Earth passed through a major swarm of particles in the comet's orbit, and nearly 100,000 meteors could be seen in an hour. The maxima of the Leonid showers occur about every 33 years, which is just the period of the comet from which they originated. As the years go by, the particles spread out more and more along their orbit, but the display was spectacular in November 1966, when Figure 10-20 was taken. We are therefore looking forward to another beautiful display around 1999.

➤ **FIGURE 10-23.** In part (a) the particles are spread around the entire orbit, but in (b) they are clustered in one area.

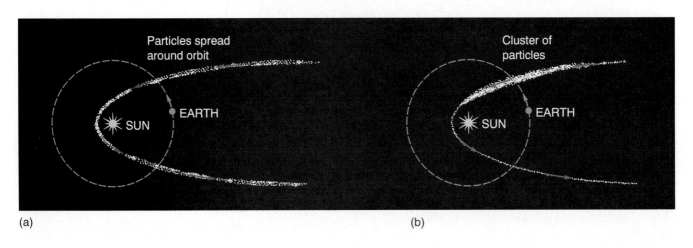

METEORITES AND CRATERS

Meteorites are classified in three categories: iron meteorites (or *irons*), which are made up of 80 to 90% iron (most of the rest being nickel); stony meteorites *(stones)*, which are just what their name implies but often contain flakes of iron and/or nickel; and *stony irons*, about half iron and half rock. ➤ Figure 10-24 shows three different meteorites.

About 90% of all meteorites are stones, but if you find a meteorite, it will likely be an iron. The reason for this apparent paradox is that stony meteorites are very similar to regular earthly stones, especially after the meteorites have weathered for a few years. Thus they are difficult to recognize. Irons, however, are found by people using metal detectors and are different in appearance from regular earthly rocks.

Some iron meteorites are found by people who realize that the rock they kicked was "too heavy."

How do we know, then, that most meteorites are stones? Meteorites can be found in the Antarctic, where they become exposed when snow is blown from them. Here, where there are no natural rocks to cause confusion, 90% of the meteorites we find are stones.

The largest meteorite ever found, the Hoba meteorite, weighs about 65 tons and is in Namibia, in southwestern Africa. Iron meteorites such as this are much stronger than stones and therefore do not break up in the atmosphere as stones do. The Hoba meteorite (so called after the name of the farm on which it lies) barely disturbed the surrounding surface of the Earth, apparently because it en-

(a)

(b)

➤ FIGURE 10-24. The iron meteorite in part (a) and the stony meteorite in (b) are part of the collection of Robert Haag, a collector who is known by some as the "Meteor Man." (c) This 14-ton iron is the largest meteorite ever found in the United States. The farmer and his son who found it in 1902 moved it three-fourths of a mile and then charged people 25 cents to view it.

(c)

CLOSE UP

Hit by a Meteorite?

Something spectacular fell in Tunguska, Siberia, on June 30, 1908. The fireball was bright enough to be seen in daylight, and the sound that resulted from its explosion some 10 kilometers above the Earth was heard as far away as 1000 kilometers (600 miles). The explosion, which was equivalent to about a 50-megaton bomb, knocked down trees as far as 30 kilometers away, and a man 80 kilometers away was knocked down by the shock wave from the explosion.

Recent work on the fate of asteroids as they plunge through the atmosphere indicates that the object that caused the Tunguska event was an asteroid. Astronomers Jack Hills and M. Patrick Goda calculate that an asteroid at least 80 meters across traveling at 22 kilometers per second would have broken up violently in the atmosphere and caused the disruption at Tunguska in 1908.

Human bodies take up a small portion of the Earth's surface, so it is unlikely that a person will be hit by a meteorite or even see one land. There is an unconfirmed report of a monk being killed in 1650 in Milan, but no good evidence that a meteorite has ever killed anyone. In 1954, a meteorite came through the roof of a house in Sylacauga, Alabama, and hit a woman on the bounce, severely bruising her hip. In 1971, a house in Wethersfield, Connecticut, was hit by a meteorite, and in an extremely unusual coincidence, a house about a mile away was hit just 11 years later. These meteorites were a few inches across and did no major damage to the houses (if puncturing the roof and ceiling is not major damage).

On August 31, 1991, in Noblesville, Indiana, two boys were standing by the sidewalk talking when they heard a whistle followed by a thud. Twelve feet from them they found a meteorite in a crater about 9 centimeters wide and 4 centimeters deep. They reported that the rock felt slightly warm when they picked it up.

Another recent close call occurred when a meteorite plunged through the rear of a car about 40 miles north of New York City on October 9, 1992 (▶Figure C10-1). The football-sized meteorite had a mass of 27 kilograms (about 60 pounds). The fireball caused by the meteorite (or from the larger one from which it broke) had been seen from as far away as Frankfort, Kentucky.

▶ **FIGURE C10-1.** Michelle Knapp of Peekskill, New York, examines the damage done by a stony meteorite in October 1992. She has been offered $69,000 for the stone.

tered the atmosphere at a very small angle and was greatly slowed down before reaching the surface.

You may be unable to travel to Africa to see the largest meteorite, but the second largest is in captivity in New York City. This 34-ton giant was hauled from Greenland in the 1890s by the explorer Robert Peary and is now in the Hall of Meteorites of the American Museum of Natural History.

What happens when a meteorite strikes the Earth? Quite naturally, it makes a hole. You can confirm that such a hole will be much larger than the meteorite by throwing a rock into mud to simulate a meteorite strike. You will see that a crater much larger than the rock is produced. ▶Figure 10-25 is a photograph of the most prominent impact crater on Earth, Meteor Crater near Winslow, Arizona. The crater is in the desert about 40 miles east of Flagstaff. The crater, nearly a mile across, is 180 meters deep and has a rim rising 45 meters above the

The crater is 600 feet deep and rises 150 feet above the surrounding ground.

➤ FIGURE 10-25. This view of Meteor Crater shows the guesthouse and entrance road at the bottom left edge of the crater.

surrounding desert ground. To appreciate the size of the crater, look at the left center of Figure 10-25 and find the parking lot and guest building.

Some 25 tons of iron meteorite fragments have been found around the crater, some as far as four miles away. The meteorite that formed Meteor Crater is estimated to have had a total mass of 300 million kilograms (300,000 tons) and to have been about 45 meters across. It struck about 25,000 years ago at a speed of about 11 kilometers/second (25,000 miles/hour), releasing an energy equivalent to 1000 Hiroshima bombs.

Many other meteorite craters have been found on Earth, but most are far less impressive than Meteor Crater because weather has worn them down or they are under the sea. Meteor Crater is not only the most recently formed major crater, but it is well preserved because of its location in the dry Arizona desert.

It is estimated that a meteorite larger than 1 kilometer in diameter strikes the Earth on the average once every 3 or 4 million years. A hit by a meteorite 1 kilometer in diameter would be equivalent to a 5000-megaton bomb and would produce a crater 10 kilometers in diameter. Such an explosion would have effects far beyond the area of impact because it could send enough dust into the atmosphere to block out a significant amount of sunlight. A hypothesis put forward nearly a decade ago suggests that such a meteorite strike was responsible for the extinction of the dinosaurs. (See the accompanying Close Up.)

The crater shown in ➤Figure 10-26 was formed on May 17, 1990, in a freshly planted wheat field in Russia. The 10-meter diameter crater was caused by an iron meteorite about 1 meter across. The impact energy is calculated to have been equivalent to $1\frac{1}{2}$ tons of TNT.

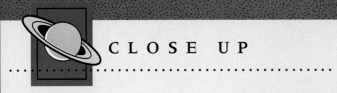

CLOSE UP

Meteors and Dinosaurs

Sixty-five million years ago, dinosaurs became extinct. In fact, nearly 75% of all species on Earth became extinct during the same short time period (as geological time is measured). The reason for this mass extinction has been debated by scientists for some time, and a number of different catastrophes have been hypothesized as the cause for the extinctions.

In 1980 astronomy entered the picture. In that year, Walter Alveraz, Luis Alveraz (Walter's father), Frank Asaro, and Helen Michel proposed a solution for which they had real evidence. At a number of sites around the Earth, they found a layer of clay that contained the elements iridium, platinum, and osmium in much greater abundance than normally found on Earth. How could this layer have been deposited? These elements are much more abundant in meteorites than in the crust of the Earth. Perhaps a giant meteorite—an asteroid—struck the Earth, exploded, and sent its debris high into the atmosphere to fall to Earth and form the layer. To account for the thickness of the layer found around the world, the meteorite must have been about 10 kilometers in diameter—not an unusual size for an asteroid. The debris, which also would have included earthly material pulverized by the impact, would spread as a cloud around the Earth and then gradually settle to form the layer found by the Alveraz team.

But would the explosion have wiped out entire species all around the Earth? No, not directly. But consider what effect a giant cloud of dust would have on vegetation on Earth. Such a cloud could remain in the atmosphere for more than a year, darkening the Earth below. Much vegetation would die, along with many animal species that depended on vegetation for food. Dinosaurs are included in this category as well as many other species. Small creatures such as rodents that could forage for seeds and nuts could survive, however. And, indeed, the fossil record shows that such creatures did survive this period.

One test of this asteroid hypothesis is to find out how long ago the clay layer was formed; that is, to see if it is 65 million years old. If you have studied geology, you know that such dating of past events is a common procedure. The result? The clay was deposited about 63 million years ago, remarkably close to the value given for the mass extinction as shown by the fossil record.

Since 1980, when the Alveraz team presented their data, their hypothesis has become more and more accepted. (In fact, it is usually referred to as a *theory*, in accordance with the practice of calling an idea a theory only after it has gained some acceptance, as was discussed in the first few chapters.)

➤ **FIGURE 10-26.** This 10-meter-wide crater was formed in Russia in 1990 by a 1500-kilogram iron meteorite.

CONCLUSION

The solar system is indeed a collection of diverse objects. As our space probes venture out to study it in more detail, we are continually surprised by the beautiful diversity we find. As the last few chapters have shown, great differences exist not only among planets, asteroids, comets, meteors, and satellites, but even between objects within these various categories. On the other hand, we cannot help but be impressed by the similarities we see—even between objects we place in different categories.

As Chapter 7 explained, the objects in our solar system formed at about the same time, and their formation and differentiation were determined by the conditions that existed at the time. We are beginning to understand why they are different and why they are the same, but we have a long way to go. That is what makes the study of the solar system exciting.

RECALL QUESTIONS

1. The connection between comets and meteors is demonstrated when one sees
   A. sporadic meteors on almost any night.
   B. a comet in one part of the sky and meteors in another part of the sky on the same night.
   C. a predictable shower of meteors.
   D. [None of the above; there is no connection.]

2. Which of the following statements about Pluto is true?
   A. It has two moons (that we know of).
   B. Its orbit is within 2 degrees of the plane of the Earth's orbit.
   C. Its orbit is as circular as the Earth's orbit.
   D. We are confident that it was once a moon of Neptune.
   E. [None of the above is true.]

3. The search for Pluto was undertaken based on
   A. Bode's law.
   B. calculations concerning the orbit of Uranus.
   C. perturbations in the orbits of asteroids.
   D. [None of the above, for it was discovered accidentally.]

4. Pluto was discovered by
   A. William Herschel in England in the last century.
   B. Percival Lowell in Arizona in 1905.
   C. Clyde Tombaugh in Arizona in 1930.
   D. James Christy in Arizona in 1978.
   E. Asaph Hall in California in 1921.

5. Pluto's mass has been calculated from
   A. its effect on passing spacecraft.
   B. its effect on passing asteroids.
   C. the motion of Charon.
   D. [All of the above.]
   E. [None of the above.]

6. How was Ceres discovered?
   A. It was discovered by accident, during a search for comets.
   B. It was observed when it fell to Earth.
   C. It has been known since antiquity, so the manner of its discovery is unknown.
   D. It was found during a search for a "missing planet."
   E. It was found accidentally as a streak on a stellar photo.

7. The discovery of asteroids depends on the fact that, compared to the stars, the asteroids
   A. look bigger.
   B. look brighter.
   C. move.
   D. vary in brightness.
   E. are a different color.

8. The largest asteroid is closest to _____ in diameter.
   A. 6 feet
   B. 600 feet
   C. 6 miles (10 km)
   D. 600 miles (1000 km)
   E. 60,000 miles (100,000 km)

9. The total radiation reflected and radiated by an asteroid depends on
   A. its size.
   B. the total radiation it receives from the Sun.
   C. its speed in orbit.
   D. [Both A and B above.]
   E. [All of the above.]

10. The Kirkwood gaps result from
    A. radiation pressure from the Sun.
    B. the previous passage of a comet through the asteroid belt.
    C. regular gravitational pull from the largest of Jupiter's moons.
    D. regular gravitational pulls from Jupiter.
    E. previous explosions of asteroids at points in the asteroid belt.

11. The masses of the largest asteroids are measured by observing
    A. their motions as they come closest to Jupiter.

B. perturbations they cause on smaller asteroids.
  C. the motion of spacecraft passing nearby.
  D. the total radiation received from them.
  E. their orbital speed and distance from the Sun.

12. Which of the following statements about the *Voyager* spacecraft that passed through the asteroid belt is true?
    A. They were guided from Earth so that they did not collide with asteroids.
    B. They had their own guidance system to prevent collisions with asteroids.
    C. They passed through the asteroid belt with no collision-avoidance system.

13. Which of the following best describes comet orbits?
    A. They are circular, lying in the plane of the ecliptic.
    B. They lie between the orbits of Mars and Jupiter.
    C. They are elongated ellipses, lying within one astronomical unit of the Sun.
    D. They are very elongated ellipses tens to hundreds of astronomical units across.
    E. They are within the orbit of Mercury.

14. When a comet is visible in the night sky, it appears to the naked eye to
    A. move so rapidly across the sky that it is easily missed if one is not looking in the right direction.
    B. stay in the sky for a week or more, with only slight shifting among the stars each night.
    C. [Either of the above, depending upon the particular comet's motion.]

15. The nucleus of most comets is
    A. much smaller than the Earth.
    B. slightly smaller than the Earth.
    C. slightly larger than the Earth.
    D. much larger than the Earth.

16. A comet's tail results when
    A. part of the comet drifts in the direction opposite to the motion of the comet.
    B. the solar wind carries some of the gas of the coma away from the Sun.
    C. gravitational force pulls loosely held material from the Sun.
    D. tidal forces tear the comet apart.

17. The best-known comet was named after Halley because
    A. he was the first to see it.
    B. he first calculated its orbit.
    C. although he had nothing to do with the comet, he was a famous astronomer and it was named to honor him.
    D. Galileo discovered it and named it after him.
    E. [Both C and D above.]

18. A comet's tail
    A. precedes its head through space.
    B. follows its head through space.
    C. is farther from the Sun than its head is.
    D. is closer to the Sun than its head is.
    E. [None of the above.]

19. The Oort cloud is located
    A. between the orbits of Mars and Jupiter.
    B. just beyond Pluto's orbit.
    C. far beyond the orbit of Pluto.
    D. at about the distance of the nearest star.

20. Why do many comets appear brighter after passing the Sun?
    A. They are more massive after the passage.
    B. The Sun starts nuclear reactions in them.
    C. They are moving faster after passing the Sun.
    D. Their head and tail are larger after passing the Sun.

21. Short-period comets are hypothesized to
    A. be permanent parts of the inner solar system similar to asteroids.
    B. be formed when the orbits of long-period comets are changed.
    C. be formed in the inner solar system and then ejected, becoming long-period comets.

22. Why do most meteroids not reach the surface of the Earth?
    A. They bounce off the atmosphere and go back into space.
    B. They burn up in the air.
    C. They are light enough that they remain suspended in the air.
    D. They land in the ocean (since oceans cover most of the Earth).
    E. [The statement is false; almost all reach the Earth, but they are not found.]

23. Meteor showers are caused by
    A. the Earth crossing the asteroid belt.
    B. a comet's nucleus striking the atmosphere.
    C. the eccentricity of the Earth's orbit.
    D. meteorite impacts on the Earth's surface.
    E. the Earth crossing a meteoroid stream.

24. Meteors are most easily seen after midnight because
    A. the sky is darker then.
    B. the Sun is closer to rising.
    C. you are then on the "leading" side of the Earth.
    D. meteor showers occur then.
    E. [The statement is false; they are seen equally well anytime.]

25. List two ways in which Pluto's orbit is unusual compared to the other planets.

26. We cannot determine the size of Pluto from the size of its image in a telescope. How, then, can we know its size?

27. What did the discovery of Charon allow us to calculate about Pluto?

28. Why have we not named all of the asteroids that have been observed?

29. What causes the Kirkwood gaps?

30. Why have the asteroids not formed into a planet between Mars and Jupiter?

31. Why do we think that it is unlikely that the asteroids are the remains of a planet that exploded?

32. Describe the three main parts of a comet. Include approximate sizes of each part for a typical comet.

33. Describe the modified "dirty snowball" model of a comet's nucleus.

34. What causes a comet to have two tails? Why are they different shapes?

35. What is the Oort cloud?
36. We cannot hope to see the Oort cloud. What makes us think that it exists?
37. Explain why the passing air heats a meteoroid, while a wind cools a person.
38. What causes a meteor shower and why do the meteors appear to come from just one part of the sky?
39. Name and describe the three main types of meteorites. Which is most common? Which is easiest to find? Why?

QUESTIONS TO PONDER

1. If a comet is seen in the west shortly after sunset, in what direction will its tail point—toward or away from the horizon? In what direction will it point if it is seen in the east in the morning sky?
2. Find out about a meteor crater in your part of the country or the world. The geology department of your college is a suggested source of information.
3. What effect on the motion of a comet would you expect to result from the jetting of material from its nucleus?
4. Compare the size and shape of Phobos to that of the nucleus of Halley's comet.
5. At the time of this writing, NASA was planning *NEAR* (Near Earth Asteroid Rendezvous), a spacecraft that would be launched in 1998 to begin orbiting an asteroid two years later. Use *Astronomy* and/or *Sky and Telescope* magazines to write a report on the status of these plans.

CALCULATIONS

1. Assume that 1000 metric tons (one million kilograms) of meteoritic material strikes the Earth each day. Calculate the area of the Earth and determine how much material strikes each square meter each day.
2. Use the answer to the previous calculation to determine how long it would take to build up a layer of meteoritic material weighing 1 kilogram on a desktop that has an area of 2 square meters.

ACTIVITY

Observing Meteors

To observe meteors, you should have a clear sky away from city lights, so your first step is to get out into the country. Second, you must avoid a night with a bright Moon that will light up the sky so that you will have difficulty viewing. Finally, the best time for observing is after midnight, preferably from around 2 A.M. until the beginning of morning twilight.

The reason for choosing a time after midnight can be seen by referring to ➤Figure A10-1. In this figure, the Sun (not shown) is to the left, and the Earth is moving toward the top of the figure. Thus the half of the Earth lined in blue is on the leading edge as the Earth moves through space. Now consider a car driving through a rain shower and you see why most meteoroids hit the Earth on this leading side. The analogy is not quite exact because different meteoroids have very different motions through space, while raindrops all fall in about the same

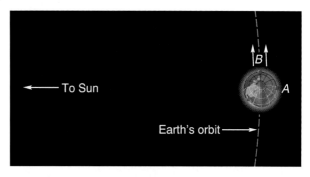

➤ **FIGURE A10-1.** The Earth is moving upward in the drawing and strikes most meteoroids on its leading edge. Those striking it from behind must catch up. It is midnight at point *A* and sunrise at point *B*.

direction. Meteoroids are able to catch up with the Earth from behind and hit the trailing half of the Earth. Still, many more strike the Earth's leading edge as the Earth sweeps them up.

Referring again to the figure, you see that it is midnight for a person standing at point *A* and that the Sun is rising for a person at *B*. Meteors, then, are better observed in the early morning hours.

Once you have found your observing location, you must wait for your eyes to adapt to the darkness. Unless you have chosen a night of a heavy meteor shower, you must be patient. Lie back and relax. (Did you bring a lawn chair?) Patience should reward you with a streak of a meteor across the sky. It is likely that you will see the meteor in some direction other than where you are looking—out of the corner of your eye. There is a good reason for this: the part of your retina that sees things out of your direct line of sight is more sensitive to motion and to changing light. An experienced observer takes advantage of this effect when looking, for example, for the motion of a satellite in the sky. The observer looks to the side of where the object is expected to be seen.

The best advice to the person looking for meteors is to pick a night when a meteor shower is expected. Table 10-1 lists the dates of the most prominent annual meteor showers.

StarLinks netQuestions

Visit the netQuestions area of StarLinks (www.jbpub.com/starlinks) to complete exercises on these topics:

1. Hale-Bopp: The Great Comet of 1997 On July 23, 1995, an unusually bright comet outside of Jupiter's orbit was discovered independently by Alan Hale of New Mexico and Thomas Bopp of Arizona. It turned into the brightest, most-viewed comet in recent memory.

2. Meteors, Meteorites and Impacts Meteors, most commonly called "shooting stars," are bright streaks of light that flash in the sky when small meteoroids enter into the Earth's atmosphere. When one of these bits of space debris lands on earth, it provides clues to the solar system.

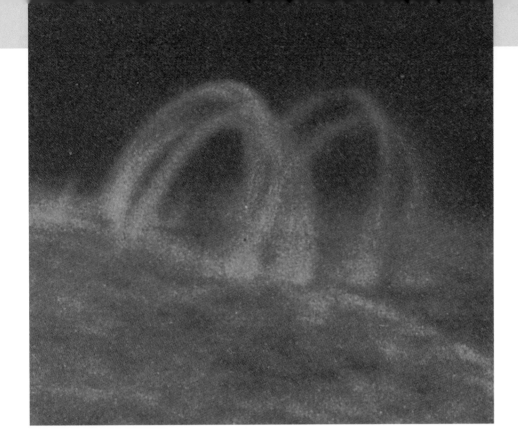

THE SUN

Solar Data
Solar Energy
CLOSE UP: The Distance to the Sun
The Source of the Sun's Energy
Solar Nuclear Reactions
CLOSE UP: Fission and Fusion Power on Earth
The Sun's Interior
Pressure, Temperature, and Density
Hydrostatic Equilibrium
Energy Transport
The Neutrino Problem

The Solar Atmosphere
The Photosphere
CLOSE UP: Data Uncertainty in the Homestake Experiment
CLOSE UP: Helioseismology
The Chromosphere and Corona
The Solar Wind
Sunspots and the Solar Activity Cycle
A Model for the Sunspot Cycle
Solar Flares
ACTIVITY: Measuring the Diameter of the Sun
ACTIVITY: Observing Sunspots

.... *THE CHAPTER-OPENING PHOTOGRAPH, which was taken from space, shows prominences on the Sun's surface, giving us a hint of the tremendous energy within. The energy comes from atomic nuclei. The following items, which appeared in* Scientific American *just one month apart in 1939, report on two breakthroughs in our knowledge about the nucleus. The first relates to the Sun's energy and is purely scientific, while the other relates to nuclear power used for a different purpose, one that led to atomic bombs and nuclear power plants.*

For the source of stellar energy we must look to some process that changes the mass of atoms and liberates a corresponding amount of energy ... in particular, to reactions between charged atomic nuclei. Here a notably successful theory has been developed by Professor Bethe of Cornell, along lines resulting from the recent work of Gamow and others. (From "What Keeps the Stars Shining?" Henry Norris Russell, *Scientific American*, June 1939, p. 369.)

For years past, scientific writers have been pointing out that if the energy known to be confined within the atom could be practically and commercially released and employed, there would be power in amounts today unknown. . . . The atomic energy of a piece of coal would exceed that of the same coal burned beneath a boiler by a ratio of a whole billion to one. . . . As this is written, physicists are at work on experiments that seem to contain the possibility if not the probability that the billion-to-one power source is at least in view. (From "Our Point of View," *Scientific American*, July 1939, p. 2.)

The Sun, the celestial object of most importance to life on Earth (➤Figure 11-1), is just an ordinary star. The cosmic importance of the Sun is limited to the fact that it is the king of the planetary system in which we live; many other stars are bigger and brighter, many are more interesting and unusual, and most will far outlive the Sun.

This chapter begins with a brief overview of the major properties of the Sun. Then the source of the Sun's energy will be examined along with various theories for the production of that tremendous energy. Finally, the Sun will be described in more detail, starting at its center, where energy production takes place, and proceeding outward.

➤ **FIGURE 11-1.** The Sun is the ultimate source of oil (including that pumped by these rigs) and of all energy on Earth except nuclear and geothermal energy.

The accompanying Close-Up explains how we know the distance to the Sun. An Activity at the end of the chapter shows how you can measure the diameter of the Sun.

sunspot. A region of the photosphere that is temporarily cool and dark compared to surrounding regions.

SOLAR DATA

As viewed from Earth, the Sun has an average angular diameter of 31'59", just barely less than 32 minutes of arc. By taking 1.50×10^8 kilometers as the average distance from Earth to Sun and by using either the method shown in the first Activity at the end of this chapter or the relationship between angular size, distance, and actual size (see Chapter 6), we can calculate that the diameter of the Sun is 1.39×10^6 kilometers. Thus the Sun's diameter is about 110 times the Earth's and about 10 times Jupiter's. ➤Figure 11-2 illustrates the great size of the Sun, and Table 11-1 lists solar data.

From Kepler's third law (as revised by Newton), we calculate that the mass of the Sun is 1.99×10^{30} kilograms. This is more than 300,000 times the mass of the Earth! The Sun's average density, then, is 1.41 gm/cm^3, about the same as the density of Jupiter. This is just the first of a number of similarities we will see between the king of the solar system and the king's subjects, the planets.

When Galileo first used a telescope to view the Sun nearly 400 years ago, he observed dark spots moving across its face. ➤Figure 11-3b is a series of photographs of the motion of such **sunspots.** Galileo concluded from sunspot motions that the Sun rotates with a period of more than a month. Today we know that the Sun exhibits differential rotation; it rotates with a period of 25.4 days at its equator and nearly 40 days near its poles. Recall that the equatorial regions of the Jovian planets also rotate faster than their polar regions.

SOLAR ENERGY

The Sun emits energy in all portions of the electromagnetic spectrum. A valuable piece of information about the Sun is the rate at which it emits its energy—its

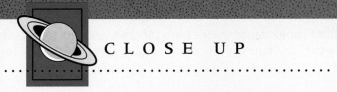

CLOSE UP

The Distance to the Sun

How do we measure the distance to the Sun? Knowledge of this distance is key to calculating the size of the Sun, and the size must be known before other solar properties—such as density and total luminosity—can be determined.

As was pointed out when the question of the heliocentric-versus-geocentric system was discussed in Chapter 2, the relative distances to the planets can be calculated by applying simple geometry to the heliocentric system. Thus Copernicus was able to calculate that Mars is 1.5 times as far from the Sun as the Earth is. We say that Mars is 1.5 astronomical units from the Sun. But how large is an astronomical unit? The situation is similar to the following scenario: You have a map of some unknown country. The map is drawn to scale, so you can tell from the map that the nation's capital is one-third of the way from the mountains to the ocean. Unless you know the scale of the map, however, you have no way of knowing the distance from the mountains to the capital. You know only relative distances. If a scale tells you that one inch on the map corresponds to 120 miles, you can make measurements on the map and calculate all distances. In the case of the solar system, Copernicus lacked the scale factor—the number of kilometers per astronomical unit.

Today, radar allows us to determine the scale factor. We can bounce radio waves from other objects and use the time of travel of the waves to determine the distance to the object. We cannot get good results bouncing radar from the Sun, both because the Sun is not a good reflector of radar and because the Sun emits its own radio waves. We can, however, reflect radar from other planets. Suppose we bounce radar from Mars when Mars is 180 degrees from the Sun in the sky. Since Earth's distance to Mars at that time is 0.5 AU, we can determine the number of kilometers in 0.5 AU and therefore determine the scale of the solar system.

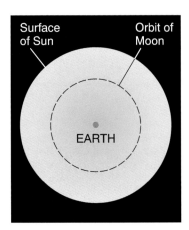

▶ FIGURE 11-2. If the Earth were at the center of the Sun, the Moon would orbit about half-way to the Sun's surface.

(a)

▶ FIGURE 11-3. (a) Sunspots appear as dark spots on the Sun's surface. (b) This photo series shows the motion of sunspots as time passes.

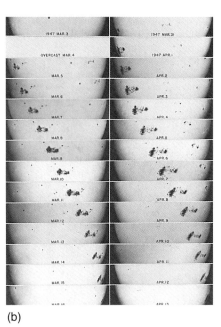

(b)

luminosity or total power output. Fortunately, this is not too difficult to determine.

We start by measuring the rate at which solar energy strikes the Earth's atmosphere. This determination was made long ago by measuring the amount of energy from the Sun that strikes an area on the Earth's surface and then correcting for the energy absorbed by the atmosphere. Today it is done most accurately from satellites above the atmosphere. Measurements show that solar energy strikes the upper atmosphere of the Earth at the rate of 1380 watts per

luminosity. The rate at which electromagnetic energy is being emitted.

Solar Energy

square meter. The following example shows how this can be used to calculate the Sun's luminosity—the "wattage" of the Sun.

EXAMPLE

We want to calculate the energy output of the Sun in watts, given that solar energy strikes the Earth at the rate of 1380 watts/m² and that the Sun is 1.50×10^8 kilometers from Earth.

Solution

First, since we have expressed the area on Earth in square meters, we should also express the Earth-Sun distance in meters:

$$1.50 \times 10^8 \text{ km} \times \frac{1000 \text{ m}}{1 \text{ km}} = 1.50 \times 10^{11} \text{ m}$$

Imagine a sphere around the Sun at the Earth's distance. We must calculate how many square meters are on that surface (see ➤Figure 11-4). To do this, use the equation for the area of a sphere, with 1.5×10^{11} meters being the radius of the sphere in the calculation:

$$\text{area of sphere} = 4\pi r^2 \quad (r = \text{radius of the sphere})$$
$$= 4(3.14)(1.50 \times 10^{11} \text{ m})^2$$
$$= 2.83 \times 10^{23} \text{ m}^2$$

Each of these square meters receives 1380 watts of solar power. Therefore,

$$\text{total solar power} = (1380 \text{ watts/m}^2) \times (2.83 \times 10^{23} \text{ m}^2)$$
$$= 3.9 \times 10^{26} \text{ watts}$$

Given the power striking the Earth's surface, you could use the inverse square law to determine this answer.

······**TRY ONE YOURSELF.** How many watts of power strike one square meter of Mars's surface? (Mars is 2.3×10^{11} meters from the Sun.) Solve this by using the total solar power calculated in the example and the area of a sphere at Mars's distance.

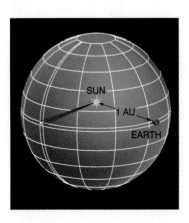

➤ **FIGURE 11-4.** An imaginary sphere is shown drawn around the Sun at the distance of the Earth. It is a simple matter to calculate the area of such a sphere and to determine the solar energy that strikes each square meter of it.

TABLE 11-1 Solar Data

	Value	Compared to Earth
Diameter	1,392,530 km	109
Mass	1.989×10^{30} kg	330,000
Density	1.41 gm/cm³	0.26
Surface gravity		28
Escape velocity	617 km/s	55
Surface temperature	6000°C	
Luminosity	3.9×10^{26} watts	
Tilt of equator to ecliptic	7.25°	
Rotation period		
Equator	25.38 days	
40° latitude	28.0 days	
80° latitude	36.4 days	

As the example shows, the Sun's luminosity is 3.9×10^{26} watts. Writing this number as 390,000,000,000,000,000,000,000,000 helps us see that it is truly an awesome amount of *power*.

The example also illustrates nicely why the inverse square law applies to radiation from the Sun (or any distant object). The area of a sphere that is centered on the Sun depends upon the square of the sphere's radius. Thus a sphere twice the distance from the Sun has an area four times as great, and only one-fourth as much energy strikes a square meter on the surface of this more distant sphere.

power. The amount of energy exchanged per unit time.

A watt is a unit of power and is a specific amount of energy each second.

The Source of the Sun's Energy

It is estimated that a 1 percent change in solar luminosity would result in a temperature change on Earth of 1 or 2 Celsius degrees (about 2 to 4 Fahrenheit degrees). When we consider that the last major ice age on Earth resulted from a temperature decrease that averaged only 5 Celsius degrees across the planet, we see how critical it is that the Sun maintain a uniform rate of energy production. What is the source of this energy that must have remained constant far into the past?

Prior to this century, a number of hypotheses had been suggested to explain the source of the Sun's energy. All have now been rejected, based on additional data. For example, it was proposed that chemical reactions (such as the burning of a fuel) are the source of the Sun's energy. We now know that this cannot be the case, simply because if the Sun were made of a fuel such as coal or oil, it would burn out in a few centuries at the rate that it is releasing energy.

In the mid-nineteenth century, Hermann von Helmholtz and William Lord Kelvin proposed that the source of the Sun's energy is a very slow gravitational contraction. We will see later how such a contraction produces energy. Their calculations showed that at the Sun's present rate of energy production, gravitational contraction could have produced the energy with a reduction in the size of the Sun so slight that it would not have been enough to notice in recorded history. The rate of energy production would not have changed appreciably over the past 100 million years. Their theory seemed to be a good one; it fit the available data.

Then in this century geologists learned that the Earth's age is not a few hundred million years but rather a few *billion* years—ten times longer. The contraction theory had to be abandoned, and the source of the Sun's energy was again an open question.

In the first decade of this century, Albert Einstein proposed, as part of his special theory of relativity, that mass and energy are interconvertible. That is, one can be changed into the other. Late in the 1920s it was hypothesized that this process could be the source of energy in the Sun. Then during the 1930s, physicists worked out the theory that today explains how the Sun has produced its tremendous power for the past 4 to 5 billion years and how it will continue this production for another similar period of time.

The conversion between mass and energy is what the equation $E = mc^2$ is about. E stands for the amount of energy that can be created from a certain amount of mass, m. The c is the conversion factor and has a value equal to the speed of light.

Solar Nuclear Reactions

Recall that the Bohr model of the atom proposed that the atom consists of a nucleus surrounded by orbiting electrons (▶Figure 11-5). That nucleus will be our focus now. An atom's nucleus makes up about 99.98% of the mass of the atom and consists of two kinds of particles: ***protons*** and ***neutrons.*** Protons have a positive electrical charge and neutrons have no electrical charge. The number of protons in the nucleus determines what element the atom is. For example, if the nucleus of an atom contains 1 proton, that atom is necessarily an atom of

proton. The massive, positively charged particle in the nucleus of an atom.

neutron. The massive nuclear particle with no electric charge.

Solar Energy 341

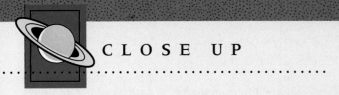

CLOSE UP

Fission and Fusion Power on Earth

The dream of unlimited energy has been with us at least since the beginning of the industrial age, and humans wondered at the tremendous power of the Sun long before that. We know now that nuclear fusion reactions are the source of the Sun's energy, and just a few decades ago humans harnessed this energy (if *harnessed* is an appropriate word here) in the hydrogen bomb. The bomb's name comes from the fact that it uses a form of hydrogen (deuterium) as its fuel. The nuclear reaction in the H-bomb is similar to that in the Sun; it produces helium from hydrogen.

Peaceful uses of fusion power have not yet been developed, although much research has taken place over the last four decades and is still in progress. A hint at the problems of controlling fusion can be seen by considering the tremendous temperatures and pressures that are necessary to produce fusion in the center of stars. On the other hand, such reactions have an essentially unlimited supply of fuel—deuterium—and therefore controlled energy production from fusion is a very attractive goal.

Although fusion is much more common than fission in the universe as a whole, the latter was the first to be developed by humans. Fission involves the release of energy when a large nucleus is broken into two medium nuclei, with mass being converted into energy in the process. The atomic bomb (poorly named, for it uses the energy of the nucleus rather than the energy of the outer atom) was the first application of fission power. Since the development of the A-bomb, we have learned to control this reaction, and today we use the energy of fission to produce electricity in nuclear power plants. Contrary to the ready availability of fuel for fusion, the uranium that must be used for fission power is definitely limited. Many people, scientists and nonscientists alike, question the wisdom of building and using fission power plants; but in any case, fission power must be viewed as only a temporary solution to the problem of finding a long-range source of energy on Earth.

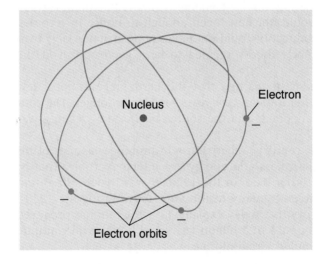

➤ FIGURE 11-5. The Bohr atom, with electrons (which have a negative charge) circling the positive nucleus. If the atom were this size, the nucleus would still be too small to see.

Do not confuse the terms *proton* and *photon*; they are very different. Refer to Chapter 4 for a description of photons.

hydrogen. If it contains 2 protons, it is helium; if 6, carbon; if 92, uranium. A nucleus of hydrogen, on the other hand, is not limited to a specific number of neutrons. Although most hydrogen nuclei contain no neutrons, some have one neutron and a few, two.

It is important to distinguish nuclear reactions from chemical reactions. The latter, which we encounter in everyday life, involve the combining of a number of atoms into a molecule or the separating of a molecule into individual atoms or into smaller molecules. When we burn paper, for example, the paper's carbon atoms combine with oxygen atoms from the air, producing carbon dioxide molecules. The nuclei of the individual atoms do not come into play. ➤Figure 11-6

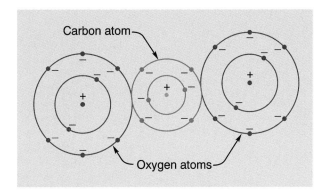

FIGURE 11-6. A carbon dioxide molecule (like all molecules) is held together by bonds between the electrons of its two oxygen atoms and one carbon atom. Nuclear forces do not come into play in the bonding.

illustrates this reaction and emphasizes that the forces between atoms involve only forces exerted by the electrons of one atom on the electrons of another, not forces exerted by the nuclei on one another.

Nuclear reactions, on the other hand, involve forces between nuclear particles; orbiting electrons are not part of these reactions. There are many types of nuclear reactions but only one will be of interest to us: the *fusion* reaction. In a nuclear fusion reaction, two nuclei combine to form a larger nucleus. They "fuse."

fusion (nuclear). The combining of two nuclei to form a different nucleus.

The primary source of energy in the Sun (and in all stars during most of their lifetimes) is a series of nuclear fusion reactions in which four hydrogen nuclei are fused to form one helium nucleus. In the process, a small fraction of the mass of the nuclei is changed into energy. This is where Einstein's theory comes into play. Let's look at the process.

Most hydrogen nuclei consist simply of one proton. Prior to the fusion reaction, we have four hydrogen nuclei, and after the reaction, there is one helium nucleus. Let us subtract the mass of one helium nucleus (6.6466×10^{-27} kilogram) from the mass of four hydrogen nuclei (each having a mass of 1.6736×10^{-27} kilogram):

$$\begin{array}{r} \text{mass of 4 hydrogen nuclei} = 6.6942 \times 10^{-27} \text{ kg} \\ \underline{\text{mass of 1 helium nucleus} = 6.6466 \times 10^{-27} \text{ kg}} \\ \text{difference} = 0.0476 \times 10^{-27} \text{ kg} \end{array}$$

The difference between the mass of the original matter and the resulting matter is very small, not only in terms of the actual amount (less than 10^{-28} kilogram) but also in that the lost mass is only seven-tenths of 1% of the original. Not even 1% of the mass is changed into energy. In fact, the energy produced by one single fusion is only enough to lift a mosquito a fraction of a millimeter. In order to produce the Sun's output of 3.9×10^{26} watts, a total of 4.3×10^9 kilograms—nearly 5 million tons—of matter must be converted into energy each second. This is about 10^{38} nuclear reactions, involving the transformation of some 610 billion kilograms of hydrogen to 606 billion kilograms of helium. Although this is a tremendous amount of matter by human standards, it is almost insignificant when compared to the Sun's total mass. If the Sun were pure hydrogen, about 100 billion years would be required to convert it entirely to helium at the present rate of consumption. (As will be discussed later in the chapter, only the inner portion of the Sun—about 10% of its mass—is involved in the reaction and will change to helium.)

In practice, the process by which hydrogen is converted into helium incorporates three steps. The reactions start with a fusion of two protons (hydrogen nuclei) and end with production of a helium nucleus containing two protons and two neutrons. During the process two other smaller nuclear particles are produced, as well as a gamma ray. The solar fusion reactions are presented in more detail in Table 11-2 and ➤Figure 11-7.

Solar Energy

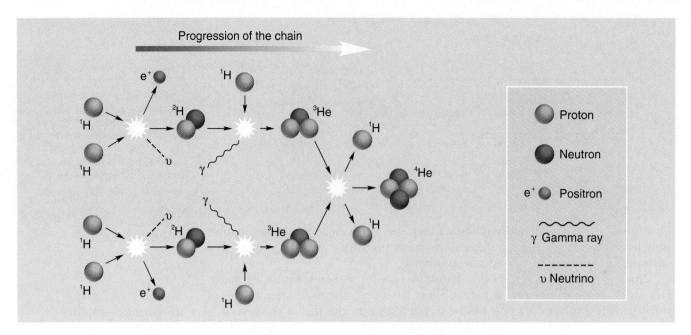

▶ **FIGURE 11-7.** The proton-proton chain begins with four protons and ends with a helium nucleus. The four protons are shown at left combining in separate reactions to produce two deuterium nuclei (each with a proton and a neutron).

proton-proton chain. The series of nuclear reactions that begins with four protons and ends with a helium nucleus.

TABLE 11-2 The Proton-Proton Chain

Reaction	Explanation
$^1_1H + {}^1_1H \rightarrow$ $^2_1H + e^+ + \nu$	Two protons combine to produce a deuterium nucleus (^2_1H; the 2 indicates the total number of protons and neutrons in the nucleus), a positron (e^+, which is a positive electron), and a neutrino (ν).
$^2_1H + {}^1_1H \rightarrow$ $^3_2He + \gamma$	A deuterium nucleus joins with another proton to produce a helium nucleus (^3_2He, containing two protons and one neutron) and a gamma ray (γ).
$^3_2He + {}^3_2He \rightarrow$ $^4_2He + 2\, {}^1_1H$	Two helium nuclei fuse to form the common type of helium (^4_2He) and two protons.

deuterium. A hydrogen nucleus that contains one neutron and one proton.

positron. A positively charged electron emitted from the nucleus in some nuclear reactions.

neutrino. An elementary particle that has little or no rest mass and no charge but carries energy from a nuclear reaction.

Look at the first reaction in the table. Here, two hydrogen nuclei fuse to form the nucleus of another type of hydrogen, called **deuterium,** which has a neutron in its nucleus along with the proton. In addition, two other particles are formed, and these two particles fly away from the reaction at great speeds. One of them, the positive electron, or *positron,* agitates nearby nuclei as it moves among them, causing these nuclei to increase in speed. If nuclei increase in speed, however, the effect we observe is a rise in temperature. It is by means of this process that the energy from fusion becomes thermal energy. The other particle, the *neutrino,* escapes from the Sun and does not cause significant heating within the Sun. This particle is discussed later in this chapter.

Hydrogen, of course, exists throughout the world. Yet it does not, on its own, fuse into helium. The reason for this is that all nuclei have the same type of electric charge (positive) and therefore they repel one another. This electrical repulsion force acts over great distances, at least compared to the very short distances over which nuclear forces act. This means that the particles are unable to get close enough together for the attractive nuclear forces to take over unless they happen to be moving toward one another at a great speed (➤Figure 11-8). Since the temperature of a gas is determined by the speed of its particles, the particles of hot hydrogen are more likely to fuse than those of cool hydrogen. Significant fusion occurs only in high-temperature matter.

In addition, as we have seen, a great number of fusions of hydrogen nuclei must occur each second in order to produce the Sun's power. Thus we know that the density of matter must be extremely high in the region of the Sun where fusion takes place. To see how and where this occurs, we will investigate the internal structure of the Sun.

THE SUN'S INTERIOR

Obviously, we cannot examine the interior of the Sun directly. Astronomers have learned much about its interior, however, by computer modeling of its interior, based on observations of its surface and our knowledge about the behavior of matter at the temperatures and pressures that are necessary to sustain fusion reactions. We know that at temperatures as high as the Sun's, matter must exist as a gas rather than as a solid or a liquid, and this fact makes the analysis easier, for gases are much simpler than liquids or solids. As will be discussed later, the temperature on the surface of the Sun is about 6000°C, and it increases greatly below the surface. At these temperatures, solids and liquids cannot exist and most electrons are stripped away from their nuclei. As a result, most of the material of the Sun's interior consists of free nuclei and free electrons. The behavior of this material, however, is similar to that of a simple gas, so we must first study properties of gases. The properties of importance to us are temperature, pressure, and *particle density*.

Pressure, Temperature, and Density

When you blow up a balloon, the gas inside exerts a pressure on the rubber of the balloon and supports it against its tendency to contract. The pressure exerted by the gas is the result of collisions of the individual molecules of the gas with the rubber surface. ➤Figure 11-9a illustrates this. Each molecule, as it strikes the rubber wall and rebounds, exerts a tiny force on the wall. Although we cannot

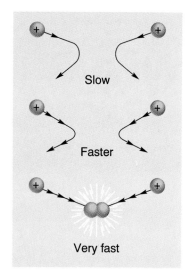

➤ **FIGURE 11-8.** Nuclei moving toward one another at too slow a speed will be repelled because of their positive charges, but if they are moving fast enough, electrical repulsion will not be strong enough to prevent them from colliding and fusing.

particle density. The number of separate atomic and/or nuclear particles per unit of volume.

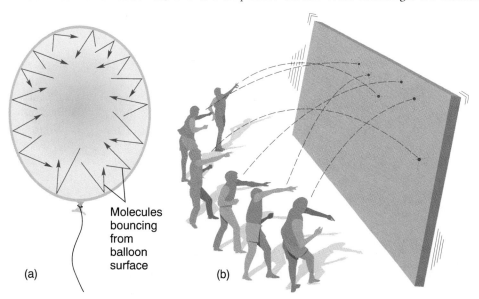

➤ **FIGURE 11-9.** The walls of a balloon are held out by numerous collisions by molecules inside the balloon in the same way that a great number of tiny sand-throwers could exert a force on a board and topple it backward.

detect each individual bounce and the force produced, the overall force exerted by the gas is simply the total of all of these individual impulses. To see this, imagine tiny grains of sand being fired at a board by a great number of sand-throwers as in Figure 11-9b. The force exerted on the board by a single grain of sand might seem negligible, but the overall result could be a force great enough to cause the board to move.

Notice that the discussion above sometimes refers to force and other times to pressure. *Pressure* is defined as the amount of force exerted per unit area and might be expressed as pounds per square inch. So when we think of a single grain of sand or a single atom rebounding from something, it is more natural to speak of the force exerted by the particle. On the other hand, when we think of many sand grains or many atoms striking over a great area, we speak of the force exerted on each unit of area, or the pressure exerted.

pressure. The force per unit of area.

What determines the pressure of a gas, then? There are two factors: the speed of the molecules of the gas and their density. To see that each of these factors is important, think again of the molecules of gas in the balloon. If this gas is heated, causing the molecules to move faster, each collision with the inside surface of the balloon will be more violent, exerting more force on the wall of the balloon. In addition, a greater speed results in more collisions. For these two reasons, the total force exerted on one square centimeter of the wall will be greater; that is, the pressure exerted by the gas will be greater. On the other hand, if the gas is cooled, it will exert less pressure. Thus we can conclude that pressure and temperature are related in a gas. More rigorous analysis shows that in fact a direct mathematical relationship exists between them so that one changes proportionately to the other.

To appreciate the effect of particle density on pressure, imagine that twice as much gas is somehow put into the balloon without allowing it to expand. If this is done, twice as many molecules will be striking the inside walls of the balloon, exerting twice as much pressure. This, of course, is what causes the balloon to expand when more gas is added. Thus we can conclude that pressure and density are related in a gas. In fact, a direct mathematical relationship exists between them so that one changes proportionately to the other.

Thus, pressure, density, and temperature are interrelated. If one changes, one or both of the others must change. Now let's turn back to the Sun and consider how these factors determine the character of the Sun's interior.

Hydrostatic Equilibrium

The Sun is held together by gravitational force. The force of gravity holds the solar material to the Sun just as the force of gravity holds the atmosphere of the Earth near its surface. In this respect, the Sun is merely a big ball of gas. Our Earth's atmosphere is more dense near the surface, not simply because the force of gravity is greater there than higher up but also because the pressure exerted by the gas above compacts the lower layers. Gases lower in the atmosphere have to support the gases above. The same applies to the Sun. At any particular depth below the Sun's surface, the pressure of the gas at that point must be enough to support the gas above. Thus it is convenient to think of the Sun as having layers, like the various layers of an onion as shown in ➤Figure 11-10a. Keep in mind, however, that in the Sun there are no distinct boundaries between layers but rather a continuous change as we move toward or away from the center.

Since the Sun is in a state of equilibrium (that is, neither noticeably contracting nor expanding), the pressure downward on any thin layer must be equal to the pressure exerted upward on that layer (Figure 11-10b). This allows us to calculate the pressure at any depth below the surface, for knowing the total mass of the Sun, we can calculate the weight of gas above any particular layer. Know-

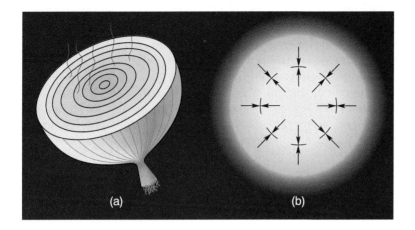

► FIGURE 11-10. (a) One can think of the Sun as consisting of multiple layers like those of an onion, except that there is no distinct boundary between the Sun's layers. (b) Within the Sun, the pressure upward on any layer must be the same as the pressure downward.

ing this, we can calculate the pressure that is needed within the layer to support the gas above.

The equilibrium conditions in the Sun are known as *hydrostatic equilibrium.* The name is almost self-explanatory: "hydro" refers to the fluid state. This is basically just a more complex case of the situation we discussed with the inflated balloon. In that case, the stretched rubber holds the air inside in a compressed state. As long as the outward pressure of the compressed air inside is enough to support the inward pressure of the rubber, equilibrium is maintained.

Since the gas at the center of the Sun is supporting the weight of the gas all the way out to the surface, we should expect great pressures at the center. In fact, the pressure there is calculated to be 1.3×10^9 times that on the surface of the Earth. This tremendous pressure pushes protons close enough together that hydrogen fusion can take place. Only near the center of the Sun are the temperature and density of hydrogen great enough to support fusion. The solar core, where fusion is taking place, extends out to perhaps 10% of the radius of the Sun.

The fusion reaction in the core provides a heat source that obviously must be taken into account when calculating conditions within the Sun. Recall that when a gas is heated, it tends to expand. A balloon expands when its temperature rises, but then it stabilizes at a (different) equilibrium condition. Likewise, the Sun exists in a state of equilibrium, with the force of gravity balanced by forces tending to expand the gas.

To see how hydrostatic equilibrium works, imagine that the Sun could somehow be compressed artificially. Under compression, the pressure within the Sun would increase. The fusion rate would then increase, pushing the Sun back out to another equilibrium position. As our discussion of the life cycle of stars in a later chapter will explain, once the energy production of a star slows and the core cools, contraction of the core begins. As long as energy production is stable, however, the Sun remains in equilibrium.

To see what happens to the energy produced in the core of the Sun, we must look at energy transport within the Sun.

hydrostatic equilibrium. In a star or a planet, the balance between pressure caused by the weight of material above and the upward pressure exerted by material below.

Energy Transport

We observe that energy is radiated from the surface of the Sun. The fusion reactions in the Sun, however, occur at its core, where the temperatures and pressures are the greatest. The heat energy produced must then be transported out to the surface.

There are three possible methods by which heat can be transferred from the center of the Sun outward: conduction, convection, and radiation. These same

The Sun's Interior

conduction (of heat). The transfer of heat in a solid by collisions between atoms and/or molecules.

convection. The transfer of heat in a gas or liquid by means of the motion of the material.

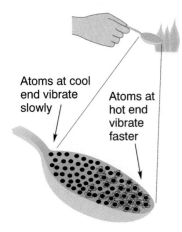

► **FIGURE 11-11.** The fast-vibrating atoms at the end of the spoon in the fire cause atoms next to them to vibrate faster. This continues until atoms at the far end are vibrating fast, meaning that that end also is hot.

► **FIGURE 11-12.** This illustrates the relative thickness of the three heat-transport zones of the Sun.

three processes occur here on Earth and, indeed, everywhere in the universe.

If you put one end of a spoon on the burner (or in the flame) of a kitchen stove and hold the other end, you'll feel your end of the spoon gradually getting warmer. Heat is being transferred by vibration through the metallic crystal structure of the spoon. The method of transfer here is called **conduction**. Imagine the atoms near the end of the spoon in the fire. As that end heats, the atoms vibrate at greater speeds (see ►Figure 11-11). These atoms exert forces on adjacent atoms of the metal, however, and cause those atoms to pick up the vibration. Gradually, the increased vibration spreads up the spoon until the atoms at the other end are also vibrating more rapidly than they were. Notice that in heat conduction, atoms do not move from one region to another but vibrational energy—heat—is transferred.

Conduction requires that the particles of the substance be in close contact, as are the atoms in a solid. In a star, this is not the case, except in some extremely dense stars, so conduction is not a significant factor in transporting heat from within the Sun.

Convection occurs when the atoms of a warm fluid (liquid or gas) move from one place to another. Put your hand about a foot above a hot stove burner. You will feel hot air rising from the burner. This takes place because heated air is less dense than cooler air and therefore the hot, less dense air rises. The result is that heat is transferred upward from the stove by the motion of the hot gas. In forced-air central heating/cooling systems, the motion of the hot (or cool) air is caused by fans.

In a star, convection between adjacent layers is significant only when the temperature difference is great compared to the pressure difference. In the case of our Sun, this condition is met only in the region within about 150,000 kilometers of the surface (►Figure 11-12). In this region, convection constantly

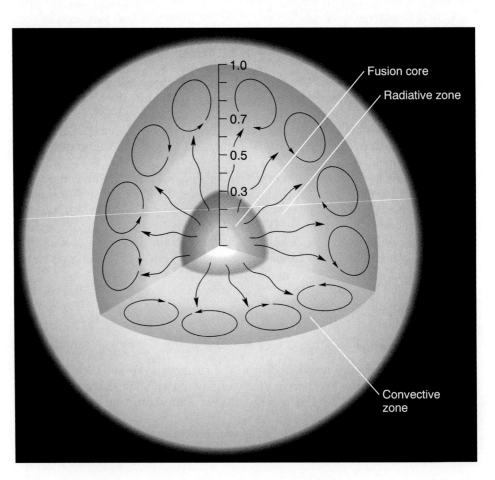

348   CHAPTER 11   The Sun

mixes the solar material as hot gas rises and cooler gas descends. Deeper within the Sun, convection is almost inconsequential and mixing does not occur to a great degree.

The final method of heat transfer is by **radiation.** If you hold the palm of your hand exposed to the stove burner, you can tell that your hand is being heated by another method besides rising hot air. To emphasize this, hold your hand off to the side, where it is not in the stream of hot air, and you will feel heat radiated to your hand. Radiation of energy occurs in all portions of the electromagnetic spectrum. Its effect on another object depends only upon whether the receiving object absorbs the particular wavelength of radiation emitted by the radiator. The air of your kitchen, for example, is transparent to most electromagnetic radiation produced by the stove burner so it does not absorb the radiation and is not heated by it directly. Your hand, however, being opaque to the radiation, absorbs it and is heated by it.

Inside the Sun, and most stars, radiation is the principal means of heat transport. If the Sun were transparent, the electromagnetic radiation produced in the core would travel outward at the speed of light and reach the surface in about 2 seconds. In actuality, the material of the Sun is nearly opaque, so the radiation that is emitted travels only a small distance before being absorbed. It is then reemitted, absorbed, reemitted, and so on, with a typical distance between successive absorptions of 1 centimeter (➤Figure 11-13). The reemissions occur in random directions, and as a result, energy that began perhaps as a gamma ray photon from the fusion of two protons travels a very circuitous path and may take on the order of millions of years to reach the surface. This seems an impossibly long time for something traveling at the speed of light, but keep in mind both the multitude of absorptions and reemissions taking place and the extreme length of the roundabout path taken by the energy.

Figure 11-12 shows the various portions of the interior of the Sun: the core, the radiative zone, and the convective zone. Once the energy of the Sun reaches the surface, it is again radiated outward. The energy from the Sun is released primarily as ultraviolet, visible, and infrared radiation, but it also comes in two less familiar forms: charged particles and neutrinos. As a later section of the chapter will explain, the charged particles flowing outward from the Sun have observable effects on the Earth. Neutrinos, however, are very difficult to detect, and they present astronomers with a major problem.

The Neutrino Problem

There is almost no doubt about the fundamental ideas of the solar model just described, for the concepts of pressure and density are well understood, and we are confident that the Sun's energy is produced by nuclear fusion. We are less sure of the details of the workings of the Sun's interior, however. The generally accepted theory of the Sun is called the ***standard solar model***, and it makes a prediction about neutrinos that can be checked to determine the validity of the model.

The standard solar model predicts that so many neutrinos flow from the Sun that 66 billion pass through every square centimeter of your body each second. Those neutrinos do not affect your body because neutrinos have a very low probability of interaction with matter they pass through, but this same low interaction rate makes them difficult for astronomers to detect. Astronomers want to measure the number of solar neutrinos that reach the Earth in order to check the standard solar model.

In order to shield neutrino detectors from cosmic rays and natural radioactivity, the detectors must be located far underground. Otherwise, the other radiation would overwhelm the few reactions caused by neutrinos. In addition, since neutrinos react so seldomly with material, a neutrino detector must contain

radiation (of heat). The transfer of energy by electromagnetic waves.

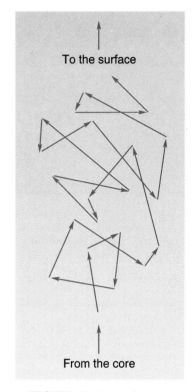

➤ **FIGURE 11-13.** A photon is absorbed and reemitted numerous times as it travels from the Sun's core. Each reemission is in a random direction.

standard solar model. Today's generally accepted theory of solar energy production.

Recall that the proton-proton chain produces two neutrinos for each helium atom produced.

FIGURE 11-14. The Homestake neutrino experiment uses 100,000 gallons of cleaning fluid in this tank located nearly a mile under the hills of South Dakota.

SAGE is an acronym for *Soviet American Gallium Experiment* (the experiment and the acronym were devised before the disappearance of the Soviet Union).

a large amount of material in order to get enough reactions to detect. The world's first solar-neutrino detector (➢Figure 11-14) began operation in the late 1960s in the Homestake Gold Mine in Lead, South Dakota. This detector, operated by a group from Brookhaven National Laboratory and led by Raymond Davis, Jr., holds 378,000 liters (100,000 gallons) of perchloroethylene, a dry cleaning fluid containing chlorine. When a solar neutrino strikes a chlorine atom with enough energy, a reaction occurs, transforming the chlorine into radioactive argon. The radioactivity of the argon can be measured, thereby revealing the neutrino reaction.

Over the past two decades, Davis's group has conducted the experiment many times. They can detect only about one-third of the number of neutrinos predicted by the standard solar model. They have checked and rechecked the apparatus and procedure, but have found no fault with them.

In an effort to confirm or improve Davis's results, other solar-neutrino experiments followed. The second such experiment, known as Kamiokande II, began in 1986 in Kamioka, Japan. This detector has the ability to provide information on the direction of the neutrinos' travel. It confirms that neutrinos do indeed come from the Sun, but like the Homestake experiment, it finds fewer neutrinos than theory predicts.

The two detectors discussed thus far were sensitive only to high-energy neutrinos that account for a very low percentage of the total neutrino production. The two latest neutrino detectors, known as SAGE and GALLEX, employ the element gallium, which is able to detect the low-energy neutrinos produced by the dominant reactions in the Sun's core. The first, SAGE, is a cooperative Russian-American experiment. It is housed in a tunnel under a mountain in the northern Caucasus, near a town called Neutrino City. The SAGE scientists have detected about 50% of the number of neutrinos predicted by the standard model.

Statistically more substantial results came from the GALLEX experiment, which is located in a tunnel in Italy. The GALLEX group reported detecting about 70% of the predicted neutrinos. This is closer to the expected value than SAGE's results, but not close enough to resolve the issue of the missing neutrinos once and for all.

It is possible that our theories concerning the detection of neutrinos are in error. Neutrinos come in three types, and today's solar-neutrino detectors detect only one type. Could it be possible that neutrinos change from one type to another during their eight-minute flight from the Sun? This is precisely what physicists Stanislaw Mikeyev, Alexei Smirnov, and Lincoln Wolfenstein proposed. The so-called MSW theory continues to be tested in experiments around the world, including the new Super-Kamiokande experiment, which is a joint Japan–U.S. collaboration to construct and operate the world's largest underground neutrino observatory.

Another possibility is that our theory of energy production within stars contains a major error. One suggestion is that a previously unknown particle is carrying off some of the energy thought to have been carried by neutrinos. The two names that have been suggested for this particle are interesting: *cosmion*, an important-sounding term, and *Weakly Interacting Massive Particle* or WIMP, a more whimsical name.

There is at least one other possibility. The prediction of the number of neutrinos produced by the Sun is based on the energy we receive from the Sun. However, as we have seen, the solar energy that comes to us as electromagnetic radiation requires millions of years to get from the core of the Sun to the photosphere where it is radiated away. Thus the energy we receive had its beginning millions of years ago. But solar neutrinos from the core reach us in only about eight minutes. Could it be that the core of the Sun has decreased its energy production since it released the electromagnetic energy we are now receiving?

If so, the change in solar activity will not have an effect on Earth until thousands or millions of years from now, when it will cause another ice age.

Astronomers eagerly seek the answer to the neutrino problem. Until we either find the missing neutrinos or find a suitable theory to replace the standard solar model, the situation will remain less than satisfactory.

THE SOLAR ATMOSPHERE

➤Figure 11-15 is a combination of six photos of the Sun, each taken by a different method and in a different portion of the spectrum. The yellow "pie wedge" at the 10 o'clock position shows the Sun in visible light—the way it appears in a "regular" photograph. In this case we see what we call the Sun's surface, although the Sun does not have a surface in the sense that the Earth does. When we speak of the surface of the Sun, we are speaking of the part of the Sun from which we receive visible light, the ***photosphere.***

Figure 11-15 shows that different information about an object is revealed, depending upon how the photograph is taken. The photo (particularly the "pie wedge" at the lower right) shows that there is material beyond the Sun's surface that is not visible to our eye. This is the solar atmosphere. It is convenient to divide the atmosphere into three regions: the photosphere, the chromosphere, and the corona. We will discuss each in turn.

photosphere. The visible "surface" of the Sun. The part of the solar atmosphere from which light is emitted into space.

Photosphere = "sphere of light."

The Photosphere

Although the photosphere is a very thin layer (about 400 kilometers thick), it does have a thickness. This means that some of the light we receive from the Sun comes from one depth within the photosphere and other light comes from

➤ FIGURE 11-15. This figure was made by combining as separate wedges six different photos of the Sun. Each photo was made using a different portion of the spectrum.

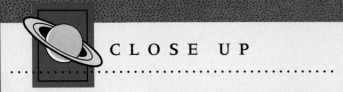

CLOSE UP

Data Uncertainty in the Homestake Experiment

The discussion of the work of Tycho Brahe in Chapter 2 pointed out that he pioneered the scientific practice of reporting not only the direct result of a measurement, but also the amount of likely error in that measurement. The data reported by the researchers in the Homestake neutrino experiment serve as a modern example of this practice in science.

➤Figure C11-1 is a graph comparing the results of the Homestake experiment through the years with the value predicted by the standard solar model. The scale on the left indicates the average number of radioactive argon atoms detected in each experiment. (Each single experiment lasted a number of days.) The dates across the bottom indicate when the experiment was performed. The pink dots across the lower part of the graph show the measured value of the number of radioactive argon atoms produced per day in each experiment.

Refer to the graph and find the first dot on the left. Its location indicates that in this particular experiment (in late 1970) about 0.2 atoms/day were detected. In the next experiment, about 0.5 atoms/day were detected. The graph tells us more about each of the experiments, however. The lines that extend above and below each of the dots are called *error bars*, and they tell us the range within which the researchers are confident the "true" measurement lies. For example, the data taken during the first experiment revealed that about 0.2 atoms/day were produced, but this value may be in error such that the true value (if the experiment were perfect) may be as little as 0 or as great as 0.5 atoms/day.

The length of the error bars in each case is not simply a guess, but is based on statistical measurements of the way that the data fluctuated day-to-day. Note that some error bars are much longer than others. Look at the report of the 1986 experiment. It shows a measurement of zero, but has an error bar that extends as high as 1.8 atoms/day.

Why are the error bars so long? This is the same as asking why the possible errors are so great. The answer lies in the difficulty in detecting neutrinos and the resulting low numbers that are detected. When few nuclear reactions are detected, researchers say that the "statistics are poor." A simple example: Suppose that we did not know that if a coin is flipped, the probability of it landing "heads" up is 50%, and that we are trying to determine this percentage by experiment. If we flip a coin only 4 times, we will have poor statistics upon which to make a judgment. On the other hand, if we flip it 10,000 times, we will have good statistics (and our result will be very close to 50%).

The line labeled "Observations" is the weighted average of all of the experiments. The line labeled "Theory" is the best theoretical prediction by John Bahcall, a leading theorist working with the standard solar model. You can see that although the error bars of a few of the experiments indicate that those experiments do not necessarily disagree with the theory, the total experimental results disagree greatly with the theory. Hence, the neutrino "problem."

---

limb (of the Sun or Moon). The apparent edge of the object as seen in the sky.

The fact that the solar spectrum peaks near the center of the visible region is not just a coincidence, for our eyes evolved so as to use the available electromagnetic radiation efficiently.

other depths. Saying that the photosphere is 400 kilometers thick means that we can see to that depth. When we look at an edge (the **limb**) of the Sun, we see to a lesser depth because we are seeing the Sun at a grazing angle. ➤Figure 11-16 shows this effect, which is important in that it allows us to analyze light from different depths within the photosphere and therefore to determine the temperature at different depths. We learn that the photosphere varies in temperature from about 8000°C at its deepest to about 4000°C near the outer edge. Overall, the light we receive from the photosphere is representative of an object about 5800°C. An intensity/wavelength graph of the radiation from the Sun (➤Figure 11-17) peaks near the center of the visible spectrum.

The pressure of the outer photosphere (calculated as for the inner layers, from knowing the gravitational force there and the amount of material above each layer) is only about 0.01 the pressure at the surface of the Earth. Knowing the temperatures and gas pressures of the photosphere, we can calculate the density of particles there. We find that the density of the matter is only 0.001 of

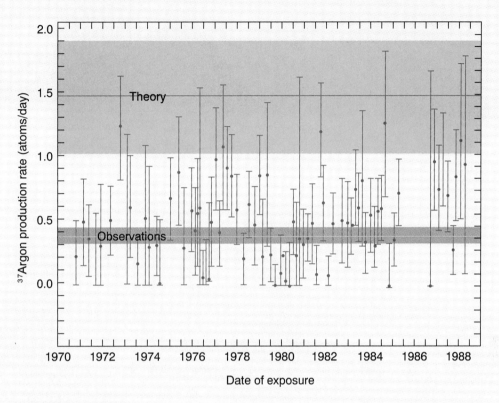

➤ **FIGURE C11-1** This graphical report of the history of the Homestake solar-neutrino experiments through the years compares their results with the theoretically expected value as calculated by John Bahcall.

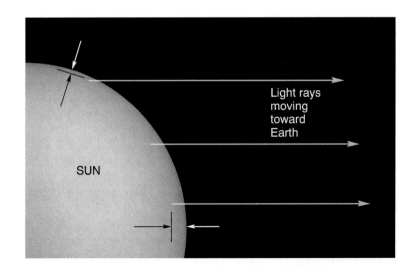

➤ **FIGURE 11-16.** The light we receive from the center of the disk of the Sun originated at a greater depth than the rays we receive from near the edge.

The Solar Atmosphere 353

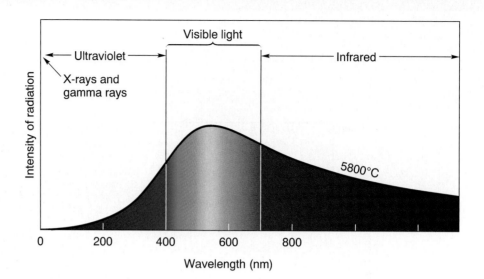

▶ **FIGURE 11-17.** The intensity/wavelength graph of light from the Sun reaches a peak about the center of the visible portion of the spectrum.

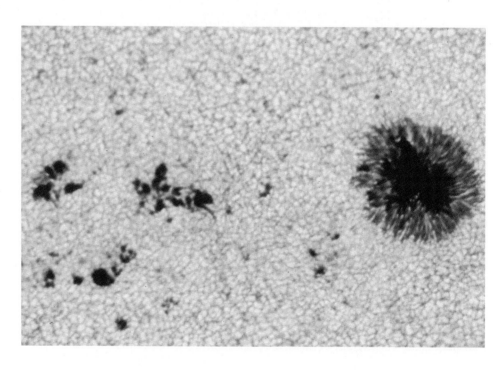

▶ **FIGURE 11-18.** A photo of the granules of the photosphere. The bright center of each granule is material that is rising from the hotter inner portion of the Sun. A granule is about 1000 km (600 miles) across, so each granule covers an area equivalent to Texas. The dark areas are sunspots of various sizes.

the density of air at sea level on Earth (even though the gravitational field there is 28 times what it is at the surface of the Earth).

▶Figure 11-18 is a photograph of part of the photosphere. It shows irregularly shaped bright areas surrounded by darker areas. Recall from the discussion of the intensity/wavelength diagram in Chapter 4 that a hotter object emits more radiation than a cooler object of the same size. The brighter areas of the Sun are brighter simply because they are hotter. This *granulation* of the Sun's surface is the result of convection. Granules are areas where hot material (the light areas) is rising from below and then descending (the dark surroundings). ▶Figure 11-19 illustrates the effect and demonstrates that the photosphere is a boiling, churning region.

granulation. Division of the Sun's surface into small convection cells.

Using methods described in Chapter 4, we can determine the chemical composition of the photosphere. We find that the mass of the photosphere is about 78% hydrogen and 20% helium. The remaining 2% consists of some 60 elements. All of these elements are known on Earth and occur in about the same proportions on Earth as in the Sun's atmosphere (with a few exceptions). The

These are percentages by mass, not by number of atoms.

CHAPTER 11 The Sun

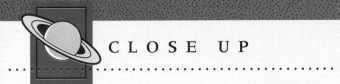

CLOSE UP

Helioseismology

In 1960 scientists discovered that the Sun is vibrating. Doppler shift measurements indicate that parts of the photosphere move up and down about 10 kilometers with a period of about 5 minutes. Since then, many other vibration frequencies have been discovered, all taking place at the same time. ▶Figure C11-2 is a computer simulation illustrating a high-frequency vibration. In this figure, the orange color represents regions where material is expanding at this instant, and blue represents areas where it is contracting. Thus, on the surface, the blue areas are moving inward and the orange areas outward. If this were a movie, in about 2 minutes the colors would reverse. The pulsations are caused by waves similar to sound waves coming from near the core of the Sun. When the waves hit the Sun's surface, they bounce back toward the interior. The combination of waves coming out and going back in produces a resonating effect, like a "gong."

Just as geologists use earthquakes to study the interior of the Earth, helioseismology—the study of vibrations of the Sun—is beginning to give us insight into the Sun's interior. The Global Oscillation Network Group (GONG) is building a network of telescopes that will observe the vibrations of the photosphere 24 hours a day.

▶ FIGURE C11-2 A computer model of solar resonance that produces the observed vibrations of the photosphere.

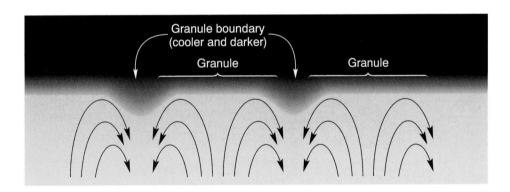

▶ FIGURE 11-19. Granules are seen where hot material from below the photosphere rises. Where it descends after cooling slightly, we see the darker edges of the granules.

exceptions are of two types: (1) elements such as helium that have masses so low that they would have escaped Earth if they were once here in abundance, and (2) elements found on Earth but whose characteristic spectra are such that they would not be detectable in the solar spectrum if the elements were as rare in the Sun as they are on Earth. As was pointed out in Chapter 7, this similarity of composition between objects as different as Earth and the Sun is not accidental, but results from the way the Sun and the solar system formed.

Notice that it is the composition of the Sun's atmosphere—not of the entire Sun—that is deduced from the solar spectrum. From our knowledge of nuclear fusion, we know that helium must be more abundant in the core of the Sun

The Solar Atmosphere 355

Again, these percentages are by mass. About 94% of the atoms of the Sun are hydrogen.

chromosphere. The region of the solar atmosphere between the photosphere and the corona.

Chromosphere = "sphere of color."

corona. The outermost portion of the Sun's atmosphere.

A telescope designed to photograph the atmosphere of the Sun (when there is no eclipse) is called a *coronagraph*.

than in the atmosphere. Overall, the Sun is theorized to be about 73% hydrogen and 25% helium; this leaves only about 2% for the remainder of the elements.

The Chromosphere and Corona

The *chromosphere,* a region some 2000 to 3000 kilometers thick lying beyond the photosphere, is not normally observable from Earth. It was first reported in the seventeenth century during a solar eclipse. It appears as a bright red flash, lasting only a few seconds, when the Moon has just covered the photosphere. During solar eclipses from 1842 to 1868, it was examined in more detail. Its spectrum was observed to be a bright line (or emission) spectrum because in viewing it we are seeing light from a hot gas with the dark sky behind it. Because the chromosphere is so much dimmer than the photosphere, it is only observed at the time of an eclipse, when the brighter portions of the Sun are blocked out (➤Figure 11-20).

Today the chromosphere and the region beyond it, the *corona,* can be observed by the use of a telescope that produces an artificial eclipse of sorts. With this instrument we can observe the Sun's atmosphere at various depths. We learn that as one moves outward from the photosphere, the temperature increases instead of diminishing as we would expect. It is as high as 100,000 degrees in the outer portions of the chromosphere and continues to increase beyond the chromosphere into the corona, where it may reach millions of degrees.

One might think that these regions would be extremely bright because of their high temperature. There are two reasons why they are dim. First, much of the radiation emitted from them is in the X-ray portion of the spectrum rather than in the visible. Second, the chromosphere and corona have a low density of matter, so hardly any matter is available to glow. The corona's density is less than one-billionth that of the Earth's atmosphere.

The reason for the high temperatures within the chromosphere and corona is not well understood. The prevailing theory has been that the high temperature is the result of sound waves that are produced within the convective regions of

➤ FIGURE 11-20. The chromosphere is seen against the dark background of space during a solar eclipse.

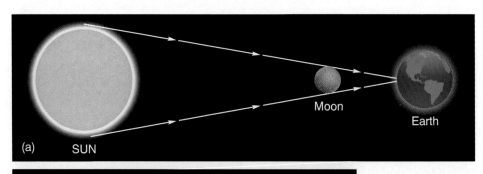
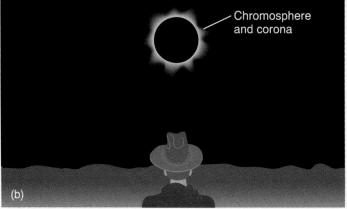

► FIGURE 11-21. The chromosphere's spicules give it an irregular and ever-changing boundary as they shoot thousands of kilometers upward into the corona and then die down within a few minutes.

the Sun and intensify as they pass outward until they are absorbed in the chromosphere and corona, heating these regions. Recent data, however, call this explanation into question, and many astronomers now believe that the heating is caused by an interaction between the Sun's magnetic field and its differential rotation. The reason for the high chromosphere and corona temperature is just one of the unanswered questions astronomers have about the Sun.

►Figure 11-21 is a photograph of the chromosphere that was taken at a wavelength that allows us to see its structure. The **spicules** that can be seen shooting upward into the corona typically reach a height of 6000 to 10,000 kilometers and last from 10 to 20 minutes.

The corona, a region extending for millions of kilometers from the Sun, has been observed during total solar eclipses for centuries although many people used to claim that it was just an optical illusion caused by the sudden dimming of light as the Sun is eclipsed. ►Figure 11-22 shows its extremely irregular appearance.

The photograph on the first page of this chapter and ►Figure 11-23 show spectacular occurrences in the Sun's atmosphere. These are **prominences,** eruptions of solar material up into the chromosphere and corona. Some of these are relatively slow-moving and remain fairly stable for as long as a few days. They may reach as high as thousands of kilometers above the photosphere. Some move much more quickly, ejecting material from the Sun at speeds up to 1500 kilometers/second and reaching heights of nearly a million kilometers. Prominences are often associated with sunspots and the solar activity cycle to be discussed in the next section.

spicule. A narrow jet of gas that is part of the chromosphere of the Sun and extends upward into the corona.

The word *corona* comes from the Latin word for crown.

prominence. The eruption of solar material beyond the disk of the Sun.

► FIGURE 11-22. This is a composite photo of the Sun and its corona taken in March 1988. The surface of the Sun is a combination of X-ray and visible-light images. The photo of the corona was taken during the March 1988 eclipse. Note the irregularity of the corona and how it streams outward to form the solar wind.

The Solar Atmosphere

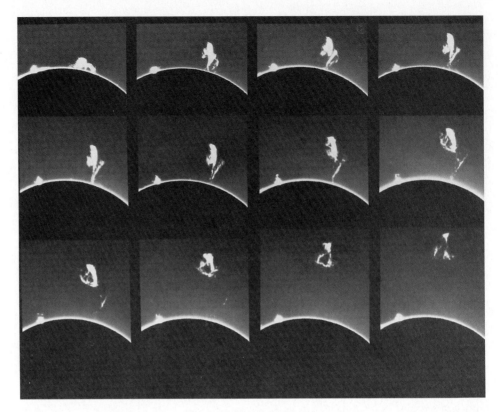

➤ **FIGURE 11-23.** This sequence of photos, taken from space, shows how this particular prominence progressed as charged particles were pushed from the Sun by its magnetic field.

When we divide the Sun into regions, we must remember that the boundaries between the regions are artificial, for we have named the various regions and distinguished between them on the basis of certain selected properties. For example, we consider that the process of heat transport is important, and we therefore talk of a radiative zone and a convective zone. If we emphasized another property, we might not make a division between these two parts of the Sun at all. In addition, remember that although the boundaries between various regions appear sharp in our drawings, in actual practice they are much less well defined. This is especially true of the outer limits of the corona, where the coronal material becomes the solar wind.

The Solar Wind

solar wind. The flow of nuclear particles from the Sun.

The *solar wind* (➤Figure 11-24) is a continuous outflow of charged particles from the Sun, mostly in the form of protons and electrons. These particles stream through space, taking mass away from the Sun. Near the Earth, the solar wind normally travels at about 400 kilometers/second and has a density of from 2 to 10 particles per cubic centimeter. Recall that one effect of the solar wind is that it causes comet tails to point away from the Sun as its particles sweep comet material along with them.

Perhaps a more dramatic effect of the solar wind is the auroras seen near the poles of the Earth. Auroras result from the solar wind being trapped by the magnetic field of the Earth. Where the Earth's magnetic field lines converge toward the surface of the Earth, the electrically charged particles strike the molecules of the upper atmosphere and cause them to emit the beautiful, eerie glow we call an aurora.

SUNSPOTS AND THE SOLAR ACTIVITY CYCLE

Observations of dark spots on the Sun were reported by the Chinese as early as the fifth century B.C. It is sometimes possible to see very large ones with the naked eye if the Sun is viewed when it is very near the horizon. Europeans did not report sunspots until Galileo saw them with his telescope, perhaps because the Europeans did not have observers as astute as the Chinese or perhaps—after Aristotelian thought was adopted—because Aristotle had proclaimed that the Sun was flawless. (We tend not to see what we disbelieve.)

In the late eighteenth century, Alexander Wilson hypothesized that sunspots were places where we were seeing through the outer surface of the Sun and into a cooler interior. William Herschel, the discoverer of Uranus (Historical Note, Chapter 9), even thought that the interior of the Sun might be cool enough to support life. Today's spectroscopic measurement of solar temperatures reveals that sunspots are indeed about 1500 degrees cooler than the surrounding photosphere. They are still very hot, however. Sunspots are temporary phenomena, lasting anywhere from a few hours to a few months.

The explanation for sunspots involves the magnetic field of the Sun, which can be measured using a technique discovered late in the last century. The magnetic field of an object causes each emission line of the object's spectrum to split into two or more lines, and the strength of the magnetic field can be determined from the extent of the splitting. The splitting can be measured in the spectrum of light from individual parts of the Sun and is an important tool in studying the Sun.

Sunspots often appear in pairs, aligned in an east-west direction. Early in this century it was found that the magnetic field in a sunspot is about 1000 times as strong as the magnetic field of the surrounding photosphere. In addition, we find that sunspot pairs have opposite magnetic polarities, one being north and the other south.

Sometimes the Sun contains a great number of sunspots, and sometimes few or no sunspots are seen. In 1851, Heinrich Schwabe, a German chemist and amateur astronomer, discovered that there is a fairly regular cycle of change in the number of sunspots and that the cycle lasts about 11 years. He found that although individual spots do not last long, about the same number are found on the Sun at any one part of its cycle. The cycle varies somewhat in period but averages about 11 years between repetitions. ➤Figure 11-25 is a graph showing how the number of sunspots has changed during this century.

A Model for the Sunspot Cycle

At a sunspot maximum, most spots occur about 35 degrees north or south of the equator. Then, as the cycle progresses, the spots are seen closer and closer to the

➤ FIGURE 11-24. Observing the area near the Sun reveals the solar wind. The color seen here is not real, of course, but indicates different intensities of the wind at the time the photo was taken.

The splitting of spectral lines by strong magnetic fields is called the *Zeeman effect*, after the Dutch physicist who discovered it.

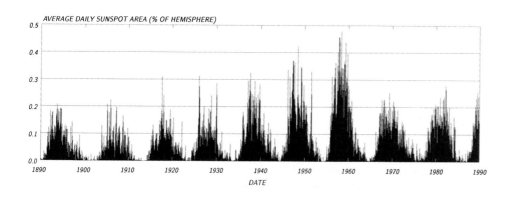

➤ FIGURE 11-25. The number of sunspots varies with a period of about 11 years, but there is a great difference in the maximum number during each cycle.

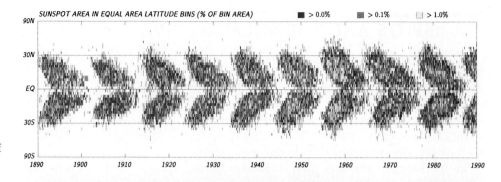

► **FIGURE 11-26.** This plot, called a butterfly diagram, shows the location and relative number of sunspots as the years pass. Notice that when each set of butterfly wings forms, the sunspots are at higher latitudes. Then they move closer to the equator. Notice also that about every 11 years there is a period of very few sunspots.

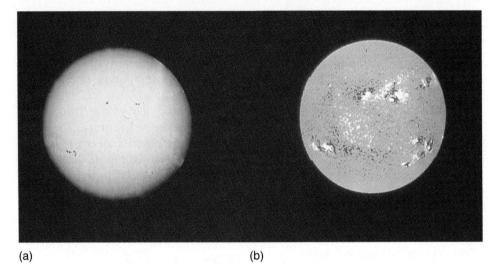

► **FIGURE 11-27.** (a) A visible-light photo of the Sun, showing sunspots. (b) A magnetic map of the Sun on the same day shows where the magnetic field is strongest on the Sun.

equator. By the time they reach the equator, the cycle is at a minimum, and new spots are beginning to form again at greater latitudes. If we plot the location of the spots as time goes by, we get the pattern shown in ►Figure 11-26, called a *butterfly diagram* for obvious reasons. It is important to point out that a given sunspot does not move from high to lower latitudes. (A sunspot's lifetime is much shorter than the 11-year solar cycle.) Instead, the diagram tells us that as time passes and old sunspots die out, new ones form closer to the Sun's equator.

Another interesting feature of the 11-year cycle is that during one cycle, the easternmost sunspot of each pair in a given hemisphere of the Sun is a north magnetic pole and the western sunspot a south pole. (In the other hemisphere, the opposite is true.) Then during the next cycle the pattern reverses, with the eastern sunspot in that hemisphere being a south pole. From one cycle to another, the general magnetic field of the Sun also reverses. Thus the entire magnetic cycle of the Sun has a 22-year period rather than an 11-year period. ►Figure 11-27 illustrates the relationship between sunspots and the Sun's magnetic field.

A modern hypothesis explains the existence of sunspots and their 11-year cycle as being due to patterns of magnetic field lines within the interior of the Sun. It is thought that groups of these lines form "tubes" threading through the Sun. When the tubes first form, they are relatively straight and buried deep within the Sun as shown in ►Figure 11-28a. The differential rotation of the Sun, however, causes the lines to wrap around the Sun as shown progressively in parts (b), (c), and (d) of Figure 11-28. As the tubes become more and more twisted around the Sun, they are forced to the surface. When they break through, we see a pair of sunspots, one with a north magnetic pole and one with a south magnetic pole. This breaking through the surface causes the lines to weaken and die out at the same time as more lines are forming deep within the

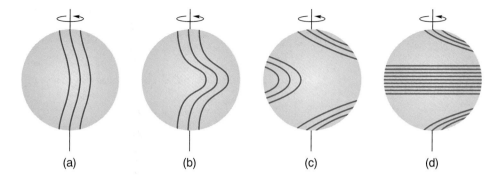

► FIGURE 11-28. It has been proposed that tubes of magnetic field lines form just below the Sun's surface. The Sun's faster rotation near its equator then twists the tubes around the Sun until they force one another through the surface.

Sun. Detailed analysis shows that such a chain of events would cause the magnetic field direction in the tubes (and therefore the magnetic field of the entire Sun) to reverse in direction when new lines are formed.

This model, though perhaps hard to imagine, does fit the observations. As with any new hypothesis, we should expect that as new observations are made, it may well have to be adjusted and refined or even discarded.

Solar Flares

The turbulent magnetic field of the Sun is responsible for the prominences discussed earlier, giving prominences their unique shapes such as those in the chapter-opening photograph and Figure 11-23. It also causes the colossal flareups called *solar flares* that normally occur during sunspot maxima. Lasting from a few minutes to a few hours, a solar flare can release the equivalent energy of a few thousand of our largest nuclear weapons. According to the model presented in the last section, these flares occur when a great number of twisted tubes of magnetic field lines release their energy at once through the photosphere. Extremely energetic particles are blasted out of the Sun, reaching the Earth in about three days. They are responsible not only for spectacular auroras but also for disruptions of earthly radio transmissions.

The radio disruption occurs when particularly energetic particles of the solar wind strike a layer of the Earth's atmosphere called the ionosphere. The ionosphere plays a part in radio transmission because it reflects radio waves back down to the surface of Earth. Normally, the Earth's magnetic field prevents particles of the solar wind from reaching the ionosphere by deflecting and trapping them, but the high-energy particles in the solar wind are able to penetrate to the ionosphere. When the ionosphere is disrupted by the solar wind, we may experience static in radio reception or even complete loss of signal.

Do solar flares have other effects on the Earth and on the lives of those of us who live on this planet? As we learn more about the Sun, we may find that what seem like small quirks in the Sun's behavior are actually of major importance to life on Earth.

solar flare. An explosion near or at the Sun's surface, seen as an increase in activity such as prominences.

As the discussion of the formation of the solar system explained, the solar wind had a major function in determining the nature of today's solar system.

CONCLUSION

The importance of the Sun to life on Earth is obvious. To astronomers, the Sun takes on even more importance because it is by far the closest star. Since astronomers must understand stars if they are to understand the workings of the universe, the Sun becomes critical in such a study.

Previous chapters of this text have been devoted to a study of the planets, and this chapter has explored the workings of the Sun. Chapter 13 will describe the formation of stars including the Sun.

RECALL QUESTIONS

1. The Sun's energy is generated by
   A. gravitational contraction.
   B. nuclear fission.
   C. hydrogen fusion.
   D. helium fusion.
   E. chemical reactions.

2. The layer of the Sun that is normally visible to us is the
   A. corona.
   B. chromosphere.
   C. photosphere.
   D. core.
   E. solar wind.

3. Sunspots are areas on the Sun that are
   A. hotter than their surroundings.
   B. cooler than their surroundings.
   C. brighter than their surroundings.
   D. [Both A and B above.]
   E. [Both B and C above.]

4. The energy produced in nuclear reactions in the Sun results from
   A. friction as the nuclei crash together.
   B. heat produced from the electrical effects of the reactions.
   C. the increase in mass of the particles due to the reactions.
   D. the decrease in mass of the particles due to the reactions.

5. Why is a high temperature needed for energy production in the core of the Sun?
   A. Hydrogen will not combine with oxygen at a low temperature.
   B. Energy is needed to overcome electrical repulsion.
   C. Electrons will not recombine at low temperatures.
   D. The force of gravity is greater at high temperatures.
   E. Speeds are less at high temperature, so there is more time for reactions between nuclei.

6. We know that the Sun's energy does not result from a chemical burning process because
   A. of the Doppler effect.
   B. of the redshift.
   C. the Sun would have burned up already.
   D. [Both B and C above.]
   E. [Both A and B above.]

7. The two forces producing hydrostatic equilibrium in the Sun to determine its size are
   A. electrical forces and gravity.
   B. nuclear forces and gravity.
   C. electrical forces and gas pressure.
   D. electrical forces and nuclear forces.
   E. gravity and gas pressure.

8. As the Sun "burns,"
   A. its total mass decreases very slightly.
   B. its total mass increases very slightly.
   C. its energy decreases, but the Sun's mass remains the same.
   D. energy is produced, but the Sun's mass remains the same.
   E. [None of the above.]

9. During a total solar eclipse, the Sun's atmosphere becomes visible. Why?
   A. It is brighter during an eclipse because of light reflected from the Moon.
   B. The light reemitted after absorption becomes visible because the brighter Sun is blocked out.
   C. The atmosphere becomes hotter during an eclipse.
   D. [The statement is not true.]

10. The total luminosity of the Sun can be calculated from its
    A. rotation period and temperature.
    B. rotation period and diameter.
    C. diameter and distance from the Earth.
    D. diameter and the solar energy at Earth's distance.
    E. distance from Earth and the solar energy detected at Earth's distance.

11. Two factors that determine the pressure of a gas are
    A. the speed of the molecules and the particle density.
    B. the speed of the molecules and the gas's temperature.
    C. nuclear reactions in the gas and its temperature.
    D. chemical reactions in the gas and its temprature.
    E. [Both C and D above.]

12. At any particular level within the Sun, the pressure outward is
    A. less than the pressure inward.
    B. equal to the pressure inward.
    C. greater than the pressure inward.
    D. [No general statement can be made.]

13. Granulation of the photosphere is a direct result of
    A. heat conduction.
    B. convection.
    C. heat radiation.

14. A prominence is
    A. a cool spot on the Sun.
    B. the ejection of material from the photosphere.
    C. a fairly permanent bulge on the photosphere's surface.
    D. a reaction within the Sun's core.

15. The solar wind extends
    A. about to Mercury's orbit.
    B. about to Venus's orbit.
    C. almost to Earth's orbit.
    D. far beyond the Earth's orbit.

16. The 11-year cycle of sunspots corresponds to
    A. the period of change in the magnetic field of the Sun.
    B. the rotation period of the Sun near the equator.
    C. the rotation period of the Sun near the poles.
    D. the revolution period of Jupiter.
    E. [None of the above.]

17. Which of the following is the thinnest layer of the Sun?
    A. Corona
    B. Chromosphere
    C. Photosphere
    D. Radiative layer
    E. Convection layer

18. Solar energy strikes the Earth at the rate of 1380 watts/m². It strikes a sphere that is 2 astronomical units from the Sun at a rate of
    A. 345 watts/m².
    B. 690 watts/m².
    C. 1380 watts/m².
    D. 2760 watts/m².
    E. 5520 watts/m².

19. The primary source of energy for the Sun is a series of nuclear reactions in which
    A. four hydrogen nuclei fuse to form a helium nucleus.
    B. a helium nucleus fissions to form four hydrogen nuclei.
    C. uranium nuclei fission to form a number of other elements.
    D. two nuclei fuse to form uranium or plutonium.
    E. oxygen nuclei combine to form more massive nuclei.

20. To begin nuclear fusion in a star, high temperatures are required in order to overcome the
    A. nuclear force between the protons.
    B. nuclear force between the electrons.
    C. electrical force between the neutrons.
    D. electrical force between the protons.

21. The photosphere is
    A. the layer of the Sun where energy is created from mass.
    B. the outermost layer of the Sun.
    C. the layer of the Sun that we see when viewing the Sun.
    D. the layer of the Sun in which we see granulation.
    E. [Both C and D above.]

22. When four hydrogen nuclei fuse to form a helium nucleus, the total mass at the end is _____ the total mass at the beginning.
    A. less than
    B. the same as
    C. more than

23. The sun emits its most intense radiation in which region of the electromagnetic spectrum?
    A. Radio
    B. Infrared
    C. Visible
    D. Ultraviolet
    E. X-ray

24. Nuclear theory predicts that we should detect
    A. fewer neutrinos than we do.
    B. just the amount of neutrinos that we do, confirming the theory.
    C. more neutrinos than we do.

25. What is meant when it is said the Sun has "differential" rotation?

26. Describe some evidence that shows that the source of solar energy cannot be chemical reactions.

27. Distinguish between chemical and nuclear reactions, giving an example of each.

28. What is it about the nucleus of an atom that distinguishes one atom from another?

29. Name the chemical element that is consumed and the element that is produced in the Sun. What produces the energy when the change occurs?

30. In what physical state is most of the material in the interior of the Sun?

31. Define and explain *hydrostatic equilibrium*.

32. How do we know how great the pressures are at certain depths below the surface of the Sun?

33. List the three methods of heat transfer giving an example of each. What method(s) is important in which region(s) of the Sun?

34. If all electromagnetic radiation travels at the speed of light, why does radiated energy take so long to get from the center of the Sun to the surface?

35. Describe the thickness and temperature of the photosphere.

36. If the chromosphere and corona are so hot, why are they not brighter than the photosphere?

37. According to present theory, what causes sunspots?

QUESTIONS TO PONDER

1. If the magnetic field of the Sun reverses every 22 years, what is meant when we refer to the "northern" hemisphere of the Sun?

2. Explain how energy produced by mass-decrease becomes heat within the Sun.

3. Distinguish between force and pressure as defined in science, and give an example of units in which each can be expressed.

4. Describe the relationship among pressure, density of particles, and temperature in a gas.

5. How would the size of the Sun change if the rate of fusion reactions increased?

6. What method of heat transport results in a room being warmer near the ceiling than lower down?

7. Explain how we measure the temperature at various depths in the photosphere.

8. Explain why hot air rises and relate this to why convection does not occur between most layers in the Sun.

9. We see the Sun not as it is now, but as it was eight minutes ago. The energy we detect, however, began millions of years ago. Discuss the implications this has on what we mean by the word "now."

CALCULATIONS

1. Calculate the angular size of a 15,000-kilometer sunspot as seen from Earth.
2. If one kilogram of coal is burned in one second, it will produce 2 million watts of power. How many kilograms would have to be burned each second to produce the Sun's energy output? If the Sun were made of coal, how much time would pass before it would burn out at its present rate of energy production? (In fact, oxygen would be needed for the burning—nearly three times more mass of oxygen than of coal—so only one-fourth of the Sun's mass would be coal.)

ACTIVITY

Measuring the Diameter of the Sun

With simple equipment one can measure the size of the Sun with a fair degree of accuracy. All that is needed is a sunny day, a piece of cardboard, a ruler, and the knowledge that the Sun is 150,000,000 kilometers from Earth. A Close Up in Chapter 6 discussed observing an eclipse by pinhole projection; we will use pinhole projection here also.

Punch a small hole (perhaps one-eighth inch) in a piece of cardboard and hold the cardboard so that the Sun shines through the hole onto a surface behind it. (Refer to Figure A6-3.) You may have to adjust the size of the hole to get an image bright enough to see clearly, and you might use additional cardboard to shield your screen from reflected sunlight.

Being sure that the screen is perpendicular to a line from the pinhole, measure the diameter of the image of the Sun and the distance from the pinhole to the screen. Now, as shown in ▶ Figure A11-1, the following ratio applies:

$$\frac{\text{diameter of Sun}}{\text{distance to Sun}} = \frac{\text{diameter of image}}{\text{distance from screen to image}}$$

Use this equation to calculate the diameter of the Sun.

To get a feel for the accuracy of your measurement, make several measurements with the screen at different distances. How closely do your various measurements agree? What is the largest source of error in this procedure? How does the value you obtained compare to that found in Table 11-1?

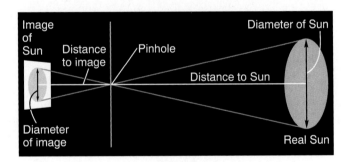

▶ FIGURE A11-1 This drawing, obviously not to scale, illustrates the relationship between distances and sizes of the Sun and its image. Since the triangle at the left is similar to that at the right, the ratio of distance to size is the same for each.

ACTIVITY

Observing Sunspots

Sunspots were first observed by the naked eye, as described in the text, but such a method is not recommended. You would probably have to search the setting Sun for long periods of time over many years before you saw your first sunspot, and staring even at the setting Sun might damage your eyesight.

A more realistic way to view sunspots is with a telescope. *Do not, however, look directly at the Sun with a telescope.* You can obtain a solar filter to put over the front of the telescope but it is even better to use the telescope to project an image of the Sun on a screen, as described below.

First, a caution: The intensity of sunlight is so great that you run a risk of damaging your telescope. One way to decrease this risk is to cover the objective lens with a piece of cardboard with a hole cut in it smaller than the lens. Tape it down so it won't fall off, and it will block out some of the light. If your telescope has a finderscope, you should cover it by taping a piece of cardboard (without a hole) over its objective lens. This will not only prevent eye damage to someone who, out of habit, looks through it, but it will prevent the Sun from burning out the finderscope's crosshairs.

In using a telescope to view the Sun, set it up by first focusing on a distant object (NOT THE SUN). Then pull the eyepiece out just slightly. Now is the time to cover the finderscope and partially cover the objective. Point the telescope exactly at the Sun. This is not as easy as it sounds, and the best way is to move it until the shadow of the telescope tube is smallest. DON'T LOOK THROUGH THE FINDERSCOPE. When this is done, you should be able to see a spot on a screen held behind the eyepiece. Focus by moving the screen and eyepiece to various positions until you get the view you want. A cardboard shield around the telescope will shadow your image from direct rays and improve your view. If you have a star diagonal (which reflects the image off to the side), use it. Trace the image of the Sun and any spots you see. Then repeat your observation in a day or two and look for motion of the sunspots.

Figure A6-2 (in the final Activity in Chapter 6) shows a small telescope being used to project the Sun's image during an eclipse.

Have fun, but be careful.

netQuestions

Visit the netQuestions area of StarLinks (www.jbpub.com/starlinks) to complete an exercise on this topic:

1. Energy and Neutrino Production in the Sun A neutrino is an elementary particle that has little or no rest mass, and no charge but carries energy from a nuclear reaction.

CHAPTER 12

Mauna Kea Observatory, Mauna Kea, Hawaii.

MEASURING THE PROPERTIES OF STARS

Stellar Luminosity
Apparent Magnitude
Distances to Stars—Parallax
CLOSE UP: Naming Stars
Absolute Magnitude
CLOSE UP: A Long-Range Proposal
Motions of Stars
Spectral Classes
CLOSE UP: Determining the Spectral Class of a Star
The Herzsprung-Russell Diagram
Spectroscopic Parallax
Luminosity Classes
Analysis of the Procedure

The Sizes of Stars
Multiple Star Systems
Visual Binaries
Spectroscopic Binaries
Eclipsing Binaries
Other Binary Classifications
Stellar Masses and Sizes from Binary Star Data
The Mass-Luminosity Relationship
Cepheid Variables as Distance Indicators
CLOSE UP: The Mathematics of the Mass-Luminosity Relationship
HISTORICAL NOTE: Henrietta Leavitt

ASTRONOMER SIDNEY WOLFF AND HER HUSBAND were involved in the building of the observatory on Mauna Kea, a 14,000-foot high mountain in the Hawaiian Islands, in the late 1960s. In an essay in The Scientist, she writes:

> We had a wonderful time in those early days, developing a site without even such basic amenities as a source of water; a site where all power had to be generated locally because there was no power line, a site where blizzards raged in

winter and where even in summer temperatures dipped to freezing every night. But we learned. We learned about altitude sickness and the best strategies for forcing our bodies to acclimate. We learned first aid so we could cope with accidents, since professional help was hours away. We learned how to handle heavy machinery and how to maintain generators. We learned more about telescope gears and worm drives and how to repair scored gears than we ever wanted to know. Nearly every one of us who was involved in those early days can tell—loves to tell—stories of being nearly trapped on the mountain during a blizzard, of hiking to the summit because the road was blocked by snow, of climbing to the top of the dome to remove snow so that not a moment of observing was lost. It was a great adventure, an adventure that surely I had not envisioned when I planned a life of research alone in my office.*

*Sidney C. Wolff, National Optical Astronomy Observatories.

Only recently in human history have we even become aware of the astonishing fact that each of the thousands of stars we see is another sun similar to the one that rules our sky. This realization makes obvious the immense distances to those stars; just imagine how far away our Sun would have to be to be as dim as a star. Can we hope to learn much about such faraway objects? As Chapter 4 explained, spectral analysis can be used to determine both the temperatures of stars and the elements of which their atmospheres are composed. Galileo and Newton would have been amazed that we can learn such things.

Temperatures and chemical compositions are only the beginning, however. This chapter will discuss how parallax allows us to calculate the distances to many stars and how, once we know their distances, we can determine other properties including their luminosities, their motions, their sizes, and their masses. We find relationships among the various properties, and these relationships give us clues as to why one star differs from another, how stars are formed, how their lives progress, and how they die. We will delay discussion of the life cycle of stars until the next chapter, turning our attention now to how we measure those properties of stars that will divulge information about their life cycles.

STELLAR LUMINOSITY

luminosity. The rate at which electromagnetic energy is being emitted.

When we speak of the brightness of a star, we must be careful to distinguish between its apparent brightness (➤Figure 12-1) and its luminosity. Chapter 11 explained how the luminosity of the Sun can be calculated if we know the Sun's distance and the amount of solar radiation striking a given area of Earth in a certain amount of time. Suppose two stars differ in apparent brightness so that different amounts of light reach the Earth from the two stars. The cause for this might be any combination of three things: (1) one star may be inherently brighter than the other (its luminosity may be greater); (2) one star may be closer, making it appear brighter; or (3) there may be more interstellar material absorbing light from one star than from the other.

power. The rate at which energy is transferred, or the amount of energy transferred per unit time.

In our discussion of the two quantities, *brightness* refers to the apparent brightness seen from Earth, and *luminosity* refers to the total amount of power emitted by a star. You may occasionally find the term *absolute luminosity*; this simply emphasizes that we are speaking of the radiation actually being emitted, not the radiation reaching Earth.

368   CHAPTER 12   Measuring the Properties of Stars

► **FIGURE 12-1.** Some of the stars of the constellation Orion appear bright because of their proximity, while others are inherently so luminous that they look bright from Earth even though they are very far away. Note that the stars appear to be different sizes in the photo. These differences are not due to the stars' actual sizes, but occur because brighter stars expose larger areas of the photographic plate.

Apparent Magnitude

One of the greatest astronomers of the pre-Christian era was Hipparchus, a Greek thinker who lived in the second century B.C. Hipparchus compiled a catalog of some 850 stars, listing each star's location in the sky along with a number that designated its brightness. To indicate stars' brightnesses, he divided all visible stars into six groups, calling the brightest stars in the heavens *magnitude one* stars and the dimmest he could see *magnitude six* stars. Other stars fell in between, with differences between magnitudes representing equal differences in brightness. Thus the brightness difference between a third-magnitude star and a fourth-magnitude star on his scale was visually the same as that between a fifth- and a sixth-magnitude star. Although it may seem odd to assign the larger number to the dimmer star, we might appreciate his reasoning by thinking of the brighter star as a first-class star and the dimmest as a sixth-class star.

Today we use a slightly revised version of Hipparchus's ***apparent magnitude*** scale, for we measure the brightnesses of stars by photographic and electronic methods. Photographic techniques began to be used for such measurements in the mid-1800s. When these measurements were made, astronomers found that when two stars differ by one magnitude, we receive 2.5 times as much light from the brighter one as from the dimmer. A fifth-magnitude star is about 2.5 times brighter than a sixth-magnitude star. Moving up from fifth to fourth magnitude means that the brightness increases another 2.5 times, making a fourth-magnitude star 6.25 times brighter than a sixth-magnitude star (2.5 × 2.5 = 6.25). A magnitude change of five would mean a change in brightness of 2.5 raised to the fifth power, 2.5^5, or about 100. Astronomers recreated the scale accordingly, defining a difference of five magnitudes as corresponding to a factor

The quantity described here is apparent magnitude rather than absolute magnitude, which refers to the luminosity of a star and will be discussed later.

apparent magnitude. A measure of the amount of light received from a celestial object.

Stellar Luminosity

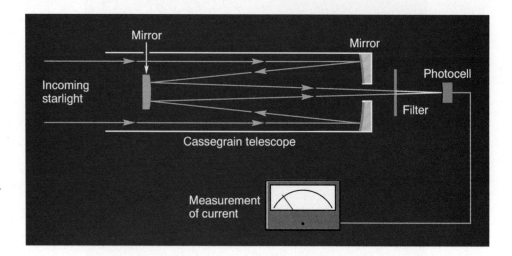

► **FIGURE 12-2** This illustrates the basic idea of photometry, in which the light from a star is focused onto a photocell that measures the amount of light. Often a filter is used to allow only light of a certain wavelength range to enter. More modern methods use CCD technology rather than photocells.

Photometry and CCDs (charge-coupled devices) were discussed in Chapter 5.

A later section in the chapter will examine stars that vary in brightness.

of exactly 100 in the light reaching us. This means that stars differing in apparent magnitude by one have a brightness ratio equal to the fifth root of 100, or 2.512.

Before electronic devices became common, astronomers measured light intensity from a star (photometry) by measuring the size and density of its image on a photograph (Figure 12-1). During the last few decades, photometry used a process similar to the way your automatic camera determines the brightness of the subject you are photographing. ►Figure 12-2 illustrates the idea. Today, however, video imaging is becoming common, and CCDs are used to detect and measure light intensity.

Although the eye is able to distinguish only a few different classes of stars (Hipparchus distinguished six), photometric methods allow us to discern the difference in brightness between two stars that may appear identical to the eye. The ability to measure magnitudes accurately has changed Hipparchus's unit-step system into a continuous one so that the magnitudes of stars are now measured to fractional values, to an accuracy of 0.001 magnitudes or better.

When the magnitude scale is defined as described, we find that some stars in Hipparchus's first-magnitude group are much brighter than others in that group. If a star were 2.5 times brighter than first magnitude, it would have to be assigned a magnitude of zero. A star 2.5 times brighter than this would have a magnitude of −1, a negative number. Sirius, the brightest star in the night sky, is about 10 times brighter than the average first-magnitude star and has an apparent magnitude of −1.47. Two other stars (and the Sun) have negative magnitudes. ►Figure 12-3 shows the approximate magnitude of a number of objects, and Appendixes E and F contain tables of the brightest and nearest stars, including magnitude values.

Table 12-1 shows the ratio of light received for a given difference in apparent magnitude. The following example shows how to use the chart.

EXAMPLE

The star W Pegasi is within the square of the constellation Pegasus (►Figure 12-4). This star varies in brightness, getting as bright as eighth magnitude. Fomalhaut (the bright star just south of Aquarius) is a first-magnitude star. Which star appears brighter, and how many times more light do we receive from that star than from the other?

Solution To answer the first question, remember that the star with the lesser magnitude is the one that appears brighter. Thus Fomalhaut is the brighter star. The difference in magnitude provides the information needed to answer the second ques-

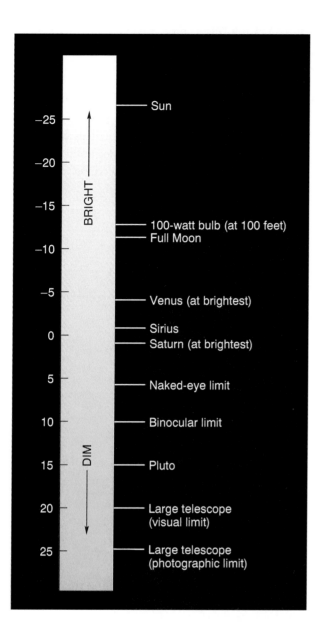

▶ FIGURE 12-3 Apparent magnitudes of a number of objects.

TABLE 12-1 Magnitude Difference versus Ratio of Brightness

Magnitude Difference*	Ratio of Light Received	
1.0	2.5	
2.0	6.3	(2.5×2.5)
3.0	16	$(2.5 \times 2.5 \times 2.5)$
4.0	40	(2.5^4)
5.0	100	(2.5^5)
10.0	10,000	(2.5^{10})

*If the difference in magnitude between two stars is 3, we receive 16 times more light from one star than from the other.

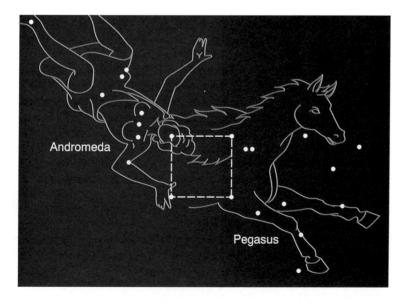

▶ FIGURE 12-4. The great square of Pegasus is an obvious feature of the fall sky. The constellations Pegasus and Andromeda share one bright star, Alpheratz, shown as an eye of Andromeda.

Stellar Luminosity

Formalhaut actually has a magnitude of 1.19, and W Pegasi gets as bright as magnitude 7.9. The magnitude values have been rounded for this example.

tion. The difference is 7. Table 12-1 does not list a difference of 7, but the values in the table can be used to determine what the light ratio is. Note that a difference in magnitude of 5 corresponds to a light ratio of 100 and a difference of 2 corresponds to 6.3. Thus a difference of 7 means that the ratio of light received is 100 × 6.3, or 630. We receive 630 times as much light from Fomalhaut as from W Pegasi (when the latter is at its brightest).

•••••**TRY ONE YOURSELF.** Barnard's star has an apparent magnitude of about 10. Suppose that on some particular night Mars is measured to have an apparent magnitude of 2. How many times more light do we receive from Mars than from Barnard's star?

As indicated in the example, telescopes allow us to see stars that are much dimmer than sixth magnitude. A telescope with a diameter of 12 centimeters (about 5 inches) might permit one to see, under perfect conditions, stars of about thirteenth magnitude. With a 5-meter telescope, one can photograph stars as dim as twenty-fifth magnitude.

It is valuable to know the brightnesses—the apparent magnitudes—of stars, but this is not a quantity that is inherent in the stars. Rather it depends partially upon their distances from Earth and therefore upon the position of the Earth in the Galaxy. We would like to know the *actual luminosity* of each star, for this would tell us something about the star itself independent of the Earth's location. To do this, however, we must know the distance to the star.

DISTANCES TO STARS—PARALLAX

At the time of Copernicus, the fact that parallax could not be observed was used as an argument *against* the heliocentric theory. Now its observation is an argument *for* the theory.

parallax angle. Half the maximum angle that a star appears to be displaced due to the Earth's motion around the Sun.

Recall from the earlier discussion of the heliocentric/geocentric debate (Chapter 2) the fact that stellar parallax had not been observed was an argument against the heliocentric system. Copernicus, however, held that it was not observed simply because the stars are too far away. Not until the mid-1800s was stellar parallax first observed, for the maximum displacement of the nearest star is only about 1.5 seconds of arc. Although the angle between lines from opposite sides of the Earth's orbit toward the star is 1.52 seconds, astronomers define the **parallax angle** as half of that, 0.76 seconds (so that there is a straightforward application of the properties of right triangles—see ➤Figure 12-5).

The formula used to determine the distance to a star by parallax can be written as follows:

➤ **FIGURE 12-5.** The nearest star has a displacement of about 1.5 arcseconds. The parallax angle is defined as half of this, using one astronomical unit as the baseline. The drawing is far out of scale.

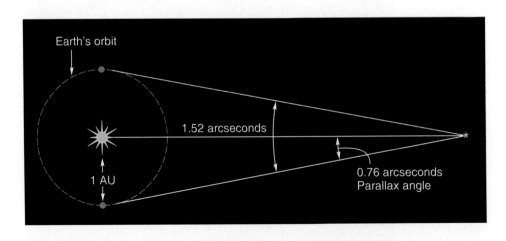

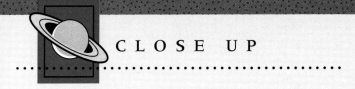

CLOSE UP

Naming Stars

Most of the names of the constellations are Latin translations of the original Greek names. Most of the names of stars, however, are of Arabic origin. Ptolemy's *Almagest* was preserved and passed on by Arab astronomers, who assigned names to numerous stars. The Arabic translation of the English article "the" is "al," and this explains why so many names of stars begin with those letters—Alcor (in the constellation Ursa Major), Aldebaran (in Taurus), and Alpheratz (in Andromeda; see Figure 12-4).

Only the brightest stars have popular names, however. Stars are assigned "official" names by a number of methods. The brightest stars in each constellation are given a Greek letter according to their brightness. Thus Elnath, the second brightest star in Taurus, is β *Tauri*. The brightest star in Taurus is Aldebaran, but since Aldebaran is part of a binary system of stars (to be discussed later in the chapter), it is named α *Tauri A*, and its dimmer companion is α *Tauri B*. Dimmer stars are given English letters followed by their constellation names (for example, *W Pegasi*, a star that was used in the example concerning stellar magnitudes).

We soon run out of letters in the Greek and English alphabets, of course, so most stars are known only by a catalog number. While the brightest star in Taurus is named *Aldebaran* and α *Tauri*, it is also known as *87 Tau*. Less distinguished stars don't have popular names, however, and don't have letter designations. One near Aldebaran just goes by the tag *75 Tau*.

$$\text{distance to star (light-years)} = \frac{3.26 \text{ light-years}}{\text{parallax angle in seconds of arc}}$$

Astronomers usually prefer, however, to express it in a different distance unit, the **parsec**. With distance in parsecs, the equation becomes

$$\text{distance to star (parsecs)} = \frac{1}{\text{parallax angle in seconds of arc}}$$

parsec. The distance of an object that has a parallax angle of one arcsecond.

In fact, this equation defines the parsec: One parsec (abbreviated pc) is the distance to a star that has a parallax angle of one arcsecond. The parsec is the unit astronomers normally use to express stellar distances. One parsec is equal to about 3.26 light-years, or 206,265 AU.

Only stars within about 120 parsecs (400 light-years) have parallax angles large enough to permit accurate parallax measurements from Earth. In 1989 the European Space Agency launched *Hipparchos* (*H*igh *P*recision *PA*Rallax *CO*llecting *S*atellite), which measured the positions and parallaxes of more than 100,000 stars to an accuracy of 0.002 arcseconds before its power failed. NASA is now planning the Space Interferometry Mission (SIM), which will be the first space mission with an optical interferometer as its primary instrument. SIM is designed to achieve an accuracy, over the whole sky, or 4 microarcseconds. Such an accuracy would revolutionize the field of astrometry—the precise measurement of the positions of stars.

Four microseconds is 0.000004 arcseconds!

Having more accurate stellar distances will be very valuable to astronomers, both because knowledge of the distance to a celestial object is often the key to determining other quantities about the object, and because it will help us to determine the distance scale of the universe more accurately. The next section explains how knowing the distance to a star allows us to calculate the star's luminosity, and a later section will explain why the accuracy of our measurements of distances to very remote objects often depends upon accurate knowledge of the distances to nearby stars.

CLOSE UP

A Long-Range Proposal

This chapter has shown the importance of knowing the distances to stellar objects and that parallax provides the most direct method of measuring those distances. As telescopes and techniques improve, astronomers are able to observe parallax at greater and greater distances. The procedure is limited, however, by the baseline of the parallax triangle: the diameter of the Earth's orbit. Scientists and engineers at the Jet Propulsion Laboratory (JPL) in Pasadena, California, have proposed that we use a specially designed space probe to extend the baseline. They propose sending a telescope 1000 AU away from Earth and using it in conjunction with earthbound telescopes to allow parallax measurements of stars much more distant than is possible with our present short baseline. The project is called TAU for Thousand Astronomical Unit.

The JPL group hopes for a launch of the telescope in 2005, but 50 years would pass before the telescope reaches its destination. Thus it would not be in operation during its designers' lifetimes. Some people find such a long-term project exciting, but others argue that it is not politically realistic to ask for funding for something that will not bear fruit for so long.

TAU probably has a potential payoff date farther into the future than any other project seriously proposed for NASA thus far, but it is only one of a number of space endeavors that, by their nature, cannot be completed during the terms of the elected officials who determine their fates. Somehow our political systems must provide for such long-range planning.

Absolute Magnitude

Chapter 11 explained how the luminosity of the Sun can be calculated using the distance from Earth to the Sun and the solar power striking a square meter of the Earth. In a similar manner, the luminosity of any light source can be calculated if one knows the distance to the light source and the brightness of the source. (The power striking a given area of Earth determines brightness.) You and I make subconscious judgments similar to this when we look at a distant light at night—a street light, for example—and decide that the lamp is dim or bright. We subconsciously take into account how bright the light appears as well as its distance from us in order to decide (qualitatively) the wattage of the bulb. If we had the tools to measure brightness and distance quantitatively, we could calculate the lamp's power in watts.

➤ FIGURE 12-6. If any two of the three quantities are known, the other can be calculated. Interstellar absorption of light is generally so little that it can be ignored.

➤ Figure 12-6 emphasizes that if we know any two of the following three factors we can calculate the third: apparent brightness (or, in astronomy, apparent magnitude), luminosity, and distance. The connection between these three will be used a number of times in the chapters to come. Brightness is most easily measured, and it is always one of the two known quantities. When the other known quantity is distance, we can calculate the luminosity; when the luminosity is known, we can calculate the distance.

The triple connection between brightness, distance, and luminosity allows astronomers to calculate the luminosity of some 1000 nearby stars; that many stars are close enough for us to determine their distance fairly accurately by parallax. If we want to know a star's luminosity in watts, we can use the procedure discussed for the Sun near the beginning of Chapter 11. Usually, however, astronomers use another method to state the intrinsic luminosity of a star. The *absolute magnitude* of a star is defined as the apparent magnitude that the star would have if it were located 10 parsecs from the Earth. Rather than calculating values of absolute magnitude from apparent magnitude and distance, one example will be used to illustrate the relationship.

absolute magnitude. The apparent magnitude a star would have if it were at a distance of 10 parsecs.

Sirius has an apparent magnitude of −1.47 and is 2.7 parsecs from Earth.

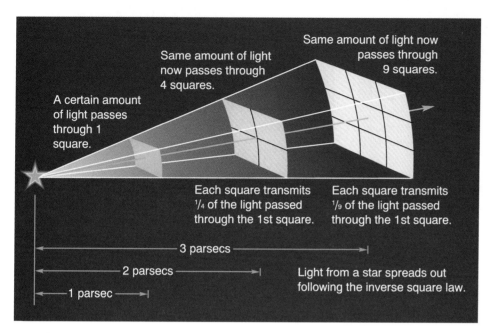

▶ **FIGURE 12-7.** The inverse square law. Light from a star spreads out in all directions, of course. This diagram shows the path of just some of the light from the star at the left, the light that hits the first square (at 1-parsec distance). The light continues to spread, and when it has gone 2 parsecs from the star, it has spread out to cover four squares. Thus less light illuminates each of the four squares, and the star's light is only one-fourth as bright. Finally, when the light has traveled 3 parsecs, it covers nine squares and is only one-ninth as bright.

To determine its absolute magnitude, imagine that it is moved to a distance of 10 parsecs. Its distance from us is thereby increased by a factor of about four ($^{10}/_{2.7}$). According to the inverse square law (▶Figure 12-7), if something is moved four times farther away, it will appear one-sixteenth, or $(1/4)^2$, times as bright. Now refer to Table 12-1 and notice that a brightness ratio of 16 corresponds to a magnitude change of 3. Therefore 3 is added to Sirius's apparent magnitude of -1.47 to yield a value of $+1.53$ for its absolute magnitude. Appendix E lists Sirius's absolute magnitude as $+1.45$, very close to the result obtained here. The error occurred because $^{10}/_{2.7}$ was rounded to 4.

The inverse square law was discussed at the end of Chapter 4.

MOTIONS OF STARS

You sometimes hear reference to the "fixed stars." In fact, the stars are not fixed but are moving relative to the Sun. This motion is obviously not visible to the naked eye, for the constellations have retained their shape fairly well over the centuries. In 1718, Edmund Halley discovered, however, that stars do move with respect to one another and therefore constellations do gradually change their shapes.

▶Figure 12-8 shows two photographs, taken 22 years apart, of a magnified portion of the sky. The arrows point to a particular star, Barnard's star, that has moved noticeably during that time.

Barnard's star is the second closest star to the Sun and shows the greatest motion as observed from Earth. It moves at the rate of 10.25 seconds of arc per year. Motion expressed as the angle through which a star moves each year (as seen from the Sun) is called the ***proper motion*** of the star. (You might speculate as to how a star could have *improper* motion, but the name comes from an old use of the word "proper" meaning "belonging to," for this is the motion that

proper motion. The angular velocity of a star as measured from the Sun.

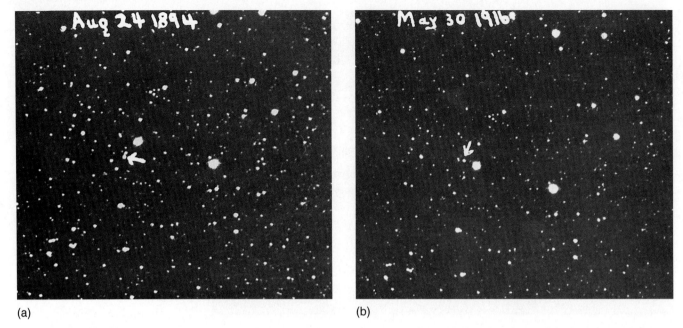

➤ **FIGURE 12-8.** (a) The arrow indicates a star named Barnard's star. (b) This photo, taken 22 years later, shows that Barnard's star has moved noticeably during that time. (The width of each photo is about 1 degree, so the full Moon would cover about half the width of the photo.)

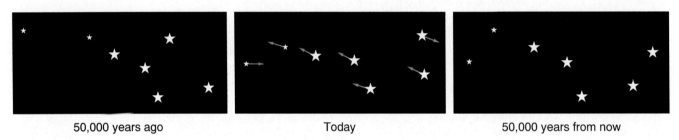

➤ **FIGURE 12-9.** The center drawing shows the Big Dipper as it is today; the arrows indicate the proper motions of its stars. From these motions, we can conclude that it once had the shape shown in the left drawing and will someday have the shape shown in the right one.

actually belongs to the star as opposed to observed motion that is due to Earth's movement.) Only relatively nearby stars show proper motion, so the background stars might still be called *fixed* stars although they only appear to be fixed because their proper motion is too small to detect. ➤Figure 12-9 shows how proper motion has changed the shape of the Big Dipper over the past 50,000 years and how its shape will continue to change in the future. We normally identify constellations by the stars that appear brightest, of course, but since these stars are often the closest ones, they have the greatest proper motion. Thus the constellations are gradually changing shape over the ages.

Proper motion does not tell us the actual velocity of a star in normal units of velocity, but the velocity can be calculated by using the small-angle formula used to determine the Moon's size in Chapter 6. (Instead of the width of the Moon, the distance the star moved is calculated.) Of course, one must know the distance to the star to make such a calculation (➤Figure 12-10). In doing the calculation, only the speed of the star *across* our line of sight—its *tangential velocity*—is being computed.

The velocity of a star toward or away from the Earth—its *radial velocity*—is easier to detect and measure than its tangential velocity. Chapter 4 discussed how

UFO reports often state the speed of the unidentified object, although the proper motion of the object is what is actually observed. This is one example of how a poor understanding of astronomy can lead to inaccurate conclusions concerning celestial objects.

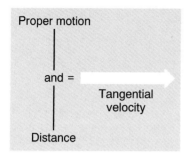

▶ FIGURE 12-10. In order to calculate a star's tangential velocity, both its proper motion and its distance must be known.

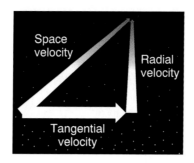

▶ FIGURE 12-11. Since radial velocity and tangential velocity are at right angles, the Pythagorean theorem is used to add them.

this measurement is made and emphasized that the Doppler effect measures only the star's radial velocity.

A star's actual motion relative to the Sun—its *space velocity*—is a combination of its radial and tangential velocities. Since these two are at right angles to one another, they must be added as shown in ▶ Figure 12-11 to calculate the star's space velocity.

Naturally, the Earth's movement in its orbit affects the observed motions of stars and must be taken into account in calculating the velocity of a star. Once this is done, we have the star's velocity relative to the Sun. We will see later that there are methods for determining the Sun's movement relative to the distant, "fixed" stars. If this motion is taken into account, we can determine a star's velocity relative to the distant stars.

space velocity. The velocity of a star relative to the Sun.

SPECTRAL CLASSES

▶Figure 12-12 is a photo of Orion taken in an unusual way. During the 30-minute exposure, the camera was held steady, but its focus was changed in steps so that the stars became more and more out-of-focus as they drifted by toward the west (the right). This causes each star to form a fan-shaped image, and it reveals the colors of the stars. Most of the brightest stars are blue. The red-orange star at upper left is Betelgeuse, and the red "star" at lower center is the Orion nebula.

As the discussion in Chapter 4 explained, the color of a star is determined by its temperature. In fact, the wavelength at which the maximum energy is emitted from a star provides an accurate way to measure the star's temperature. Refer back to the intensity/wavelength graph of the Sun in Figure 11-17, which shows that the Sun's energy output peaks near the center of the visible region and indicates a temperature of about 5800°C for the surface of the Sun.

Another method of analyzing the spectrum provides an independent measure of temperature. This method depends upon the absorption of radiation at various wavelengths—the absorption spectrum. As we have seen, the absorption of certain wavelengths is what allows us to determine the chemical elements in a star's atmosphere, but the absorption by an atom depends not only upon what element it is, but also upon the state of its electrons (that is, whether some have been moved to higher energy levels or stripped from the atom). In a gas at higher temperature, more atoms will be at higher energy levels, and the transitions that occur from these atoms will be different from the transitions that take place in atoms of a cool gas (whose atoms are at lower energy levels). This provides a method of determining temperature other than by the intensity/wavelength graph. The accompanying Close Up explains this in greater detail.

▶FIGURE 12-12. The colors of the stars are obvious in this photo of Orion. The photo was made by successively de-focusing the camera as the stars moved by toward the right.

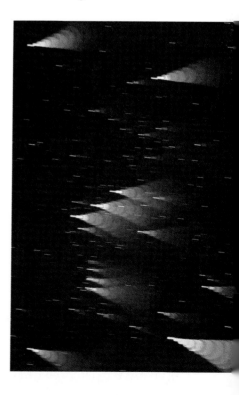

CLOSE UP

Determining the Spectral Class of a Star

Before studying this Close Up, you may find it helpful to reread a related Close Up in Chapter 4, "The Balmer Series."

Hydrogen is the most common gas in the atmosphere of stars, so it will serve as our example here. Recall that each hydrogen atom has one electron in orbit around its nucleus. Suppose that the hydrogen gas near a star is cool enough that most of its electrons are in the ground state ($n = 1$ in Figure ➤C12-1). When an atom with its electron in the ground state absorbs a photon of light, the photon will be of an energy that will move the electron from the ground state to one of the other energy levels. These lines form the Lyman series, which is in the ultraviolet region of the spectrum; therefore they do not appear in the visible part of the spectrum and cannot be seen in Figure 12-14.

Now consider a hotter star, one with a significant number of its atmospheric hydrogen atoms with electrons in the second energy level. For one of these atoms to absorb a photon, that photon must be of the right energy to move the electron from level 2 to higher levels. Figure C12-1 shows that the wavelengths of these photons will be part of the Balmer series and will be in the visible range. Look at the spectra of Figure 12-14. The hydrogen-α line is indicated at 656 nm, in the red part of the spectrum. The spectrum of the coolest star, the K5 star, does not show a pronounced absorption line for this energy. The reason is that the star is not hot enough for many of its atmospheric hydrogen atoms to be in the level 2 state.

Look at the hydrogen-α absorption line in the hotter stars. As one moves to hotter and hotter stars, the absorption becomes more intense. In fact, the line is most pronounced in A-type stars. The figure does not include spectra of the hottest stars, O- and B-type. In these stars, the hydrogen-α absorption line is again less pronounced. The reason in this case is that in these ultrahot stars most hydrogen atoms have so much energy that their electrons are at level 3. Thus there are few electrons at the second level to absorb photons that correspond to the hydrogen-α absorption line.

Only one line of the Balmer series of hydrogen has been considered here. A complete analysis would require an examination of a number of lines of several different chemical elements. The spectral class of a star is determined by the relative intensities of these lines.

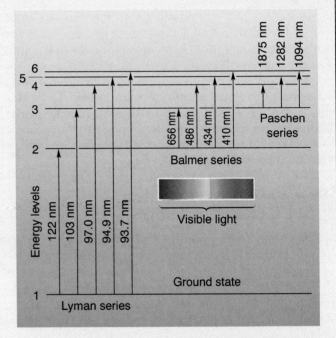

➤ **FIGURE C12-1** The energy levels of the hydrogen atom. If many atoms are in the $n = 2$ state, their absorption of photons will cause a dark line in the visible portion of the spectrum.

Annie Jump Cannon (➤Figure 12-13), an astronomer at Harvard College Observatory, devised a classification scheme for spectra and separated several hundred thousand stars according to their spectral type. The classes were first labeled alphabetically, but later their order was rearranged by temperature and some classes were dropped. The spectral types used today are designated, from hottest to coolest, as O B A F G K M (➤Figure 12-14).

The hottest stars, O stars, range in temperature from 30,000°C to 60,000°C. The coolest stars are the M stars, with temperatures less than 3500°C.

Within each spectral class, stars are subdivided into 10 categories by number, so our Sun is listed as a G2 star. You need not be concerned about these subdivisions, but you should remember the classification scheme from hottest to

► FIGURE 12-13. Annie J. Cannon, a member of the Harvard College Observatory for almost 50 years, was the founder of the spectral classification scheme in use today.

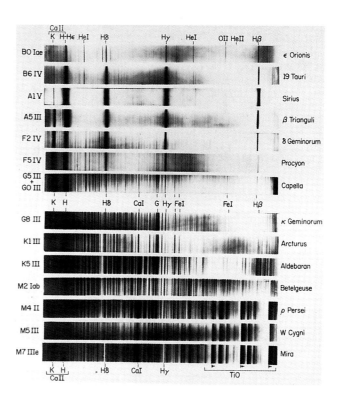

► FIGURE 12-14. The spectra of stars of various classes, from B to M, with their classes and subclasses shown at left. The name of each star is listed at right, and a few of the major absorption lines are identified at the bottom.

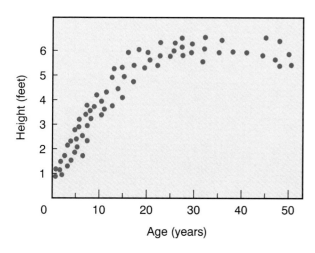

► FIGURE 12-15. A plot of the age versus height of people in a neighborhood might look like this. Each dot represents one person.

coolest as being O B A F G K M. The traditional mnemonic is "Oh, Be A Fine Girl, Kiss Me" although you may want to devise your own less sexist memory aid (or substitute *guy* for *girl*).

The Hertzsprung-Russell Diagram

Suppose that an extraterrestrial being, unfamiliar with humans, somehow learned the age and height of each person in your neighborhood and then used these values to plot a graph such as that shown in ►Figure 12-15. The first thing he would notice is that a pattern exists: there is some relationship between height and age for humans. Then if our imaginary alien were very analytical, he could learn an important fact about the life cycle of humans even though no one

Hertzsprung-Russell diagram. A plot of absolute magnitude (or luminosity) versus temperature (or spectral class) for stars.

main sequence. The part of the H-R diagram containing the great majority of stars; it forms a diagonal line across the diagram.

person had been analyzed through his or her entire life. The alien could logically hypothesize that an individual spends most of his or her life at about the same height; that height being the tallest achieved by the person. In our study of stars, we have a similar chart, the *Hertzsprung-Russell diagram.*

Early in this century, Ejnar Hertzsprung, a Danish astronomer, and Henry Norris Russell, an American astronomer at Princeton University, independently developed the diagram now named for them and often called simply the H-R diagram. The diagram shows that a pattern exists when stars are plotted by two properties: their temperature (or spectral class) and their absolute magnitude (or luminosity). ➤Figure 12-16 is similar to the diagram Russell plotted in 1913. The stars are not evenly distributed over the entire chart, but seem to group along a diagonal line. If we include many more stars than those on Russell's first diagram, we obtain a plot as shown in ➤Figure 12-17. (Notice the Sun's position on the diagram.)

About 90% of all stars fall into a group running diagonally across the diagram called *main sequence* stars. Other groups are named as shown in the diagram. The significance of the groups will be discussed later in this chapter and in the next few chapters, along with the life cycle of stars.

➤ **FIGURE 12-16.** This is similar to the first H-R diagram, plotted by Henry Norris Russell in 1913. He plotted absolute magnitude versus spectral class and saw an obvious pattern in the distribution of stars. Notice that his category labeled N contained no stars.

➤ **FIGURE 12-17.** A modern H-R diagram shows that stars fall into various categories, including main sequence stars, white dwarfs, red giants, and supergiants.

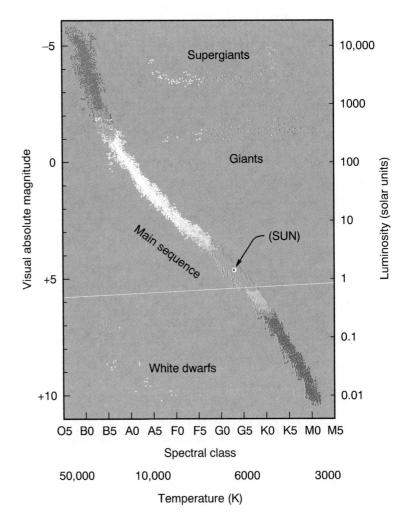

380   CHAPTER 12   Measuring the Properties of Stars

Spectroscopic Parallax

As we have seen, the distances to the nearest stars can be measured using stellar parallax. The H-R diagram provides another method of measuring such distances, one that is not confined to neighboring stars. Although this method does not involve parallax, it is called *spectroscopic parallax.*

As will be shown later, it is possible to determine from the spectrum of a star whether that star is a main sequence star, a giant (or a supergiant; the term *giant* is often used for both), or a white dwarf. Suppose that we observe a particular star and determine that it is a main sequence star. As we have seen, it is a fairly routine procedure to ascertain the temperature of a star. Let's suppose that our star has a temperature of 10,000 K. Refer to the H-R diagram of ➤Figure 12-18. The region of the main sequence where stars have a temperature of 10,000 K is marked on the diagram. It is now a simple matter to determine that our chosen star has an absolute magnitude between about +1 and +2.

Recall the connection between absolute magnitude, apparent magnitude, and distance. If we know two of these, we can calculate the third. Use of the H-R diagram has allowed us to determine, within a small range, the absolute magnitude of the star. Since apparent magnitude is directly measurable, we can calculate the star's distance.

As indicated, we know the absolute magnitude only within a certain range, and therefore our precision in determining the distance is limited. Keep in mind, however, that this is *always* the case with a measurement. In using spectroscopic parallax, the source of the error is obvious, but as we have seen, the determi-

spectroscopic parallax. The method of measuring the distance to a star by comparing its absolute magnitude to its apparent magnitude.

In practice the temperature cannot be measured with absolute precision, so some error is introduced in this manner also.

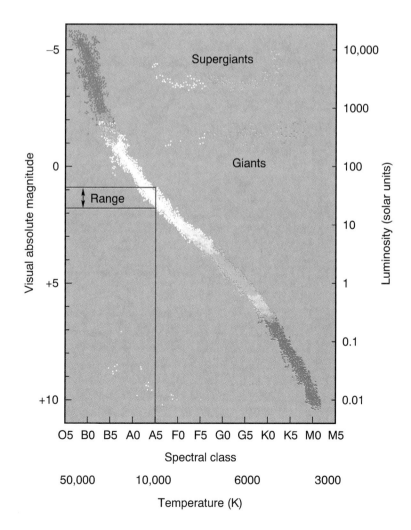

➤ FIGURE 12-18. The absolute magnitude of a main sequence star with a temperature of 10,000 K can be determined from the H-R diagram.

The Hertzsprung-Russell Diagram

nation of distance by trigonometric (stellar) parallax is also limited in accuracy, both because of the difficulty in measuring the small angles involved and because other motions must be taken into account, such as the proper motion of the star being measured. All measurements contain error. The important thing in science is to be cognizant of how great the likely error is.

Because of the limits in the accuracy of spectroscopic parallax, it is most useful when applied to groups of stars that are nearly the same distance from Earth. As the next chapter will show, it is common for stars to be grouped with many other stars in large clusters. The stars of each cluster are very close to one another compared to their distance from Earth, and we use statistical methods to combine individual measurements of spectroscopic parallax to determine the distance to the cluster. The resulting value has much less error than the distance we calculate to any individual star.

Such averaging over a number of cases to improve accuracy is a valuable and common technique in science.

A simplified example will illustrate the method of spectroscopic parallax.

EXAMPLE

Spica, the brightest star in the constellation Virgo, is a B1-type main sequence star with a temperature of about 20,000 K. Its apparent magnitude is 0.91. From this data, determine whether its distance from Earth is less than 10 pc, approximately 10 pc, greater than 10 pc, or much greater than 10 pc.

Solution

Referring to Figure 12-18 and using the fact that Spica is a B1-type star, we determine that its absolute magnitude is between about -3 and -4.5 (although a safer range might be from -2.5 to -5.0).

Next, let's compare our absolute magnitude values with Spica's observed apparent magnitude. Spica's apparent magnitude is 0.91, and we can reason that since its apparent magnitude has a value greater than its absolute magnitude, it is farther than 10 pc away. (Remember, greater magnitude means a dimmer star, so its apparent brightness is less than its brightness would be at 10 pc.) In fact, from the difference between $+0.91$ and -3 or -4, we can conclude that it would have to be moved fairly far to bring it as close as 10 pc. So it is much farther away than 10 pc.

Actually, Spica is about 75 parsecs, or 240 light-years, from us.

TRY ONE YOURSELF. The star 40 Eridani is a main sequence star of spectral type K0, which means that its temperature is 5100 K. Its apparent magnitude is $+4.4$ (Actually, this is the magnitude of the brightest star of a triple-star system. Multiple stars are discussed later.) What can you conclude about the distance to 40 Eridani?

Luminosity Classes

Recall from Chapter 4 that a solid object emits a continuous spectrum because its atoms interact with one another and thereby distort their energy levels. This smears what would be separate, discrete wavelengths into a continuous range of wavelengths. Recall also that an absorption spectrum is produced when light passes through a star's atmosphere. In the 1880s, Antonia Maury discovered that absorption lines are also subject to a smearing effect and that, in general, they are not fine lines. This discovery has become very valuable in classifying stars.

Maury worked at Harvard College Observatory with Annie Jump Cannon and Henrietta Leavitt, who will be discussed later in this chapter.

The atmosphere of a main sequence star is fairly dense, and therefore its atoms collide frequently and stretch what would have been a thin absorption line into a broader line. Red giant stars have thinner atmospheres, so the broadening does not occur to the extent seen in a main sequence star. Their spectral lines are narrower. Through the examination of the extent of line broadening,

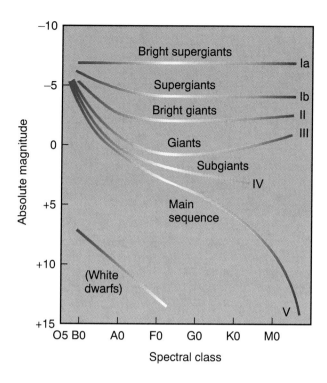

► FIGURE 12-19. Stars can be classified according to luminosity class, which is determined by analysis of their spectra.

as well as other subtle differences in spectra, stars are classified into various *luminosity classes* that are located on the H-R diagram as shown in ►Figure 12-19.

The example of spectroscopic parallax in the last section involved only main sequence stars. The classification of stars into luminosity classes allows spectroscopic parallax to be used with any star, for instead of reading the star's absolute magnitude from the main sequence, one reads it from the appropriate location on the H-R diagram.

luminosity class. One of several groups into which stars can be classified according to characteristics of their spectra.

Do not confuse *luminosity class* with *spectral class* discussed earlier.

Analysis of the Procedure

Observe that in using the method of spectroscopic parallax to find a star's distance from us, the following two triple connections are used (►Figure 12-20): By knowing the temperature of a star and the star's type, its absolute magnitude is determined; then, by knowing its absolute magnitude and its apparent magnitude, its distance is calculated. This last triple connection is used in a different way than it was before. For nearby stars, distance and apparent magnitude were used to determine absolute magnitude.

Absolute magnitudes calculated for nearby stars enabled astronomers to draw the H-R diagram. Once the patterns of that diagram were known, they could be

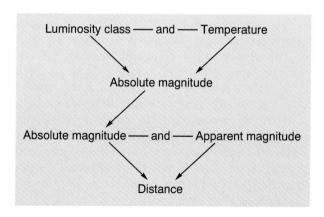

► FIGURE 12-20. From the luminosity class and temperature of a star, its absolute magnitude can be determined. From this and its apparent magnitude, its distance can be determined.

The Hertzsprung-Russell Diagram

www.jbpub.com/starlinks

applied to stars that are too far away to permit distance measurement by parallax. It is reasonably assumed that these stars fit the same H-R diagram pattern as do nearby stars, so the diagram is then used to determine their absolute magnitude.

The basic procedure outlined here is often used in astronomy (and in other sciences). By observing familiar objects (nearby stars in this case), we see patterns and formulate laws (or statements of relationships). We then assume that these patterns and laws hold for more distant objects of the same type. This allows us to use the same patterns to learn more about those distant objects. At times it appears that the whole system is a house of cards ready to fall down, but usually cross-checks are available; measurements can be taken in a number of ways, thereby permitting us to verify theories by independent measurements.

The Sizes of Stars

Although the sizes of a few of the largest stars have now been directly detected by interferometry methods, in general a star is observed as only a point of light. It shows no size. How, then, can we determine the size of an object so distant that it appears to have no size?

Consider the group of stars in the lower left corner of the H-R diagram, the **white dwarfs.** These stars are hot, but their location in the lower part of the diagram indicates that they are intrinsically dim. One might think that a hot star would be bright, for this is not only the pattern seen in the main sequence but it also follows from the intensity/wavelength graph: As an object becomes hotter, the wavelength of the maximum radiation emitted becomes shorter, and the *total amount of radiation from the star increases*. How, then, can these stars be dim? The answer is that they are small. Being hot, each square meter of their surface emits more energy per second than a cooler star, but they simply have small total surface areas. The name "white dwarf" indicates their temperature (white-hot) and their size.

An everyday example might help here. You can adjust the burner of an electric stove to various temperatures. As the burner gets hotter, its color changes from dull red to orange-red (and perhaps to orange). At the same time, its overall brightness increases. Imagine, though, that you have a very small burner and you make it orange-hot. If it is small enough, it will emit less total radiation—it will be less bright—than a very large burner that is not as hot. The small, hot burner corresponds to a white dwarf in that its small size causes its overall luminosity to be less than average.

On the other hand, consider the **giants** and **supergiants.** How do we know that these are large stars? Simply because they are very bright in spite of their low temperatures. The giants are often called *red giants*, paralleling the name given the white dwarfs. (A single stove burner set to "warm" does not yield much light, but if enough of these relatively cool burners are lit, they provide quite a lot of red light.)

In these examples of white dwarfs, giants, and supergiants, qualitative reasoning was used to learn something about their sizes. Determination of stellar sizes is not limited to those classes of stars, however; nor is it limited to qualitative methods. Knowing the temperature of an object, we know the amount of energy emitted each second per square meter of its surface. Then if we know the total energy emitted by the object each second (by knowing the absolute magnitude), it is easy to calculate the area of its surface and therefore its diameter. ➤Figure 12-21 illustrates the triple connection in the chain of reasoning. We find that stars on the main sequence range from about 0.1 times the Sun's diameter up to as much as 20 times the Sun's diameter. Non-main sequence stars, however, differ in size even more; white dwarfs may be as small as 0.01 of the Sun's size, and supergiants may be 100 times larger in diameter than the Sun.

white dwarf. A very small, hot star. (Typical diameter: 0.01 that of the Sun.)

giant star. A star of great luminosity and large size. (10 to 100 times the Sun's diameter.)

supergiant. A star of very great luminosity and size. (100 to more than 1000 times the Sun's diameter.)

The Stefan-Boltzmann law, $E = \sigma T^4$ (discussed in a Close Up in Chapter 4), is used to calculate the energy emitted per square meter of a star's surface each second.

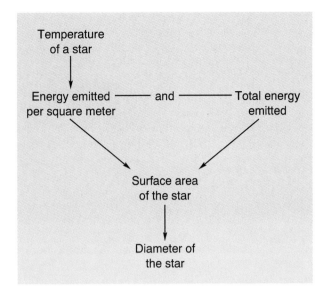

▶ **FIGURE 12-21.** If the temperature of a star and the total power emitted by it are known, the triple connection shown here can be used to calculate its diameter.

It is always desirable to have a second, separate method of measuring a quantity. This allows us not only to check our theories regarding the object being measured but also to check our measurement techniques. Fortunately, there is a second method for measuring the sizes of stars, although only a few stars can be measured by this technique. To see how it works, some discussion of multiple star systems is necessary.

MULTIPLE STAR SYSTEMS

A few decades ago astronomy books reported that about one-fourth of the objects that appear to be single stars really contain two or more stars in a close grouping. Books published a decade ago reported this as about one-third. A few years ago, it was considered to be about half. Now we can safely say that *more than half* of what appear as single stars are in fact multiple star systems. These systems must be distinguished from pairs of stars that appear close together as a result of being nearly in the same line of sight from Earth (▶Figure 12-22). These *optical doubles*, as they are called, are merely chance alignments of stars. The stars in multiple star systems are gravitationally bound so that they revolve around one another. When two stars are gravitationally linked, they are said to be a **binary**

optical double. Two stars that have small angular separation as seen from Earth but are not gravitationally linked.

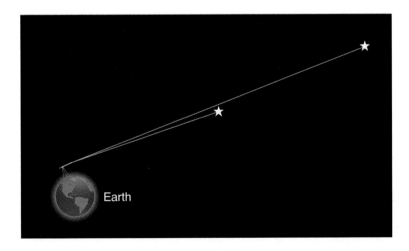

▶ **FIGURE 12-22.** Two stars are said to form an optical double if they appear close together but have no actual relationship to one another.

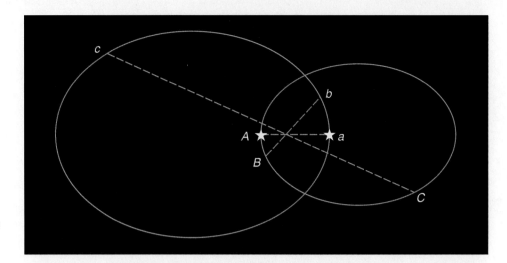

▶ FIGURE 12-23. Binary stars orbit their common center of mass. When one star is at *A*, the other is at *a*, then *B* and *b*, and so forth. Can you tell which is the more massive star?

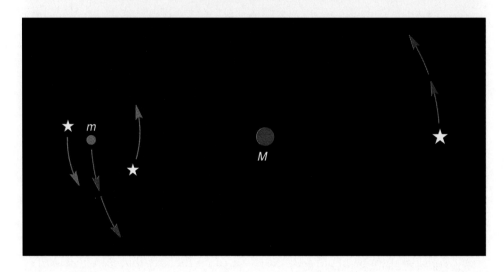

▶ FIGURE 12-24. A triple-star system. The two stars at the left orbit their center of mass (point *m*). This pair and the star at right orbit the overall center of mass at *M*.

binary star system. A system of two stars that are gravitationally linked so that they orbit one another.

visual binary. An orbiting pair of stars that can be resolved (normally with a telescope) as two stars.

star system. ▶Figure 12-23 illustrates how the two stars of a binary pair revolve about their common center of mass, in the same manner as discussed in a previous chapter for a star and a planet. Although groups of more than two stars are common, emphasis here will be on binary systems since multiple star systems are simply combinations of a binary system and a single star or of two binary systems (▶Figure 12-24).

In 1802 William Herschel obtained the first observational evidence that some double stars orbit one another. (This was also the first direct evidence of gravitational force at work outside our solar system.) Herschel had taken an interest in double stars in hopes of using them for parallax measurements, but his discovery of binary star systems turned out to be much more important to astronomy, for it is by such systems that we are able to determine the mass of stars.

Binary systems are classified into a number of categories according to how they are detected. We will discuss each in turn and then explore what can be learned from binary stars.

Visual Binaries

A *visual binary* is a system in which the pair of stars can be resolved as two in a telescope. Using the largest telescopes, perhaps 10% of the stars in the sky are visual binaries. In a small telescope, only a small fraction of these can be resolved,

386 CHAPTER 12 Measuring the Properties of Stars

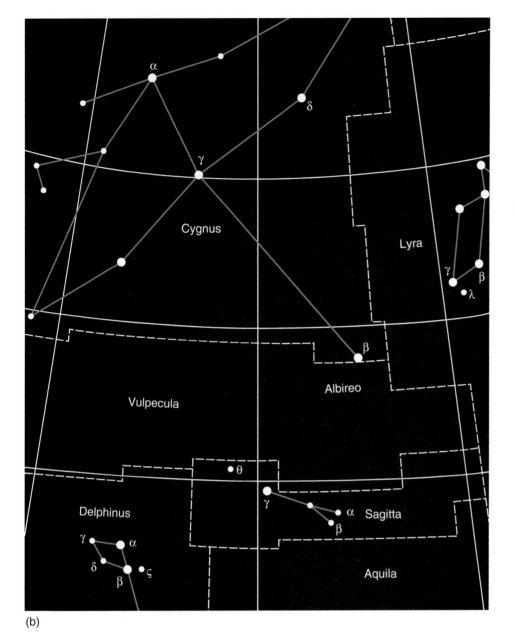

▶ **FIGURE 12-25.** (a) Albireo is a binary pair that shows an obvious color difference between the two stars. (b) Albireo is the bottom star of the cross of Cygnus. (Normally, the designation β would mean that the star is the second brightest in the constellation (after α), but labeling in Cygnus does not follow the rule strictly and β Cygni is the fourth brightest star in Cygnus.)

but some are beautiful sights, particularly when the two stars are very different in color. ▶Figure 12-25 is a photograph of Albireo, a visual binary in Cygnus, taken by an amateur astronomer. If you have access to a telescope, find Albireo some clear summer night.

If two stars appear close together in the sky, how can astronomers tell if they are a binary system? The surest way is simply to observe the pair over a period of time and look for signs of revolution. ▶Figure 12-26 shows a binary (at the right in the photograph) that reveals obvious orbital motion over a number of years. Things are usually not this easy, however, for in order for us to be able to resolve a binary pair, the stars must be either very close to Earth or very far apart. The Albireo pair, for example, is separated by some 4500 AU. As indicated by Kepler's laws, a great distance of separation indicates a long period of revolution, and the Albireo binary is thought to have a period of many tens of thousands of years. Since no detectable orbital motion has occurred in the relatively short time over which we have photographic records, it is difficult to confirm that such a system is indeed a binary pair. We do know, however, that the two

Multiple Star Systems

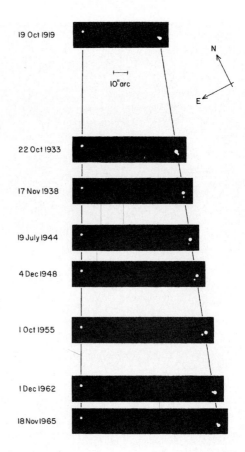

▶ **FIGURE 12-26.** The visual binary Kruger 60 can be seen during its 45-year period. In addition, its proper motion away from the star at the left can be seen.

stars are about the same distance from us, and it is therefore thought that they are gravitationally linked.

A second method of determining whether a pair is indeed a binary system employs that powerful astronomical tool, the Doppler effect, which leads to the next type of binary system.

Spectroscopic Binaries

▶Figure 12-27 shows two spectra of the star κ (the Greek letter, kappa) Arietis, taken at different times. The upper spectrum contains more lines than the lower. Look closely and you will see that this is the result of each of the lines in the lower spectrum having broken into two in the upper. Continuing observation of this star reveals that the spectrum repeatedly goes through a cycle in which each line gradually breaks into two, which spread until they reach a maximum separation and then come together again. The explanation for this is that we are seeing not one, but two stars and that they are revolving around one another so that the Doppler effect causes us to see separate spectral lines where one star is moving toward us and the other away. Such a binary system is called a *spectroscopic binary*.

spectroscopic binary. An orbiting pair of stars that can be distinguished as two due to the changing Doppler shifts in their spectra.

▶ **FIGURE 12-27.** Two spectra of the spectroscopic binary κ Arietis are shown here. Notice that lines that are single in the bottom spectrum are split in the upper one. This is particularly evident for the lines to the right of center. When single lines appear, the two stars are moving at right angles to the line of sight, and where they are double, one star is moving toward us and the other away.

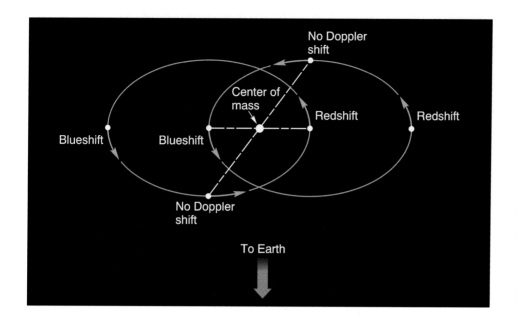

► **FIGURE 12-28.** When the binary stars whose spectra are shown in Figure 12-27 are moving perpendicular to the line of sight, no Doppler shift is observed. The Doppler effect is seen only when the stars are moving toward or away from us.

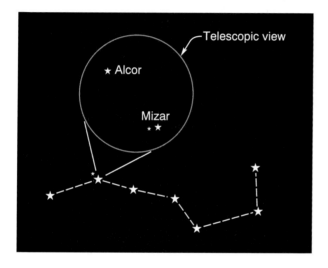

► **FIGURE 12-29.** Even a small telescope reveals that Alcor and Mizar are actually three stars.

The Doppler effect shows only the radial components of the stars' motion; that is, the motion toward or away from us. If a binary pair is oriented so that its plane of revolution is perpendicular to a line from Earth, no Doppler shift is observed in its spectral lines (►Figure 12-28). On the other hand, if the plane of revolution is tilted directly toward the Earth, the Doppler effect allows measurement of the stars' actual velocities during the part of their orbits when one is moving directly toward us and the other directly away. At any other orientation of the plane of revolution, we detect only the radial component of the stars' motions.

In the handle of the Big Dipper is a particularly interesting example of binary stars (►Figure 12-29). As pointed out in Chapter 5, good eyesight reveals two stars at that location, the brighter one named Mizar and the dimmer Alcor. Although it had been thought that Alcor and Mizar form an optical double rather than a gravitationally linked binary, more recent observations of their radial velocities indicate that they are orbiting one another. In any case, if you view Mizar and Alcor through a telescope, you will see that Mizar itself is two stars (Figure 5-12). Following the standard practice for naming double stars, we call the brighter Mizar A and the other Mizar B. Mizar A and B are a widely separated visual binary and have a period of at least 3000 years.

In 1889 a Harvard College astronomer, Edward Pickering, found that Mizar A is a spectroscopic binary with a period of only 104 days. When it was discovered

Multiple Star Systems

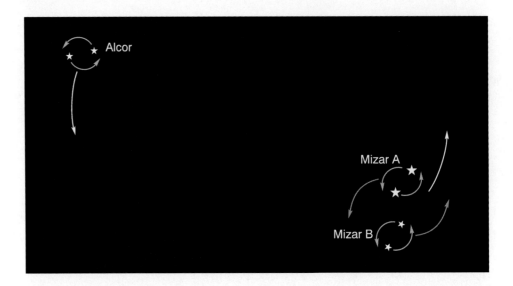

► **FIGURE 12-30.** Spectroscopic evidence reveals that Alcor and Mizar actually form a six-star system.

Mizar B is called a *single-line spectroscopic binary system* rather than a double-line system such as Alcor.

that Mizar A was binary, Mizar B was scrutinized. In this case, the spectral lines did not separate into two parts but instead moved back and forth; first they redshifted and then blueshifted. Mizar B is indeed a binary star, but the spectrum of its companion is not bright enough to be observable. The shifting spectrum that is observed is the spectrum of the brighter star of the pair. We deduce the existence of the companion from that motion.

Finally, it has been found that Alcor is a spectroscopic binary. Thus the dot in the handle of the Big Dipper is in fact six stars (►Figure 12-30)!

Eclipsing Binaries

A special case of binary stars occurs when their plane of revolution is along a line from Earth so that one star moves in front of the other as they orbit. The star Algol, in the constellation Perseus, is a good example of an eclipsing binary. Its brightness changes periodically, and in 1783, John Goodricke, an amateur astronomer, explained the brightness change as being due to a dimmer companion passing in front of the brighter. ►Figure 12-31a shows a graph of the light received from Algol as time passes, the system's **light curve.** Notice that every 69 hours the apparent magnitude of Algol changes from 2.3 to 3.5. This happens when Algol A is partially eclipsed—see Figure 12-31b. Midway between these dips in the light curve we see smaller dips, caused by the dimmer companion being eclipsed.

John Goodricke was hearing impaired.

light curve. A graph of the numerical measure of the light received from a star versus time.

Since Goodricke's discovery that Algol is a binary star, its nature has been confirmed from spectroscopic evidence. Like Mizar B, its companion is too dim to show its own spectrum, but the spectral lines of Algol A move back and forth in rhythm with its cycle of brightness changes. Like most eclipsing binaries, it can now be classed as both an eclipsing binary and a spectroscopic binary.

Other Binary Classifications

There are at least two other ways to detect binary stars. Sometimes a star is seen to shift back and forth in its position among the other stars, indicating that it is revolving around an unseen companion. A number of such systems, known as **astrometric binaries,** have been found.

astrometric binary. An orbiting pair of stars in which the motion of one of the stars reveals the presence of the other.

If one star of a binary system is much hotter than the other, their spectra differ enough from each other that it is possible to ascertain that the spectrum we see is not from a single star but is a composite of two spectra. Such a system

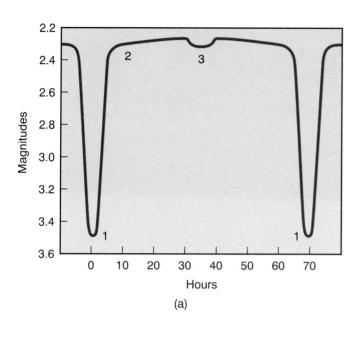
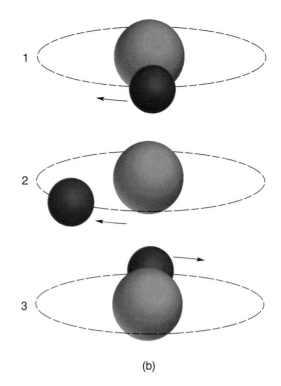

► FIGURE 12-31. The light curve of Algol (a) is explained by the eclipsing of its components, one of which is much darker than the other. Positions 1, 2, and 3 in the light curve correspond to the numbers in part (b).

is called a *composite spectrum binary.* This provides another method of detecting binary stars, but nothing can be learned of the motions of the stars in such a system.

composite spectrum binary. A binary star system with stars having spectra different enough to distinguish them from one another.

STELLAR MASSES AND SIZES FROM BINARY STAR DATA

Binary stars are interesting in themselves. For example, astronomers speculate on the stability of a planetary system around a star that is part of a binary system and on how conditions would be different on such planets because of the extra sun. The major importance of binary stars, however, is that they allow us to measure masses of stars—something that can be measured in no other way. Recall from Chapter 7 that the mass of a planet can be calculated from the orbit of one of its moons and that the mass of the Sun can likewise be calculated from the orbits of the objects circling it. The calculation involves Kepler's third law as revised by Isaac Newton. Stellar masses of binary systems are calculated in the same way.

In order to do such a calculation, we must know two things about the orbit of one (or both) of the stars: the size of the ellipse (its semimajor axis) and the period of revolution. The latter is easy to ascertain in all cases except when the period is extremely long. The former is a little more complicated.

►Figure 12-32 shows the orbit of a visual binary as it might be observed. The center of mass, or *barycenter,* of the two stars can be determined from the fact that it must always lie along a line connecting them (as was illustrated in Figure 12-23). Each of the orbits is an ellipse, but notice that the center of mass of the two stars does not lie symmetrically in either ellipse—it is not at the foci of the ellipses. The explanation for this is that we are not seeing the ellipses straight-on. A circle, when viewed at an angle, yields an elliptical shape (►Figure

barycenter. The center of mass of a binary pair of stars.

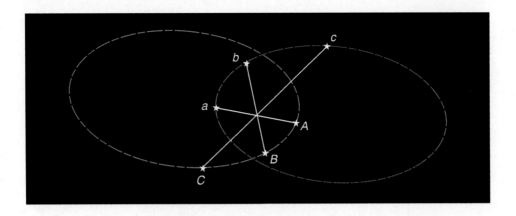

▶ FIGURE 12-32. This shows the observed orbits of a binary pair along with the locations of the stars at three times. The center of mass must be located where the lines that connect the stars cross. Notice, however, that the center of mass is not at the focus of either ellipse. (Compare this to Figure 12-23.)

12-33), and an ellipse viewed at an angle also yields an ellipse but one of a different shape than the original. The fact that the center of mass is not at one of the ellipses' foci can be used to determine how much the ellipses are tilted with respect to our line of sight (▶Figure 12-34).

Once the true shape of the ellipse is determined, its size can be calculated using the small-angle formula. (The distance to the pair must be known to do this, of course.) Knowledge of the size of one of the stars' ellipses, along with knowledge of the period of its motion, allows us to calculate the *total mass* of the two stars. To determine how the mass is distributed between the two, we need only consider the ratio of the two stars' distances to the center of mass. This is analogous to the way that we can calculate the weight of each of the people on a seesaw if we know the total weight of the two people and know how far each is sitting from the center if they are balanced (see ▶Figure 12-35).

▶FIGURE 12-33. A circular shape, such as the top of a trash can or a basketball rim, appears elliptical when viewed at an angle.

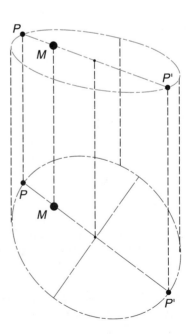

▶FIGURE 12-34. This figure shows the apparent path (above) and the true path (below) of a binary star. Although both shapes are ellipses, if you turn your book, you can see that point *M* is at a focus of the lower ellipse.

392   CHAPTER 12   Measuring the Properties of Stars

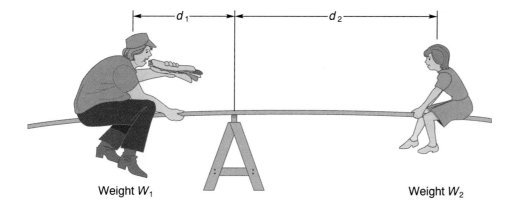

► **FIGURE 12-35.** Starting from the simple relationship $w_1 d_1 = w_2 d_2$, it can be shown that, if W is the total weight of the two people,

$$w_1 = \frac{W \times d_2}{d_1 + d_2}$$

This same method allows us to determine the mass of each star in a binary system if we know the total mass and the distance of each star from the center of mass.

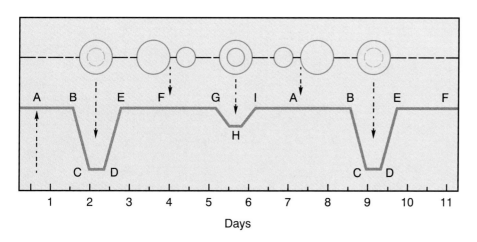

► **FIGURE 12-36.** Suppose Doppler effect data tell us that the relative velocity of these two stars as they pass one another is 8.0×10^2 km/s (which is 6.9×10^7 km/day). Assume that in the leftmost drawing of the stars, the small one is moving to the right. At point B on the light curve, it began to be hidden by the large star. At point D it starts to emerge. From B to D is about 0.8 days, so it took the small star this long to cross the large one. The diameter of the larger star must then be (0.8 days) $\times$ (6.9×10^7 km/day), or 5.8×10^7 km. In a similar manner you can calculate the diameter of the smaller star.

In the case of a spectroscopic binary, if we can be confident that the plane of the stars' revolution lies very close to a line from Earth, we can do a similar calculation, but instead of calculating the size of the orbit with the small-angle formula, we calculate it from a knowledge of the maximum speed of the star in orbit and the period of the orbit. We are usually not able to know the inclination of the orbit, however, so a mass calculation cannot be done. Valuable information about *average* masses of stars in a great number of spectroscopic systems can be obtained, however, by assuming an average inclination of the orbits. (Since any orbit inclination from 0 degrees to 90 degrees is equally likely, if enough systems are included in the analysis, the average inclination will be 45 degrees.)

Eclipsing binaries that are also spectroscopic binaries provide us with a way of measuring not only the masses of the two stars but also their sizes. Since the fact that they eclipse one another means that the inclination of their orbit with our line of view is zero (or very nearly so), their Doppler shift tells us their velocities. Knowing their velocities and the time it takes to complete an eclipse, we can calculate the size of each star as well as the luminosity of each. ►Figure 12-36 shows a simplified case that illustrates this calculation. Recall that we can also determine the size of a star if we know its luminosity, distance, and apparent

Stellar Masses and Sizes from Binary Star Data

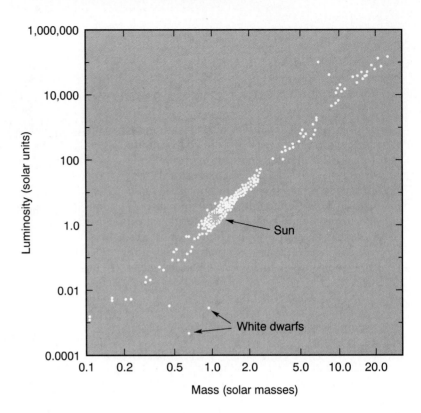

► **FIGURE 12-37.** When the luminosities of stars are plotted against their masses, an obvious relationship is seen for most stars—main sequence stars. The three at the bottom that do not fit here are white dwarfs. (Notice that the scale at the bottom is a logarithmic scale.)

magnitude. Eclipsing binary systems give us an independent method of measuring star sizes.

THE MASS-LUMINOSITY RELATIONSHIP

Suppose the masses of a number of stars are compared to their luminosities by plotting a graph. ►Figure 12-37 shows such a graph, plotting mass in solar masses along the *x*-axis and luminosity in solar units along the *y*-axis. There is an obvious correspondence for main sequence stars. More massive stars are more luminous.

This ***mass-luminosity diagram*** was produced from knowledge of binary stars that are close enough to Earth to yield the necessary data, but it is reasonable to assume that more distant stars also follow this pattern. (Otherwise, we would be claiming that we live in some special place in the universe where stars behave differently.) The mass-luminosity relationship is valuable to astronomers in investigating less accessible stars and in constructing and evaluating hypotheses concerning the life cycle of stars. This will be discussed in the next chapter.

mass-luminosity diagram. A plot of the mass versus the luminosity of a number of stars.

The mass-luminosity relationship holds only for main sequence stars. The mathematics of the relationship is in the Close Up on page 395.

CEPHEID VARIABLES AS DISTANCE INDICATORS

Eclipsing binary stars are sometimes called eclipsing *variables,* since their light intensity varies over time. Other types of variable stars are found in the heavens, and one particular type is of importance to us here because it provides a method of measuring distances.

In 1784 John Goodricke, the same astronomer who explained Algol's variations, discovered that the star δ Cephei varies in luminosity in a regular way but that its variations cannot be explained by an eclipse of a binary companion. This star changes its apparent magnitude between 4.4 and 3.7 every 5.4 days,

Goodricke made his discovery when he was 19 years old and just two years before his death.

δ is the Greek letter, delta.

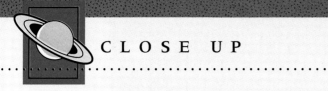

CLOSE UP

The Mathematics of the Mass-Luminosity Relationship

The graph of Figure 12-37 is empirical. That is, it is plotted from values that result directly from measurements rather than from a theory of how mass and luminosity are expected to be related. Since the points that represent main sequence stars lie along a straight line, one can determine the mathematical equation for that line. The equation is found to be

$$L = M^{3.5}$$

where L is the luminosity of a star on the main sequence and M is its mass. (Raising a number to a power of 0.5 means taking its square root. Raising it to a power of 3.5 means cubing it and then multiplying by its square root.)

Here this equation is used to calculate the luminosity of a star that has three times the mass of the Sun:

$$L = M^{3.5} = (3)^{3.5} = 3 \times 3 \times 3 \times \sqrt{3} = 46.8$$

Thus this star has a luminosity about 47 times that of the Sun.

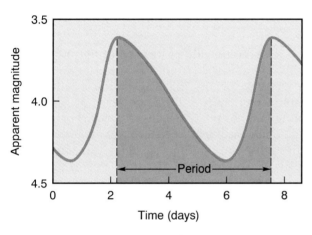

▶ **FIGURE 12-38.** This is the light curve of δ Cephei, the prototype of a class of stars that have light curves with this shape and are therefore called Cepheid variables.

▶ **FIGURE 12-39.** Henrietta Leavitt.

with a light curve as shown in ▶Figure 12-38. Soon thereafter, other stars exhibiting this characteristic light curve—brightening quickly and then dimming more slowly—were seen. Doppler effect data show that such stars are actually pulsating—changing size—in rhythm with their changes in luminosity, and it is fairly easy to identify this class of stars, now called *Cepheid variables,* or *Cepheids.* Each Cepheid has a very constant period of variation; the periods range from about one day to about three months for different Cepheids.

In 1908 a Harvard College astronomer, Henrietta S. Leavitt (▶Figure 12-39), published data on variable stars in the Small and Large Magellanic Clouds (▶Figure 12-40). The Magellanic Clouds appear to the naked eye as hazy cloudlike patches in the sky of the Southern Hemisphere and were first reported to Europeans by Magellan after his voyage around the world. Telescopes, however, reveal that the clouds are gigantic groups of stars now known to be separate galaxies. Leavitt discovered that in the case of Cepheid variables, the brighter variables have the longer periods.

It may seem strange that period and *apparent* magnitude are related, for apparent magnitude is not a quantity intrinsic to the star; rather, it depends on our

Cepheid (pronounced (SEF-e-id) variable. One of a particular class of pulsating stars.

FIGURE 12-40. The Large Magellanic Cloud is on the left of this photo, and the Small Magellanic Cloud is on the right. (The bright star Achernar is at the top right.)

FIGURE 12-41. The period-luminosity diagram for Cepheid variables.

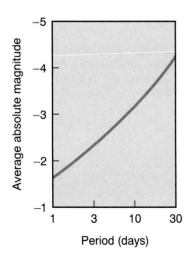

distance from it. The reason the relationship exists is that all of the stars of the Magellanic Clouds are about the same distance from us—at least compared to the distance to the clouds. To see this, suppose that you are on a hill outside your town at night, observing the street lights of the town (each of which, we will assume, has the same absolute luminosity). The lights will *appear* to be many different brightnesses, depending upon each light's distance from you. Now suppose that the hill is high enough so that you can use a small telescope to observe the lights of a town 200 miles away. You will find that the lights of this town all appear about equally bright, since each of them is 200 miles away, give or take only a few miles.

The stars of the Magellanic Clouds do not all have the same intrinsic luminosity, but since they are all at about the same distance, their apparent magnitudes are related directly to their absolute magnitudes. For example, if the absolute magnitude of a certain star in the Large Magellanic Cloud is five magnitudes less than its apparent magnitude, the absolute magnitude of each star in the cloud will be five magnitudes less than its apparent magnitude.

Astronomers quickly realized the importance of the relationship between the magnitude and the period of Cepheids: the period (which is easy to determine) allows the absolute magnitude to be determined, and this quantity can be used along with apparent magnitude to determine the distance to the variable star. The problem was that only the *apparent* magnitudes of the Cepheids in the Magellanic Clouds were known. The absolute magnitudes of these stars could not be determined because the clouds are too far away for parallax to be observed. The apparent magnitudes of Cepheids in the clouds are around 15; they appear dim. It was obvious from Cepheid variables nearer the Earth, however, that Cepheids, as a class, are very luminous stars. This provided qualitative evidence that the Magellanic Clouds are at a great distance from Earth.

All that was needed was to find one Cepheid variable near enough to the Earth that its distance could be measured by parallax. But there is none. Beginning in 1917, Harlow Shapley (to be discussed more fully when galaxies are examined) worked out a complex statistical method to determine distances to Cepheids within our galaxy. Using such methods, astronomers by the late 1930s were confident that they had determined correct distances to a number of Cepheids. This allowed them to plot a graph of period against *absolute* magnitude, as shown in ➤Figure 12-41.

HISTORICAL NOTE

Henrietta Leavitt

Henrietta Swan Leavitt, one of seven children, was born on July 4, 1868. She graduated from Radcliffe College (then known as the Society for the Collegiate Instruction of Women) in 1892. In her senior year she took a course in astronomy, and after teaching and traveling for a few years, she joined the Harvard College Observatory in 1895 as a student and volunteer research assistant. Her work caught the attention of those in charge, and in 1902 she was given a permanent position on the staff. Soon she was named chief of the photographic photometry department.

Leavitt's primary work was with variable stars, and she discovered large numbers of them. In 1908 she published a list of 1777 variable stars in the Magellanic Clouds; the list included a table of 16 Cepheid variables on which she had very precise data. Concerning these she inserted a comment: "It is worthy of notice that in Table VI the brighter variables have the longer periods."[1] This brief comment went unnoticed by other astronomers. Later that year, because of ill health, she was forced to return to her family home in Wisconsin.

During her years of illness in Wisconsin, Leavitt continued her work (the observatory sent stellar photographs to her). When she returned to Cambridge in 1912, she published a report on 25 Cepheid variables. In it she stated, "A remarkable relation between the brightness of these variables and the length of their periods will be noticed."[2] She included a graph showing the relationship between period and brightness, but although she recognized the importance of the discovery, her duties at the observatory (and the nature of the work being done there) prevented her from following up on it.

Henrietta Leavitt was one of a group of some 40 women hired by Edward Pickering, starting in the 1880s, when it was unusual for women to work outside the home. Although women had made contributions to astronomy prior to this time, many still considered science an inappropriate field for them. Pickering was not really a progressive thinker in hiring the women; he did so because they would work for less money than men. Nonetheless, he was rewarded by the significant advances made by many members of the group, particularly Annie Jump Cannon and Henrietta Leavitt.

Although her health was never good and her hearing was impaired from childhood, Leavitt worked at the Harvard Observatory until her death from cancer in 1921.

---

1. Henrietta Leavitt, "1777 Variables in the Magellanic Clouds," *Annals of the Harvard College Observatory* 60 (1908): 107.
2. Henrietta Leavitt, *Periods of 25 Variables in the Small Magellanic Cloud*, Harvard College Observatory Circular no. 173 (March 3, 1912).

To show how the period-luminosity chart can be used to determine distances, suppose that a Cepheid variable has a period of 10 days and an apparent magnitude of 15. Figure 12-41 shows that a Cepheid with a period of 10 days has an absolute magnitude of about −3. Now that both apparent and absolute magnitude are known, distance can be calculated. (By our nonmathematical method, you can see that the star must be extremely far away, for its apparent magnitude is greater than its absolute magnitude by 18.) Use of Cepheid variables tells us that the Large and Small Magellanic Clouds are, in fact, 48,000 and 56,000 parsecs away, respectively.

One of the reasons that it took from 1912 to the late 1930s for astronomers to determine the correct relationship between Cepheids' periods and absolute magnitudes accurately is that there are actually two different types of Cepheid variables, differing from each other in luminosity by about two magnitudes (for the same period). We can now detect the difference between the two, however, so the problem no longer exists.

Although astronomers have detected fewer than a thousand Cepheid variables, their importance greatly outweighs their number. Because they are among the very brightest of stars, we can see them not only in distant parts of our own galaxy but in other galaxies, giving us a method of measuring distances to these faraway worlds.

Corresponding to normal metric usage, 1000 parsecs is one kiloparsec, abbreviated kpc, so the Large Magellanic Cloud is 48 kpc away.

CONCLUSION

As this chapter has shown, once we know the distance to stars, many other avenues open up to us. Knowing distance and proper motion, we calculate a star's tangential velocity. The Doppler effect allows us to measure radial velocity, and from these two velocities, we can determine a star's actual motion relative to the Earth and to other stars. From distance and apparent magnitude, we can calculate a star's absolute magnitude. This knowledge can be combined with knowledge of the star's temperature to calculate its size.

On the other hand, the patterns in the H-R diagram allow us to determine a star's absolute magnitude without first knowing its distance, so that we can use the distance versus absolute magnitude versus apparent magnitude relationship in reverse to determine stellar distances.

The discovery and analysis of binary stars have provided a second method of learning a star's size and also a method of measuring a very fundamental property of a star, its mass.

Finally, a particularly useful type of star, the Cepheid variable, provides a tool to measure even greater distances, opening up the entire field of galactic astronomy.

RECALL QUESTIONS

1. Suppose you observe a previously uninvestigated star and find its apparent magnitude. To determine its absolute magnitude, you need to know the star's
   A. distance.
   B. color.
   C. velocity.
   D. brightness as seen from Earth.
   E. Doppler shift.

2. The absolute magnitude of a star is
   A. the same as the apparent magnitude.
   B. equal to the greatest the apparent magnitude can be, in the case of a variable star.
   C. equal to the apparent magnitude if the star is 10 parsecs away.
   D. the size of a star from one side to the other.
   E. equal to the brightness of the star on the clearest night.

3. A magnitude change of 1 corresponds to a brightness change of about 2.5 times. A magnitude change of 2 corresponds to a brightness change of about
   A. 1.3 times.   D. 6.3 times.
   B. 2.5 times.   E. 15.9 times.
   C. 5 times.

For the next three items, refer to the following table:

Star	Apparent Magnitude	Absolute Magnitude	Distance (parsecs)
Sirius	−1.5	—	2.7
Vega	0.0	0.5	—
Antares	0.9	−5.1	160
Fomalhaut	1.15	—	6.9

4. Which of the following is about the correct distance for Vega?
   A. 4.2 parsecs   C. 11.7 parsecs
   B. 8.1 parsecs   D. 26.7 parsecs
   E. [The answer cannot be determined from the information given.]

5. Which star appears dimmest to an observer on Earth?
   A. Sirius
   B. Vega
   C. Antares
   D. Fomalhaut
   E. [The answer cannot be determined from the information given.]

6. Fomalhaut's absolute magnitude is about
   A. 0.75.   D. −2.
   B. 2.   E. −0.75.
   C. 12.

7. Star X is twice as far away as Star Y. The parallax angle of Star X is
   A. half that of Star Y.
   B. the same as that of Star Y.
   C. twice that of Star Y.
   D. four times that of Star Y.
   E. [The answer cannot be determined from the information given.]

8. The distances to nearby stars can be measured by
   A. comparing the apparent magnitudes of several stars.
   B. bouncing radar pulses from their surfaces.
   C. measuring the time it takes light to get here from them.
   D. measuring their shifting motion against background stars through the year.
   E. [Both B and C above.]

9. Star S and Star K have the same tangential velocity, but Star S is closer to Earth. Which has the larger proper motion?
   A. Star S
   B. Star K

C. They both have the same proper motion.
D. [The answer cannot be determined from the information given.]

10. Star S and Star K have the same tangential velocity, but Star S is closer to Earth. Which has the larger radial velocity?
    A. Star S
    B. Star K
    C. They both have the same radial velocity.
    D. [The answer cannot be determined from the information given.]

11. If a star is 100 light-years away, what is its approximate distance in parsecs?
    A. 3000 parsecs   D. 9 parsecs
    B. 900 parsecs   E. ⅓ parsec
    C. 30 parsecs

12. To use spectroscopic parallax, we must know
    A. the star's diameter.
    B. the star's temperature.
    C. the distance from the Earth to the Sun.
    D. the star's luminosity class.
    E. [Both B and D above.]

13. If the temperature of a star increased without a change in the star's size, its point on the H-R diagram would move
    A. up and to the left.   C. down and to the left.
    B. up and to the right.   D. down and to the right.

14. Stars on the main sequence that have a small mass are
    A. bright and hot.
    B. dim and hot.
    C. dim and cool.
    D. bright and cool.
    E. [Any of the above; there is no regular relationship.]

15. To observe spectroscopic binaries, we rely on
    A. knowing the composition of the individual stars.
    B. our knowledge of the distance separating the stars.
    C. our knowledge of the distances from Earth to the stars.
    D. the change in light intensity as the stars orbit.
    E. the Doppler effect.

16. The stars of binary star systems
    A. revolve around a point midway between their centers.
    B. revolve around a point somewhere between their centers (but not necessarily midway).
    C. revolve such that one star remains still, and the other revolves around it.
    D. do not revolve around each other.

17. Binary star systems are especially important to us because they allow us to calculate the _____ of stars.
    A. compositions   D. temperatures
    B. proper motions   E. masses
    C. radial velocities

18. Which type of binary system provides the most information about its component stars' masses and sizes?
    A. Eclipsing binaries
    B. Spectroscopic binaries
    C. Visual binaries
    D. Composite spectrum binaries
    E. [All of the above about equally.]

19. We can determine the size of stars by
    A. measuring the size of their image in a telescope.
    B. a measurement using a spectroscope.
    C. calculations based on temperature and absolute magnitude.
    D. Doppler effect measurements.
    E. mass measurements.

20. Cepheid variables can be used to find distances because their
    A. luminosity is related to their period.
    B. radial velocity is related to their mass.
    C. distance is related to their mass.
    D. magnitude is related to their color.
    E. period is related to their radial velocity.

21. If the orbital plane of a certain binary star system is not parallel to a line from Earth, the maximum radial velocity of one of its stars measured by the Doppler effect will be
    A. less than its true radial velocity.
    B. greater than its true radial velocity.
    C. either greater or less than its true radial velocity, depending upon other factors.

22. If a star has a parallax angle of 0.20 arcseconds, how far away is it?
    A. 0.20 parsec
    B. 1.0 parsec
    C. 2.0 parsecs
    D. 5.0 parsecs
    E. [The answer cannot be determined from the information given.]

23. When a star's spectrum is redshifted as a result of the Doppler effect, we know that the star is
    A. much cooler than average.
    B. slightly cooler than average.
    C. about average temperature.
    D. hotter than average.
    E. moving away from Earth.

24. Spectroscopic parallax allows us to measure
    A. the distances to stars using the Earth's orbital motion.
    B. the distances to stars using the Earth's rotational motion.
    C. the distances to stars using the H-R diagram.
    D. the temperatures of stars using their spectra.
    E. the radial speed of stars using their spectra.

25. Considering only stars on the main sequence, the most massive stars are the
    A. hottest and brightest.
    B. hottest and dimmest.
    C. coolest and brightest.
    D. coolest and dimmest.
    E. [No general statement can be made.]

26. In order to calculate (or judge) the velocity of an object moving in the sky, what quantities must we know?

27. Define and distinguish between radial velocity and tangential velocity.

Recall Questions

QUESTIONS TO PONDER

1. Why do only nearby stars show measurable proper motion? How would the tangential velocity of nearby and distant stars compare?
2. In Figure 12-26 (Kruger 60), the binary pair is getting farther from the single star. From this series of photographs it would be impossible to tell which has the proper motion. Still, the binary pair would be taken as the most likely candidate. Why?
3. It is possible for a binary star system to fall into two or more classifications (such as visual binary and eclipsing binary). Describe some situations in which this might be the case.
4. Explain why there is such a simple relationship between the apparent and absolute magnitudes of stars in the Magellanic Clouds.
5. Suppose that two given stars differ by three in apparent magnitude. Is their difference in absolute magnitude less than three, equal to three, greater than three, or can the difference even be determined from the information given? Explain your answer.
6. Do you think that it is reasonable to use the H-R diagram, which is plotted from data on nearby stars, for stars much more distant from us? Why or why not?
7. Explain how an extraterrestrial could logically make conclusions about a typical person's lifetime from a chart such as Figure 12-15.
8. Would a mass-luminosity diagram (Figure 12-37) show the same relationship if the scales were independent of the Sun instead of being multiples of the Sun's values?
9. Why was it important to find the distance to a Cepheid variable after Leavitt had plotted the data for Cepheids in the Magellanic Clouds?
10. Why is spectroscopic parallax called "parallax" even though no angle measurement is involved?

CALCULATIONS

1. What is the ratio of light received from stars that differ in magnitude by 15?
2. About how many times as much light reaches us from Antares, the brightest star in Scorpius (apparent magnitude −1.0), than from τ (the Greek letter, tau) Ceti (apparent magnitude 3.5)?
3. The star Ross 128 has a parallax angle of 0.30 seconds. How far away is it in parsecs? In light-years?
4. α (the Greek letter alpha) Centauri is the nearest naked-eye star to the Sun. Its apparent magnitude is −0.01 and its distance is 1.35 parsecs. Will its absolute magnitude be greater or less than 0?
5. Use data from the last question to calculate the parallax angle of α Centauri.
6. Altair, one of the summer triangle stars, is on the main sequence, is spectral type A7 (8000°C), and has an apparent magnitude of 0.77. Is Altair closer than 10 parsecs, about 10 pc away, somewhat farther than 10 pc, or much farther than 10 pc?
7. Rigel, a bright supergiant, is a B8Ia star with a temperature of about 15,000°C. Its apparent magnitude is +0.14. What can you conclude about its distance from us?
8. A certain star has a mass four times greater than the Sun. How does its luminosity compare to the Sun's?

StarLinks netQuestions

Visit the netQuestions area of StarLinks (www.jbpub.com/starlinks) to complete exercises on these topics:

1. Distances to the Stars
If you observe something from one position and then change your position, nearby objects will appear to move with respect to distant objects. This observation, or method of judging distance, is called parallax.

2. Multiple Star Systems
More than half of what appear as single stars in the night sky are really multiple star systems.

CHAPTER 13

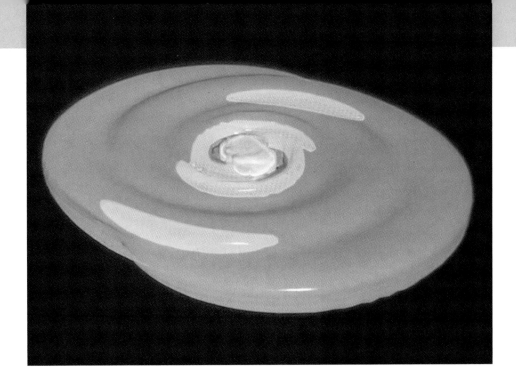

INTERSTELLAR MATTER AND STAR FORMATION

The Interstellar Medium
Interstellar Dust
CLOSE UP: Holes in the Heavens?
CLOSE UP: Blue Skies and Red Sunsets
Interstellar Gas
Clouds and Nebulae
A Brief Woodland Visit
Star Birth

The Collapse of Interstellar Clouds
CLOSE UP: A Celestial Godzilla?
Protostars
Evolution toward the Main Sequence
Star Clusters
CLOSE UP: ET Life VI—The Life Equation
ACTIVITY: Deep Sky Objects with a Small Telescope

MANY OF THE DETAILS of our theories of the life cycles of stars come from computer modeling. The chapter-opening image is a frame from a videotape made by Richard H. Durisen (Indiana University) and his colleagues using a Cray supercomputer. It represents one stage in the formation of a star. As material collapses to form the star, it flattens into a pancake shape. Then the rapid rotation of the star causes it to shed its outer layers. Some of the material shown in blue will eventually collapse into the blue-green area, forming a planetary system. The material shown in red will collapse further to become the star. Astronomers believe our solar system formed in a manner very similar to this.

Previous chapters, particularly the last one, discussed the tools and techniques by which astronomers measure the properties of stars and discern relationships among these properties. The next few chapters will show how these tools and techniques have enabled us to learn about the universe beyond the solar system. Astronomers study the dust and gas that exist in interstellar space and are learning *how this material forms into new stars, how these stars live out their lives, and how they eventually "die."*

I hope that you will be impressed with the tremendous amount of information astronomers have amassed about the life cycles of stars. You may not be surprised to learn that many of our theories concerning stars' lives are somewhat tentative and that many questions remain.

THE INTERSTELLAR MEDIUM

An everyday definition of *medium* is "an intervening substance through which something is transmitted." The interstellar *medium* transmits light.

There are two seemingly contradictory truths about the space between stars. First, a large amount of gas and dust exists in that space. Second, the space between stars is very nearly a perfect vacuum. The secret to reconciling these statements is to realize the vastness of interstellar space. Although the dust and gas of space are so thinly distributed that their molecular density (the number of molecules per cubic centimeter) is less than in the best vacuum in a laboratory on Earth, space is so vast that the total amount of material between stars is enough to affect the light that passes through it and—as will be explained later—is enough to provide the material for the formation of new stars.

Interstellar Dust

A nearby Close Up explains why this hypothesis for the dark areas was abandoned.

Astronomers have been aware of dark areas in the sky (➤Figure 13-1) for ages. Although it was once proposed that these dark patches are simply spaces between the stars that allow us to see into the darkness beyond, we now know that the dark areas are caused by giant clouds of interstellar dust that block light from stars behind them. In the 1930s, astronomers became aware that grains of dust

➤ **FIGURE 13-1.** Interstellar clouds of dust hide the stars behind them. (a) The dark C-shaped nebula is an interstellar dust cloud in the constellation Aquila. (b) The Horsehead nebula results from clouds of dust blocking light from a bright nebula behind. Find the dark profile of a horse's head at the right center of the photo.

(a)   (b)

CLOSE UP

Holes in the Heavens?

Sir William Herschel, the discoverer of Uranus, proposed that the dark patches we see in the sky are simply large spaces between the stars that allow us to see into the dark void beyond. How do we know that this is not the case? Today, of course, infrared imaging allows us to detect the dust that makes up the "holes in space." But before we ever made infrared images of the dust, an argument based on geometry gave astronomers reason to doubt Herschel's idea. Recall that stars are not all at the same distance from us; they are not on a "celestial sphere." Instead, they are spread out so that some are relatively nearby and others are very far away. Thus, for us to be able to see through gaps between the stars, the gaps would have to be similar to tunnels, and the tunnels would have to be perfectly aligned with Earth. Suppose you are deep in a forest. Even with open clearings among the trees, you might not be able to see beyond the forest. For you to see outside, there would have to be straight, open trails. In the case of stars, it is extremely unlikely that so many open trails ("tunnels") would be perfectly aligned with Earth.

---

exist throughout interstellar space, and not just in clouds. The *Infrared Astronomical Satellite* (*IRAS,* launched in 1983) and the *Hubble Space Telescope* have enabled astronomers to photograph the diffuse interstellar dust. ➤Figure 13-2 is an image of a region of Orion (taken by the HST) as it appears in infrared. Note the wispy appearance of the material in the infrared image. Its similarity to cirrus clouds on Earth leads to its name, **interstellar cirrus.** Cirrus emits infrared radiation because it is warmed slightly by light that it absorbs. Wispy interstellar cirrus clouds span huge volumes of space, from parsecs to tens of parsecs across. To

interstellar cirrus. Faint, diffuse dust clouds found throughout interstellar space.

➤ **FIGURE 13-2.** The Hubble Space Telescope obtained this infrared image of the edge of the Orion Nebula. The image reveals the diffuse, patchy clouds of the nebula.

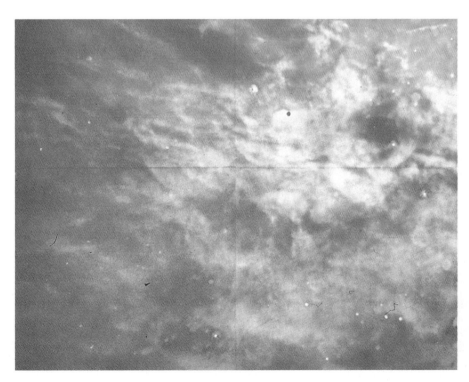

The Interstellar Medium 403

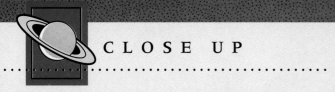

CLOSE UP

Blue Skies and Red Sunsets

The same phenomenon that explains the reddening of starlight by interstellar dust also explains why the sky is blue and sunsets are red (or at least orange). As the discussion of interstellar reddening pointed out, dust scatters blue light more than it scatters red. A similar effect happens in the Earth's atmosphere, where the scattering of sunlight by the atmosphere is what gives the daytime sky its color. The sky appears blue simply because particles in the air (and molecules of the air) scatter higher-frequency light—blue—more than lower-frequency light. It follows from this that light that reaches us directly from the Sun without scattering is somewhat deficient in these higher frequencies.

Consequently, the Sun seen from Earth's surface appears redder than it appears when viewed from a vantage point in space. From Earth, the Sun appears somewhat yellow, but from space it is closer to white.

When we look at the Sun as it is setting, the sunlight that reaches our eyes has traveled even farther through the atmosphere than sunlight at midday. Thus even more of the high-frequency light is scattered away, and the Sun appears red. This effect is especially noticeable if the atmosphere contains significant amounts of dust. The dust that remained in the air from major volcanic eruptions in 1991 caused beautiful red sunsets for the next few years.

---

appreciate this size, recall that a parsec is somewhat greater than 3 light-years and that the solar system is only about 0.001 light-years across.

Interstellar dust grains make up only a tiny fraction of the mass of the interstellar medium, but dust has very definite effects on light. We have seen one of the effects, the reduction in the amount of light from distant stars—and sometimes the complete extinction of that light. Another effect results from the fact that while dust grains may absorb some light, they scatter much of it, and this scattering is more efficient for light of shorter wavelengths. As a result, blue light is scattered more than red light (►Figure 13-3). Thus the light from distant stars is *reddened* by the dust through which it passes.

The reddening caused by scattering should not be confused with the redshift of the Doppler effect. In the case of reddening by scattering, the positions of the spectral lines are not changed, but with the Doppler redshift, the spectral lines actually shift. ►Figure 13-4 illustrates the difference.

How large are the interstellar dust grains? We know that they must be smaller than the wavelength of light, because they would not scatter blue light more than red if they were not. This means that the dust grains are the size of

> The effect by which starlight is blocked completely by interstellar material is called *interstellar extinction*.

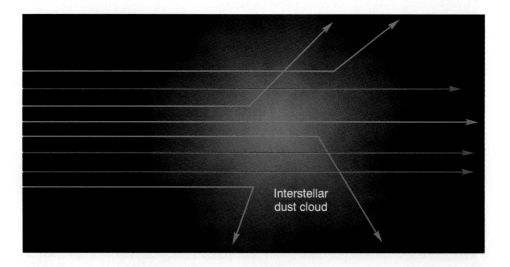

►FIGURE 13-3. Grains of dust scatter blue light more efficiently than red light, so a star seen through a cloud of dust appears redder than it would if the dust were not present.

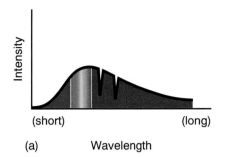 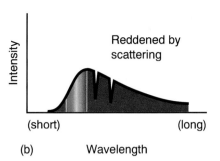 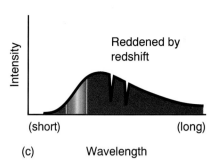

▶ **FIGURE 13-4.** (a) An intensity-wavelength curve showing two (exaggerated) absorption lines. (b) This is the same source with its light reddened by scattering. Notice that the absorption lines remain in the same place. (c) Again, the same source but with its light reddened by the Doppler shift. The entire curve is shifted to the right in this case.

particles of cigarette smoke and smaller. (You may have noticed that a cloud of cigarette smoke has a blue tint.)

What is the dust made of? Spectral analysis indicates that it contains carbon in the form of graphite, similar to the carbon that is found in comets, but beyond this we know little about its chemical makeup. Chapter 15 will discuss the origin of the interstellar material and the related hypotheses about the composition of the dust.

Interstellar Gas

Dust in space accounts for only about 10% of the total mass of the interstellar material. The remainder is simply gas, which in most cases is extremely diffuse (so that for most purposes the volume it occupies may be considered a vacuum), but which in some cases clusters together into giant clouds. The gas between the stars reveals its presence in a number of ways, including the following:

- Clouds of gas can be seen in photographs of the sky. The red glow behind the Horsehead nebula in Figure 13-1b is a giant cloud of gas that is glowing due to ultraviolet light from hot stars within it. The glow is a *fluorescence* process; that is, the atoms of the gas absorb ultraviolet light from nearby hot stars, and this causes the atoms to become energized. The atoms lose their extra energy by emitting light—*fluorescing*. ▶Figure 13-5 shows other examples of such bright **emission nebulae.** The term *nebula* (plural *nebulae*) has its origin in the adjective *nebulous,* which means "lacking definite form or limits." This is an apt description of an interstellar cloud, for such clouds have no definite boundaries. A cloud is called a nebula if it is dense or bright enough to show up in a photograph.

- Interstellar gas causes absorption lines in the stellar spectra. Astronomers can distinguish these absorption lines from the absorption lines of a stellar atmosphere in three ways. First, absorption lines due to interstellar gas tend to be narrower then those produced by a star's atmosphere (▶Figure 13-6). This difference occurs because the atoms of a star's atmosphere are generally moving fast in random directions, causing a range of Doppler shifts. Second, since the star and the interstellar gas probably have different velocities, the lines caused by the stellar atmosphere will have a different Doppler shift than those caused by the interstellar gas. (The velocities referred to here are the overall velocities of the star and of the interstellar gas cloud, not the random molecular speeds within each gas.) Finally, the interstellar gas will generally be much cooler than the gas of the stellar atmosphere. Thus the atoms will be in their ground state, and most of the absorption will take place in the ultraviolet region of the spectrum. (For a

fluorescence. The process of absorbing radiation of one frequency and reemitting it at a lower frequency.

emission nebula. Interstellar gas that fluoresces due to ultraviolet light from a star near or within the nebula.

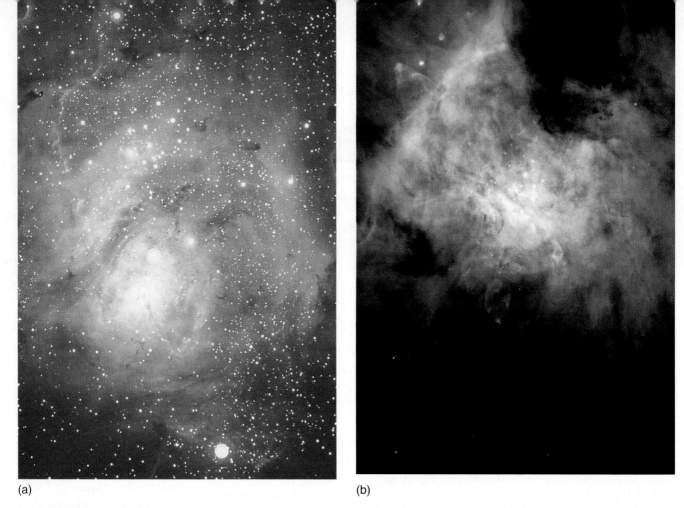

(a) (b)

➤ **FIGURE 13-5.** (a) The Lagoon nebula in Sagittarius is an emission nebula. (b) This is a mosaic of 45 images of the Orion nebula taken by the HST. Light emitted by oxygen is shown as blue, hydrogen emission is shown as green, and nitrogen as red. The overall color balance is close to what an observer near the nebula would see.

discussion of this phenomenon, see "Determining the Spectral Class of a Star," a Close Up in Chapter 12.)

■ In 1951, two American astronomers, E. M. Purcell and H. I. Ewen, used a specially built radio telescope to detect radiation emitted by interstellar hydrogen. Scientists had predicted that hydrogen atoms should emit radiation with a wavelength of 21 centimeters, but Purcell and Ewen's telescope was the first to reveal it. Hydrogen is the most common element in stars, and astronomers had expected it would also be the most common element in

As Chapter 16 will explain, the 21-centimeter radiation is very valuable to us in our efforts to map the Milky Way Galaxy.

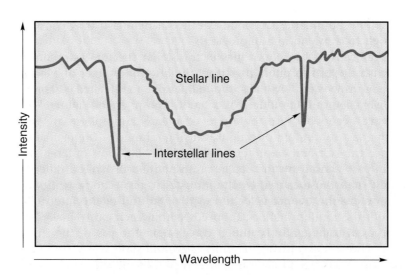

➤ **FIGURE 13-6.** This graph is a highly magnified portion of the intensity-wavelength graph of light from a star. It shows three absorption lines. The outer two are much finer than the central one, indicating that they were caused by interstellar gas.

➤ **FIGURE 13-7.** The Pleiades is an open cluster 410 light-years from Earth. Photographed with a telescope, the reflection nebula surrounding the Pleiades is visible. The dimmer stars are part of the cluster and are the same distance from us as the bright O- and B-type stars. The spikes from each star are caused by diffraction around secondary mirror supports in the telescope. (Subaru is the Japanese name for the Pleiades. You may recognize this pattern of stars from the logo on the back of a car.)

interstellar space. The 21-centimeter radiation confirmed this expectation and allows us to spot interstellar hydrogen at great distances, for the interstellar medium absorbs very little of the 21-centimeter radiation. In addition, we are able to detect radio emission lines from interstellar gases—including water (H_2O), carbon monoxide (CO), ammonia (NH_3), and formaldehyde (H_2CO)—at other wavelengths. These gases are much less abundant than hydrogen, however.

Clouds and Nebulae

As indicated earlier, interstellar material is not uniformly distributed; it tends to clump together into what we call *interstellar clouds*. The interstellar cirrus clouds are made up mostly of gas with its molecules widely dispersed. At sea level on Earth, there are about 2×10^{19} molecules of air per cubic centimeter. The interstellar cirrus contains fewer than 1000 molecules per cubic centimeter. It is difficult to determine the shape and size of these diffuse clouds, but most of them seem to be wispy sheets rather than spherical clumps. The sheets are probably less than one parsec thick, but many parsecs wide and long. The total mass of a cirrus cloud might equal the mass of a small-to-average star.

If a cloud is near a hot star, the ultraviolet radiation from the star will cause the cloud to fluoresce (as in Figure 13-5), and the cloud is then known as an emission nebula. A diffuse cloud of dust that happens to be behind or beside a bright star may be visible to us because it reflects the star's light back toward us. ➤Figure 13-7 is a photograph of such a ***reflection nebula.*** When we see a re-

reflection nebula. Interstellar dust that is visible due to reflected light from a nearby star.

The Interstellar Medium

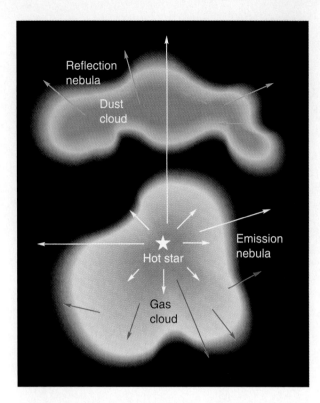

▶ FIGURE 13-8. A hot star is shown inside a cloud of gas. The star causes the cloud to glow, and the cloud is therefore known as an emission nebula. When the star's light passes through the dust cloud, some of the blue light is scattered. If the Earth happens to be located to the left of the page, this dust could appear as a blue reflection nebula. If the Earth is located at the top of the page, and if the cloud is very thick, the star's light may not be able to penetrate it, and the cloud will appear as a dark nebula. In the case shown, however, some light gets through the cloud. It is reddened in the process, making the star appear redder than it really is.

▶ FIGURE 13-9. The Trifid nebula in Sagittarius is a red (or pink) emission nebula that appears to be divided into three parts by dark nebula that form lanes blocking the light from behind. Above the Trifid nebula is a blue reflection nebula, unconnected to the Trifid.

flection nebula, we are seeing light that has been scattered by dust in the cloud, so the nebula appears blue. ▶Figure 13-8 is a diagram of a star surrounded by an emission nebula with a reflection nebula nearby. ▶Figure 13-9 is a photograph of a real example that is similar to the diagram. In the photograph the pink emission nebula (the Trifid nebula, in Sagittarius) surrounds hot central stars that provide the energy to make it fluoresce. Above it is an unconnected reflection nebula.

The Trifid nebula also contains absorbing dust lanes that extinguish the light from the emission nebula. Dust clouds such as this may have densities ranging from as low as 1000 particles per cubic centimeter to perhaps 1,000,000 particles per cubic centimeter. (By earthly standards, this is not very dense, however, for air at sea level is still 20 million-million times more dense than the densest interstellar cloud.) A dust cloud is able to extinguish light from a star behind it only if the cloud is extremely large, perhaps hundreds of parsecs across. The photographs in Figures 13-1 and 13-9 contain other examples of **dark nebulae.** Although it is the dust of a dark nebula that is responsible for blocking starlight, most of the particles in such a nebula are atoms, about 75% hydrogen and 23% helium with the remaining 2% being heavier elements.

If we take an inventory of the dust, gas, and stars within a radius of about

dark nebula. A cloud of interstellar dust that blocks light from stars on the other side of it.

a thousand parsecs from Earth, we find that interstellar matter contributes about 15–30% of the total mass. Even though astronomers are able to take an inventory only of our "neighborhood," they are unable to make this figure more precise primarily because determining the mass of the interstellar material is so difficult. Does interstellar matter make up this same percentage of total matter elsewhere in the Galaxy? We would like to know the answer to this question, but we are unable to make the necessary measurements.

As will be discussed later, the interstellar material is constantly in a state of flux, for it is being used up as new stars are formed and being replenished as old stars die.

A BRIEF WOODLAND VISIT

Imagine that you are a visitor from another star system who has landed in a forest on Earth for a two-day visit. What could you learn about the life cycles of the woodlands during your short visit? You would see small and large trees, but would you be able to determine that trees progress from small to large? In two days you would have no chance to see any growth. You would see decayed material on the forest floor, but it might not be obvious that this is what remains of once-standing trees. Perhaps you might be lucky enough to see a change take place. You might see a limb fall from a tree. Would you conclude that trees get smaller by dropping their branches? Or perhaps you might see an entire tree fall. What would that tell you?

Like an extraterrestrial forest visitor, we humans are brief tourists in the universe. During our short lifetimes, we see very little change in the heavens. As we measure time, the stars change slowly. We do have an advantage over the forest visitor in that previous generations have handed down their observations to us so that our learning cycle is longer than one lifetime. The earliest reports of astronomical observations are only a few thousands years old, however; this is almost nothing compared to the total life of the heavens.

It is possible, though, to learn about the life cycle of stars. We are fortunate that a few stellar events happen quickly enough for us to observe them over human lifetimes or (in some cases) over much shorter times. Although astronomers cannot experiment directly with their subject matter, they can observe tremendous numbers of stars in various stages of development. This, along with earthbound experiments on the material of which stars are made, allows scientists to develop theories of stellar life cycles in which we can have reasonable confidence.

STAR BIRTH

Like other scientific theories, today's theories of star birth and star death developed slowly. Successful theories do not spring fully developed into a scientist's mind. Theories in astronomy result from long struggles with data from the heavens, often accompanied by experimentation here on Earth.

The story of today's understanding of the life cycles of stars starts early in this century with astrophysicist Henry Norris Russell playing the major role. The H-R diagram is the key to understanding the lives of stars, but it is not easy to read. Russell's early theories held that stars begin their lives as red giants, become O- and B-type main sequence stars, and then move down the main sequence, gradually dimming as they live out their lives. When new evidence arrived, particularly from H-R diagrams of clusters and from new knowledge of nuclear energy processes, Russell realized that stars live most of their lives on the main sequence with very little change in their positions and that the red giant stage is not the beginning of a star's life but is near the end. We will start at the beginning.

➤ **FIGURE 13-10.** This HST image shows material from a very small part of the Eagle nebula being swept away from dense areas where new stars are forming.

➤ **FIGURE 13-11.** This is the Eagle nebula, about 6000 ly distant and 20 ly across. (See if you can see the nebula as a landing eagle.)

The Collapse of Interstellar Clouds

Stars are born in the coldest place in the Galaxy, the giant molecular clouds (GMCs) of interstellar space. These clouds have a temperature of about 10 Kelvins (−440°F) and are composed mainly of molecular hydrogen. The average density of a GMC is about 200 molecules per cubic centimeter, but the cloud may be as large as 100 parsecs across. In all, a GMC may contain as much as a million solar masses of material.

➤Figure 13-10 is a picture of a very small portion of M-16, the Eagle nebula. The dark pillar-like structures are part of a GMC, so they are made up of gas and dust. Intense ultraviolet radiation from hot, massive newborn stars that lie at upper right beyond the photo cause the columns. ➤Figure 13-12 illustrates the process. Prior to the sequence shown in the figure, a molecular cloud was fairly evenly spread through this region of space. In part (a) of the figure, ultraviolet radiation from upper right is blowing the cloud back and evaporating gas outward from the cloud's surface. At the same time, the radiation illuminates the surface of the cloud, causing it to glow.

Certain parts of the cloud are more dense than average. In part (b) a dense globule of gas is about to be uncovered. Globules like this are called "EGGS," for "Evaporating Gaseous Globules." Radiation pushes the surrounding gas away from the EGG in much the same way that a strong wind on a beach blows sand along, uncovering shells and leaving trails of sand behind the shells. Part (c) of Figure 13-12 shows most of the nebula blown away, but a snake of gas and dust remains that has been protected from radiation by the EGG. Later, the nebula will blow still farther away, leaving behind a teardrop shape—part (d). Even later, all of the nebula will disappear, leaving behind only the EGG.

What is the EGG? At least in some cases, it is the beginning of a star. Within the globule, gravitational forces between individual molecules and dust particles are sufficient to draw the particles together. The globule was in the process of this contraction when the ultraviolet light began blowing away the surrounding nebula. The contraction continues within the globule, and when the center of the EGG gets hot enough to give off its own radiation, it will itself blow away any remaining parts of the nebula that have not fallen inward and a star will remain.

Astronomers estimate that there are 5000 giant molecular clouds in the Galaxy.

The name EGGs also fits the globules because they are embryonic stars.

410   CHAPTER 13   Interstellar Matter and Star Formation

Stars born in this manner probably do not have planets around them, for the radiation that causes the pillars probably disperses the material that would have ended up as planets.

Some stars form in more isolated conditions, without nearby hot stars blowing their raw material away. In these cases, parts of the nebula that remain near the embryonic star will continue to fall into it, perhaps resulting in a very large star, or perhaps revolving around the star to become a planetary system.

One question remains before we consider how the EGG becomes a star. What triggers the collapse of parts of a giant molecular cloud to form globules? The answer may involve collisions between GMCs. Along the boundary of such a collision, molecules would be forced closer together, perhaps close enough that gravitation would take over and continue the compression. Another source of the waves may be interstellar shock waves. As such a wave strikes the GMC, it forces parts of the cloud together enough that gravity becomes dominant.

There are at least four possible sources of interstellar shock waves. We have seen one—radiation from hot, newly forming stars. Second, as the next section will explain, very massive stars undergo a period during which they release enormous quantities of material at great speeds. These bursts are a source of shock in the surrounding GMC to trigger the birth of new stars.

The material flowing from stars is commonly called the *stellar wind*. The *solar wind* that we detect (and that results in auroras on Earth) is far weaker than the bursts of stellar wind described here.

A third source of interstellar shock waves is a supernova—the explosion of a star as it nears the end of its life. Supernovae will be discussed later. Finally, we know that tremendous shock waves move around the entire galaxy, forming its galactic arms (Chapter 16). These waves may be the most common trigger of star formation.

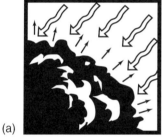

(a) The surface of a molecular cloud is illuminated by intense ultraviolet radiation from nearby hot stars. The radiation evaporates material off of the surface of the cloud.

(c) The EGG has now been largely uncovered. The shadow of the EGG protects a column of gas behind it, giving it a finger-like appearance.

➢ FIGURE 13-12. Pillars like those in Figure 13-10 are the result of strong ultraviolet radiation striking a giant molecular cloud.

(b) As the cloud is slowly eaten away by the ultraviolet radiation, a denser than average globule of gas begins to be uncovered. Because this globule of gas—dubbed an "EGG"—is denser than its surroundings it is not evaporated as quickly and so is left behind. Forming within at least some of the EGGs are young stellar objects.

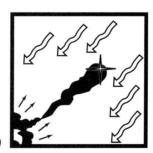

(d) Eventually the EGG may become totally separated from the molecular cloud in which it formed. As the EGG itself slowly evaporates the star within is uncovered, and may appear sitting on the front surface of the EGG.

Star Birth

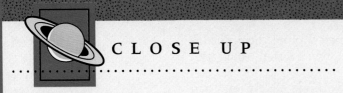

CLOSE UP

A Celestial Godzilla?

Figure C13-1 is a photograph taken by David Malin with the 3.9-meter Anglo-Australian Telescope in Coonabarabran, Australia. The unusual photograph appears to show a giant celestial beast about to devour a white french fry. In fact, the "monster" is an interstellar cloud about 30 parsecs (100 light-years) away, and the "french fry" is a distant spiral galaxy, far beyond the Milky Way Galaxy.

The molecular cloud, named *CG4*, is one of a number of objects that are called *cometary globules* because of what appears to be a cometlike tail. The explanation for CG4's appearance has nothing to do with comets (or monsters), however. The red glow in the mouth of the monster is due to an emission nebula. This nebula is contained within the green portion of the cloud, and radiation from stars within it causes its glow. The nebula receives so much radiation from these clouds that it is rapidly expanding, blowing the outer portions of the cloud with it. The blue region, the tail of the monster, is a reflection nebula, dust that appears blue to us because of scattered light. Millions of years from now, the entire cloud will be dispersed, revealing the stars that are the cause of its destruction.

CG4, like other cometary globules, is smaller than most interstellar clouds, which are typically more than 10 parsecs across. This beautiful monster is perhaps one-fourth of a parsec—less than a light-year—across. It lies in the constellation Puppis, which is south of Canis Major and appears low on the southern horizon for a viewer in the southern United States.

➤ **FIGURE C13-1.** A molecular cloud can have an unusual shape and contain many colors, as is apparent in this true-color photograph of the cometary globule CG4.

Protostars

Whatever the cause of the beginning of the collapse of material within the core of a giant molecular cloud, once the collapse begins, the increased gravitational force causes the molecules of the core to pick up speed as they fall inward. The condition of fast-moving molecules is synonymous with high temperature, however, so material near the center of the collapse becomes hot. Soon the pressure caused by this increased temperature prevents further collapse of the central core. Material farther from the core continues falling inward, crashing down on the central core and causing a further increase in temperature.

The hot core is now called a **protostar.** It receives its energy from the infall of material, so the ultimate source of its energy is gravitational. This is similar to the way a hammer that is repeatedly lifted and dropped onto a piece of metal heats both itself and the piece of metal. Each molecule that falls onto the core converts gravitational energy into thermal energy.

As the cloud contracts and heats, about half of its gravitational energy is radiated away from the heating center, and as the center gets hotter and hotter,

protostar. A star in the process of formation before it reaches the main sequence.

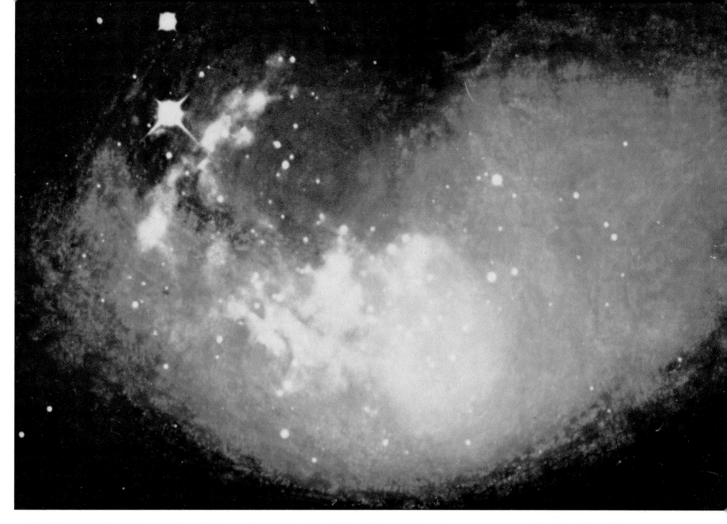

▶ **FIGURE 13-13.** This is an infrared photo of a small region of the Orion nebula where new stars are forming. The colors represent radiation from 30 to 100 microns. A visible-light photo has been superimposed on the infrared image in order to show the stars.

it radiates more and more energy. In only a few thousand years, the center is as bright as the Sun. We cannot see the center directly, however, because the outer portion, which continues to fall relatively slowly toward the center, blocks most of the radiation. This *cocoon,* or *cocoon nebula,* absorbs the radiation from the center, becoming warmer as it does so.

A warm object emits infrared (heat) radiation, and it is the infrared radiation emitted from the cocoon that we see from Earth and that gives evidence for the existence of protostars. The *Infrared Astronomy Satellite (IRAS)* has observed many such small infrared sources. ▶Figure 13-13 is an infrared photograph and is therefore shown in false color, with each color representing a different intensity of infrared radiation. Such false-color photographs of nonvisible sources are very useful tools in astronomy, but do not be fooled into thinking that the objects have colors like the photographs.

cocoon nebula. The dust and gas that surround a protostar and block much of its radiation.

Evolution toward the Main Sequence

▶Figure 13-14 shows the progress of a protostar of one solar mass on an expanded H-R diagram. (Notice that the main sequence is now compressed into the left side of the figure.) The diagram had to be expanded beyond the coolest stars on the main sequence because a protostar begins its life much cooler than any main sequence star. The details of this part of the star's *evolutionary track* are not known, but the protostar begins as a cool, dim object, warms up by gravitational contraction, and moves toward the main sequence. Finally, as the

evolutionary track. The path on the H-R diagram taken by a star as its luminosity and color change.

Star Birth

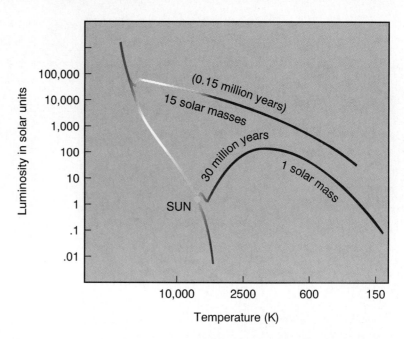

FIGURE 13-14. This H-R diagram shows temperatures much lower than on previous diagrams in order to include the early, cool stages of a star's formation. A more massive star ends up at a higher location on the main sequence because it is hotter and brighter. Since gravitational forces are greater on the more massive star, it spends much less time in its protostar stage.

In speaking of evolutionary tracks and stellar evolution, astronomers are using the term *evolution* in a very different way than biologists do. One star is said to "evolve" as it lives its life. We are not speaking of an entire species evolving through generations.

T Tauri stars. A certain class of stars that show rapid and erratic changes in brightness.

The Horsehead nebula of Figure 13-1 is a smaller part of this same Orion nebula. New stars are thought to be forming in the dark horsehead.

cocoon continues to contract, its smaller size causes it to appear dimmer so the star begins to move downward on the diagram as it nears the main sequence.

Motion on the H-R diagram does not mean the star is actually moving in space, of course. Recall from the last chapter the diagram of people's height versus age. As a person ages, his or her dot moves on this diagram. This is analogous to the motion of a star on the H-R diagram.

As time passes, the center of the protostar continues to shrink and become hotter, emitting more radiation all the time. This radiation gradually blows away the outer portions of the cocoon. Some of the cocoon may not be blown away but may condense to form planets, as discussed in Chapter 7. In some stars the cocoon is apparently blown away by sudden bursts of stellar winds. Certain stars, named **T Tauri stars** after the variable star T in the constellation Taurus, appear to be young stars undergoing some sort of instability that causes enormous flares. These flares are thought to play a part in blowing away the cocoon of newly forming stars, particularly K- and M-type stars, which—being at the bottom of the main sequence—are of low mass.

The most massive stars, the O- and B-type stars, follow a track on the H-R diagram toward the top of the main sequence (Figure 13-14). These stars are much fewer in number than stars of lesser mass, but they are so energetic that they signal their presence by emitting ultraviolet light that stimulates the hydrogen of their cocoon (and the nebula beyond the cocoon) to emit its own light, forming an emission nebula. ➤Figure 13-15 is a photograph of the Orion nebula, an emission nebula lit by energy from new stars within it.

The infalling particles of massive stars experience much greater gravitational force than do particles in low-mass stars, and massive stars reach the main sequence much faster. M-type stars remain protostars for hundreds of millions, perhaps billions, of years. The Sun, with more mass, spent about 30 million years in this phase, but the most massive stars remain protostars for only tens of thousands of years. Recall that stars of low mass may undergo a period of instability just before joining the main sequence. Massive stars do likewise, but their insta-

FIGURE 13-15. The Orion nebula is an emission nebula that contains a dark nebula where new stars are forming. The Orion nebula is easily visible in small telescopes or even binoculars.

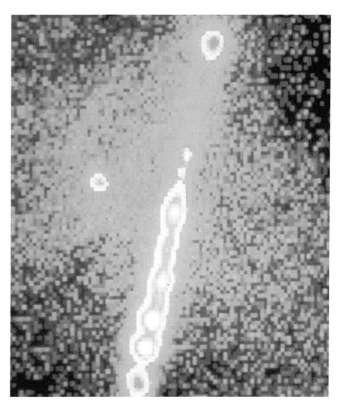

FIGURE 13-16. A jet of material is seen emerging from the star at the top of the photograph. Such a jet can extend to distances of about 1 light-year, and the gas that makes it up has speeds of hundreds of kilometers per second.

bility is more violent; O and B stars blow material out at supersonic speeds during this time, creating a shock wave in the surrounding material. This shock wave may be one of the triggers that starts the collapse of other portions of the interstellar cloud to form more stars.

The great amount of radiation from massive stars is what limits how massive a star can be. Astronomers calculate that a star with a mass greater than about 100 solar masses will emit radiation so intense that it will prevent more material from falling into the star, thereby limiting the star's size.

On the other hand, protostars with masses of less than about 0.08 solar masses do not have enough internal pressure to ignite hydrogen fusion. They heat up due to gravitational contraction but never become main sequence stars. Eventually, they contract as far as they can and then begin to cool, becoming planetlike objects—cold cinders in space. Recall that Jupiter and Saturn emit more energy than they receive from the Sun and that this comes from gravitational energy that remains from their formation.

Recent infrared observations have revealed that protostars are very commonly surrounded by disks of gas and dust. These observations fit well with the theory of the formation of the solar system and lead us to believe that planetary systems are common around stars.

In addition, streams of material have been observed flowing from the poles of many protostars. ▶Figure 13-16 shows a single jet of material extending downward on the image from a protostar at the top. In other cases, two jets emerge from the star, one from each pole. Astronomers hypothesize that this bipolar outflow plays an important part in reducing the angular momentum of stars. Recall from the earlier discussion of the formation of the solar system that theory had predicted that the Sun should be spinning faster than it is. If, early in the Sun's formation, material was ejected from its poles, that ejection would have resulted in reducing the Sun's rotational speed to what is observed. Much remains to be learned about the part played by the ejection of matter from the poles of protostars.

Star Clusters

galactic (or open) cluster. A group of stars that share a common origin and are located relatively close to one another.

globular cluster. A spherical group of up to hundreds of thousands of stars, found primarily in the halo of the Galaxy.

Globular is pronounced "glob" (as of mud) rather than "globe."

The Pleiades (Figure 13-7) is a small cluster of stars in the constellation Taurus (Figure 13-11). The brighter stars of the cluster are visible to the naked eye and are a beautiful sight even in a small telescope. ➤Figure 13-17 shows two other clusters of the same type, called *galactic clusters*, or *open clusters*. Galactic clusters are found primarily in the disk of the Galaxy. ➤Figure 13-18 shows another type of star cluster, a *globular cluster*. Globular clusters contain hundreds of thousands of stars. They are not confined to the galactic disk, but orbit the center of the Galaxy in all directions so that they spend most of their time outside the disk. ➤Figure 13-19 shows the location of the two types of clusters relative to the Galaxy.

➤**FIGURE 13-17.** Many clusters are named by their number in the New General Catalog (NGC), which was compiled between 1864 and 1908. (a) The cluster NGC 457, in Cassiopeia, is called the Owl cluster and contains about 80 stars. The bright star is φ (Phi) Cassiopeiae. (b) A pair of open clusters (NGC 869 and NGC 884) in Perseus.

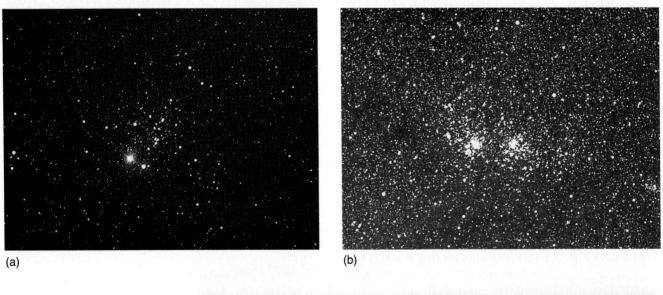

(a) (b)

➤**FIGURE 13-18.** This is a globular cluster (M13). Globular clusters are gravitationally bound groups of hundreds of thousands of stars.

CHAPTER 13 Interstellar Matter and Star Formation

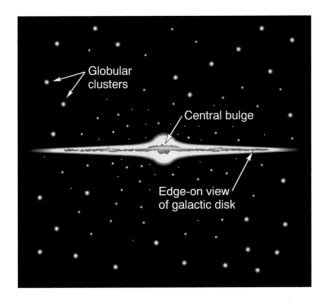
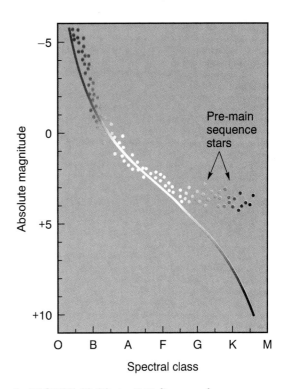

► FIGURE 13-19. Galactic (open) clusters are found within the disk of the Galaxy, and globular clusters are found primarily in the halo surrounding the disk.

► FIGURE 13-20. An H-R diagram of a very young cluster shows that stars of low mass have not yet reached the main sequence.

Clusters are important to astronomers for two reasons. First, all of the stars in a given cluster are at about the same distance from us. This means that their apparent magnitude is a direct indication of their absolute magnitude—the ones that look the brightest really are the brightest. Second, all the stars within a cluster formed at about the same time. (We will see that there is good evidence for this seemingly wild conjecture.) This means that they formed out of the same giant molecular cloud, and therefore it is probably safe to assume that all of them have about the same chemical composition.

Much of what we know about star formation has come from examination of clusters. The fact that all the stars in a cluster began forming at about the same time allows us to learn how stars of different mass have progressed at different rates. ►Figure 13-20 is a representative H-R diagram of a very young cluster. The diagram indicates that stars of low mass have not yet reached the main sequence. All of the stars of the cluster began their formation at essentially the same time, when their interstellar cloud began its collapse. H-R diagrams of clusters provide evidence that stars of low mass spend much more time in the protostar stage than do more massive stars. As will be explained a little later, H-R diagrams of older galactic clusters can be used in a similar manner to reveal when stars end the main sequence part of their lives.

Thus far, the evolutionary path of stars has been tracked through the protostar stage. When the core of a protostar becomes hot and dense enough, nuclear fusion begins. The star becomes stable and begins the main sequence portion of its life. The discussion of this stage of a star's life will be put off until the next chapter.

As was observed earlier, there is evidence that all stars within a cluster began their lives at about the same time. The regularities we see in H-R diagrams such as Figure 13-20 prove the validity of this assumption, for what other explanation could there be for the regularities? The next chapter will continue to examine the H-R diagrams of star clusters, for they are very valuable tools in our quest to understand stellar life cycles.

Star Birth

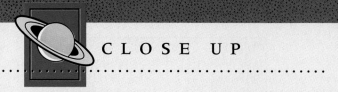

CLOSE UP

ET Life VI—The Life Equation

Astronomer Frank Drake of Cornell University proposed an equation for calculating the number of communicative civilizations in the Galaxy. The equation has been written in a number of forms, including this one:

$$N = R f_p n_p f_l f_i f_c L$$

where R is the rate at which stars form in our galaxy, f_p is the fraction of these stars that have planetary systems, n_p is the average number of planets per system that are Earth-like enough to support life, f_l is the fraction of those on which life develops, f_i is the fraction of life-forms that evolve to intelligence, f_c is the fraction of the intelligent races who are interested in interstellar communication, and L is the lifetime of the typical communicative civilization.

The factors in the equation, as you might surmise, are far from well known. Let's take a quick look at each. Based on the average lifetime of stars and on the number of stars in the Galaxy, we have a pretty good idea of the rate of star formation; the first term on the right is known to be in the range of 1 to 10 stars per year of the Sun's type. The more we learn about how planetary systems are formed, the better we can estimate what fraction of stars have planets circling them. Better knowledge of the origin of our own system will help in this regard, and as Chapter 7 explained, we are beginning to understand the solar system's formation. In addition, evidence is accumulating that planetary systems are common. We might suspect that if this is so, many systems probably have life-supporting planets. (After all, our system *almost* has three planets capable of supporting Earth-type life.) Thus we can make reasonably confident estimates of the values for the first three terms in the equation. Beyond this, things get fuzzier.

As a Close Up in Chapter 7 described, an experiment has been performed that indicates that the most basic molecules of which life is made are formed easily and naturally under conditions expected to be prevalent on some young planets. Many biologists believe that, given the right conditions and enough time, life will develop from these components, but they have no way of testing this hypothesis. The last three terms are even less well known. Does life evolve easily to intelligence, and if so, is it common for intelligent races to be interested in communication with aliens? (We seem to enjoy fiction on this topic, but we hesitate to spend money for a search.) Finally, how long does the typical civilization last? How long will ours last?

Various astronomers have inserted their best guesses into the equation, and—depending upon whether they are optimists or pessimists—have come up with answers anywhere from "only a few" to "millions." The equation's value, however, lies less in its mathematical answer than in showing us what we must learn in order to calculate an answer. It shows that we have a lot to learn.

CONCLUSION

Interstellar space is so barren of matter that for many purposes it can be considered a vacuum. Vast quantities of dust and gas do exist between the stars, however, and this material is important to astronomers for two principal reasons. First, radiation from stars passes through the interstellar medium on its journey to Earth, and if astronomers are to interpret the message of starlight correctly, they must understand how the interstellar medium affects radiation. Second, interstellar matter is the stuff of which stars are made. If astronomers want to understand the life cycles of stars, they must first understand the substance from which the stars form.

Clouds of dust and gas in interstellar space differ in density, composition, and size. Interstellar cirrus is extremely diffuse, but some dust clouds are dense and large enough to extinguish light from stars on the other side of them. In some clouds, the processes of fluorescence and reflection provide beautiful images that can be appreciated by astronomers and non-astronomers alike.

Technological developments during the past few decades—particularly in microwave and radio telescopes—have provided many answers to questions about the interstellar medium. But, as is usual in science, they have also provided additional questions.

The quest to understand how stars are formed from the interstellar medium is of special interest to astronomers, and many mysteries remain, particularly about the initial steps in the formation of a star.

Whatever serves as the trigger to begin the collapse of portions of a giant molecular cloud, mutual gravitational forces between the particles continue and accelerate the compression. And as the core of the cloud collapses, gravitational energy is transformed into thermal energy, which raises the temperature of the protostar. Finally, the inner portion of the collapsing matter becomes so hot and dense that nuclear fusion starts. At that time, the cloud begins its life on the main sequence of the H-R diagram. Examination of star clusters has shown the importance of a star's mass in determining how the star develops, for more massive stars become hotter and more luminous than less massive stars. The next chapter will show that mass is not only important in determining the early development of a star, but that it is the single most important property in determining the fate of stars.

RECALL QUESTIONS

1. In the densest interstellar cloud, the particle density is _____ the particle density of air at sea level on Earth.
   A. much less than
   B. about the same as
   C. much greater than

2. The reddening of starlight as it passes through interstellar dust clouds is due to
   A. the Doppler effect.
   B. fluorescence.
   C. scattering.

3. A nebula that glows due to fluorescence is called
   A. a fluorescence nebula.
   B. a reflection nebula.
   C. an emission nebula.

4. Absorption lines due to interstellar gas can be distinguished from spectral absorption lines caused by a star's atmosphere because
   A. their Doppler shifts are different.
   B. different chemical elements are found in the two different places.
   C. [The question is misleading; interstellar gas does not cause absorption lines.]

5. A protostar is
   A. a newly forming star.
   B. a main sequence star.
   C. a star nearing the red giant stage.
   D. a black dwarf.

6. H-R diagrams of young star clusters show
   A. fewer than about 10 stars.
   B. very few of the most luminous stars.
   C. that massive stars have not reached the main sequence.
   D. that low-mass stars have not yet reached the main sequence.

7. What is the primary difference between a star and a planet as each forms?
   A. age   D. chemical composition
   B. mass   E. velocity
   C. volume

8. Which of the following is true?
   A. Compared to the length of a star's main sequence life, its protostar stage is relatively short in duration.
   B. Protostars are surrounded by cocoons of gas and dust.
   C. Protostars radiate mainly in the infrared.
   D. [All of the above.]

9. Prior to reaching the main sequence, a star's energy comes from
   A. gravitation.
   B. nuclear fusion.
   C. nuclear fission.
   D. hydrogen conversion to helium.
   E. [Both B and D above.]

10. While a star is on the main sequence, its energy comes primarily from
    A. gravitational shrinking.
    B. nuclear fusion.
    C. nuclear fission.
    D. proton instability.
    E. chemical reactions.

11. As a star is forming by the condensing of gases, the gases
    A. cool as they fall.
    B. heat up as they fall.
    C. stay about the same temperature.
    D. [Any of the above, depending upon the mass involved.]

12. A star is considered to begin its main sequence life when
    A. it starts to collapse.
    B. its protostar life begins.
    C. nuclear reactions start.
    D. it begins to move off the main sequence.
    E. its planetary system has formed.

13. The contraction of an interstellar cloud to become a star is caused by
    A. magnetic forces.
    B. electric forces.
    C. nuclear forces.

D. gravitational forces.
E. [Both A and B above, for they are closely related.]

14. An evolutionary track on an H-R diagram reveals
    A. the changes that occur in a star's life.
    B. the changes that occur as one star evolves into another.
    C. the motion of a star relative to others in its cluster.
    D. the motion of a star relative to others in the entire galaxy.
    E. [Both C and D above.]

15. Which of the following is *not* considered a possible trigger to begin the collapse of an interstellar gas cloud?
    A. a shock wave from a supernova
    B. a shock wave occurring during the formation of very massive stars
    C. a shock wave resulting from radiation from nearby emission nebula
    D. a shock wave passing around the galaxy
    E. [All of the above are considered likely triggers.]

16. Clusters help us to learn about the evolution of stars because stars in a cluster
    A. are all about the same temperature
    B. are all about the same mass.
    C. interact with one another and affect each other's evolution.
    D. are all about the same age.
    E. can be observed more easily than individual stars.

17. The H-R diagram for a single star cluster is different from the usual H-R diagram because all stars in the cluster have the same
    A. mass.
    B. temperature.
    C. diameter.
    D. rotational speed.
    E. age.

18. What is a protostar?

19. List three possible events that may be responsible for triggering the collapse of interstellar clouds.

20. Which stars take longest to complete their protostar stage? Why?

21. What is the source of the energy that powers protostars?

22. Why is there a lower limit to the possible mass of a star? An upper limit?

23. Explain how galactic clusters provide observational evidence for theories of stellar evolution.

QUESTIONS TO PONDER

1. Starlight that has passed through a cloud of dust is reddened. So is starlight from a star that is moving away from us. In each case, how can we tell that the light from a star has been reddened? Perhaps what we see is the actual color of the star.

2. Refer to Figure 13-9, the Trifid nebula. Describe the process that produces each of the following effects: the pink nebula, the dark lanes in the pink, and the blue nebula.

3. Explain how an examination of the H-R diagram for a large group of stars can tell us the relative length of the various stages of a star's life.

4. Why do astronomers feel that there must be a triggering mechanism for the beginning of an interstellar cloud's collapse? Why couldn't it happen on its own?

5. Why do we think that most stars form from a *rotating* disk of gas? Couldn't it just as well be nonrotating?

6. Massive stars have much more hydrogen in their cores than less massive stars. Why do they run out of hydrogen faster than stars of low mass?

7. Which factors in the Drake equation (found in the Close Up "The Life Equation") are known with most certainty and which are least known?

CALCULATION

1. Insert your own best guess into the Drake equation (found in the Close Up "The Life Equation") and calculate the corresponding number of communicative civilizations.

ACTIVITY

Deep Sky Objects with a Small Telescope

Many beginning telescope users limit themselves to viewing the Moon and planets, but a number of interesting objects outside the solar system are accessible with a small telescope or even with binoculars. The star maps in this book can help you find the objects listed below. At the end of each listing is the time of year when the object is highest in the sky in the evening, but each may be seen in roughly the same place later at night earlier in the year.

Orion nebula—A cloud of dust and gas 1500 LY away. Its diameter is some 15 LY. Find the Trapezium, four stars in a small dark area of the nebula (sometimes called the Fish Mouth). (January through March)

Alcor and Mizar—A naked-eye double star in Ursa Major. Your telescope will reveal that Mizar is itself a double star. (March through July)

Sagittarius—Such a rich area of the sky for interesting objects that if you slowly scan it with your telescope, you will find a number of clusters and nebulae. (July and August)

Albireo—A binary pair with obviously different colors. (Figure 12-25) It is the "beak" star of the swan Cygnus, or the bottom star of the Northern Cross. (August through October)

Andromeda galaxy—A spiral galaxy similar to the Milky Way. Can you detect an oval shape? (October through December)

Pleiades—An open cluster visible to the naked eye. It contains a few hundred stars (although only six or seven are visible with the naked eye) and is about 400 LY away. See if you can detect the nebula around the stars. (November through February)

h and χ (the Greek letter, chi) Persei—A pair of open clusters visible to the naked eye (though not as easily as the Pleiades). In the sword handle of Perseus, they are an excellent view in binoculars. (November through February)

To fully enjoy your telescope (or binoculars), a good star atlas is almost a necessity. The following are recommended:

The Night Sky. Ian Ridpath and Wil Tirion. London: Collins Gem Guides, 1985. (about $5)

Mag 5 Star Atlas. Barrington, N.J.: Edmund Scientific Co. ($6.95)

Whitney's Star Finder (4th ed.). C. Whitney. New York: Alfred A. Knopf, 1985. ($6.95)

A Field Guide to the Stars and Planets (2d ed.). D. Menzel and J. Pasachoff. Boston: Houghton Mifflin, 1983. (about $10)

Star Gazing through Binoculars: A Complete Guide to Binocular Astronomy. S. Mensing. Blue Ridge Summit, Pa.: TAB Books, 1986. ($14.95).

To keep up with developments in astronomy, you might write to the nonprofit Astronomical Society of the Pacific (390 Ashton Avenue, San Francisco, CA 94112), whose aim is to share the excitement of astronomy with a wider public. Its members include scientists, teachers, hobbyists, and thousands of people from around the world who enjoy reading about the universe.

StarLinks netQuestions

Visit the netQuestions area of StarLinks (www.jbpub.com/starlinks) to complete an exercise on this topic:

1. The Interstellar Medium (ISM) Interstellar medium is the name for the stuff that exists in the space between the stars. But what's the stuff?

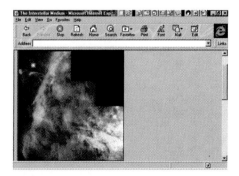

CHAPTER 14

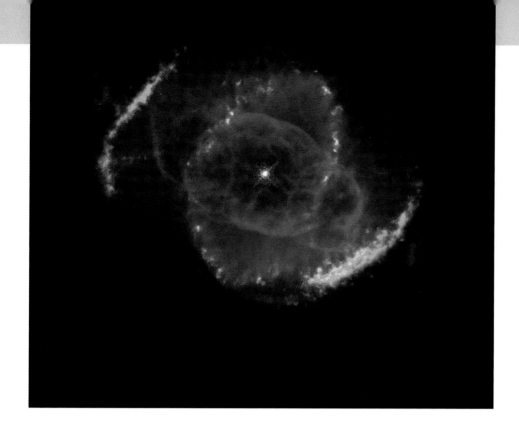

THE LIVES AND DEATHS OF STARS

Brown Dwarfs
Stellar Maturity
Stellar Nuclear Fusion
The Stellar Thermostat
Main Sequence Life
CLOSE UP: Lifetimes on the Main Sequence
Star Death
Flyweight Stars
Heavier than Flyweights—The Red Giant Stage
Electron Degeneracy

Lightweight Stars
The Helium Flash
Mass Loss from Red Giants
Planetary Nebulae
White Dwarfs
HISTORICAL NOTE: Tycho Brahe's Nova
Novae
The Chandrasekhar Limit
Type I Supernovae

·······   THIS BEAUTIFUL HUBBLE SPACE TELESCOPE IMAGE shows the "Cat's Eye nebula," officially named NGC6543. The image reveals that the nebula contains concentric gas shells, jets of high-speed gas and unusual shock-induced knots of gas. Estimated to be 1000 years old, the nebula is a visual "fossil record" of the dynamics and late evolution of a dying star.

Astronomers hypothesize that the star might be a double-star system. The effects of two stars orbiting one another most easily explains the intricate structures that are much more complicated than features seen in most nebulae that surround dying stars. The two stars are too close together to be individually resolved by Hubble, and *so they appear as a single point of light at the center of the nebula. According to the most generally accepted model, a fast "stellar wind" of gas blew off the central star and created the elongated shell of dense, glowing gas.*

The color picture, taken with the Wide Field Planetary Camera-2, is a composite of three images taken at different wavelengths. Such photographs reveal the beauty of distant dying stars, but I hope that as we study the lives and deaths of stars, you will also admire the beauty of the theory that underlies these events.

Stars form from the material of the interstellar medium. As a portion of a giant molecular cloud collapses, the matter in its core becomes hotter and more dense until nuclear fusion begins. When this happens, the object joins the family of stars, for it is then on the main sequence of the H-R diagram, where it will spend most of its life. This chapter describes stars of low mass, those toward the bottom of the main sequence (including our Sun). It will show that although all stars live their main sequence lives in much the same manner, they differ greatly in the way they die. The lightest ones just seem to fade away, "giving up the ghost" gently. Others expand to become red giants and then puff away their outer layers, thereby revealing their previously hidden white-hot cores.

As we study the amazing events of stars' lives, keep in mind that we are describing the results of various models of stellar interiors We cannot examine the inside of a star directly. Instead, we observe the properties of the surfaces of stars and make mathematical models of what must be occurring inside them to produce our data. Such models have improved dramatically in the last 20 years as computer capabilities have increased; the calculations that produce today's models are numerous and complex.

Theoretical models cannot be divorced from reality, of course. They must correspond to observations. For example, the major supernova of 1987 served as a case history for testing models of supernovae. The supernova excited astronomers not only because it provided another opportunity to test their theories, but—as we will see—because it confirmed most of their predictions.

BROWN DWARFS

www.jbpub.com/starlinks

As the previous chapter explained, the mass of a collapsing object determines how long the object remains in the protostar stage. Once a star has completed its protostar stage and reached the main sequence, its mass continues to be the dominant property determining the steps it passes through during its life. Before discussing stellar life cycles, therefore, the question of what limits the mass of a star should be examined.

The maximum mass that a star can have is probably about 100 solar masses. If a prostar has more mass than this, it collapses quickly and develops tremendous amounts of energy, so much energy, in fact, that it blows itself apart before it can establish equilibrium on the main sequence.

Suppose that a very small portion of an interstellar cloud becomes compressed enough for gravitational force to take over and continue the collapse. Calculations show that if this protostar has a mass greater than about 8% of the Sun's mass, the object's core will become hot and dense enough for nuclear fusion to begin. In that case, the object joins the main sequence as a full-fledged star. But what if the protostar has less than enough mass for fusion to begin? In this case, the object will still heat up due to gravitational contraction. ➤Figure 14-1(a) shows hypothesized evolutionary trails of five protostars on an H-R diagram. The two with the most mass become stars on the main sequence. Two other stars on the diagram—those with masses of 0.07 and 0.02 times the mass of the Sun, heat up so that they glow with a dull red color, but nuclear fusion never starts in their core. Their dull color gives them their name: ***brown dwarfs***. The final object has the mass of the planet Jupiter, which may once have emitted a small amount of visible light, but now is too cool to do so.

The question of how small an object can be and still be classified as a brown dwarf is open for discussion. Even an object as small as the planet Jupiter (0.001 solar masses) heats up as it forms. Such a small object probably does not deserve to be classified in a stellar category at all, so we will arbitrarily set the lower mass limit of brown dwarfs at twice the mass of Jupiter, or about 0.002 solar masses.

Brown dwarfs have been predicted since the mid-1900s, but until one was found, astronomers were not able to confirm that it is possible for such a small portion of an interstellar cloud to collapse. Until 1994, the only objects whose existence was confirmed that were near the required size were the planets of our solar system. In October 1994, using adaptive optics with a 60-inch telescope on Palomar Mountain, the first brown dwarf was seen. It appears very close to another star, a red dwarf, and a year after the discovery astronomers confirmed that the brown dwarf is actually gravitationally bound to the nearby star.

Figure 14-1(b) is an HST photograph of the brown dwarf, called GL229B. This is a false-color image, so the dwarf appears white rather than dull red. Its much brighter binary companion, GL229, is beyond the picture to the left, but GL229 is so bright that it floods the detector and causes the blur and the diffraction spike from the optical system. The brown dwarf is about 20 to 25 times the mass of Jupiter, but is about the same diameter as Jupiter. The two stars are about 18 light-years away in the constellation Lepus.

brown dwarf. A starlike object whose mass is too small to sustain nuclear fusion Probable limits of mass are from 0.002 to .08 solar masses.

➢ **FIGURE 14-1.** (a) Protostars with masses less than about 0.08 solar masses never reach the main sequence. Of the five stars whose evolutionary path is shown here, two become main-sequence stars, two become brown dwarfs, and one becomes a planet. (b) The small object just right of center is GL229B, a brown dwarf, photographed by the HST in far red light. The dwarf is a companion to a much brighter star, Gliese 229, that is off the left edge of the image. The two stars are about 18 light-years away from us and are separated from one another by at least 4 billion miles.

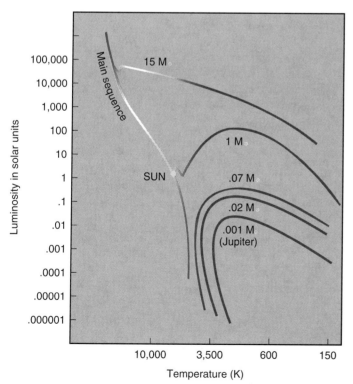

(a)   (b)

Notice that I have avoided referring to a brown dwarf as a star. The reason for this is that the word *star* is reserved for an object massive enough to sustain nuclear fusion in its core—that is, to have reached stellar maturity on the main sequence.

STELLAR MATURITY

When the center of a collapsing star becomes hot and dense enough, hydrogen fusion begins. Gravitation serves as the energy source for protostars, heating them and causing the emission of radiation, but main sequence stars—including our Sun—have as their energy source the fusion of hydrogen into helium.

Stellar Nuclear Fusion

Chapter 11 described the fusion reactions that occur in the Sun. That series of reactions, known as the proton-proton chain, is the predominant reaction that provides the energy for stars of low mass, up to about 1.5 solar masses. For the sake of completeness, it is repeated in ➢Figure 14-2.

Stars that have masses greater than about 1.5 solar masses have higher temperatures in their cores, and a different chain of nuclear reactions predominates. This series of reactions involves carbon, nitrogen, and oxygen and is called the *carbon cycle* or the *CNO cycle*. Its steps are shown in ➢Figure 14-3. If you examine the steps of the CNO cycle, you will see that its overall effect is to transform four hydrogen nuclei into one helium nucleus, just like the proton-proton chain. Carbon-12 is one of the nuclei that participate in the reaction, but for each carbon-12 nucleus that enters the reaction, one is produced at the end. Therefore there is no net change in the carbon; it participates only as a facilitator.

carbon (or CNO) cycle. A series of nuclear reactions that results in the fusion of hydrogen into helium, using carbon-12 in the process.

Chemical reactions sometimes involve a catalyst, a chemical that stimulates the reaction but is not used up in the reaction. Carbon acts in this manner in the CNO cycle.

➢ **FIGURE 14-2.** In the proton-proton chain, four hydrogen nuclei are changed to one helium nucleus, releasing energy in the process. The symbol e^+ denotes a positive electron, a positron. Dashed arrows indicate that that nucleus takes part in the next reaction. In the third reaction, the second hydrogen-3 nucleus comes from a second occurrence of the two previous steps.

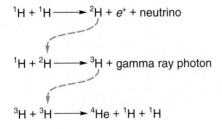

$$^1H + {}^1H \longrightarrow {}^2H + e^+ + \text{neutrino}$$

$$^1H + {}^2H \longrightarrow {}^3H + \text{gamma ray photon}$$

$$^3H + {}^3H \longrightarrow {}^4He + {}^1H + {}^1H$$

➢ **FIGURE 14-3.** The CNO cycle changes four hydrogen nuclei into one helium nucleus, with an attendant release of energy.

$$^1H + {}^{12}C \longrightarrow {}^{13}N + \text{gamma ray photon}$$

$$^{13}N \longrightarrow {}^{13}C + e^+ + \text{neutrino}$$

$$^1H + {}^{13}C \longrightarrow {}^{14}N + \text{gamma ray photon}$$

$$^1H + {}^{14}N \longrightarrow {}^{15}O + \text{gamma ray photon}$$

$$^{15}O \longrightarrow {}^{15}N + e^+ + \text{neutrino}$$

$$^1H + {}^{15}N \longrightarrow {}^{12}C + {}^4He$$

The Stellar Thermostat

Thermostats on the walls in our homes have two features: a thermometer to read the temperature and a switch to turn on and off the furnace or air conditioner. If the thermometer indicates that the temperature is below some preset level, the switch turns on the furnace until the thermometer indicates that the temperature is high enough, whereupon the switch turns off the furnace. The core of a main sequence star has an analogous regulating mechanism, which controls the rate of consumption of the hydrogen fuel.

Suppose that somehow the rate of hydrogen fusion in a star begins to increase. This causes the temperature of the core to increase. When the temperature of a gas increases, however, the gas expands and the expansion serves as a switch to decrease the rate of fusion. This occurs because in the expanded gas, the average distance between hydrogen nuclei has increased, and the increase in distance means fewer collisions and therefore fewer fusions. The result is that the expansion causes the rate of energy production to decrease, which causes the expansion to stop. In this way equilibrium is reached, and fusion reactions occur at a uniform rate.

On the other hand, if the rate of fusion in a star's core somehow decreases, the core will contract, decreasing the distance between nuclei. This causes the fusion rate to increase. Another effect that occurs when the core contracts is the conversion of gravitational energy into heat, just as occurred when the original interstellar cloud collapsed. Therefore, energy comes both from an increase in fusion and from gravitational energy, and both of these energies contribute to the thermostat that brings the star once again to equilibrium. The overall effect of the stellar thermostat is that nuclear fusion proceeds at a rate that is just enough to balance the force of gravity that tends to compress the star.

The analogy of a thermostat is an alternative description of hydrostatic equilibrium, discussed in Chapter 11.

Main Sequence Life

The stellar thermostat is important during the main sequence life of a star. In a main sequence star, hydrogen in the core of the star is continually being converted to helium as the nuclear reaction occurs. This causes the number of particles in the core to gradually decrease (since four hydrogen nuclei are converted to one helium nucleus), and this results in the core shrinking slightly. In turn, the contraction causes the temperature to rise within the core, and this increases the rate of fusion. Therefore, more energy is released by the core, and as this energy flows outward, it causes the outer portion of the star to expand somewhat. These steps are outlined in ➤ Figure 14-4. We see the effect of these

```
The number of nuclei
decreases due to fusion.
            ↓
       The core shrinks.
            ↓
      Gravitational energy
        heats the core.
            ↓
      The fusion rate increases.
         ↙        ↓
     Additional energy is
      released by the core.
       ↙              ↘
The star becomes    The outer portions
 more luminous.      of the star expand.
                         ↓
                    The surface cools.
```

➤ **FIGURE 14-4.** This diagram shows why a star becomes more luminous and cooler as it ages on the main sequence. Each arrow indicates that one event causes the next.

Stellar Maturity

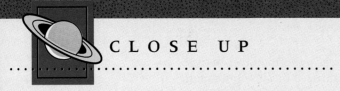

CLOSE UP

Lifetimes on the Main Sequence

The amount of time that a star spends on the main sequence depends on two things: the amount of hydrogen in its core and its rate of hydrogen consumption. The relationship can be expressed as a proportionality as follows:

$$T \sim \frac{\text{amount of hydrogen}}{\text{rate of hydrogen consumption}}$$

where T represents the star's lifetime and the symbol "$\sim$" means "proportional to." Since all stars have roughly the same proportion of core hydrogen at equivalent points in their lives, the amount of hydrogen in a star is proportional to its mass. The rate at which the hydrogen is consumed depends upon the star's luminosity. Therefore

$$T \sim \frac{M}{L}$$

A Close Up in Chapter 12 explained that a star's luminosity (L) is proportional to its mass (M) raised to the power of 3.5. Thus

$$T \sim \frac{M}{M^{3.5}} \sim \frac{1}{M^{2.5}}$$

Using the lifetime of the Sun as a standard, so that T_S is the lifetime of the Sun on the main sequence and M is measured in units of the Sun's mass, the expression can be written as an equation:

$$T = \frac{T_S}{M^{2.5}}$$

Then the main sequence lifetime of a star with a mass of 3 solar masses can be calculated as follows:

$$T = \frac{T_S}{3^{2.5}} = \frac{T_S}{3 \times 3 \times \sqrt{3}} = 0.064 T_S$$

Thus a star with three times the Sun's mass will live on the main sequence only about 6% as long as the Sun. Since the Sun's main sequence life (T_S) is about 10 billion years, this star's life will be about 640 million years.

The cooling of the outer layers occurs because as a gas expands, it cools.

changes on the H-R diagram (►Figure 14-5), for the increase in energy from the star means that it has a greater luminosity, so it moves upward on the diagram. In addition, the expansion of the star causes its outer layers to cool a little, and this causes the star to move to the right on the diagram. These changes are the reason the main sequence has a perceptible width. Stars start their main sequence lives on the left side of the strip, then move up and to the right as they age.

The changes that the Sun is undergoing will have drastic effects on the Earth. The Sun began its main sequence life about 5 billion years ago, when it was only about one-third as luminous as it is now. It will continue on the main sequence for another 5 billion years, at which time it will be twice as luminous as it is now. This will raise the average temperature of the Earth by about 20 Celsius degrees, enough to melt the polar caps and drastically change the climate of Earth.

Twenty Celsius degrees equal 36 Fahrenheit degrees.

Stars that are more massive than the Sun have a much greater fusion rate because their cores have greater pressures and higher temperatures—the thermostat is set higher in massive stars. This means they are more luminous. (Recall the mass-luminosity relationship for main sequence stars: the more massive, the more luminous.) The greater fusion rate also causes these massive stars to use up their core hydrogen in a much shorter time. The most massive stars fuse hydrogen so quickly that their cores run out of hydrogen in only a few million years. Calculations for the least massive stars, on the other hand, show that they will continue hydrogen fusion on the main sequence for hundreds of billions of years.

The name M11 means that this cluster is number eleven in the Messier Catalog of Nebulae and Star Clusters, originally developed in 1781 by the French astronomer Charles Messier (1730–1817).

Observational evidence for the shorter lifetimes of massive stars comes from galactic clusters. ►Figure 14-6a is the H-R diagram of the Pleiades. Although this is considered a young cluster, its most massive stars are already leaving the main sequence. Arrows on the figure indicate their evolutionary paths after they end

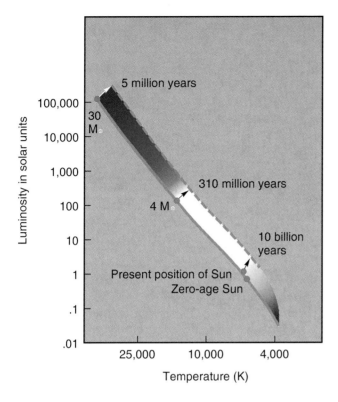

▶ FIGURE 14-5. As stars age on the main sequence, they change slightly in temperature and luminosity. The nearby Close Up shows how main sequence lifetimes are calculated.

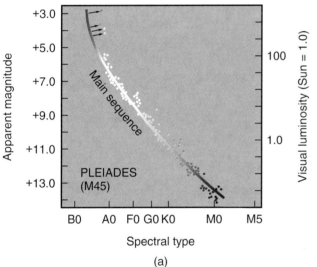
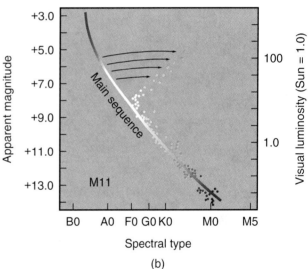

▶ FIGURE 14-6. (a) An H-R diagram of the Pleiades reveals that its most massive stars have started leaving the main sequence. The arrows indicate the evolutionary paths they must have taken. (b) The cluster M11 is older than the Pleiades, and stars are turning off the main sequence at a lower point.

their main sequence lives. Figure 14-6b is a similar diagram for the cluster M11, a cluster older than the Pleiades. Notice that the most massive stars have moved even farther to the right and that stars of less mass are now leaving the main sequence.

STAR DEATH

Until the end of their lives on the main sequence, the primary difference between the evolution of stars of various masses is in the amount of time they spend as protostars and as main sequence stars. From this point on, however, the mass of a star determines which of a number of very different paths its life will take. Each of these will be discussed in turn, beginning with stars of very low mass. Since mass classifications are fairly arbitrary and astronomers have not agreed on official names for the different classes, we will group them like boxers, calling them—in order from least massive to most massive—flyweights, lightweights, middleweights, and heavyweights. Remember, though, that the classification is really based on mass rather than weight.

Flyweight Stars

Recall from the discussion of energy transport within the Sun in Chapter 11 that little convection takes place in the Sun except in the outer layers. That is why, once hydrogen is used up in the core of a star like the Sun, the core is not replenished with fresh hydrogen from outside. This causes the star to move from the main sequence.

In the least massive stars (those with a mass of less than about 0.4 solar masses), however, convection occurs throughout most or all of the volume of the star (►Figure 14-7). Therefore, hydrogen from throughout the star is cycled through the core, and the entire star runs low on hydrogen at the same time. When this happens, the rate of fusion in the core will decrease. As described earlier, this will cause the core to contract and heat as gravitational energy is converted into thermal energy. In a more massive star, this heat is carried outward by radiation, but a flyweight distributes its heat by convection. The result is that the entire star will contract and heat up. On the H-R diagram, the star will move toward the lower left (►Figure 14-8) and will become a white dwarf.

Computer models indicate that a star in the flyweight class will require 20 billion years or more to complete the burning of its hydrogen fuel and end its main sequence life. Since this is greater than the age of the universe, the white dwarfs that astronomers see could not have originated in the manner described above. For this reason, the discussion of white dwarfs will be delayed until later.

The lowest mass stars may live as main sequence stars for as long as 100 billion years.

► **FIGURE 14-7.** The Sun (at left) contains a large radiative zone. A star of very low mass, on the other hand, consists of one convective zone, meaning that all of its material mixes.

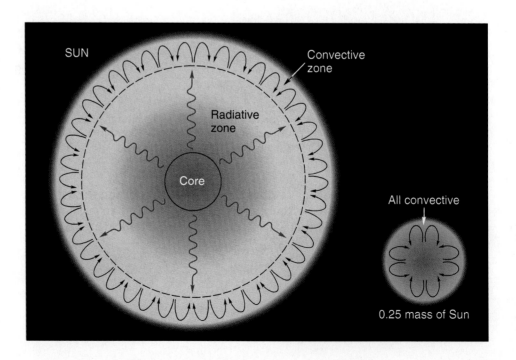

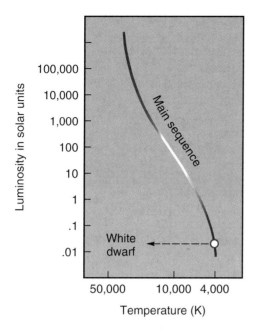

► FIGURE 14-8. Flyweights spend their main sequence lifetimes at the very bottom of the main sequence. Although the universe is not old enough for any to have completed that stage of their life, when they do, they will slowly become hotter and join the category of white dwarfs.

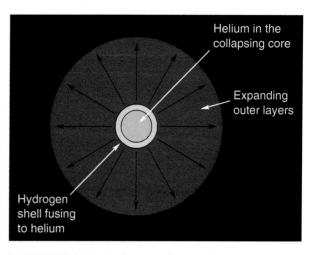

► FIGURE 14-9. As the core of a star shrinks, it heats up, causing additional hydrogen fusion in a shell surrounding the core. (The core and hydrogen-fusing shell are actually a much smaller fraction of the star than is shown here.)

HEAVIER THAN FLYWEIGHTS—THE RED GIANT STAGE

About 90% of the stars in the sky are on the main sequence. This tells us that a typical star spends 90% of its luminous lifetime there. In all cases except for stars of very low mass (the flyweights, with mass less than about 0.4 solar masses), the next step for stars leaving the main sequence is essentially the same. They become red giants.

The process of becoming a red giant starts when the core begins to run low on hydrogen fuel. Until then, the nearly constant production of fusion energy kept the core from collapsing. Now, lacking a source of energy to fight gravitational collapse, the core starts to shrink dramatically. Just as in the case of the protostar, however, the contraction converts gravitational energy into heat and radiation. In fact, the energy produced in the core actually *increases* over what it was when the energy source was fusion. The resulting increase of radiation from the core causes the shell of material around the core to heat up enough that hydrogen fusion begins there (►Figure 14-9).

Now gravitational energy is producing heat in the core, and fusion is producing heat in a shell surrounding the core. These two sources of energy in the star's center result in a great increase in radiation from the center and cause the outer part of the star to expand. When a gas expands, it cools, and the outer portion of the star does just that. Thus the star moves to the right on the H-R diagram, toward the red giant region. At the same time it moves upward because its total luminosity is increasing due to the additional energy being produced in the core.

It may seem strange that a star can decrease its surface temperature and at the same time increase its luminosity, for when an object cools, it emits *less* radiation per square meter of its surface. The key to understanding this seeming contradiction is to realize just how large the red giant becomes. The Sun, at this

The cooling of a gas as it expands is what causes the gas escaping from an aerosol spray can to feel so cold.

stage in its life as a giant, will have expanded so that it encompasses the orbit of Mercury! It is true that each square meter of the Sun's cooler surface will emit less radiation, but the surface will have become so tremendously large that the total radiation emitted will be greater than before. ➤Figure 14-10 illustrates how the Sun will change its position on the H-R diagram. As the Sun expands it will move from point A, its turn-off point from the main sequence, toward point B.

The core of the red giant at this time consists of helium nuclei intermingled with electrons, a mixture that has properties similar to a regular gas. A simple relationship exists between pressure, temperature, and volume. For example, an increase in pressure exerted on a normal gas causes its volume to decrease and its temperature to increase. That is why gravitational pressure causes a star's core to contract and heat up when nuclear fusion shuts down at the end of main sequence life. Table 14-1 lists some properties of a typical red giant.

Electron Degeneracy

The core of a red giant does not continue to contract indefinitely, however. Once the density of the core has increased beyond a certain value, the material of the core changes to a different state of being, and the normal relationship between pressure, temperature, and volume no longer holds. At the great density achieved at this point, only one thing places a limit on how much the core contracts—the

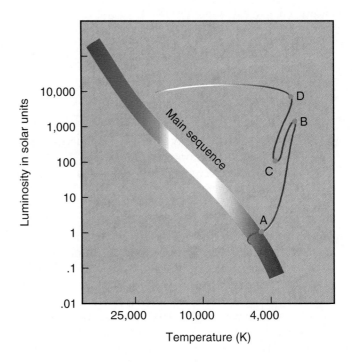

➤ **FIGURE 14-10.** The Sun will enter the first stage of its death when it leaves the main sequence at point A. The helium flash will occur when it reaches the luminosity and temperature of point B, and this will cause a quick decrease in its luminosity until it reaches point C. Helium fusion around the Sun's inner carbon core will then move it up to point D, where it will reach its greatest luminosity.

TABLE 14-1 A Typical Red Giant	
Absolute magnitude	−1
Luminosity (solar units)	500
Mass (solar units)	1
Diameter (solar units)	200
Average density	10^{-7} gm/cm^3 (10^{-7} Sun's density)
Surface temperature	3600 K

electrons' resistance to being squeezed together. The matter of the core is now said to be *degenerate,* or *electron degenerate.*

When a normal gas is heated, its pressure increases and the gas expands. In degenerate matter, pressure does not depend on temperature; it depends only upon density. The properties of degenerate matter lead to a strange phenomenon. Refer to the graph in ▶Figure 14-11a, which shows the relationship between the radius and the mass of a main sequence star. As one would expect, the more massive the star is, the larger its radius. The degenerate core of a red giant does not act this way, however. Figure 14-11b shows what happens in this case. The more massive the core, the smaller its radius is! This is very different from our experience with nondegenerate—regular—matter, and its has a profound influence on the next event of the star's death struggle.

electron degeneracy. The state of a gas in which its electrons are packed as densely as nature permits. The temperature of such a high-density gas is not dependent on its pressure as it is in a "normal" gas.

LIGHTWEIGHT STARS

All main sequence stars become red giants (or supergiants, which will be discussed in the next chapter) when they leave the main sequence. How stars change during the red giant stage, however, depends upon how massive they are. First to be considered are the lightest stars on the main sequence, the *lightweight* stars. This group includes stars from about 0.4 solar masses up to about 6 solar masses, and therefore it includes our Sun.

The Helium Flash

As the hot, degenerate core of a red giant radiates its heat outward, hydrogen continues to fuse to helium in a shell around the core. The hydrogen-burning shell gradually works its way outward, all the time dumping its helium "ashes" onto the core. Because of the strange relationship between the mass and the radius of a degenerate core, the increased mass falling onto the core causes the core to shrink further and further. This, in turn, causes its temperature to continue to rise. Finally, the temperature of the red giant's core reaches a critical temperature, calculated to be about 100,000,000 K. At this point, shown as point B on Figure 14-10, helium nuclei begin to combine, forming carbon. The three-

▶ **FIGURE 14-11.** (a) In the case of normal, main sequence stars, the more massive the star is, the larger its radius. (b) In the case of the degenerate core of a red giant, however, the more massive the core is, the *smaller* its radius.

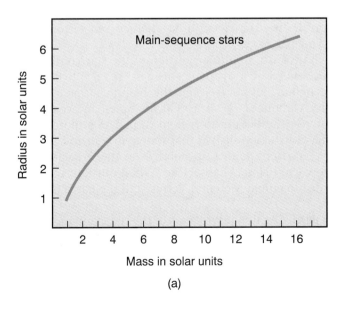
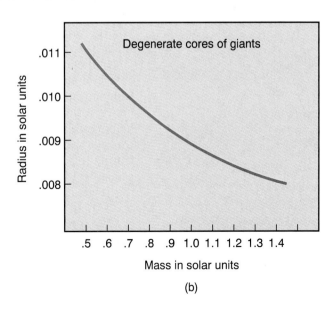

(a) (b)

A helium nucleus is also known as an *alpha particle*, and the conversion of three helium nuclei into a carbon nucleus is called the *triple-alpha process*.

TABLE 14-2 Helium Fusion Reactions That Occur in Red Giants

Reaction	Explanation
$^4_2He + {}^4_2He \rightarrow {}^8_4Be + \gamma$	Two helium nuclei combine to produce a beryllium-8 nucleus and a gamma ray (γ).
$^4_2He + {}^8_4Be \rightarrow {}^{12}_6C + \gamma$	Another helium nucleus combines with the beryllium nucleus to form a carbon-12 nucleus and another gamma ray.
$^4_2He + {}^{12}_6C \rightarrow {}^{16}_8O + \gamma$	Carbon in the core fuses with helium and produces oxygen and another gamma ray (and more energy).

➤ **FIGURE 14-12.** This diagram shows schematically the steps that occur as helium fuses into carbon in the core of a red giant.

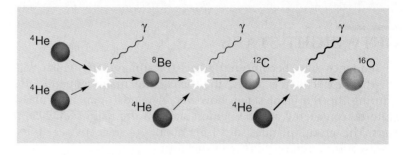

helium flash. The runaway helium fusion reaction that occurs during the evolution of a red giant.

step fusion reaction that produces carbon from helium is shown in more detail in Table 14-2 and ➤Figure 14-12.

The reaction does not proceed smoothly, however. Degenerate matter is a good conductor of heat, so the additional heat produced by the new nuclear reaction spreads rapidly throughout the core. The core ignites quickly and violently, a process known as the **helium flash.** The speed of the reaction is aided by the fact that the core is degenerate, so that it does not expand as it heats. (Expansion of the core would slow the reaction.) The helium flash is so violent in some stars that it consumes the core's available helium in only a few seconds. When the helium flash occurs, the surface of the star contracts slightly and becomes hotter. This stage in the evolution of the star is shown in Figure 14-10 as the movement from point B to point C.

One might think that the violence of the helium flash would have a drastic effect on the outer portion of the star, but the core is very small and the energy released by the flash dissipates slowly toward the surface, taking thousands of years to be released. Nevertheless, computer models show that the helium flash changes the core drastically. One effect is that the sudden increase in temperature and the production of carbon destroy the electron degeneracy and restore the stellar thermostat. Thus the violence is very short-lived.

Immediately after the helium flash, the inner portion of the star consists of a degenerate carbon core surrounded by a hydrogen-fusing shell. The hydrogen fusion continues to produce helium, of course, and the helium forms a layer between the hydrogen and the carbon. Temperatures at the inner surface of the helium are great enough now for the helium to fuse to carbon in a more controlled fashion, and the star has a structure such as that shown in ➤Figure 14-13. During this time, the star evolves from point C to point D on the H-R diagram of Figure 14-10. Its size is now even larger than before. When our Sun reaches this stage, it will have expanded to enfold Mercury, Venus, and the Earth (➤Figure 14-14).

What will happen to the Earth? As the outer layers of the red giant approach Earth, our planet will find itself speeding through the hot gas of the Sun's atmosphere. Earth's atmosphere will be ripped away. The planet will be slowed by friction and will begin falling toward the Sun's center. Meanwhile the Earth's crust will melt and then vaporize. Our planet will become one with the Sun.

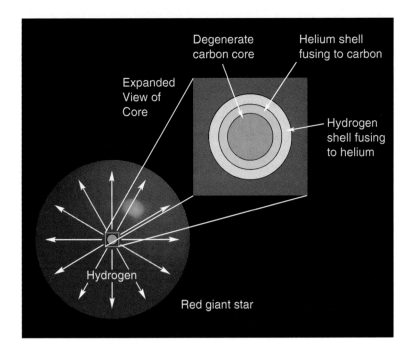

▶ FIGURE 14-13. When a red giant reaches the stage where it has a carbon core, the heat from that shrinking core ignites helium fusion in a shell around it. At the same time, hydrogen is fusing in a second shell beyond the first. This activity occupies very little of the volume of the star.

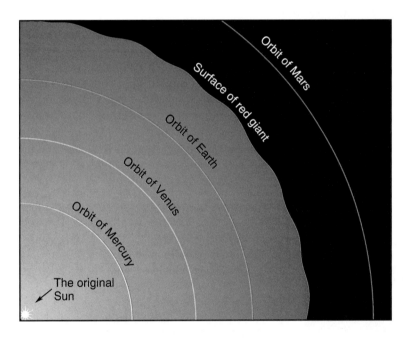

▶ FIGURE 14-14. When the Sun becomes a red giant, it will expand until its surface is somewhere between Earth and Mars. Compare this to its present size.

The discussion just presented actually applies only to the lightest of the lightweights. Computer simulations indicate that stars more massive than about 2 solar masses do not experience a helium flash. The cores of these stars get hot enough toward the end of hydrogen burning that they make a smooth transition to helium burning without becoming degenerate. When their supply of helium is exhausted, however, their carbon core shrinks and becomes degenerate just as it does for lighter, Sun-like stars. Therefore, in both cases the interior of the new red giant has the structure of Figure 14-13.

Mass Loss from Red Giants

During its main sequence life, our Sun continually sheds material as the solar wind. Orbiting X-ray and ultraviolet telescopes such as the *High Energy Astro-*

Lightweight Stars   435

physical Observatories and the *International Ultraviolet Explorer* have detected emission spectra from around other main sequence stars, indicating that stars commonly have hot chromospheres and coronas just as our Sun does. It seems safe to assume that they also have stellar winds like the Sun. The solar wind carries away about 10^{-14} of the Sun's mass each year. This is an extremely small amount, for at this rate, the Sun will lose only 0.01% of its mass in 10 billion years. Mass loss is much more important for red giants.

We are able to detect a slight Doppler shift in emission lines from gas near red giants. Based on the amount of the Doppler shift, astronomers conclude that the gas is being blown off red giants at speeds between 10 and 20 kilometers/second, and that a typical red giant loses about 10^{-7} solar masses per year. The reason for the greater mass loss in red giants is that these stars are so much larger that they exert less gravitational force on material near their surface. In addition, the helium-burning process that provides much of the energy for red giants is unstable and varies greatly in response to small changes in temperature. As a result, a red giant undergoes instabilities and pulsations as helium fusion increases and decreases. In fact, during a portion of a red giant's life, it is common for the star to pulsate with periods less than one day. Astronomers think that the core instabilities and pulsations of red giants are the reason for the great mass loss, although the process is not well understood.

Planetary Nebulae

planetary nebula. The shell of gas that is expelled by a red giant near the end of its life.

➤Figure 14-15 shows photographs of two objects called **planetary nebulae.** Planetary nebulae were given that name decades ago because, when viewed in a small telescope, many of them have a blue-green color that resembles the color of Uranus and Neptune viewed through similar telescopes. In fact, planetary nebulae have nothing to do with planets. Spectroscopic analysis reveals—again by the Doppler effect—that these nebulae are made up of material moving out-

➤ **FIGURE 14-15.** Planetary nebulae consist of a sphere of gas ejected from a red giant. Part (a) is the Ring nebula in the constellation Vega, and (b) is the Helix nebula (NGC 7293) in Aquarius. The Ring nebula can be seen with a small telescope (Figure 14-16).

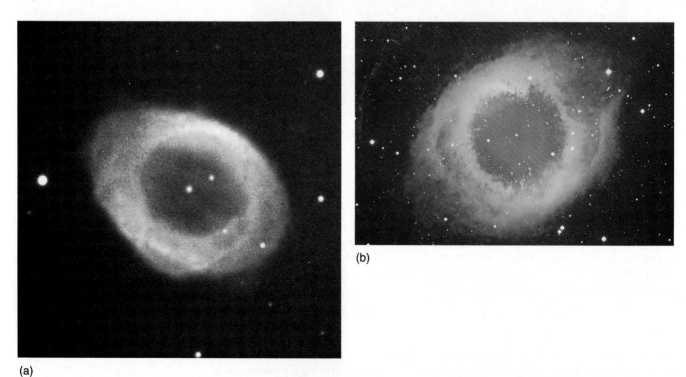

➤ **FIGURE 14-16.** (a) In a small telescope, the Ring nebula appears as a tiny circle. (b) It can be found in Lyra, which is a small constellation located adjacent to Cygnus.

ward from a very hot central star. Many appear to be doughnut-shaped, but this is because we are looking through a shell of material when we view them, as explained in ➤Figure 14-17. The material glows because ultraviolet radiation from the central hot star causes it to fluoresce.

At least two models attempt to explain planetary nebulae:

- One model holds that the pulsations in the core of a red giant continue to increase in intensity until the outer layers of the star also become unstable. With each pulse, part of the star's envelope is blown away. After perhaps 1000 years—a short time by stellar standards—the entire outer portion of the star will have been ejected, forming a shell around the original core.
- A second model, now gaining in favor, holds that the stellar winds emitted from a dying star occur in definite stages, and that at the earliest stage, the outer layers of the star are blown off as a slow, dense wind at speeds of perhaps 90,000 kilometers per hour. This finally leaves behind a hot, dense star that emits high-energy radiation. Now the character of the wind changes, so that it becomes a low-density, but very fast wind, with speeds up to 18 million kilometers per hour. When this second wind catches the first, it pushes the first forward, causing a region where there is a very dense wave. (To picture this, imagine that a group of walkers is being overtaken and run into by a group of runners. Where the two groups are colliding, the people are bunched very close together as the runners push the walkers into one another.) In addition, the intense radiation causes that dense area to glow, and this is the shell that we see as a planetary nebula. In the next chapter, when we discuss supernovae, we will see that some hour glass-shaped planetary nebulae are being discovered that can be explained using this model.

The fluorescence produces an emission spectrum, and the green color is due primarily to ionized oxygen. (Recall the fluorescence of a comet's coma and tail.)

Whichever model is correct (or perhaps a combination of the two will be found to be the best explanation for the phenomenon), not all planetary nebulae

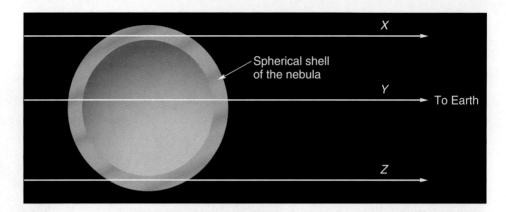

➤ FIGURE 14-17. Three lines of sight through the spherical shell of the Ring nebula. Lines *X* and *Z* pass through much more of the glowing nebula than does line *Y*; thus the nebula will appear brighter here.

appear as rings. The Dumbbell nebula of ➤Figure 14-18(a) shows just a hint of a spherical structure. The HST image of a planetary nebula in part (b) reveals that the initial ejections from the star were periodic and fairly uniform (as evidenced by the concentric spheres). The bright inner regions of the nebula must have been produced by irregular ejection of material, and dense clouds of dust condensed from the ejected material. The dense clouds are shown here in yellow.

Only a few thousand planetary nebulae have been observed. Although this may seem a large number, it is very small compared to the number of stars, and this fact indicates that planetary nebulae do not last long. The material that we see around the central star quickly dissipates into space and becomes part of the interstellar medium.

As the nebula dissipates, the hot, bright core of the star begins to peek through. This core may have a temperature of 100,000 K, and its appearance causes the star to shift its position on the H-R diagram. To include this stage on an H-R diagram, the diagram must be extended toward higher temperatures than have been included in previous diagrams. ➤Figure 14-19 shows that the star moves far to the left. The core remains at its very luminous stage for a relatively short time; then it quickly moves down the H-R diagram and collapses to become a white dwarf.

➤ FIGURE 14-18. (a) This is the Dumbbell Nebula (M27) imaged by traditional methods with an Earth-bound telescope. (b) This HST image of NGC 7027 is a composite of images in the visible and infrared. The central part is not solid, as it may appear in the false-color image.

(a)

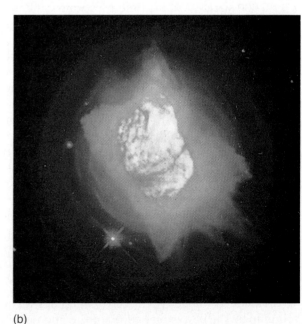

(b)

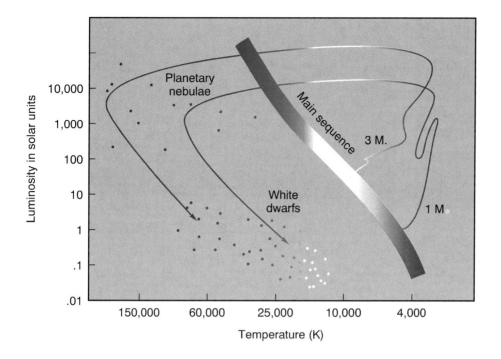

► **FIGURE 14-19.** This H-R diagram has been expanded to the left—to higher temperatures—in order to include stars that are passing through the planetary nebula stage. The evolutionary tracks of one-solar-mass star and a three-solar-mass star are shown.

WHITE DWARFS

White dwarfs were discovered long before astronomers had predicted their structure. In 1844 the German astronomer Friedrich Bessel (1784–1846) noticed that the star Sirius (►Figure 14-20) wobbled back-and-forth slightly. He hypothesized that an invisible companion orbiting Sirius causes the wobble; that is, Sirius and another star form a binary pair. Bessel's calculations showed that the unseen

Recall that a binary system in which the motion of one star reveals the presence of the other is called an *astrometric binary*. Bessel discovered, also in 1844, that Procyon is an astrometric binary. Procyon is the brightest star in Canis Minor.

► **FIGURE 14-20.** Sirius (in the constellation Canis Major) is the brightest star in the sky, except for the Sun, of course. Bessel found that Sirius wobbles, causing him to predict that it is one of a binary pair and has an unseen companion.

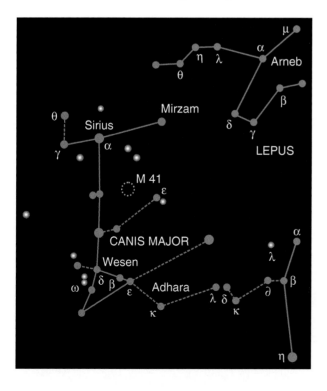

White Dwarfs

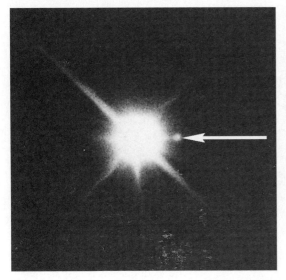
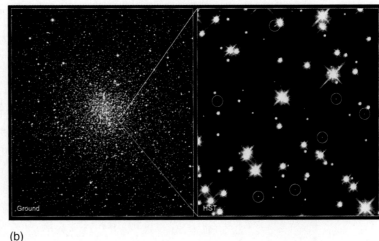

➤ FIGURE 14-21. (a) Sirius is revealed to be two stars in this telescopic view. The spikes that radiate from Sirius A are caused by diffraction around mirror supports within the telescope. (b) The image at left is a cluster of stars 7000 light-years away. The HST image at right is a small part of this, only 0.63 light-years across. It contains seven white dwarfs (inside the blue circles) among the other much brighter stars.

The bright star is now called *Sirius A*.

The next chapter will explain why there is an upper limit to the mass that a white dwarf can have.

companion has a mass about the same as Sirius. Eighteen years later—and sixteen years after Bessel had died—the American telescope maker Alvan G. Clark (1804-1877) discovered the companion of Sirius and, in doing so, became the first person to see a white dwarf. ➤Figure 14-21(a) shows Sirius B, as the companion is called, as it appears in a much larger telescope. Part (b) shows seven white dwarfs in M4, a compact cluster of stars.

More than 300 stars have been identified as white dwarfs, but none is bright enough to be seen by the unaided eye. Because they are so difficult to see, it is hard to know what percentage of stars are white dwarfs. Astronomers estimate about 10%.

White dwarfs are the cores of red giants that remain after the outer parts of the original stars have blown away. Lighter ones are made up of helium and carbon, but white dwarfs that were formed from more massive stars contain oxygen, neon, and sodium and may contain atomic nuclei as massive as iron. Because white dwarfs are composed of degenerate matter, the more massive a white dwarf is, the smaller it is. (Refer to the graph of Figure 14-11b.) As a white dwarf gets older, its temperature and luminosity decline without a significant change in its size, for electron degeneracy prevents it from contracting further.

White dwarfs have been observed with surface temperatures from 4000 K to 85,000 K, but computer models predict that it is possible for them to have even higher temperatures. The masses of white dwarfs range from perhaps 0.02 solar masses up to 1.4 solar masses. Since the typical white dwarf is comparable in size to the Earth, the material of these stars is tremendously dense. The density of a white dwarf is about 10^6 grams per cubic centimeter. A teaspoon of white dwarf material would weigh two tons! Table 14-3 lists the properties of a "typical" white dwarf.

TABLE 14-3 A Typical White Dwarf

Absolute magnitude	+11
Mass	0.8 solar mass
Diameter	10,000 km (¾ Earth diameter)
Density	10^6 gm/cm^3
Surface temperature	15,000 K
Surface gravity	400,000 times Earth's

When the Sun becomes a white dwarf, its mass will be about 0.6 of its present mass. The remainder of its material will have been puffed away during its red giant stage and blown away during its planetary nebula stage. It will be nearly as small as the Earth (➤Figure 14-22).

To get to the white dwarf stage, Sun-like stars have gone through the following stages: protostar, main sequence star, red giant, and planetary nebula. Protostars produce energy from gravitational contraction. Main sequence stars produce energy from nuclear fusion. Red giants produce energy from gravitational contraction of their cores (until the cores become degenerate) and from fusion within shells around the cores. White dwarfs, however, do not produce energy. They are hot because of leftover energy, and as they radiate this energy away, they get cooler. And cooler. After billions of years, a white dwarf will have cooled enough that it no longer radiates in the visible region of the spectrum. It will appear on the H-R diagram at a position below and to the right of the bottom of the main sequence. As billions more years pass, it will cool further to become a ***black dwarf***, the burned-out cinder of a once-proud star. It is unlikely that the universe is old enough for many, if any, black dwarfs to have formed.

➤Figure 14-23 reviews the evolutionary steps taken by flyweight and lightweight stars. All begin as protostars and end up as black dwarfs. The end state of more massive stars—the middleweight and heavyweight categories—is even more exotic, but must await the next chapter for discussion.

black dwarf. The theorized final state of a star with a main sequence mass less than about 8 solar masses, in which all of its energy sources have been depleted so that it emits no radiation.

➤ FIGURE 14-22. When the Sun becomes a white dwarf, it will be slightly larger than today's Earth and therefore much smaller than the present Sun. Its mass, however, will be 0.6 of what it is now.

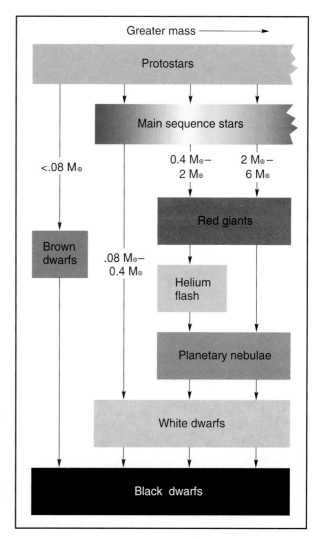

➤ FIGURE 14-23. This diagram shows the evolutionary steps taken by the stars discussed in this chapter. Unless we include planets in the category, it is probable that no black dwarfs have ever formed, so their box could have been left off the chart. The boxes for the top two categories are open at the right to indicate that more massive stars also fall into these categories. Keep in mind that the limiting masses in each case are known only approximately.

HISTORICAL NOTE

Tycho Brahe's Nova

Tycho Brahe had been taught that the heavens were perfect and unchanging. Therefore, when he saw a "new star"—a nova—he was very excited. His writings reveal his feelings:

> In the evening, after sunset, when, according to my habit, I was contemplating the stars in a clear sky, I noticed that a new and unusual star, surpassing all the other stars in brilliancy, was shining almost directly above my head; and since I had, almost from boyhood, known all the stars of the heavens perfectly (there is no great difficulty in attaining that knowledge), it was quite evident to me that there had never before been any star in that place in the sky, even the smallest, to say nothing of a star so conspicuously bright as this. I was so astonished at this sight that I was not ashamed to doubt the trustworthiness of my own eyes. But when I observed that others, too, on having the place pointed out to them, could see that there really was a star there, I had no further doubts. A miracle indeed, either the greatest of all that have occurred in the whole range of nature since the beginning of the world, or one certainly that is to be classed with those attested by the Holy Oracles.

Novae

Ancient astronomers believed that the sky never changed. Chapter 12 revealed, however, that stars move and that the constellations gradually change shape. Other changes also occur in the heavens. In 1572, Tycho Brahe observed a "new star." His excitement at the discovery is revealed in his words in a nearby Historical Note. We know now that Tycho's observation, and others since his time, are not actually new stars, but are a phenomenon associated with white dwarfs.

In A.D. 1054, Chinese scholars recorded what they called a "guest star," a new star never before seen. The next chapter will discuss this star, which was bright enough that it could be seen in the daytime.

Consider a binary star system in which one star is more massive than the other. The more massive star will end its main sequence life before its smaller companion. The massive star will then become a red giant, and, if it is in our lightweight category, it will shed its outer portion and become a white dwarf. This will leave a white dwarf in orbit with a main sequence star. The latter star then ends its fusion-burning life and begins to grow into a red giant.

At some point, the material of the outer portion of the new red giant is attracted to the white dwarf with more force than is exerted from its own core. This material will be pulled from the giant. Although some might fall directly into the white dwarf, analysis shows that most of the material pulled from the red giant will go into orbit around the white dwarf, forming an **accretion disk** (▶ Figure 14-24). In the accretion disk, matter swirling around the white dwarf is heated as it falls inward, another example of gravitational energy being changed into heat. Collisions within the disk cause its material—mostly hydrogen from the outer portions of the red giant—to fall inward onto the surface of the white dwarf. The hydrogen builds up there, becoming denser and hotter until it

accretion disk. A rotating disk of gas orbiting a star, formed by material falling toward the star.

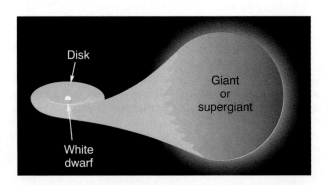

▶ FIGURE 14-24. In a binary system composed of a white dwarf and a red giant, material from the growing giant will fall toward the dwarf and go into orbit around it.

► FIGURE 14-25. These photos of Nova Cygni 1975 show its dimming from magnitude 2 at maximum light to magnitude 15.

TABLE 14-4 A Typical Nova

Luminosity increase	10,000 times
Absolute magnitude	-8, or 10^5 Suns
Time to brighten	A few days
Time to dim	6 months to one year

ignites in an explosive fusion reaction. The explosion blows off a shell of gas from the white dwarf, and although the shell contains only a tiny amount of mass (perhaps 0.0001 solar masses), it can cause the white dwarf to become 10,000 times brighter (10 magnitudes) within a few days.

People of Brahe's time thought such a newly brightened star was a new star and called it a **nova,** the Latin word for *new*. In actuality, the star is not new, but is simply the sudden brightening of an old star. ►Figure 14-25 shows one of the brightest of recent novae, and Table 14-4 contains data for a typical nova.

Because so little material is blown off during a nova, the explosion does not disrupt the binary system. The companion star soon resumes transferring matter to the white dwarf. Depending on how fast the hydrogen is transferred, fusion can be re-ignited as quickly as a few months later, or 10,000 to 100,000 years may be required for a recurrence of the nova. Because of their repeating nature, these novae are called recurrent novae.

The Chandrasekhar Limit

When the material from its companion falls onto the surface of a white dwarf, the white dwarf gains mass. Recall, however, that as an electron-degenerate object gains mass, it *decreases* in size and therefore becomes even more compacted. As it becomes more compacted, the gravitational forces tending to crush it even smaller become greater and greater. Electron degeneracy supports the white dwarf against collapsing completely, but there is a limit to the amount of pressure that degenerate electrons can withstand. The limit is reached when the white dwarf achieves a mass of 1.4 solar masses. In 1930, a 19-year-old student on his way from his native India to college in England calculated the limit. Subrahmanyan Chandrasekhar (or, as he is known to astronomers, "Chandra") was

nova (plural **novae**). A star that suddenly and temporarily brightens, thought to be due to new material being deposited on the surface of a white dwarf.

There are various types of novae, including *classical novae* and *dwarf novae*, but these differences are not important here.

Many major breakthroughs in science have been made by young persons. Newton's and Einstein's achievements are famous examples.

White Dwarfs

Chandrasekhar limit. The limit to the mass of a white dwarf star above which it cannot exist as a white dwarf. Above that limit it will not be supported by electron degeneracy.

awarded a share of the 1983 Nobel Prize in Physics for his discovery, and the limit is known as the ***Chandrasekhar limit.***

Chandra's calculations showed that if a white dwarf became more massive than 1.4 solar masses, the pressure due to gravity at its surface would be greater than the maximum pressure a degenerate gas can support. This means that white dwarfs must have masses less than 1.4 times the Sun's mass. Main sequence stars with masses of up to about 6 solar masses can end up as white dwarfs only because they lose mass during the red giant and planetary nebula stages. Stars more massive than these stars (which we have classified as lightweight stars) retain cores more massive than the Chandrasekhar limit and cannot form white dwarfs. Such middleweight and heavyweight stars end their lives in a dramatically different way than do less massive stars. The next chapter will discuss the more massive stars.

TYPE I SUPERNOVAE

White dwarfs are composed mostly of degenerate carbon (but they probably contain heavier elements as well). Imagine a white dwarf in a binary system accreting material from its companion. When the accretion brings the mass of the white dwarf above the Chandrasekhar limit, electron degeneracy can no longer support the star, and it collapses. The collapse raises the core temperature enough that carbon fusion suddenly begins. This is somewhat similar to the helium flash during the red giant stage, except that now carbon rather than helium is the fuel. Runaway fusion reactions result in carbon fusing to elements as massive as iron. The heat produced destroys electron degeneracy, and the white dwarf explodes. Whereas a red giant undergoing a helium flash has a surrounding envelope of gases to control its explosion, the white dwarf doesn't. The star explodes completely.

supernova. The catastrophic explosion of a star during which the star becomes billions of times brighter.

The destruction of a white dwarf in this manner is called a **supernova.** The name is unfortunate, for a supernova is not just a "super nova." A nova might reach an absolute magnitude of −8 (about 100,000 Suns), but a supernova attains a magnitude of −19 (10 *billion* Suns). When a supernova is observed in another galaxy, it may shine brighter than all the rest of the galaxy combined.

Astronomers classify supernovae into two categories, *Type I* and *Type II*, depending upon their spectra. The spectrum of a Type II supernova contains prominent hydrogen lines, but the spectrum of a Type I supernova contains none. Since white dwarfs contain no hydrogen, they are thought to be responsible for Type I supernovae. Type II supernovae result from the explosion of a single star and will be discussed in the next chapter.

We know now that the "nova" observed by Tycho was actually a supernova.

A Type I supernova reaches maximum brightness after a few days. It fades quickly for about a month and then declines in brightness more gradually until it dissipates in about a year. ➤Figure 14-26 shows the light curve of a Type I supernova. Theoretical models indicate that its energy (after the initial explosion) comes from radioactive decay of nuclei produced in the explosion.

What happens to the remains of a supernova? They disperse into space. ➤Figure 14-27 is a photograph of the Gum nebula, the remains of a supernova that exploded some 11,000 years ago. To calculate its age, astronomers determined the Doppler shift of the part of the remnants that is moving toward us and calculated when the explosion had to have occurred to produce the present speed. Chapter 15 will discuss Type II supernova and describe examples of past and recent supernovae.

Photos of other supernova remnants appear in Chapter 15.

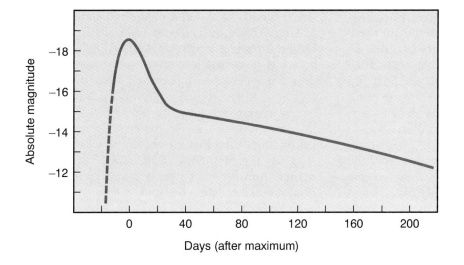

► **FIGURE 14-26.** A Type I supernova, thought to be produced by the destructive explosion of a white dwarf, reaches peak brightness in a few days and then declines in brightness, first quickly and then more slowly. The gradual decrease is due to the decay of radioactive elements in the products of the explosion.

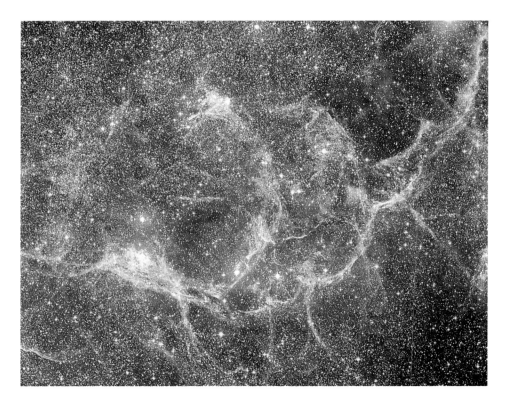

► **FIGURE 14-27.** The Gum nebula is a supernova remnant. The nearest part of it is about 300 light-years from us.

CONCLUSION

Mass is the single most important factor that determines a star's properties and how the star evolves. The least massive protostars never get hot and dense enough to join the main sequence and become hydrogen-fusing stars. These brown dwarfs heat up during their protostar stage, but after that they have no heat source, so they cool to become dark, planet-size cinders. More massive protostars become stars on the main sequence, where they fuse hydrogen to helium (and, in massive stars, to heavier elements). During their main sequence lives, they become slightly more luminous and cooler, resulting in a small change in their position on the H-R diagram. The more massive a star is, the less time it takes to exhaust its fuel and end its main sequence life.

In stars massive enough to support fusion but with masses less than about 0.4 solar masses, their entire supply of hydrogen mixes as fusion takes place. Although the universe is not old enough for any of these stars to have had time to complete their main sequence lives, computer models show that when they run low on hydrogen, they will shrink and heat up until electron degeneracy sets a lower limit to their size. They will then live out their lives as white dwarfs until they, too, cool to be burned-out cinders.

More massive stars fuse hydrogen in their cores without any intermixing with the outer portions of the stars, and when they run low on fuel, they become brighter, larger, and cooler. That is, they become red giants. Their electron-degenerate cores heat up until they become hot enough to begin helium fusion. When the Sun (and other stars about its mass) reach this point, a helium flash occurs, and they consume their helium quickly, getting hotter and dimmer for a time.

During the red giant stage, stars lose great amounts of matter from their outer layers because of intense stellar winds. At the end of the red giant stage, Sun-like stars blow away their outer shells and become beautiful planetary nebulae. Left behind when the nebula has dispersed into space is the star's white-hot degenerate core—a white dwarf. White dwarfs are dim and hard to see, but a white dwarf that is part of a binary system will advertise its presence as a nova if its companion, as it swells to become a red giant, leaks material onto the hot surface of the white dwarf.

When the mass of a white dwarf that is accreting matter from a companion becomes greater than the Chandrasekhar limit, the star explodes in a brilliant supernova. This completely destroys the star, dispersing its radioactive remnants through space.

Stars with masses greater than about 6 solar masses end the red giant portion of their lives in a different way than stars of lower mass. The next chapter will describe the behavior of these stars, including the supernovae that they produce.

RECALL QUESTIONS

1. The length of a star's main sequence lifetime is determined by the star's
    A. carbon content.
    B. distance from the center of the galaxy.
    C. surface temperature.
    D. mass.
    E. spectral type.

2. Why do stars of great mass live longer on the main sequence than stars of lesser mass?
    A. The massive stars have more hydrogen fuel.
    B. The massive stars burn their fuel more slowly.
    C. The massive stars go through many stages of fusion.
    D. [More than one of the above.]
    E. [None of the above; the statement in the question is false.]

3. If the rate of hydrogen fusion within the Sun were somehow to increase, the core would
    A. collapse and the Sun would grow cool.
    B. collapse and heat up further.
    C. expand and therefore tend to slow the fusion.
    D. expand and therefore increase in temperature.
    E. stay the same size but become hotter.

4. Why does hydrogen fusion occur only in a star's center?
    A. Only near the center is there enough heat and pressure.
    B. Only near the center is there enough hydrogen that is not mixed with other elements.
    C. Only near the center is the speed of light favorable for the reaction to occur.
    D. Heat is transferred down to the center during the main sequence life of the star.
    E. [The statement is false; fusion occurs throughout the star's volume. This is what causes the surface to be bright.]

5. Why do massive stars run out of hydrogen in their cores faster than less massive stars?
    A. Their hydrogen fuses faster because of greater pressure.
    B. There is less hydrogen in their cores.
    C. The cores of less massive stars contain a greater percentage of helium, which slows hydrogen fusion.
    D. The cores of less massive stars contain a lesser percentage of helium, which slows hydrogen fusion.
    E. [The statement is false; more massive stars do not run out of hydrogen faster than stars of less mass.]

6. The Sun will at some time in the future become
    A. a red giant.
    B. a white dwarf.
    C. a black dwarf.
    D. [All of the above.]
    E. [None of the above.]

7. Which of the following lists the stages in a star's life in correct order?
    A. Main sequence, protostar, white dwarf, red giant
    B. Protostar, red giant, main sequence, white dwarf
    C. Protostar, main sequence, white dwarf, red giant
    D. White dwarf, protostar, main sequence, red giant
    E. Protostar, main sequence, red giant, white dwarf

8. Red giants are more luminous than white dwarfs because
    A. red giants are hotter.
    B. red giants are closer.

C. red giants are larger.
D. [All of the above.]
E. [None of the above; the statement is not true.]

9. Why are all known white dwarfs relatively close to the Sun?
   A. White dwarfs are formed only in our neighborhood of the Galaxy.
   B. Light from distant white dwarfs has not yet reached the Earth.
   C. No white dwarfs are bright enough to be seen at great distances.
   D. Light from distant white dwarfs is too redshifted to be seen.
   E. [The statement is false; white dwarfs are seen at all distances from the Sun.]

10. What is the difference between the Sun and a one-solar-mass white dwarf?
    A. The Sun is larger.
    B. The Sun has more hydrogen.
    C. They have different energy sources.
    D. [All of the above.]
    E. [None of the above.]

11. One of the causes for the phenomenon called a nova is
    A. the fusion of iron in the core of a massive star.
    B. the infall of material onto a neutron star from a white dwarf.
    C. the transfer of material onto a white dwarf in a double star system.
    D. the collapse of a protostar.
    E. the death of a massive star and the formation of a black hole.

12. The nuclear reactions in a star's core are kept under control so long as
    A. the star's luminosity depends on its mass.
    B. the pressure of the gas in the core depends on its temperature.
    C. the star's density depends on its mass.
    D. the star's mass depends on its temperature.

13. A planetary nebula is
    A. the vastly expanded shell of a dying star.
    B. a cloud of gas out of which stars form.
    C. a cloud of cold dust in space.
    D. the same as a white dwarf.
    E. a circular ring around a black hole.

14. Compared to a young star of the same mass, an older star contains
    A. more hydrogen.
    B. more elements heavier than hydrogen.
    C. about an equal amount of each element.
    D. more of every element.
    E. [No general statement can be made.]

15. The mass of a white dwarf is
    A. always greater than 4 solar masses.
    B. always less than 0.5 solar masses.
    C. less than the mass of its original main sequence star.
    D. greater than the mass of its original main sequence star.
    E. [Two of the above.]

16. A brown dwarf is
    A. the final fate of all stars.
    B. the final fate of stars like the Sun, but not of less massive stars.
    C. a stage of a star's life prior to the white dwarf stage.
    D. a stage of a star's life after the white dwarf stage.
    E. a warm starlike object that has too little mass to support fusion in its core.

17. When more material falls into a white dwarf, its size
    A. increases.
    B. remains the same.
    C. decreases.
    D. [Any of the above, depending upon the nature of the white dwarf.]

18. When the core of a star shrinks after hydrogen fusion stops,
    A. the core cools and the star expands.
    B. the core cools and the star contracts.
    C. the core heats and the star expands.
    D. the core heats and the star contracts.

19. The Chandrasekhar limit is a limit to a star's
    A. size.
    B. volume.
    C. density.
    D. mass.
    E. [Both A and B above.]

20. Some supernovae occur when the Chandrasekhar limit is exceeded in a white dwarf. In this case, the energy producing the explosion is
    A. chemical in nature.
    B. nuclear fusion.
    C. nuclear fission.
    D. degenerate electrons.
    E. gravity.

21. What property of a star is most important in determining the stages of its evolution?

22. Why do stars of the lowest mass not become red giants?

23. What causes the core of a star to heat up after its hydrogen fusion ceases?

24. List the stages in the life of a one-solar-mass star.

25. What will happen to the Earth when the Sun becomes a red giant? How large will the Sun become?

26. Starting at the center and going outward, list the various layers of a red giant that contains a carbon core.

27. Describe two hypotheses designed to explain what causes a planetary nebula. Describe the appearance of a planetary nebula. What causes it to glow?

28. Describe a white dwarf with regard to size, mass, luminosity, and temperature.

29. What supports a white dwarf from further collapse?

30. What happens to the size of a main sequence star when mass is added to it? What happens to the size of a white dwarf when mass is added to it?

31. Why does a recurrent nova continue to flare up time-after-time?

QUESTIONS TO PONDER

1. Massive stars have much more hydrogen in their cores than do less massive stars. Why, then, do they run out of hydrogen faster than stars of low mass?
2. We see relatively few white dwarfs compared to main sequence stars, but astronomers are confident that white dwarfs are very common. Explain this discrepancy.
3. What determines whether the carbon cycle will occur in a particular star?
4. Explain why a star becomes cooler as it ages on the main sequence.
5. Explain how star clusters provide evidence that supports our theories of what happens to stars after their main sequence life ends.
6. What causes a star to get larger as it becomes a red giant? What causes its surface to cool?
7. Explain why a Sun-like star, once it becomes a red giant, undergoes a helium flash, but a star a few times more massive than the Sun begins helium fusion gently.
8. Why are novae and Type I supernovae not included in Figure 14-23, which shows the evolutionary steps of normal stars?

CALCULATIONS

1. Compared to the Sun's life expectancy, how long would a star of 10 solar masses remain on the main sequence? (Hint: See the Close Up "Lifetimes on the Main Sequence.")
2. The Sun's life expectancy on the main sequence is expected to be 10 billion years. What is the life expectancy of a star with a mass half that of the Sun? (Hint: See the Close Up "Lifetimes on the Main Sequence.")

Visit the netQuestions area of StarLinks (www.jbpub.com/starlinks) to complete an exercise on this topic:

1. **Star Evolution** When our Sun dies, it will expand into a red giant or super-giant and then collapse to form a white dwarf. For stars more massive than the Sun, the end is much more dramatic.

CHAPTER 15

The great supernova of 1987 (right of center) occurred near the Tarantula nebula (above center).

THE DEATHS OF MASSIVE STARS

Middleweight and Heavyweight Stars
Type II Supernovae
Detecting Supernovae
CLOSE UP: *Supernovae from Lightweight Stars?*
SN1987A
Theory: The Neutron Star
Observation—The Discovery of Pulsars
CLOSE UP: *The Pulsar in SN1987A?*
Theory: The Lighthouse Model of Neutron Stars/Pulsars
Observation—The Crab Pulsar and Others

Middleweight Conclusion
General Relativity
A Binary Pulsar
CLOSE UP: *The Distance/Dispersion Relationship*
The Heavyweights
Black Holes
Properties of Black Holes
Detecting Black Holes
Our Relatives—The Stars
CLOSE UP: *Black Holes in Science, Science Fiction, and Nonsense*

•••••••• THE FOLLOWING QUOTATION *from Stephen Maran indicates the excitement SN1987A generated among astronomers:*

> February 25, 1987, was a quiet day at the Goddard Space Flight Center outside Washington. I was walking down a hallway toward the Coke® machine when I overheard the excited comment, "It's the worst one in 300 years!" A group of scientists were milling around, handing a telegram back and forth, and talking excitedly as new arrivals joined them. "My God," I thought, "there must have

been a horrible volcanic eruption." But I had misheard. It was not an Earthly disaster but a cosmic cataclysm, spotted the day before from South America, that was arousing normally unflappable astronomers to a level of excitement that, for some, approached ecstasy. It was not the "worst" but the "first" and not a volcanic eruption but the explosion of a star in our own backyard."

*From Stephen P. Maran, "A Blue Supergiant Dissects Itself in a Cosmic Explosion," Smithsonian, April 1988, pp. 46–47.

T he previous chapter described how stars of low mass, those with masses less than about six solar masses, evolve from protostars to white dwarfs. In general, we might say that such stars end their lives in a fairly uneventful manner. Although all but the lightest become red giants and some put on beautiful displays as planetary nebulae, all-in-all, lightweight stars die not with a bang, but with a whimper. If a white dwarf happens to be part of a binary system, things may get explosive, however. The resulting novae and particularly supernovae give a preview of what happens when massive stars die.

You might say that the smallest stars don't die—they just fade away.

The death of a massive star is truly a major event. Instead of gently puffing their outer layers away, some stars blow their outer portions away in cataclysmic supernova explosions such as the one Stephen Maran describes in the opening vignette. Many end up as superdense neutron stars, sending out powerful lighthouse beacons across space. The most massive, however, end their lives by swallowing themselves—becoming black holes.

MIDDLEWEIGHT AND HEAVYWEIGHT STARS

Although the final destinies of middleweights (perhaps 6 to 12 solar masses) and heavyweights (greater than about 12 solar masses) are far different, they behave similarly during most of their lives, and so they will be considered together for now.

The least massive lightweight stars experience a helium flash rather than a gradual change.

The stars that have the most mass live very short main sequence lives. A star 15 times as massive as the Sun takes only about 10 million years to burn up its hydrogen. Like a less massive star, it then expands to become a red giant. During its red giant stage, the 15-solar-mass star changes gradually from a hydrogen-fusing central core to one that fuses helium. Because of its greater mass and corresponding greater core temperature, it becomes brighter than the standard red giant, and we call it a **supergiant.** The most prominent supergiant is Betelgeuse, the bright star in Orion's right shoulder (➤Figure 15-1). A supergiant may be a million times brighter than the Sun and have an absolute magnitude of −10. ➤Figure 15-2 shows the position of red supergiants on the H-R diagram, and Table 15-1 lists some typical data for a supergiant.

supergiant. The evolutionary stage of a massive star after it leaves the main sequence.

Because of the greater temperatures and pressures inside red supergiants, fusion continues to produce heavier elements, including neon, silicon, and even iron. As each of these elements is produced, energy is released, the core heats, and the star expands and shifts on the H-R diagram. The various reactions and their approximate longevity are shown in Table 15-2 for a 15-solar-mass star. The core of the red supergiant now contains a number of layers, as illustrated in ➤Figure 15-3.

TYPE II SUPERNOVAE

Based on their spectra, we divide supernovae into two types. A Type I supernova has no hydrogen lines in its spectrum, and most supernovae of this type are

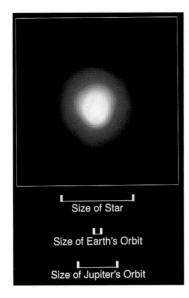

➤ **FIGURE 15-1** This HST image of Betelgeuse is the first direct image of a star (other than the Sun). It reveals a huge atmosphere with a mysterious hot spot on the star's surface. In the constellation Orion (Figure 1-4 on page 16), Betelgeuse is the bright star in the hunter's right shoulder.

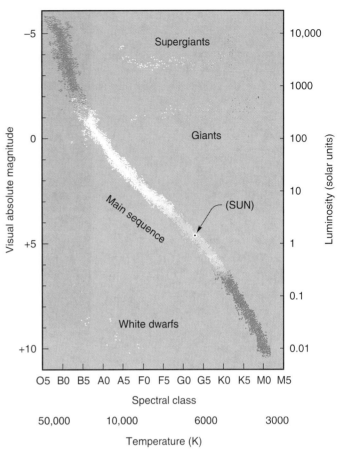

➤ **FIGURE 15-2** Supergiants have about the same surface temperatures as red giants, but they are much brighter.

TABLE 15-1 A Typical Supergiant

Absolute magnitude	−5
Luminosity (solar units)	10,000
Mass (solar units)	12
Diameter (solar units)	1200
Average density	10^{-8} gm/cm^3 (10^{-8} Sun's density)
Surface temperature	3600 K

TABLE 15-2 The Evolution of a 15-Solar-Mass Star*

Element Fused	Fusion Products	Time	Temperature
Hydrogen	Helium	10,000,000 years	4,000,000 K
Helium	Carbon	>1,000,000 years	100,000,000 K
Carbon	Oxygen, neon, magnesium	1000 years	600,000,000 K
Neon	Oxygen, magnesium	A few years	1,000,000,000 K
Oxygen	Silicon, sulfur	1 year	2,000,000,000 K
Silicon	Iron	A few days	3,000,000,000 K

*Adapted from Philip Flower, *Understanding the Universe* (St. Paul, Minn.: West Publishing Co., 1990).

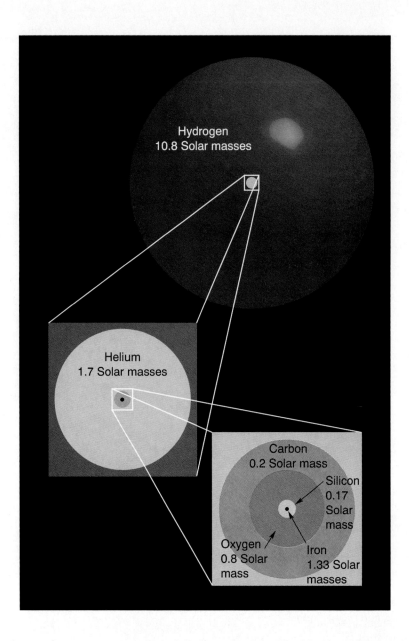

► **FIGURE 15-3** Most of the volume of a 15-solar-mass red giant is hydrogen, but the central part is multilayered. The hypothesized mass of each layer is shown.

TABLE 15-3 Typical Supernovae

	Type I	Type II
Spectrum	No hydrogen lines	Prominent hydrogen lines
Magnitude at peak	−19	−17
Light curve	Sharp peak	Broader peak
Expansion rate	10,000 km/sec	5000 km/sec
Mass ejected	0.5 solar masses	5 solar masses

thought to be produced by white dwarfs in binary systems, as described in the previous chapter. The spectra of Type II supernovae, on the other hand, reveal prominent hydrogen lines, indicating that these supernovae result from stars that still contain hydrogen in their outer layers. Differences between the two types of supernovae are outlined in Table 15-3 and typical light curves are shown in ►Figure 15-4.

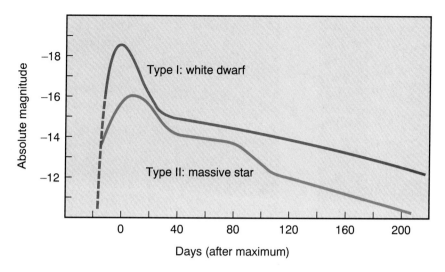

➤ **FIGURE 15-4** Typical light curves from the two types of supernovae are shown. Type I supernovae get brighter and have a sharper peak. The curve is dashed at the beginning, during the time that the luminosity of the star is usually not observed.

The process by which Type II supernovae occur is not well understood, and various models have been proposed. According to the leading model, once the core of a supergiant has reached the stage where silicon has formed and begins fusing to iron, things start to happen quickly. Computer modeling indicates that the silicon layer will fuse to iron in only a few days. Considering that stellar lifetimes are measured in millions and billions of years, a few days is a remarkably short time. But after this, things change even faster!

Silicon fuses to iron in the center of the star. Previously, when a new central core (first of helium, then carbon, then oxygen, and finally silicon) was formed, the material of that core began to fuse to heavier elements when it got hot enough. Iron is different. All of the fusion reactions in Table 15-2 produce energy when they occur. When silicon nuclei fuse to form iron, for example, energy is produced. If iron were to fuse together to form even more massive nuclei, however, energy would be *absorbed* instead of released. In other words, if iron is to fuse, it must have a supply of energy. Therefore, as the core iron shrinks and heats up, it does not fuse to something more massive. Instead, once its mass reaches the Chandrasekhar limit, the core collapses violently. In much less than a second, the tremendous pressure generated by gravitation pushes pairs of electrons and protons together to form a neutron and a neutrino in each case.

Neutrinos were described in Chapter 11 during the discussion of fusion reactions in the Sun.

Computer simulation indicates that when the core reaches the minimum size that is possible, it rebounds—its surface bounces outward. Meanwhile the outer layers of the core are falling inward to fill the gap left when the iron core shrank. The rebounding surface collides violently with the infalling material, and produces two effects: First, the collision is energetic enough to cause iron to fuse into heavier elements. (Recall that this requires an *input* of energy. The necessary energy comes from the collision.) Second, the collision sends shock waves outward that throw off the outer layers of the supergiant. Some astronomers question whether the shock wave is energetic enough to cause the star to explode. One theory holds that as the shock wave begins to lose its energy, it is heated by great numbers of neutrinos escaping the core. The heating action takes less than a second, and at the end of it, the newly energized shock wave has enough energy to blow away the outer portion of the star.

Supernova theory is still young, and advances will have to be made before it can be considered well founded.

We know that a major fraction of the energy of a supernova appears in neutrinos.

Some Type II supernovae are more luminous than others, but a typical one reaches a peak absolute magnitude of −17. This is nearly a billion times brighter than the Sun! ➤Figure 15-5 shows a supernova that occurred in the galaxy M81 in March, 1993. The other dots in the photograph are not part of M81, but are

Type II Supernovae   453

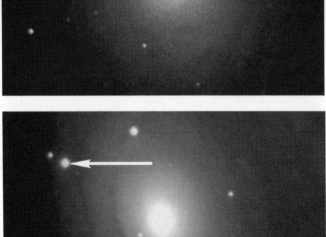

(a)

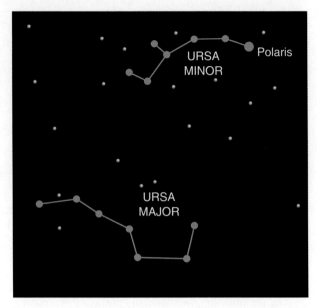

(b)

> **FIGURE 15-5** (a) Supernova 1993J was discovered in a spiral arm of the galaxy M81 on March 28, 1993. It is identified by an arrow in the lower photo, taken March 30, when it had an apparent magnitude between 10 and 11. Notice that nothing appears at that spot in the upper photo, which was taken earlier. The other dots in the image are stars in our Galaxy and are not associated with M81. (b) M81 is visible in binoculars at this location in Ursa Major.

stars in our galaxy. Except for the supernova, stars in M81 show up only as a haze.

Physicists have long recognized the special place that iron holds in the list of chemical elements. When lighter nuclei are fused to form heavier nuclei, energy is produced, but only so long as the newly formed nucleus is not more massive than iron. Elements heavier than iron cannot be produced without some source of energy, so they are not formed spontaneously in nature. What, then, is the source of the heavy elements that we find here on Earth? Supernovae provide the answer. Supernova explosions in the distant past produced the heavy elements and blasted them away into space to become part of the interstellar material from which the Earth was formed.

Detecting Supernovae

Three supernovae have been seen with the naked eye in our galaxy; they occurred in the years 1054, 1572, and 1604. The most spectacular on record occurred in the constellation Taurus on July 4, 1054, and was observed by Chinese astronomers who reported that it was bright enough to be seen in daylight and to read by at night. It remained bright for a few weeks, then gradually faded until it disappeared from view after about two years. Invention of the telescope was centuries away, of course, so the Chinese of the eleventh century had no way to continue observing their "guest star" (as they called the object). In 1731, an amateur astronomer reported a small nebula in Taurus, and two hundred years later Edwin Hubble discovered that the nebula consists of material expanding at a rate such that it must have begun its expansion at the time the

Since the Crab nebula is 7000 light-years from us, the supernova actually occurred about 8000 years ago (7000 + the 950 years since 1054), but we speak of it as if it happened about 940 years ago.

Edwin Hubble (1889-1953) is an important American astronomer whose name will appear often in the next two chapters.

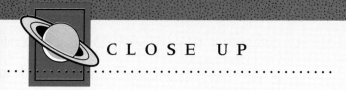

CLOSE UP

Supernovae from Lightweight Stars?

A white dwarf that is accreting material from a binary companion continues to do so until its mass reaches the Chandrasekhar limit, whereupon it collapses and suddenly initiates carbon fusion. A Type I supernovae is the result. Some stellar models indicate that the same process might be responsible for Type II supernovae in red giants of 5 to 6 solar masses. Consider the point in its evolution when such a star is a red giant with a carbon core. Surrounding the carbon core is a shell of helium that is fusing together into carbon and dropping the carbon onto the core. If this model is correct, a 5-solar-mass star would form enough carbon that the mass of its core could reach the Chandrasekhar limit. When this occurs, the carbon core collapses, heats up, and undergoes violent carbon fusion.

According to the model, the sudden detonation of carbon is powerful enough to blast the remainder of the star away in a supernova explosion. Since the outer portions of the red giant are made up primarily of hydrogen, the spectrum of the hot expanding shell of gas will contain hydrogen lines and will have the characteristics of a Type II supernova.

The mass of the star discussed here is only 5 to 6 solar masses, which puts the star in the category of lightweight stars, a group described in the last chapter as evolving into planetary nebulae and then ending up as white dwarfs. If the model presented here is correct, some may die in a much more explosive way.

Chinese reported the guest star. ➤Figure 15-6 is a photograph of the supernova remnant, called the Crab nebula because of its shape. Telescopic observations taken over the last half-century show its growth; they reveal that its outer portions are still expanding outward at about 1400 kilometers per second and that it is now about 4.4 light-years in diameter.

We have also been able to find remnants of the supernova seen by Tycho Brahe in 1572 (which provided evidence to him that the heavens were not unchanging) and the one seen by Johannes Kepler in 1604. Kepler's supernova

➤ FIGURE 15-6 The Crab nebula is the remnant of a supernova that was seen in the year 1054.

occurred just before the invention of the telescope, and since that time there have been no visible supernovae in our galaxy.

Table 15-4 lists some major supernova remnants. Images of two parts of one of these—the Cygnus loop—appear in ➤Figure 15-7.

TABLE 15-4 Some Supernova Remnants

Remnant	Distance (Light-Years)	Diameter (Light-Years)	Age (Years)
Crab nebula	6000	4.4	940
Cygnus loop	2500	100	20,000
Gum nebula	300*	2300	11,000
Tycho's supernova	9800		420
Kepler's supernova	20,000		390

*This is the distance to the nearest part of the nebula.

➤ FIGURE 15-7 (a) The Veil nebula, shown here, is part of the Cygnus loop, a supernova remnant 2500 light-years away. (b) This is a false-color image from the Hubble telescope of the eastern edge of the Cygnus loop.

▶ FIGURE 15-8 The photo at the left was taken before SN1987A occurred. The supernova is obvious in the other photo.

SN1987A

When a new astronomical object is observed, a telephone call or telegram is sent to the International Astronomical Union's Central Bureau for Astronomical Telegrams in order to establish priority of discovery. Telegram number 4316, received on February 24, 1987, read:

> W. Kunkel and B. Madore, Las Campanas Observatory, report the discovery by Ian Shelton, University of Toronto, of a mag 5 object, ostensibly a supernova, in the Large Magellanic Cloud. . . .

The opening of the Prologue describes how Shelton made his discovery.

The supernova (▶Figure 15-8) is officially known as SN1987A (*A* because it was the first discovered that year), but it is sometimes called *Supernova Shelton* after its discoverer.

One of the first things that astronomers did after the discovery of SN1987A was to examine recent photographs of the region where it occurred to determine which star was its source. It is fortunate that the supernova occurred in a region of the Large Magellanic Cloud where new stars were known to be forming, for this meant that astronomers already had an interest in the region and many photographs of it could be found. Two very luminous blue stars were found very close together at the location where the supernova occurred. This caused confusion, because it was thought that red giants, not blue stars, explode as supernovae. The explanation appeared when a shell of gas was detected (by the *International Ultraviolet Explorer* satellite) about a light-year from the central star. It appears that the shell consists of material that was shed from what was originally a red supergiant with a mass about 20 times that of the Sun. The star ejected this material about 20,000 years ago, leaving behind its hotter, blue surface. Since 20,000 years is little more than an instant in a star's life, we can say that the red supergiant ejected its surface material shortly before it exploded as a supernova.

The star that formed SN1987A (the "progenitor" star) was Sanduleak—69°202, so named because it is the 202nd star in the 69th degree south of the celestial equator ($-69°$ declination) in a catalog compiled by Nicholas Sanduleak.

Supernova theory had predicted that a burst of neutrinos would be emitted by the explosion. As discussed in Chapter 11, neutrinos are nuclear particles produced within stars and are the subject of research at various neutrino detectors around the world. Such devices make a record when they detect neutrinos. Our theory of supernovae was confirmed when neutrino researchers in the United States and Japan checked their records and reported that the number of neutrinos had increased a full three hours before the supernova was seen.

We continue to observe and learn from SN1987A, but there are still many mysteries about supernovae. Whatever their details, such events are singular ones in the lives of stars. In some cases, it appears that the entire star is blown apart, including the core; but in others, the core is left behind as a tiny remnant of the once-mighty star. The nature of this leftover core depends upon whether the star was originally in our middleweight or our heavyweight class. In either case, a unique, peculiar object is formed.

Besides its mass, other factors, such as the star's angular momentum, play a role in determining what the star leaves behind.

THEORY: THE NEUTRON STAR

The last few chapters have described much of the theory concerning stars' lives but have only occasionally cited evidence for it. (As explained in the previous chapter, much of the evidence of pre– and post–main sequence life comes from examining star clusters.) In addition, the many steps that accompanied the development of the theory have not been detailed. There is an interesting story, however, that illustrates a portion of the theory of stellar life cycles. It not only shows how the theory was confirmed, but also how the confirmation provided further knowledge—a common occurrence in science.

A hypothesis worked out in the 1930s predicted that after the mass of a star's core increases beyond the Chandrasekhar limit, the star will collapse further and that its electrons and protons will combine to form neutrons, resulting in a ***neutron star***. Just as electron degeneracy prohibits a white dwarf from collapsing under gravity, neutron degeneracy does the same for a neutron star. The hypothesis predicted that the remains of a middleweight star's collapsed core would become a neutron star—a tremendously compressed star with a mass between 1.4 and about 3 solar masses. Table 15-5 shows the properties of such a star. To try to imagine it, picture the entire mass of a star larger than the Sun compressed into a ball the size of a small city, about 20 kilometers across.

neutron star. A star that has collapsed to the point at which it is supported by neutron degeneracy.

The diameter of a typical neutron star is only 0.2% of the diameter of a white dwarf and the neutron star is a billion times more dense.

TABLE 15-5 A Typical Neutron Star

Mass	1.5 solar masses
Diameter	20 km (width of a small city)
Density	10^{15} gm/cm^3
Temperature	10,000,000 K

Astronomers had little hope of finding such a star because, despite its high temperature, its extremely small size results in it being very dim. Thus the idea was put on astronomy's back burner as an interesting hypothesis but one that seemed beyond our ability to confirm or deny. In 1967, an accidental discovery changed this.

Observation—The Discovery of Pulsars

In 1967, Jocelyn Bell (now Jocelyn Burnell), a graduate student in astronomy at Cambridge University in England, was working with Antony Hewish and a group of researchers who were searching for quasars, energetic stellar sources to be discussed in a later chapter. The radio telescope she was using for her research did not look at all like the giant radio telescope dishes normally associated with radio astronomy. Instead, it looked like a field of clothesline covering a total of 4½ acres (➤Figure 15-9). It was designed to detect faint radio sources and to see quick changes in their energy. In the course of research for her dissertation, Bell found a new, unexpected, and unexplained source of radio waves. The signal from it pulsed rapidly, about once every 1.3 seconds. This was a much more rapid pulsation than had ever been observed from a stellar energy source.

The researchers' first thought was that the waves Bell had received were of terrestrial rather than celestial origin. A check of local radio transmitters, however, failed to indicate a source. In addition, the signal was detected about four minutes earlier each night than the previous night. Recall that a given star sets four minutes earlier each night as a result of the Earth's moving around the Sun. The researchers concluded therefore that the source was in the sky and was not of human origin.

Their next thought was that a signal from an extraterrestrial race had been detected. In fact, the source was referred to for a short while as an LGM (the

➤ FIGURE 15-9 Part of the radio telescope that first detected a pulsar.

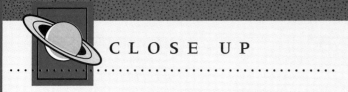

CLOSE UP

The Pulsar in SN1987A?

After the discovery of SN1987A in February 1987, astronomers began to look for a pulsar at its center. They were not particularly surprised that the pulsar had not been visible immediately because debris from the explosion would block our view of it.

On January 18, 1989, nearly two years after the supernova was discovered, an international team of astronomers headed by John Middleditch (of Los Alamos National Laboratory) observed the supernova from Cerro Tololo Inter-American Observatory in Chile for nearly seven hours, using a detector that is sensitive to visible and near-infrared light. The signals contained pulses! In order to check their procedure and their instruments, the experimenters turned the telescope to another object that they knew did not pulse. Indeed, no pulses were detected, and measurements of that object's brightness corresponded to previous brightness measurements. They announced that the pulsar in SN1987A had been discovered!

Astronomers were excited about the new pulsar for at least three reasons: First, the discovery confirmed the theory that links pulsars and supernovae. Second, the pulsar was pulsing 1968.629 times per second, indicating a tremendously high rotation rate. The star had to be spinning every 0.0005 seconds, much faster than any other known pulsar. Although astronomers expected that a new pulsar would spin faster than an old one, this great rotation rate was difficult to explain, and it indicated that more theoretical work was necessary. Third, the pulses varied slightly in frequency over the seven-hour observation. This also could not be explained using present supernova theories.

Unfortunately, when the astronomers repeated their observations a week later in an attempt to confirm the discovery, they could find no pulses in the signal. Continued observation of the supernova through the remainder of 1989 failed to reveal pulsations. The only explanation seemed to be that, during the January 18 observation, light from the pulsar had reached Earth through a temporary gap in the debris cloud around the pulsar.

As 1989 passed, theoreticians worked to adapt pulsar models to the new findings, and experimental astronomers continued to look for the pulsar's reappearance. As time passed, concern rose that perhaps the initial observation was flawed. The data and methods of the observation were examined over and over, but no flaw could be found. Then at the February 1990 meeting of the American Association for the Advancement of Science, John Middleditch reported that he and Tim Sassen, a graduate student at the University of California at Berkeley, had solved the problem. They found an old video camera in the observatory and discovered that if the camera was left on during an observation, it sometimes caused an electronic pulse with a frequency of 1968.629 cycles per second—exactly that found for the pulsar. Apparently the camera had been on the night of the pulsar's "discovery," and conditions had been just right for its signals to join the signals received from the sky.

There are lessons to be learned from the would-be discovery of the SN1987A pulsar. Although skeptics might point to this as a failure of science, it actually is an example of a successful use of the scientific method, for it illustrates how the method corrects itself. Although Middleditch showed courage in reporting the mistake, his embarrassment would have been greater if someone else had found it. The supposed discovery prompted re-analysis of our theories of pulsars, and we should be better prepared for whatever we find when—and if—the pulsar of SN1987A is finally discovered.

---

initials for "Little Green Men," a reference to science-fiction-type extraterrestrials). This speculation was abandoned for a couple of reasons. First, the pulsations continued in a very regular fashion instead of changing as they would if they contained a message. More convincing, however, was the discovery of three more such sources in other directions in the sky, each with its own characteristic rate of pulsation. It was highly unlikely that a number of different civilizations were sending such signals toward us at the same time, so the sources had to be natural ones. They were renamed ***pulsars***.

The first pulsar detected had a period of 1.3373011 seconds. Such great precision is possible in measuring the rate of a regularly repeating cycle because one can measure over a great number of the cycles and divide by that number to obtain the time for a single one. ▶Figure 15-10 shows a record of pulses from

pulsar. A celestial object of small angular size that emits pulses of radio waves with a regular period between about 0.03 and 5 seconds.

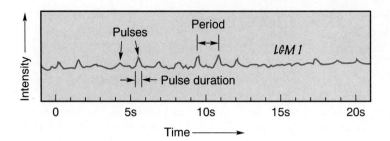

► **FIGURE 15-10** The chart of pulses from the first pulsar indicates their regularity. The difference between pulse duration and pulse period is illustrated.

Jocelyn Bell also found the next three pulsars that were discovered.

this pulsar and indicates that although they are extremely regular, they do vary in intensity.

All of the pulsars that were found had a pulse with a duration of about 0.001 second. (Figure 15-10 also illustrates the difference between pulse duration and period.) This immediately revealed to the astronomers an upper limit to the size of the object emitting the signals. The objects could be no greater than about 0.001 light-seconds in diameter—a few hundred kilometers. To see how such a conclusion could be reached even before the nature of the pulsation was known, consider what we would observe if the Sun were to brighten instantaneously. We would not see this instantaneous brightening as being instantaneous at all, for light from the part of the Sun closest to us would reach us about two seconds before light from its limb (see ►Figure 15-11). Since the Sun is about two light-seconds in radius, we would see the intensity of light build up over two seconds. Likewise, if the Sun suddenly shut off, the dimming would appear to take two seconds, rather than appearing to happen all at once.

The observed stretching out of any sudden change in the Sun would not be due to our distance from it but rather to its size. Be sure to understand this idea; it is a useful tool in astronomy that tells us the maximum size of some objects.

The smallest stars known in 1967 were the white dwarfs, but these are Earth-size objects, not small enough to emit pulsations that last only 0.001 second, at least not if the pulsation is caused by a change in the light emitted from the entire object. Thus the pulsar must be even smaller than a white dwarf. How could a star be so small? Enter the hypothesized neutron star.

THEORY: THE LIGHTHOUSE MODEL OF NEUTRON STARS/PULSARS

Let us consider how an object might emit pulses of radiation. One way would be for its surface to vibrate up and down. (A number of variable stellar objects are known to do this, including Cepheid variables.) Not only did the short duration of flashes from pulsars seem to indicate that they were not white dwarfs,

► **FIGURE 15-11** The arrows indicate light that left the Sun at the same time. Since point *X* is about two light-seconds closer to Earth than points *Y* and *Z*, its light reaches us two seconds sooner. Thus if the Sun were to increase in brightness instantaneously, we would see the light increase gradually over two seconds.

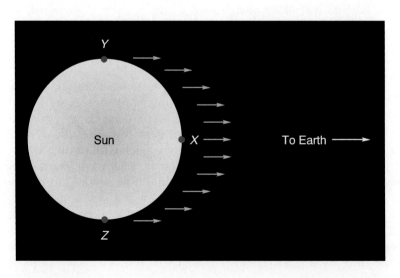

but when astronomers considered the nature of the material of a white dwarf and the gravitational force on its surface, calculations showed that the surface of a white dwarf could not vibrate as quickly as once per second. Neutron stars are much more dense than white dwarfs and have a much greater gravitational force on their surface, so their surface should beat up and down more quickly. In fact, calculations showed that they should be unable to oscillate as *slowly* as once a second.

A second possible way for an object to emit radiation in pulses is by an eclipsing binary process, with a bright object orbiting a dark one. The radiation curve of pulsars is different in nature from one that can be explained by eclipses, however. Eclipsing binaries must be ruled out.

A final mechanism for producing pulses is by radiation that comes from a small part of the surface of a rotating object. This is the mechanism that causes sailors at sea to observe pulses of light from a lighthouse. On a foggy night, they might see the lighthouse beam sweeping through the fog, but on a clear night the sailors see the light only when it shines directly at them. This makes it appear to blink on and off. Could a star rotate with a period as short as that observed for pulsars? A star the size of the Sun would be torn apart by such fast rotation, but white dwarfs or neutron stars would have two advantages in this regard: Their smaller size would mean that less force would be needed to retain their surfaces under fast rotation, and their small size and great mass would result in a much greater gravitational force on their surfaces than is experienced on the surface of the Sun. Calculations showed that a white dwarf might be able to withstand the forces involved in rotating with a period of one second, and perhaps with a period of one-fourth second, but certainly no faster than that. A neutron star, on the other hand, would have no difficulty rotating with a period of a fraction of a second.

It is easy to see what would cause either a white dwarf or a neutron star to rotate so fast. Recall that as a spinning object decreases in size, its rotation rate increases. White dwarfs, and especially neutron stars, are so small that they would be expected to be spinning very fast.

Logic seemed to be pointing more and more to the neutron star as the explanation for pulsars. In analogy with a sailor's lighthouse, the model developed to explain how neutron stars create pulses is called the ***lighthouse model.***

Recall that the Sun has a magnetic field. When the theory of the existence of neutron stars was developed back in the 1930s, it was suggested that such a star might have an extremely strong magnetic field since the star is the compacted core of a main sequence star that presumably would have had a magnetic field. The neutron star's strong magnetic field is a necessary part of the lighthouse model.

As we discussed when describing the solar system, it is common for the magnetic poles of a planet to be out of alignment with the axis of the planet's rotation. Refer back to Figure 6-28 to see the location of the Earth's magnetic pole in the Northern Hemisphere, some 1400 kilometers from the pole of rotation. Recall also that the magnetic field of the Earth traps charged particles and that these particles result in the auroras seen near the magnetic poles of the Earth. In the case of the theorized neutron star's extreme magnetic field, the energy associated with trapped charged particles would be much more intense, and a beam of radio waves and other radiation would be emitted near each magnetic pole. If the star's magnetic poles were located off the rotation axis, this beam would sweep through space as the star spins (▶Figure 15-12). Then, if the Earth were located in the path of the beam, we would see a pulse of radio waves each time the beam sweeps by us.

Notice that the lighthouse model would also predict the existence of many pulsars we could not observe on Earth, for we would see only those whose lighthouse beam happens to sweep by us. Since the length of every flash from a

---

Compare Figure 15-10 to a light curve for an eclipsing binary—Figure 12-36, for example.

As discussed earlier, the lack of a shorter rotation period for the Sun posed problems for astronomers in understanding the formation of the solar system from a nebula.

lighthouse model. The theory that explains pulsar behavior as being due to a spinning neutron star whose beam of radiation we see as it sweeps by.

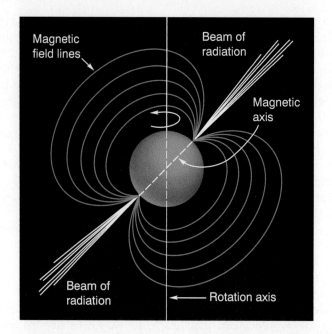

► **FIGURE 15-12** A beam of radiation is emitted near each magnetic pole of the pulsar. As the star rotates, the beam sweeps through space.

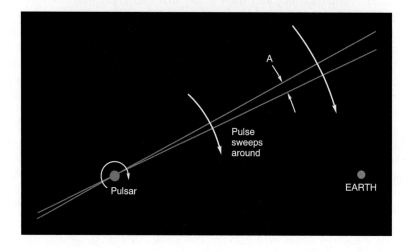

► **FIGURE 15-13** The angular width (A) of the beam from a pulsar determines how long we see its pulse. No pulsar has a long-duration signal, which indicates that pulsars' angular beam widths are narrow and therefore that we are seeing only a small fraction of the pulsars that exist.

pulsar is very short compared to its period, we can conclude that the angular size of the beam is small (►Figure 15-13). This means that from Earth we would see only a small percentage of the pulsars that exist. It also means that we may never see a pulsar in SN1987A.

In 1967, the lighthouse model seemed to be a logical explanation for the pulsar observations, but in order to test it, more pulsars would have to be found, perhaps including one related to the remnants of a supernova, where neutron stars are theorized to be.

Observation—The Crab Pulsar and Others

It was only a matter of months after the discovery of the first pulsar that one was found in the Crab nebula. It might have been found sooner, but its rate of pulsation is much faster than was expected; it has a period of 0.033 second so that it blinks 30 times per second. This short period finally ruled out completely the possibility of a spinning white dwarf as the source of pulses; a white dwarf would tear apart if it rotated 30 times per second.

Not only was the Crab pulsar flashing more frequently than others yet discovered, but it was emitting great amounts of energy in the radio region of the spectrum—100 times more than the total energy emitted by the Sun. Astronomers at the University of Arizona then observed that it also pulses in visible light. Since that time, astronomers have found that the Crab pulsar emits radiation in all regions of the spectrum, from radio waves to X-rays (which were observed by the orbiting *High-Energy Astronomy Observatory—HEAO-2*, the *Einstein Observatory*). Adding up the energy emitted by the Crab in all the various regions of the spectrum, it was found that the total energy from the Crab pulsar is more than 25,000 times the energy from the Sun.

Astronomers wondered what the source of the Crab pulsar's energy was. This question had been asked even before its pulsar had been discovered. Astronomers had long been puzzled by the luminosity of the nebula itself; it emits more energy than was thought to have been possible. Both of these energy problems were solved when it was discovered that the Crab pulsar is slowing down; its pulses were found to be growing slightly less frequent. It was hypothesized that the source of the energy that powers the nebula is the rotational energy of the pulsar. As the pulsar spins, its magnetic field propels electrons out into the nebula. These electrons are the cause of the nebula's great luminosity, but in being swept from the pulsar, the electrons in turn slow down the pulsar. If the hypothesis is correct, the amount of rotational energy lost as the object slows its spinning should correspond to the amount of energy emitted by the nebula. Calculations showed that the two amounts of energy did indeed correspond. The theory was confirmed quantitatively.

The pulsar in the Crab nebula is spinning faster than most others because the supernova in which it had its start was so recent; it occurred only 940 years ago. As time passes, this pulsar will gradually slow down. As it loses its rotational energy, the intensity of its pulses will also decrease, and it will no longer emit X-rays. Finally, in tens of thousands of years, it will be just another radio pulsar with its nebula spread so far that it can no longer be seen.

More than 440 pulsars have been discovered; a few have periods less than 0.1 second, but most have periods between 0.1 second and 4 seconds. Normally, no nebula is found surrounding pulsars for the nebula has long since dispersed. To further confirm that pulsars are indeed the neutron stars predicted to be left behind in supernovae, astronomers looked for more instances of pulsars associated with expanding nebulae. Since the Crab's pulsar was found, another has been found in the Gum nebula (Figure 14-27). Astronomers are now confident that they have found the neutron stars that theory predicted 50 years ago. We are searching for the pulsar in SN1987A. A Close Up in this chapter describes what was thought to be its discovery in 1989.

The pulse rate for a pulsar is sometimes observed to increase suddenly (this is called a *glitch*). This occurs because as the neutron star slows its spinning, forces on its surface change, causing "starquakes" and sudden changes in its shape.

MIDDLEWEIGHT CONCLUSION

▷Figure 15-14 reviews the steps a middleweight star takes from protostar to pulsar/neutron star. Only a small fraction of all stars are in the middleweight category, however, for two reasons. First, middleweights' lifetimes are short compared to lifetimes of stars with lower mass. Thus even if they are being formed at the same rate, not as many will be in existence at any given time. Secondly, only stars in a small range of mass end up with a core too massive to be supported by electron degeneracy and of low enough mass to be supported by neutron degeneracy. We know that neutron stars, which are the final ends of middleweight stars, are more massive than the Chandrasekhar limit of 1.4 solar masses, but we are not sure of the upper limit to the mass of a neutron star. (We certainly do not have any neutron-degenerate matter on Earth with which to experiment.) Theory indicates, however, that the limit falls somewhere between 2 and 4 solar masses. Thus only stars that end up with masses greater than—but not much

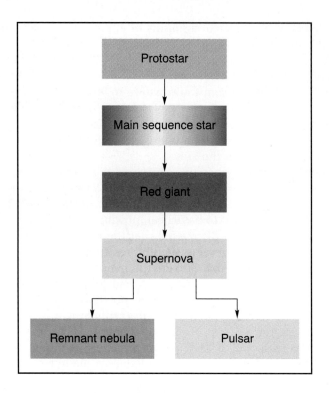

FIGURE 15-14 These are the steps a middleweight star takes from protostar to its final stages.

greater than—1.4 solar masses can ever become neutron stars and send out their characteristic lighthouse beam of radiation.

GENERAL RELATIVITY

Albert Einstein stated that in his youth he wondered whether, if he were moving at the speed of light, he could see himself in a mirror. Such questions led him in 1905 to develop the *special theory of relativity.* This theory makes the perception of electromagnetic waves independent of the motion of the observer. It allows us to answer Einstein's mirror question with a "yes," but it reveals a link between the nature of space and time that does not appear in our everyday perception of the universe. Special relativity has some interesting effects, but it is Einstein's expansion of the theory to the *general theory of relativity,* or *general relativity,* that must be used to describe the fate of massive stars.

Chapter 3 described how the general theory of relativity provided a different way of explaining what we normally call gravitation. The discussion began with a "thought experiment" involving a woman in a spaceship far from Earth. The conclusion was reached that if the spaceship were accelerating with an acceleration equal to the acceleration of gravity on Earth, the woman could not distinguish her situation in the accelerating spaceship from one in a stationary spaceship on Earth. That is, an object would seem to fall in the accelerating spaceship just as if the ship were in a gravitational field. General relativity leads to the *principle of equivalence,* which states that there is no way to distinguish between an accelerating reference frame in a location where there is no gravitational force and a nonaccelerating reference frame that is in a gravitational field.

Now we extend the thought experiment to include electromagnetic radiation. Suppose that the woman is in a spacecraft on the surface of the Earth, not accelerating. The spacecraft has a window on one side, and through the window a beam of light enters parallel to the floor. The principle of equivalence leads to a prediction that the beam will bend downward as it crosses the room. To see why this is so, imagine that the room is not on Earth but is in deep space with

special theory of relativity. A theory developed by Einstein that predicts the observed behavior of matter due to its speed relative to the person who makes the observation.

A Close Up in Chapter 3 discussed special relativity.

general theory of relativity. A theory developed by Einstein that expands special relativity to accelerated systems and presents an alternative way of explaining the phenomenon of gravitation.

principle of equivalence. The statement that effects of the force of gravity are indistinguishable from those of acceleration.

464   CHAPTER 15   The Deaths of Massive Stars

an acceleration upward equal to the acceleration of gravity. A quick burst of light enters the room as shown in ➤Figure 15-15. As the room accelerates upward, gaining speed all the time, the light takes the path shown in parts (b) and (c) of the figure. Notice that from our point of view, in the unaccelerated frame of the page in the book, the light continues in a straight line. From the point of view of the woman in the accelerating craft, however, the light will have bent downward, as shown in ➤Figure 15-16.

The principle of equivalence says that the same bending should happen when the spacecraft is stationary on the surface of the Earth. In this case, however, the bending will appear to be due to gravity.

The drawings that show the bending of light exaggerate the amount of bending, of course, for in the time required for a beam of light to cross a room, its acceleration would not result in enough curvature of the light's path to make it measurable. Perhaps if the woman's room were in an extremely strong gravitational field, where the acceleration of gravity is 10 billion times greater than on Earth, the bending would be appreciable.

If the principle of equivalence is valid, light should bend in the presence of a massive object. This prediction was made by Einstein's theory in 1907, but the predicted amount of bending near the Earth was very small, and no experimental check of the prediction was done until a solar eclipse in 1919. When the Sun is totally eclipsed by the Moon, stars can be seen in the sky, and this provides an opportunity to observe the bending of light that originated at a distant star as the light passes near the Sun.

Suppose a total eclipse occurs while the Sun—as seen from Earth—is between two bright stars. ➤Figure 15-17a shows the stars as they normally appear. During the eclipse, shown in part (b) of the figure, light from the stars must pass near the Sun before reaching Earth. When it does, the theory predicts that the light will be bent and will therefore make the stars appear slightly farther apart as shown in part (c). In practice, the bending was predicted to be very little, only 1.75 seconds of arc. This would produce very little change in the apparent po-

➤ FIGURE 15-15 A burst of light entering the accelerating spaceship (a) continues in a straight line but since the spaceship is accelerating upward, the light hits the floor. (c).

➤ FIGURE 15-16 As seen by a person in the spacecraft, the light beam bends downward as if falling to the floor. The acceleration of the beam will be measured to be exactly equal to the acceleration of the ship.

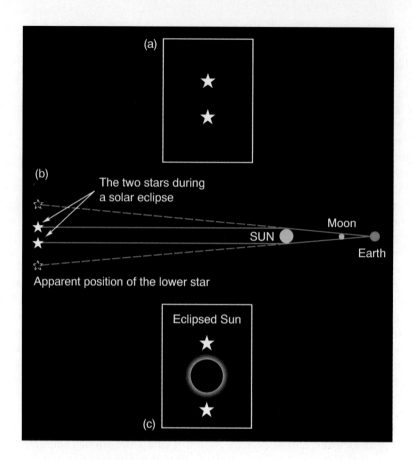

► **FIGURE 15-17** Light from the two stars is bent as it passes near the Sun, causing the stars to appear farther apart.

sition of a star close to the Sun, much less than indicated in the figure. The eclipse of 1919 did produce results in agreement with the general theory, however, and it provided the first experimental confirmation of the theory. Since then, similar measurements have been taken during other eclipses and have always confirmed the predictions of Einstein's theory.

Newton's theory predicts no gravitational effect on light since light has no mass. And, indeed, light is not observed to respond to gravity in our everyday world. Only when very strong gravitational fields are involved is the bending of light observed, and since Newton's theory had never been checked in such strong fields, no one realized that it makes incorrect predictions in those cases.

Thus the general theory of relativity presents a different way of looking at the phenomenon we call gravitation. Chapter 3 described how the theory explains gravitation not as a force but as the result of the curvature of space. The particular result of interest to us here is that the theory predicts that electromagnetic radiation will respond to this curvature in a way that will make it seem to be responding to the force of gravity. The present discussion will continue to speak of the force of gravity in Newtonian terms, but we will accept the Einsteinian prediction that electromagnetic radiation, including light, appears to respond to this force.

General relativity has survived every test to which it has been put (some of which were described in Chapter 3), and it is a well-accepted theory—not a hypothesis as is sometimes suggested by the media.

A Binary Pulsar

In 1974, Joe Taylor, a professor at the University of Massachusetts, and Russell Hulse, a graduate student working with him, were using the giant radio telescope at Arecibo to study pulsars. They discovered a very unusual pulsar, which seemed

Joe Taylor and Russell Hulse won the 1993 Nobel Prize in Physics for this work.

CLOSE UP

The Distance/Dispersion Relationship

Chapter 12 described various methods to measure distances to stellar objects. Pulsars provide another method, which relies on two phenomena: each burst of radiation from a pulsar contains an entire spectrum of wavelengths, and different wavelengths of electromagnetic radiation travel through space at slightly different speeds. For many purposes, it can be assumed that all wavelengths travel at exactly the same speed in space—the so-called *speed of light*—but in practice, since space is not a perfect vacuum, longer wavelengths travel slightly slower than shorter wavelengths. Recall from Chapter 5 that this difference in speed—called *dispersion*—is the same property that causes chromatic aberration in lenses. Its effect here is that as each pulse of radiation from a pulsar travels through space, the longer wavelengths get slightly behind.

When it was stated that the pulses from a pulsar are typically only 0.001 second in length, this referred to pulses of a single wavelength. If we consider, for example, a pulsar that emits visible light as well as radio waves, we will find that the visible-light portion of each pulse reaches us before the radio portion, although for any one wavelength of visible light or any one wavelength of radio energy the pulse length is the same—perhaps 0.001 second. The relative amount of dispersion might be stated as the time that elapses between the detection of a given wavelength of visible light and the detection of a given wavelength of radio energy in the same pulse.

Two factors determine the amount of dispersion: the distance the pulse travels through the interstellar medium and the dispersion properties of that medium. We have here another triple connection: distance to the pulsar, dispersion of the pulse, and the dispersive nature of the interstellar material (➤Figure C15-1). The amount of dispersion can be measured directly. Thus, to the degree that one of the other quantities is known, the final one can be calculated. If the dispersion properties of the interstellar matter between us and a pulsar are known, the distance to the pulsar can be determined. On the other hand, if distances can be determined by another method, this provides a means of learning more about the interstellar material.

➤ **FIGURE C15-1.** If any two of the three quantities are known, the third can be calculated.

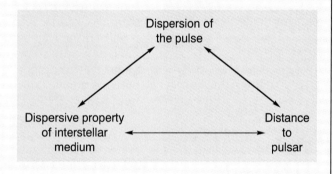

to be changing its pulsation rate regularly. After nearly a month studying the object, they deduced that the changes were due to the Doppler effect as this pulsar revolved around a companion in a binary system. The companion, however, could not be detected. As was explained in previous chapters, binary star systems are useful in determining the masses of stars. Recall, however, that when the spectrum of only one star is observed, the object's mass cannot be calculated. Astronomers were very interested in determining the masses of the objects in this case, however, for it would provide the first measurement of the mass of a pulsar and thereby serve as a check on the neutron star/pulsar theory.

Fortunately, the binary system had a short period of revolution (less than eight hours) so that many orbits could be observed over a few weeks. This permitted Hulse and Taylor to determine that the pulsar's orbit was precessing at the rate of four degrees per century, a precession much greater than Mercury's because of the great mass of the pulsar and its nearby companion.

The amount or precession predicted by general relativity depends only on the masses of the two objects in orbit and their distance of separation. If the masses of the objects had been known, the binary pulsar would have served as one more experimental check on general relativity. On the other hand, if the

The precession of Mercury was discussed near the end of Chapter 3.

General Relativity

applicability of the theory is assumed, general relativity can be used to calculate the masses of the two objects. After nearly two decades of observing the system, the orbits of the two objects are known very precisely, allowing a precise calculation of their masses. The pulsar has a mass of 1.441 solar masses, and its companion's mass is 1.3874 solar masses. The system has been a perfect one for studying the predictions of general relativity, and it was the first instance of the use of general relativity to calculate a stellar property and the first precise determination of the mass of a neutron star.

THE HEAVYWEIGHTS

A heavyweight star proceeds through its life in basically the same manner as a middleweight, although each stage occurs more quickly. Heavyweights differ from middleweights primarily in what happens to them when their core is compressed to a greater density than electron degeneracy can support. When this occurs in a middleweight, the resulting supernova leaves a neutron star at its center. In a heavyweight, an even more spectacular event happens: the core swallows itself as a black hole. The general theory of relativity plays a very important part in understanding black holes.

BLACK HOLES

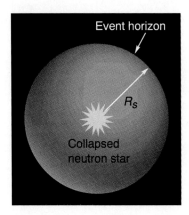

▶ FIGURE 15-18 Once a star has shrunk to inside the Schwarzschild radius, light can no longer escape its surface. The Schwarzschild radius forms a sphere whose surface is called the event horizon.

Schwarzschild radius. The radius of the sphere around a black hole from within which no light can escape.

black hole. An object whose escape velocity exceeds the speed of light.

Neutron degeneracy can no longer support a neutron star whose mass is greater than about three solar masses. Such a star will collapse. How far it will collapse is still an open question, but there is no known force capable of keeping the force of gravity from collapsing a star to zero size. This is an unimaginable situation, for we cannot envision matter having no size whatsoever, especially an object a number of times more massive than the Sun. Fortunately, we do not have to answer the question of how far the star collapses in order to predict how it will appear from outside.

After Einstein introduced general relativity, Karl Schwarzschild calculated that when a star collapses to a dimension equal to or less than what is called the **Schwarzschild radius,** light will be unable to escape the object. Recall the escape velocities of the various solar system objects that were noted earlier. The escape velocity of an object is the minimum velocity a missile must have in order to escape the gravitational field of the object. (The escape velocity from the Earth's surface is about 11 kilometers/second.) As a star decreases in size, the escape velocity from its surface becomes greater. When its radius becomes so small that its escape velocity is greater than the velocity of light, the star has reached the Schwarzschild radius (▶Figure 15-18).

The size of the Schwarzschild radius depends upon the mass of the star, for it is this mass that determines the force of gravity. The approximate radius (R_S) is given in kilometers by the formula

$$R_S = 3M \qquad (R_S \text{ in km; } M \text{ in solar masses})$$

where M is the mass of the star expressed in multiples (or fractions) of the Sun's mass. Thus a star with a mass five times greater than the Sun's will have a Schwarzschild radius equal to about 15 kilometers.

A star whose radius is less than the Schwarzschild radius is called a ***black hole;*** "black" because no light escapes from it, and "hole" because matter—or light—that falls into it can never be retrieved. A black hole exists if the star's radius is equal to or less than the Schwarzschild radius, but the radius of the black hole is considered to be the Schwarzschild radius.

The equation indicates that if matter falls into a black hole, the Schwarzschild radius becomes larger. This is reasonable, for as matter falls into the black hole, more mass is inside exerting more gravitational force and increasing the escape velocity at any given distance.

A name given to the surface of the sphere formed at the Schwarzschild radius illustrates an important and interesting feature of black holes. The surface is called the ***event horizon.*** Just as we cannot see beyond the Earth's horizon, we cannot see inside the event horizon. More than that, there is no way we can know about an event inside that sphere. Nothing that happens there is accessible to us and may just as well be in another universe, since we can have no experience of it.

event horizon. The surface of the sphere around a black hole from within which nothing can escape. Its radius is the Schwarzschild radius.

Properties of Black Holes

Although the theory of black holes involves general relativity and therefore might be considered complicated, black holes themselves are very simple. A black hole can be described completely by only three numbers, one for its mass, one for its electric charge, and one for its angular momentum. A full description of any other object in the universe would involve a long list of properties, including size, color, chemical composition, texture, and temperature (in addition to the three named above). For a black hole, these other properties have no meaning. For example, chemical composition is meaningless because it does not matter what original material condensed to form the black hole. That matter no longer exists as a chemical element in the black hole. Whatever the properties of the material that formed the black hole, that information is forever removed from the universe.

The mass of a black hole is, in principle, easy to measure. It would be measured the same way as the mass of any other celestial object. That is, we would observe an object that is in orbit around the black hole and apply Kepler's third law to calculate the total mass of the black hole and the object.

A black hole might have an electric charge, either positive or negative. In principle, this charge could be measured by detecting its effect on charged objects near the black hole. In reality, a black hole is not expected to have an electric charge, because if it had a positive charge, for example, the black hole would draw in negative charges until it became neutral. For this reason, electric charge is not normally considered in discussing black holes.

The final property that a black hole may have is angular momentum. Recall that as an object shrinks in size, it increases its rotation rate, and therefore a black hole would be expected to spin rapidly. Results of the theory of general relativity show that this rapid rotation would change spacetime around the black hole. In effect, the spinning black hole would drag nearby spacetime around with it. Because of this, light passing near a rotating black hole on one side behaves differently than light passing by the other side. On one side the light is moving along with spacetime, and on the other side it is moving against the motion of spacetime. This, of course, is not easy to imagine, for motions of spacetime are not part of our everyday experience.

Because a black hole can be described completely by just three properties, astronomers say that "a black hole has no hair." Any properties of objects that fall into the black hole are forever gone from the universe. They leave behind only the properties of mass, electric charge, and angular momentum. A black hole is a simple, "hairless" thing.

Detecting Black Holes

Astronomers predicted the existence of black holes in the 1930s when they realized that a star's mass may cause it to collapse beyond neutron degeneracy. A prediction of something as unusual as a black hole certainly calls for observational verification. But how?

Actually, Pierre Simon Laplace proposed in 1798 that a very massive star might have such a strong gravitational force that light could not escape it. This was 100 years before Einstein's theory linked gravitation with the travel of light!

➤ **FIGURE 15-19** If a black hole and red giant or supergiant form a binary system, material will be pulled from the giant (left) and will swirl around the black hole, causing X-rays to be released from the heated material in the disk.

There is, of course, no hope of seeing a black hole directly, for we cannot see something from which no light escapes. If, however, matter is falling into a black hole, we should expect some of that matter to orbit the black hole in a manner similar to the way matter falling into a white dwarf orbits it (causing a nova as it falls into the white dwarf). Since the gravitational field near a black hole is so strong, the orbital speed of nearby matter would be extremely great, and as collisions among particles turned the regular orbital motion to random thermal motion, the matter would reach temperatures of hundreds of millions of degrees. ➤Figure 15-19 illustrates material being pulled into orbit around a black hole from a companion binary star. Such hot material would radiate great amounts of energy and since it is not yet inside the event horizon, we should be able to detect it. We can predict the characteristics of the radiation, and we know that the object should appear as an X-ray source.

Numerous X-ray sources have been found in the heavens, particularly by NASA's HEAOs and by the Japanese and the European orbiting X-ray observatories. Are all of these black holes? Probably not. Only if one of these sources is found to be associated with a particularly massive star can we hope that it is a black hole. When we wish to know the mass of a star, we search for a binary system. Then if we find a binary system in which one of the stars is invisible with a mass greater than four or five solar masses, we can conclude that the star must be collapsed (or otherwise it would be visible). Finally, if the star emits X-rays characteristic of those predicted for a black hole, we would have good evidence for claiming to have found a black hole.

In the 1960s, astronomers discovered an X-ray source in the constellation Cygnus. Because it was the first X-ray source found in that constellation, it was named Cygnus X-1. Then in 1971, they discovered that the location of Cygnus X-1 corresponds with a ninth-magnitude star named HDE 226868, a blue supergiant of spectral type O. A periodic Doppler shift of the spectrum of the supergiant indicates that it is part of a binary system with a period of 5.6 days, but its companion is invisible. An O-type supergiant is expected to be a very massive star. Calculations reveal that if HDE 226868 has the mass expected of an O-type star—about 30 solar masses—its companion must have a mass greater than 3 solar masses. If so, the companion is probably a black hole.

Cygnus X-1 was the first candidate for a black hole. For years astronomers sought other explanations for the behavior of the object. For example, if the mass of HDE 226868 is less than normal for an O-type supergiant, its companion

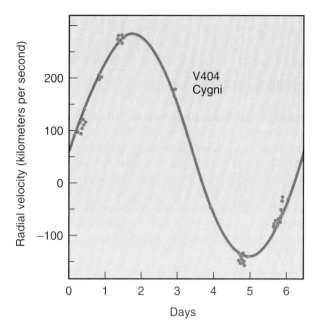

➤ **FIGURE 15-20** Applying Doppler shift analysis to the spectrum of V404 Cygni shows that its radial velocity changes with a period of 6.47 days. The dots indicate measured velocities (by the Doppler shift) at different times, and the curve is drawn to fit the data.

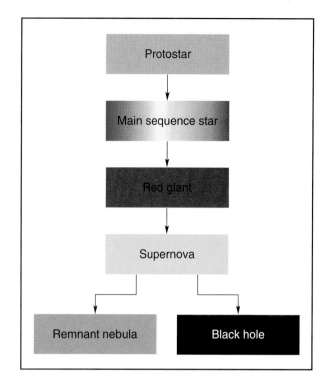

➤ **FIGURE 15-21** The steps in a heavyweight's life.

could have a mass less than the Sun. If this is so, the companion may be a neutron star instead of a black hole. Such explanations cannot be made to fit all the data, however, and it appears that Cygnus X-1 is indeed a black hole.

Since the discovery of Cygnus X-1, other black hole candidates have been found. A few are even part of eclipsing binary systems, thus providing us with further information about them.

In 1989, the star V404 Cygni, A G- or K-type star, called attention to itself by erupting with a powerful X-ray flare. In 1992, three European astronomers reported that Doppler shift analysis of V404 Cygni shows that it orbits an unseen companion with a period of 6.47 days (➤Figure 15-20). In this case, the mass of the visible star does not present a problem, for even if the calculation is based on its least possible mass, the dark companion must have a mass of at least 6.3 solar masses, and probably 8 to 12. V404 Cygni is now considered almost surely to be a black hole. It further convinces astronomers that black holes are a reality. The discovery of black holes not only confirms the theory of the death of the most massive stars but also serves as another confirmation of the theory of general relativity. Although the general public may still think that black holes are on the fringes of science, these objects are in fact firmly entrenched in astronomical theory.

➤Figure 15-21 reviews the steps taken by heavyweight stars as they progress from protostars to black holes.

The astronomers: Jorge Casares, Phil Charles, and Tim Naylor.

Chapter 17 discusses black hole candidates that astronomers are finding in the centers of some galaxies.

OUR RELATIVES—THE STARS

The idea that the stars are our relatives might be surprising, but in fact astronomers can show that humans and stars are related—albeit distantly.

Astronomers divide the stars into two classes, called *population I* stars and *population II* stars. The two populations are distinguished by the amount of heavy elements they contain. Population II stars contain very little material in their

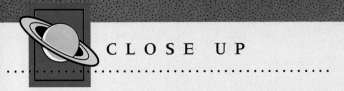

CLOSE UP

Black Holes in Science, Science Fiction, and Nonsense

Black holes are fantastic objects and are fruitful subjects for science fiction as well as for a lot of nonsense. We will try to separate the science from the nonsense and the science fiction.

Nonsense

There is a common misconception that a black hole is a giant vacuum cleaner, sweeping up matter across wide portions of space. In fact, the gravitational force of a black hole is unusually great only near the black hole. To make this clear, suppose that the Sun could magically become a black hole without losing any mass. If it did so, the gravitational force it exerts on the Earth would not change. The force exerted on the Earth would remain the same, and the Earth would continue in its same orbit. The only difference to us on Earth would be that the lights would go out: no radiation would arrive from the Sun. Newton's law of gravity states that the force of gravitational attraction between two objects depends only upon the masses of the two objects and the distance between them. Although predictions from Einstein's theory differ from those of Newton's, these differences show up only where the forces are extremely great, so Newton's theory can still be used to discuss Earth's orbit. The strength of Newton's gravitational force does not depend upon the Sun's size, only its mass.

Where is the increased gravitational field, then? To understand this, note that with the Sun in its present state, the closest one can get to it (and still be outside) is its surface—some 700,000 kilometers from the center. If a person could go down inside the Sun, the force of gravity on him would become *smaller*, for there would then be a gravitational force back toward the matter near the surface (➤Figure C15-2a). In fact, at the center of the Sun, the person would be weightless, for he would be attracted equally in all directions.

The difference in the case of the solar-mass black hole is that now one can get closer to the center of the

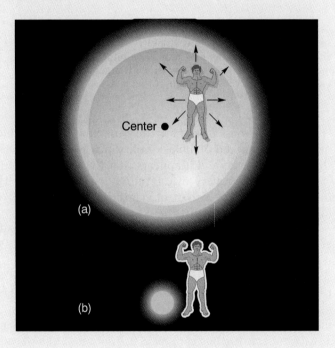

➤ **FIGURE C15-2.** (a) If a person could exist inside the Sun, he would weigh less than he did on the surface, for gravitational forces would be exerted on him in all directions by parts of the Sun. (b) If the entire Sun could be shrunk to a small enough ball, a person could be the same distance from its center as in part (a) and still be outside its surface. He would then weigh much more than in part (a).

---

atmospheres other than hydrogen and helium. Heavier elements are produced in the cores of stars as fusion takes place, so heavy elements do exist in their cores, but little or none is seen in their atmospheres.

The spectra of population I stars, on the other hand, reveal that their atmospheres contain heavier elements. The separation of stars into these two groups is somewhat arbitrary since a continuum actually exists from stars that have no elements beyond hydrogen and helium in their atmospheres to those that contain the greatest amounts of heavy elements; nonetheless, the distinction is convenient.

The existence of different amounts of the heavy elements in different stars can be easily explained. Recall that stars are formed from interstellar clouds of star while remaining outside its surface (Figure C15-2b). For such a black hole (only six kilometers across) one could get within a few kilometers of the center, and—as predicted by Newton's law of gravity—the force of gravity would be very great at these small distances. (Actually, Newton's law would not make accurate predictions this close to the black hole, but it correctly predicts that the force would be extremely large.)

Predictions from Science

Stephen Hawking, whose brief biography appears in Chapter 18, made some interesting predictions concerning black holes. Based on calculations of the density of the universe at its beginning (Chapter 18 again), he predicted that it may be possible that *mini* black holes formed at the time. These might have been the size of a pinhead and have had the mass of an asteroid. Alas, later theorizing by Hawking showed that if such mini black holes ever existed, they would have long since evaporated by radiation.

Unusual phenomena—by earthly standards—occur as an object falls into a black hole. For one thing, as object nears the black hole, it is pulled apart by tidal forces. Recall that tidal forces result because of the difference between gravitational forces on one side of an object and the other. An object approaching a black hole would feel a much stronger gravitational force pulling on its side nearer the hole than on the other side. The force difference would pull the object apart. It would be impossible for a person to fall into a black hole and remain intact.

Even more strange would be the observation of something falling into a black hole. Forget for now the destruction caused by tidal forces. Einstein's theories of relativity tell us that if we could watch the object fall, we would never see it reach the event horizon. We would see it getting closer and closer to the event horizon and getting redder and redder (Doppler shift-like), but because of the distortion of time that is predicted by the theory of relativity, it would take forever—according to our observation—for the object to reach the event horizon. As time is reckoned on the object, however, it would fall into the hole very quickly.

Science Fiction

Where does an object go when it falls into a black hole? It has been speculated that it may appear elsewhere or "elsewhen"—at another place or another time. Such travel through "hyperspace" or through time has lent itself to numerous science fiction plots. If it indeed occurs, we should see "white holes" where matter and energy are appearing out of nowhere. (In the language used, the matter and energy enter a black hole, pass through a "worm hole," and emerge from a white hole.) No such phenomenon has been observed.

Even more speculative is the idea that the matter may come out in another universe—not in another galaxy, but in a parallel universe. Since by definition we have no contact with such a universe, we have no way of verifying such speculation. Thus it is not in the realm of science at all. (Recall that a hypothesis must be verifiable to be classified as scientific.) While the hypothesis of white holes in our universe might perhaps qualify as a scientific hypothesis, the speculation of a parallel universe must remain pure science fiction.

---

gas and dust and that most stars end their lives by blowing (or exploding) much of their mass back out into space. The material they expel into space therefore contains some of the heavier elements produced within the star. Thus population II stars are those old stars that were formed from interstellar material long ago in the history of the universe, before the interstellar material became enriched in heavy elements. Population I stars, on the other hand, are young stars formed from material that contained the remains of previous generations of stars.

Which population is our Sun? We can answer this question without knowing anything about the spectrum of the Sun. Recall that the solar system was formed from material that did not fall all the way into the Sun as the interstellar cloud collapsed; it was formed from the same material that made the Sun. The fact that the Earth (and the other planets) contains heavy elements means that the cloud from which the planets formed contained those materials. The Sun is a population I star (although it contains less heavy material than some other stars).

Recall that fusion in the core of stars continues to release energy as heavier and heavier elements are formed until iron is produced. In order for iron nuclei to fuse with other nuclei to form heavier elements, there must be an *input* of energy. The reaction that forms heavier elements from iron does not release energy; it absorbs it. Although most of the matter of which the Earth is made is less massive than iron, many elements are more massive than iron. Such matter could not have been formed by fusion within the core of a star. It was instead formed during a supernova explosion, when tremendous amounts of energy were available. Most of the energy of a supernova is used up in releasing radiation (especially neutrinos) and in blasting away the outer parts of the star, but a small fraction is absorbed by the material of the star, where light elements are fused into heavy elements.

This discussion has implications for us humans. The material that makes up our bodies is from the Earth. Where was it before it was part of the Earth? In an interstellar cloud. And before that? In a star!

Harlow Shapley, former director of the Harvard College Observatory whose work will be discussed later, lists cosmic evolution as one of the 10 revelations that have most affected modern humans' life and thought. He says, "Nothing seems to be more important philosophically than the revelation that the evolutionary drive, which has in recent years swept over the whole field of biology, also includes in its sweep the evolution of galaxies, and stars, and comets, and atoms, and indeed all things material."*

---
*Harlow Shapley, *Beyond the Observatory* (New York: Scribner, 1967), pp. 15–16.

CONCLUSION

All stars get their starts when shock waves compress parts of cold interstellar dust clouds. Most differences between stars, from the various stages through which each star will progress as it lives and dies to how long each of those stages lasts, result entirely from differences in mass. The universe itself has not existed long enough for the least massive stars to have ended their lives, but the most massive ones have short lives, and at the end they become the most astonishing things in nature: black holes. ➤Figure 15-22 summarizes the life cycles of all of the stars in our boxing stable: flyweights, lightweights, middleweights, and heavyweights.

This chapter completes our examination of the stars. Just as the horizons expanded many times over when our focus changed from the solar system to the stars, they must expand again in the next chapter, which begins the examination of the largest class of objects in the universe: the giant galaxies of stars.

RECALL QUESTIONS

1. The core of a mature supergiant
   A. is layered, with heavier elements in the center.
   B. is layered, with lighter elements in the center.
   C. is uniformly composed of hydrogen and helium.
   D. is uniformly composed of helium.
   E. is composed primarily of iron and carbon.

2. Which type of supernovae have hydrogen lines in their spectra?
   A. Type I (from binary systems)
   B. Type II (from single stars)
   C. [Neither type.]

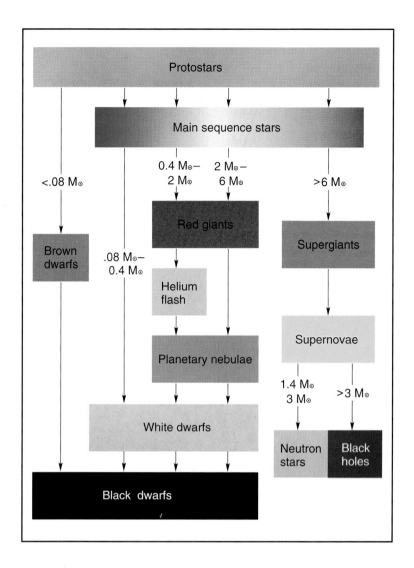

▶ **FIGURE 15-22** This diagram summarizes the steps in the life cycles of stars of various masses. The limits in mass are not well known. (The symbol "<" means *less than*, and ">" means *more than*.

3. The most massive element that can be formed by nuclear fusion with the liberation of energy is
   A. helium. D. iron.
   B. carbon. E. lead.
   C. oxygen.

4. The most recent naked-eye supernova was observed
   A. in 1054 by the Chinese.
   B. in the 1500s by Tycho Brahe.
   C. around 1600 by Kepler.
   D. in 1929 by Hertzsprung and Russell.
   E. in 1987 by Ian Shelton.

5. SN1987A was the explosion of a
   A. red giant. C. white dwarf.
   B. blue supergiant. D. pair of stars.

6. The pulse rate of pulsars is slowed because
   A. they convert rotational energy into radiation.
   B. they drag companion stars around.
   C. of friction with the interstellar medium.
   D. of the conservation of angular momentum.

7. The first candidate for a black hole was
   A. Cygnus X-1. D. Centaurus A.
   B. Algol (the "Demon Star"). E. Xi Ursa Majoris.
   C. Sigma Xi.

8. Theory predicts that a neutron star should spin fast because
   A. it was given increased speed by the supernova explosion.
   B. it was given increased speed by a companion star.
   C. it conserved mass as it collapsed.
   D. it conserved angular momentum as it collapsed.
   E. [The statement is false; neutron stars are not predicted to spin fast.]

9. Which of the following is the same as a neutron star?
   A. The Crab nebula D. A pulsar
   B. A Cepheid variable E. [None of the above.]
   C. A dark nebula

10. Pulsars are
    A. larger and more massive than neutron stars.
    B. larger but less massive than neutron stars.
    C. smaller but more massive than neutron stars.
    D. smaller and less massive than neutron stars.
    E. the same as neutron stars.

11. Pulsars emit sharp bursts of pulses, each lasting for less than one second. From this we can conclude that pulsars

A. have a great amount of energy.
B. have little energy.
C. are small in size.
D. are moving rapidly away from us.
E. are far away from Earth.

12. There is good evidence that black holes have been detected
    A. by use of data concerning their chemical compositions.
    B. in binary systems.
    C. by observing planets falling into them.
    D. in planetary nebulae.
    E. in galactic clusters.

13. The first pulsar was discovered by
    A. a professional American astronomer using a radio telescope.
    B. a professional American astronomer using a visible-light telescope.
    C. a professional British astronomer using a radio telescope.
    D. a professional British astronomer using a visible-light telescope.
    E. a British graduate student using a radio telescope.

14. According to present theory, the pulses of radiation from a pulsar are due to
    A. pulsations of the surface of the star.
    B. pulsations from within the core of the star.
    C. eclipses of the star by a binary companion.
    D. rotation of the star.
    E. [Any of the above, depending upon the particular pulsar.]

15. What remains after a supernova?
    A. A main sequence star
    B. A white dwarf
    C. A neutron star
    D. A black hole
    E. [Either C or D above, depending upon the mass of the star.]

16. Which of the following lists the stages in the life of very massive stars after they leave the main sequence?
    A. Supergiant, supernova, black hole
    B. Supergiant, supernova, neutron star
    C. Supergiant, white dwarf, black dwarf
    D. Supergiant, planetary nebula, white dwarf
    E. Planetary nebula, supergiant, black hole

17. When more material falls into a black hole, the diameter of its event horizon
    A. increases.
    B. remains the same.
    C. decreases.
    D. [Any of the above, depending upon the nature of the black hole.]

18. In which of the following categories are the objects the least dense?
    A. Main sequence stars
    B. Nebulae
    C. Pulsars
    D. Red giants
    E. White dwarfs

19. Which of the following have the greatest absolute magnitude when they are at their brightest?
    A. Black holes
    B. Protostars
    C. Pulsars
    D. Supernovae
    E. White dwarfs

20. Which of the following were first detected by Jocelyn Bell, a graduate student at Cambridge University in England?
    A. Black holes
    B. Protostars
    C. Pulsars
    D. Supernovae
    E. White dwarfs

21. What do Cygnus X-1 and V404 Cygni have in common (other than that they are in the same constellation)?
    A. Both are neutron stars.
    B. Both are planetary nebulae.
    C. Both are white dwarfs
    D. Both are probably black holes.
    E. Both are supergiants.

The following question is from a Close Up:

22. If the Sun could magically and suddenly become a black hole (of the same mass), the Earth would
    A. continue in its same orbit.
    B. be pulled closer, but not necessarily into the black hole.
    C. be pulled into the black hole.
    D. fly off into space.

23. Compare the luminosity of a supernova to that of a massive main sequence star.

24. What happens to the size of a main sequence star when mass is added to it? A white dwarf? A neutron star? A black hole?

25. Outline the stages in the lives of middleweight and heavyweight stars. What causes them to end their lives differently?

26. What leads us to believe that the object seen by the Chinese in 1054 was a supernova?

27. Describe how pulsars were discovered. What led astronomers to conclude that they were neutron stars?

28. When astronomers were searching for the nature of pulsars, why, if one assumes that the pulses from a pulsar are caused by vibrations in their surfaces, did the observed rates of pulsation seem to rule out both white dwarfs and neutron stars?

29. Describe the lighthouse model of pulsars. What observation(s) led astronomers to select this model from among other suggested models?

30. What is a black hole? How can we expect to observe one?

31. Define *Schwarzschild radius* and *event horizon*.

32. What general statement can be made about the escape velocity from a black hole?

33. If the Sun (magically) became a black hole, what would be the effect on the Earth? Explain.

34. What is the observational evidence that black holes exist?

QUESTIONS TO PONDER

1. The fusion of lightweight nuclei is said to produce energy, yet fusion cannot occur until a high temperature is reached. Doesn't the high temperature indicate that energy is being supplied to the reaction rather than being produced by it? Explain.

2. Consult an astronomy magazine, such as *Astronomy, Mercury,* or *Sky and Telescope,* and report on the latest findings concerning SN1987A.

3. When astronomers were looking for pulsars in supernova remnants, they automatically searched the Crab nebula, for the result of a recent supernova. If the lighthouse model is correct, why was it unlikely that we would find a pulsar there (and therefore lucky that one was found)?

4. The Schwarzschild radius is not the radius of the matter that makes up the black hole. Of what is it the radius?

5. Explain why, if HDE 226868 has less mass than expected, Cygnus X-1 would have too little mass to be a black hole. Why is this not a problem with V404 Cygni?

6. Explain why main sequence stars, white dwarfs, neutron stars, and black holes respond differently with regard to changes in their sizes when matter is added to them.

7. Explain why the observation that radiation from the first pulsar was appearing four minutes earlier each night led astronomers to conclude that the pulses were not of earthly origin (or, if they were, that astronomers had produced them).

8. Compare the mechanism that has slowed the rotation of the Sun (Chapter 7) to the mechanism that slows pulsars.

9. In order to calculate how fast a white dwarf (or neutron star) would pulse by vibrating its surface, we must know the strength of the gravitational field at its surface. What other property of the object must be known?

10. It is predicted that supernovae explode at the rate of one per second over the entire universe. Why don't we see more of them?

CALCULATIONS

1. What is the value of the Schwarzschild radius of a star of 7 solar masses?

2. If the mass of a black hole is doubled, how does the size of its event horizon change?

3. Betelgeuse is a red giant 650 light-years away. Tycho's supernova was 9800 light-years away. If Betelgeuse were to become a supernova, how many times brighter than Tycho's supernova would it appear?

StarLinks netQuestions

Visit the netQuestions area of StarLinks (www.jbpub.com/starlinks) to complete exercises on these topics:

1. Supernovae Three supernovae have been seen with the naked eye in our galaxy. The most brilliant occurred in the constellation Taurus, and was observed in 1054 by Chinese astronomers.

2. Schwarzchild Radius The Schwarzchild Radius is the radius of the sphere around a black hole from within which no light can escape.

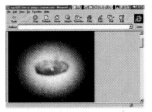

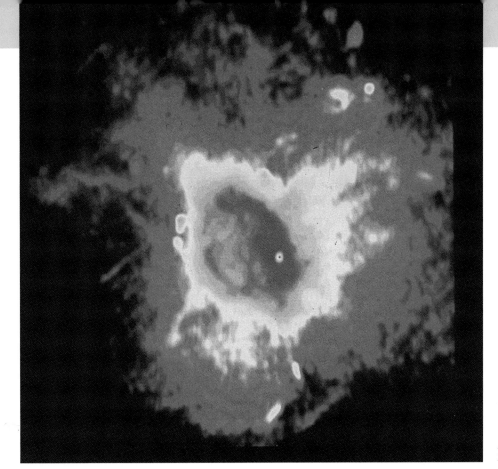

A radio map of the center of the Milky Way Galaxy.

CHAPTER 16

THE MILKY WAY GALAXY

Our Galaxy
Globular Clusters
Components of the Galaxy
HISTORICAL NOTE: The Shapley-Curtis Debate
Galactic Motions
CLOSE UP: ET Life VII—Where Are They?
The Mass of the Galaxy
The Spiral Arms
CLOSE UP: Calculating the Mass of the Inner Galaxy

CLOSE UP: The Milky Way: A Barred Spiral Galaxy?
Spiral Arm Theories
The Density Wave Theory
The Chain Reaction Theory
The Galactic Nucleus
The Evolution of the Galaxy
Age and Composition of the Galaxy
The Galaxy's History
ACTIVITY: The Scale of the Galaxy

SERENDIPITY IS DEFINED AS "the faculty of making fortunate and unexpected discoveries by accident."* In 1933, Karl Jansky of Bell Telephone Laboratories was using an antenna he had built for studying the direction of arrival of static from thunderstorms in order to improve radio telephone service to Europe. He detected radio waves he could not account for, and after much work he determined that they were coming from the center of the Galaxy. Years later, in a letter to a colleague, Jansky wrote:

> As is quite obvious, the actual discovery, that is, the first recording made of galactic radio noise, was purely accidental and no doubt would have been made sooner or later by others. If there is any credit due me, it is probably for a stub-

born curiosity that demanded an explanation for the unknown interference and led me to the long series of recordings necessary for the determination of the actual direction of arrival.**

Serendipitous discoveries are fairly common in science, and analysis has shown that almost all of them have certain features in common. First, as Jansky wrote in his letter, the discoverer has a "stubborn curiosity that demand(s) an explanation for the unknown...." The data that led to Jansky's discovery were only a series of tiny blips on a chart. Jansky could easily have put the chart aside and ascribed the unexpected blips to equipment problems or to any number of possible "unexplainable" sources. Instead, his stubborn curiosity led him to investigate.

A second feature common to serendipitous discoveries is that the discoverer is extremely knowledgeable about the subject he or she is investigating. The fact that the discoveries are unexpected does not mean that just anyone will make them, for important discoveries are not made by the unprepared. Instead, they are made by competent observers engaged in serious scientific investigation.

*The American Heritage Dictionary, 1976.
**Quoted by W. T. Sullivan III in Serendipitous Discoveries in Radio Astronomy, ed. K. Kellermann and B. Sheets (Green Bank, WV: National Radio Astronomy Observatory, 1983).

Throughout this text we have seen how our perception of the universe has grown from a small universe centered on our immediate environment to a larger universe whose center was thought to be located farther and farther away. As the next chapter will show, today it appears that the center simply does not exist: the universe has no center. Hand-in-hand with this receding center has come a tremendous expansion of our realization of the size of the universe. Galileo caused controversy not only because he placed Earth off-center, but because the Earth played such a small part in his universe.

Galileo's contemporary, Giordano Bruno, did propose that the stars were suns, and he paid for his belief with his life; he was burned at the stake for heresy in 1601.

Yet Galileo did not realize that the Sun that rules our sky is just one of countless suns—that each of the stars he saw in the sky is another sun. When astronomers shift their focus from the solar system to the stars, their entire scale of thinking must expand. For example, as we have seen, the unit of measure used in the solar system, the astronomical unit, is very inconvenient for stellar distances; therefore astronomers use the light-year and parsec to describe such distances.

Another mental leap in thinking about distances and sizes is required when stars are considered not as individuals but in galaxies. The leap is so great that it may be impossible to truly understand the distances involved. It is fun to try, however. We'll see that our Galaxy contains about a million-million solar masses. Give some thought to what such a number means.

When objects are studied at greater and greater distances from us, our knowledge of their nature and properties is more recent and less certain. This chapter will discuss the galaxy of stars of which the Sun is such an insignificant part and describe how astronomers measure its properties. It will then compare two conflicting theories that attempt to explain the structure of galaxies like ours. Finally, it will describe how our Galaxy formed and evolved to its present state. We'll see that in the case of galaxies, measurements are much less precise than we might be comfortable with.

OUR GALAXY

▶Figure 16-1 shows part of the faint band of light that stretches around the sky, encircling the Earth. The ancient Greeks named this hazy band *galaxies kuklos*, meaning "the milky circle." The Romans called it *via lactae*, "milky way." You

▶ FIGURE 16-1. This wide-angle photo shows the Milky Way in Sagittarius (left) and Scorpius (right). Comet Halley is visible at left center in this photo taken March 13, 1986.

are encouraged to find a good clear night sky away from city lights and look at the beautiful Milky Way, a sight many of us in the modern world never get a chance to experience. The Milky Way completely encircles the Earth, passing through Sagittarius, Aquila, Cygnus, Cassiopeia, and Auriga and between Gemini and Orion on the northern hemisphere of the celestial sphere. Then it passes through Monoceros, near Canis Major, and through Vela and Crux on the southern hemisphere.

We know today that the haze of the Milky Way results from the many stars in the disk of the galaxy in which we live. The Sun is one of about 200 billion stars that make up what we call the **Milky Way Galaxy,** the Milky Way, or simply the Galaxy. Most stars in the Galaxy are arranged in a wheel-shaped disk that circles around a bulging center (▶Figure 16-2). The diameter of the Galaxy is

Milky Way Galaxy. The galaxy of which the Sun is a part. From Earth, it appears as a band of light around the sky.

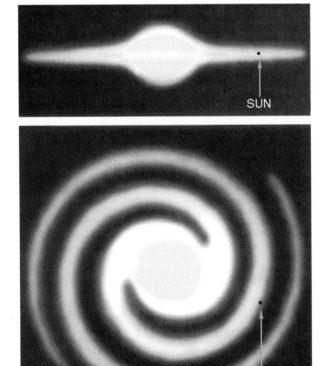

▶ FIGURE 16-2. The Sun is located about two-thirds of the way out along the disk of the Galaxy, shown here both edge-on (top) and face-on (bottom). Recent evidence indicates that our Galaxy may not have a symmetric nucleus as shown here. See the Close Up "The Milky Way: A Barred Spiral Galaxy?"

Our Galaxy

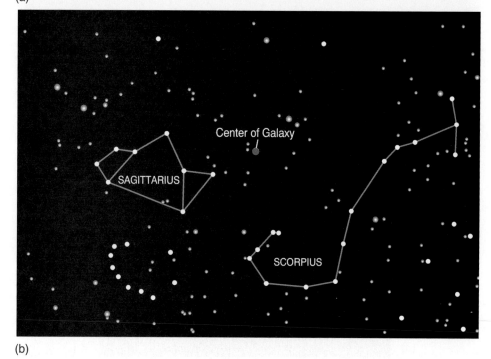

▶ FIGURE 16-3. (a) A photo of the region of Sagittarius and Scorpius. (b) The center of the Galaxy is within the boundaries of Sagittarius, but it appears to be between Sagittarius and Scorpius at about the location indicated.

SN1987A (Chapter 15) occurred in the Large Magellanic Cloud.

Notice that the term *Milky Way* is being used here to refer to the effect seen in the sky. Whether the term refers to this phenomenon or to the galaxy of stars itself should be clear from the context.

about 30,000 parsecs (100,000 light-years), and the Sun is a little more than halfway out from center. The galactic center is in the direction of Sagittarius (▶Figure 16-3) in our sky.

With only a few exceptions, every naked-eye object in our sky is part of the Galaxy. The Magellanic Clouds (Figure 12-40) are exceptions that are visible from the Southern Hemisphere. These small galaxies are close to the Milky Way Galaxy, but not part of it. An exception that those of us in the Northern Hemisphere can view in the fall sky is the Andromeda galaxy (▶Figure 16-4) in the constellation Andromeda. The Andromeda galaxy is the oval patch on the September sky chart at the end of this book. Our Galaxy is probably similar to the Andromeda galaxy.

The discovery that we live in a galaxy of hundreds of billions of stars was made fairly recently. It began, though, four hundred years ago when Galileo turned his telescope to the Milky Way and found that it is not made up of haze (or even milk!), but of stars far too numerous to count. The Milky Way appears misty because so many of the stars are so far away that the naked eye cannot distinguish individual stars and sees only the overall illumination from them.

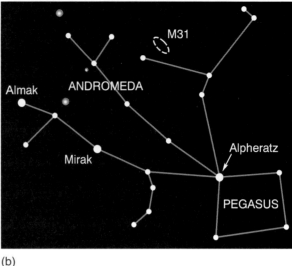

▶ **FIGURE 16-4.** (a) This is a photo of the Andromeda galaxy, a spiral galaxy more than 2 million light-years from our Galaxy. (b) The galaxy can be seen with the naked eye as a fuzzy spot at the location indicated as M31, which is the Messier designation for the Adromeda galaxy.

When Galileo looked through his telescope, he saw more haze behind the many stars his telescope revealed, and he concluded that it was caused by even more stars too faint to see individually.

William Herschel, discoverer of Uranus, wrote that through a telescope

> We find [the stars] crowded beyond imagination along the extent of [the Milky Way]; . . . so that, in fact, its whole light is composed of nothing but stars of every magnitude from such as are visible to the naked eye down to the smallest points of light perceptible with the best telescope.

The telescopic view of the Milky Way led astronomers to conclude that we live in a disk of stars. When we view the Milky Way in the sky, we are looking along that disk (▶Figure 16-5), and when we are looking in other directions, we are looking out of the disk.

As the Milky Way is viewed from Earth, it appears at first glance that we are at its center, for to the naked eye, despite some local variations, the Milky Way seems to be about as bright in one direction as in another. In the 1780s, in order to determine where the Sun lies relative to the disk, William Herschel and his sister Caroline made star counts in nearly 700 selected regions distributed around the sky. They reasoned that if more stars were found in one direction than in another, that direction could be assumed to be toward the center of the disk. Their conclusions not only confirmed the disklike shape of the Galaxy but indicated that the Sun is indeed at the center, for they saw about the same number of stars in all directions in the Milky Way. ▶Figure 16-6 shows the shape they arrived at for the Galaxy. Notice that the Sun is nearly at its center.

▶ **FIGURE 16-5.** When we see the Milky Way in the sky, we are looking along the disk of the Galaxy. Otherwise we are looking out of the disk.

▶ **FIGURE 16-6.** The Herschels' counting of stars led them to conclude that the Galaxy is shaped like this. The Sun is located at the bright spot within the Galaxy.

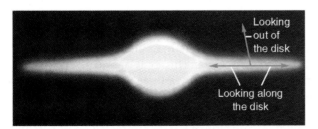
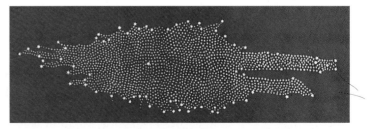

Our Galaxy

Jacobus Kapteyn (1851–1922) was one of 15 children. From the age of 27 until he retired at 70, he held a position at the University of Groningen in the Netherlands.

In the early part of this century, Jacobus C. Kapteyn sought to find the Sun's location by analyzing the density of stars in various directions from the Sun. He did this by measuring not only the number of stars in each direction but their distances from us. He found that the density of stars decreases in every direction from the Sun, and it was logical for him to conclude, like William and Caroline Herschel, that the Sun is at the center of the disk.

The conclusion that the Galaxy centers on the Sun was viewed with skepticism, as you might expect. After all, we once thought that the Earth was the center of the universe only to find that it circles the Sun. Were we now discovering that the Sun is the center? The evidence from the Herschels and from Kapteyn pointed to an affirmative answer, but the finding seemed to be contrary to the trend that began before written history and continued through Copernicus and Galileo.

Today we know that the Sun is not at the center. To understand why the Herschels and Kapteyn obtained their erroneous results, imagine yourself standing in a large forest. Suppose you try to decide whether or not you are at the center of the forest by counting trees in all directions. Unless you are very near an edge, you will see the same number of trees in all directions even though you are nowhere near the center. The reason is that the trees themselves prevent you from seeing beyond a certain distance. If you cannot see to the edge of the forest, this method of determining your location is not valid.

Another analogy: If you are surrounded by dense fog, the fact that you can see the same distance in every direction is not evidence that you are in the center of anything.

The situation in the case of the stars is somewhat different, for the stars do not fill our view as do the trees in a forest. When we look out among the stars, though, interstellar dust and gas place a limit on how far we can see. The Herschels were unaware of the existence of this material. They assumed that they could see all the way to the edge of the group of stars within which our Sun lies. Since they could only see a limited distance and since the Sun is not near an edge, they concluded that it was at the center.

Likewise, interstellar dust kept Kapteyn from counting the stars correctly. The density of stars at greater and greater distances from the Sun appeared to decrease because when Kapteyn counted stars at great distances, the interstellar dust prohibited him from seeing them all. In Figure 16-1, the dark areas stretching along the Milky Way are dust and gas clouds.

These investigators reached erroneous conclusions simply because one of their assumptions—that there is nothing in space to block the view of distant stars—was wrong. Incorrect assumptions cause trouble not only in science but in everyday life. Often we are not even aware of what our assumptions are, and this prevents us from even accepting the possibility that our conclusion may be in error.

Globular Clusters

As a previous chapter showed, some stars begin their lives in galactic clusters. These clusters are called "galactic" because they are found within the disk of the Galaxy. The Pleiades (Figure 13-7) is the most prominent example of this type of cluster, which may typically contain hundreds of stars. ➤Figure 16-7 shows two other galactic or open clusters. ➤Figure 16-8 shows an example of a much larger type of cluster, the **globular cluster.** These beautiful, symmetrical clusters may look as though they have solid centers, but they are actually groups of hundreds of thousands of stars. The stars are so densely packed in the center of the cluster that we simply see a white area, not the individual stars. The average separation of stars near the center of a globular cluster is about 0.5 light-year. (In the Sun's region of space, stars are separated by an average distance of about 4 or 5 light-years.) Globular clusters are not confined to the disk of the Galaxy but are seen outside the disk.

globular cluster. *A spherical group of up to hundreds of thousands of stars, found primarily in the halo of the Galaxy.*

Globular is pronounced "glob" (as of mud) rather than "globe."

While Kapteyn was seeking to determine the Sun's location in the Milky

▶ FIGURE 16-7. The photo shows two open clusters in Perseus.

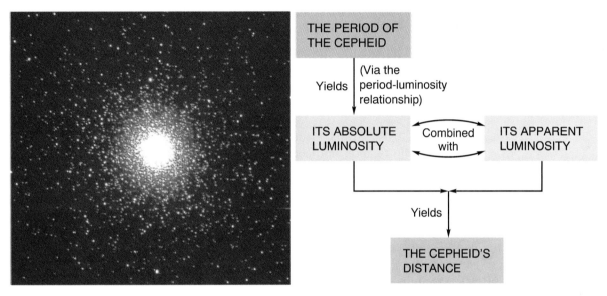

▶ FIGURE 16-8. This is a photo of a globular cluster, known as 47 Tucanae. It is not solid, but stars near its center are so closely spaced that they look as though they are touching.

▶ FIGURE 16-9. Once Shapley had determined the relationship between Cepheids' periods and absolute luminosities, he could use a chain like the one shown here to determine the distance to any Cepheid variable.

Way by studying star locations, Harlow Shapley was trying to do the same using globular clusters. He had a problem determining the distances to globular clusters, however, for they are much farther away than the stars Kapteyn was analyzing. Just a few years earlier Henrietta Leavitt had discovered the relationship between the periods and the apparent magnitudes of Cepheid variables in the Magellanic Clouds. Shapley observed Cepheid variables in globular clusters but could not use them as distance indicators until he had determined the relationship between their periods and absolute magnitudes (see ▶ Figure 16-9). He used a statistical method with Cepheid variables within the galactic disk to do so and then turned his attention to Cepheids in globular clusters.

In 1917 Shapley published results of his survey of distances and directions to the then-known 93 globulars. He showed that they are not distributed evenly around the sky but tend to be located more on one side, centered about the constellation Sagittarius. In fact, they seemed to be distributed in a sphere cen-

Harlow Shapley (1885–1972) quit school after the fifth grade, returned to school at age 16, and earned a Ph.D. from Princeton in astronomy at age 27.

Our Galaxy   485

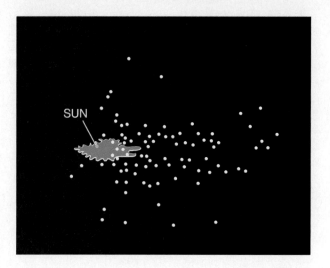

► FIGURE 16-10. The distribution of the globular clusters is shown relative to the Sun and to the Herschels' model of the Galaxy.

tered on a point thousands of light-years away from the Sun. ►Figure 16-10 shows the approximate distribution of globular clusters compared to the Herschels' model of the Galaxy. Shapley assumed that the clusters revolve around the center of the Galaxy and therefore concluded that the Galaxy's center lies at the middle of the group of globular clusters. This meant that the Galaxy is much larger than indicated by the Herschels' model. It also meant that the Galaxy is not centered about the solar system; that is, our Sun is not at the center of the Galaxy.

In the 1920s, further evidence indicated that the Sun is not at a unique position in the Galaxy. Jan Oort (who proposed the comet cloud that bears his name) and Bertil Lindblad studied the motions of great numbers of stars near the Sun. They found that there is a pattern in the velocities of stars, depending upon their directions from the Sun. Kepler's third law, when applied to stars revolving around the center of the Galaxy, predicts that stars closer to center should move faster and those farther from center should move slower. This is the reason for the pattern of velocities shown in the Oort-Lindblad analysis. They concluded, as had Shapley, that the center of the Galaxy was thousands of light-years away in the direction of Sagittarius.

Bertil Lindblad (1895–1965) was a Swedish astronomer who was president of the Royal Swedish Academy of Sciences for 22 years and was chair of the Nobel Foundation when he died in 1965.

It should be pointed out that Oort and Lindblad saw a pattern only after analyzing very great numbers of stars, for stars have random motions along with their pattern of motion around the galactic center. This is similar to the way that cars on a multilane freeway have a pattern of motion in one direction though at any given time certain cars may be changing lanes or otherwise deviating from the pattern. Oort and Lindblad ignored individual stellar motions and concentrated on patterns of average motions.

Finally, in 1930 the interstellar dust was discovered. This resolved the conflict between the conclusions of Herschel and Kapteyn on the one hand and of Shapley, Oort, and Lindblad on the other for the interstellar dust had prevented Herschel and Kapteyn from seeing to the edge of the disk. Although Shapley's original values for the size of the Galaxy had to be revised when it was discovered that there are two types of Cepheid variables, his basic deductions were correct. He had shown that the Galaxy was much larger than previously thought. The boundaries of the universe were again pushed back.

COMPONENTS OF THE GALAXY

The Galaxy can be described as having four components, the disk (which contains the Sun), the nuclear bulge, the halo, and the galactic corona (►Figure 16-11).

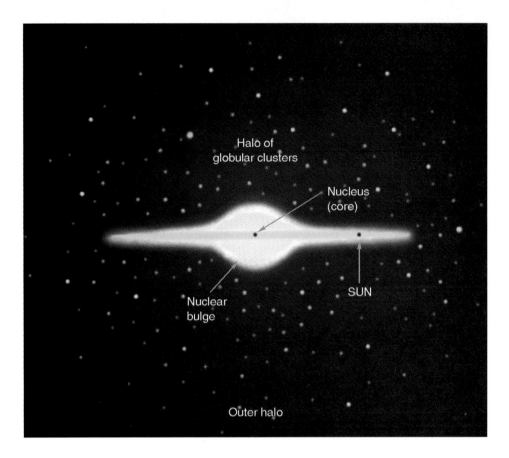

► FIGURE 16-11. The Galaxy consists of a nuclear bulge, a rotating disk, and a halo. The Sun is one of a few hundred billion stars in the galactic disk. The halo probably extends farther than indicated here.

Each of the four components will be described here, and then some will be discussed in more detail in sections of their own.

The *disk* is the large, flat part of the Galaxy that rotates in a plane around its center. The disk contains individual stars, clusters of stars—particularly open clusters—and almost all of the gas and dust found in the Galaxy. The edges of the disk are not well-defined, and therefore its width and thickness cannot be stated exactly. As noted earlier, its diameter is about 30,000 parsecs. Stars are most crowded near the plane of the Galaxy and become less crowded as we move from that plane. The disk is generally considered to be about 1000 parsecs thick, which makes its thickness about 3% of its diameter. Thus, it has the shape of about three compact disks stacked on one another.

Since star formation is more likely to occur where the interstellar material is most dense, stars are born with greatest frequency near the galactic plane. Gradually, their motions result in them wandering away from the plane. Very massive stars have short lifetimes, so we would expect to find few of them far from where they were formed. This is indeed the case, for almost all of the O-type stars lie within about 100 parsecs of the galactic plane. Only less massive stars live long enough to move very far from the plane.

It is difficult to determine the structure of the galactic disk from our position inside it because interstellar dust obstructs our view of distant locations. This dust is what limited early investigators' attempts to determine the Sun's position in the Galaxy. A later section will describe how radio astronomy has provided evidence for the spiral arms, and a Close Up outlines new evidence that the Galaxy may not be a standard spiral galaxy after all.

The *nuclear bulge* of the Galaxy is somewhat less than 10,000 parsecs (30,000 light-years) in diameter. Thus, if the disk is imagined to be a stack of three compact disks, a ping pong ball must be added at their center to represent the bulge. The ball should be squashed somewhat to represent the Galaxy correctly, because the bulge is wider along the disk than it is across its other di-

disk (of a galaxy). The flat, dense portion of a spiral galaxy that rotates in a plane around the nucleus.

O-type stars lie within about 300 light-years of the center of a disk that is about 3000 light-years thick.

nuclear bulge. The central region of a spiral galaxy.

Components of the Galaxy 487

HISTORICAL NOTE

The Shapley-Curtis Debate

In the late 1910s, there was considerable controversy about the size of the universe and the nature of the spiral nebulae. A number of scientific papers were published on these topics, and the controversy attracted national attention. Harlow Shapley of Mount Wilson Observatories and Heber Curtis of Lick Observatory became the primary spokesmen for two opposing views. In April 1920, the National Academy of Sciences invited these two astronomers to present papers before a meeting of that group in Washington, D.C., and to engage in a public debate concerning their research and conclusions.

At that time, Shapley had calculated the diameter of the Galaxy as about 300,000 light-years. The reason that his result was too large is that he did not know about the interstellar medium. Another error led him to calculate the distance to the Magellanic Clouds to be only 75,000 light-years—less than his calculated diameter of the Galaxy: he was using Cepheid variables to determine the distance to the clouds, but he did not know that there are two types of Cepheids and that he was seeing the unknown type in the Magellanic Clouds.

Adriaan van Maanen, a Dutch astronomer who was a friend of Shapley, had published results showing that he could observe rotation of the Andromeda nebula. In seeing the rotation, he was observing proper motion of parts of the nebula. Since we cannot observe proper motion within an object unless it is relatively close to us, this indicated that the nebula is nearby. (Van Maanen's observation was simply in error.) Because of his own observations and those of van Maanen, Shapley concluded that not only are the Magellanic Clouds and the Andromeda nebula part of our own system of stars, but that other spiral nebulae are also part of that system.

Curtis held the opposite view. He alleged that the group of stars that includes the Sun is much smaller than Shapley claimed and that the Andromeda nebula and other spiral nebulae are outside that group and are other "island universes." One bit of evidence he used to back his view was that Vesto Slipher of Lowell Observatory had investigated 15 spiral nebulae and found that 11 of them have significant redshifts. The redshifts indicated that they are moving away from us at great speeds, and this would make it unlikely that they are nearby. (These redshifts will be of major importance in the next chapter.)

It is interesting that although Shapley thought that spiral nebulae are part of our Galaxy, he held that the universe is larger than Curtis envisioned. In his recollections of the debate, Shapley says that the subject assigned to the debaters was the size of the universe, but that Curtis turned the topic to the nature of the spiral nebulae.

Most of the two or three hundred people present at the debate were members of the National Academy of Sciences, but the debate made headlines in the *New York Times* and excited public interest. When, just a few years later, Hubble discovered Cepheids in the Andromeda nebula and showed conclusively that it was indeed another "island universe" (as galaxies were called), it looked as though Shapley had been entirely wrong in the debate. In his own mind, however, Shapley felt that the score was even, for he had been more correct in predicting the overall size of the universe.

---

mension. Stars, dust, and gas are much more densely packed inside the nuclear bulge than they are anywhere else in the Galaxy. The bulge and its mysterious core will be discussed in a later section.

Figure 16-11 shows the galactic **halo,** which contains the globular clusters that caused Harlow Shapley to conclude that the Sun is not at the center of the Galaxy. Observe that a person viewing the globular clusters from the position of the Sun would see many more in some directions than in others, and that their distribution has its center at the nuclear bulge. Besides the globular clusters, the halo contains small amounts of gas and dust.

Globular clusters feel a gravitational force toward the center of the Galaxy and thus cannot simply hover at their positions in the halo. Instead, they orbit the galactic nucleus, passing through the disk twice during each orbit. ➤ Figure 16-12 indicates their motion. About 150 globular clusters are known to be associated with the Milky Way, and the orbits of 75 of them have been calculated. Of these, 63 are in the halo and 12 are confined to the disk.

halo (around a galaxy). The outermost part of a spiral galaxy; fairly spherical in shape, it lies beyond the spiral component.

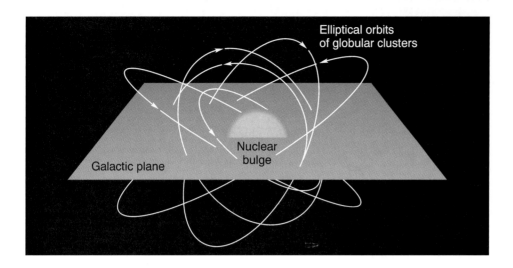

► FIGURE 16-12. Globular clusters orbit the galactic nucleus, passing through the disk twice during each orbit.

TABLE 16-1 Galactic Data

Radius of disk	50,000 light-years
Radius of nuclear bulge	15,000 light-years
Total radius of halo	200,000 light years?
Sun's distance from center	27,000 light-years
Sun's orbital period	250,000,000 years
Thickness of disk	3000 light-years
Number of stars	200 to 400 billion

The next chapter will discuss an unresolved problem in galactic astronomy: the motions of galaxies indicate that much more gravitational force is exerted on galaxies than can be explained by the amount of mass that we can see. Astronomers conclude that there is more invisible mass in the universe than visible mass—by a factor of 5 to 10! It is now thought that much of this mass exists in *extended halos* that surround galaxies. Therefore an outer halo was included in Figure 16-11 even though it has never been detected directly. We do not even know what the corona consists of—small black holes and/or great numbers of neutrinos have been hypothesized. It is thought to extend to perhaps two or three times the radius of the disk and halo.

The outer halo is sometimes called the *galactic corona*.

Table 16-1 shows today's values for various Milky Way properties. Keep in mind, though, that the numbers are approximate, not only because every measurement is uncertain to some extent, but because there are no specific boundaries for the various parts of the Galaxy. The total diameter of the halo is little more than a guess.

Galactic Motions

One method for determining the motion of the Sun around the galactic center relies on the observation that the orbits of globular clusters seem to be randomly distributed around the center of the Galaxy. If we assume that the average speed of all of the clusters, relative to the nucleus, is zero, we can measure their speeds relative to the Sun and attribute the average motion that is observed to the motion of the Sun. An analogy would be a person in a boat drifting on the ocean far from land and trying to measure the boat's speed. Suppose that the person

The Doppler effect is used to measure the speeds of the globular clusters.

Components of the Galaxy

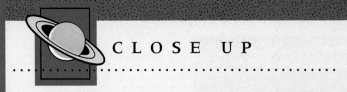

CLOSE UP

ET Life VII—Where Are They?

I hope that in studying the material in this chapter you have paused to contemplate the great size of the Galaxy and the tremendous number of stars in it. Perhaps your thoughts have turned to the implications these numbers have for the question of the likelihood of the existence of extraterrestrial life. A previous ET Life Close Up discussed the life equation, which uses the number of stars as its starting point. Even though eligible stars would have to be limited to population I stars (because they contain heavy elements) and perhaps to types F, G, and K, the number of stars that could support a planet with intelligent life is immense. A previous chapter pointed out that astronomers are discovering that the formation of planetary systems seems to be commonplace, and scientists have gone to great efforts to search for signals from intelligent extraterrestrial beings.

There is an opposing point of view. The argument goes as follows: If the formation of life is usual in the Milky Way, there should be a large number of civilizations like ours. We would expect many of the civilizations to be millions of years older than ours, because the Sun was formed fairly recently in the Galaxy's history. The older civilizations would be scientifically advanced enough to be able to travel to other stars. In fact, some would have been forced to leave their home planet when their star reached the end of its main sequence life and moved toward the red giant stage. These civilizations would have moved to hospitable planets near other stars and would have colonized space. (Aren't we likely to do so within the next million years?)

Therefore, if life begins and evolves as readily as some thinkers would want us to accept, the Galaxy should be teeming with life. We should have encountered not only radio signals from extraterrestrial beings, but Earth should have been visited. The question is, "Where are they?" Although some people say that UFOs of extraterrestrial origin have visited Earth on a number of occasions, no report of UFOs controlled by extraterrestrial beings has been substantiated to the degree demanded for acceptance in science. Few astronomers consider the reports credible.

The argument presented here leads to either of two conclusions: (1) Earth has been (and continues to be) visited by extraterrestrials who for some reason do not want to reveal themselves, or (2) intelligent life is rare in the Galaxy. If we refuse to accept the first, we are left with the second, and the money we spend searching is wasted. Or so the argument goes.

There is another completely different argument against spending money on the search for extraterrestrial intelligence (SETI). Recall that a good theory in science must have within itself the seeds of its own destruction. The theory must be capable of being proved wrong. Some scientists point out that no matter how much we spend over the next decade in searching for extraterrestrial radio signals, it is very possible that we will find nothing. This will not prove that extraterrestrials are not broadcasting radio signals, however, but only that we have not found them. Will those in favor of the search continue to ask for more money? When do we stop? This line of thinking leads to the conclusion that SETI is bad science and is not worth the considerable amounts of money being proposed for it. What do you think?

---

is able to measure the speed of planes flying overhead, and that equal numbers of planes are seen flying eastward and westward. If the person can assume that planes fly eastward at the same speed as planes fly westward, she would know that the average speed of all the planes is zero. But if her measurements show that relative to her the planes' average speed is 10 miles per hour toward the east, she could conclude that her boat is moving 10 miles per hour toward the west.

In practice, wind speed would affect the speed of the planes, and her assumption would not be valid.

Scientists always look for different methods to measure a quantity, and in this case another method exists. Recall that Oort and Lindblad studied the motions of great numbers of stars relative to the Sun and applied statistical methods to determine the motion of stars in the Sun's neighborhood. Similar studies yield a value for the speed of the Sun. The best measurements show that the Sun is traveling in a nearly circular path around the galactic center at a speed of about 220 kilometers per second. It is now moving toward the constellation Cygnus. Knowing that the radius of the Sun's orbit is 8500 parsecs (27,000 light-years), we can calculate the circumference of the Sun's path and find that it takes about

Saying that the Sun is moving toward Cygnus does not mean that it is getting closer to Cygnus; it means that the Sun is moving in that direction through space. Besides, each star of the constellation Cygnus has its own motion—some toward us and some away from us.

250 million years to complete one revolution. Although this seems a tremendously long time, the Sun has completed some 20 orbits during its 5-billion-year lifetime.

If the Galaxy had almost all of its mass concentrated at its center, its parts would rotate according to Kepler's third law, just as the planets do in the solar system. ➤Figure 16-13a shows the *galactic rotation curve* that would be expected in this case. If the mass were distributed so that most of it is fairly near the center, but not in the center, the curve would look more like that of Figure 16-13b. In fact, the galactic rotation curve is somewhat like Figure 16-13c. This indicates that the galactic mass is unevenly distributed, and the fact that the

galactic rotation curve. A graph of the orbital speed of objects in the galactic disk as a function of their distance from the center.

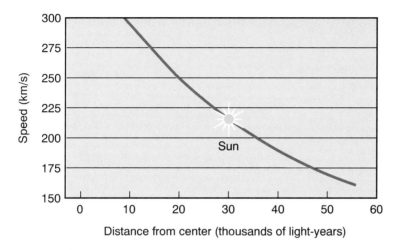
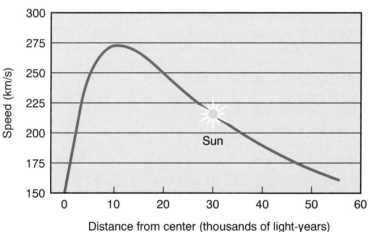
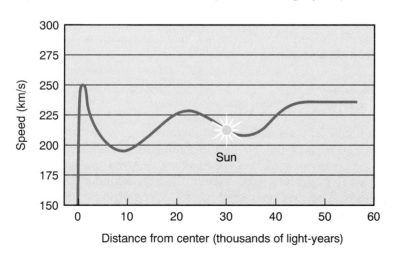

➤ **FIGURE 16-13.** (a) If all of the mass of the Galaxy were concentrated at its center, the galactic rotation curve would follow Kepler's laws, so that the speeds of objects farther and farther from the center move at lower and lower speeds, as shown in this graph. (b) If the mass were distributed so that most of it is somewhat near—but not at the center—a rotation curve such as this would result. (c) This is the observed rotation curve of the Galaxy. It indicates that large amounts of mass orbit the center far beyond the Sun's orbit.

Components of the Galaxy

speeds of objects even far out from the Sun are not lower than the Sun's speed indicates that large amounts of mass lie beyond the Sun's distance.

As we might expect, however, the orbits of stars are not perfectly circular. Along with the general orbiting motion of each portion of the Galaxy, each star has its own peculiar motion—perhaps having a component of motion from one side of the disk to the other or perhaps moving closer to the galactic center at one time and moving farther away at another time as it follows an elliptical path. In general, however, the paths of stars seem to be nearly circular.

The Mass of the Galaxy

Chapter 3 showed that Isaac Newton revised Kepler's third law to read

$$\frac{a^3}{P^2} = K(m_1 + m_2)$$

a = average radius of the orbit
P = period of the orbit
K = a constant whose value depends on the units used in the equation
m_1 and m_2 = the masses of each of the objects

Later it was shown that the relationship holds not only for objects in orbit about the Sun but also for satellites in orbit around the planets. Finally, Chapter 12 showed the law's application to binary star systems.

The discovery (in 1927) by Oort and Lindblad that the Galaxy in the Sun's neighborhood undergoes differential rotation meant that Kepler's third law might be applied to it to calculate the masses involved. In the case of a star revolving around the Galaxy's center, one of the masses in the equation is the mass of the star, and the other is the mass of the entire inner Galaxy, including all objects in the Galaxy that are closer to the center than that star is. This may seem strange since the inner portion of the Galaxy is made up of a number of objects rather than one as in the case of the Earth's orbit around the Sun. It can be shown, however, that this is the correct application of the equation.

In the accompanying Close Up, Kepler's law is used to calculate the mass of the inner galaxy. The value obtained, 110 billion solar masses, must be taken as approximate. Also keep in mind that this is not the mass of the entire Galaxy but simply the mass of the inner portion. Recent analysis of the pattern of rotation in the outer parts of the Galaxy indicates that the total mass of the Galaxy is about 10^{12} (a thousand billion) solar masses, or about 10 times more mass than is calculated for the inner Galaxy assuming the applicability of Kepler's third law. At present, the nature of this additional mass is unknown for there are not enough stars to account for that much mass. We will see later that a similar problem exists in the case of other galaxies.

Refer to the discussion of dark matter in Chapter 17 and again in Chapter 18.

THE SPIRAL ARMS

spiral galaxy. A disk-shaped galaxy with arms in a spiral pattern.

The **spiral** nature of the Galaxy is not obvious from observations in visible light because we can see only limited distances along the plane of the Galaxy. In 1951, however, astronomers at Yerkes Observatory detected that as one looks either toward or away from the galactic center, the distribution of O- and B-type stars is not uniform. They seem to be clustered at certain distances. ➤Figure 16-14 illustrates the concentrations of this type. This was the first hint of the spiral nature of the Galaxy. More evidence came with the discovery of the 21-centimeter line of hydrogen during the same year.

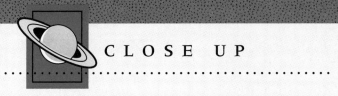

CLOSE UP

Calculating the Mass of the Inner Galaxy

The equation for Kepler's third law, along with the Sun's motion, can be used to calculate the mass of the inner part of the Galaxy. In the equation, the constant K is equal to one if a is expressed in astronomical units, P in years, and m_1 and m_2 in solar masses. The first step, then, is to express the Sun's 30,000 light-year distance from the Galaxy's center in astronomical units. Appendix A shows that one light-year is equivalent to 63,000 astronomical units. Thus,

$$30{,}000 \text{ LY} \times \frac{63{,}000 \text{ AU}}{1 \text{ LY}} = 1.9 \times 10^9 \text{ AU}$$

Using this value, along with the Sun's period of 250,000,000 years, in Kepler's equation,

$$\begin{aligned} m_1 + m_2 &= a^3/P^2 \\ &= \frac{(1.9 \times 10^9 \text{ AU})^3}{(2.5 \times 10^8 \text{ yr})^2} \\ &= 1.1 \times 10^{11} \text{ solar masses} \end{aligned}$$

The value obtained is the total of the mass of the Sun and the mass of the inner Galaxy. The mass of the Sun, of course, is negligible compared to the value obtained, so this answer is simply the mass of the part of the Galaxy that lies within the Sun's orbit.

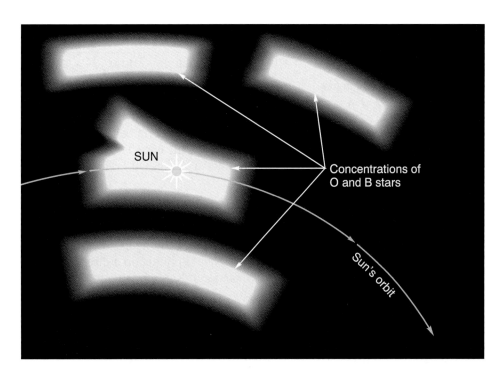

▶ FIGURE 16-14. O- and B-type stars are found at greatest concentrations in the shaded areas.

Cool hydrogen gas emits radiation of a particular wavelength: 21.1 centimeters. This is part of the radio portion of the spectrum and is not as readily absorbed by the interstellar medium as is light. Radio telescopes can therefore use **21-centimeter radiation** to detect high concentrations of cool hydrogen such as exists in interstellar clouds. Since hydrogen is the main component of interstellar material—and the entire universe—this is an ideal method of detecting cool hydrogen clouds at great distances.

As we have seen, it is gas and dust from which new stars arise and to which some of the material of a massive star returns at the end of the star's lifetime. Therefore hydrogen gas clouds detected by 21-centimeter radiation are located at the same place as newly forming stars. Recall that O and B stars are very

twenty-one-centimeter radiation. Radiation from atomic hydrogen, with a wavelength of 21.1 centimeters.

The Spiral Arms

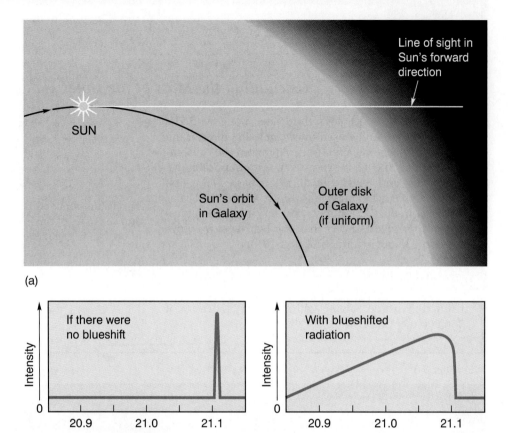

FIGURE 16-15. If hydrogen gas were distributed uniformly in the galactic disk, 21-centimeter radiation from the Sun's forward direction would show a fairly regular blueshift (c).

massive and are therefore short-lived. This means that they will be found only where stars have recently formed and that we might expect to find hydrogen clouds at locations identified by the astronomers at Yerkes Observatory as having high concentrations of O- and B-type stars.

It may seem that it would be impossible to determine the distance of a source of 21-centimeter radiation, for a radio telescope would seem to reveal only the direction of the source of the waves. In a sense this is true. The Doppler effect, however, allows us to determine the radial motion of the hydrogen with respect to us. Suppose, for example, that a radio telescope is pointed in a direction across the Galaxy as shown in ➤Figure 16-15a. Because of Keplerian motion, we would be moving toward the hydrogen along this direction, so a blueshift of the 21-centimeter radiation would be seen. If the hydrogen were distributed uniformly, the radiation would be Doppler shifted so that its wavelength would range from just less than 21.1 cm to quite a bit less. A graph of radio intensity versus wavelength might then appear as shown in part (c) of this figure.

On the other hand, consider how the radiation would appear if the hydrogen is located in spiral arms as shown in ➤Figure 16-16a. In this case, each spiral arm would have a fixed value for its Doppler shift, and the graph would appear as shown in Figure 16-16b. This is how it appears in an actual case, so 21-centimeter radiation provides us further evidence for the spiral nature of the Galaxy.

Mapping the Galaxy in 21-centimeter radiation contains plenty of room for error, but there is no doubt that ours is a spiral galaxy. As the next chapter will show, a spiral appearance is common for galaxies. ➤Figure 16-17a shows a spiral galaxy that may be similar to the Milky Way, although recent work indicates that our Galaxy may have an elongated nucleus, more like the galaxy of Figure 16-17b. A nearby Close Up discusses this idea.

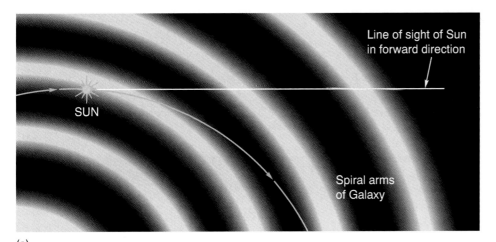
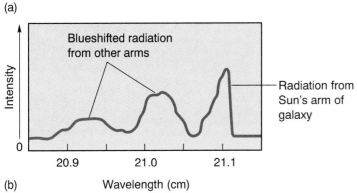

▶ FIGURE 16-16. Twenty-one-centimeter radiation observed in the Sun's forward direction actually shows peaks of blueshifted radiation (b). The peaks occur because the hydrogen that emits the radiation is concentrated in spiral arms (a).

▶ FIGURE 16-17. If we could get outside our Galaxy and view it face-on, it would probably look like one of these galaxies. The bright stars in each image are part of our Galaxy. Because they are much closer to us than the other galaxy, we see them as we see flies on a car windshield when we look through it. (a) NGC 2997 is a spiral galaxy in the southern constellation Antlia. (b) The galaxy M83 in Hydra (also a southern constellation) has a slightly elongated nucleus. Our Galaxy probably has two primary arms like this one, with irregular branches extending from each.

The Spiral Arms

CLOSE UP

The Milky Way: A Barred Spiral Galaxy?

The next chapter will show that spiral galaxies are of two types. About half are standard spiral galaxies, like the one shown in Figure 16-17a. The other half have an elongated central bulge that appears as a bar across the center, like the one in ▶Figure C16-1. Until recently astronomers thought that the Milky Way is a standard spiral galaxy, but evidence is accumulating that our Galaxy is a barred spiral galaxy with an elongated nucleus that is probably shorter in proportion to the rest of the galaxy than the one in Figure C16-1.

The new evidence comes primarily from two sources. We are learning more and more by analysis of 21-centimeter radiation, which penetrates the clouds of interstellar space and allows us to detect the motion of clouds of cool hydrogen gas near the nuclear bulge. Second, in 1989 NASA launched the *Cosmic Background Explorer (COBE)*. It was designed to measure short-wavelength radio waves from space (for reasons to be discussed in Chapter 18).

Suppose the central bulge is elongated with its long axis oriented at some oblique angle to the line of sight, as in ▶Figure C16-2. The Sun (and Earth) is located at the point indicated in the figure. Line C extends to the center of the Galaxy, and lines L and R are drawn at equal angles on each side of line C. Notice that the line of sight on the near side of the nucleus (the left side here) passes through part of the nucleus, but the equivalent line of sight on the other side does not. If the nucleus is shaped like this and has this orientation, astronomers should detect an asymmetry in the radiation from near the nucleus. New data from *COBE*, as well as data from Japanese astronomers, correspond to what would be expected from such an elongated nucleus.

Further information comes from analyzing motions of the gas near the nucleus. The gas moves around the nucleus in a noncircular pattern, corresponding to what would be expected from a barred nucleus. The data indicate that the bar is a fairly short one and that it is oriented almost in line with the Sun, so that measurements like those illustrated in Figure C16-2 are very difficult to make.

Astronomers are not in complete agreement on the interpretation of the new data, but because of the importance of knowing the basic shape of our Galaxy, the subject has become a hot topic of research.

▶ FIGURE C16-1. The Milky Way may be a barred spiral galaxy similar to this one.

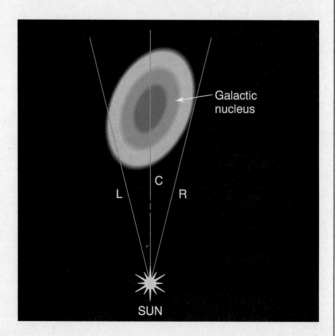

▶ FIGURE C16-2. Opposite sides of an elongated nucleus—a bar—do not appear the same when viewed from the Sun, because lines that make equal angles with the central line pass through different volumes of the bar. Data indicate that the nucleus of the Milky Way is slightly elongated with its long axis almost aligned with the Sun.

SPIRAL ARM THEORIES

It may seem at first glance that spiral arms could be explained by the fact that stars near the center of the Galaxy complete a circle in less time than those farther out. This would occur even if the stars do not orbit according to Kepler's third law simply because stars near the center have less distance to cover to complete their orbits. Consider a group of stars that at a given time are in a straight line across the center of the Galaxy as shown in ➤Figure 16-18a. As the stars move around the center (even if they all move at the same speed), the line of stars will wind up into a spiral. The problem is that they would wind up too much. Recall that the Sun has made some 20 revolutions around the Galaxy. As Figures 16-18b, c, and d show, the line of stars would wind up so much that it would not be distinguishable as a line. The Galaxy would appear as a fairly uniform disk. Yet we do perceive a spiral pattern in our Galaxy and in many other galaxies. Therefore this simple differential rotation hypothesis cannot be correct.

Currently, there are two competing theories to explain the spiral nature of galaxies: the density wave theory and the chain reaction theory.

The Density Wave Theory

The *density wave theory* was first proposed in 1960 by Bertil Lindblad and holds that what we see in a spiral arm of a distant galaxy is not a simple fixed line of

density wave theory. A model for spiral galaxies that proposes that the arms are the result of density waves sweeping around the galaxy.

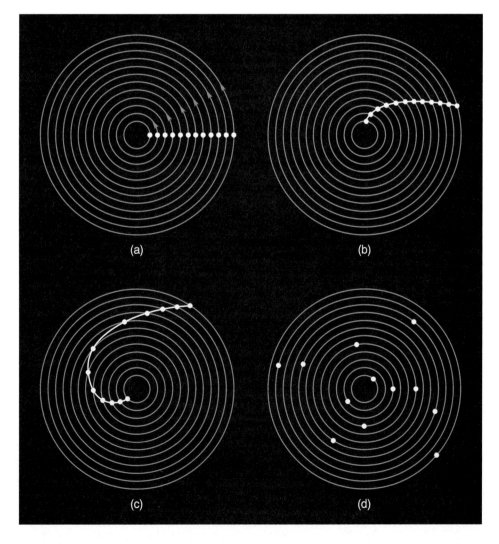

➤ **FIGURE 16-18.** Stars that start out in a straight line from the Galaxy's center (a) would begin to show a spiral pattern (b) and (c), but after a few revolutions, the pattern would be lost (d).

Spiral Arm Theories

density wave. A wave in which areas of high and low pressure move through the medium.

stars but a line formed by the brightest stars and the glowing nebulae surrounding them. The theory holds that stars revolve around the Galaxy independent of the spiral arms and that the arms are simply areas where the density of gas is greater than at other places. According to this theory, there are almost as many stars per volume between the arms as in the arms, but the arms contain more of the brightest stars and a higher density of gas and dust. The areas of denser gas move around the Galaxy in *density waves,* causing the formation of new stars and glowing emission nebulae. This is best explained by an analogy.

Suppose cars are traveling on a long superhighway at a speed exceeding the speed limit—perhaps leaving campuses for spring break. Also traveling along the road, at less than the speed limit, is a traffic patrol car with its radar on. We are observers in a helicopter high in the sky and we see the cars fairly evenly distributed along the highway except around the police car. For a short distance behind and in front of the police vehicle, the cars are bunched up. As a given car approaches the police cruiser from behind, the car slows down, slowly passing the feared patrol. Then when the driver feels that his or her car is safely in front of the police cruiser, it again speeds up. As a result, there is constantly a high density of cars around the police car even though the cars making up that group change all the time. ➤Figure 16-19 illustrates the situation. The police car moves along, seeming to carry its high-density group along with it in what we see from the helicopter to be a density wave.

In a galaxy, the density wave consists of a region of gas and dust that is more dense than normal. Density waves are common here on Earth, for every sound wave is a density wave with regions of high and low density making up the wave. The major difference between sound waves and the density waves of a spiral galaxy is that in the atmosphere of Earth, a sound wave travels faster than the particles of the gas itself but in the near-vacuum of a galaxy, the wave travels more slowly than the particles. The gas and dust particles—as well as stars—catch the wave from behind and pass through it in much the same way that cars in the analogy passed through the density pulse around the police car. ➤Figure 16-20 illustrates the idea.

➤ FIGURE 16-19. This represents an overhead view of a highway seen as time passes. Car X is a police car, moving just slower than the speed limit. Bunched up around that car are other slow-moving cars that only gradually pass the police car. Thus, as time goes by, the cars near X are different, but a high density of cars remains there.

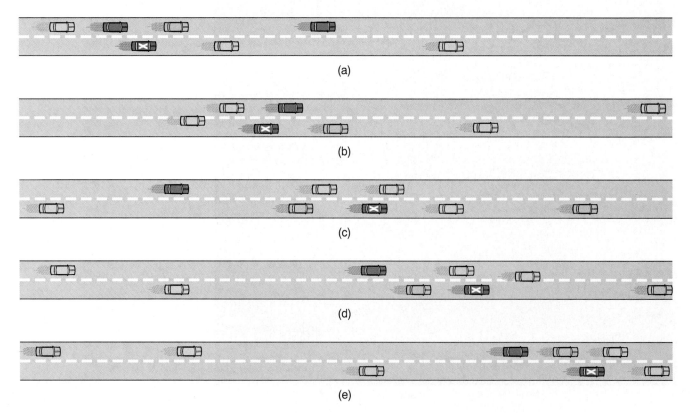

498   CHAPTER 16   The Milky Way Galaxy

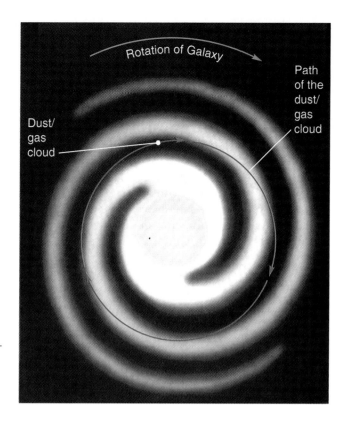

► FIGURE 16-20. The orbit of objects in the Galaxy is very nearly a circle. Since their orbital speed is greater than the speed of the density wave that forms the spiral pattern, they pass in and out of the spiral arms. In the figure, a cloud of dust and gas is about to enter a spiral arm. This will cause the cloud to compress, resulting in star formation.

Now recall that new stars are formed when an interstellar cloud of gas collapses. The density wave theory holds that the trigger for this collapse is the wave. As interstellar clouds approach the density wave from behind, they are compressed and stars are formed. Although stars of all masses are formed along the edge of the density wave, the brightest, most massive stars have ended their lives before they pass far from this edge. This means that when we look at a spiral galaxy, the spiral arms are obvious to us because they are the areas containing the bright stars.

One problem with the density wave theory is the question of how the density wave is sustained through the life of the galaxy. Computer simulations indicate that it would die out. In addition, a recent infrared image of the center of the Whirlpool galaxy (Figure 16-21b) indicates that spiral arms penetrate much further into the nucleus of a galaxy than had previously been thought. The density

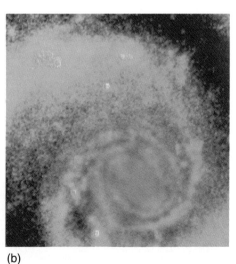

► FIGURE 16-21. (a) The Whirlpool galaxy (M-51), seen here in visible light, is an excellent example of a spiral galaxy. (b) This infrared image of the central portion of the Whirlpool galaxy indicates that the spiral arms penetrate farther into the nucleus than was previously thought possible. The finding conflicts with predictions made by the density wave theory.

(a)   (b)

Spiral Arm Theories

wave theory had predicted that a galactic nucleus absorbs density waves, preventing the waves from extending into it. The new information from the Whirlpool galaxy is causing a reexamination of the density wave theory. Even before the discovery, however, another theory had been proposed: the chain reaction theory.

The Chain Reaction Theory

According to the **chain reaction theory** of galactic spiral arms, the triggers that start the collapse of most interstellar clouds are nearby supernova explosions. Then, as the more massive stars finish their lives and become supernovae, they trigger more star formation, and so on. Thus, the formation of new stars is confined to areas where this process is taking place. Now differential rotation enters the picture. Computer analysis shows that at the rate at which massive stars would be formed, Keplerian motion would cause spiral arms to be formed and sustained.

The chain reaction model is a much more recent proposal than the density wave model. Time will tell which theory best fits further analysis and new observations. (Or perhaps parts of each will have to be combined into a new theory.) At present both theories are able to explain the spiral structure of this type of galaxy, but the chain reaction theory has the advantage of being able to explain how a spiral arm would begin since the chain reaction starts with a single supernova.

THE GALACTIC NUCLEUS

Observations by Shapley, Oort, and Lindblad in the early part of this century revealed that the center of the Galaxy lies in the direction of Sagittarius. Because visible light from the galactic nucleus does not reach us, astronomers had to await the development of nonoptical telescopes to learn more about that nucleus.

In 1960, observations of the 21-centimeter radiation pinpointed the galactic nucleus and revealed that hydrogen clouds there are in a very turbulent state, moving at very high speeds. In addition, the radiation indicated that two great expanding arms of hydrogen are moving away from the center: one on this side moving toward us and another on the other side moving away. The Doppler effect cannot tell us about material moving across our line of sight, but one hypothesis is that a ring or a shell of hydrogen has been blown out of the center.

The orbiting infrared observatories have revealed that the galactic nucleus is a very bright infrared source—equivalent to hundreds of Suns. The size of the source (judged from its angular size and its distance) seems to be only a few light-years across, which is very small for the amount of infrared radiation emitted.

chain reaction theory. A model for spiral galaxies that explains the arms as resulting from a series of supernovae, each triggering the formation of new stars.

➤ FIGURE 16-22. The *Cosmic Background Explorer (COBE)* satellite provided this false-color image of the inner part of the Galaxy. The wide-angle view shows the nuclear bulge and the dense central plane near the bulge.

➤Figure 16-22 is an infrared photograph of the inner part of the Galaxy. It may look like a photograph of the entire galaxy, but we certainly have not been outside the Galaxy to take such a photo. This one, taken by the *Cosmic Background Explorer* satellite, shows only the Galaxy's central portion, including the nuclear bulge. The reddest part is where intervening interstellar dust absorbs much of the infrared. The white dots are stars near us.

Finally, gamma rays detected by satellite receivers reveal that an even smaller source in the nucleus emits tremendous amounts of energy in this region of the spectrum. What is the source of energy in the nucleus of the Galaxy? The only source that seems feasible is gravity, and the current hypothesis holds that the center of the Galaxy is a gigantic black hole, orbited nearby by a great number of stars. As matter falls into the black hole, losing its orbital motion due to collisions, energy is released. The release of energy here is similar to the energy released as matter falls into stellar black holes, discussed in the previous chapter. In the case of the galactic nucleus, however, a *super*massive black hole is at the center, and it is surrounded by a great amount of matter, causing enormous quantities of energy to be released.

Recent discoveries of fast-moving stars and gas near the center of the Galaxy lend support for the black hole hypothesis, for the Doppler shift indicates that the gas is orbiting the center with a speed that would require a central mass of perhaps 5 million solar masses in order to hold the gas in orbit. It seems that only a black hole could be this massive.

Some other spiral galaxies also have tremendously powerful energy sources at their center, and we see signs of violent activity near those centers. The supermassive black hole hypothesis is rapidly gaining support among astronomers and seems to be the best explanation for the energy source in our Galaxy and in similar galaxies. As the next chapter will show, there are galaxies that produce even more energy in their nuclei. We are far from understanding the nucleus of our Galaxy, and further knowledge must come from observations of other galaxies as well as our own.

www.jbpub.com/starlinks

THE EVOLUTION OF THE GALAXY

The first step in developing a theory of the formation of the Galaxy is to consider the chemical content and the age of the Galaxy's components. Once that is done, the pieces can be put together to give us a picture of how the Galaxy developed.

Age and Composition of the Galaxy

A previous chapter discussed the division of stars into population I and population II according to their percentage of heavy elements. Although in reality there is a continuum from the lowest percentage of heavy elements in extreme population II to the greatest percentage in extreme population I, the grouping is still convenient. On the average, population II stars contain only about 1% of the heavy elements of population I stars.

Population I stars are stars that have formed from the remains of previous generations of stars, and most of them are found in the disk of the Galaxy. In fact, stars with the greatest percentage of heavy elements lie close to the nuclear bulge. The farther from the bulge, the lower the fraction of heavy elements. Most globular clusters, on the other hand, are made up of population II stars. Of the known globular clusters in the Milky Way, about 70% have an average heavy-element content of about one-twentieth of that of the Sun. The remaining 30% contain about one-third the heavy-element content of the Sun.

Star clusters provide a convenient method of measuring the age of stars. Recall that all stars in a cluster were formed at about the same time. Knowing this, astronomers can determine a cluster's age by analyzing its H-R diagram.

Astronomers say that a star with a large abundance of heavy elements is *rich in metals*, or *metal rich*.

The abundance of heavy elements decreases by a factor of 0.8 for each thousand parsecs from the center of the disk.

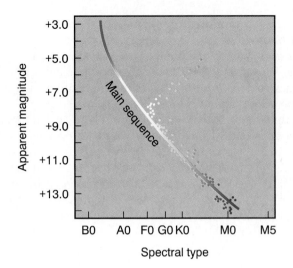

► **FIGURE 16-23.** The H-R diagram of a galactic cluster shows a definite main sequence turn-off point. From the location of the turn-off point on the main sequence, astronomers can determine the age of the cluster.

Consult the beginning of Chapter 14 for a discussion of the determination of the age of a star cluster.

►Figure 16-23 is the H-R diagram of a cluster old enough that its heaviest stars have left the main sequence. The position of the main sequence "turn-off point" permits a calculation of the age of the cluster.

The classification of stars in a specific portion of the Galaxy can provide a clue to the age of that portion. For example, in the part of the galactic disk near the Sun, there are no white dwarfs. Since white dwarfs do not form until after stars have gone through the main sequence and red giant stages, this sets a lower limit of about 10 billion years for the age of the Sun's portion of the disk.

Within the disk, O- and B-type stars are found primarily near the galactic plane. Since these stars have short lifetimes, the fact that they are not found at great distances from the Galaxy's plane tells us that star formation does not occur to any great extent except along that plane.

Stars in the halo of the Galaxy are older than those in the disk. While the oldest stars in the portion of the disk where the Sun is located are about 10 billion years old, the *youngest* stars in globular clusters are about 11 billion years old. Thus, globular clusters formed before the disk portion of the Galaxy. The oldest globular clusters are about 15 to 17 billion years old, and since these are the oldest stars we find, we take this to be the minimum age of the Galaxy.

The age of stars in the nuclear bulge is difficult to determine because the bulge is obscured by dust. It can be studied only in the X-ray, infrared, and radio portions of the spectrum. Observations through a few gaps in the dust have allowed astronomers to sample the outermost bulge population, however, and they have learned that the stars of the nuclear bulge are very rich in heavy elements. The stars of the bulge all appear to be red giants of types K and M. This tells us that the nuclear bulge must be very old, according to the following logic: First, the stars themselves are old, for K- and M-type stars live long lives on the main sequence before becoming red giants, and secondly, they were formed from the remains of previous generations of stars.

The Galaxy's History

Although astronomers disagree about the specifics, a general picture of the evolution of our Galaxy is emerging. The process is somewhat parallel to the way stars form. The Galaxy began as a tremendous cloud of gas and dust bigger than the present galactic halo (►Figure 16-24). Mutual gravitation between the cloud's parts gradually pulled it together. The center portion was the first to become dense enough for stars to form, and it soon became a hotbed of star formation. Soon thereafter, stars began forming in dense pockets, hundreds of

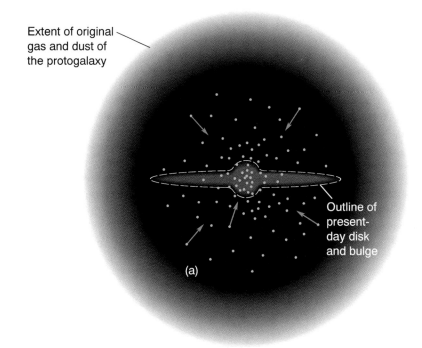
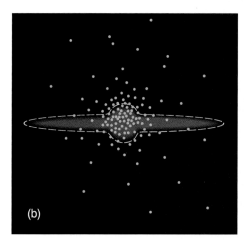
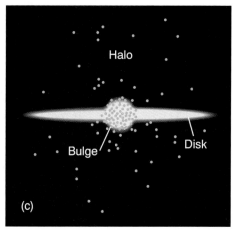

► **FIGURE 16-24.** (a) According to a leading theory of galactic evolution, the Galaxy began as a sphere of gas and dust about 15 to 17 billion years ago. (b) As the sphere collapsed toward the center, stars and clusters formed, so that the Galaxy appeared like this some 12 to 14 billion years ago. (c) The Galaxy reached its present state after the remaining gas and dust formed into a disk and spiral arms developed.

light-years across, that were in orbit around the center. They formed what are now the globular clusters.

The initial giant cloud had some rotation, and as it contracted, it spun faster. Just as occurs during a star's protostar stage, the rotating matter formed into a disk. In the case of a star, almost all orbiting material becomes caught in the disk, but in the Galaxy, the density of matter in the disk is low enough that globular clusters pass through it without being captured. Finally, density waves formed in the Galaxy's disk, creating the spiral arms where star formation continues today.

Even though this scenario is very sketchy, not all astronomers agree with all of it. For example, in some models, several separate clouds of gas merge to form the Galaxy rather than one. Our knowledge of the Galaxy has only recently progressed to the point where we can formulate meaningful models of its formation, and today's models will have to be developed, and perhaps combined, before we can be confident of their validity.

As discussed earlier, the density wave theory is not the only possible explanation for the spiral arms.

The formation of galaxies is discussed further in the next chapter.

The Evolution of the Galaxy

CONCLUSION

Our Sun is one of some 200,000,000,000 stars in an enormous spiral galaxy that is itself just one of countless galaxies in the observable universe. The nature of our Galaxy was discovered only rather recently because our vision is obscured by gas and dust along the Galaxy's disk. Although measurements of the Galaxy's characteristics are necessarily imprecise, some of its properties can be measured using techniques dating from the time of Isaac Newton, such as the calculation of the Galaxy's mass using Kepler's third law. Detection and measurement of many other features of the Galaxy must rely on more modern methods and equipment, such as radio astronomy, which has opened many doors to understanding the Milky Way.

Analysis of 21-centimeter radiation leads us to conclude that our Galaxy is a spiral one, like many others we observe. Two principal theories have been proposed to explain the spiral arms of a galaxy, the density wave theory and the chain reaction theory.

The nuclear bulge of our Galaxy contains a high density of stars. The tremendous energy that pours from the nucleus, as well as the rapid motions of objects near it, are consistent with the hypothesis that a massive black hole is at the center of the Galaxy.

Data that are accumulating about the Galaxy permit us to develop models of galactic formation, but these models, like many of astronomy's theories about the Milky Way, are very tentative and await further research by today's and tomorrow's scientists.

RECALL QUESTIONS

1. Our Sun is near the center of the Milky Way Galaxy.
   A. True.
   B. False.
   C. We don't know if this is true or false.

2. Astronomers of the eighteenth and nineteenth centuries thought that the Sun was near the center of the Milky Way Galaxy because they counted the same number of stars in the disk of the Galaxy in every direction. The reason they were not correct is that the Galaxy
   A. is an irregular galaxy with a chaotic shape.
   B. contains dust that obscures its distant regions.
   C. has the shape of a tube with the Sun near one end.
   D. has two kinds of Cepheid variables, so that all distance measurements until recently were incorrect.
   E. has a giant black hole at its center.

3. Shapley first discovered the direction to the center of the Milky Way by observing the distances and directions to
   A. globular clusters.
   B. other galaxies.
   C. the galactic center itself.
   D. open clusters.
   E. [None of the above.]

4. The Sun is one of about how many stars in the Milky Way Galaxy?
   A. 100 million to 200 million
   B. One billion
   C. 100 to 400 billion
   D. One trillion
   E. [We don't have any idea how many stars are in the Galaxy.]

5. Evidence for the spiral nature of the Galaxy comes primarily from
   A. Cepheid variable data.
   B. Doppler shift data of stars.
   C. 21-centimeter data.
   D. observations of globular clusters.
   E. [Both A and D above.]

6. We have learned that the Galaxy is a spiral galaxy by
   A. the reflection of light from nebulae.
   B. the Doppler shift of radio waves.
   C. observations of globular clusters.
   D. comparing ages of various clusters.
   E. plotting the stars of various clusters on the H-R diagram.

7. Why is observation of distant parts of the Galaxy limited in the visible region?
   A. Interstellar dust blocks visible light.
   B. The redshift of distant stars makes them difficult to detect.
   C. Opacity of the atmosphere blocks visible light.
   D. Relative motion of the spiral arms makes them difficult to detect.

8. Radio observations can provide information about the Galaxy that we cannot learn from observations in visible light because
   A. radio waves are able to pass through clouds of dust.
   B. stars emit a greater amount of energy in radio wavelengths than in visible wavelengths.
   C. the velocity of radio waves is greater than that of light.
   D. our Galaxy is spiral in nature.

9. William and Caroline Herschel counted stars in various directions in an attempt to determine the Sun's position in the Galaxy. Later, Jacobus Kapteyn tried to im-

prove on their determination by taking into account
- A. the interstellar dust.
- B. radio waves from stars.
- C. bright nebulae between the stars.
- D. redshift data on stars in various directions.
- E. the density of stars in various distances.

10. Oort and Lindblad determined the motion of the Sun relative to other stars by analyzing
    - A. the interstellar dust.
    - B. radio waves from stars.
    - C. bright nebulae between the stars.
    - D. redshift data on stars in various directions.
    - E. the density of stars in various distances.

11. The most dependable estimates of the mass of the Galaxy are based on
    - A. Kepler's laws.
    - B. observations of nebulae in spiral arms.
    - C. star counts.
    - D. distances to Cepheid variables.

12. According to the density wave theory of the spiral arms,
    - A. the number of stars per volume is about the same between the spiral arms as in them.
    - B. the number of stars per volume is much greater between the spiral arms than in them.
    - C. the number of stars per volume is much less between the spiral arms than in them.
    - D. [The theory predicts nothing about the number of stars per volume of space.]

13. How does the density wave theory of spiral arms explain why they are brighter than the areas between the arms?
    - A. There are more stars per volume in the arms than between them.
    - B. Brighter stars are in the spiral arms because they follow along with the arms as they move.
    - C. Brighter stars are in the spiral arms because they are formed there and are short-lived.

14. According to the chain reaction theory of spiral galaxies' arms,
    - A. stars revolve around the galactic center in perfect circles.
    - B. stars obey Kepler's third law as they move around the galactic center.
    - C. stars are about evenly spread around the galaxy.
    - D. [Two of the above.]
    - E. [None of the above.]

15. A black hole is thought to be at the center of the Galaxy because
    - A. no light comes from that area of the Galaxy.
    - B. large amounts of energy come from a small source there.
    - C. motions of objects near the center indicate a large mass there.
    - D. [Both A and C above.]
    - E. [Both B and C above.]

QUESTIONS TO PONDER

1. If stars revolve around the galactic center independent of the spiral arms, it would seem that there should be as many stars per unit volume between the arms as inside them. Yet the text states that there is *almost* the same density of stars between the arms as inside them. Why would there be a difference at all?

2. Explain how Kepler's law is used to calculate the mass of the Galaxy. Discuss the limitations of this method.

3. Explain why the hypothesized galactic rotation curve of Figure 16-12b extends down to very low speeds near the zero point on the bottom axis while the curve of Figure 16-12a does not.

4. If a radio telescope cannot reveal directly the distance to a source of 21-centimeter radiation, how can this radiation be used to reveal spiral arms in the Galaxy?

5. Write a report on various models of the formation of the Galaxy based on the article "How the Milky Way Formed," In *Scientific American*, January 1993, pp. 72-78.

6. What is the evidence that stars of the nuclear bulge formed before stars in other parts of the Galaxy?

CALCULATIONS

1. Show how one can calculate the period of the Sun's motion around the Galaxy, given that the Sun's distance to the center is 30,000 LY and its speed is 250 km/sec.

2. Suppose that a portion of another galaxy is observed to move with a speed of 250 km/sec and to be at a distance of 40,000 LY from the center of that galaxy. What is the period of revolution of that portion of the galaxy?

3. Use the data from the last question to calculate the mass of that galaxy that is inside the radius of the portion described.

4. Suppose we wish to construct a physical model of the Galaxy. If the Sun is represented by a small grain of sand 0.1 millimeter across, what will be the diameter of the model? (Hint: See the Activity)

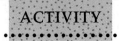

ACTIVITY

The Scale of the Galaxy

Suppose that you wish to construct a scale drawing of our region of the universe. The Sun has an actual diameter of about 1,500,000 kilometers, and you represent it by a dot the size of a period on this page, about 0.5 mm across. The average distance between stars in our region of the Galaxy is about 5 LY. Therefore, you must calculate what the corresponding distance would be on your drawing, given that one light-year is equal to 9.5×10^{12} km.

First, set up the ratio of the distance on your scale to the actual distance:

$$\frac{\text{distance on scale}}{\text{actual distance}} = \frac{0.5 \text{ mm}}{1.5 \times 10^6 \text{ km}}$$

Next, calculate the distance between stars in kilometers:

$$5 \text{ LY} \times \frac{9.5 \times 10^{12} \text{ km}}{1 \text{ LY}} = 4.8 \times 10^{13} \text{ km}$$

Now, use that value as the actual distance and solve for the distance in the drawing that represents the separation of stars:

$$\frac{\text{distance on scale}}{4.8 \times 10^{13} \text{ km}} = \frac{0.5 \text{ mm}}{1.5 \times 10^6 \text{ km}}$$

$$\text{distance on scale} = 1.6 \times 10^7 \text{ mm}$$

Finally, change this to more appropriate units:

$$1.6 \times 10^7 \text{ mm} \times \frac{1 \text{ m}}{10^3 \text{ mm}} = 1.6 \times 10^4 \text{ m}$$

This is 16 kilometers! It means that on your drawing stars must be 16 kilometers, or 10 miles, apart.

The Galaxy is about 100,000 LY in diameter, or 20,000 times the average distance between stars. This means that the Galaxy on your scale would be 320,000 kilometers across. This is beyond imagining, and distances to other galaxies have not yet been calculated. Therefore you will have to try another scale.

Suppose you take the average distance between stars on your scale to be 5 centimeters. Although the stars would be too small to see on an actual drawing to this scale, you might cheat and make each star a very tiny speck.

On your drawing, 5 centimeters will correspond to 5 LY. Now calculate the size of the Galaxy, the distance to the Andromeda galaxy (which is actually about 2 MLY away), and the distance to the most distant objects we can see (which, as discussed in the next chapter, are about 15 billion LY away).

StarLinks netQuestions

Visit the netQuestions area of StarLinks (www.jbpub.com/starlinks) to complete exercises on these topics:

1. A View of the Milky Way The English word "galaxy" is derived from the Greek root *gala* meaning *milk*. The name *Milky Way* is a literal translation of the Latin *via galactica*, and is used to identify the band of light observed in the sky.

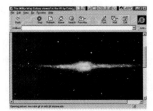

2. The Shape of the Milky Way In 1610 Galileo Galilei, the first astronomer with a telescope, discovered that the Milky Way was the light of countless stars. In the almost 400 years since Galileo's discovery, we've learned much more.

CHAPTER 17

Several hundred galaxies are visible in this Hubble Space Telescope photo. The image covers a speck of sky 1/30th the diameter of the full Moon.

THE DIVERSE GALAXIES

The Hubble Classification
Spiral Galaxies
Elliptical Galaxies
Irregular Galaxies
HISTORICAL NOTE: Edwin Hubble
Hubble's Tuning Fork Diagram
Measuring Galaxies
Distances Measured by Distance Indicators
The Hubble Law
HISTORICAL NOTE: Milton Humason, Mule Driver/Astronomer
CLOSE UP: Observations, Assumptions, and Conclusions
The Hubble Law Used to Measure Distance
The Tully-Fisher Relation

CLOSE UP: The Precision of Science
The Masses of Galaxies
Clusters of Galaxies—Missing Mass
The Origin of Galactic Types
The Cloud Density Theory
The Merger Theory
Look-Back Time
Active Galaxies
Quasars
Competing Theories for the Quasar Redshift
Seyfert Galaxies
Quasars and the Gravitational Lens
The Nature of Active Galaxies and Quasars

•••••• *SPIRAL NEBULAE WERE DISCOVERED IN THE LATE 1700s, but their nature remained a mystery for nearly 150 years. Early in this century, Adriaan van Maanen, an astronomer at Mount Wilson Observatory, was convinced that they were clouds of gas rather than galaxies similar to the Milky Way. He determined to settle the question based on evidence, and between 1916 and 1923 he made careful measurements of pairs of photographs taken up to 12 years apart. He found that the nebulae rotated with periods of tens of thousands of years. Such a* great rotation rate for a galaxy would mean that stars on its outer edge would have to be moving faster than the speed of light. Thus his data proved that the nebulae were not galaxies.

Other astronomers obtained conflicting data, but van Maanen was using the largest telescope in existence, and he found a great rotation speed for each of the seven nebulae for which he had the necessary pairs of photographs.

In 1924, Edwin Hubble (also working at Mount Wilson Observatory) proved conclusively, by the use of Cepheid variables, that the nebulae were separate galaxies.

Why did van Maanen obtain incorrect results? In Science and Objectivity (Iowa State University Press, 1988), Norriss Hetherington argues that although van Maanen was attempting to be impartial, his preconceived conclusion distorted his findings. Hetherington cites other examples of unintentional nonobjectivity in astronomy (including Lowell's Martian canals) and warns that scientists must be constantly on guard against allowing their anticipated results to cloud their data.

Viewed with the naked eye in a good dark sky on a clear night, the Andromeda galaxy (▶Figure 16-4) appears as a fuzzy little patch of light. Through binoculars, the fuzzy patch looks larger but is still only a fuzzy patch. Back through the ages this spot must have been a great source of wonder to curious people. It was called the Andromeda *nebula* because of its nebulous, or fuzzy appearance.

After the invention of the telescope, many more such nebulous objects were found but their nature remained a mystery. In 1924, Edwin Hubble found Cepheid variables in three of what had been called *spiral nebulae*, and he thereby showed that they in fact were spiral *galaxies*.

The importance of Hubble's discovery was tremendous, for it greatly expanded our appreciation of the size of the universe. Like the realization centuries earlier that our planet is just one of many planets, the realization that our Milky Way Galaxy is just one of a number of galaxies was a giant leap in understanding

▶ FIGURE 17-1. Edwin Hubble showed conclusively that the Andromeda nebula is in fact made up of stars and is therefore another galaxy. His discovery expanded our concept of the universe tremendously.

our place in the universe. We know now that the Andromeda galaxy is a spiral galaxy much like our own and that it is about 2.2 MLY (million light-years) away. When you view the Andromeda galaxy, give some thought to the fact that the light reaching your eye has been traveling for more than 2 million years.

This chapter will first describe how astronomers classify and measure properties of "normal" galaxies. Then it will turn to more unusual objects, including active galaxies and quasars.

THE HUBBLE CLASSIFICATION

The serious study of galaxies began with their observation and classification. This is a common starting place in science. When we come upon an assortment of objects about which we know little or nothing, we begin by grouping them into classes according to their observable properties.

Edwin Hubble provided the classification scheme that is the basis for the one still in use today. He divided galaxies into three types: spiral, *elliptical,* and *irregular,* with subdivisions within each category. More recently, astronomers have discovered objects in intergalactic space that fit none of Hubble's categories, not even irregular. This chapter will discuss each category in turn, then look at how properties of galaxies are measured. The next chapter will examine the objects that are too peculiar to be called "irregular."

elliptical galaxy. One of a class of galaxies that have smooth spheroidal shapes.

irregular galaxy. A galaxy of irregular shape that cannot be classified as spiral or elliptical.

Spiral Galaxies

Hubble divided spiral galaxies into two groups, normal ones like that of ►Figure 17-2 and galaxies with bars—called *barred spirals* (►Figure 17-3). Regular spirals are designated by the capital letter *S*, and barred spirals by *SB*. Each type is then further subdivided into categories a, b, and c, depending upon how tightly the spiral arms are wound around the nucleus. Galaxies with the most tightly wound arms (type a) also have the most prominent nuclear bulges. ►Figure 17-4 shows an Sa galaxy and two Sb galaxies. The galaxy in Figure 17-2 is an Sc

barred spiral galaxy. A spiral galaxy in which the spiral arms come from the ends of a bar through the nucleus rather than from the nucleus itself.

► FIGURE 17-2. The galaxy M100 is type Sc. Note its loosely wound arms.

► FIGURE 17-3. This galaxy, called NGC 1365, is a barred spiral galaxy. It is part of a cluster of galaxies about 60 million light-years from our Galaxy.

► **FIGURE 17-4.** (a) The Sombrero galaxy (M104, in Virgo) is a type Sa galaxy, so classified because of its large nuclear bulge. (b) NGC 4565 is an Sb galaxy. Note the prominent dust lane cutting across the center of each of these galaxies. (c) M88 is also an Sb galaxy, but we see it at an oblique angle.

Galaxies (and other objects) are named by their appearance in the Messier catalog (prefix *M*) or in the New General Catalog (prefix *NGC*).

The correct name for the three-dimensional shape of an elliptical galaxy is an ellipsoid.

galaxy. When a spiral galaxy is seen edge-on, it often displays the lane of dust and gas clouds that we see in our own Galaxy.

Barred spiral galaxies are also classified according to how tightly wound their arms are. Figure 17-3 is type SBb. A few galaxies seem to have the nuclear bulge and disk of a spiral galaxy, but no arms. Hubble called those *S0* galaxies.

Most spiral galaxies are from 50,000 to 2,000,000 light-years across and contain from 10^9 to 10^{12} stars. (We will see later how such measurements are made.) For comparison, recall that the Milky Way is about 100,000 light-years across and contains about 2×10^{11} stars.

Elliptical Galaxies

►Figure 17-5 shows some elliptical galaxies and makes clear why they were given that name. Various elliptical galaxies show different eccentricities in their elliptical shape, depending in part on their orientation to Earth. (A football appears round if viewed end-on, for example.) The actual eccentricity of an elliptical galaxy is difficult to determine because its orientation is unknown. Elliptical galaxies are classified from round (E0) to very elongated (E7).

Most of the galaxies in existence are ellipticals, but most galaxies listed in catalogs are spirals. The reason for this is that although a few giant elliptical galaxies are larger than any spiral galaxy (having 100 times more stars than the Milky Way), most are small and dim (with one-millionth as many stars as the Milky Way). Dwarf elliptical galaxies are visible from Earth only if they are relatively nearby.

▶ **FIGURE 17-5.** Elliptical galaxies show no spiral structure.

Irregular Galaxies

Hubble found that some galaxies did not fit into either of the above categories, nor did they exhibit other common characteristics. Reference to ▶Figure 17-6 indicates why these were called irregular galaxies. Fewer than 20% of all galaxies fall in the category of irregulars, and they are all small, normally having fewer than 25% of the number of stars in the Milky Way.

The Magellanic Clouds are usually classed as irregular galaxies, although some astronomers think that the Large Magellanic Cloud is a barred spiral that has been disrupted by its proximity to the Milky Way and perhaps by a past collision with the Small Magellanic Cloud. Collisions between galaxies should not be unusual, because on the average they are separated by distances only about 20 times their diameter. Stars within a galaxy, on the other hand, are separated by millions of times their diameter and therefore collide infrequently. Because of their great distances from Earth, however, galaxies exhibit no proper

▶ **FIGURE 17-6.** (a) The Large Magellanic Cloud, the nearest galaxy to the Milky Way at only 160,000 light-years away, is an irregular galaxy. (b) This is another irregular galaxy, NGC 1313, that is visible from the Southern Hemisphere.

(a) (b)

The Hubble Classification

HISTORICAL NOTE

Edwin Hubble

Like many famous persons, Edwin Hubble has been the subject of a number of myths. He is said to have been a professional quality boxer, a wounded World War I hero, and a lawyer, but no records exist to support any of these claims. Embellishments aside, one thing that we know for certain is that Hubble was a very capable and energetic scientist.

Hubble was born on November 20, 1889, in Marshfield, Missouri, where he was one of seven children. In high school he showed promise in both academics and athletics, becoming a track star and placing in the top quarter of his class. When he was sixteen, he entered the University of Chicago, where he continued to excel at science and math while lettering in track and basketball. After graduation he received a Rhodes scholarship and enrolled at the University of Oxford where he studied law (his family hoped he would become a lawyer). By 1914 Hubble was bored with the study of law and decided instead to do graduate work in astronomy at Yerkes Observatory in Wisconsin.

Early in his graduate career, Hubble attended a presentation by Vesto M. Slipher. A hot topic in astronomy at the time was whether spiral nebulae are part of our Galaxy or are galaxies in their own right. Slipher's presentation included observational data that supported the latter hypothesis, but not everyone was convinced. Possibly inspired by that talk, Hubble began photographing nebulae using the Yerkes's 24-inch reflecting telescope. His Ph.D. thesis, entitled "Photographic Investigations of Faint Nebulae," grew out of this earlier work. Although this work was not scientifically rigorous, it led Hubble closer to the conclusion that the nebulae were extragalactic objects separate from the Milky Way.

Before Hubble could reach any final conclusions, World War I began. Three days after receiving his Ph.D., Hubble enlisted and eventually rose to the rank of major in the 86th "Black Hawk" Division. Despite his success in the military, Hubble never saw any fighting because the Armistice had been declared by the time his division reached Europe. Hubble was disappointed but went instead to Mount Wilson Observatory where he returned to his study of the nature of "nongalactic nebulae."

Using the 100-inch reflector at Mount Wilson, Hubble began observing NGC 6822. By 1923 he had found several smaller nebulae and 12 variable stars within it. In 1924 he married Grace Burk Leib, who later became responsible for some of the myths that now surround Hubble. Upon his return to work, Hubble began observing M31 (a.k.a. the Andromeda nebula). Hubble distinctly resolved six variable stars in the Andromeda "nebula." This discovery convinced him that M31 is a galaxy outside our own, and he concluded that other spiral nebulae must also be separate, distinct galaxies. In the course of this work, Hubble observed several Cepheid variables in the galaxies M31 and M33. By comparing each star's luminosity with his observations of the star's apparent brightness, Hubble was able to deduce the distance to the star and its surrounding galaxy. He deduced that M31 and M33 are about 930,000 light-years distant. This distance placed them far outside the known boundaries of the Milky Way, so Hubble had even more proof to support his hypothesis that there are other galaxies outside our own. Hubble also attempted to classify the galaxies that he observed, which led to his so-called tuning fork diagram that is still widely used today.

In the mid-1920s, Hubble started investigating the expanding universe hypothesis. As he observed galaxies such as M31 and M33, his colleague Humason measured their radial velocities. By combining these radial velocities with the measured distances to the observed galaxies, the two men deduced what is called the Hubble law of redshifts: $v = H_0 d$. The law was published in a 1929 paper on the expansion of the universe. It sent shock waves through the astronomical community. The Hubble law indicates that the universe is expanding because the law predicts that velocities of galaxies increase at increasing distances from any chosen point. From 1931 to 1936 Hubble concentrated on extending this law to increasingly greater distances.

As if all this weren't enough, Hubble next began researching how the density of galaxies changes with distance. This work wasn't as groundbreaking as his earlier work, but he still strongly influenced the direction of astronomical research and wrote books on astronomical subjects for the general public. In the last few years of his life, he pushed for the construction of a 200-inch reflecting telescope at Mount Palomar in California. During World War II, he joined the staff of the U.S. Army's Ballistics Research Laboratory where he used his early astronomical training to lead a group calculating artillery-shell trajectories. After the war, he continued his work at the now-completed Palomar Observatory until he died of a stroke in 1953. His memory lives on with the many researchers who have built upon his earlier work.

Much of the material for this Historical Note comes from Osterbrock, Gwinn, and Brashear, "Edwin Hubble and the Expanding Universe," *Scientific American*, July 1993. That article is highly recommended.

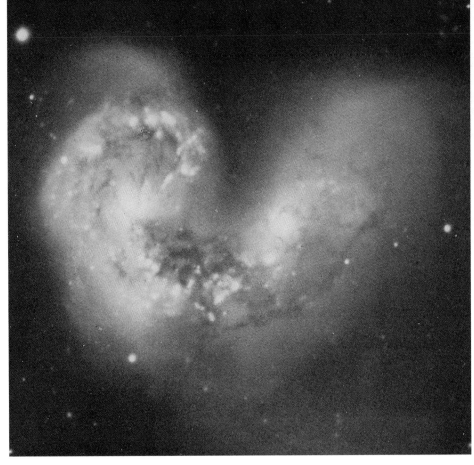

➤ **FIGURE 17-7.** NGC 4038-9 is known as "the Antennae." It is two galaxies in the process of colliding.

motion (motion across our line of sight), so we have to deduce past collisions from their present appearance. ➤Figure 17-7 shows two galaxies, called "the Antennae," that are in the process of colliding. Computer simulations show that gravitational forces between the galaxies produce odd formations such as the two "antennae."

Computer simulations also show that colliding galaxies actually pass through one another with few collisions between individual stars, although the interstellar dust and gas from the two galaxies intermingle. The collisions between clouds of interstellar gas result in greater gas density and therefore increased star formation. In addition, gravitational forces drastically alter the shapes of the two galaxies.

Sometimes galaxies may merge, especially in the case of a collision between a large and a small galaxy. In this case, the larger one swallows up the smaller, cannibalizing it.

Hubble's Tuning Fork Diagram

Edwin Hubble linked the three types of galaxies—ellipticals, normal spirals, and barred spirals—in what is called his ***tuning fork diagram*** (➤Figure 17-8). In his plan, S0 galaxies form the connecting link, because they have characteristics of both elliptical galaxies and spiral galaxies. Astronomers once thought that the diagram represents an evolutionary sequence, with galaxies changing from one form to another, but old stars are found in all three types, indicating that one type does not evolve to another, at least not as simply as indicated by the diagram. Table 17-1 summarizes the properties of the various galactic types.

tuning fork diagram. A diagram developed by Edwin Hubble to relate the various types of galaxies.

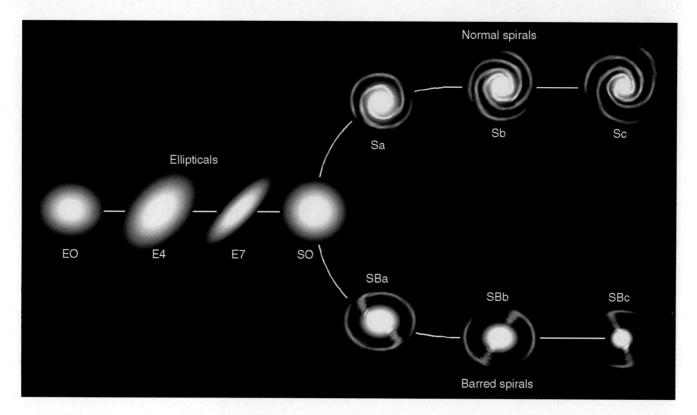

▶ **FIGURE 17-8.** The tuning fork diagram developed by Edwin Hubble provides a convenient way to categorize galaxies. It does not represent an evolutionary sequence, however.

TABLE 17-1 Types of Galaxies

Type	Designation	Description
Elliptical	E0–E7	Galaxies that appear circular (E0) to very elongated (E7)
Spiral	Sa–Sc	Disklike galaxies with a nuclear bulge and spiral arms. Sa type have large nuclei and tightly wound arms. Sc have small nuclei and open arms.
Barred spiral	SBa–SBc	Spiral galaxies with elongated nuclei. Subtypes are determined as with spiral galaxies.
S0	S0	Disklike galaxies, but with no spiral structure
Irregular	Irr	Galaxies that do not fit into any of the other types

MEASURING GALAXIES

The most important properties of a galaxy we can measure are its distance, mass, and motion. As we have seen on a number of occasions, if the distance to an extended object is known, its size and absolute luminosity can be calculated from its angular size and apparent magnitude, respectively.

Distances Measured by Distance Indicators

Distances to galaxies are measured using a number of different distance indicators. One of these, the Cepheid variable, has already been described. The calcu-

▶ **FIGURE 17-9.** The photo at the left is the Andromeda galaxy. The rectangle in the photo is enlarged at the right, and the lines indicate two Cepheid variables near the edge of the galaxy.

lation of the distances to the Magellanic Clouds was the first use of Cepheid variables as distance indicators on an extragalactic (outside the Milky Way) object. ▶Figure 17-9 shows two Cepheid variables in the Andromeda galaxy, the galaxy Hubble used to show that the so-called spiral nebulae were indeed galaxies.

Although Cepheid variables are very bright stars, we can distinguish them only in relatively nearby galaxies, out to perhaps 20 MLY. If we wish to measure the distance to a galaxy in which we are unable to see Cepheids, we must find other distance indicators. By analysis of giants, supergiants, and novae in our Galaxy and nearby galaxies, astronomers have learned that all have approximately the same range of luminosity. In addition, large globular clusters and supernovae are of consistent brightness from one galaxy to another. This allows us to use these bright stars and clusters as distance indicators in more distant galaxies.

An object used to measure distance because its absolute magnitude is known is often called a *standard candle*.

The logic here is interesting. By determining the luminosities of nearby Cepheid variables, astronomers determined the period-luminosity relationship for Cepheids. Then they assume that Cepheids in other galaxies are basically the same as the ones in our Galaxy, so they use that same period-luminosity relationship in reverse to calculate the absolute luminosity and thereby the distances to Cepheids in the nearer galaxies. Knowing these distances, they learn that the brightest stars and globular clusters in galaxies of each type have approximately the same luminosity. That is, the brightest 50 stars in one spiral galaxy have approximately the same average luminosity as the brightest 50 in every other spiral galaxy. Likewise for globular clusters. Now astronomers can use these stars and clusters as distance indicators to learn the distances to galaxies in which they can see the individual objects. ▶Figure 17-10 summarizes this chain, which allows the measurement of distances as far as about 100 MLY.

Here again we assume—with successful results—that the same laws of science that apply here on Earth also apply in distant galaxies.

At distances at which we can no longer see individual objects within a galaxy,

Measuring Galaxies 515

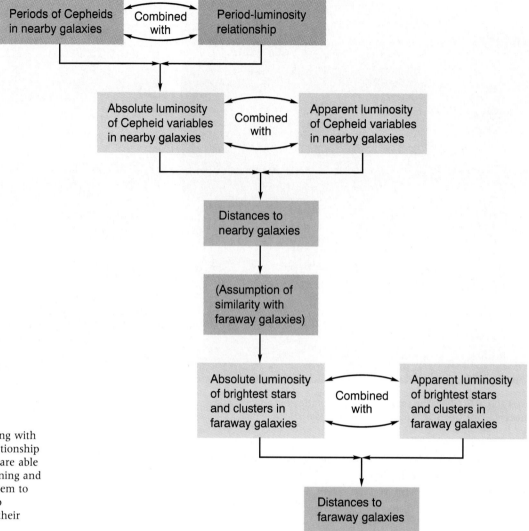

► **FIGURE 17-10.** Starting with the period-luminosity relationship of Cepheids, astronomers are able to follow a chain of reasoning and observation that allows them to determine the distances to galaxies too far away for their Cepheids to be visible.

the distance indicator becomes the galaxy itself. This method, as you might suspect, is extremely imprecise. Suppose, for example, that we see a very distant spiral galaxy. Since we know the range of luminosities spiral galaxies have, we can make some judgment as to that galaxy's distance by assuming that it is an average spiral galaxy. If we assume that it is among the dimmest galaxies of its type, we find its nearest probable distance. On the other hand, if we assume that it is among the brightest, we find its farthest probable distance. In this way, we find the range of distances within which we can be fairly certain the galaxy falls.

When applied to an individual galaxy, the whole-galaxy method of assessing distance is very imprecise. Fortunately, the method need not be restricted to individual galaxies. As will be discussed later, galaxies exist in clusters, which contain from a few galaxies to thousands of galaxies per cluster. If we consider a cluster that contains a great number of spiral galaxies, we can logically assume that the brightest spiral galaxy in that cluster has about the same luminosity as the brightest spiral in another cluster. Thus, we use the brightest galaxies as a distance indicator to the cluster.

Notice that one measurement builds on another in a series of steps. As a result, if there is an error in a beginning step, the error will be transmitted up through the chain of steps. For example, if somehow there is an error in our understanding of the period-luminosity relationship for Cepheids, the stated distances to the farthest galaxies will have to be adjusted. Fortunately, constant checks are made as new data arrive.

The above analysis makes an important assumption: Galaxies in our neighborhood of the universe are basically the same as those farther away. This may seem reasonable, but remember that we are seeing distant galaxies not as they are today but as they were in the past. Light coming from a galaxy 100 million light-years away has been traveling 100 million years, and we cannot rule out the possibility that galaxies have changed in that time.

The Hubble Law

In 1912, Vesto M. Slipher, an astronomer working in Percival Lowell's observatory in Flagstaff, Arizona, was assigned to examine the spectrum of some of the spiral nebulae to learn about their chemical composition. A then-current hypothesis regarding the nature of these objects was that they were planet systems in formation, and Lowell was seeking to discover life on other planets. Instead of finding evidence for life, Slipher found something else that had profound implications. He found a redshift in the spectra of most of the nebulae he examined. If the redshift was due to the Doppler effect, it meant that most of the other nebulae were moving away from us—as fast as 1800 kilometers/second. There seemed to be no reasonable explanation for this strange finding.

In 1924, Edwin Hubble showed that the nebulae are in fact galaxies. Then, he and Milton Humason used the 100-inch telescope on Mount Wilson to photograph the spectra of these objects. Their work not only confirmed the findings of Slipher but showed that there is a pattern in the speeds with which galaxies are receding from us.

➤Figure 17-11 is a diagram by Hubble and Humason on which they plotted the distances to a number of galaxies against the galaxies' recessional velocities. It indicates that the more distant a galaxy is, the faster it is moving away from us. These data, taken during the 1920s, used distance indicators that relied on incorrect Cepheid variable values, so the data had to be adjusted when more was learned about Cepheids. As ➤Figure 17-12 indicates, however, the relationship still holds when more and more galaxies are added using today's distance values.

The implications of Hubble's findings for our understanding of the past and

Lowell's telescope, on which Slipher worked, had only a 24-inch aperture. The 100-inch telescope used by Hubble and Humason therefore had 16 times more light-gathering power.

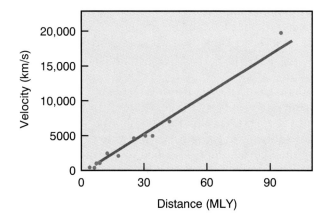

➤ FIGURE 17-11. This is similar to a diagram prepared by Hubble and Humason in 1931. It shows a relationship between the recessional velocities of galaxies and their distances.

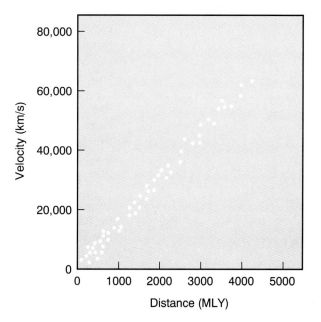

➤ FIGURE 17-12. The relationship found by Hubble and Humason is still valid using more recent data, but reevaluations of distances have caused the slope of the line to change.

HISTORICAL NOTE

Milton Humason, Mule Driver/Astronomer

Milton Humason (1891–1972) was born in Minnesota. He grew up in southern California, however, where he liked the mountain on which he lived, Mount Wilson, more than he did school, so when he finished the eighth grade, he dropped out of high school to wander the mountain. Later, when George Hale chose that mountain to construct a large observatory with a 60-inch telescope, Humason was hired as a mule driver to transport equipment up the steep slopes.

Humason learned about the observatory by talking to the construction workers, and when it was completed, he was hired as the first janitor for Mount Wilson Observatory. A young man with great curiosity, he constantly talked to the astronomers about their work, and eventually he was allowed to help them develop photographic plates in the darkroom. He then progressed to helping them take photographs with the large telescope.

Humason was aware of the inadequacy of his education and persuaded Harlow Shapley and Seth Nicholson, another astronomer at the observatory, to teach him math, including calculus. In a few years he was hired as a full-time night assistant and gave up his janitorial duties. Finally, in 1930, when he was 29 years old, he was appointed to the professional staff of the observatory, which by then had built a 100-inch telescope. Humason worked closely with Edwin Hubble using the 100-inch telescope on Mount Wilson, but since Hubble was the leader in their research, some astronomy books fail to even mention the former mule driver who worked with him.

When the 200-inch Hale telescope was completed in 1948 on nearby Mount Palomar, Humason used that instrument in his research. His contributions were so significant that he was awarded an honorary doctor's degree from the University of Lund, Sweden.

It is interesting that Milton Humason almost discovered Pluto. At the same time that Percival Lowell was attempting to find a planet beyond Neptune, W. H. Pickering was doing similar calculations in an attempt to predict its position. He asked Humason to take photographs of certain areas of the sky where he predicted the planet to be, but Humason was unable to locate the planet. Later examination of his plates showed that he had photographed Pluto twice, but in one case it was too close to a star to be visible, and in the other it happened to coincide with a flaw on the photographic plate.

cosmology. The study of the nature and evolution of the universe as a whole.

future of the universe are very important, for they imply that the universe is expanding. In fact, Hubble's work has become the foundation for today's theories of *cosmology*, which is the subject of the next chapter. That chapter will discuss the evidence for—and the implications of—the idea that the universe is expanding. It will show that the redshift that Hubble observed is not really due to the Doppler effect. The difference is not important here, however, because we are interested in using Hubble's findings only as a tool to measure distances to galaxies. What is important is that astronomers observe a regular relationship between redshift and distance, a pattern that is valuable to them in determining distances to faraway galaxies, as will be shown. Because it is convenient (and common) to think of the redshift as being due to the Doppler effect, that practice will be followed here.

➤Figure 17-13 shows photographs, all taken with the Palomar telescope, of five different galaxies and their spectra. Compare the location of the two prominent absorption lines in each of the spectra, noting that the dimmer the galaxy, the farther to the right (toward longer wavelengths) is the pair of lines. The distances to the galaxies, determined by methods described in the previous section, show that—as we would expect—the dimmer galaxies are farther away.

Using the distance values given and interpreting the redshift as due to the Doppler effect, ➤Figure 17-14 shows a "Hubble plot" of the galaxies' distances versus their recessional velocities. A dot on the diagram corresponds to the recessional velocity and distance of each of the five galaxies. The dots appear to be in a straight line (allowing for errors in measurement), so a straight line has been drawn to represent the trend they show. This graph confirms the relationship shown in the graphs in Figures 17-11 and 17-12.

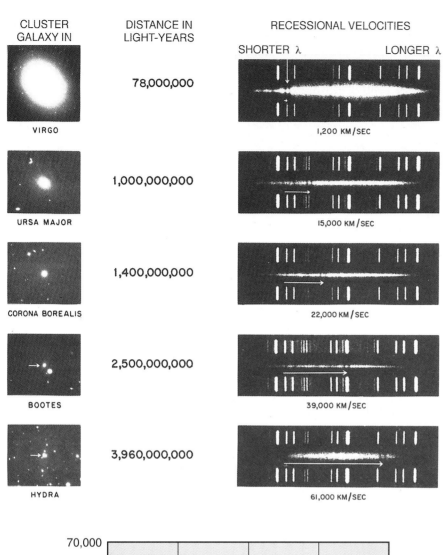

► **FIGURE 17-13.** The photos at the left in the figure are of galaxies in clusters at various distances from us. At the right, the spectrum of each is shown. (The spectrum is the white, horizontal streak between the reference lines.) Arrows indicate how far the two darkest absorption lines have shifted in each case. (λ = wavelength)

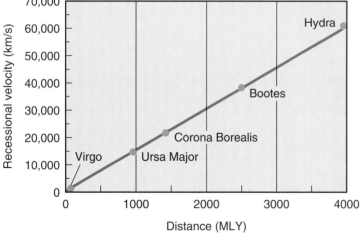

► **FIGURE 17-14.** A graph of recessional velocity versus distance for the five galaxies of Figure 17-13 shows that their recessional velocities are proportional to their distances. Each galaxy is identified by the constellation it is in.

Whenever data can be represented by a straight line on a graph, there is a direct proportionality between the quantities on the two axes. Thus, the relationship between the two quantities, radial velocity, v, and distance, d, can be written as

$$v = H_0 d$$

where H_0 is the constant of proportionality, called the **Hubble constant** in honor

Hubble constant. The proportionality constant in the Hubble law; the ratio of recessional velocities of galaxies to their distances.

Measuring Galaxies 519

Hubble law. The relationship that states that a galaxy's recessional velocity is directly proportional to its distance.

of the man who discovered this **Hubble law.** The value of the constant is simply the slope of the line on the graph. The slope of a line on a graph is defined as the ratio of the change in the quantity plotted on the y axis to the change in the quantity plotted on the x axis. The following example calculates the Hubble constant as an example of calculating the slope of a line.

EXAMPLE

Calculate the slope of the line in the graph of Figure 17-14, thereby calculating the Hubble constant.

Solution

Calculating the slope requires a comparison of the changes in the two quantities represented on the graph. Thus, two points must be chosen on the data line of Figure 17-14. Observe that the line happens to cross the point where the velocity is 30,000 km/s and the distance is 2000 MLY. This will be one of the chosen points. The line also passes through the point where both quantities are zero; this will be the second point. Now the slope is calculated as follows:

$$\text{slope} = \frac{30{,}000 \text{ km/s} - 0}{2{,}000 \text{ MLY} - 0}$$

$$= 15 \text{ km}/(s \times \text{MLY})$$

····· **TRY ONE YOURSELF.** Calculate the slope of the line in the 1931 Hubble-Humason graph (Figure 17-11) to find the value of the Hubble constant as determined with that data.

Astronomers usually use megaparsecs (million parsecs) as their distance unit. Fifteen km/(s × MLY) is 50 km/(s × Mpc).

Thus, according to the graph, the Hubble constant is about 15 km/(s × MLY) (read as "kilometers per second per million light-years"). It is important to see what Hubble's constant means and what its units represent. A value of 15 km/(s × MLY) means that for each million light-years a galaxy is from us, its speed is greater by 15 kilometers per second. Refer again to the graph to check that a galaxy at a distance of 1000 MLY has a speed of 15,000 km/s. In like manner, a galaxy 3000 MLY away has a speed of 45,000 km/s.

The slope of the graph of Figure 17-12, which represents a great amount of modern data, also yields a Hubble constant of 15 km/(s × MLY). The slope of the 1931 Hubble-Humason graph is much greater than this. Their calculation for the constant was about 180 km/(s × MLY), about 10 times the more modern value. This was due to their misunderstanding of the luminosity of Cepheids, as discussed earlier.

The data shown in Figure 17-12 and referred to as more modern data correspond to data compiled by Allan Sandage of the Mount Wilson Observatory and Gustav Tammann of the University of Basel, who have worked for years on the question. Another group of researchers uses slightly different methods for determining distances and obtains values close to 25 km/(s × MLY). Today, astronomers are fairly confident that the true value lies somewhere between these two findings, between 15 and 25 km/(s × MLY) [or 50 and 80 km/(s × Mpc)].

The Hubble constant is very difficult to determine with accuracy. We can calculate the distance to galaxies in nearby clusters fairly accurately using established distance indicators, but the motion of a galaxy within a nearby cluster is significant compared to the motion of the cluster (so that the galaxy may even be approaching us). Thus, the velocity of a nearby galaxy is due in large part to random motions within its cluster. It therefore does not fit the Hubble law. A distant galaxy, on the other hand, is in a cluster whose motion away from us is much greater than the galaxy's individual random motion, so its redshift follows the Hubble law. For these galaxies, however, determination of the distance to

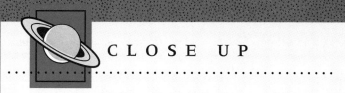

CLOSE UP

Observations, Assumptions, and Conclusions

It is always important in science to separate observations from the conclusions that are based on the observations. The situation illustrated in Figure 17-13 serves as a good example of this rule. The figure shows simply that the dimmer a galaxy is, the greater the redshift of its spectrum. Although the figure only indicates this relationship for five galaxies, the relationship has been found to be true in innumerable cases and has been confirmed by a number of researchers. It is based directly on many *observations* and is considered indisputable.

The middle column of the figure shows the distance of each galaxy from Earth and allows us to conclude that the farther away a galaxy is, the greater its redshift is. This conclusion, however, is not based directly on observations, but upon a conclusion about the distances to galaxies. Its validity depends upon how accurate the distance determinations are. One would suppose that the decreasing brightnesses of the galaxies in the figure are due to greater distance, but that is not proven by the figure. Actually, each of the galaxies in the figure is part of a cluster of galaxies, and the distance to the cluster has been measured using a variety of distance indicators. We are confident that the distances given are fairly reliable, although no one would pretend that they are exact.

Even though the distances to the galaxies are known with some accuracy, it is important to remember that the relationship between redshift and distance is a *conclusion*, not a direct observation. Now we take the next step: we assume that the redshift is caused by the Doppler effect, and therefore that it is caused by the galaxies' motions relative to the Earth. The Hubble relationship between recessional velocity and distance depends upon this *assumption*. Is there any other possible explanation for the redshift? The next chapter will show that there is, although the explanation is similar to that of the Doppler effect and does not negate the Hubble law.

Scientists must be on guard to remember the difference between what is observed—the evidence—and the conclusions that they reach based on those observations. Between an observation and a conclusion lies one or more assumptions, and the validity of the conclusion is based not only on the accuracy of the observations, but upon the validity of the assumptions.

---

the cluster is much less precise. So in cases where we can measure the distance accurately, the value obtained for velocity is less meaningful, and where we can have faith in our velocity measurement, the distance measurement is imprecise.

It is important to know the value of the Hubble constant because knowledge of the expansion rate of the universe is fundamental to our understanding of the universe as a whole. This idea will be discussed again in the next chapter.

The Hubble Law Used to Measure Distance

Distances to nearby galaxies can be measured using Cepheid variables as distance indicators, but as we progress to more distant galaxies, Cepheids can no longer be seen, and we must use globular clusters and the brightest stars as distance indicators. At even greater distances, these cannot be seen, and the brightest galaxies serve as distance indicators to clusters of galaxies. Finally, at the farthest distances, even this method cannot be used, and we turn to the Hubble law to indicate a galaxy's distance. This idea is illustrated with an example.

EXAMPLE

Suppose a very faint source is observed to have a redshifted spectrum that indicates a recessional speed of 120,000 km/s. Assume that the Hubble law applies to this object and calculate its distance.

Solution

First, one must decide which value of the Hubble constant to use. Here the distance is calculated using both 15 km/(s × MLY) and 25 km/(s × MLY), to give a range of distance within which one can be fairly confident the object lies.

Using 15 km/(s × MLY):

$$v = H_0 d$$
$$120{,}000 \text{ km/s} = 15 \text{ km/(s} \times \text{MLY)} \times d$$
$$d = 8000 \text{ MLY}$$

Now using 25 km/(s × MLY):

$$120{,}000 \text{ km/s} = 25 \text{ km/(s} \times \text{MLY)} \times d$$
$$d = 5000 \text{ MLY}$$

Thus, the object is between 5000 MLY and 8000 MLY away. This is 5 to 8 billion light-years!

You may be uncomfortable with such great uncertainty in an answer, but remember that without the Hubble law and this calculation, we would have no idea whatsoever of the distance to the object. The Hubble law does provide us with an idea of its distance.

······ TRY ONE YOURSELF. Suppose an object is observed with a redshift that indicates a speed of 90,000 km/s. What is the range of distance within which we can feel reasonably sure that the object lies?

The Tully-Fisher Relation

Recently, two astronomers, Brent Tully of the University of Hawaii and J. Richard Fisher of the National Radio Astronomy Observatory, discovered that spiral galaxies with wider 21-centimeter lines have greater absolute luminosities. The reason for the relationship is that a galaxy's spectral line is wide if the galaxy is rotating fast, because the Doppler effect causes a redshift from one side of the galaxy and a blueshift from the other side. Thus, the faster the rotation, the wider the line. In addition, more massive galaxies would be expected (1) to rotate faster and (2) to be brighter. ➤Figure 17-15 shows the logic that indicates that a wide spectral line would be associated with a bright galaxy.

Using the **Tully-Fisher relation,** astronomers can determine the absolute magnitude of a galaxy and use it as a distance indicator. The new relationship is

Tully-Fisher relation. A relation that holds that the wider the 21-centimeter spectral line, the greater the absolute luminosity of a spiral galaxy.

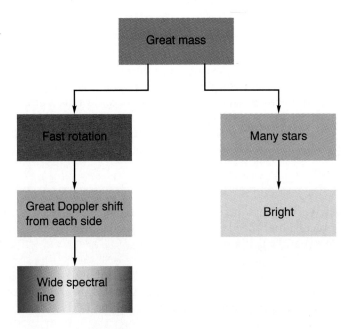

➤ FIGURE 17-15. A more massive galaxy would be expected to be bright and to have a wide 21-centimeter spectral line. Similarly, a galaxy of low mass is dim and has a narrow line. Tully and Fisher discovered this relationship experimentally.

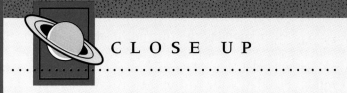

CLOSE UP

The Precision of Science

It is sometimes said that science is an exact study. If those who make such statements mean that the measurements of science are exact, they are wrong. It is obvious that measurements of distances to galaxies are not exact. What is not so obvious is that *every* measurement in science is to some degree an approximation. No measurement is exact. What scientists attempt to do is to be aware of how *inexact* their measurements are. For example, in calculating the distance to a galaxy, parallel calculations are made to determine the probable error in the measurement. Measurements may show that a galaxy is 200 MLY away, with a likely error of 50 MLY. This means that the galaxy is measured to be between 150 and 250 MLY away. Calculating the likely error is a common practice in natural sciences. Thus, scientists attempt to be specific about their inexactness.

In addition, the previous Close Up stated that scientists must take into account assumptions that are consciously or unconsciously included in the measurements. In the case of determining distances to faraway galaxies, for example, astronomers assume that there is negligible intergalactic material that might diminish the light from those galaxies. Early measurements of the size of the Galaxy were in error because they did not take into account the interstellar dust and gas. A scientist tries to be aware of the assumptions involved in each measurement. There is no claim that a measurement is exact. In fact, there is little room in science for the word "exact."

being used to check other methods of distance measurement and to improve our knowledge of distance to galaxies.

➤Figure 17-16 reviews the various methods used to measure distances in astronomy, from radar within the solar system to the Hubble law for the most distant objects.

THE MASSES OF GALAXIES

A galaxy's mass can be determined in a number of ways, all of them limited in precision primarily because of the tremendous distances involved. One method of measuring the mass of a galaxy is by observing the rotation periods of some parts of it. Naturally, we cannot wait for part of a galaxy to complete a revolution, for this takes millions of years. Instead, we use Doppler shift data to measure the

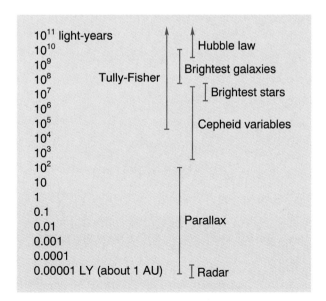

➤ FIGURE 17-16. A review of the various methods of measuring distance in astronomy. Approximate limits are indicated for each method.

velocity of part of a galaxy. Then knowing the distance of that part from the center, we can determine the period of revolution of the stars located there. As in the case of our own Galaxy, Kepler's third law then allows us to calculate the mass of the material within the orbit of the part of the galaxy being studied.

Another method of measuring galactic masses is similar to that used to measure stellar masses. There are many cases of a pair of galaxies revolving around one another. Again, we cannot wait the long times necessary to measure the period of revolution, but we can use the Doppler shift to gauge the speeds of the galaxies. The problem with this method is that it is difficult to determine the angle of the plane of revolution to our line of sight (➤Figure 17-17), and knowledge of this is necessary for an accurate measurement. Making measurements for great numbers of such binary galaxies allows us to determine an average mass for a given type of galaxy, however.

Clusters of Galaxies—Missing Mass

Most galaxies are part of clusters rather than lonely nomads drifting through space. ➤Figure 17-18 shows a cluster of galaxies. Thousands of such clusters are

Be careful with the terminology here. A *galactic cluster* is an open cluster of stars. It is not a *cluster of galaxies*.

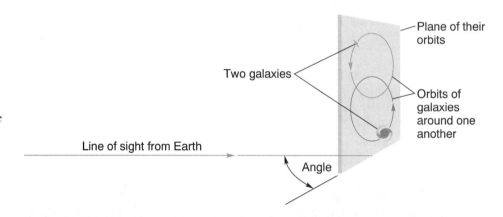

➤ **FIGURE 17-17.** In order to use the Doppler shift to measure the speeds of two galaxies that orbit one another, we would have to know the tilt of their plane of revolution to our line of sight.

➤ **FIGURE 17-18.** A number of galaxies are visible in this cluster, called the Fornax cluster. It is in the constellation Fornax in the Southern Hemisphere.

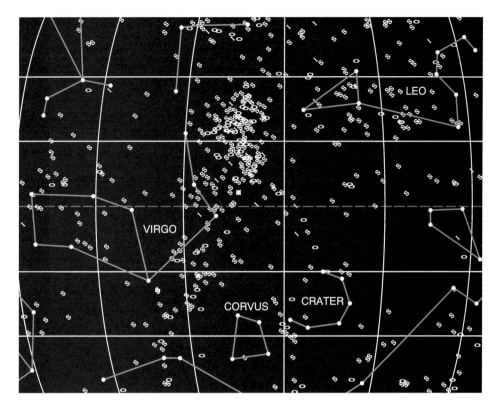

► **FIGURE 17-19.** This figure shows galaxies to the sixteenth magnitude. The dashed line passing through the center of the map is the celestial equator. Spiral galaxies are designated by a slanted *S*, and ellipticals by an oval. The Virgo cluster of galaxies is in the center, just north of the celestial equator. (This map was produced by the Voyager software for the Macintosh.)

known, and the largest clusters may contain as many as 10,000 galaxies. ►Figure 17-19 is a sky chart that includes the constellations Virgo and Leo. Instead of showing stars, the chart shows galaxies to the sixteenth magnitude. The large group of galaxies just above the center of the chart includes the Virgo cluster. It contains some 2500 galaxies! Not all of the galaxies in the clump on the diagram are part of the Virgo cluster, however, for some are nearer to us and some farther from us than the cluster is. To determine whether a particular galaxy is part of a given cluster, astronomers must determine the galaxy's distance, because a cluster of galaxies is not just a group of galaxies that are near each other on the celestial sphere (like a constellation) but a gravitationally linked assemblage.

Our Milky Way Galaxy is part of a group of about two dozen galaxies, including the Magellanic Clouds and the Andromeda galaxy, that form a cluster called the *local group.* The Andromeda galaxy and the Milky Way are by far the largest members of this cluster, which appears to contain only one other spiral galaxy.

A third method of measuring the masses of galaxies takes advantage of their clustering. There is only one force that could hold a galaxy within its cluster: the gravitational force. If we observe a galaxy near the outside of a large cluster and assume that it is in a relatively circular orbit, we can use the Doppler effect to measure its speed and therefore its period. This allows us to calculate the total mass of the galaxies within its orbit. Calculating the mass of the cluster based on measurements for only one galaxy would not be meaningful because there would be too many uncertainties, but the same measurement, repeated for a number of galaxies on the outer portion of the cluster, gives a useful value for the mass of the cluster and therefore an average value for the masses of the galaxies within the cluster.

Notice on Figure 17-19 that galaxies are not evenly distributed at all.

Recall that all galaxies of a cluster have nearly the same redshift.

local group (of galaxies). The cluster of 20 or so galaxies that includes the Milky Way Galaxy.

In April 1997, a new member of the local group was announced. Named Antlia, it has only about one million stars.

The Masses of Galaxies 525

When measurements are made by the various methods, we find that the cluster method reveals that clusters have much more mass than is accounted for by the visible stars within the galaxy. Within our own Galaxy, the interstellar gas and dust account for at least 10 to 20% of the total mass, and the rotation curve of the Milky Way indicates that even considering the interstellar matter, we can account for as little as one-tenth of the total mass of the Galaxy. (Recall that the rotation curve leads us to hypothesize a massive galactic corona.) Assuming that other galaxies have similar amounts of interstellar matter, there is still a large amount of **missing mass**. There must be a very large amount of **dark matter** in the universe—matter that is too cool to emit enough radiation for us to detect it. We know that intergalactic dust and gas exist between galaxies in a cluster, but this is also not nearly enough to account for the calculated mass. There are recent indications that the halo of our own Galaxy contains more material than previously thought, and perhaps galaxies' halos account for much or all of the dark matter that makes up the missing mass. The quest for an understanding of the nature of dark matter is a very active field in astronomy today, because—as the next chapter will discuss—it is fundamentally important to cosmology.

If stars are grouped into galaxies and galaxies are grouped into clusters, do clusters group into something bigger? Yes, we call them **superclusters**. Our local supercluster includes the local group and the Virgo cluster (Figure 17-19) as well as other clusters. It has a diameter of perhaps 100 MLY and contains some 10^{15} solar masses. Between superclusters are great voids with no galaxies.

> **missing mass.** The difference between the mass of clusters of galaxies as calculated from Keplerian motions and the amount of visible mass.
>
> **dark matter.** Matter that is too cool to emit sufficient radiation to allow us to detect it.
>
> **supercluster.** A group of clusters of galaxies.
>
> 10^{15} is an unimaginable number, a thousand million million.

THE ORIGIN OF GALACTIC TYPES

Astronomers once thought that Hubble's tuning fork diagram might represent the evolutionary sequence of galaxies. Perhaps spirals evolve into ellipticals or vice versa. This hypothesis can be dismissed quickly. First, we know that ellipticals do not evolve into spirals because elliptical galaxies contain very little gas and dust, while spirals contain significant amounts. Ellipticals have already used up their gas and dust forming stars. What about evolution the other way? This proposal presents mechanical problems. Once a disk has formed in a galaxy, there is no way that the galaxy, on its own, could disperse its disk and acquire a symmetric elliptical shape.

Another discarded theory held that the difference is caused simply by rotation, that spiral galaxies evolved from dust clouds that were spinning fast and ellipticals came from slowly spinning clouds. This theory cannot account for the fact that elliptical galaxies have completed their star formation and have little interstellar material, however.

Today, there are two leading theories that explain why galaxies exist in various types. Each will be discussed in turn.

> Evidence for the lack of interstellar material in ellipticals comes from the *Infrared Astronomy Observatory*, which detects very little infrared radiation from ellipticals. Interstellar material radiates in the infrared region of the spectrum.

The Cloud Density Theory

Early in the history of the universe, before galaxies, some gas/dust clouds must have been more dense than others. One theory of galactic formation holds that elliptical galaxies are those that formed from the most dense clouds. Because of the great density of these clouds, star formation would have proceeded quickly and would have used up all dust and gas before a disk had a chance to form. Once stars have formed, they act as individual particles that orbit the center of the galaxy and have little interaction with one another. This is why galactic clusters are able to pass through the disk of our Galaxy without being captured. So, according to this theory, early star formation explains the lack of a disk in elliptical galaxies.

> These theories are not listed in the margin glossary because they have no generally accepted names. The names here are unofficial.

Clouds that had a lower density of gas and dust would have formed stars less frequently when the cloud contracted, and the dust and gas would have collapsed into a disk before star formation used it all up. The reason that a rotating cloud of gas and dust forms into a disk is that gas acts like a fluid, where the particles interact a great deal. When two fluids collide, energy is dissipated and the fluids do not pass through one another. (Recall that in the case of star birth, a disk is a common occurrence.) Within a galactic disk, star formation proceeds slowly. Instabilities cause density waves that assist star formation, so that new stars are still being born in spiral galaxies today.

Various hypotheses exist to explain why some spiral galaxies have bars and others do not. One idea is that faster-spinning clouds form elongated nuclei, or bars. Another hypothesis notes that barred spiral galaxies generally do not have a halo of globular clusters. It proposes that the lack of a halo results in instabilities that cause a bar to form.

A puff of wind cannot pass through still air without carrying the still air along (and reducing the wind speed in the process).

The Merger Theory

A more recent theory proposes that spiral galaxies formed before elliptical galaxies, and that ellipticals are the result of mergers of spirals. The direction of spin of galaxies is random, so merged galaxies would often have little or no overall rotation (which is what is observed for ellipticals). The lack of interstellar matter in ellipticals is explained by the fact that the merger would compress interstellar clouds and would cause strong density waves. During the merger, then, star formation would be frequent and interstellar matter would be used up.

Notice that the word theory is being used here for what we should probably call a hypothesis. There is no sharp dividing line between the terms.

This theory is supported by the observation that ellipticals make up a greater percentage of the galaxies in large clusters where galaxies are packed close together. In these clusters, mergers would have been frequent and would have produced numerous elliptical galaxies. In loosely packed clusters of galaxies, on the other hand, ellipticals are fairly rare.

You may have noticed that nothing has been said about irregular galaxies. Neither theory explains these oddballs well. Astronomers know that a number of galaxies that were once classified as irregulars are actually pairs of galaxies that are in the process of collision and that others are spiral galaxies with large, dense dust clouds that hide their spiral patterns. Perhaps all irregular galaxies can be explained like this. In any case, irregulars make up a small fraction of all galaxies.

It should be obvious that the question of why some galaxies are of one type and others are of another type is far from settled. This is just one of the unanswered questions that make extragalactic astronomy a lively source of research today.

Look-Back Time

As new techniques are developed, we are seeing objects farther and farther away. We have now detected objects that may be as far away as 18 billion light-years. If these objects are truly this far, the light we see left them 18 billion years ago. We are seeing far into the past. Astronomers speak of **look-back time**, a term that emphasizes this idea. An object 18 billion light-years away has a look-back time of 18 billion years.

We are referring now to distant objects because, as will be explained later, we cannot be sure that they are galaxies.

look-back time. The time light from a distant object has traveled to reach us.

One of the problems with using large galaxies as distance indicators relates to the idea of look-back time. When we see these galaxies, we are not seeing them as they are today but as they were in the past. There is some evidence that within large clusters of galaxies, some galaxies combine to form supergalaxies. That is, some galaxies are "cannibalized." If this occurs, the largest galaxies in nearby clusters are not the same size as the largest galaxies in distant clusters because we are seeing nearby clusters at a later stage of their life than we see distant ones. In the distant ones, not enough of their life has passed for galaxies to have gobbled up one another. This would invalidate the assumption that distant clusters are similar to nearby clusters and would call into question the practice of using galaxies as distance indicators. Although this is a possible problem, the Hubble law seems to apply equally as well to faraway clusters where galaxies are used as distance indicators as it does to nearby galaxies where other methods are used for determining distance. This indicates that the dilemma posed by look-back time is not a major obstacle in this case.

Among the objects with the greatest look-back times, we find a class of objects not yet discussed: active galaxies. These objects play a major part in helping us to determine the scale of the universe.

ACTIVE GALAXIES

Until now the discussion of this chapter has centered on what are called *normal galaxies*, those that fit the Hubble classification. Now we turn to objects that do not fit into those categories, objects that were discovered in the last 50 years by radio astronomy.

All galaxies emit some radio waves; the Milky Way Galaxy radiates them from its nucleus. The radio waves from a normal galaxy constitute only about 1% of that galaxy's total luminosity, however. In the late 1940s, astronomers began observing strong extragalactic radio sources, and in 1951 it was discovered that a radio source in the constellation Cygnus is actually a double source associated with a galaxy. Cygnus A, as the source is called, emits about a million times more radio energy than does the Milky Way.

radio galaxy. A galaxy having greatest luminosity at radio wavelengths.

Since the discovery of Cygnus A, many more such **radio galaxies** have been discovered, typically emitting millions of times more radio waves than a normal galaxy. ➤Figure 17-20 shows Centaurus A. Observe that the visible galaxy in this case is an elliptical galaxy with a prominent lane of interstellar gas and dust. The double-lobed feature seen in Centaurus A is common in radio galaxies, and most of the galaxies associated with double-lobed radio sources are either giant ellipticals or spirals. The radio lobes are enormous, sometimes extending 15 MLY from the visible galaxy.

Recall that the Milky Way is about 100,000 LY, or 0.1 MLY in diameter. From end to end, large double lobes are 300 times this size!

➤**FIGURE 17-20.** Centaurus A is an elliptical galaxy, shown in (a) in visible light. In (b) we see its radio image. The radio sources are primarily in two lobes at opposite sides of the galaxy.

(a)

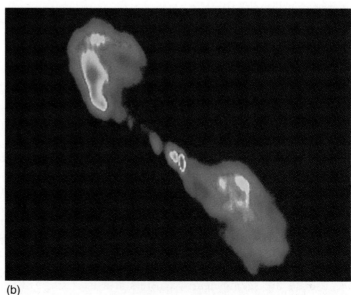

(b)

528   CHAPTER 17   The Diverse Galaxies

(a) (b)

▶ **FIGURE 17-21.** (a) This visible-light photo shows Virgo A (M87), a giant elliptical radio galaxy. (b) A short-exposure infrared image by the Hubble Space Telescope shows a jet of material coming from the nucleus of the galaxy. The nucleus is the bright white spot toward the left.

Radio galaxies often appear unusual when viewed in visible light. ▶Figure 17-21 shows a jet of hot gas being emitted from Virgo A. This feature is seen in a number of cases. More recently, galaxies have been observed that have properties like those of radio galaxies but have their primary emission at wavelengths other than the radio region of the spectrum. Astronomers therefore classify radio galaxies as one type of a group of high-energy galaxies called ***active galaxies.***

During the first decade of observing active galaxies, many types of active galaxies were discovered, and various hypotheses were put forward to explain them. Then, less than 10 years after the discovery of active galaxies, the mystery deepened with the discovery of what appeared to be an entirely new type of object.

active galaxy. A galaxy with an unusually luminous nucleus.

Because the energy of an active galaxy comes from its nucleus, astronomers often refer to active galactic nuclei (or *AGN*s) rather than *active galaxies.*

Quasars

Prior to 1960, astronomers thought that intense radio waves like those from active galaxies necessarily come from large objects such as galaxies. Individual stars are such weak sources of radio waves that before 1960 the Sun was the only star from which radio waves had been detected. In that year, however, two radio sources were found in the sky that were so small that they appeared to be stars. Recall that planets' sizes can be measured when they occult a star. The same phenomenon can be used in reverse to measure the size of a distant object. In the case of one of these radio sources, 3C 273 (so named because it is the 273rd object listed in the third Cambridge catalog), its visual and radio images were observed as it was occulted by the Moon. By observing 3C 273 as its light and radio waves were blocked out by the Moon and again as they reappeared on the other side, details of the source could be determined. The object appeared to be very small—more like a star than a galaxy. In addition, it had a small jet protruding from it like the jets that have been observed from some radio galaxies.

The radio waves from 3C 273 seemed to have two sources: the jet and the

quasar (quasi-stellar source, or QSS). A small, intense celestial source of radiation with a very large redshift.

main body of the object. We have seen that double-lobed radio sources are common for active galaxies.

The spectra of both 3C 273 and 3C 48 (the other source mentioned above) were found to be extremely unusual. Although spectral lines were prominent, the lines could not be identified with any known chemical element. Were we seeing objects that had entirely different chemical elements than are known to us? Because of their unusual nature, the objects were called *quasi-stellar radio sources,* or *quasars.*

In 1963, Maarten Schmidt of the California Institute of Technology found the solution to the puzzle of the unusual spectrum of 3C 273. He found that the prominent spectral lines that had been seen are simply hydrogen spectral lines that are very greatly redshifted. Each wavelength is redshifted to a value 16% greater. (That is, the ratio of the change in wavelength to the nonshifted wavelength from a stationary source is 16%. $\Delta\lambda/\lambda = 0.16$.) ➤Figure 17-22 illustrates the situation. Schmidt's colleague Jesse Greenstein examined the spectrum of 3C 48 and found that its spectral lines are shifted even farther—by about 37%.

If the redshifts of the two quasars are caused by the Doppler effect, one quasar is moving at 15% the speed of light, and the other is moving at 30% the speed of light (or 90,000 kilometers per second). These were speeds far greater than any yet encountered for celestial objects (or for terrestrial objects, other than nuclear particles).

If the great redshifts follow the Hubble law, the closest of the two quasars must be as far away as the farthest galaxies. Yet it is brighter than those galaxies, even though in size it is more like a star than a galaxy.

Recall that red giants are typically light-minutes in diameter but that galaxies are thousands, or hundreds of thousands, of light-years in diameter.

Since the early 1960s, about 7000 quasars have been discovered. Unlike the first two, most are not sources of radio waves. Quasars are bluish-white objects and are X-ray emitters. In addition, many vary in intensity in an irregular way, typically changing intensity in weeks or months. This latter observation confirms their small size; they cannot be larger than a few light-weeks or light-months in diameter. A few have been found with intensity variations of less than a day.

The first two quasars seen, with redshifts of 16% and 37%, have smaller redshifts than most; the 16% redshift is the least yet observed. The greatest redshift ($\Delta\lambda/\lambda = 4.73$) observed to date indicates that the speed of its quasar is about 94% the speed of light.

If quasars' redshifts fit the Hubble law, the objects are extremely far away. The Hubble constant is known with limited precision, but if we calculate distances to quasars using the smallest and largest likely values for the constant, we

➤ **FIGURE 17-22** (a) An "at rest" spectrum. The wavelengths of two lines are indicated. (b) The same spectrum shifted 16%. Observe that the pattern of lines is the same.

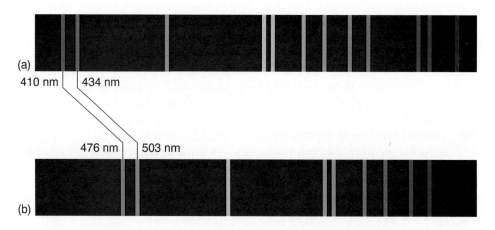

obtain distances of 11 and 18 billion light-years, respectively, for the most distant quasar.

Now consider what the luminosity of such an object must be for us to be able to detect it at that distance. It must be far more luminous than an entire galaxy. Finally, remember that we are speaking about a small object—far smaller than a galaxy.

Competing Theories for the Quasar Redshift

When astronomers were faced with the prospect of an object being so small and yet so luminous, they were forced to reexamine their assumptions. As was pointed out earlier, strictly speaking, the Hubble law is not due to the Doppler effect, though the basis for the law is mathematically equivalent to that effect. The next chapter will discuss the theorized cosmological reason for the validity of the Hubble relationship. Nevertheless, the Hubble law indicates that quasars are tremendous distances from us. Is there any other possible explanation for the redshift?

Einstein's general theory of relativity predicts that light leaving a massive object will be redshifted due to the mass of the object. Calculations show that in order to produce redshifts such as those seen in quasars, the gravitational field near the object must be far greater than that near the most massive neutron star, and well-grounded nuclear theory tells us that there is a limit to the mass of a neutron star. After a certain mass is reached, a neutron star cannot exist. A black hole is formed. The theory of relativity simply cannot be used to explain the redshift.

Perhaps quasars are nearby objects with enough local motions that they do not fit the Hubble law. (Recall that galaxies within the local group do not fit it.) ➤Figure 17-23 shows fast-moving objects ejected from our Galaxy; some are moving at nearly the speed of light. If this is what quasars are, then they do not have such tremendous luminosities after all.

The problem with this *local hypothesis* is not only that we can imagine no source for such objects but that we must ask why we do not see similar objects from other galaxies. We cannot expect our Galaxy to be unique in this regard. Yet if other galaxies emit such fast-moving objects, we would see some of them with tremendous *blue*shifts as they move toward us. In fact, however, such blueshifted objects are not seen, and the local hypothesis is not considered a likely explanation.

Perhaps there is some other explanation. Such an explanation might require drastically different laws of physics than those of today, however. We never can disregard such a possibility, but we must progress with what we know until it is

local hypothesis. A proposal stating that quasars are much nearer than a cosmological interpretation of their redshifts would indicate.

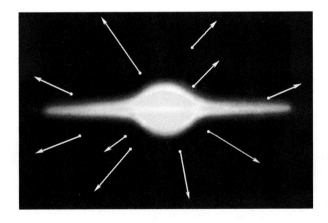

➤ **FIGURE 17-23.** The local hypothesis holds that quasars are relatively nearby objects that have been ejected from the Galaxy at great speeds. If quasars are emitted by galaxies, however, some should be moving toward us from nearby galaxies, and these would show blueshifts. No quasar has a blueshift.

Active Galaxies

definitely ruled out. The redshift *can* be explained by the Hubble law, even though that law leads us to conclude that quasars are at distances greater than we are accustomed to and are therefore much more energetic than we are accustomed to. Almost all astronomers now agree that quasars' redshifts do indeed fit the Hubble law.

Seyfert Galaxies

Recall that one of the first two quasars seen has a double-lobed radio source and a jet from its center. In this regard, it resembles an active galaxy. As astronomers found more and more quasars, they discovered that most of the nearer ones are associated with clusters of galaxies. The quasars are more luminous than the galaxies, but the association causes us to look for more similarities.

One particular type of spiral galaxy, a **Seyfert galaxy** (➤Figure 17-24), has a very luminous nucleus that—although it is not as luminous as a quasar—in some cases varies in intensity in time periods even shorter than those of quasars. Similarities in the spectra of Seyfert galaxies and quasars point to a further link between galaxies and quasars, indicating that perhaps a quasar is a galactic nucleus.

Recent high-resolution photos of nearer quasars show a fuzz on the image near the quasar. Again, it appears that quasars may be at the nuclei of some type(s) of galaxies and that the fuzz is caused by the stars of the galaxies.

Seyfert galaxy. One of a class of galaxies having active nuclei and spectra containing emission lines.

➤ **FIGURE 17-24.** The Seyfert galaxy NGC 4151 has a redshift of 0.001.

gravitational lens. The phenomenon in which a massive body between another object and the viewer causes the distant object to be seen as two or more objects.

Quasars and the Gravitational Lens

Twin quasars were discovered in 1979 in Ursa Major. The two quasars are very close together, separated by only 6 seconds of arc. They have the same luminosity, the same redshift ($\Delta\lambda/\lambda = 1.4$), and identical spectra. The explanation for these identical twins lies with the theory of general relativity. That theory predicts the possibility of a large mass bending light from a more distant object so that two images of the distant object appear. The predicted phenomenon is called a **gravitational lens**, but until the discovery of the twin quasars, not much attention was paid to the prediction. Since the discovery of the twin quasars, the intervening galaxy has been found, and we are now confident that the twins are actually one quasar that has been made to appear double by the galaxy that lies between it and us (➤Figure 17-25). Only a few dozen examples of gravitational lensing have been found since 1979. One interesting case is shown in ➤Figure 17-26(a). The two yellow-orange spots in the image are twin images of a single

➤ **FIGURE 17-25.** Light from the quasar at the left is bent as it passes by the galaxy at the center. This causes the light to appear to come from twin quasars, one on either side of the galaxy. Such gravitational lenses are predicted by the theory of general relativity.

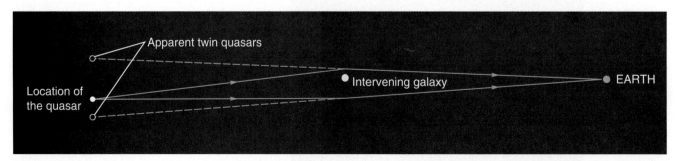

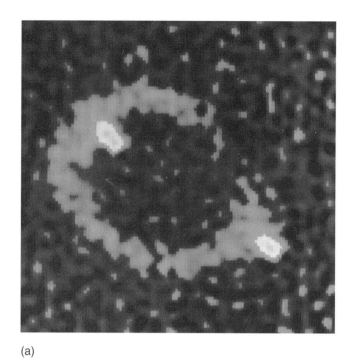

> FIGURE 17-26. (a) This false-color image was made by radio interferometry with the Very Large Array. The ring is called an "Einstein Ring". (b) In 1995 the Hubble Space Telescope was used to discover this case of quadruple gravitational lensing.

quasar behind a massive object that acts like a gravitational lens. The ring is caused by another object behind whatever object causes the lensing effect. This object forms a ring because it is so well aligned that light passing on every side of the intervening massive object is bent toward Earth, making the distant quasar appear to be on all sides of it at once.

Arcs, such as in Figure 17-26(a), are the most common type of lensing, although most cases are not as complete as that arc. Pairs of lensed objects are the next most common. Figure 17-26(b) shows a much rarer case, one with a quadruple pattern. This was found by the HST in 1995.

The importance of the discovery of gravitational lenses is not only that they provide another confirmation of the general theory of relativity but that they indicate that quasars are indeed very distant—that their redshift fits the Hubble law. They really are the most distant objects yet observed.

Suppose that we plot the number of known quasars versus their distances from our Galaxy. The graph would not be very instructive because as the distance from our Galaxy (or from any point) increases, the volume of space in a shell increases. This happens because the area of a large sphere is greater than the area of a small sphere (➤Figure 17-27). Therefore, even if quasars are evenly distributed throughout the universe, we would expect to see more of them at greater distances.

To truly illustrate the distribution of quasars, the graph should show the *density* of quasars—the number of quasars per unit of volume of space—versus their distance from us. ➤Figure 17-28 is such a graph. The distance shown on its horizontal axis depends on the value chosen for the Hubble constant, of course, and other choices are possible. Nevertheless, even if a different value were chosen for the Hubble constant, the shape of the graph would remain the same. The graph leaves no doubt that quasars appear at a fairly specific distance from the Milky Way. Since distance is equivalent to time, this indicates that quasars existed during a relatively short period in the distant past.

Active Galaxies

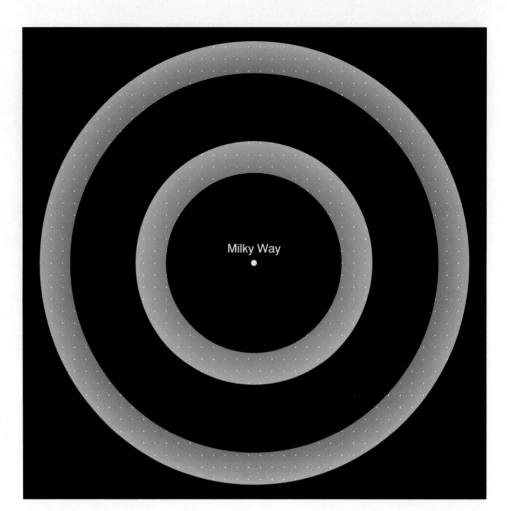

▶ **FIGURE 17-27.** If two equally thick shells are drawn at different distances from us, the more distant shell will contain more space. Even if quasars (dots in the figure) were equally spaced throughout the universe, more would be seen in the faraway shell.

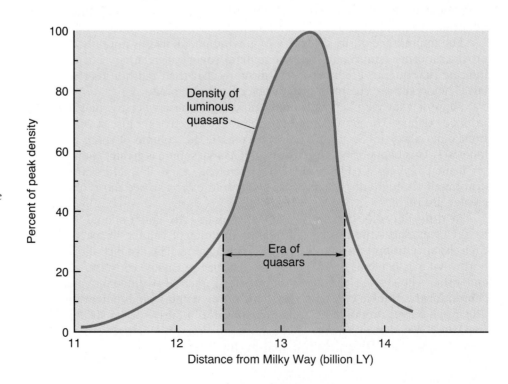

▶ **FIGURE 17-28.** The graph shows the relative abundance of quasars at various distances from the Milky Way. The distance scale is based on a Hubble constant of 20 km / (s × MLY).

THE NATURE OF ACTIVE GALAXIES AND QUASARS

The first question that has to be answered in trying to solve the puzzle of the nature of active galaxies and quasars is, "Why are these strange objects not found among nearby galaxies?" To answer, we invoke the idea of look-back time: active galaxies and quasars are far away because they existed only in the distant past and are not part of today's universe. Therefore, any theory about the nature of these mysterious objects must explain why they existed long ago but not today.

Be aware of the tremendous scale of galactic astronomy; a few million light-years is a very small distance.

According to today's prevalent theories, the tremendous amount of energy that comes from active galaxies and quasars is caused by an immense black hole at the nucleus of the galaxy. The black hole is surrounded by an accretion disk that is extremely hot because material falls into it with a very great speed. The temperature of the accretion disk explains the enormous energy emitted by the objects. What remains to be explained is why some active galaxies and quasars are so much more luminous than others, why some emit intense radio waves and others do not, why the luminosity of some of them changes so rapidly, and, in general, why there appear to be so many different types of these peculiar objects.

Various theories attempt to explain the many observations about active galaxies and quasars, but none is completely satisfactory. The leading theory provides a link between the different types of objects. It holds that they are all basically the same, and that they appear different depending upon their orientation with respect to us. ➤Figure 17-29 illustrates the hypothesized object, which is an active galactic nucleus. At its center is a supermassive black hole of

An active nucleus, of course, is the nucleus of an active galaxy.

➤ FIGURE 17-29. The leading model for active galaxies (including quasars) holds that they are basically the same type of object, but that they appear different because they are oriented at different angles to our line of sight. We see this one obliquely, neither in line with its disk (in which case we could not see its accretion disk) nor along one axis, in which case we would see an intense radio source.

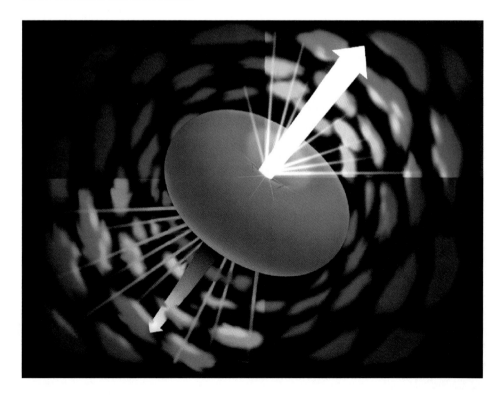

Recall from Chapter 13 (and from Figure 13-16) that jets of material have been observed coming from the poles of protostars. The galactic case might be a similar phenomenon on a much larger scale.

There are at least two theories for the visibility of the jets or the lack thereof. Both depend upon the amount of material that lies along the trail of the jet.

➤ **FIGURE 17-30.** (a) A false-color radio image of the galaxy Cygnus A shows two radio lobes some 200,000 light-years apart. A jet is visible extending from the source (an active galactic nucleus) toward one of the radio lobes. (b) A false-color image (red) is superimposed on a visible-light image (blue-white) of the radio galaxy NGC 1316 in the constellation Fornax. Each radio lobe is about 600,000 light-years across.

many million (or even a few billion) solar masses. It is surrounded by an accretion disk, perhaps a few light-months across.

A much larger ring shaped like a fat doughnut (a *torus*) surrounds the accretion disk in the same plane. The torus is composed of relatively cool gas and dust. It may be several light-years across and is dense enough that light from the accretion disk cannot penetrate it. Light from the disk does reach the outside world, however, for it shines out of the "doughnut hole" in two directions, forming two cones of light. Radiation in all other portions of the electromagnetic spectrum, including high-energy wavelengths, pours out along with the light. A jet of material is ejected by some unexplained process along the axis of each of the cones. Finally, irregular clouds of gas and dust move in orbit around the center.

Now suppose that the active nucleus in the figure is oriented so that we see it edge-on, so that the torus hides the accretion disk from our view. Telescopes would reveal infrared energy coming from the clouds, but would not show the high-energy radiation from the accretion disk. In some cases the jets cause emission of visible light, but in other cases they do not. If the active nucleus has energetic jets, astronomers would observe large radio-emitting areas at a great distance on either side of the galaxy. ➤Figure 17-30 (and 17-20b) shows the radio lobes of some active galaxies.

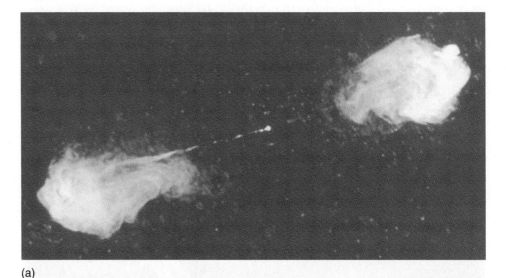
(a)

(b)

536   CHAPTER 17   The Diverse Galaxies

When the torus is seen obliquely, as it is drawn in Figure 17-29, radiation from the accretion disk reaches us, and the object appears as a very energetic source that emits radiation across the spectrum, including X-rays. These are the objects we call quasars. On the other hand, suppose the jet is aimed directly toward us. Again it would appear as a small, energetic source, but in this case it might vary quickly in intensity as clouds of dust move across our line of sight. We call such objects **BL Lac objects** after their prototype, *BL Lacertae*.

How does this model explain the observation that active galaxies are not found in our neighborhood? There is a limited amount of material near the black hole in the galactic center, and after that material spirals into the black hole, the galaxy calms down to become a more standard galaxy. In today's universe, no galaxy remains with a nucleus that is still in the quasar stage. This means that previous quasars and active galaxies are the ancestors of today's galaxies.

➤Figure 17-30c provides further evidence that galaxies have progressed from having quasars at their centers, to active galaxies, to normal galaxies. The three galaxies in the figure are classed as normal galaxies, but they all show characteristics of having black holes at their centers. A census of 27 nearby galaxies carried out by the Hubble Space Telescope and ground-based telescopes in Hawaii suggests that nearly all galaxies may harbor supermassive black holes that once powered quasars. It seems supermassive black holes are so common that nearly every large galaxy has one.

In addition, the black hole's mass is proportional to the mass of the host galaxy. This suggests that the growth of the black hole is linked to the formation of the galaxy in which it is located. Finally, the number and masses of the black holes found are consistent with what would have been required to power the quasars. Perhaps a giant step has been taken in our understanding of the evolution of galaxies. Time will tell.

BL Lac objects. Especially luminous active galactic nuclei that vary in luminosity by a factor of up to 100 in just a few months.

BL Lac objects are also called *blazars*.

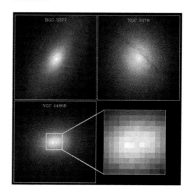

➤ **FIGURE 17-30c.** The three galaxies are believed to contain supermassive black holes in their nuclei. One has a double nucleus.

CONCLUSION

This chapter has progressed from knowledge about which astronomers are fairly confident to knowledge that is very tentative. Galaxies can be classified with confidence, and Hubble's classification is based simply on appearance. Although the tuning fork diagram may appear to indicate an evolutionary sequence, it does not. At least two competing theories are available to explain why galaxies exhibit such different appearances.

Not much can be known about a galaxy unless we first know its distance, so the study of galactic distances is important, and various distance indicators are in use today. To measure distances to the farthest galaxies, a new tool, the Tully-Fisher relation, has been added to the Hubble law.

Astronomers can calculate the mass of a galaxy by applying gravitational theory to the measured motions of parts of the galaxy. They measure the mass of a cluster of galaxies in a similar manner. Both calculations show that we can account for only a small fraction of the mass that must exist, and the mystery of the missing mass has not been solved.

A busy area of research today is the study of active galaxies and quasars. These distant, energetic objects that have been discovered by the methods of radio astronomy have brought about major changes in galactic astronomy. They seem to produce new questions almost as prodigiously as they produce energy.

RECALL QUESTIONS

1. Cepheid variables are important in calculating
   A. the distances to galaxies.
   B. the compositions of stars.
   C. the compositions of the interstellar medium.
   D. the ages of galaxies.
   E. the temperatures of stars.

2. About how many galaxies are within range of the largest telescopes?
   A. 20
   B. 103
   C. 1000 to 2000
   D. 1 to 4 million
   E. More than 100 billion

3. The Magellanic Clouds are
   A. galaxies.
   B. extremely high atmospheric clouds.
   C. globular clusters.
   D. supernovae.
   E. of unknown nature.

4. The local group consists of
   A. about 100 nearby stars.
   B. about 15 to 30 nearby stars.
   C. about 100 nearby galaxies.
   D. about 15 to 30 nearby galaxies.
   E. the closest planetary nebulae.

5. Which of the following is used as a distance indicator to galaxies?
   A. Cepheid variables
   B. Globular clusters
   C. The brightest supernovae
   D. [All of the above.]
   E. [None of the above.]

6. The masses of galaxies are determined by using the Doppler effect to measure the speeds
   A. of parts of individual galaxies.
   B. of individual galaxies in a cluster.
   C. of galaxies that are part of a binary pair.
   D. [All of the above.]
   E. [None of the above; we cannot determine the masses of galaxies.]

7. Hubble found that the objects that were first called spiral *nebulae* are instead spiral *galaxies* by observing
   A. supernovae in them.
   B. Cepheid variables in them.
   C. radiation characteristic of black holes coming from them.
   D. black holes in their centers.
   E. that their spectra are characteristic of spectra of stars.

8. If galaxy X is four times more distant than galaxy Y, then according to the Hubble law, galaxy Y is receding
   A. 16 times faster.
   B. 4 times faster.
   C. 2 times faster.
   D. 1.6 times faster.
   E. [No general statement can be made.]

9. Measurements of the total mass in clusters of galaxies indicate that they contain far more mass than is visible to us. The "missing mass" is now known to be
   A. tiny black holes.
   B. cool gas and dust between the galaxies.
   C. black dwarfs within the galaxies.
   D. [None of the above; we don't know what it is.]

10. Which of the following is the largest?
    A. A galactic cluster
    B. A cluster of galaxies
    C. The Milky Way
    D. The Large Magellanic Cloud
    E. [Both A and B above; they are the same.]

11. The Tully-Fisher method of measuring distances to galaxies depends on the relationship between a galaxy's absolute magnitude and
    A. the galaxy's size.
    B. the galaxy's redshift.
    C. the galaxy's apparent color.
    D. the width of a line in the galaxy's spectrum.
    E. [Both A and C above.]

12. The Hubble law relates
    A. absolute magnitude and temperature.
    B. apparent magnitude and temperature.
    C. proper motion and radial velocity.
    D. proper motion and distance.
    E. distance and radial velocity.

13. The Hubble constant refers to
    A. the fact that the speed of light never changes.
    B. a number in a formula that will never change its value.
    C. the rate at which the universe is expanding at the present time.
    D. the amount of mass in the universe.
    E. the amount of mass in galactic cores.

14. Suppose that we know that a galaxy 5 million light-years away is receding at 20 miles per second. What would be the Hubble constant based on this single galaxy?
    A. 4 mi/(s × MLY)
    B. 15 mi/(s × MLY)
    C. 25 mi/(s × MLY)
    D. 100 mi/(s × MLY)
    E. [None of the above.]

15. The look-back time of an object is
    A. how long light from the object takes to reach Earth.
    B. numerically equal to the distance to the object in light-years.
    C. larger for more distant objects.
    D. [All of the above.]
    E. [None of the above.]

16. The Hubble law is based primarily upon
    A. calculations of galaxy masses.
    B. calculations of galaxy rotational speeds.
    C. Doppler shift data.
    D. knowledge of galactic evolution.
    E. data concerning the number of stars in galaxies.

17. Which of the following is true about quasars?
    A. They seem to be extremely large compared to most galaxies.
    B. They seem to be very small for the energy released.
    C. If they are as far away as they seem, they are very energetic.
    D. [Both A and C above.]
    E. [Both B and C above.]

18. The evidence for small size of quasars comes from
    A. the amount of energy they release.
    B. their distance from us.
    C. the rapidity of their luminosity changes.
    D. comparison with Cepheid variables.
    E. the magnitude of their blueshift.

19. We cannot judge quasars' distance by using their absolute luminosity as we do in the case of galaxies because
    A. they have no absolute luminosity.
    B. there are no nearby quasars with which to compare distant ones.
    C. their luminosity is too great.
    D. their recessional speeds are unknown.

20. The radio waves from a radio galaxy can come from an area many times bigger than the visible object.
    A. True
    B. False
    C. No general statement can be made concerning this.

21. Which of the following is an observation rather than the result of theory?
    A. Light from most galaxies is redshifted.
    B. Most galaxies are receding from us.
    C. The rate of expansion of the universe is slowing.
    D. Galaxies farther away are moving away faster.
    E. Quasars emit more energy than most galaxies.

22. If the local hypothesis concerning quasars is true, we would expect
    A. that quasars would be less luminous.
    B. that quasars would be more luminous.
    C. to observe quasars with blueshifted spectra.
    D. quasars to be emitted from galaxies other than our own.
    E. [Both C and D above.]

23. What is meant by a distance indicator, and what serves as a distance indicator for nearby galaxies? For more distant galaxies?

24. Describe three methods for measuring the mass of a galaxy other than the Milky Way.

25. What is the local group?

26. What is meant by *dark matter* in galactic astronomy?

27. The chapter stated that we know that elliptical galaxies do not evolve into spiral galaxies. The reason given was that we observe that spirals contain gas and dust, while ellipticals do not. Explain why the observation proves the statement.

28. State the Hubble law. What is the currently accepted range of the Hubble constant? Include units in your answer.

29. Describe some difficulties encountered in determining the Hubble constant with accuracy.

30. Explain how the Hubble law can be used to determine the distance to a galaxy. Discuss the limitations of this method of determining distance.

31. What is meant by *look-back time?*

32. In what way are active galaxies "active"?

33. After the spectra of quasars were understood, why were the objects still so difficult to explain?

34. What do we mean when we say that a quasar has a redshift of 25%? Which quasar is moving away faster, one with a redshift of 25% or one with a redshift of 30%?

35. What is the present theory of the origin of the energy of quasars?

36. Why are there no nearby quasars?

QUESTIONS TO PONDER

1. Explain how we determine the distances to galaxies that are too far away to allow us to see Cepheids in them.

2. The difficulty in determining the orientation of the plane of revolution of binary galaxies limits the usefulness of Doppler shift measurements of velocity. Why does the same problem not exist in using Doppler shift data to determine the speed of part of an individual spiral galaxy?

3. Discuss the limitations on the accuracy of measurements of distances to galaxies. Doesn't the lack of precision in these measurements make galactic astronomy less a science than other sciences? Discuss.

4. Compare the use of bright galaxies as distance indicators to grading a class "on the curve."

5. If the slope of the Hubble graph were greater (with distance plotted on the *x* axis), would the Hubble constant be larger or smaller? Explain.

6. What is a gravitational lens and what does the fact that they have been observed tell us about quasars?

7. Why is a spectral line of a galaxy broadened by the galaxy's rotation?

8. In describing what was called the *merger theory* of galaxy formation, no explanation was given for why some spiral galaxies have bars and others do not. Propose an explanation.

9. List similarities and differences between the collapse of an interstellar cloud to form a star and the formation of a galaxy.

CALCULATIONS

1. Determine the slope of the graph shown in ➤Figure 17-31. Be sure to include units in your answer.

➤ **FIGURE 17-31.** This graph might represent the changing speed of a race car. Determine the slope of the graph. (Calculation #1)

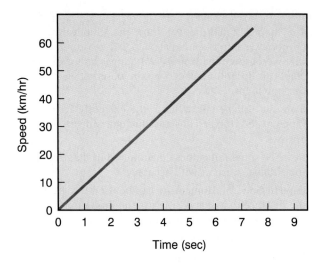

2. If the data for the Hubble expansion were as shown in ➤Figure 17-32, what would be the value of the Hubble constant?

➤ **FIGURE 17-32.** If these were correct data for the Hubble graph, what would be the Hubble constant? (Calculation #2)

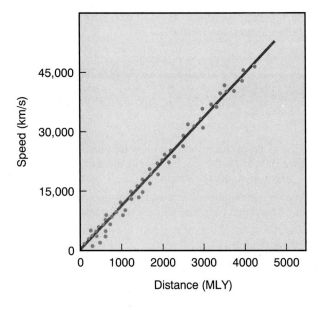

3. Suppose that an object is seen with a redshift that indicates a speed of 100,000 km/s away from us. If this redshift fits the Hubble law, how far away is the object? (Assume some reasonable value for the Hubble constant.)

4. Suppose that someone is moving away from you at a constant speed of 15 mi/hr and is at the moment 60 miles away. If he has maintained this speed since leaving you, how long ago were you together?

5. For the motion in question 4 above, calculate the "Hubble constant".

6. The photo below shows the Large Magellanic Cloud, a satellite galaxy of the Milky Way. It is 50 kiloparsecs from us, and has an angular size of 650 arcminutes (') by 550'. Use the small angle formula (in Chapter 6) to calculate the dimensions of the Large Magellanic Cloud in kiloparsecs.

7. The photo below shows NGC 4535. Its angular size is 6.8′ by 5′. Assume that its longest dimension is the same as the diameter of the Milky Way (about 100,000 light-years), and use the small angle formula to calculate its distance from us. (Actually, NGC 4535 is probably smaller than the Milky Way.)

Visit the netQuestions area of StarLinks (www.jbpub.com/starlinks) to complete exercises on these topics:

1. **Estimating Distances in Space** There are various methods to measure distances to stellar objects.

2. **Distance Scaling** To measure the distance to far-away galaxies, scientists can use certain stars called Cepheids as cosmic yardsticks.

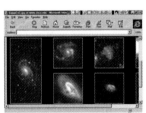

CHAPTER 18

This is a photo of Centaurus A, a radio galaxy, whose understanding involves Einstein's theories of relativity.

COSMOLOGY: THE NATURE OF THE UNIVERSE

The Search for Centers and Edges
Einstein's Universe
The Expanding Universe
What Is Expanding and What Is Not?
The Cosmological Redshift
CLOSE UP: Wrong Explanation: The Doppler Effect
Olbers's Paradox
Cosmological Assumptions
The Big Bang
Evidence: Background Radiation
CLOSE UP: The Steady State Theory

The Age of the Universe
CLOSE UP: Science, Cosmology, and Faith
The Future: Will Expansion Stop?
Evidence: Distant Galaxies
The Density of Matter in the Universe
The Inflationary Universe
The Flatness Problem
The Horizon Problem
CLOSE UP: Stephen Hawking, the Ultimate Theoretician
The Grand Scale Structure of the Universe

······· ALBERT EINSTEIN: *"The existence and validity of human rights are not written in the stars."*
"The unleashed power of the atom has changed everything save our modes of thinking, and we thus drift toward unparalleled catastrophes."
(Asked by the press what would happen in the event of nuclear war) "Alas, we will no longer be able to listen to the music of Mozart."

"If A is success in life, then A equals X plus Y plus Z. Work is X, Y is play and Z is keeping your mouth shut."

"One thing I have learned in a long life—that all our science, measured against reality, is primitive and childlike—and yet it is the most precious thing we have."

"The most beautiful thing to be felt by man is the mysterious side of life. There is the cradle of Art and real Science."

"The most incomprehensible thing about the universe is that it is comprehensible."

This final chapter asks three simple questions: What is the nature of the universe, what was its past, and what will be its future? The questions are those of the subject of cosmology, which studies the universe as a whole rather than its individual parts. Many of the ideas of the previous chapter will be useful in answering cosmology's questions, but—as you might suspect—the questions are not easy to answer and today's answers must be regarded as tentative.

Nature has provided a number of clues to be read by cosmologists, and amazingly they have been able to reach very meaningful answers to the questions—questions that must have been among the first asked by intelligent humans.

THE SEARCH FOR CENTERS AND EDGES

Throughout the first 17 chapters of this book, a story has unfolded about changes in our basic ideas of the nature and extent of the universe. These changes have produced different answers to the question of whether or not humans are located at the center of the universe. Ptolemy's model placed the Earth at the center. ➤Figure 18-1 is a representation of the idea of Ptolemy's universe (although the planets and their complicated epicycles are not included). The stars in Ptolemy's geocentric model were hypothesized to be on a sphere surrounding the Earth.

Copernicus moved the Earth from center and replaced it with the Sun, but the stars remained on a sphere surrounding the solar system. The sphere of stars was generally considered to lie just beyond the most distant planet, and this was the entire physical universe. Some important observations did not fit Copernicus's heliocentric model, however. The model could not explain why objects

Dates are given in margin notes for review and comparison.

Ptolemy: c. A.D. 150.

Copernicus: 1473–1543.

➤ **FIGURE 18-1.** Ptolemy's universe placed the Earth at the center of a sphere of stars. This drawing does not include the planets, which would move between the Earth and the celestial sphere.

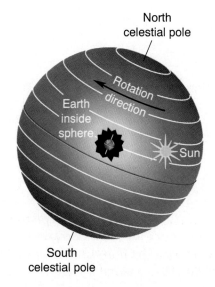

544 CHAPTER 18 Cosmology: The Nature of the Universe

thrown upward were not left behind by the spinning Earth, nor could it explain the lack of stellar parallax as the Earth moved around the Sun. A half century later, Galileo introduced the concept of inertia to explain motion, and he argued that stellar parallax was not observed because the stars are simply too far away. For this to be true, the universe had to be much larger than people had suspected.

Isaac Newton proposed that the universe is infinite in extent. He argued that if it is not, the force of gravity would cause it to collapse to its center. An infinite universe would not be in danger of collapse because each object in it would be pulled equally hard in all directions. Newton's idea did not catch on, however. Others claimed that gravitational force has limited range and therefore cannot cause the universe to collapse. An infinite universe simply conflicted too strongly with "common sense."

During the eighteenth century, William Herschel's discoveries led to a better understanding of the Milky Way, again enlarging the boundaries of the known universe. Herschel's data, however, indicated that our solar system is at the center of the Galaxy, thereby returning the human race to center. Herschel proposed that the many fuzzy nebulae visible in telescopes are not part of the Milky Way but are other galaxies like ours. Like Newton's idea of an infinite universe, Herschel's proposal was not accepted, and by the end of the nineteenth century, most astronomers were convinced that what we call the Milky Way Galaxy comprised the entire universe.

In 1917, Harlow Shapley used globular clusters to conclude that the Sun is not at the center of the Galaxy, and in the next decade other astronomers calculated the Sun's motion. This showed finally that the Sun is not "special," either in nature or in position.

In 1923, Edwin Hubble discovered Cepheid variables in the Andromeda "nebula." This pushed back the limits of the universe tremendously, for astronomers recognized that our Galaxy of hundreds of billions of stars is just one of numerous galaxies. The universe continued to grow in people's minds, and humans were far removed from any central position.

Galileo: 1564–1642.

Newton: 1642–1727.

Herschel: 1738–1822.

At the time, galaxies were called island universes, *the name suggested by Immanuel Kant.*

Shapley: 1885–1972.

Hubble: 1889–1953.

Einstein's Universe

While Hubble and others were using observations to advance our knowledge of the universe, a theoretician was approaching the problem from a different perspective. Einstein's general theory of relativity (1916) substituted curved space for gravitational force. The theory holds that space itself is curved near a massive object and that the curvature causes the acceleration that previously had been attributed to the force of gravity. Chapters 3 and 15 describe several observations that confirm Einstein's theory and show that it explains many phenomena better than Newton's does.

Newton hypothesized that his gravitational force would cause a finite universe to collapse. The same question appears in general relativity, but it is couched in different language. If the mass of the universe will eventually result in the collapse of the universe, space is curved enough that the universe is said to be ***closed.*** Curvature of space is difficult to imagine because it involves a fourth dimension. The world of our direct experience is a three-dimensional one, and the best we can do to imagine a closed universe is to think of a two-dimensional plane being curved so that it forms a sphere (▶Figure 18-2a).

Chapter 3 compared our existence in a three-dimensional world that is curved into a fourth dimension to the situation of imaginary "flatfleas" on a two-dimensional surface that is curved (Figure 18-2b). If the creatures in the figure travel far enough in what they perceive as a straight line, they will return to their original positions. Likewise, in a closed universe, a beam of light sent in one direction will travel around the universe and return to its starting point (if the universe lasts long enough). Like the finite two-dimensional surface of a

Einstein: 1879–1955.

closed universe. The state of the universe if it stops expanding and begins contracting.

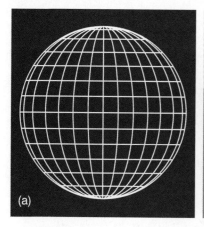

▶ **FIGURE 18-2.** (a) A closed universe can be thought of as a two-dimensional plane curved into the third dimension so that it becomes a sphere. (b) Imagine that fleas on the surface of the sphere think that they live in only two dimensions. That is, the surface appears flat to them. Yet, if they travel far enough in one direction they will end up where they began.

sphere, such a universe would have no boundaries even though it would have a limited volume and a limited number of stars.

The alternative to a closed universe is an **open** one that does not curve back onto itself. This can be visualized in a two-dimensional analogy as a saddle shape (▶Figure 18-3). An open universe is an infinite one that continues on forever.

Einstein believed that the universe is closed and finite. He would not accept the possibility that the universe was anything but static and unchanging. When he applied his equations of general relativity to the universe, he concluded that such a universe would collapse on itself, just as Newton had concluded. This seemed an unacceptable situation to Einstein, and he decided to adjust his theory to eliminate that conclusion. To do so, he had to insert into the equations a term that adds a cosmological force to support the universe against collapse. The force would cause one mass to repel another, but the force would be very unusual, for instead of weakening with distance, it would become greater with distance. This explains why we do not experience the force at the small distances we normally deal with. The hypothesized cosmological force is significant only at great distances.

Einstein regretted having to adjust his theory in such an arbitrary, ad hoc manner. Adding the term violated the principle that a theory should be aesthetically simple. Nevertheless, he felt that the adjustment was necessary if his equations were to apply to a closed universe. If he had rejected the cosmological force, he probably would have concluded from his theoretical work that the universe is expanding. Instead, the expansion was discovered observationally.

open universe. The state of the universe if it continues expanding without stopping.

"Unchanging" does not mean that we observe no changes in parts of the universe. The term applies only to the universe as a whole.

Something is *ad hoc* if it is designed for one specific purpose. An ad hoc committee, for example, is one that is created for a specific job.

▶ **FIGURE 18-3.** An open universe can be represented as a plane curved into the shape of a saddle. The figure shows only a portion of the shape, for it extends forever in every direction.

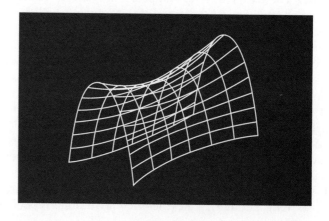

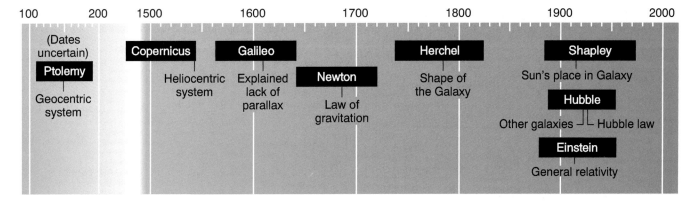

► **FIGURE 18-4.** The life spans of the scientists whose cosmological contributions were described in the last few pages are shown here.

THE EXPANDING UNIVERSE

In 1929, Edwin Hubble announced that the pattern of redshifts of distant galaxies indicates that they are moving away from us and that the more distant the galaxy, the faster it is moving away. The previous chapter explained how the resulting *Hubble law* has become a valuable tool to measure the distances to far-away galaxies. The law also has tremendous implications concerning the nature of the universe, however.

First, if other galaxies are moving away from ours, does this not mean that our Galaxy must be at the center? Have we finally discovered that we are at a special location after all? Hubble published his findings in 1929, 12 years after Shapley's work with globular clusters had taken our Sun out of the central position within the Galaxy. By this time it had become a working premise that our location within the cosmos is not central. To see why Hubble's discovery does not in fact conflict with this basic premise, consider the following analogy.

Imagine that you are a trainer in a flea circus and that you put a number of educated fleas on a balloon that your assistant is blowing up. You have instructed these fleas to hold their positions on the balloon. ►Figure 18-5 shows the balloon being blown up with the fleas in place.

Now imagine what a particularly intelligent flea sees when it looks out toward neighboring fleas. (Assume either that the flea can see around the curvature of the balloon, or that the fleas are close enough together that the balloon seems flat to them.) The intelligent flea will see every other flea getting farther and farther away as the balloon is blown up. In addition, the fleas more distant from him will be moving away at a greater speed than the ones nearby. The more distant a given flea is from the observer, the faster it is moving.

The important point is that the same result would be obtained no matter which of the fleas is the intelligent flea. While the balloon is being blown up, every flea sees every other flea moving away with a velocity that depends upon the flea's distance from the observer.

The analogy shows that if galaxies are part of an expanding universe, we will see exactly what Hubble and Humason observed: Every galaxy will be seen to be moving away from us at a speed that depends upon how far away it is. A number of astronomers since Hubble have surveyed galaxies at greater and greater distances and have come to the same conclusion: The universe is expanding.

The previous chapter included a Close Up concerning the differences between observations and conclusions. The distinction is important in this case. The pattern of redshifts is an *observation*. All spectral lines of distant (dim) galaxies show the redshift, not just the lines of the visible spectrum. The observations are 100%

►Figure 18-4 shows when the scientists who are mentioned here lived.

► **FIGURE 18-5.** Fleas on a balloon get farther apart when the balloon is blown up.

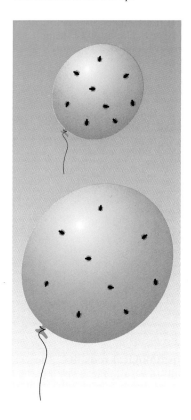

in agreement with what would be predicted for light from a receding object. Thus, we *conclude* that the galaxies are getting farther away from us. We do not, however, *observe* this.

What Is Expanding and What Is Not?

Saying that the universe is expanding certainly does not mean that the solar system is expanding. Nor does it mean that stars within our Galaxy are getting farther apart. In fact, it does not even mean that all galaxies are getting farther apart.

The observed redshift leads us to conclude that other *clusters* of galaxies are moving away from ours. Some individual galaxies are actually moving toward us. This occurs in two ways. First, galaxies in the local group are moving randomly within that group as each responds to the overall gravitational force of the others. Thus, the Andromeda galaxy and a half-dozen others in the local group are at this time moving toward the Milky Way Galaxy.

Second, in nearby clusters, the same random motion of individual galaxies results in some of the galaxies moving toward us at the present time, even though the cluster in which each galaxy exists is moving away from the local group. No individual cluster—as far as we can tell—is expanding.

It is the clusters of galaxies that are moving farther apart. Thus, although we often say that other galaxies are moving away from us, we should really say that other clusters are moving away from our cluster.

The Cosmological Redshift

After the redshift of light from distant galaxies was discovered, astronomers assumed that the Doppler effect was the cause. Using the Doppler effect, they calculated the velocities of the galaxies. The Hubble law is a statement of the relationship between a galaxy's distance and its (Doppler effect) velocity, as the previous chapter showed. Modern cosmology explains the redshift in an entirely different manner.

This does not mean, of course, that the Doppler shift is not the correct explanation of other observations in astronomy.

An expanding universe does not mean that clusters of galaxies are rushing through space. Instead, space itself is expanding. ➤Figure 18-6 illustrates the difference using the balloon analogy again. This time, tiny seeds have been glued to the balloon, each seed representing a cluster of galaxies. Part (a) shows the Doppler interpretation of the expansion of the universe. The balloon is expanding against a background grid that represents space, indicating that the galaxies move through space.

Part (b) shows the modern cosmological interpretation of the expanding universe. The grid that represents space is part of the surface of the balloon here. As the balloon expands, space itself expands.

Why don't individual galaxies expand along with space? General relativity shows that objects that are held together by their own gravity, such as the Earth, the solar system, the Galaxy, and the local group, do not expand as space expands. The force of gravity (or—in terms of general relativity—the local curvature of space) holds them together just as each individual seed on the balloon holds itself together and does not expand with the balloon.

This interpretation of the expansion of the universe means that clusters of galaxies do not actually have a velocity through space and therefore that the Doppler effect does not explain the redshift in their spectra. A Doppler redshift is caused by the relative motion of an object that is emitting a wave. The ***cosmological redshift*** has a different cause. Long ago when a distant galaxy emitted some electromagnetic waves, those waves were not redshifted. But as space expanded, the waves expanded—that is, they lengthened—along with space. The more time the radiation has traveled, the longer its waves have become. ➤Figure 18-7 illustrates the difference between the two explanations.

cosmological redshift. The shift toward longer wavelengths that is due to the expansion of the universe.

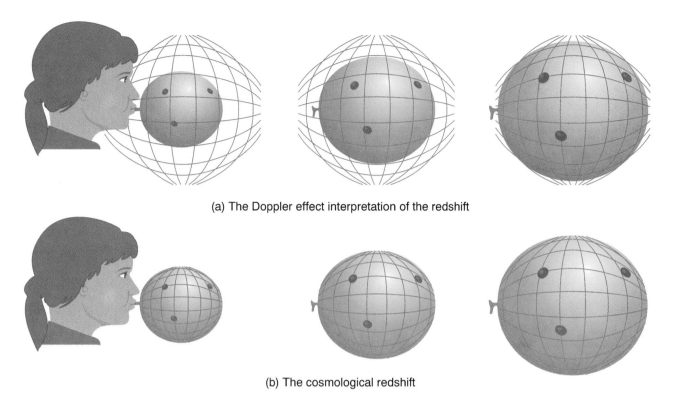

(a) The Doppler effect interpretation of the redshift

(b) The cosmological redshift

▶ **FIGURE 18-6.** Each tiny seed glued to the balloon represents a cluster of galaxies. (a) This is the old, incorrect interpretation of an expanding universe in which the universe expands through space. (b) This is the correct cosmological interpretation. Space expands, dragging the clusters of galaxies along with it.

▶ **FIGURE 18-7.** (a) This shows a Doppler redshift that results from the motion of the source. The wave traveling backward toward the radio telescope is lengthened by the motion of the source. (b) When a wave is emitted by a nonmoving object, it has the same wavelength in all directions. As space expands, that wave lengthens. If the wave takes long enough to get from the source to the telescope, it will have lengthened significantly. This is the case for galaxies that exhibit the cosmological redshift.

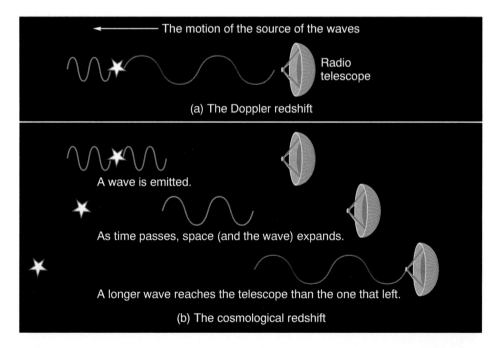

The Expanding Universe 549

CLOSE UP

Wrong Explanation: The Doppler Effect

Why is the Doppler effect the incorrect explanation for the cosmological redshift? The answer lies in one of the criteria for a good scientific theory: aesthetics.

Quasars have been discovered that have a redshift greater than 4. A redshift of 4 means that the ratio of the shift in a particular wavelength ($\Delta\lambda$) to the unshifted wavelength (λ) is equal to 4. In the standard application of the Doppler effect, this means that the velocity of the object (v) is equal to four times the velocity of light (c), because

$$\frac{\Delta\lambda}{\lambda} = \frac{v}{c}$$

This cannot happen! An object cannot travel at a speed greater than the speed of light. To resolve this dilemma, astronomers apply the law of special relativity to the problem because that law must be used when objects move with speeds near the speed of light. When the rules of special relativity are used, the calculated speed of a receding cluster of galaxies is always less than the speed of light.

The fallacy here is that special relativity is not applicable in a universe where general relativity is the controlling rule. Special relativity can be applied only where space can be considered flat. In the case of distant clusters, the curvature of space must be taken into account, so special relativity cannot be used.

To review this relationship between laws, recall that Newton's law of gravitation should not be considered *wrong*. Rather, it is a law with limited applicability that must be replaced by general relativity in certain cases. It is accurate enough for most everyday uses, however. Likewise our standard concept of velocity (where we do not worry about objects moving at or near the speed of light) works in everyday situations. Special relativity needs to be used only when speeds approach the speed of light. Special relativity is also a law with limited applicability, however. It can only be used for nonaccelerated objects, and it cannot be used when space curvature is a factor.

When we try to combine physical laws that are not consistent with one another, we violate the principle that a theory should be aesthetically pleasing. Consistency is important in applying scientific principles.

Thus the Doppler effect cannot be used as the explanation for the cosmological redshift. The correct explanation lies in the idea of the expansion of space, as discussed in the text.

In equation form the Hubble law is $v = H_0 d$ where v = the velocity of the object, d = the distance to the object, and H_0 is the Hubble constant.

When the Hubble law was written, astronomers assumed that the observed redshift was caused by the now-rejected Doppler explanation. The law, which states the connection between a galaxy's distance and its velocity, is still valuable, however, if we think of the velocity in the Hubble equation as being caused by the expansion of space rather than velocity *through* space. Although astronomers sometimes use more modern ways of stating the redshift/expansion-rate relationship without using the concept of velocity, this text will continue to use the traditional Hubble relationship.

Olbers's Paradox

Why is the sky dark at night? This appears to be a very simple question, but it actually has important cosmological implications. If stars (or galaxies or clusters of galaxies) are distributed throughout space, the sky should not be dark, because no matter in which direction one looks, the line of sight should end on a star (▶Figure 18-8). It is true that distant stars appear dimmer than nearby ones, but in a universe of evenly distributed stars, there are more stars at great distances, and this eliminates the dimming effect. If you are deep within a forest, you see a tree no matter which direction you look. Distant trees look smaller, but there are more of them.

Perhaps the absorption of light by dust clouds prevents the sky from being bright. But dust clouds heat up as they absorb light, and since they absorb light from an infinite number of stars, they should get hot enough to glow. Dust clouds cannot be used to explain the black sky.

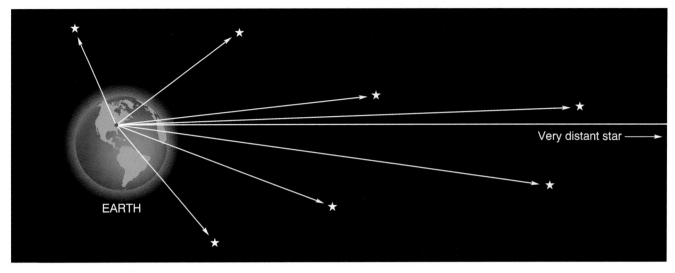

> **FIGURE 18-8.** In an infinite, static universe, your line of sight would end at a star no matter in which direction you looked.

The paradox between what is observed and the argument that the sky should not be black was pointed out by Wilhelm Olbers in 1823. It is called **Olbers's paradox**. The problem exists whether the universe is open or closed. An open universe extends without limit, so the line of sight should extend through space until it hits a star. In a closed universe, the line of sight could possibly extend all the way around until the surface of a star is encountered.

The solution to Olbers's paradox—the reason that the sky is dark—is that the universe is expanding. An expanding universe solves the problem by two mechanisms. First, when light from distant galaxies is redshifted, it also loses energy. This occurs because the energy of a photon of light depends on the light's frequency, so a photon of lower frequency (due to redshift) has less energy. Second, an expanding universe cannot have existed forever, for in the past it must have been smaller than it is now. And before that it was even smaller. At one time in the past, it must have had a beginning. As our line of sight goes farther and farther across space, it goes further and further back in time. It cannot go past the beginning of the universe, however. We cannot see beyond that.

These two explanations eliminate the paradox, but they depend upon the fact that the universe is expanding and that it had a beginning. This latter point is at the heart of the major theory of modern cosmology, the *big bang theory*. Before that discussion, however, the fundamental assumptions of modern cosmology must be described.

Olbers's paradox. An argument showing that the sky in a static universe could not be dark.

Photons were discussed in Chapter 4.

COSMOLOGICAL ASSUMPTIONS

The assumptions of cosmology cannot be proved, but evidence indicates that they are reasonable. The first assumption is that the universe is *homogeneous*. This means that no matter where an observer is positioned in the universe, the universe will look essentially the same. This does not mean, of course, that there is another Earth from which an observer can look to see another Mars in the sky. It does not even mean that there is another local group of galaxies similar to our local group. It means, rather, that on the largest scale, the universe has about the same density and composition of matter at one location as it has at another. As this text has described, humans have come to learn that they are not in a favored place in the universe. In cosmology, this becomes an underlying assumption.

homogeneous. Having uniform properties throughout.

► **FIGURE 18-9.** Although the ocean surface is homogeneous, you see something different when you look in different directions because the surface is not isotropic.

isotropy. The property of being the same in all directions.

The second assumption is somewhat related to the first. The assumption of *isotropy* (I-SOT-rah-pee) states that the universe looks the same in all directions. Although this sounds like homogeneity, it is actually a different quality. The ocean in ►Figure 18-9 is homogeneous, for no matter where a person is located on the surface of the ocean, the view is the same. However, someone on the surface sees a different view when she looks across the waves than she sees looking along the waves. This means that the ocean is not isotropic.

Like homogeneity, isotropy applies only on the largest scale. Recall that when we look out from Earth, we can see farther when we look out of the disk than if we look along the disk. The universe is not isotropic on this small scale. To apply the idea of isotropy, one must imagine being outside the Galaxy and even outside the local group of galaxies. In this case, the assumption is reasonable.

cosmological principle. The basic assumption of cosmology that holds that on a large scale the universe is the same everywhere.

The two assumptions of cosmology are so much at the heart of the subject that they form what is called the *cosmological principle.* The principle formalizes the idea that we are not located at a special place in the universe. It brings to final completion the Copernican revolution that proposed that the Earth is not central. The cosmological principle not only states that the Earth is at no special place in the universe but that there *is* no special place in the universe.

universality. The property of obeying the same physical laws throughout.

An additional assumption is *universality.* This means that the same physical laws apply everywhere in the universe. Although this may seem so obvious that it need not be stated, recall that it was fairly recently in human history that Isaac Newton found the first law of nature that could be shown to apply beyond the Earth. Now we have vastly more evidence for this assumption as we analyze radiation from the most distant objects, but we must remember that it is still an *assumption* that this principle holds throughout the universe.

THE BIG BANG

The present motion of galaxies leads us to believe that at one time in the past, they were all together. Suppose a large firecracker is placed at the center of a small stack of sand. A moment after the explosion of the firecracker, we analyze the situation. ►Figure 18-10 shows what might appear. Soon after the explosion, grains of sand are at various distances from their original position, and—here is the important point—grains that are farther from their respective starting points are moving faster. That is how they got farther from one another. No matter which grain one considers, all others are moving farther from it.

big bang. The theorized initial explosion that began the expansion of the universe.

Calvin, of the cartoon Calvin and Hobbes, prefers the name the Horrendous Space Kablooie.

The explosion that started the expansion of the universe is called the *big bang,* an irreverently trivial name for the event that began the universe. In its most basic sense, the standard big bang model is simply the idea that every bit of the matter and energy in the universe was once compressed to an unimaginable density. In the big bang, the material exploded apart.

We must be careful not to think of the material of the big bang as existing at a certain place, at one point in the universe. It *was* the universe—the entire universe. It did not exist as part of the universe.

After the initial explosion, the nuclear particles formed into atoms of low mass (hydrogen and helium) then clustered under the force of gravity into stars, clusters of stars, galaxies, and clusters of galaxies. The clusters of galaxies are still moving apart. Again, we must be careful not to think of the universe as expanding into empty space around it, for space does not exist apart from the universe. The universe is all there is, and its parts are moving away from one another. The example of the explosion in a pile of sand is misleading if one does not remember that in the case of cosmological expansion, galaxies do not move through space as the grains of sand do. We have no experience with the expansion of space in our everyday lives, so Earthly examples must involve motions through space.

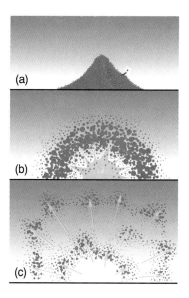

▶ FIGURE 18-10. Part (a) shows a firecracker about to explode inside a small pile of sand. Parts (b) and (c) show the sand later, with a few arrows included to illustrate that the sand grains that are farthest from the initial explosion are moving fastest.

Evidence: Background Radiation

The idea of the big bang arose from the evidence for the expansion of the universe. We must ask, however, if there is other evidence to support the theory. The first evidence was found quite accidentally when a prediction made by the big bang theory was found to be correct. As scientists learned more about nuclear processes in the 1940s, they were able to make predictions about the character of the early universe shortly after the big bang. The material of the big bang was originally extremely hot, but it cooled as it expanded outward, in much the same way that a gas cools when it expands. The hot matter of the very early universe was opaque to radiation, but when it cooled to about 3000 K, it became transparent. At this point, the gas that made up the universe was emitting radiation that had the characteristics of radiation from a 3000-degree object, and this radiation existed over the entire universe.

The universe has remained transparent to radiation since that time, so it was predicted that this radiation should still be in existence. Remember that as we look at objects at greater and greater distance, we are looking back in time. If we were to detect the radiation left over from the big bang, it would be coming from far back in time and therefore from a great distance. According to the Hubble law, then, it would be greatly redshifted. The original 3000-degree radiation peaked in the infrared region of the spectrum, but calculations showed that it would by now have been redshifted to the microwave region, with wavelengths of a millimeter or so. Such radiation would not be characteristic of radiation from a 3000-degree object, but it would appear the same as radiation from a very cold object—one at about 3 K. It was predicted that this **background radiation** should be striking Earth from all directions but that it should be very faint.

In 1948, when the prediction of background radiation was made, there was no way to detect such weak waves so the prediction was laid aside and forgotten. Then in the mid-1960s, a group of physicists at Princeton University were studying the big bang and again predicted the existence of background radiation, not realizing that the prediction had been made some 15 years before. They were confident that they could build a radio receiver that would be able to detect the radiation, and they set about building one on the roof of the Princeton biology building. They were too late.

At the same time the predictions were being made at Princeton, only a few miles away two employees of Bell Telephone Laboratories, Arno Penzias and Robert Wilson (▶Figure 18-11), were doing applied research on microwave

3 K corresponds to −270°C.

background radiation (or cosmic background radiation). Long-wavelength radiation observed from all directions; believed to be the remnants of radiation from the big bang.

The Big Bang 553

▶ **FIGURE 18-11.** Arno Penzias (right) and Robert Wilson discovered the cosmic background radiation using Bell Laboratories' horn-shaped radio antenna, seen behind them.

transmission in hopes of improving the transmission of messages. They were frustrated, however, by some low-intensity radio waves they observed to be coming from all directions toward their receiver.

As you might suspect, the radiation they had found turned out to be the cosmic background radiation predicted by the physicists at Princeton and by astronomers years before—the prediction that was based on the big bang theory. Penzias and Wilson were awarded the Nobel Prize in 1978 for their discovery of the radiation—radiation for which they had not been searching and the cosmological implications of which they were unaware.

Recall that at the time of the big bang, the condensed material made up the entire universe. This is why the background radiation now comes to us from all directions. The radiation fills the universe just as it did at the beginning. The matter and energy comprise the entire universe, and no space is left over.

Until the discovery of the background radiation, there was another popular cosmological theory (see the Close Up "The Steady State Theory"), but that discovery established the big bang theory as the accepted explanation for the beginning of the universe.

In 1989 NASA launched the *Cosmic Background Explorer* (*COBE*, pronounced "KOH-bee"). Its purpose is to measure the background radiation at various wavelengths and to map the sky in those wavelengths. *COBE*'s results provide an astounding corroboration of the accuracy of predictions concerning cosmic background radiation. ▶ Figure 18-12 is an intensity-wavelength graph like those

▶ **FIGURE 18-12.** The theoretical intensity-wavelength graph (of the type discussed in Chapter 12) for cosmic background radiation is shown in red. The tiny black lines represent data from *COBE*. The fit is remarkable!

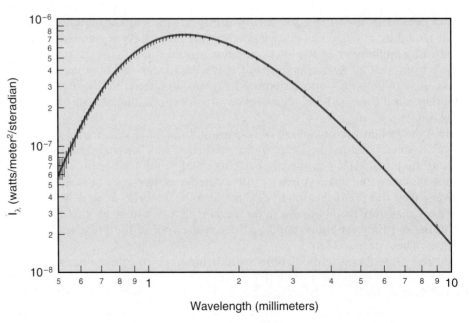

554   CHAPTER 18   Cosmology: The Nature of the Universe

CLOSE UP

The Steady State Theory

The cosmological principle holds that we are at no special place in space. It does, however, allow for us being at a special place in time; for at the present moment, the universe is less dense than it was in the past and more dense than it will be in the future because expansion reduces its density. In 1948, a *perfect cosmological principle* was proposed by astronomers Fred Hoyle, Hermann Bondi, and Thomas Gold. It states that our location in time is no more special than our location in space, and that the universe of the past was the same as it is now and the same as it will be in the future. Since the overall universe does not change in this cosmology, it is called the *steady state theory.*

Before looking at the implications of the perfect cosmological principle, consider why such a thing might be proposed. Remember that a theory should be aesthetically pleasing. The perfect cosmological principle is neater because it expands the nonspecialness to include time as well as space. The perfect cosmological principle carries the nonspecialness idea through both space and time.

It seems to be almost undeniable that the universe is expanding. How can this be squared with the steady state theory? It seems that if clusters of galaxies are getting farther apart, the universe will necessarily be less dense in the future than it is now. Hoyle, Bondi, and Gold proposed that matter is created in the space between galaxies and that this new matter is sufficient for the universe to maintain its density (➤Figure C18-1). Such spontaneous creation of matter would violate a long-established principle of physics: the conservation of mass/energy, which holds that the total amount of mass and energy in the universe cannot change.

The principle of the conservation of mass/energy is dear to scientists, but it has changed in the past. Before this century, there were two such conservation principles: the conservation of mass and the conservation of energy. It was thought that neither matter nor energy could be created or destroyed. Then in 1905, Einstein's theory proposed that mass could be changed into energy—a prediction that was dramatically demonstrated in the atomic bomb. Einstein combined the two theories into one, the conservation of mass/energy, which sees mass and energy as two forms of the same thing and holds that the total mass and energy in the universe remains the same.

The amount of mass that would have to be created in order to fill the gaps between galaxies as they move apart would be very small—only one new hydrogen atom in each cubic meter of space every 10 billion years or so. This means that if one new atom were created every 10,000 years in the volume of the Houston Astrodome, the density of the universe would remain the same. The theory holds that as millions of years pass, newly formed atoms collect to form new stars and galaxies, and that the expanded space is thereby refilled, so that the universe is no different now than it was in the past and than it will be in the future.

The fact that the steady state theory violates the principle of conservation of mass/energy should not be considered a strike against the theory. First, according to the predictions, the necessary creation events are so rare that they would be far below any possible detection limit and therefore far below the accuracy to which the principle has been verified. Secondly, the big bang theory may also involve a violation of the principle. In this case it is a gigantic one-time violation, as all matter and energy are created in one big burst at the beginning.

After the discovery of the cosmic background radiation, many attempts were made to reconcile its existence with the steady state theory. None were successful. Today the steady state theory is considered a relic of the past, another example of a theory destroyed by data it cannot explain.

➤ **FIGURE C18-1.** If the universe does not decrease in density as galaxies move apart, matter must be created in the space between galaxies.

used to describe the temperature of stars in Chapters 12 through 15. The red line on the figure is the predicted intensity of radiation at each wavelength. The tiny black lines represent data points measured by *COBE*. Each data point is depicted as a line, which takes into account possible errors in the measurement. The longer lines at the left indicate that the measurements there—those at shorter wavelengths—are less accurate. The data fit is extraordinary.

The latest results show that the temperature of the background radiation from *COBE* data is 2.726 K (with an uncertainty of 0.01 K). This matches theory to within 0.03%, which is 1000 times better than the best data before *COBE*.

The Age of the Universe

If the big bang theory is correct, we can use the Hubble constant to determine how long ago the big bang occurred, for the constant tells us how fast galaxies are spreading apart. Instead of writing v for velocity in the Hubble law, substitute the definition of velocity: the ratio of distance to time.

$$v = H_0 d$$

$$\frac{d}{t} = H_0 d$$

Now choose two widely separated galaxies and let d be the distance they are apart. Thus, t is the time taken to go that distance, or the age of the universe. Solving the equation for t gives

$$t = 1/H_0$$

This indicates that the age of the universe is simply the reciprocal of the Hubble constant. This constant is normally expressed as km/(s × MLY) or km/(s × Mpc). Before we can substitute numbers to calculate the age of the universe, both distances must be expressed in the same units rather than one in kilometers and the other in millions of light-years. One million light-years is about 9.5×10^{18} kilometers, so a Hubble constant of 15 km/(s × MLY) corresponds to 1.6×10^{-18} km/(s × km). The age of the universe is the reciprocal of this, or

$$t = \frac{1}{\frac{1.6 \times 10^{-18}}{s}} = 6.3 \times 10^{17} \text{ s} = 2 \times 10^{10} \text{ yr}$$

This is 20 billion years. On the other hand, using 25 km/(s × MLY) as the value of the Hubble constant gives an age for the universe of 12 billion years.

This calculation assumes that galaxies have continued to move apart at the same rate back through their history; that is, that the Hubble constant does not change over time. In fact, we would expect the speeds of galaxies to change. Because of the gravitational force they exert on each other, the recessional speeds of galaxies should be decreasing. This would mean that in the past, they were moving apart faster than they are now and that the age of the universe is therefore less than calculated above. For this reason, and because the Hubble constant is not known precisely, 20 billion years is taken as the *upper limit* to the age of the universe. The universe is this age or younger.

As Chapter 17 noted, the most distant quasar yet found has a redshift of 4.73. This indicates a look-back time of 94% of the age of the universe. Although we know the age of the universe fairly imprecisely (12 to 20 billion years), Doppler shift data are accurate enough that we can be confident that the farthest quasar thus found has a look-back time of 94% of whatever is the correct value for the age.

The most recent studies yield a value of 26 km/(s × MLY) for the Hubble constant, putting the age of the universe at about 12 billion years.

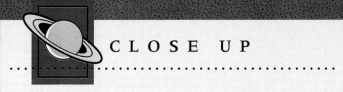

CLOSE UP

Science, Cosmology, and Faith

Where did the matter and energy of the big bang originate? Science cannot answer that question at present. In fact, it seems possible—perhaps likely—that it will never be able to answer that question. For if the matter existed in atomic form before the big bang, it did not carry information with it through the big bang. Thus the matter itself can tell us nothing about its pre–big bang history.

Does this then mean that science can say that there must be a creator, and that science has, after all, found a need for God? An individual scientist may believe this, but science itself cannot use God as the explanation for the big bang. Science cannot use God for any explanation. It has been said that science avoids God. It does, indeed. The reason that science does not use God for explanations is basically that science has been successful in explaining the material world without reference to a God. Science, by intention, uses natural causes to explain natural effects.

We say that science is successful in its method because scientific explanations of the workings of the material world have led us to further understanding of that world. The fact that the success has come without reference to God indicates that the material universe seems to be describable by completely natural principles.

What about cases where science is unable to find an answer? If science, when it comes to something that it cannot explain at the time, were to explain it by reference to God, the search for an explanation would end. If Newton had used God as an explanation for why things fall to Earth, he would never have developed the theory of gravitation. If we use God as an explanation for the big bang, there would be no reason to look further for a natural explanation. Use of supernatural explanations would shut down science. History, however, tells us that it is profitable to look for natural explanations. Supernatural explanations cannot be used in natural science.

Testability

There is another reason that science cannot use God for an explanation, and this relates to the reason that traditional science does not accept creationism as a science: A theory of science must be able to be shown to be wrong. A theory must be testable. Every theory must be regarded as tentative, as being only the best theory we have at present. It must contain within itself its own possibility of destruction. The 1948 prediction concerning cosmic background radiation was such a case. If the background radiation had not been found, the big bang theory would have had to have been adjusted, or—if enough such contradictions appeared—the theory would have had to be dropped and replaced with another. This has happened a number of times in science. It happened to the steady state theory and to other theories presented in this text.

On the other hand, if science relied on a creator to explain the inexplicable, there would be nowhere to go, no way to prove that explanation wrong. The question would have already been settled.

This is not to say that God might not be the explanation. Many people believe that a creator is the ultimate explanation of everything. Perhaps some questions, like the origin of the material for the big bang, simply cannot be answered without reference to a creator. But in that case, science simply cannot answer the question. The question is beyond the realm of science. Science does not deny the existence of God. God is simply outside its realm.

THE FUTURE: WILL EXPANSION STOP?

What came before the big bang? As discussed in the accompanying Close Up, that question may not be a scientific one. But it might. To see how scientists might get a hint at what came before the big bang, we look into the *future*.

If galaxies are moving apart, they will obviously be farther apart in the future than they are now. Where does this stop? There *is* an agent to stop it: gravity. The force of gravity must be slowing the expansion of the universe. The question is, will it slow it enough to stop it and bring the galaxies back together?

There are two possible answers: Either the expansion stops or it doesn't. First, suppose the expansion does not stop. This will mean, simply, that the clusters continue to get farther apart. Figure 18-13a illustrates a universe that continues to grow. In this universe, the stars use up the hydrogen fuel that powers them,

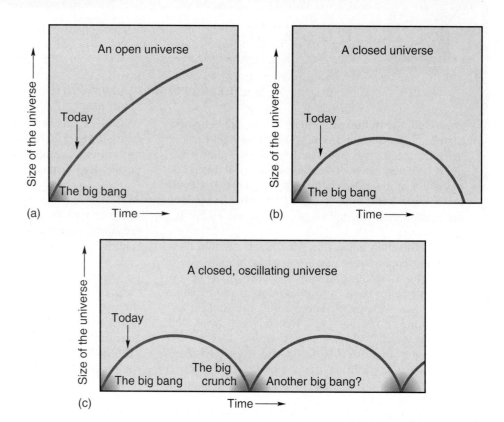

► **FIGURE 18-13.** (a) An open universe would continue to grow in size indefinitely. (b) A closed universe would someday stop growing and begin to contract. (c) Could it be possible that a closed universe could oscillate, going through consecutive big bangs and big crunches?

A closed universe will stop its expansion. An open one won't.

oscillating universe theory. A big bang theory that holds that the universe goes through repeating cycles of explosion, expansion, and contraction.

the glowing stars become fewer, the glow fades, and the universe fizzles out. Not a very attractive idea, but a possibility.

On the other hand, suppose the expansion stops, as shown in Figure 18-13b. The gravity will begin to pull the galaxies back toward one another. Intelligent beings (our descendants?) living on some other planet will be able to use the Doppler effect to observe that the galaxies are then getting closer together. The infall will continue until all matter in the universe is condensed into a tremendously dense ball—the "big crunch." What might happen then? Perhaps another big bang, another universe.

If this is the future, perhaps it was the past. Perhaps this is what preceded the big bang: a previous universe containing the same matter and energy as ours. If so, then today's universe may be just one phase of a number of oscillations, each having its own big bang. Such a scenario is called an *oscillating universe* (Figure 18-13c). It is a much more exciting prospect than a fizzling-out universe.

Evidence: Distant Galaxies

There are two ways to search for an answer to the question of whether the universe will stop expanding. One way is to see if we can measure how much the expansion is slowing with time. Of course, we cannot use the few years of our observations and expect to see a change. However, we are seeing distant galaxies as they were in the past, not the present. When we look at the most distant galaxies—at active galaxies and quasars—we are detecting light that left them billions of years ago, and the Doppler shift of that light tells us their speed then, not now. Therefore we should be able to tell how much they are slowing down by comparing their speeds to the speeds of galaxies nearer to us. If we can determine this, we can calculate whether their rate of slowing is enough to bring them to a stop.

In practice, the observation is tough to make. The problem is that in order to look far enough back in time to get a significant change in speed, we are looking at objects so distant that the light is extremely dim. The Doppler effect is easily observable, but we cannot get accurate measurements of the distances to these galaxies. The primary reason for this is that distance is determined from magnitude, and we cannot be certain how stellar populations within galaxies have changed over time. If there were significantly more brighter (or dimmer) stars in galaxies when they were younger, our calculations of distances to the faraway galaxies will be incorrect.

➤Figure 18-14 is a graph of velocity versus distance for galaxies. On it one line is drawn to show the relationship as predicted by the steady state theory. This is the straight line, for the theory holds that the expansion is exactly the same now as it was in the past (and will be in the future). On the other hand, if gravity is slowing the expansion, the actual case will be a line that curves, perhaps like the other line. Along this line, the velocities of the most distant galaxies are faster than the steady state theory would predict, for the velocities were greater in the distant past.

The curved line in ➤Figure 18-15 represents the borderline case between a universe that will expand forever (an open universe) and one that will stop its expansion and start falling inward on itself (a closed universe). Suppose we plot measured values on the graph. If they fall on the straight line, they will indicate that we live in a steady state universe. If they fall in the green area of the graph, it will indicate an open universe, one that will expand forever. On the other hand, if the data indicate a curvature greater than the curved line (in the red area), the universe is closed.

The borderline case between an open and a closed universe is called a *flat universe.* Inflationary universe theories, discussed in the next section, predict that the universe is exactly flat and that the data should fall on the curved line. If the universe is flat, it will gradually slow its expansion but will stop only after an infinite amount of time; therefore it will never fall back inward. While a closed

flat universe. The condition of the universe if gravity just balances its expansion so that it stops expanding only in an infinite amount of time.

➤ FIGURE 18-14. If the Hubble data fall along the straight line, a steady state universe will be indicated. If the data fall along a curve such as the one shown, it will indicate that the universe was expanding faster in the past and has slowed its expansion.

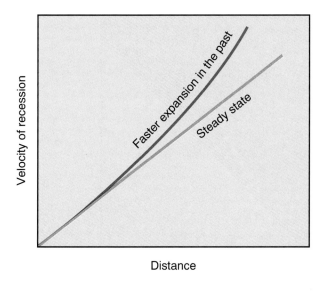

➤ FIGURE 18-15. If the data fall in the green area, the universe is expanding too fast to stop, and we live in an open universe. If the true case is in the red area, this indicates that the expansion speed changes greatly as time passes and that the expansion will stop.

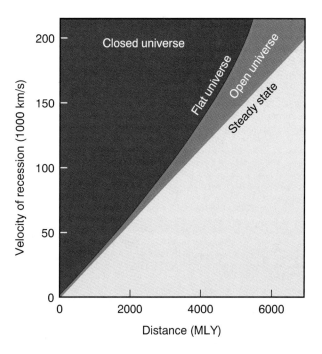

The Future: Will Expansion Stop?

universe is represented in three dimensions by a sphere and an open universe by the shape of a saddle, the corresponding representation of a flat universe is simply a plane, like the top of a desk. If the universe is flat, it is one that Newton would understand. In order for the universe to be flat, however, it must meet the exact conditions to be on the borderline of being open and closed.

➤Figure 18-16 shows present data plotted on the graph. Notice that the data indicate a line curving upward, showing that galaxies are moving more slowly now than in the past. The data are not accurate enough, however, to tell us whether or not the expansion will stop, although one might say that there is a hint that the slowdown is not great enough to stop the expansion. That is, there are more points plotted to the right of the borderline case than to the left. So although the conclusion must be considered very tentative, indications are that galaxies are not losing speed fast enough to ever stop moving apart. It is important to emphasize, however, that this is an extremely tentative conclusion.

The Density of Matter in the Universe

The second method of seeking an answer to the question is to determine the overall density of matter (and energy) in the universe. If there is enough mass/energy in each volume of space, there will be enough gravitational force to stop the expansion. Chapter 17 described evidence, based on the amount of mass/energy necessary for galaxies and clusters to be gravitationally bound, that dark matter makes up about 90% of the matter of the universe.

If astronomers are incorrect in their prediction that great amounts of dark matter exist throughout the universe, then the universe is open and will continue to expand forever. If dark matter does exist in the amount predicted by gravitational studies, however, then the universe is just barely dense enough to lie on the boundary between open and closed. The density that the universe would have to have to be perfectly flat is called the ***critical density.*** Today's best calculations indicate that the universe's density is somewhere between 0.1 of the critical density and 2 times that density. Therefore the universe appears to be flat or almost flat.

The theory of general relativity (Chapter 3) predicts that both matter and energy cause what we call gravitational force.

critical density. The density of a perfectly flat universe.

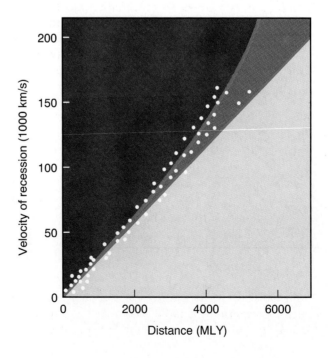

➤ **FIGURE 18-16.** When present data are plotted on the graph, we see that recessional speeds seem to have been greater in the past, so that the steady state theory does not fit. Velocity data such as these are not accurate enough, however, to determine whether the universe is open or flat.

THE INFLATIONARY UNIVERSE

Scientists do not like coincidences, and the standard big bang model is unable to explain two cosmological observations that appear to be coincidences. Attempts to solve these two problems, as well as others, led astronomers to propose a modification of the standard model.

The Flatness Problem

The universe seems to be either flat or very nearly flat. This would be an extreme coincidence because a flat universe is a special case—a universe that has a density that puts it exactly on the border between open and closed. Of all the possible densities that the universe might have, why would it have this particular one? The universe could be a little more dense or billions of times more dense, and in either of these cases it would be closed. (If it were much more dense than it is, it would be short-lived, so that planets—and life—would not have had time to form.) On the other hand, it could be slightly less dense or billions of times less dense, in which case it would be open. (If it were much less dense than it is, gravity would not have gathered the diffuse matter into stars and planets, so again life would not exist.) The *flatness problem* asks, "Of all the possible conditions of the universe, why is it so nearly flat?"

The flatness problem is even worse than just described, however, because if the universe's density had not been exactly equal to the critical density in the beginning, the curvature of the universe would have increased as time passed, and the curvature would be even greater now. Consider the ratio of the actual density of the universe to the critical density. If, immediately after the big bang, that ratio was equal to 1 (that is, the universe's density was equal to the critical density), then the ratio would still be equal to 1 today. If, however, the ratio at the big bang was less than 1, calculations show that it would be *much* less than 1 today. And if the ratio just after the big bang had a value greater than 1, the ratio would have become even greater today. For today's ratio to be somewhere between 0.1 and 2, it must have been extremely close to 1 at the beginning. Could this be coincidence?

The *inflationary universe model* has an explanation for the flatness of the universe. The model provides details of the first fractions of a second after the big bang, and according to its predictions, during the short time interval between 10^{-35} second and 10^{-32} second after the big bang, the universe suddenly expanded at an extremely fast rate. In fact, it grew by a factor of 10^{50} during that time. To understand how this solves the flatness problem, we must return to the analogy of fleas on a balloon. If the fleas are on a small balloon, the curvature will be obvious to them. If the balloon is very large, however, the fleas will not be able to detect its curvature (►Figure 18-17). The inflationary universe model holds that an "inflation" of the universe took place so quickly that any portion of it that had a great curvature before the inflation became flat as a result of the inflation.

Like any analogy, the balloon story does not work perfectly, for even a large balloon has some curvature. The inflationary model predicts that curvature was completely removed during the quick expansion. Therefore the universe's flatness is not a coincidence but a necessary consequence of the brief period of inflation very early in history.

The Horizon Problem

Cosmic background radiation is remarkably uniform no matter in what direction one looks. *COBE* data reveal that the radiation varies in temperature by less than one part in 10,000. This would indicate that the temperature of the universe was

the flatness problem. The inability of the standard big bang model to account for the apparent flatness of the universe.

inflationary universe model. A modification of the big bang model that holds that the early universe experienced a brief period of extremely fast expansion.

If the diameter of a helium nucleus expanded by a factor of 10^{50}, the nucleus would be 40 billion billion light-years across.

►Figure 18-18 is a drawing of the *COBE* satellite.

The Inflationary Universe 561

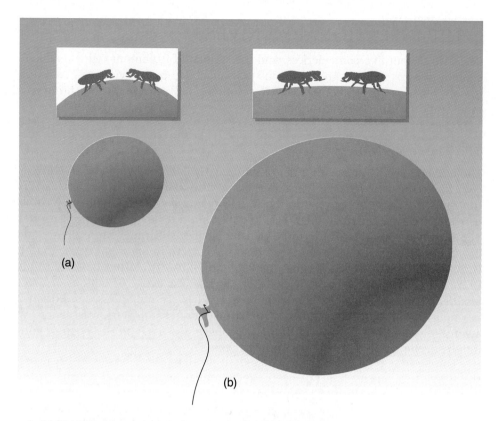

► **FIGURE 18-17.** The fleas on the small balloon (a) can tell that their universe is curved, but if the balloon is large enough, it will appear flat to the fleas (b).

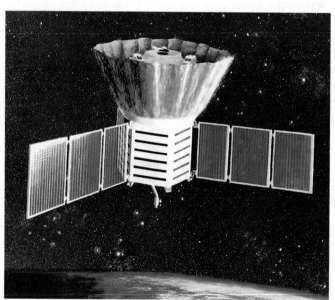

► **FIGURE 18-18.** The *Cosmic Background Explorer (COBE)* was launched in 1989 and is providing important data regarding background radiation. Figures 18-20 and 18-21 were produced from *COBE* images.

extremely uniform at the time that the background radiation was emitted. This does not seem possible in the standard model of the big bang because in that model the various portions of the universe could not have been in "communication" with one another when the radiation was emitted. To see why this is so, return to the balloon analogy one more time. First, imagine a balloon that is not expanding, and suppose that energy in the form of heat is added to one portion of the balloon. That energy will spread around the balloon, but this will not happen instantaneously. The amount of time required for the energy to distribute all over the balloon depends on how fast the energy travels. Nevertheless, the energy will eventually distribute uniformly around a nonexpanding balloon.

562   CHAPTER 18   Cosmology: The Nature of the Universe

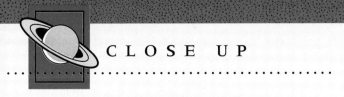

CLOSE UP

Stephen Hawking, the Ultimate Theoretician

Scientists are sometimes categorized as either experimentalists or theoreticians, depending upon whether they are primarily involved in the experimental or theoretical aspects of their science. William Herschel, who was highlighted in a Historical Note in Chapter 9, was primarily an experimental astronomer—an observer—while Albert Einstein is commonly agreed to be the premier theoretical physical scientist in the first half of this century. This half-century's answer to Einstein is generally considered to be Stephen Hawking, a professor of mathematics at Cambridge University (➤Figure C18-2).

Soon after entering Cambridge, Hawking began having physical problems; he stumbled and slurred his speech. His condition was diagnosed as amyotrophic lateral sclerosis, commonly called Lou Gehrig's disease. This disease, usually fatal, disables the voluntary muscles of the body. Hawking became depressed, abandoned his studies, and began drinking heavily. Then in January 1963 his life took another turn when he met Jane Wilde, a student of language who was more attracted to Stephen's mind than she was bothered by his deteriorating physical condition. They were married in 1965 and he became a father.

His marriage gave Hawking the determination he needed to continue with his life and work, and his brilliance soon came to the fore. He wrote his dissertation on "Properties of the Expanding Universe" and was awarded the Ph.D in 1966.

Hawking continued working on the relationship of general relativity to black holes and to the big bang and wrote papers on these subjects even as his physical condition deteriorated. By the early 1970s, he was confined to a wheelchair. The disease does not affect the mind, however, and the fact that it is not particularly painful left Hawking free to use his mind fully. His inability to use a pencil required him to work with long, complicated equations in his mind. He reports that it is impossible to handle the very complicated equations, so he has developed geometrical ways of thinking instead, giving problems a diagrammatic interpretation.

New discoveries continued to pour from his mind, including the seemingly absurd idea that the laws of quantum mechanics require that black holes emit a stream of radiation. This concept, slow to be accepted by other astronomers, has since become accepted, and the radiation is now known as "Hawking radiation."

Hawking's disease affects not only his skeletal muscles, but also the muscles used in speech. Until 1986, he often used an interpreter who was able to understand his distorted language. That year, Hawking developed pneumonia with severe complications. In order to save his life, surgeons removed his trachea. Now he has no voice at all. Instead, he uses three fingers (which are all he can control) to call up words on a computer screen. A voice synthesizer delivers a sentence once he has completed it. Using this method of communication, he has completed a book intended to explain his ideas to the general public: *A Brief History of Time: From the Big Bang to Black Holes* (Bantam Books, 1988).

After learning of his disabilities, one understands why Stephen Hawking is a theoretician rather than an experimentalist. His brilliance was apparent even in his youth, but it may be that his lack of physical abilities has served to make him even more brilliant—a nearly perfect cerebral being, a mind contemplating the universe.

➤ **FIGURE C18-2.** Stephen Hawking in front of a statue of Isaac Newton. (Hawking holds the same position at Cambridge University that Newton held.)

Now suppose that the balloon is expanding when energy is added to one portion of it. If the balloon is expanding rapidly enough, the energy will not be able to reach distant portions of the balloon because those portions are moving away too fast. There is a "horizon" beyond which the change in energy cannot be transmitted. In this rapid-expansion case, any portion of the balloon that experiences a change will not communicate this change to the rest of the balloon, and therefore the properties of the balloon will not be uniform.

In the case of the real universe, a change in one portion can be communicated to another portion with a maximum speed equal to the speed of light. The standard big bang model predicts that at the time that the cosmic background radiation was emitted, the universe was expanding at such a rate that the "horizon" distance was much less than the distance across the universe. Therefore any little change in one part of the universe could not be communicated to faraway parts. This means that there would be no reason for the entire universe to be at the same temperature and astronomers should not observe the same temperature in the background radiation in various directions. Hence, the *horizon problem*.

According to the inflationary universe model, before inflation occurred, the universe was smaller than the horizon distance. Therefore, any change in one portion of the universe was transmitted everywhere, and it was possible for the entire universe to be at a uniform temperature. Then inflation occurred. Two parts of the universe that were originally within each other's horizons were outside those horizons after inflation. They were still at the same temperature, however, as a result of their condition before inflation. This is why all parts of today's observable universe show a uniform background radiation.

The inflationary universe model was proposed in the early 1980s by Alan H. Guth of MIT. The model has been further developed by Guth and other cosmologists, including Stephen Hawking, so that now, along with dark matter, it is the best explanation we have for the development of the universe.

➤Figure 18-19 is a NASA drawing that summarizes the inflationary big bang theory. At upper left is the universe at the big bang and then just after the big bang. At the latter time, the universe contained extremely small irregularities. The universe then expanded rapidly during the inflationary period. The central image is from the COBE satellite and its production is explained in the next section. It pictures the universe 300,000 to 500,000 years after the big bang. By that time, the universe had cooled down enough to become transparent to radiation. Galaxies and stars began to form. The bottom right part of the image

This is called a *horizon* by analogy with the *horizon* that places a limit on how far we can see on Earth.

the horizon problem. The inability of the standard big bang model to account for directional uniformity of the background radiation.

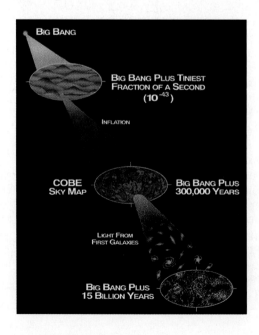

➤ FIGURE 18-19. This is a summary of the big bang theory. The central image—the COBE Sky Map—is not a drawing, but is a map of the cosmic background radiation. Its mottled appearance is caused by minute temperature variations in the universe at the time when the background radiation was released.

shows today's universe with each dot representing a cluster of galaxies. As we have emphasized previously, times such as those given in the figure are not well known at all.

THE GRAND SCALE STRUCTURE OF THE UNIVERSE

The success of the inflationary theory at first caused a problem in explaining the formation of galaxies, for in a completely uniform universe, there would be no irregularities around which matter could group into galaxies and clusters of galaxies. Not only did galaxies form, however, but the discovery of extremely old quasars indicates that matter clumped together relatively soon after the big bang.

Discoveries made during the 1980s added to the problem. As we accumulated data on the distribution of clusters of galaxies (➢Figure 18-20), we were able to construct three-dimensional drawings to show their locations. The drawings showed that clusters of galaxies are not distributed uniformly at all, but are arranged as if to form bubbles, with a void between the walls of the bubbles. To visualize this, think of soap bubbles in a sink. Galaxies make up the surfaces of the bubbles, and where the bubbles join, we find the largest clusters of galaxies.

A possible solution to the problem of the formation of galaxies and their bubblelike distribution has recently been found. If theorists assume that the universe at the time of the great inflation contained large amounts of cold dark matter, they find that gravity waves will be generated in this matter, and the gravity waves will lead to the formation of galaxies in walls and bubble-type configurations. If the dark matter that makes up most of the universe is not the normal matter of our experience but some exotic matter that cannot be perceived directly, it would not have emitted radiation as part of the cosmic background radiation. The curvature of space caused by its mass would affect the distribution of that radiation, however, and would prevent the radiation from being perfectly uniform.

In 1992, George Smoot of the Lawrence Berkeley Laboratory and his *COBE* team reported finding tiny irregularities in the cosmic background radiation. The irregularities amounted to a deviation of only 0.00003 K! In order to detect these tiny irregularities, thought to be due to tiny differences in the temperature of the early universe, the *COBE* team first had to subtract irregularities caused by known

Recall the mention of WIMPs (Weakly Interacting Massive Particles) in Chapter 11 in connection with the neutrino problem.

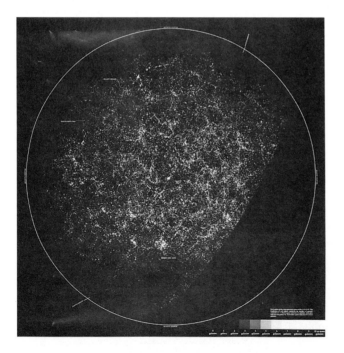

➢ FIGURE 18-20. Each dot on the diagram is a cluster of galaxies. The figure shows that clusters are clumped together in an irregular way.

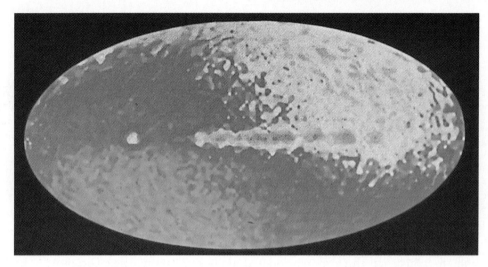

> FIGURE 18-21. The Earth moves around the Sun as the Sun moves around the Galaxy's center and the Galaxy moves in the local group. Each motion causes a Doppler shift in the background radiation. The blue and red portions of this *COBE* infrared map of the sky indicate the part of the sky we are moving toward and away from, respectively.

sources of radiation. These include the effect of the Sun's radiation, the effect caused by known radiators within the Milky Way, and a slight Doppler shift in the radiation that is caused by our motion through space (➤Figure 18-21). The central oval of ➤Figure 18-19 shows the resulting infrared map of the sky. The colors show slight differences in the temperature of portions of the long-past universe, the universe before the formation of stars. These tiny irregularities are the roots of the structures we see in today's universe. Without the irregularities in the early universe, there would be no Galaxy, no star, no Earth, no us.

As you probably conclude from the above discussion, there is great uncertainty in this scenario. Do not think, though, that it is pure speculation, for each of the various modifications of the model makes predictions about today's universe, and the predictions can be tested against observations. Cosmologists are in fairly general agreement with the basic idea of the inflationary theory. The details, however, are tentative. As new observatories bring us a flood of new data, we will undoubtedly have to revise the models. Some aspects of them will be discarded, but some will live on to be tested again and again until they join the body of firmly established science.

A simple example of a testable prediction of inflationary theories is the relative fractions of the chemical elements that would be produced and that should be found today.

CONCLUSION

In *The Nature of Reality*,* Richard Morris writes, "The scientific conception of the universe has changed so much that about the only thing that has remained the same is the word itself." This chapter—and this book—has shown how much our perception of the universe has changed. Nevertheless, we might pose the following question to a cosmologist: "What practical use can be made of the study of cosmology?" The answer is simply that not all pursuits need to have practical or useful consequences. We engage is some, like music, art, philosophy (and cosmology), for the sheer intellectual and artistic pleasure they provide. Curiosity is part of human nature, and questions like those raised in this chapter are among the most fundamental that can be asked—and perhaps answered—by science.

Our astronomical quest began 18 chapters ago in our home—the solar system. As you look back on the geocentric and heliocentric theories presented in the first few chapters, I hope that you see them as small, beginning steps in the development of today's view of the universe. After studying the solar system, we traveled outward to the stars and then much farther to galaxies. Finally, our study of cosmology considered the universe as a whole. I trust that this has given you a broader perspective on the natural world and a better appreciation of how the process of science helps us to understand the magnificent universe of which we are a part. I trust also that your quest for the universe will continue through your life.

*New York: The Noonday Press, 1987.

RECALL QUESTIONS

1. An open universe is one that
   A. will eventually collapse to form "a big crunch."
   B. will never stop expanding.
   C. [Either A or B above, depending upon its density.]

2. Einstein's conviction that the universe is closed and finite
   A. resulted directly from his general theory of relativity.
   B. resulted directly from his special theory of relativity.
   C. was a belief that caused him to make a change in his general theory of relativity.
   D. was a belief that caused him to make a change in his special theory of relativity.

3. What *evidence* leads us to believe that the universe is expanding?
   A. The big bang theory
   B. The inflationary universe theory
   C. The expansion of the solar system
   D. The redshift of most galaxies
   E. [All of the above.]

4. The assumption of isotropy states that the universe looks the same
   A. at all times.
   B. in all locations.
   C. in all directions.
   D. [All of the above.]

5. The cosmological principle refers to
   A. a comparison of our location in the universe to other locations in the universe.
   B. the redshift used to determine distances.
   C. Cepheid variables used to determine magnitude and therefore distance.
   D. the Doppler effect used to determine velocities.
   E. [Both B and C above.]

6. The existence of 3-degree background radiation supports
   A. the big bang theory.
   B. the steady state theory.
   C. the big bang theory and the steady state theory about equally.
   D. neither the big bang theory nor the steady state theory, for neither can explain it.

7. Big bang theories predict a _____ speed for the most distant galaxies than do steady state theories.
   A. a lesser
   B. a greater
   C. the same

8. The Hubble law is a relationship between galaxies'
   A. redshifts and colors.
   B. distances and redshifts.
   C. redshifts and spectral types.
   D. colors and spectral types.

9. If present data are correct, the universe is
   A. closed, but not oscillating.
   B. closed and oscillating.
   C. flat.
   D. open.

10. A better knowledge of which of the following would most aid us in determining whether the universe is open or closed?
    A. The age of the Earth
    B. The age of the solar system
    C. The size of the Galaxy
    D. The average density of the universe
    E. [All of the above would help about equally.]

11. What *evidence* indicates that an explosion took place at the beginning of the universe?
    A. The density of matter in the universe
    B. Low-intensity radiation from all directions
    C. The big bang theory
    D. The inflationary universe theory
    E. The existence of supernovae

12. If the average density of the universe is less than the critical density, the universe
    A. will expand forever.
    B. will eventually collapse.
    C. was produced in a big bang.
    D. is an inflationary universe.
    E. violates the laws of energy conservation.

13. Suppose that you read that astronomers calculate the age of the universe to be 15 billion years. Their calculation is based on
    A. calculations of the density of matter in the universe.
    B. the Hubble law.
    C. the age of rocks on Earth and Moon.
    D. the inflationary universe theory.
    E. [All of the above.]

14. The Hubble law is based on
    A. the big bang theory.
    B. the oscillating theory.
    C. the Doppler effect.
    D. inflationary theories.
    E. [None of the above.]

15. Which of the following forces may halt the expansion of the universe?
    A. Nuclear forces
    B. Electrical forces
    C. Gravitational forces

16. The cosmological redshift indicates that
    A. most stars in the Galaxy are moving away from the Sun.
    B. most galaxies are moving away from ours.
    C. stars at greater distances are hotter.
    D. space is expanding.
    E. [All of the above.]

17. The horizon problem and the flatness problem
    A. are solved by the standard big bang theory.
    B. are solved by the inflationary universe theory.
    C. have not yet been solved by modern cosmology.

18. Describe at least six steps in the historical progression of our ideas of the universe from Ptolemy's geocentric universe to Einstein's relativistic space curvature.

19. It is said that the universe is expanding. Explain just what this means.

20. Explain how, if almost all galaxies are moving away from ours, we are not necessarily at or near the center.

21. Name and explain three basic assumptions of cosmology.
22. What is background radiation and how does it enter the discussion of cosmological theories?
23. How is the age of the universe calculated?
24. Describe two methods of determining how quickly the expansion of the universe is slowing.
25. What is the oscillating universe?
26. Explain how the *COBE* results of Figure 18-21 relate to the universe of today.

QUESTIONS TO PONDER

1. Quasars do not exist in our part of the universe. Why is this not a violation of the homogeneity principle?
2. Give an example from nature (other than the one given in the chapter) of something that is homogeneous but not isotropic, and then give an example of something that is isotropic but not homogeneous.
3. Discuss the observation of background radiation as an example of the confirmation of a theory, comparing it to the prediction of stellar parallax in the 1600s.
4. The age of the universe calculated from the Hubble constant assumes a constant velocity of recession for galaxies. If in fact galaxies were moving faster at an earlier time, is the universe older or younger than calculated? Show your reasoning.
5. Where in the universe did the big bang occur? Explain.
6. It has been said—perhaps only half in jest—that when astronomers come upon a problem they cannot solve, they use a black hole. Give two examples that may have led someone to say this.
7. Explain why the horizon problem exists in the standard big bang model.
8. The horizon problem results from difficulties explaining how the background radiation can be so uniform in all directions. Figures 18-20 and 18-21 show that it is not perfectly uniform, however. Explain.

CALCULATIONS

1. If the Hubble constant is 20 km/(s × MLY), what is the maximum age of the universe)?
2. If the Hubble constant is 17 km/(s × MLY), what is the maximum age of the universe?

StarLinks netQuestions

Visit the netQuestions area of StarLinks (www.jbpub.com/starlinks) to complete an exercise on this topic:

1. **The Big Bang Theory** It all began with a really Big Bang.

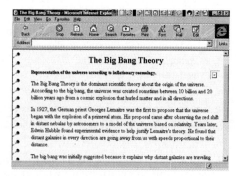

Appendixes

- A Units and constants
- B Solar data
- C Planetary data
- D Planetary satellites
- E The brightest stars
- F The nearest stars
- G The constellations
- H Answers to selected questions and try one yourself exercises

Units and Constants

APPENDIX A

Metric Prefixes

nano (n) = 10^{-9}
micro (μ) = 10^{-6}
milli (m) = 10^{-3}
centi (c) = 10^{-2}
kilo (k) = 10^{3}
mega (M) = 10^{6}

Metric-English Conversion

1 kilometer (km) = 0.6214 miles
1 meter (m) = 39.37 inches
2.54 centimeter (cm) = 1 inch
1 kilogram (kg) = 1000 g (weighs 2.2 pounds)
1 gram (g) = 0.001 kg (weighs 0.035 oz)

Units of Length

1 astronomical unit (AU) = 1.49598×10^{11} m
= 92.96×10^{6} miles
1 light-year (LY) = 6.324×10^{4} AU
= 9.461×10^{15} m
= 5.879×10^{12} miles
1 parsec (pc) = 2.063×10^{5} AU
= 3.086×10^{16} m
= 3.262 LY
1 nanometer (nm) = 10 Angstroms (Å)
= 10^{-9} m

Constants

Speed of light = 2.9979×10^{8} m/s
Electron mass = 9.1095×10^{-31} kg
Proton mass = 1.6726×10^{-27} kg

Temperature

	Kelvin (K)	Celsius (°C)	Fahrenheit (°F)
Absolute zero	0	−273	−459
Freezing point of water	273	0	32
Boiling point of water	373	100	212

$$K = °C + 273$$

$$\frac{°C}{°F - 32} = \frac{5}{9}$$

Appendix B

Solar Data

	Value	Ratio to Earth
Diameter	1,392,530 km	109
Mass	1.989×10^{30} kg	330,000
Average density	1.41 gm/cm^3	0.26
Surface gravity	270 m/s^2	28
Escape velocity	617 km/s	55
Surface temperature	6000°C	
Luminosity	3.9×10^{26} watts	
Absolute magnitude	4.83	
Tilt of equator to ecliptic	7.25 °	
Rotation period		
Equator	25.38 days	
40° latitude	28.0 days	
80° latitude	36.4 days	

PLANETARY DATA

APPENDIX C

Physical Data

Planet	Equatorial Diameter* (km)	(Earth = 1)	Oblateness	Mass** (Earth = 1)	Density (gm/cm³)	Surface Gravity (Earth = 1)	Escape Velocity (km/s)
Mercury	4,878	0.382	0	0.055	5.43	0.38	4.25
Venus	12,104	0.95	0	0.82	5.24	0.90	10.36
Earth	12,756	1	0.0034	1	5.52	1	11.18
Mars	6,794	0.53	0.009	0.107	3.93	0.38	5.02
Jupiter	142,800	11.2	0.064	318	1.32	2.69	59.6
Saturn	120,660	9.5	0.102	95.1	0.70	1.19	35.6
Uranus	51,000	4.0	0.024	14.5	1.2	0.93	21.1
Neptune	49,500	3.9	0.027	17.2	1.76	1.22	24.6
Pluto	2,290	0.2	?	0.002	2	0.05	1

*Equatorial diameter of Earth = 12,756 km.
**Mass of Earth = 5.9742×10^{24} kg.

Axis Tilt and Rotation Period

Planet	Equatorial Tilt to Orbital Plane (degrees)	Sidereal Rotation Period* (days or hours)
Mercury	0	58.65 d
Venus	178	−243 d
Earth	23.44	23.934 h
Mars	25.20	24.623 h
Jupiter	3.12	9.842 h
Saturn	26.73	10.65 h
Uranus	97.86	−17.2 h
Neptune	28.8	16.05 h
Pluto	119	−6.39 d

*The "−" signs indicate retrograde rotation.

Orbit Data

Planet	Semimajor Axis (10^6 km)	(AU)	Orbital Period	Orbital Eccentricity	Inclination to Ecliptic (degrees)
Mercury	57.9	0.387	87.97 d	0.2056	7.0
Venus	108.2	0.723	224.7 d	0.0068	3.39
Earth	149.6	1	365.26 d	0.0167	0
Mars	227.9	1.524	1.881 y	0.0934	1.85
Jupiter	778.3	5.203	11.86 y	0.0485	1.3
Saturn	1427	9.539	29.46 y	0.0556	2.49
Uranus	2870	19.19	84.01 y	0.0472	0.77
Neptune	4497	30.06	164.79 y	0.0086	1.77
Pluto	5900	39.44	248.5 y	0.250	17.2

Planetary Satellites

APPENDIX D

The Moon

	Value	Compared to Earth
Diameter	3476 km	0.27
Mass	7.35×10^{22} kg	0.0123
Density	3.34 gm/cm^3	0.61
Surface gravity		0.165
Escape velocity	2.4 km/s	0.21
Sidereal rotation period	27.322 days	
Synodic period (phases)	29.531 days	
Surface temperature	$-170°$C to 130°C	
Albedo	0.07	0.2
Tilt of equator to orbital plane	6.68°	
Orbit		
Average distance from Earth	384,400 km	
Closest distance	363,000 km	
Farthest distance	405,500 km	
Eccentricity	0.055	

Satellite Data

Planet Satellite	Average Distance (1000 km)	Sidereal Period (days)	Diameter (km)	Mass (10^{18} kg)
Earth				
Moon	384.4	27.322	3476	73,500
Mars				
Phobos	9.38	0.319	$28 \times 22 \times 18$	0.011
Deimos	23.46	1.262	$16 \times 12 \times 10$	0.002
Jupiter				
Metis	127.96	0.295	40?	0.095
Adrastea	128.90	0.298	$25 \times 20 \times 16$	0.019
Amalthea	181.3	0.498	$270 \times 170 \times 150$	7.2
Thebe	221.9	0.675	110×90	0.76
Io	421.6	1.769	3630	89,200
Europa	670.9	3.551	3138	48,700
Ganymede	1,070	7.155	5262	149,000
Callisto	1,880	16.689	4800	108,000
Leda	11,094	238.7	16?	0.0057

continued on next page

Planet Satellite	Average Distance (1000 km)	Sidereal Period (days)	Diameter (km)	Mass (10^{18} kg)
continued from previous page				
Himalia	11,480	250.6	186?	9.5
Lysithea	11,720	259.2	36?	0.076
Elara	11,737	259.7	76?	0.76
Anake	21,200	631	30?	0.038
Carme	22,600	692	40?	0.095
Pasiphae	23,500	735	50?	0.19
Sinope	23,700	758	36?	0.076
Saturn				
Pan	133.57	0.573	20	
Atlas	137.67	0.602	40 × 30 × 30	
Prometheus	139.35	0.613	140 × 100 × 80	
Pandora	141.70	0.629	110 × 100 × 70	
Epimetheus	151.42	0.694	140 × 100 × 100	
Janus	151.47	0.695	220 × 200 × 160	3.7
Mimas	185.54	0.942	392	45
Enceladus	238.04	1.370	500	74
Tethys	294.67	1.888	1060	740
Telesto	294.67	1.888	34 × 28 × 26	
Calypso	294.67	1.888	34 × 22 × 22	
Dione	377.42	2.737	1120	1,050
Helene	378.1	2.737	36 × 32 × 30	
Rhea	527.04	4.518	1530	2,500
Titan	1,221.86	15.945	5150	135,000
Hyperion	1,481.1	21.277	410 × 260 × 220	17.1
Iapetus	3,561.3	79.330	1460	1,880
Phoebe	12,954	550.0	220	
Uranus				
Cordelia	49.7	0.33	~40	
Ophelia	53.8	0.38	~50	
Bianca	59.2	0.43	~50	
Cressida	61.8	0.46	~60	
Desdemona	62.7	0.48	~60	
Juliet	64.6	0.49	~80	
Portia	66.1	0.51	~80	
Rosalind	69.9	0.56	~60	
Belinda	75.3	0.62	~60	
Puck	86.0	0.76	170	
Miranda	129.7	1.413	484	75
Ariel	191.2	2.520	1160	1,400
Umbriel	266.0	4.144	1190	1,300
Titania	435.8	8.706	1610	3,500
Oberon	582.6	13.463	1550	2,900
Neptune				
Naiad	48.2	0.30	60	
Thalassa	50	0.31	80	
Despoina	52.5	0.33	150	
Galatea	62	0.36	160	
Larissa	73.6	0.55	200	
Proteus	117.6	1.12	415	
Triton	354.6	5.877	2760	21,400
Nereid	5,513	359.4	340	21
Pluto				
Charon	19.7	6.387	1200	

The Brightest Stars

APPENDIX E

Star	Popular Name	Apparent Magnitude	Apparent Brightness Compared*	Absolute Magnitude	Absolute Luminosity Compared**	Distance (LY)	Spectral Type
	Sun	−26.8	1.4×10^{10}	4.83	1	0.000015	G2
α CMa A	Sirius	−1.47	1	1.4	24	8.7	A1
α Car	Canopus	−0.72	0.50	−3.1	1,500	98	F0
α Boo	Arcturus	−0.06	0.27	−0.3	110	36	K2
α Cen	Rigel Kentaurus	0.01	0.26	4.4	1.5	4.2	G2
α Lyr	Vega	0.04	0.25	0.5	54	26.5	A0
α Aur	Capella	0.05	0.25	−0.6	150	45	G8
β Ori A	Rigel	0.14	0.23	−7.1	59,000	900	B8
α CMi A	Procyon	0.37	0.18	2.7	7.1	11.4	F5
α Ori	Betelgeuse	0.41	0.18	−5.6	15,000	520	M2
α Eri	Achernar	0.51	0.16	−2.3	710	118	B3
β Cen AB	Hadar	0.63	0.14	−5.2	10,000	490	B1
α Aql	Altair	0.77	0.13	2.2	11	16.5	A7
α Tau A	Aldebaran	0.86	0.12	−0.7	160	68	K5
α Vir	Spica	0.91	0.11	−3.3	1,800	220	B1
α Sco A	Antares	0.92	0.11	−5.1	9,400	520	M1
α PsA	Fomalhaut	1.15	0.090	2.0	14	22.6	A3
β Gem	Pollux	1.16	0.089	1.0	34	35	K0
α Cyg	Deneb	1.26	0.081	−7.1	59,000	1600	A2
β Cru	(Beta Crucis)	1.28	0.079	−4.6	5,900	490	B0.5
α Leo A	Regulus	1.36	0.074	−0.7	160	84	B7

*Compared to Sirius, the brightest star other than the Sun.
**Compared to the Sun.

The nearest stars

APPENDIX F

Star	Apparent Magnitude	Apparent Brightness Compared*	Absolute Magnitude	Absolute Luminosity Compared**	Distance (LY)	Spectral Type
Sun	−26.8	1.4×10^{10}	4.83	1	0.000015	G2
Proxima Centauri	11.5	0.0000065	15.5	0.000054	4.2	M5
α Centauri A	0.01	0.26	4.4	1.5	4.2	G2
α Centauri B	1.5	0.065	5.8	0.41	4.2	K5
Barnard's star	9.5	0.000041	13.2	0.00045	5.9	M5
Wolf 359	13.5	0.0000010	16.8	0.000016	7.6	M6
Lalande 21185	7.5	0.00026	10.4	0.0059	8.1	M2
Luyten 726-8A	12.5	0.0000026	15.4	0.000059	8.2	M5
Sirius A	−1.47	1	1.4	24	8.7	A1
Sirius B	7.2	0.00034	11.5	0.0021	8.7	White dwarf
Ross 154	10.6	0.000015	13.3	0.00041	9.4	M4
Ross 248	12.2	0.0000034	14.8	0.00010	10.3	M5
ε Eridani	3.7	0.0086	6.1	0.31	10.7	K2
Ross 128	11.1	0.0000094	13.5	0.00034	10.8	M4
Luyten 789-6	12.2	0.0000034	14.6	0.00012	11.0	M6
61 Cygni A	5.2	0.0021	7.6	0.078	11.2	K5
61 Cygni B	6.0	0.0010	8.4	0.037	11.2	K7
ε Indi	4.7	0.0034	7.0	0.14	11.2	K5
τ Ceti	3.5	0.010	5.7	0.45	11.4	G5
Lacaille 9352	7.4	0.00028	9.6	0.012	11.4	M1
Procyon A	0.4	0.18	2.7	7.1	11.4	F5
Procyon G	10.8	0.000012	13.1	0.00049	11.4	White dwarf

*Compared to Sirius, the brightest star other than the Sun.
**Compared to the Sun.

THE CONSTELLATIONS

APPENDIX G

Name	Genitive	Abbreviation	Approximate Position		English Meaning
			Right Ascension	Declination	
Andromeda	Andromedae	And	01^h	+40	Andromeda*
Antlia	Antliae	Ant	10	−35	Air pump
Apus	Apodis	Aps	16	−75	Bird of paradise
Aquarius	Aquarii	Aqr	23	−15	Water bearer
Aquila	Aquilae	Aql	20	+05	Eagle
Ara	Arae	Ara	17	−55	Altar
Aries	Arietis	Ari	03	+20	Ram
Auriga	Aurigae	Aur	06	+40	Charioteer
Bootes	Bootis	Boo	15	+30	Herdsman
Caelum	Caeli	Cae	05	−40	Chisel
Camelopardus	Camelopardis	Cam	06	−70	Giraffe
Cancer	Cancri	Cnc	09	+20	Crab
Canes Venatici	Canum Venaticorum	CVn	13	+40	Hunting dogs
Canis Major	Canis Majoris	CMa	07	−20	Big dog
Canis Minor	Canis Minoris	CMi	08	+05	Little dog
Capricornus	Capricorni	Cap	21	−20	Sea goat
Carina	Carinae	Car	09	−60	Keel of ship
Cassiopeia	Cassiopeiae	Cas	01	+60	Cassiopeia* (Queen of Ethiopia)
Centaurus	Centauri	Cen	13	−50	Centaur*
Cepheus	Cephei	Cep	22	+70	Cepheus* (King of Ethiopia)
Cetus	Ceti	Cet	02	−10	Whale
Chamaeleon	Chamaeleonis	Cha	11	−80	Chameleon
Circinis	Cirini	Cir	15	−60	Compass
Columba	Columbae	Col	06	−35	Dove
Coma Berenices	Comae Berenices	Com	13	+20	Berenice's hair*
Corona Australis	Coronae Australis	CrA	19	−40	Southern crown
Corona Borealis	Coronae Borealis	CrB	16	+30	Northern crown
Corvus	Corvi	Crv	12	−20	Crow
Crater	Crateris	Crt	11	−15	Cup
Crux	Crucis	Cru	12	−60	Southern cross
Cygnus	Cygni	Cyg	21	+40	Swan (or northern cross)
Delphinus	Delphini	Del	21^h	+10	Dolphin or porpoise
Dorado	Doradus	Dor	05	−65	Swordfish
Draco	Draconis	Dra	17	+65	Dragon
Equuleus	Equulei	Equ	21	+10	Little horse
Eridanus	Eridani	Eri	03	−20	River Eridanus*
Fornax	Fornacis	For	03	−30	Furnace
Gemini	Geminorum	Gem	07	+20	Twins
Grus	Gruis	Gru	22	−45	Crane

*These are proper names.

Name	Genitive	Abbreviation	Approximate Position		English Meaning
			Right Ascension	Declination	
Hercules	Herculis	Her	17	+30	Hercules*
Horologium	Horologii	Hor	03	−60	Clock
Hydra	Hydrae	Hya	10	−20	Hydra* (water monster)
Hydrus	Hydri	Hyi	02	−75	Sea serpent
Indus	Indi	Ind	21	−55	Indian
Lacerta	Lacertae	Lac	22	+45	Lizard
Leo	Leonis	Leo	11	+15	Lion
Leo Minor	Leonis Minoris	LMi	10	+35	Little lion
Lepus	Leporis	Lep	06	−20	Hare
Libra	Librae	Lib	15	−15	Scales of justice
Lupus	Lupi	Lup	15	−45	Wolf
Lynx	Lincis	Lyn	08	+45	Lynx
Lyra	Lyrae	Lyr	19	+40	Harp or lyre
Mensa	Mensae	Men	05	−80	Table (or mountain)
Microscopium	Microscopii	Mic	21	−35	Microscope
Monoceros	Monocerotis	Mon	07	−05	Unicorn
Musca	Muscae	Mus	12	−70	Fly
Norma	Normae	Nor	16	−50	Carpenter's level
Octans	Octantis	Oct	22	−85	Octant
Ophiuchus	Ophiuchi	Oph	17	00	Ophiuchus* (serpent bearer)
Orion	Orionis	Ori	05	+05	Orion* (hunter)
Pavo	Pavonis	Pav	20	−65	Peacock
Pegasus	Pegasi	Peg	22	+20	Pegasus* (winged horse)
Perseus	Persei	Per	03^h	+45	Perseus*
Phoenix	Phoenicis	Phe	01	−50	Phoenix
Pictor	Pictoris	Pic	06	−55	Easel
Pisces	Piscium	Psc	01	+15	Fishes
Piscis Austrinus	Piscis Austrini	PsA	22	−30	Southern fish
Puppis	Puppis	Pup	08	−40	Stern of ship
Pyxis	Pyxidis	Pyx	09	−30	Compass of ship
Reticulum	Reticuli	Ret	04	−60	Net
Sagitta	Sagittae	Sge	20	+10	Arrow
Sagittarius	Sagittarii	Sgr	19	−25	Archer
Scorpius	Scorpii	Sco	17	−40	Scorpion
Sculptor	Sculptoris	Scl	00	−30	Sculptor
Scutum	Scuti	Sct	19	−10	Shield
Serpens	Serpentis	Ser	17	00	Serpent
Sextans	Sextantis	Sex	10	00	Sextant
Taurus	Tauri	Tau	04	+15	Bull
Telescopium	Telescopii	Tel	19	−50	Telescope
Triangulum	Trianguli	Tri	02	+30	Triangle
Triangulum Australe	Trianguli Australi	TrA	16	−65	Southern triangle
Tucana	Tucanae	Tuc	00	−65	Toucan
Ursa Major	Ursae Majoris	UMa	11	+50	Big bear
Ursa Minor	Ursae Minoris	UMi	15	+70	Little bear
Vela	Velorum	Vel	09	−50	Sails of ship
Virgo	Virginis	Vir	13	00	Virgin
Volans	Volantis	Vol	08	−70	Flying fish
Vulpecula	Vulpeculae	Vul	20	+25	Fox

*These are proper names.

Answers to Selected Questions and Try One Yourself Exercises

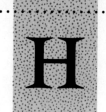

PROLOGUE

Recall Questions

2. D 4. D 6. B 8. A
10. Nebula, planet, galaxy, star, Sun, Moon, comet, meteor.
12. The great number of stars in our Galaxy.
14. Clusters of billions of stars.

CHAPTER 1

Try One Yourself

$$4.5" \times \frac{1}{60"} = 0.075'$$

$$0.075' \times \frac{1°}{60'} = 0.00125°$$

Recall Questions

2. D 4. C 6. D 8. B 10. D 12. D 14. E 16. E
18. A 20. B 22. E
24. Ptolemy lived in the second century A.D.
26. Stars near Polaris appear to move in circles around the north celestial pole.
28. The shadow of the Earth on the Moon always appears as part of a circle during a lunar eclipse.
30. Summer solstice occurs around June 21 and winter solstice occurs around December 22.
32. Normally, the planets move eastward among the stars.
34. Mercury and Venus never appear far from the Sun in the sky.
36. They can be found in the east shortly before sunrise or in the west shortly after sunset.
38. (a) The theory must fit the data. (b) The theory must make falsifiable predictions. (c) The theory should be aesthetically pleasing.

Calculations

2. 240 arcminutes

CHAPTER 2

Try One Yourself

$$\frac{a^3}{p^2} = 1 \text{ AU}^3/\text{year}^2$$

$$\frac{a^3}{0.615^2} = 1$$

$$a^3 = 0.615^2$$

$$a = 0.723 \text{ AU}$$

Recall Questions

2. A 4. B 6. C 8. C 10. B 12. E 14. C 16. A
18. A 20. B 22. B
24. The theory held that the Earth rotates on its axis, causing the entire sky to appear to rotate around the Earth.
26. He used epicycles in order to improve accuracy. They were necessary because he held that planets move in circular orbits (rather than elliptical orbits as proposed by Kepler).
28. The farther a planet is from the Sun, the slower it moves.
30. The geocentric model proposed that the center of these planets' epicycles stayed between the Earth and the Sun. The heliocentric model proposes that these planets circle the Sun in orbits that are inside the Earth's orbit and thus never appear far from the Sun.
32. Place two tacks in a board. Use a loop of string that is larger than needed to reach around the tacks. Stretch the string with a pencil and draw around. To make the ellipse more eccentric, use a shorter string or move the tacks farther apart.
34. From slowest to fastest: Saturn, Jupiter, Mars, Earth, Venus, Mercury. From closest to farthest from the Sun: Mercury, Venus, Earth, Mars, Jupiter, Saturn. (Notice that the two lists are in reverse order from one another.)

Calculations

2. 0.39 AU 4. 3600 quarters

CHAPTER 3

Recall Questions

2. D 4. D 6. A 8. C 10. D 12. B 14. A 16. B
18. D 20. A 22. A

Calculations

2. One-fourth of 150 pounds, or 37.5 pounds.

CHAPTER 4

Try One Yourself (Wave Motion in General)

$$V = \lambda \times f$$
$$336 \text{ m/s} = \lambda \times 4000 \text{ cycles/s}$$
$$\lambda = 0.08375 \text{ m}$$

(Notice that the resulting units are actually *meters/cycle*. That is appropriate, however, for one wave is one cycle.)

Try One Yourself (The Doppler Effect as a Measurement Technique)

The difference between the two wavelengths is 0.048 nm. Substituting the values in the Doppler equation:

$$V = c(\Delta\lambda/\lambda)$$
$$= (3.0 \times 10^8 \text{m/s})(0.048/656.285)$$
$$= 2.2 \times 10^4 \text{ m/s}$$

Since the shifted wavelength is shorter than the unshifted one, the star is moving toward the Sun.

Recall Questions

2. B **4.** A **6.** B **8.** A **10.** A **12.** C **14.** B **16.** E
18. B **20.** D **22.** D **24.** D
26. We know that the 400 nm light has the higher frequency because its wavelength is shorter.
28. Radio, infrared, visible, ultraviolet, X-rays, gamma rays.
30. In the Bohr model of the atom, the electron is the negatively charged object that orbits the nucleus. The nucleus is the central, massive core of an atom. A photon is the least possible amount of electromagnetic energy of a given wavelength.
32. The amount of energy of a photon is directly proportional to the frequency of the light. (Light of double the frequency has photons of double the energy, for example.)
34. A continuous spectrum is produced by a hot solid (or a very dense hot gas). An emission spectrum is produced by a hot gas with a density less than that which produces a continuous spectrum. An absorption spectrum is produced when light having a continuous spectrum passes through a gas that is too cool to emit light.

CHAPTER 5

Try One Yourself (Angular Size and Magnifying Power)

1.5 meters is 1500 millimeters.

$$M = \frac{F_{obj.}}{F_{eye.}}$$
$$= \frac{1500 \text{ mm}}{12 \text{ mm}}$$
$$= 125$$

Try One Yourself (Light-Gathering Power)

Light-gathering power depends upon the square of the diameter of the objective lens:

$$\frac{5^2}{3^3} = \frac{25}{9}$$
$$= 2.8$$

Thus, the light-gathering power is 2.8 times greater with the larger telescope.

Recall Questions

2. A **4.** D **6.** C **8.** A **10.** B **12.** C **14.** E **16.** B
18. A **20.** C

Calculations

2. The eyes gather only 1/200 as much light as the telescope.
4. Eight meters. The light gathering power is proportional to the square of the diameter. 8 squared is 64. 64 × 4 telescopes is 256, which is also 16 squared.

CHAPTER 6

Try One Yourself

$$W = \frac{\theta d}{57.3}$$
$$= \frac{0.52 \times 384{,}000 \text{ km}}{57.3}$$
$$= 3480 \text{ km}$$

Recall Questions

2. E **4.** A **6.** C **8.** B **10.** D **12.** E **14.** D **16.** B
18. E **20.** C **22.** D
24. The angular size of an object is the angle between two lines that start at the observer and go to each side of the object. The zenith is the point in the heavens directly over one's head.
26. The angular size of the Moon is about ½ degree, which is about the same as that of the Sun, although the Moon's angular size is sometimes slightly larger or slightly smaller.
28. New, waxing crescent, first quarter, waxing gibbous, full, waning gibbous, third (or last) quarter, waning crescent.
30. A lunar eclipse is visible to everyone on Earth who can see the Moon at the time the eclipse occurs. An annular or total solar eclipse can be seen only along a narrow strip along the surface of the Earth. The reason for this difference is the difference in size between the Moon's shadow and the Earth's shadow.
32. A lunar eclipse can occur only at full moon, and a solar eclipse can occur only at new moon.
34. A solar eclipse is annular if the Moon is at too great a distance from the Earth for its umbral shadow to reach the Earth's surface. If that shadow reaches the surface, the eclipse is total.
36. A magnetic field is a region where magnetic forces can be perceived. A magnetic field can be detected by placing a magnet in the suspected location and determining whether there is a magnetic force on the magnet. The magnet used is commonly a small one that is free to turn, that is, a compass.

38. (a) Continental drift has been measured. (b) Rift zones, from which spreading is occurring, have been found. (c) Plate motion explains a number of earthquake zones, mountain ranges, and ocean trenches.
40. Craters were caused by impacts of meteorites. Material thrown out by the impact caused the rays that radiate from some craters.

Calculations

2. No, the Moon would have been more than 7 degrees from the zenith at Alexandria because it is close enough to the Earth that parallax is observed in its case.

CHAPTER 7

Try One Yourself

The one-way travel time to Venus is half of 4.8 minutes, or 2.4 minutes. This is 144 seconds. Using the velocity of light:

$$\text{distance} = \text{velocity} \times \text{time}$$
$$= (3.0 \times 10^5 \text{ km/s}) \times 144 \text{ s}$$
$$= 4.32 \times 10^7 \text{ km}$$

Since one AU is 1.5×10^8 km,

$$4.32 \times 10^7 \times \frac{1 \text{ AU}}{1.5 \times 10^8 \text{ km}} = 0.288 \text{ AU}.$$

Venus is 0.72 AU from the Sun. Since $(1 - 0.72)$ is 0.28, the above answer is reasonable.

Recall Questions

2. C **4.** A **6.** A **8.** A **10.** A **12.** B **14.** D **16.** A
18. C **20.** B **22.** E **24.** C
26. Newton expanded Kepler's law to include the total mass of the two objects.
28. Jupiter is the largest planet. Its diameter is 11 times Earth's, and its mass is 318 times Earth's.
30. The masses of the planets that have moons were calculated using Kepler's third law. Mercury's and Venus's masses were calculated based on their gravitational effect on passing asteroids and comets. After spacecraft were sent near them, more accurate values were obtained based on the planets' effects on the spacecraft.
32. All of the planets revolve around the Sun in the same direction. All except Venus, Uranus, and Pluto rotate on their axes in the same direction as they revolve.
34. Compared to the Jovians, the terrestrials are nearer the Sun, smaller, less massive, more solid, slower rotating, and more dense and have thinner atmospheres and fewer moons. No terrestrial planet has a ring, while all Jovian planets have rings.
36. Earth is just slightly more dense than Mercury and Venus, somewhat more dense than Mars, and much more dense than the Jovian planets and Pluto.
38. The temperature of a gas is related to the speed of its molecules, and at the same temperature, the molecules of a more massive gas move more slowly.
40. Nonvolatile elements condensed in the inner solar system, but volatile elements were swept outward by the solar wind.

42. The asteroids are planetesimals that were prohibited from forming a planet by the effects of Jupiter's gravitational force. The Oort cloud is hypothesized to have resulted when small objects in the outer solar system were thrown outward by gravitational forces as Jupiter or Saturn passed near them.

Calculations

2. The predicted planet would be 77.2 AU from the Sun. (See the Close Up, The Titius-Bode Law.)
4. 9×10^8 meters, or 6 AU.

CHAPTER 8

Recall Questions

2. B **4.** B **6.** D **8.** C **10.** E **12.** D **14.** C **16.** B
18. D **20.** B **22.** D **24.** A
26. Venus is closest to Earth in size; Mars in rotation period.
28. Mercury's rotation period is exactly ⅔ of its period of revolution. This has occurred because the mass of the planet is not evenly distributed through its volume. The planet has an eccentric orbit, and gravitational forces toward the Sun have changed its rotation rate so that resonance occurs between its rotation and revolution.
30. There are two reasons for the extremes in temperature. First, day and night periods are very long (88 Earth days each) providing long periods of time for the surface to heat and cool. Second, there is no atmosphere to block sunlight during the day and to retain heat at night.
32. The greenhouse effect is the phenomenon of infrared radiation being prohibited from escaping through a planet's atmosphere, thereby causing a high temperature on the planet. In a florist's greenhouse an additional effect occurs: the hot air is trapped inside the greenhouse by the walls and ceiling.
34. The tilt of Mars's axis is very similar to that of Earth's. Mars's equator is tilted 25.2° from its orbital plane, while Earth's is tilted 23.44°. (Tables normally show equatorial tilt rather than axis tilt: equatorial tilt is the same as the tilt of the axis with respect to a line perpendicular to the orbital plane.)
36. The polar caps of Mars are made of frozen water and frozen carbon dioxide.

Calculations

2. The volume of a planet depends upon the cube of the diameter, so Jupiter's volume is 11^3, or more than 1300 times as great as Earth's.

CHAPTER 9

Recall Questions

2. B **4.** C **6.** A **8.** C **10.** D **12.** D **14.** B **16.** B **18.** C
20. The band at Jupiter's equator has a rotation rate of 9 hours, 50 minutes.
22. The outer atmosphere is primarily hydrogen and helium at a fairly low pressure. As one goes deeper, pressure increases until the material would be judged a liquid. About 15,000

kilometers below the cloudtops, liquid metallic hydrogen is found, and it extends down to whatever core exists.

24. Saturn's rings are aligned with its equator, which is tilted at about 27° to the planet's orbital plane. Thus, as Saturn orbits the Sun, we on Earth (which is relatively near the Sun) sometimes see the southern side of Saturn and its rings and sometimes the northern side. Midway between these views, the rings are edge-on to our view and are nearly invisible.

26. Knowing the distance to a planet and its angular size, we can calculate its actual size (as explained in an earlier chapter). When a planet passes in front of a star, observation of the amount of time that the planet blocks the star's light, along with knowledge of the planet's speed, gives us another method to calculate its size.

28. Uranus's axis tilt (98°) is such that each of its poles nearly points to the Sun at one time during its revolution period.

30. When a moon changes its distance from its planet, the amount of tidal force exerted on the moon changes. A change in tidal force causes the moon to flex, and this produces heat. The more massive the planet, the greater the effect on its moon, and the closer the moon is to the planet, the greater the effect. Thus, the inner moons of Jupiter that have a non-circular orbit experience tidal heating the most. Io is closest to Jupiter and has an eccentric orbit. Thus, it is heated the most by tidal forces.

Calculations

2. The object would have ⅗ the density of the Earth.

• • • • • • • • • • • • • • • •

CHAPTER 10

Recall Questions

2. E **4.** C **6.** D **8.** D **10.** D **12.** C **14.** B **16.** B
18. C **20.** D **22.** B **24.** C

26. The most accurate values of Pluto's size come from observations of eclipses of its moon, Charon.

28. Before an asteroid is named, its orbit must have been determined accurately. The orbits of most observed asteroids have not been calculated.

30. Gravitational pull from Jupiter disrupts the orbits of asteroids so that the weaker gravitational forces between them cannot pull them together.

32. The nucleus of a comet is its solid core. Comet nuclei are irregular in shape and are typically a few kilometers across. The coma is made up of gas and dust surrounding the nucleus. It has a very low density and may be as large as 100,000 km in diameter. A comet has two tails, one made up of ions and one of dust. Tails are typically from 10^7 to 10^8 km long.

34. One tail is made up of ions swept away by the solar wind. It is straight because the ions move away from the comet at great speeds. The other tail consists of dust pushed away slowly by radiation pressure. The curvature results from the fact that the material moves away from the coma very slowly compared to the coma's speed.

36. First, the vast majority of long-period comets are in elongated elliptical orbits around the Sun. They must therefore spend most of their time far from the Sun. Second, comets are regularly being captured in the inner solar system. The existence of the cloud explains their origin.

38. A meteor shower results when the Earth encounters a swarm of meteoroids. As the Earth moves into the swarm, the resulting meteors seem to originate in the direction in which the Earth is moving.

Calculations

2. (The answer to #1 is 1.96×10^{-9} kg/m²/day.) #2: 255,000,000 days, or about 700,000 years.

• • • • • • • • • • • • • • • •

CHAPTER 11

Recall Questions

2. C **4.** D **6.** C **8.** A **10.** E **12.** B **14.** B **16.** A
18. A **20.** D **22.** A **24.** C

26. The most convincing evidence is that if a chemical process were the source of the Sun's power, the Sun would burn out over a relatively few centuries, and we know that it has released energy fairly uniformly for (at least) many millions of years.

28. The number of protons in nuclei distinguish one element from another.

30. Most of the Sun is made up of gas.

32. The pressure at any depth within the Sun is caused by the weight (and therefore by the mass) of gas above that layer. To calculate the mass of gas above a given point within the Sun, one needs to know the density at levels above that point. This density is calculated from knowledge of pressures and temperatures. (Since the pressures and densities at each layer depend upon pressures and densities of layers above, the calculations are best done by a computer.)

34. Photons from the Sun's core are continuously absorbed and reemitted by material within the Sun. Since a photon is absorbed after traveling only about one centimeter and since photons are emitted in random directions, the radiation travels a very great distance before arriving at the photosphere.

36. First, most of their radiation is in the X-ray portion of the spectrum. Second, they have a very low density, and therefore there is not much material in them to emit light.

Calculations

2. 2×10^{20} kg. (This is about 200,000,000,000,000,000,000 tons.) It would burn out in 10^{10} seconds, or about 7600 years.

• • • • • • • • • • • • • • • •

CHAPTER 12

Try One Yourself (Apparent Magnitude)

The difference in apparent magnitude between the two objects is 8. Referring to Table 12-1, we see that a difference of 5 corresponds to a ratio of 100 and a difference of 3 corresponds to a ratio of 16. Therefore we receive 100×16, or 1600 times more light from Mars than from Barnard's star.

Try One Yourself (Spectroscopic Parallax)

Refer to Figure 12-18. It shows that a K0-type star has an absolute magnitude between about +5.5 and +6.5. Since its apparent magnitude is +4.4, we can conclude that 40 Eridani is closer to us than 10 pc. (In fact, it is 4.8 pc from the Sun.)

Recall Questions

2. C 4. B 6. B 8. D 10. D 12. E 14. C 16. B
18. A 20. A 22. D 24. C
26. We must know the object's distance as well as its angular velocity.

Calculations

2. They differ in magnitude by 4.5, so we receive between 40 (which corresponds to a difference of 4) and 100 (which corresponds to 5) times as much light from Antares than from τ Ceti. (The actual ratio is about 63.)
4. Moving α Centauri to a distance of 10 parsecs would entail moving it more than seven times farther away than it is now, thereby making it appear much dimmer. Therefore its absolute magnitude must be greater than zero. (Its actual absolute magnitude is 4.4.)
6. It is closer than 10 pc. (Actual distance: 5.1 pc.)
8. Using Figure 12-37, we see that it is more than 100 times more luminous than the Sun.

CHAPTER 13

Recall Questions

2. C 4. A 6. D 8. D 10. B 12. C 14. A 16. D
18. A protostar is a star in the process of formation; it is heated by gravitational energy as its material falls inward.
20. Less massive stars take longer, because the lesser mass means there is less gravitational force pulling each particle toward the center.
22. More massive stars release more radiation as they join the main sequence, and stars above about 100 solar masses probably emit radiation so intense that no more material can fall onto them.

CHAPTER 14

Recall Questions

2. E 4. A 6. D 8. C 10. D 12. B 14. B 16. E
18. C 20. B
22. The material of a star of very low mass completely mixes by convection during the star's main sequence life. Therefore, there is little or no material outside the core to expand at the end of that life.
24. A one-solar-mass star goes through the following stages: protostar, main sequence star, red giant, (helium flash), planetary nebula, white dwarf, black dwarf.
26. Carbon core, helium shell (fusing to carbon), hydrogen shell (fusing to helium), main body of hydrogen.
28. A typical white dwarf has a diameter less than that of the Earth, a mass somewhat less than that of the Sun, an absolute magnitude of 11 (which is about 250 times more luminous than the Sun), and a temperature of about 15,000 K.

30. When mass is added to a main sequence star, the star becomes larger. White dwarfs become smaller when mass is added, however.

Calculations

2. $T = \dfrac{1}{M^{2.5}} = \dfrac{1}{0.5^{2.5}} = 5.7$

Thus the star will live 5.7 times as long as the Sun, or 57 billion years.

CHAPTER 15

Recall Questions

2. B 4. E 6. A 8. D 10. E 12. B 14. D 16. A
18. B 20. C 22. A
24. When mass is added to a main sequence star, the star becomes larger. White dwarfs and neutron stars become smaller when mass is added, however. If the size of a black hole is considered to extend to its event horizon, added mass makes a black hole larger.
26. The fact that the Chinese "guest star" was very bright and that today we see an expanding cloud of debris at its location leads us to conclude that it was a supernova.
28. Pulsars vibrate faster than the surface of a white dwarf should be able to vibrate. On the other hand, they vibrate more slowly than neutron stars should vibrate.
30. A black hole is an object that is so dense that its gravitational force is great enough that light cannot escape. We might observe a black hole when material falls toward it, for that material would heat up as it falls and emit radiation.
32. The escape velocity from inside the event horizon of a black hole is greater than the velocity of light.
34. Radiation has been observed that has the characteristics predicted to occur when material falls toward the event horizon of a black hole.

Calculations

2. If the mass of a black hole is doubled, the radius of its event horizon (the Schwarzschild radius) will also double.

CHAPTER 16

Recall Questions

2. B 4. C 6. B 8. A 10. D 12. A 14. B

Calculations

2. 250 km/s is 8.3×10^{-4} LY/yr. The circumference of the path is 2.5×10^5 LY. The time for one revolution is therefore 3.0×10^8 years.
4. If 0.1 mm corresponds to 1.4×10^6 km (the diameter of the Sun) and the Galaxy's diameter is 10^{18} km, then the Galaxy on this scale is about 70,000 km across. We can't build it; the Earth's radius is only 13,000 km!

CHAPTER 17

Try One Yourself (The Hubble Law)

According to figure 17-11, a distance of 90 MLY corresponds to a speed of about 17,000 km/s. Thus

$$\text{slope} = \frac{17,000 \text{ km/s} - 0}{90 \text{ MLY}}$$

$$= 190 \text{ km/(s} \times \text{MLY)}$$

Try One Yourself (The Hubble Law Used to Measure Distance)

Using a Hubble constant of 15 km/(s × MLY):

$$V = Hd$$
$$90,000 \text{ km/s} = 15 \text{ km/(s} \times \text{MLY)}$$
$$d = 6000 \text{ MLY}$$

Using 25 km/(s × MLY) in the same manner, we obtain 3600 MLY; thus the distance is between 3600 MLY and 6000 MLY.

Recall Questions

2. E 4. D 6. D 8. B 10. B 12. E 14. A 16. C
18. C 20. A 22. C
24. (a) Application of Kepler's third law to measured values of rotation rates of various parts of a galaxy. (b) Application of Kepler's third law to binary pairs of galaxies. (c) Calculations of the mass necessary to prevent a galaxy from leaving the cluster of galaxies in which it is observed.
26. Not enough mass is directly observed in clusters of galaxies to hold the clusters together. Since they obviously hold together, more mass must exist than we can observe, and since it is not easily observed, it must be dark matter.
28. The Hubble law states that the velocity of recession of a distant galaxy is proportional to the galaxy's distance. The Hubble constant seems to be between 15 and 25 km/(s × MLY), which is between 50 and 80 km/(s × Mpc).
30. To use the Hubble law to determine the distance to a galaxy, one measures the recessional velocity of the galaxy and then calculates its distance based on a value of the Hubble constant. Although recessional velocity can be accurately measured, some of that velocity may not be due to the Hubble expansion. The greatest problem is that the Hubble constant is not accurately known, however.
32. An active galaxy emits more than normal amounts of radiation from its nucleus.
34. For a quasar with a redshift of 25%, the ratio of the amount of shift in wavelengths of its spectral lines to the unshifted wavelengths is 0.25. The quasar with a redshift of 30% is moving away faster.
36. If they were nearby and moving at the speeds indicated by their redshifts, they would have had to have been produced recently. It is thought, instead, that quasars do not exist in the universe of today, and we see them only because they are very distant and therefore have a great look-back time.

Calculations

2. The slope of the line in Figure 17-32 is 11 km/(s × MLY), so that is the value of the Hubble constant from that graph.
4. 4 hours.
6. 0.80 kpc × 0.95 kpc

CHAPTER 18

Recall Questions

2. C 4. C 6. A 8. B 10. D 12. A 14. C 16. B
18. Ptolemy described the Greek geocentric system. Copernicus proposed a heliocentric system that used circles for the planets' orbits. Kepler developed three laws to explain planetary motions better. Galileo found evidence for the heliocentric system, and Newton linked heavenly and earthly laws. Newton also stated the relative nature of position and velocity. Herschel determined that we live in a galaxy of stars, and Shapley located the Sun's place in the Galaxy. Hubble found that our galaxy is just one of many. Einstein's work provided the idea of curvature of space, an important idea in modern cosmology.
20. If all objects are moving apart from one another, it will appear from each of the objects that all of the others are moving away from it.
22. Background radiation is radiation that comes from all directions in space. Its existence and characteristics must be explained by any successful cosmological theory.
24. One method is to determine the speed of the very distant galaxies and compare it to the speed of nearer galaxies. Since we are seeing the distant ones in the past, we can determine how much their speed must have changed. The other method is to determine whether there is enough mass in the universe for gravitational attraction to stop the expansion.
26. The *COBE* results show that the background radiation was almost uniform, but not perfectly so. The lack of perfect uniformity is what resulted in the clustering of matter to form galaxies.

Calculations

2. 5.5×10^{17} seconds, or 1.8×10^{10} years.

Glossary

Angstrom (abbreviated Å). A unit of length equal to 10^{-10} m. There are 10 Angstroms in a nanometer.

Apollo asteroids. Asteroids that cross the Earth's orbit.

absolute magnitude. The apparent magnitude a star would have if it were at a distance of 10 parsecs.

absorption spectrum. A spectrum that is continuous except for certain discrete wavelengths.

accelerate. To change the speed or direction of motion of an object.

acceleration. A measure of how rapidly the speed or direction of motion of an object is changing.

accretion disk. A rotating disk of gas orbiting a star, formed by material falling toward the star.

achromatic lens (or achromat). An optical element that has been corrected so that it is free of chromatic abberation.

active galaxy. A galaxy with an unusually luminous nucleus.

adaptive optics. A system that monitors and changes the shape of a telescope's objective to produce the best image.

albedo. The fraction of incident sunlight that an object reflects.

altitude. The height of a celestial object measured as an angle above the horizon.

angular momentum. The tendency of a rotating or revolving object to continue its motion.

angular separation. Measured from the observer, the angle between lines toward two objects.

angular size (of an object). The angle between two lines drawn from the viewer to opposite sides of the object.

annular eclipse. An eclipse in which the Moon is too far from Earth for its disk to cover that of the Sun completely, so the outer edge of the Sun is seen as a ring.

aphelion. The point in its orbit when a planet (or other object) is farthest from the Sun.

apogee. The point in the orbit of an Earth satellite where it is farthest from Earth.

apparent magnitude. A measure of the amount of light received from a celestial object.

asteroid belt. The region between Mars and Jupiter where most asteroids orbit.

asteroid. Any of the thousands of minor planets that orbit the Sun.

astrometric binary. An orbiting pair of stars in which the motion of one of the stars reveals the presence of the other.

astronomical unit (AU). A unit of distance equal to the average distance between the Earth and the Sun.

astronomical unit. A distance equal to the average distance between the Earth and the Sun.

astrophysics. Physics applied to extraterrestrial objects.

aurora. Light radiated in the upper atmosphere due to impact from charged particles.

autumnal equinox. The point on the celestial sphere where the sun crosses the celestial equator moving south.

BL Lac objects. Especially luminous active galactic nuclei that vary in luminosity by a factor of up to 100 in just a few months.

Bohr atom. The model of the atom proposed by Niels Bohr; it contains electrons in orbit around a central nucleus and explains the emission of light.

background radiation (or **cosmic background radiation**). Long-wavelength radiation observed from all directions; believed to be the remnants of radiation from the big bang.

barred spiral galaxy. A spiral galaxy in which the spiral arms come from the ends of a bar through the nucleus rather than from the nucleus itself.

barycenter. The center of mass of two astronomical objects revolving around one another.

big bang. The theorized initial explosion that began the expansion of the universe.

binary star system. A system of two stars that are gravitationally linked so that they orbit one another.

black dwarf. The theorized final state of a star with a main sequence mass less than about 8 solar masses, in which all of its energy sources have been depleted so that it emits no radiation.

black hole. An object whose escape velocity exceeds the speed of light.

blueshift. A change in wavelength toward shorter wavelengths.

brown dwarf. A starlike object whose mass is too small to sustain nuclear fusion. Probable limits of mass are from 0.002 to .08 solar masses.

Cassegrain focus. The optical arrangement of a reflecting telescope in which a mirror is mounted so that it intercepts the light from the objective mirror and reflects the light back through a hole in the center of the primary.

Cepheid (pronounced SEF-e-id) variable. One of a particular class of pulsating stars.

Chandrasekhar limit. The limit to the mass of a white dwarf star above which it cannot exist as a white dwarf. Above that limit it will not be supported by electron degeneracy.

capture theory. A theory that holds that the Moon was originally solar system debris that was captured by Earth.

carbon (or CNO) cycle. A series of nuclear reactions that results in the fusion of hydrogen into helium, using carbon-12 in the process.

catastrophic theory. A theory of the formation of the solar system that involves an unusual incident such as the collision of the Sun with another star.

celestial equator. A line on the celestial sphere directly above the Earth's equator.

celestial pole. The point on the celestial sphere directly above a pole of the Earth.

celestial sphere. The sphere of heavenly objects that seems to center on the observer.

center of mass. The average location of the various masses in a system, weighted according to how far each is from that point.

centripetal force. The force directed toward the center of the curve along which the object is moving.

chain reaction theory. A model for spiral galaxies that explains the arms as resulting from a series of supernovae, each triggering the formation of new stars.

charge-coupled device (CCD). A small semiconductor that serves as a light detector by emitting electrons when it is struck by light. A computer uses the pattern of electron emission to form images.

chromatic aberration. The defect of optical systems that results in light of different colors being focused at different places.

chromosphere. The region of the solar atmosphere between the photosphere and the corona.

closed universe. The state of the universe if it stops expanding and begins contracting.

cocoon nebula. The dust and gas that surround a protostar and block much of its radiation.

coma (KOH-mah). The part of a comet's head made up of diffuse gas and dust.

composite spectrum binary. A binary star system with stars having spectra different enough to distinguish them from one another.

conduction (of heat). The transfer of heat in a solid by collisions between atoms and/or molecules.

conservation of angular momentum. A law that states that the angular momentum of a system will not change unless an outside force is exerted on the system.

constellation. An area of the sky containing a pattern of stars named for a particular object, animal, or person.

continental drift. The gradual motion of the continents relative to one another.

continuous spectrum. A spectrum containing an entire range of wavelengths rather than separate, discrete wavelengths.

convection. The transfer of heat in a gas or liquid by means of the motion of the material.

core (of the Earth). The central part of the Earth, probably consisting of a solid inner core surrounded by a liquid outer core.

corona. The outermost portion of the Sun's atmosphere.

correspondence principle. The idea that predictions of a new theory must agree with the theory it replaces in cases where the previous theory has been found to be correct.

cosmological principle. The basic assumption of cosmology that holds that on a large scale the universe is the same everywhere.

cosmological redshift. The shift toward longer wavelengths that is due to the expansion of the universe.

cosmology. The study of the nature and evolution of the universe as a whole.

crescent (phase). The phase of a celestial object when less than half of its sunlit hemisphere is visible.

critical density. The density of a perfectly flat universe.

crust (of the Earth). The thin, outermost layer of the Earth.

Doppler effect. The observed change in wavelength from a source moving toward or away from the observer.

dark matter. Matter that is too cool to emit sufficient radiation to allow us to detect it.

dark nebula. A cloud of interstellar dust that blocks light from stars on the other side of it.

density wave theory. A model for spiral galaxies that proposes that the arms are the result of density waves sweeping around the galaxy.

density wave. A wave in which areas of high and low pressure move through the medium.

density. The ratio of mass to volume.

deuterium. A hydrogen nucleus that contains one neutron and one proton.

differential rotation. Rotation of an object in which different parts have different periods of rotation.

differentiation. The sinking of denser materials toward the center of planets or other objects.

diffraction grating. A device that uses the wave properties of electromagnetic radiation to separate the radiation into its various wavelengths.

diffraction. The spreading of light upon passing the edge of an object.

disk (of a galaxy). The flat, dense portion of a spiral galaxy that rotates in a plane around the nucleus.

dispersion. The separation of light into its various wavelengths upon refraction.

double planet theory. A theory that holds that the Moon was formed at the same time as the Earth.

dynamo effect. The generation of magnetic fields due to circulating electric charges, such as in an electric generator.

eccentricity of an ellipse. The result obtained by dividing the distance between the foci by the longest distance across an ellipse (the *major axis*).

eclipse season. A time of the year during which a solar or lunar eclipse is possible.

ecliptic. The apparent path of the Sun on the celestial sphere.

electromagnetic spectrum. The entire array of electromagnetic waves.

electron degeneracy. The state of a gas in which its electrons are packed as densely as nature permits. The temperature of such a high-density gas is not dependent on its pressure as it is in a "normal" gas.

electron. One of the negatively charged particles that orbit the nucleus of an atom.

ellipse. A geometrical shape every point of which is the same total distance from two fixed points (the foci).

elliptical galaxy. One of a class of galaxies that have smooth spheroidal shapes.

elongation. The angle of the Moon or a planet from the Sun in the sky.

emission nebula. Interstellar gas that fluoresces due to ultraviolet light from a star near or within the nebula.

emission spectrum. A spectrum made up of discrete wavelengths rather than a continuous band of wavelengths.

epicycle. The circular orbit of a planet in the Ptolemaic model, the center of which revolves around the Earth in another circle.

escape velocity. The minimum velocity an object must have in order to escape the gravitational attraction of an object such as a planet.

event horizon. The surface of the sphere around a black hole from within which nothing can escape. Its radius is the Schwarzschild radius.

evolutionary track. The path on the H-R diagram taken by a star as its luminosity and color change.

eyepiece. The magnifying lens (or combination of lenses) used to view the image formed by the objective of a telescope.

field of view. The actual angular width of the scene viewed by an optical instrument.

fireball. An extremely bright meteor.

fission theory. A theory that holds that the Moon formed when material was spun off from the Earth.

flat universe. The condition of the universe if gravity just balances its expansion so that it stops expanding only in an infinite amount of time.

flatness problem. The inability of the standard big bang model to account for the apparent flatness of the universe.

fluorescence. The process of absorbing radiation of one frequency and reemitting it at a lower frequency.

focal length. The distance from the center of a lens or a mirror to its focal point.

focal point (of a converging lens or mirror). The point at which light from a very distant object converges after being refracted or reflected.

focus of an ellipse. One of the two fixed points that define an ellipse. (See the definition of *ellipse*.)

frequency. The number of repetitions per unit time.

full (phase). The phase of a celestial object when the entire sunlit hemisphere is visible.

fusion (nuclear). The combining of two nuclei to form a different nucleus.

Galilean moons. The four natural satellites of Jupiter that were discovered by Galileo.

galactic rotation curve. A graph of the orbital speed of objects in the galactic disk as a function of their distance from the center.

galactic (or open) cluster. A group of stars that share a common origin and are located relatively close to one another.

general theory of relativity. A theory developed by Einstein that expands special relativity to accelerated systems and presents an alternative way of explaining the phenomenon of gravitation.

geocentric model. A model of the universe with the Earth at its center.

giant star. A star of greater luminosity and larger size than a main sequence star of the same temperature.

gibbous (phase). The phase of a celestial object when between half and all of its sunlit hemisphere is visible.

globular cluster. A spherical group of up to hundreds of thousands of stars, found primarily in the halo of a spiral galaxy.

granulation. Division of the Sun's surface into small convection cells.

gravitational lens. The phenomenon in which a massive body between another object and the viewer causes the distant object to be seen as two.

greenhouse effect. The effect by which infrared radiation is trapped within a planet's atmosphere through the action of particles, such as carbon dioxide molecules, within that atmosphere.

Hertzsprung-Russell diagram. A plot of absolute magnitude (or luminosity) versus temperature (or spectral class) for stars.

Hubble constant. The proportionality constant in the Hubble law; the ratio of recessional velocities of galaxies to their distances.

Hubble law. The relationship that states that a galaxy's recessional velocity is directly proportional to its distance.

halo (around a galaxy). The outermost part of a spiral galaxy; fairly spherical in shape, it lies beyond the spiral component.

heliocentric. Centered on the Sun.

helium flash. The runaway helium fusion reaction that occurs during the evolution of a red giant.

hertz (abbreviated Hz). The unit of frequency equal to one cycle per second.

horizon problem. The inability of the standard big bang model to account for directional uniformity of the background radiation.

homogeneous. Having uniform properties throughout.

hydrostatic equilibrium. In a star or a planet, the balance between pressure caused by the weight of material above and the upward pressure exerted by material below.

hypothesis. A tentative explanation.

image. The visual counterpart of an object, formed by refraction or reflection of light that came from the object.

inclination (of a planet's orbit). The angle between the plane of a planet's orbit and the ecliptic plane.

inertia. The property of an object whereby it tends to maintain whatever velocity it has.

inflationary universe model. A modification of the big bang model that holds that the early universe experienced a brief period of extremely fast expansion.

interferometry. A procedure that allows a number of telescopes to be used as one by taking into account the time at which individual waves from an object strike each telescope.

interstellar cirrus. Faint, diffuse dust clouds found throughout interstellar space.

inverse square law. Any relationship in which some factor decreases as the square of the distance from its source.

ion. A charged atom or molecule resulting from the atom's loss or gain of an electron.

irregular galaxy. A galaxy of irregular shape that cannot be classified as spiral or elliptical.

isotropy. The property of being the same in all directions.

Kelvin temperature scale. A temperature scale with its zero point at the coldest possible temperature ("absolute zero") and a degree that is the same size (same temperature difference) as the Celsius degree.

Kuiper belt. A band of comets hypothesized to exist closer to the solar system than the Oort cloud.

large impact theory. A theory that holds that the Moon formed as the result of an impact between a large object and the Earth.

lighthouse model. The theory that explains pulsar behavior as being due to a spinning neutron star whose beam of radiation we see as it sweeps by.

light curve. A graph of the numerical measure of the light received from a star versus time.

light-gathering power. A measure of the amount of light collected by an optical instrument.

light-year. The distance light travels in a year.

limb (of the Sun or Moon). The apparent edge of the object as seen in the sky.

local group (of galaxies). The cluster of 20 or so galaxies that includes the Milky Way Galaxy.

local hypothesis. A proposal stating that quasars are much nearer than a cosmological interpretation of their redshifts would indicate.

look-back time. The time light from a distant object has traveled to reach us.

luminosity class. One of several groups into which stars can be classified according to characteristics of their spectra.

luminosity. The rate at which electromagnetic energy is being emitted.

lunar eclipse. An eclipse in which the Moon passes into the shadow of the Earth.

lunar month. The Moon's synodic period, or the time between successive similar phases.

lunar ray. A bright streak on the Moon caused by material ejected from a crater.

magnetic field. A region of space where magnetic forces can be detected.

magnetosphere. The volume of space in which the motion of charged particles is controlled by the magnetic field of the planet rather than by the solar wind.

magnifying power (or magnification). The ratio of the angular size of an object when it is seen through the instrument to its angular size when seen with the naked eye.

main sequence. The part of the H-R diagram containing the great majority of stars; it forms a diagonal line across the diagram.

mantle (of the Earth). The thick, solid layer between the crust and the core of the Earth.

mare (plural **maria**). Any of the lowlands of the Moon or Mars that resemble a sea when viewed from Earth.

mass-luminosity diagram. A plot of the mass versus the luminosity of a number of stars.

mass. The quantity of inertia possessed by an object.

meridian. An imaginary line that runs from north to south, passing through the observer's zenith.

meteorite. An interplanetary chunk of matter that has struck a planet or moon.

meteoroid. An interplanetary chunk of matter smaller than an asteroid.

meteor shower. The phenomenon of a large group of meteors seeming to come from a particular area of the celestial sphere.

meteor. The phenomenon of a streak in the sky caused by the burning of a rock or dust particle as it falls.

micrometeorite. A tiny meteorite.

Milky Way. Historically, the diffuse band of light that stretches across the sky. Today the term refers to the Milky Way Galaxy.

Milky Way Galaxy. The galaxy of which the Sun is a part.

minute of arc. One-sixtieth of a degree of arc.

missing mass. The difference between the mass of clusters of galaxies as calculated from Keplerian motions and the amount of visible mass.

Newtonian focus. The optical arrangement of a reflecting telescope in which a plane mirror is mounted along the axis of the telescope so that the mirror intercepts the light from the objective mirror and reflects it to the side.

nanometer (abbreviated nm). A unit of length equal to 10^{-9} meters.

neap tide. The least difference between high and low tide, occurring when the solar tide partly cancels the lunar tide.

nebula (plural **nebulae**). An interstellar region of dust and/or gas.

neutrino. An elementary particle that has little or no rest mass and no charge but carries energy from a nuclear reaction.

neutron star. A star that has collapsed to the point at which it is supported by neutron degeneracy.

neutron. The massive nuclear particle with no electric charge.

nonvolatile element. An element that is gaseous only at a high temperature and condenses to liquid or solid when the temperature decreases.

nova (plural **novae**). A star that suddenly and temporarily brightens, thought to be due to new material being deposited on the surface of a white dwarf.

nuclear bulge. The central region of a spiral galaxy.

nucleus (of atom). The central, massive part of an atom.

nucleus (of comet). The solid chunk of a comet, located in the head.

Occam's razor. The principle that the best explanation is the one that requires the fewest unverifiable assumptions.

Olbers's paradox. An argument showing that the sky in a static universe could not be dark.

Oort cloud. The theorized spherical shell lying between 10,000 and 100,000 AU from the Sun containing billions of comet nuclei.

objective lens (or objective). The main light-gathering element—lens or mirror—of a telescope. It is also called the primary lens.

oblateness. A measure of the "flatness" of a planet, calculated by dividing the difference between the largest and smallest diameter by the largest diameter.

$$\text{oblateness} = \frac{(d_{\text{large}} - d_{\text{small}})}{d_{\text{large}}}$$

occultation. The passing of one astronomical object in front of another.

open universe. The state of the universe if it continues expanding without stopping.

opposition. The configuration of a planet when it is opposite the Sun in our sky. That is, the objects are aligned as follows: Sun—Earth—planet.

optical double. Two stars that have small angular separation as seen from Earth but are not gravitationally linked.

oscillating universe theory. A big bang theory that holds that the universe goes through repeating cycles of explosion, expansion, and contraction.

Ptolemaic model. The theory of the heavens devised by Claudius Ptolemy.

parallax angle. Half the maximum angle that a star appears to be displaced due to the Earth's motion around the Sun.

parallax. The apparent shifting of nearby objects with respect to distant ones as the position of the observer changes.

parsec. The distance of an object that has a parallax angle of one arcsecond.

partial lunar eclipse. An eclipse of the Moon in which only part of the Moon passes through the umbra of the Earth's shadow.

partial solar eclipse. An eclipse in which only part of the Sun's disk is covered by the Moon.

particle density. The number of separate atomic and/or nuclear particles per unit of volume.

penumbral lunar eclipse. An eclipse of the Moon in which the Moon passes through the Earth's penumbra but not through its umbra.

penumbra. The portion of a shadow that receives direct light from only part of the light source.

perigee. The point in the orbit of an Earth satellite where it is closest to Earth.

perihelion. The point where an object in orbit about the Sun is closest to the Sun.

phases (of the Moon). The changing appearance of the Moon during its cycle, caused by the relative positions of the Earth, Moon, and Sun.

photometry. The measurement of light intensity from a source, either the total intensity or the intensity at each of various wavelengths.

photon. The smallest possible amount of electromagnetic energy of a particular wavelength.

photosphere. The visible "surface" of the Sun. The part of the solar atmosphere from which light is emitted into space.

planetary nebula. The shell of gas that is expelled by a red giant near the end of its life.

planetesimal. One of the small objects that formed from the original material of the solar system and from which a planet developed.

planet. Any of the nine (so far known) large objects that revolve around the Sun.

plate tectonics. The motion of sections of the Earth's crust across the underlying mantle.

positron. A positively charged electron emitted from the nucleus in some nuclear reactions.

power. The rate at which energy is transferred, or the amount of energy transferred per unit time.

precession. The conical shifting of the axis of a rotating object.

precession (of an elliptical orbit). The change in orientation of the major axis of the elliptical path of an object.

pressure. The force per unit of area.

prime focus. The point in a telescope where the light from the objective is focused. This is the focal point of the objective.

principle of equivalence. The statement that effects of the force of gravity are indistinguishable from those of acceleration.

prominence. The eruption of solar material beyond the disk of the Sun.

proper motion. The angular velocity of a star as measured from the Sun.

proton. The massive, positively charged particle in the nucleus of an atom.

proton-proton chain. The series of nuclear reactions that begins with four protons and ends with a helium nucleus.

protostar. A star in the process of formation before it reaches the main sequence.

pulsar. A celestial object of small angular size that emits pulses of radio waves with a regular period between about 0.03 and 5 seconds.

quarter (phase). The phase of a celestial object when half of its sunlit hemisphere is visible.

quasar (quasi-stellar source, or QSS). A small, intense celestial source of radiation with a very large redshift.

Roche limit. The minimum radius at which a satellite (held together by gravitational forces) may orbit without being broken apart by tidal forces.

radial velocity. Velocity along the line of sight, toward or away from the observer.

radiant (of a meteor shower). The point in the sky from which the meteors of a shower appear to radiate.

radiation (of heat). The transfer of energy by electromagnetic waves.

radioactive dating. Any of a number of procedures that examine the radioactivity of a substance to determine its age.

radio galaxy. A galaxy having greatest luminosity at radio wavelengths.

redshift. A change in wavelength toward longer wavelengths.

reflection nebula. Interstellar dust that is visible due to reflected light from a nearby star.

refraction. The bending of light as it crosses the boundary between two materials in which it travels at different speeds.

resolving power (or resolution). The smallest angular separation detectable with an instrument. Thus it is a measure of an instrument's ability to see detail.

retrograde motion. The east-to-west motion of a planet against the background of stars.

revolution. The orbiting of one object around another.

rift zone. A place where tectonic plates are being pushed apart, normally by molten material being forced up out of the mantle.

rotation. The spinning of an object about an axis that passes through it.

Schwarzschild radius. The radius of the sphere around a black hole from within which no light can escape.

Seyfert galaxy. One of a class of galaxies having active nuclei and spectra containing emission lines.

scarps. Cliffs in a line.

scientific model. A theory that accounts for a set of observations in nature.

second of arc. One-sixtieth of a minute of arc.

sidereal day. The amount of time that passes between successive passages of a given star across the meridian.

sidereal period. The amount of time required for one revolution (or rotation) of a celestial object with respect to the distant stars.

solar day. The amount of time that elapses between successive passages of the Sun across the meridian.

solar eclipse (or eclipse of the Sun). An eclipse in which light from the Sun is blocked by the Moon.

solar flare. An explosion near or at the Sun's surface, seen as an increase in activity such as prominences.

solar wind. The flow of nuclear particles from the Sun.

space velocity. The velocity of a star relative to the Sun.

special theory of relativity. A theory developed by Einstein that predicts the observed behavior of matter due to its speed relative to the person who makes the observation.

spectrometer. An instrument that separates electromagnetic radiation according to wavelength. (A spectrograph is a spectrometer that produces a photograph of the spectrum.)

spectroscopic binary. An orbiting pair of stars that can be distinguished as two due to the changing Doppler shifts in their spectra.

spectroscopic parallax. The method of measuring the distance to a star by comparing its absolute magnitude to its apparent magnitude.

spectrum. The order of colors or wavelengths produced when light is dispersed.

spicule. A narrow jet of gas that is part of the chromosphere of the Sun and extends upward into the corona.

spiral galaxy. A disk-shaped galaxy with arms in a spiral pattern.

spring tide. The greatest difference between high and low tide, occurring about twice a month when the lunar and solar tides correspond.

stadium. An ancient Greek unit of length, perhaps equal to 0.15 to 0.2 kilometers. Various stadia were in use.

standard solar model. Today's generally accepted theory of solar energy production.

star. A self-luminous celestial object.

stellar parallax. The apparent annual shifting of nearby stars with respect to background stars. (Later this term will be used to refer to the angle of shift.)

stellar wind. The flow of nuclear particles from a star.

summer and winter solstice. The points on the celestial sphere where the Sun reaches its northernmost and southernmost positions, respectively.

sunspot. A region of the photosphere that is temporarily cool and dark compared to surrounding regions.

supercluster. A group of clusters of galaxies.

supergiant. A star of very great luminosity and size.

supergiant. The evolutionary stage of a massive star after it leaves the main sequence.

supernova. The catastrophic explosion of a star during which the star becomes billions of times brighter.

synodic period. The time interval between successive similar alignments of a celestial object with respect to the Sun.

syzygy. A straight line arrangement of three celestial objects.

Tully-Fisher relation. A relation that holds that the wider the 21-centimeter spectral line, the greater the absolute luminosity of a spiral galaxy.

T Tauri stars. A certain class of stars that show rapid and erratic changes in brightness.

tail (of comet). The gas and/or dust swept away from a comet's head.

tangential velocity. Velocity perpendicular to the line of sight.

theory. A hypothesis or set of hypotheses that have been well tested and verified.

tidal friction. Friction forces that result from tides on a rotating object.

total lunar eclipse. An eclipse of the Moon in which the Moon is completely in the umbra of the Earth's shadow.

total solar eclipse. An eclipse in which light from the normally visible portion of the Sun (the photosphere) is completely blocked by the Moon.

troposphere. The lowest level of the Earth's (and some other planets') atmosphere.

tuning fork diagram. A diagram developed by Edwin Hubble to relate the various types of galaxies.

twenty-one-centimeter radiation. Radiation from atomic hydrogen, with a wavelength of 21.1 centimeters.

umbra. The portion of a shadow that receives no direct light from the light source.

universality. The property of obeying the same physical laws throughout.

vernal and autumnal equinoxes. The points on the celestial sphere where the Sun crosses the celestial equator while moving north and south, respectively.

visual binary. An orbiting pair of stars that can be resolved (normally with a telescope) as two stars.

volatile element. A chemical element that exists in a gaseous state at a relatively low temperature.

volatile. Capable of being vaporized at a relatively low temperature.

wavelength. The distance from a point on a wave to the next corresponding point.

weight. The gravitational force between an object and the planetary body on which the object is located.

white dwarf. A very small, hot star that is the final stage of evolution of a star like the Sun. A star supported by electron degeneracy.

winter solstice. The point on the celestial sphere where the Sun reaches its southernmost position.

zenith. The point in the sky located directly overhead.

zodiac. The band that lies 9° on either side of the ecliptic on the celestial sphere.

Index

References to illustrations are in bold type.

Aberration, chromatic, 124–**125**
Absolute luminosity, 368
Absolute magnitude, 374–375, 383–384, 396–397, 417
Absorption spectrum, 106, 110–112
Acceleration
   defined, 70
   for various motions, 79
Accessories, telescope, 135–138
Accretion disk, **442**–443
Accuracy
   as a criterion of scientific models, 46–47
   in measurement, 51–52, 381–382
Active galaxies, 528–537
   nature of, 535–537
Active optics telescopes, 135
Activities:
   Circular Motion, 94
   Deep Sky Objects with a Small Telescope, 421
   Do-It-Yourself Phases, 193–194
   Measuring the Diameter of the Sun, 364
   Observing a Solar Eclipse, 195
   Observing Jupiter and Saturn, 304
   Observing Meteors, 335–336
   Observing Sunspots, 364–365
   Observing the Moon's Phases, 194
   The Radius of Mars's Orbit, 61–62
   The Rotating Earth, 61
   The Scale of the Galaxy, 506
   Viewing Mercury, Venus, and Mars, 263–264
Adams, John C., 91
Adaptive optics telescopes, 135, 143
Aesthetic criterion, 32–33, 49–51
Age of Earth and Moon, measuring, 189
Age of the universe, 556
Airy, George, 300
Albedo, 234, **387**
Albireo, 105, **387**, 421
Alcor, **389–390**, 421
Aldren, Buzz, 64, 77
Algol, 390
Alkaid, **17**
Almagest, 27, 373
Alpha Centauri, 117
Alpha particle, 434
Altitude, defined, 23
Alveraz, Luis, 332
Alveraz, Walter, 332
Andromeda (constellation), **371**
Andromeda Galaxy, 421, 482, **483**, **508**–509, **515**

Angstrom, defined, 99
Anglo-Australian Telescope, 412
Angular momentum, conservation of, 214, 216
Angular separation, 18, 126, 130
Angular size, 126
   of the Moon, 157
Annular eclipse, **169–170**
Answers to questions and Try-One-Yourself, A-12–A-18
Antennae, the (colliding galaxies), **513**
Aphelion, 233
Aphrodite Terra (on Venus), **240**
Apogee, 160
Apollo asteroids, 313
Apollo missions, 73–64, 76–77, 184, 186, 189, 198
Apparent magnitude, 369–372, 383, 396, 417
Aquinas, Saint Thomas, 40, 41, 65
Arcminutes, 18
Arcseconds, 18
Arecibo telescope, **140**, 148, 466
Ariel, **293**
Aristarchus, 33, 42
Aristotle, 40–41, 75
Armstrong, Neil, 63–64, 77, 198
Asaro, Frank, 332
Assumptions in science, 521
Assumptions of cosmology, 551–552
Asteroid belt, 199, 310
Asteroids, 199, **203**, 309–310, 312–314
   Apollo, 313
   detection of, 309–310
   discovery of, 203
   naming of, 310, 312
   number of, 309–310
   orbits of, 310, 312–**313**
   origin of, 313–314
Astrology, as scientific theory, 34
Astrometric binaries, 390, 439
Astronomical constants, A-2
Astronomical Society of the Pacific, 421
Astronomical unit, 7, 48, A-2
Astronomy
   modern, 10
   studying, xix–xxiii
   value of, 11, 566
Astrophysics, 10
Atmosphere
   of Earth, **179**, 181

   of planets, 209–212
Atmospheric turbulence, 131
Atom
   Bohr model of, 106–111
   energy levels of, 107–111
Autumnal Equinox, **21**, 26
Auroras, 174, **176**, 358
Axis tilt of planets, A-4

Background radiation, 553–556
Bahcall, John, 352
Balmer series, 112, **113**, 378
Barnard, Edward, 248
Barnard's star, 220, 372, 375–**376**
Barred spiral galaxy, **509**, 510, 514
   Milky Way as a, **496**
Barycenter, 80, 391
Bell, Jocelyn, 458–460
Bessel, Friedrich, 439–440
Beta Cygni, 105, **387**, 421
Beta Pictoris, **220**, **221**
Betelgeuse, 450
Bethe, Hans, 338
Bible, The, 65
Big bang, 552–554, 556–560
Big Dipper, **13**, 14, **17**, 376
Binary stars, 219, 385–394
   and planetary systems, 219–220
   astrometric, 390, 439
   calculating stellar mass and size from, 391–394
   composite spectrum, 391
   eclipsing, 390, **391**, 393–394
   novae, in, 442–444
   rotation determined by the Doppler effect, 117
   spectroscopic, **388–390**
   visual, 386–**388**
Biot, J. B., 324
BL Lac objects, 537
Black dwarf stars, 441
Black holes, 468–471, 472–473
   detecting, 469–471
   in binary systems, **470**–471
   in fiction, 473
   in the center of galaxies, 501
   misconceptions about, 472–473
   properties of, 469
Blackbody radiation, 105
Blink comparator, 306
Blue color of sky, 404
Blueshift, 114
Bode, Johanne, 200, 203

Bode's Law, 200–201
Bohr model of the atom, 106–111, **342**
Bohr, Niels, 107, 108
Bondi, Hermann, 555
Brahe, Tycho, **51**–52
    Supernova discovered by, 442, 455, 456
Brightest stars, table of, A-8
Brightness of stars, 368
Brown dwarfs, 424–426
Bruno, Giordano, 480
Buffon, Georges Louis de, 215
Burnell, Jocelyn, 458–460
Butterfly diagram, 360

Caesar, Augustus, 25
Caesar, Julius, 25
Calendar, 21, 25
Callisto, **232**, **275**, 277–**278**
Caloris Basin, 230–**231**, 233
Calvin, John, 69
Camera, 123–**124**
Canali of Mars, 248
Canis Major, **439**
Cannon, Annie Jump, 378, **379**
Capture theory of Moon's origin, 186
Carbon cycle, **426**
Carbon dioxide, and greenhouse effect, 243, 245
Cassegrain focus, 132–**133**
Cassini spacecraft, 283
Cassini's division, 285–286
Catastrophic theories of solar system formation, 215, 221
Catholic Church, Roman, 25, 68–69
CCD (Charge-coupled device), **137**, 370
Celestial coordinates, 20
Celestial equator, 20, 22
Celestial objects, measuring positions of, 18–19
Celestial photographs, 30
Celestial pole, 14, **16**
Celestial Police, 201
Celestial sphere, 14, **16**, 28, 43
    and Sun's motion, 20–26
    definition of, 14
Centaurus A, **528**, **543**
Center for High Angular Resolution Astronomy, 145
Center of mass, 79–81
    of binary star system, 389, 391
    of Moon and Earth, 80
Centripetal force, 74
Cepheid variables, 394–397
    used to measure galactic distances, 485–486, **515**
Ceres, 203, 310, **312**
Cerro Tololo Inter-American Observatory, 459
CETI (Communication with Extraterrestrial Intelligence), 148
Chain reaction theory, of spiral arm formation, 500
Challenger spacecraft, 312
Chandrasekhar limit, 443–444, 453, 455, 458
Chandrasekhar, Subrahmanya, 443–444
Chaos theory, 315

CHARA (Center for High Angular Resolution Astronomy), 145
Charge-coupled device (CCD), **137**
Charon, **308**–309
Chemical reactions vs. nuclear reactions, 343
Chi Persei, 421
Christianity and science, 40–41
Christy, James W., 308
Chromatic aberration, 124–**125**
Chromosphere, **356–357**
Circular motion, 73–74
Clark, Alvan G., 440
Closed universe, 545–546, 558–560
Cloud density theory of galaxy formation, 526–527
Clouds, interstellar, 403–405, 407–409
    collapse of, 410–411
Cluster
    globular, **416**–417
    H-R diagram of, **417**, **429**, 501–**502**
    of galaxies, 516, 524–526
    open (or galactic), **416**–417
CNO cycle, **426**
COBE, 496, 500, 501, 554, 556, 561, **562**, 565–566
Cocoon nebula, 413
Collins, Peter, 423
Color
    and wavelength, 99–**100**
    as measure of temperature, 102–105, 377
    from reflection, 101–102
    of planets, 101–102
Columbus, Christopher, 153–154
Coma, of a comet, 316–318
Comet Halley, 315, **316**, **317**–319, **481**
Comet Mrkos, **320**
Comet West, **314**
Cometary globule, **412**
Comets, **314**–323
    as part of evolution of solar system, 213
    nature of, 316–319
    Oort cloud of, **321**–322
    orbits of, 315–**316**
    parts of, 316–321
    short-period, 322–323
    tails of, 316, 318, **319**–321
    Whipple's model of, 317–318
Communication with Extraterrestrial Intelligence (CETI), 148
Composite spectrum binaries, 390–391
Conclusions in science, 521
Conduction (of heat), 348
Conservation of angular momentum, 214, 216
Conservation of mass/energy, law of, 555
Constants, astronomical, A-2
Constellations, 15–17
    definition of, 16
    of the zodiac, 22
    table of, A-10–A-11
Continental drift, 176–177
Continuous spectrum, 105
Convection (of heat), 348–349
Coordinates, celestial, 20
Copernican model, 41–51, 68, 544–545

accuracy of, 46–47
and calculation of planetary distances, 48–49
and epicycles, 47, 50
and Galileo's observations, 64–69
compared to Ptolemaic model, 41, 46–51
epicycles and, 47, 50
Copernicus, Nicolaus, **41**–42, 49, 69, 544–545, 547
    biographical sketch, 42
    heliocentric system of, 41–51, 544–545
Core, of Earth, 171–**172**
Corona, of Sun, 167–**168**, 356–358
Correspondence principle, 91, 202
Cosmic Background Explorer (COBE), 496, 500, 501, 554, 556, 561, **562**, 565–566
Cosmic background radiation, 553–556
Cosmic Connection, The, 258
Cosmion, 350
Cosmological principle, 552
Cosmological redshift, 548–550
Cosmological space curvature, 87–89, 545–546
Cosmology, 518, 543–566
    and general relativity, 547–551
    and religious faith, 557
    and the future, 557–560
    assumptions of, 551–552
    Bing Bang theories of, 552–554, 556–560
    history of, 544–547
    inflationary theory of, 561–564
    reason for studying, 566
    steady state theory of, 555
Crab Nebula, 454–**455**, 456
Crab pulsar, 463–464
Craters, meteorite, 329–**332**
Creationism, 557
Crescent phase (of moon or planet), 67, **160**–162
Criteria for scientific models, 31–33, 46–51
Critical density, 560
Crust, of Earth, 171–**172**
Curtis, Heber, 488
Curvature of space, 87–89, 545–546
Cygnus, **18**, 105, **387**
Cygnus A, 528, **536**
Cygnus loop, **456**
Cygnus X-1, 470–471

Dark Matter, **402**, **408**, 526, 560
Dark nebula, **402**, **408**
Dark spot (on Neptune), **295**
Darwin, Sir George Howard, 186
Data-fitting criterion, 32, 46–48
Davis, Raymond, Jr., 350
Day, sidereal vs. solar, 208–209
De Revolutionibus, 41, 51
Declination, 20
Deep sky objects with a small telescope, 421
Deimos, **227–228**, **257**–258, 260
    discovery of, 258
Degeneracy
    electron, 432–433, 434

neutron, 458
Density
 critical, 560
 definition of, 171
 of a gas, 346
 of Earth and Moon, 171
 of matter in the universe, 560
 of planets, 206–**208**
 particle density, 345–346
Density wave theory of spiral arm formation, 497–500
Descartes, Rene, 214
Deuterium, 344
Devens, R. M., 305
Differential rotation, 267
Differentiation of planetary materials, 171–172
Diffraction, 129–**130**
 of radio waves, 139
Diffraction grating, 138
Dinosaurs, extinction of, 332
Dione, **284**
Disk, galactic, **487**
Dispersion, 124
Distance indicators, 514–517, 527–528
Distances
 measured by parallax, 159
 to planets by the heliocentric model, 48–49
 to stars, 372, 373, 381–382, 383–384
Distance/dispersion relationship, 467
Doppler, Christian, 113
Doppler effect, 112–117
 and binary stars, 117, **388–390**
 and cosmological redshift, 548–550
 and galactic spiral arms, 494–495
 and Hubble law, 517–522
 and quasars, 530–531
 and rotation of the Sun and planets, 117
 as a measurement technique, 115–117
 as cause of cosmological redshift, 548–550
 defined, 113
 in light, 114–117
 in sound, 112–114
 in water waves, 112–113
 misconceptions about, 114
 used to measure the mass of galaxies, 523–525
Double planet theory of Moon's origin, 186–187
Drake, Frank, 147, 418
Dumbbell nebula, **438**
Dust disks around stars, 220
Dwarf star, **380, 383,** 384
 black, 441
 brown, 424–426
Dynamo effect, 173, 232, 271

Earth, 171–181
 age of, 189
 atmosphere of, **179,** 181
 compared to Venus, 242–243, 245
 data page, 180
 data with other planets, A-4–A-5
 greenhouse effect on, 245
 interior of, 171–172
 magnetic field of, 172–176
 mass of, determining, 171
 plate techtonics on, 176–179
 precession of, 84–85
 size of, 154, **155,** 198–**199**
 shadow of, **163**–165
 tides on, 81–83
 viewed from space, **174–175**
Eccentricity
 of ellipse, **53**
 of planetary orbit, 206
Eclipse
 dates of lunar, 166
 dates of solar, 168, **170**
 lunar, 27, 162–166
 solar, 166–170, 195
Eclipse season, 164–165
Eclipsing binaries, 390, **391,** 393–394
Ecliptic, **21–22**
Einstein, Albert, 90, 107, 341, 464, 545–546, 563
 (See also relativity)
 biographical sketch, 90
 quotes by, 543–544
Einstein Observatory, 147
Electromagnetic spectrum, 100–101
Electron degeneracy, 432–433, 434
Electron
 in Bohr atom, 107–110
 mass of, A-2
Element, Chemical, Identification of, 109, 111
Elements, volatile and nonvolatile, 213
Ellipse, **53–55,** 391–**392**
 semimajor axis of, **53,** 55
Elliptical galaxy, 509, 510, **511,** 513
Elongation, definition of, 30, 161
Emission nebula, 405–**406, 412**
Emission spectra, 106, 109–111, 437
Enceladus, **284**
Energy transport in the Sun, 347–**349**
Epicycles
 defined, 31
 in Copernican model, 47, 50
 in Ptolemaic model, 31–32, 50
Equal areas, law of, 54–55, 79
Equatorial coordinate system, 20
Equinox, 26, **44**
Equivalence principle, 87, 464–466
Eratosthenes, 154
Escape velocity, 210–**212,** 468
 planetary, table of, A-4
Europa, **275,** 277
European Infrared Space Observatory, 146
European Southern Observatory, **134–**135
European Space Agency, 318
European X-ray Observatory Satellite (EXOSAT), 147
Evening star, 49
Event horizon, **468,** 469
Evolutionary theories of solar system formation, 214–215, 216–219, 221
Evolutionary track on H-R diagram, 413–**414,** 439
Expansion of the universe, 517–518, 547–550
Exponential notation, 8

Extraterrestrial life
 and the inflationary universe, 561
 and the life equation, 418
 and the origin of life, 222
 messages to, 148
 on Mars, 252–253
 search for, 147
 why has there been no contact by, 490
Eye, resolving power of, 130
Eyepiece (of telescope), 124, 127–**128**

Faith, religion, and cosmology, 557
False color image, **121, 138**
Feynman, Richard, 96
Field of view, 127–**128**
Fireball, 324
First-quarter moon, 160–**162**
Fisher, J. Richard, 522
Fission, 342
Fission theory of Moon's origin, 186–187
Flat universe, 87–89, 559–560
Flatness problem (of cosmology), 561
Fluorescence, 405
Flyweight stars, 430
Focal length, 123
 and magnification, 126–127
Focal point, 123
Focus, of an ellipse, 53
Force, 70–77
 (See also: Newton's Laws)
 centripetal, 74
 direction of, for various motions (table), 79
 gravitational, 75–77
Frequency of waves, 97–98
 and the Doppler effect, 113–116
Full phase (of moon or planet), 66, **161–162**
Fusion, 337, 338, 426
 and fission, 342
 in a supergiant, 451, 453
 of helium, 433–434
 of hydrogen, 341–345, 426–427
 rate of, in stars, 427

Galactic clusters, 484, **485**
Galactic halo, **487**–489
Galactic rotation curve, 491–492
Galaxies, 6, 507–537
 active, 528–537
 as part of expanding universe, 548
 barred spiral, **496, 509,** 510, 514
 clusters of, 516, 524–526
 distances to, 514–523
 elliptical, 509–**511,** 514
 Hubble classification of, 509–511, 513, **514**
 Hubble law applied to, 517–522
 irregular, 509, **511,** 513, 514
 local group of, 525
 masses of, 523–526
 measuring, 514–522
 origin of, 526–528
 radio, 528–529
 Seyfert, **532**
 spiral, **7, 509–510,** 513, 514

INDEX I-3

superclusters of, 526
theories of the formation of, 526–527
Galaxy, the, 6–**7, 479**–503
   age of, 501–502
   as a barred spiral galaxy, 496
   as part of the local group, 525
   as seen from Earth, 3–**4,** 480–483
   black hole in the center of, 501
   components of, 486–488
   data table, 489
   definition of, 3
   discovery of radio waves from, 140
   evolution of, 501–503
   globular clusters in, 484–486, 488
   mass of, 492, 493
   nucleus of, **487**–488, **500**–501
   position of Sun in, **481**–482
   radio waves from the center of, 140
   rotation of, 489–492
   scale model of, 9, 506
   size of, 481–482, 506
   spiral arms of, 492–500
   star formation in, 487
Galilean moons of Jupiter, 65, **275–278**
Galilean relativity, 117
Galilei, Galileo, 64–70, 545, 547
   and Kepler, 64
   and Saturn's rings, 280
   biographical sketch, 68–69
   celestial observations of, 64–69
   sun observed by, 64, 338
   telescope observations by, 64–69
   view of the universe of, 480
Galileo spacecraft, **10,** 185, 265–266, 278, 310
Gallex, 350
Gamma ray telescopes, 147
Gamma rays, 100
Gamow, George, 338
Ganymede, **232, 275,** 277
Gases
   and escape velocity, 211–212
   pressure, temperature, and density of, 345–346
Gaspra, **310**
Gassendi crater, **188**
Gauss, Karl Frederick, 248
Gemini, 28–**29**
General theory of relativity, 86–91, 464–468, 545–546
   and quasars, 531, 532–533
Geocentric model, 26–31, 33
   and Galileo's observations, 64–69
   defined, 27
   Ptolemaic model, 28, 30–33
   compared to heliocentric model, 46–51
Giant molecular clouds, **410**–411
Giant star, **380,** 383, 384
Gibbous phase (of moon or planet), 67, **161**–162
Giotto space problem, 317, 318
Global Oscillation Network (GONG), 355
Globular clusters, 484–**486,** 502
   motions in the Galaxy of, 488
God, 557
Goddard Space Flight Center, 449
Gold, Thomas, 555

Goldreich, Peter, 296
Golubkina crater on Venus, **239**
Goodricke, John, 390, 394
Granulation (of Sun's surface), **354**
Gravitation, law of, 75–77
Gravitational lens, 532–**533**
Gravitational mass vs. inertial mass, 86
Gravity
   and Einstein, 89–91
   and general relativity, 89–91
   and light, 87
   and tides, 81–83
   equivalence to acceleration, 87, 89–91
   center of, 80
   law of, 75–77
Great dark spot (on Neptune), **295**
Greek celestial model, 27–30
Green Bank observatory, **139**
Green Bank Telescope, **142**–143
Greenhouse effect, 243, 245
Gregorian calendar, 25
Gum Nebula, 444, **445**
Guth, Alan H., 564

H Persei, 421
H-R Diagram, 379–384
   (See also, Stars)
   and star formation, 409
   evolutionary path of stars on, 413–**414**
   of a cluster, **417, 429,** 501–**502**
Haag, Robert, 329
Haldane, J. B. S., 222
Hale telescope, **133,** 134, 135, 518
Half-life, radioactive, 189
Hall, Asaph, 258
Halley, Edmund, 315, 375
Halley's Comet, 315, **316, 317**–319, 481
Halo (of the Galaxy), **487**–489
Harvard College Observatory, 378, 382, 395, 397
Hawking, Stephen, **563,** 564
Heavyweight stars, 450–456, 468–471
Head (of a comet), 316–**317**
HEAOs (High-Energy Astronomy Observatories), 147, 435–436
Heliocentric, definition of, 41
Heliocentric model, 41–58
   (See also Copernical Model)
   and Galileo's observations, 64–69
   as developed by Copernicus, 41–51
   compared to geocentric model, 46–51
Helieoseismology, 355
Helium, fusion reaction of, 433–434
Helium flash, 433–435
Helix nebula, **436**
Helmholtz, Hermann von, 341
Herschel, Caroline, 483, 484
Herschel, John, 290
Herschel, William, 288, 290, 359, 386, 403, 483, 484, 486, 545, 547
   biographical sketch, 290
Hertz, defined, 99
Hertzsprung-Russel (H-R) Diagram, 379–384
   (See also, Stars)
   and star formation, 409
   evolutionary path of stars on, 413–**414**

of a cluster, **417, 429,** 501–**502**
Hetherington, Norriss, 508
Hewish, Antony, 458–460
High-Energy Astronomy Observatories (HEAOs), 147, 435–436
Hipparchos (spacecraft), 373
Hipparchus, 369, 370
Hoba meteorite, 329
Homestake neutrino experiment, the, **350,** 352–353
Homogeniety of the universe, 551–552
Horizon problem (of cosmology), 561–562, 564
Horsehead nebula, **5, 402,** 405, 414
Hoyle, Fred, 555
Hubble classification of galaxies, 509–511, 513, **514**
Hubble constant, 519–522
   change in, 558–559
   to determine age of universe, 556
Hubble, Edwin, 455, **508,** 509, 511, 512, 517, 545, 547
   biographical sketch of, 512
Hubble law, 517–522, 547–550
   and quasars, 530–532
   defined, 520
Hubble Space Telescope (HST), 131, **149**–150, 220, 283, 308, 403, 456
Hubble tuning fork diagram, 513–**514**
Humason, Milton, 517, 518
Huygens, Christian, 280
Hydrogen
   fusion of, 343–345, 426–427
   liquid metallic, 270
   spectrum of, **106**
Hydrostatic equilibrium, 346–347
Hypotheses, definition of, 35
Hyperion, **284,** 315

Image formation, 123
Inclination of a planet's orbit, **205**
Inertia, 70–71
Inflationary universe theories, 561–564
Infrared Astronomy Satellite (IRAS), 146, 403, 413, 526
Infrared radiation, 100, 103, 145–146
Intensity/wavelength graph, 102–104
Interferometer, **143**–145
International Ultraviolet Explorer, 147, 436, 457
Interstellar cirrus, **403,** 407
Interstellar clouds, 403–405, 407–409
   collapse of, 410–411
Interstellar dust, **402**–405, 414, 486
Interstellar extinction, 404
Interstellar matter, 402–409
Interstellar shock waves, 411
Inverse square law, 117–118, 273, **375**
Io, **10, 275–276**
Ion, 217–218, 320
Iron meteorites, **329**
Irregular galaxy, 509, **511,** 513, 514
Isotropy, 552

Janus, 288
Jansky, Karl, 140, 479–480
Jovian planets, 265–301

classification of, 206–209, 212
Julian calendar, 25
Jupiter, 266–279
   and the Galileo mission, 265–266, 278
   as seen from Earth, 266–**267**
   asteroid belt and, 313–314
   composition of, 269–271
   data page, 279
   data, with other planets, A-4–A-5
   differential rotation of, 267
   energy from, 272–274
   interior of, 271
   magnetic field of, 271
   red spot of, 268, **269**
   ring of, **278**
   rising and setting times of, 304
   satellites of, 65, **274–278**

Kamiokande II experiment, 350
Kant, Immanuel, 214
Kapteyn, Jacobus C., 484, 486
Keck telescope, 135–**136**
Kelvin, William Lord, 341
Kelvin temperature scale, 96–**97**
Kennedy, John F., 76
Kepler, Johannes, **52**
   biographical sketch, 56
   contributions of, 58
   theory on spacing of planets, 56
Kepler's laws, **54–55,** 57–58
   and center of mass, 81
   and path of Moon around Earth, 81
   applied to the Galaxy, 491–492
   Newton's laws and, 77–79
   used to measure the mass of galaxies, 524
Kepler's supernova, 455–456
Kepler's third law, 55, 57
   and binary stars, 387, 391
   and rings of Saturn, 286–287
   and table of planetary data, 57
   revised by Newton, 79
   used to calculate mass of the Sun and planets, 202–204
   used to calculate mass of galaxies, 524
Kilogram, 71
Kirkwood, Daniel, 313
Kirkwood's gaps, **313**
Kirchhoff, Gustav, 106
Kirchhoff's laws, 106
Koppernigk, Mikolaj, (See Copernicus, Nicolaus)
Kuiper Airborne Observatory, 146
Kuipter belt, 32, **323**
Kunkel, W., 457

Lagoon nebula, **406**
Laplace, Pierre Simon de, 469
Large impact theory of Moon's origin, 187
Large Magellanic Cloud, **1, 396**–397, 457, **511, 540**
Las Campanas Observatory, 1, 457
Law of Inertia, 70–71
Law of Universal Gravitation, 75–77
Leap year, 21, 25
Leavitt, Henrietta S., 395, 485
   biographical sketch of, 397

Lens, **123**–125
   achromatic, 125
   focal length of, 123
Leonid meteor shower, **305**–306, **326,** 327, 328
Letters on Sunspots, 68
Le Verrier, Joseph, 89, 91
Life equation, 418
Life, Extraterrestrial
   and the inflationary universe, 561
   and the life equation, 418
   and the origin of life, 222
   messages to, 148
   on Mars, 252–253
   search for, 147
   why has there been no contact by, 490
Life, origin of, 222
Light
   and the equivalence principle, 87
   and gravity, 464–466
   and special relativity, 88
   and the Doppler effect, 114–117
   and the electromagnetic spectrum, 95–111
   diffraction of, 129–130
   dispersion of, 124
   nature of, 97, 110
   refraction of, 122–123
   speed of, 88, 99, 123
   wave motion of, 99
Light curve, 390, **391**
Light-gathering power (of a telescope), 128–129
Lighthouse model, of pulsars and neutron stars, 461–462
Lightweight stars, 433–439
   Supernovae from, 455
Light year, defined, 8
Limb (of Sun or Moon), 352
Lindblad, Bertil, 486, 497, 500
Liquid metallic hydrogen, 270
Local group, of galaxies, 525
Local hypothesis, for distance of quasars, 531
Look-back time, 527–528
Lowell Observatory, 248, 281, 306, 517
Lowell, Percival, 248, 249, 306–307
Luminosity, 339, 368
Luminosity classes, 382–383
Lunar craters, 181, 183
Lunar eclipse, 27, 162–166
   dates of, 166
   types of, **153, 165**–166
Lunar module, 77
Lunar month, 162, **163**
Luther, Martin, 69
Lyman series, 112, **113**
Lyra, **437**

Maanen, Adriann van., 488, 507–508
Maat Mons (on Venus), **241**
Madore, B., 457
Magellan spacecraft, 239, 241
Magellanic Clouds, **1, 396**–397, 482, **511**
Magellanic supernova (SN1987A), **1**–2
Magnetic field, **172**

and the formation of the solar system, 217–218
   and dynamo effect, 232
   of Earth, 172–176
   of planets, (See individual planets)
Magnetosphere, 271
Magnification (of a telescope), 125–128
Magnitude,
   absolute, 374–375, 383–384, 396–397
   apparent, 369–372, 383, 396
   difference, 371
   of clusters, 417
Main sequence stars, 380, 427–429, 431
Malin, David, 412
Mantle, of Earth, 171–**172**
Maran, Stephen, 449, 450
Mare, 181, 183–184
Mare Orientale, 230–**231**
Mariner spacecraft, 185, 229–231, 249, 251, 252
Mars, 245–260
   as seen from Earth, 246
   atmosphere of, 255–257
   data page, 259
   data, with other planets, A-4–A-5
   Kepler's third law applied to, 57
   life on, 248, 252–253
   moons of, 227–228, 257–258, 260
   motion of, 247–248
   polar caps of, **246,** 252
   radius of orbit, calculating, 57
   retrograde motion of, 28–**29,** 30–31, **45**–46
   rising and setting times for, 264
   size, mass, and density of, 247
   voyages to, 248–249, 251–252, 256, 257, 260
   water on, 253–255
Mass,
   and space curvature, 89–91
   definition of, 71
   distribution of in solar system, 204, **207**
   in Newton's second law, 72–73
   inertial vs. gravitational, 86
   measuring, of solar system objects, 202–204
   missing, 526
   of electron and proton, A-2
   of galaxies, 523–526
   of stars, 391–394, 415, 425
Mass spectrometer, 253
Mass-luminosity relationship, 394, 395
Mauna Kea Observatory, **134,** 135–**136,** 145
Mauri, Antonia, 382
Maxwell, James Clerk, 95–96
Measurement, uncertainty in, 51–52, 381–382
Mercury, 228–236
   as seen from Earth, 29–32, **228**–229
   atmosphere of, 231
   data page, 235
   data, with other planets, A-4–A-5
   favorable dates for viewing, 263
   in Copernican model, 49–50
   in Ptolemaic model, 32
   interior of, **233**

magnetic field of, 232
motions of, 233–234
precession of orbit of, 89–91
sidereal vs. solar day on, 233–**234**
size, mass, and density of, 231–232
surface of, **228–231**
temperatures on, 234
visited by *Mariner*, **229–232**
Merger theory of galaxy formation, 527
Meridian, 42
Messier number, 428
Meteor Crater, 181, 330–**331**
Meteor showers, **305**–306, **326–328**
  table of,
Meteorites, 324, **329–332**
  striking the Moon, 181, 183
Meteoroids, 324, 325–326
Meteors, 3, 323–328
  and extinction of dinosaurs, 332
  observing, 335–336
Metric units, A-2

Michel, Helen, 332
Mid-Atlantic rift, 176–**177**
Middleditch, John, 459
Middleweight stars, 450–456, 463–464
Mikeyev, Stanislaw, 350
Milky Way Galaxy, 6–7, **479**–503
  age of, 501–502
  as a barred spiral galaxy, 496
  as part of cluster of galaxies,
  as seen from Earth, 3–**4**, 480–483
  black hole in the center of, 501
  components of, 486–488
  data table, 489
  definition of, 3
  discovery of radio waves from, 140
  evolution of, 501–503
  globular clusters in, 484–486, 488
  mass of, 492, 493
  nucleus of, **487**–488, **500**–501
  position of Sun in, **481**–482
  radio waves from the center of, 140
  rotation of, 489–492
  scale model of, 9
  size of, 481–482, 506
  spiral arms of, 492–500
  star formation in, 487
Miller, Stanley, 222
Miller-Urey Experiment, 222
Mimas, **284**, 285–286
Minor planets, 199, **203**, 309–310, 312–314
  (See also, Asteroids)
Minute of arc, defined, 18
Miranda, **293–294**
Missing mass, 526
Mitchell, Maria, 318
Mizar, **17**, **389–390**, 421
Model, 35
Molecular clouds, **410**–411
Month, 162, **163**
Moon
  age of, measuring the, 189
  and solar eclipses, 166–170
  and tides on earth, 81–83
  angular size of, 157

Apollo missions to, 63–64, 76–77, 184, 186, 189
back side of, **138**
changes in apparent size of, 159–160
data page, 182
diameter of, by the small angle formula, 158
distance to, measuring, 155–158
eclipses of, 162–166
gravitational aspects of travel to, 76–77
history of, 187–189
interior of, 183–**184**
observations of, by Galileo, 64
orbit of, 80–81, 160, 164
origin of, 186–187
parallax of, 156
partial eclipse of, 165
phases of, **160**, 162
phases of, observing,
rotation and revolution of, 83–84
surface of, 181, **182–185**
surface of, compared to Mercury, 229
theories of the origin of,
total eclipse, **153**, 165–166
Moribito, Linda, 275
Morning star, 49
Morris, Richard, 566
Motion
  circular, 73–74
  natural (Aristotelian), 27, 65
  of stars, 20–26, 43–44
  proper, 375–376
  retrograde, 29–31, 45–46
  of the sky, **6**, 14–**15**, 42–43
  types of (table), 79
Mount Wilson Observatory, 512, 517, 520
Multiple Mirror Telescope, 145
Multiple Star Systems, 385–394
  (See also, Binary stars)
Murmurs of Earth, 273

Nanometer, 99
National Aeronautics and Space Administration (NASA), 146, 374
National Radio Astronomy Observatory, 139, 522
Nature of Reality, The, 566
Neap tide, 83
Nearest stars, table of, A-9
Nebula, 6, 405, 407–409
  cocoon, 413
  Crab, 454–456
  dark, **402**, 408
  defined, 3
  Dumbbell, **438**
  emission, 405–**406**, **412**
  Gum, 444, **445**, 456, 463
  Helix, **436**
  Horsehead, **5**, **402**, 405, 414
  Laboon, **406**
  planetary, **436–439**
  reflection, **407–408**, **412**
  Ring, **436**, 437
  Rosette, **406**
  spiral, 507–508
  Tarantula, **1**
  Trifid, **408**

Veil, **456**
Neptune, **294–301**
  composition of, 294–297, **298**
  data page, 299
  data, with other planets, A-4–A-5
  discovery of, 89, 91, 300
  Great Dark Spot, **295**
  moons of, 297–298, 300–301
  orbit of, 204–**205**
  Pluto as a former moon of, 309
  rings of, 301
  rotation of, 296
  Voyager explorations of, 294, 295, 297–298, 300–301
Nereid, 297, 300
Neutrinos, 344, 349–351, 352–353, 453, 457
Neutrino problem, The, 349–351, 352–353
Neutron, 341
Neutron degeneracy, 458
Neutron star, 458–468
  (See also, Pulsar)
  determining the mass of, 468
  typical, 458
New Astronomy, The, 54
New General Catalog, The (NGC), 416
New moon, 161–**162**
Newton, Isaac, 70, 79, 545, 547
  and light spectrum, 97
  and nature of light, 110
  and orbits of comets, 315
  biographical sketch, 72
  universal law of gravitation of, 75–79
Newton's laws of motion, 70–74
  and the law of gravitation, 75
  and Kepler's laws, 77–79, 202–204
  applied to circular motion, 73–74
  first, 70
  importance of, 85–86
  second, 70, 71–73
  third, 73
Newtonian focal arrangement, **131**, 132
Newtonian relativity, 117
NGC (New General Catalog) classification, 416
Noblesville, IN meteorite fall, 330
Nonvolatile elements, 213
North Star, **15**
Nova Cygni 1975, **443**
Nova Cygni 1992, **423**
Northern Cross, **18**
Novae, 423–424, 442–444
  in binary systems, 442–444
  typical, 443
Nuclear bulge, galactic, **487**–488
Nuclear fission, 342
Nuclear fusion, 337, 338, 426
  and fission, 342
  in a supergiant, 451, 453
  of helium, 433–434
  of hydrogen, 341–345, 426–427
  rate of, in stars, 427
Nuclear reactions, solar, 341–345
Nuclear reactions vs. chemical reactions, 343
Nucleus
  of a comet, 316–318

of the atom, 107
of the Galaxy, **487**–488, **500**–501

Oberon, **293**
Objective lens, 124
Oblateness, 85
Occam's Razor, 33
Occultation, 290
Olbers, Heinrich, 203
Olbers's paradox, 550–551
Olympus Mons, 249–**251**
On the Revolutions of the Heavenly Spheres, 41, 51
Oort cloud, **321**–322, 323
Oort, Jan, 322, 486, 500
Oparin, A. P., 222
Open clusters, **484**–485
Open universe, 546, 557–560
Ophiuchus, 22
Opposition of planets, 246
Optical double, **385**
Orion, **16, 369, 377, 410, 451**
Orion A molecular cloud, **410**
Orion nebula, **403, 413, 414**–**415,** 421
Oscillating universe theory, 558
Our First Century (by R. M. Devens), 305
Owl cluster, **416**
Ozma, Project, 147
Ozone
  on Earth, 179, 181
  on Earth and Mars, 255

Pallas, 310
Palomar Mountain telescope (Hale Telescope), **133,** 134, 135, 518
Parallax, **47, 48,** 156, **372**–373
  and the distance to the Moon, **156**
  defined, 47
  spectroscopic, 381–**382,** 383–384
  stellar, **48,** 372–373
Parsec, 373
Partial lunar eclipse, 165–166
Partial solar eclipse, 168–**169**
Particle density, 345–346
Paschen series, 112, **113**
Peekskill, NY meteorite fall, **330**
Pegasus, 370–**371**
Penumbra, **163**–165
Penumbral lunar eclipse, 165
Penzias, Arno, 553–**554**
Perigee, 160
Perihelion, 89, 233
Period
  in Kepler's third law, 55, 57
  sidereal vs. synodic, 162–**163,** 208–**209**
Perseid meteor shower, 327, 328
Phases of the Moon, **160**–162
Phases of Venus, 66–69
Phobos, 227, **257**–**258,** 260
  discovery of, 258
  escape velocity of, 211
Photographs of sky, 30
Photometry, 137–138, **370**
Photon, 108
Photon model of light vs. wave model, 110
Photosphere, 110, 351–356

Piazzi, Giuseppe, 203, 310
Pickering, Edward, 389
Pioneer spacecraft, 267–268, 270, 272, 282
Pioneer Venus, 239
Planet X,
Planetary data, A-4–A-5
Planetary distances, and Titius-Bode Law, 200–201
Planetary exploration, reason for, 260
Planetary motion, Kepler's laws of, **54**–**55,** 57–58, 77–79
Planetary nebulae, **436**–**439**
Planetary rings, 285–286, 288
Planetesimals, 271, **218**
Planets
  (See also, individual planets)
  apparent motion of, 28–33
  around other stars, 219–221
  atmospheres of, 209–212
  calculating distances to by heliocentric model, 48–49
  classifying, 206–209, 212
  color of, 101–102
  data tables, 199, 204, 206, 208, 210, A-4–A-5
  definition of, 3
  diameters of, 198–**199,** 206–**207**
  distances to, 48–49, 198–**199**
  eccentricities of orbits, 206
  escape velocities of, 210–**212,** A-4
  in Copernican model, 43, 45–46, 49–50
  in Ptolemaic model, 28–32
  Kepler's laws and 54–55, 57–58
  masses and densities of, 202–204, A-4
  measuring rotation rates of, 117
  minor, 199, **203,** 309–310, 312–314
  motions of, 29–31, 45–46, 54–55, 57, 204–206
  rotations of, 117, 208–**210**
  satellites and rings of, 204–205, 208
  sidereal periods of, 55, 57
  speeds of, in orbit, 50, 54–55, 57
Plate tectonics, 176–179, 250–251
Pleiades, **407, 410,** 421, 484
Pluto, 306–309, 311
  and Charon, **308**–309
  as a possible moon of Neptune, 309
  as seen from Earth, **307**
  classification as a planet, 309
  data page, 311
  discovery of, 306–**307**
  mass of, 308–309
  orbit of, 204–**205,** 307
  size of, 198, **232**
Polaris, **15,** 85
Pope, Alexander, 72
Pope Gregory XIII, 25
Population I and II stars, 471–474
Positron, 344
Power,
  definition of, 341
  of reflected radar signal, 200
Powers of ten notation, 8
Precession,
  of Earth, 84–85

of Mercury's orbit, 89–91
Prediction criterion, 48–49
Prefixes, metric, A-2
Pressure, 345–346
  of a gas, 211
Prime focus, 133
Principia, The, 72, 315
Principle of equivalence, 87, 464–466
Project Cyclops,
Project Ozma, 147
Prominences, **337, 357**–**358**
Proper motion, of star, 375–376
Proton, 341
  mass of, A-2
Proton-proton chain, 344, **426**
Protostars, 412–416, 424–425
Ptolemaic model, 28, 30–33, 40
  compared to Copernican model, 41, 46–51
  criteria for judging, 31–33
Ptolemy, Claudius, 27–**28,** 31, 373, 544, 547
Pulsars, 458–468
  and planetary systems, 221
  determining the mass of, 468
  discovery of, 458–460
  distance/dispersion relationship of light from, 467
  in binary system, 466–468
  in SN1987A, 459
  lighthouse model of, 461–462
  pulse duration of, 460
  typical, 458
Pythagoras, 27, 32

Quadrant, Tycho's, **52**
Quandrature, 61
Quarter phase (of moon or planet), 68, **160**–162
Quasars, 529–537
  competing theories for, 531–532
  nature of, 535–537
  redshift of, 530–532
  twin, 532–**533**
Quasi-stellar radio sources, (See quasars)

Radar, used to measure distances to planets, 49, 200–201
Radial velocity, 115–**116,** 376–377
Radiant (of a meteor shower), **327**
Radiation
  cosmic background, 553–556
  electromagnetic, 99–101
  of heat, 349
  21-centimeter, 407
Radio galaxies, 528–529
  See also Active galaxies
Radio telescopes, 139–143
  resolution of, 139, 143–145
Radio waves, 100
  diffraction and, 139
  from space, discovery of, 139, 140
Radioactive dating of lunar materials, 188, 189
Radioactive decay, 189
Rays, lunar, **183**

INDEX I-7

Red color of sunset, 404
Red giants, **380, 383,** 384, 431–433
    mass loss from, 435–436
Red spot (on Jupiter), 268–**269**
Redshift
    cosmological, 548–550
    defined, 114
    of quasars, 530–532
Reflecting telescopes, 131–132
    optical arrangements of, 132–133
    versus refractors, 132
Reflection nebula, **407–408,** 412
Refracting telescopes, 123–125
    chromatic aberration in, 124–125
    versus reflectors, 132
Refraction, 122–123
    by atmosphere, 166, 229
Regulus, 116
Relativity,
    and gravitation, 89–91, 532–533
    general theory of, 86–91, 464–468, 531, 532, 545–546
    Newtonian (Galilean), 117
    special theory of, 88, 341, 464
Religion, and science, 40–41, 50, 68, 557
Resolving power
    of a telescope, 129–131
    of radio telescopes, 139, 143–145
    of the human eye, 130
Retrograde motion, 29–31
    definition of, 29
    epicycles used to explain, 30–31
    in Copernican model, 45–46
    in Ptolemaic model, 30–31, 50
Revolution,
    Copernicus's, 51
    definition of, 160
    sidereal, 162–**163**
Rift zone, 176–**177**
Right ascension, 20
Rings, planetary, 285–286, 288
    and the chaos theory, 315
Ring nebula, **436,** 437
Roche limit, 288
Roentgen Satellite (ROSAT), 147–148
Roman Catholic Church, 25, 68–69
Rosette nebula, 216, **217,** 406
Rotation
    definition of, 160
    differential, 267
    of planets, 117, 208–**210**
    of the Galaxy, 491–492
    of the Sun, 117
    sidereal vs. synodic, 163
Russell, Henry Norris, 380, 409

Sagan, Carl, 258, 272–273
Sagan, Linda, 272–273
SAGE, 350
Sagittarius, **4,** 421, **482**
Sandage, Allan, 520
Satellites of planets, 204–205, 208
Saturn, **280**–288
    as seen from Earth, **281**
    composition of, **282**–283
    data page, 287
    data, with other planets, A-4–A-5
    energy radiated from, 283
    interior of, **282**–283
    magnetic field of, 282
    measuring rotation of rings around, 117
    moons of, **284–285**
    motions of, 280–282
    oblateness of, 282
    Pioneer, Voyager, Cassini, and, 282–283
    rings of, 285–**286,** 288
    rising and setting times of, 304
    size, mass, and density of, 280
Scale model,
    of the Earth-Moon system, 158, 163
    of the solar system, 8, 198–**199**
    of the Galaxy, 506
    of the Universe, 8–**9**
Scarps, 230
Schiaparelli, Father Giovanni, 248
Schmidt, Maarten, 530
Schwabe, Heinrich, 359
Schwarzschild radius, **468**–469
Science
    and religion, 40–41, 557
    data and conclusions in, 521
    nature of, 2–3, 5, 40
    observations, assumptions, and conclusions in, 521
    precision of, 50–51, 523
Scientific model, 26–27, 30–33, 35, 384
    and astrology, 34
    definition of, 26
    criteria for judging, 31–33, 46–51
Scientific notation, 8
Scorpius, **4,** 482
Sea of Tranquility, 77
Search for Extraterrestrial Intelligence (SETI), 147
Seasons, and movement of Sun, 22–26
Secchi, Father Angelo, 248
Second of arc, 18
Serendipity, 479–480
Seyfert galaxy, **532**
Shapley, Harlow, 396, 474, 486–487, 488, 500, 518, 545, 547
Shapley-Curtis debate, 488
Shelton, Ian, 1–2
Shepherd moons, 286, **296–297**
Shock waves, 411
Shu, Frank, 411
Sidereal day, 209, 233–**234**
Sidereal period, 162–**163**
Sirius, 371, 374–375, **439–440**
Skills, study, xix–xxiii
Slipher, Vesto, 488
Sky,
    blue color of, 404
    motion of, **6,** 14–**15**
    naked eye, 6
Small angle formula, 158
Small Magellanic Cloud, **396**–397, 511
Smoot, George, 565
SN1987A, **449–450, 457,** 459
SN1993J, 453–**454**
Solar activity cycle, **359–361**
Solar data, table of, 340
Solar day, 208–209, 233–**234**
Solar eclipse, 166–170, 195
    annular, 169–**170**
    as a test of relativity, 465–466
    dates and locations of, 168, **170**
    observing, 195
    partial, 168–**169**
    total, 167–**168**
Solar energy, 337–345
    (See also, Sun)
Solar flares, 361
Solar spectrum, **106,** 352, **354,** 359
Solar system
    debris in, 309
    distances in, 198–202
    distribution of mass in, 204
    formation of, 213–219, 221
    formation of, catastrophic theories, 215, 221
    formation of, evolutionary theories, 214–215, 216–219, 221
    measuring distances in, 199–202
    measuring mass of objects in, 202–204
    scale model of, 8, 198–**199**
Solar wind, 219, 358–**359,** 436
Solstice, 24, **44**
    defined, 24
Sombrero galaxy, **510**
Soviet American Gallium Experiment, 350
Space, curvature of, 87–89, 545–546
Space Infrared Telescope Facility, 146
Space Telescope, 131, **149**–150
Space velocity, 377
Space warp, 87–**89,** 545–**546**
Special relativity, 88, 341, 464
Spectral class, of stars, 377–379
Spectrometer, 138
Spectroscopic binaries, **388,** 390
Spectroscopic parallax, 381–**382,** 383–384
Spectroscopic Survey Telescope (SST), 138–**139**
Spectrum, **97**–98
    absorption, 110–112, 377
    continuous, 105
    electromagnetic, **100**
    emission, 106, 109–110
    of reflected light, 101–102
    of various gases, **106**
    shifted by the Doppler effect, 114–117
    solar, 106
    thermal, 103–104
Speed
    of light, 88, 99
    relative versus "real," 117
Sphere, as normal shape of planet, 260
Spicules (of the chromosphere), 357
Spiral arms
    of the Galaxy, 492–500
    theories of, 497–500
Spiral galaxy, **7, 509–510,** 513, 514
Spiral nebulae, 507–508
Spring tide, 83
Sputnik, 197
Stadium, as unit of measure, 154–155
Standard solar model, the, 349–350
Star clusters, **416**–417, 429, 501–502
Star Date magazine, 318

Stars
    angular separation of, 18, 126, 130
    black dwarf, 441
    brightest, A-8
    color and temperature of, 102–105, 337
    composition of, 111–112
    death of, 429–438
    distances to, 372, 373, 381–382, 383–384
    energy source of, 341–345
    evolutionary track of, 413–**414**, 439, 451
    flyweight, 430
    formation of, 401, 402, 409–417, 487
    fusion in, 426–427
    giant, **380, 383,** 384
    H-R diagram of, 379–384
    heavyweight, 450–456, 468–471
    intensity/wavelength graph of, 102–104
    lightweight, 433–439
    luminosity classes of, 382–383
    luminosity of, 367–372, 274, 396–397
    magnitude of, 369–372, 374–375, 383–384, 396
    main sequence, 380, 427–429, 431
    main sequence, lifetime on, 427–429, 431
    mass of, determining 391–394
    mass of, maximum and minimum, 415, 425
    mass-luminosity relationship of, 394, 395
    measuring the properties of, 367–397
    middleweight, 450–456, 463–464
    motions of, 375–377
    multiple systems of, 385–394
    naked eye, 6
    naming, 373
    nearest, 9, 48, A-9
    neutron, 458–468
    planetary systems around, 219–221
    population I and population II, 471–474, 501
    quasars, 529–537
    red giant, 431–433, 435–436
    size of, 384–385, 390–394
    spectral classes of, 377–379
    supernovae, 444–445
    temperatures of, 102–105, 377–379, 380–382, 384–385, 427–428
    thermal spectrum of, 102–104
    thermostat of, 427
    white dwarf, 430, 439–444, 460–461
Steady state theories of cosmology, 555, 559
Stefan-Boltzmann Law, 105, 384
Stellar luminosity, 368
Stellar mass, 391–394, 415, 425
Stellar parallax, **48**
Stellar thermostat, 427
Stellar wind, 411, 436
Steward Observatory Mirror Lab, 135
Stickney (crater on Phobos), **258**
Stony irons, 329
Stony meteorites, **329**
Study skills, xix–xxiii

Sudbury Neutrino Observatory, 350
Sumerians, 17
Sun, 337–361
    and the evolution of solar system, 213, 214–215, 217–219
    as a red giant, 431–**432**, 434–**435**
    as a population I star, 473–474
    as a white dwarf, 440–**441**
    atmosphere of, 110–**111**, 351–358
    chromosphere of, **356**
    composition of, 354–356
    corona of, 167–**168**, 356–**358**
    diameter of, measuring, 338, 364
    distance to, measuring, 339
    data table, 340, inside back cover
    eclipses of, 166–170
    energy of, source of, 337–338, 341–345
    energy transport in, 347–349
    granulation of surface of, **354–355**
    hydrostatic equilibrium of, 346–347
    interior of, 345–351
    lifetime of, 428–429
    luminosity of, measuring, 338–341
    magnetic field of, and formation of the solar system, 217–218
    magnetic field of, and sunspots, 359–**361**
    mass of, 204, 338
    motion of, 20–26, 43–44, 490–491
    neutrino problem of, 349–351, 352–353
    photosphere of, 110, 351–356
    position in the Galaxy of, **481**–482
    prominences on, **337**, 357–**358**
    rotation of, 117, 415–416
    size of, 198–**199**, 338
    spectrum of, **106**, 352, **354**, 359
    standard model of, 349–351, 352–353
    tides on Earth produced by, 83
    vibrations within, 355
Sunset, red color of, 404
Sunspots, 338, **339**, 359–361
    cycle of, **359–361**
    Galileo's observation of, 64, 338
    observing, 364–365
Superclusters of galaxies, 526
Supergiant stars, **380, 383,** 384, 450, 451
    typical, 451
Supernova, Crab, 454–455
Supernova, Magellanic, **449–450**, **457**, 459
Supernova 1987A, **449–450**, **457**, 459
Supernova 1993J, 453–**454**
Supernovae
    detecting, 454–456
    light curves of, **453**
    remnants of, **445**, 456
    Type I, 444–**445**, 450, **452**–454
    Type II, 444, 450, 452
    typical, 452
Swift, Jonathan, 258
Synodic period, 162–**163**
Syzygy, 246

T Tauri stars, 220, 414
Tail, of a comet, 316, 318, **319**–321
Tammann, Gustav, 520
Tangential velocity, 116–**116**, 376, 377
Tarantula nebula, **1**, 540, **541**

Taurus, **16, 410**
TAU Project, 374
Taylor, Joe, 466, 467
Tectonics, plate, 176–179, 250–251
Telescope
    Arecibo, **140**, 148, 466
    Hale, **133**, 134, 135, 518
    Hubble Space, 131, **149**–150, 220, 283, 308, 403, 456
    Keck, 135–**136**
    Multiple Mirror, 145
    Spectroscopic Survey, 138–**139**
    Very Large, 134–135
    Uhuru, 147
    Yerkes, **132**
Telescopes
    accessories, 135–138
    adaptive optics, 135, 143
    Cassegrain focus, 132–**133**
    chromatic aberration in, 124–**125**
    field of view of, 127–**128**
    focal arrangements of, 132–133
    gamma ray, 147–149
    infrared, 145–146
    large optical, 132–135
    light-gathering power of, 128–129
    magnifying power of, 125–127
    Newtonian focus, **131**, 132
    powers of, 125–131
    radio, 139–143
    reflecting, 131–132
    refracting, 123–125
    resolving power of, 129–131
    small, to observe deep sky objects, 421
    ultraviolet, 147
    x-ray, 148
Temperature
    and the H-R diagram, 380–381
    determined by color, 102–105
    of a gas, 211–212
    of a star, 102–105, 377–379, 380–382, 384–385, 427–428
    scales, 96–**97**
Terrestrial planets, 227–260
    classification of, 206–209, 212
Theory, definition of, 35
Thermal spectrum, 102–104
Third quarter moon, 160–**162**
3-degree radiation, 553–556
Tidal friction, 84
Tides, 81–85
    and precession of Earth, 84–85
    due to the Sun, 83
    friction caused by, 84
Titan, **232**, 284–**285**
    escape velocity from, 285
Titania, **293**
Titius, Johann, 290
Titius-Bode Law, 200–201
Tomgaugh, Clyde W., 306–307
Total lunar eclipse, **153**, 165–166
Total solar eclipse, 167–**168**
Transit of meridian, 42
Tremaine, Scott, 296
Trifid nebula, **408**
Triple-alpha process, 434
Triton, **298**, 300

Troposphere of Earth's atmosphere, 179
Try-One-Yourself exercises, answers to, A-12–A-18
Tully, Brent, 522
Tully-Fisher relation, 522–523
Tunguska event, 330
Tuning fork diagram, 513–**514**
21-centimeter radiation, 407, 493–495
Tycho crater, **183**

UFOs, 376
U.S., viewed from space, **175**
U.S. Naval Observatory, 258
Uhuru Telescope, 147
Ultraviolet telescopes, 147
Ultraviolet radiation, 100
Ulysses spacecraft, 270
Umbra, **163**–165
Umbriel, **293**
Uncertainty in measurement, 51–52
Units
  of distance in astronomy, 5–8
  of temperature, 96–97, A-2
  metric, A-2
Universal gravitation, law of, 75–77
Universality principle, 552
Universe, (See also cosmology)
  age of, 556
  closed, 545–546, 558–560
  density of matter in, 560
  expansion of, 517–518, 547–550
  flat, 559–560
  homogeniety of, 551–552
  inflationary, 561–564
  isotropy of, 552
  open, 546, 557–560
  oscillating, 558
  scale of, 8–**9**, 480
  structure of, 564–566
Uranus, 288–294
  and discovery of Pluto, 306–307
  as seen from Earth, **288**
  composition of, **291**
  data page, 289
  data, with other planets, A-4–A-5
  discovery of, 288
  interior of, **291**
  magnetic field of, 293
  moons of, **293**, **294**
  orientation and motion of, 292–**293**
  rings of, 291–**292**

Urey, Harold, 222
Ursa Major, **13**, **17**, **454**
Ursa Minor, **454**

V404 Cygni, **471**
Vallis Marineris, **251**
Van Allen Belts, 174–**175**
van Maanen, Andriaan, 488, 507–508
variable stars
  Cepheid, 394–397
  eclipsing binary, 390, **391**, 393–394
Vega, 85
Veil nebula, **456**
Velocity
  of light, 88, 99
  radial, 376–377
  relative versus "real," 117
  space, 377
  tangential, 376, 377
Venera spacecraft, 238–239, 240
Venus, 236–245
  as morning and evening star, 29–32, 49
  as seen from Earth, 29–32, 236, **238**
  atmosphere of, 240–243, 245
  axis tilt of, **237**
  compared to Earth, 242–243, 245
  data page, 244
  data, with other planets, A-4–A-5
  favorable dates for viewing, 264
  greenhouse effect on, 243, 245
  in Copernican model, 49–50
  in Ptolemaic model, 32
  magnetic field of, 236
  motions of, 236–237
  phases of, 66–69, **67**, 238
  rotation of, 236–237
  size, mass, and density of, 236
  surface of, 237–**240**
  visits to, 237, 239
Vernal equinox, **21**, 26
Very Large Array, 144, **145**, 533
Very Large Telescope, 134–135
Very Long Baseline Array, 144
Vesta, 310
Viking I and II, 251–253, 256, 257
Virgo A, **529**
Virgo cluster of galaxies, **525**
Visual binaries, 386–**388**
Volatile elements, 213
Voyager spacecraft,
  and Jupiter, 268, 270, 275

  and Neptune, 294, 296–298, 300
  and Uranus, 290–291, 293
  and Saturn, 283–283
  messages to extraterrestrials on, 273
Vulcan, 90

Waning moon phases, **161**–162
War of the Worlds, 248
Warp of space, 87–**89**, 545–**546**
Wave motion, 98–99
Wave theory of light, 97, 99, 110
Wavelength, **98**–99
  and Doppler effect, 113–116
  of various colors, 99–**100**
Waxing moon phases, **160**–162
Weakly Interacting Massive Particle, 350
Wegener, Alfred, 176
Weight
  and law of universal gravitation, 75
  compared to mass, 71, 75
  definition of, 75
Weizacker, Carl von, 216
Welles, Orson, 248
Wells, H. G., 248
Whipple, Fred L., 317
Whirlpool galaxy, **499**–500
White dwarf star, **380**, **383**, 384, 430, 439–444
  as explanation for pulsars, 460–461
  typical, 440
Wien's Law, 105
Wilson, Robert, 553–**554**
Wilson, Alexander, 359
WIMP, 350
Windows in atmosphere, 101
Winter solstice, 24, **44**
Wolff, Sidney, 367–368

Yerkes Observatory, 131–**132**, 492, 512

X-rays, 100
X-ray telescopes, 147

Year, 21
Yerkes Observatory, **132**

Zach, Francis von, 203
Zeeman effect, 359
Zenith, 154
Zodiac, 22

CREDITS

Cover: Courtesy of Richard Wainscoat, Robert Williams and the Hubble Deep Field Team (STCScI)/NASA. **pii:** Courtesy Anglo-Australian Observatory. **pvi:** © 1980 Anglo-Australian Telescope Board. **pvii:** NASA. **pviii:** University of Texas McDonald Observatory. **pix:** NASA. **px:** NASA. **pxi:** NASA. **pxii:** Courtesy Anglo-Australian Observatory. **pxix:** NASA. **Prologue Chapter Opener:** Chris Floyd. **Fig. P-1:** Fred Espenak/Science Photo Library, Photo Researchers, Inc. **Fig. P-2:** Dennis Milon/Photo by Allan E. Morton. **Fig. P-3:** Copyright © 1980, Royal Observatory, Edinburgh, Anglo-Australian Telescope Board. **Fig. P-4:** Jim Baumgardt, Burlingame, CA. **Fig. P-6:** © 1980, Anglo-Australian Telescope Board. **Fig. P-8:** Author. **Fig. P-10:** Painting by David Seals.

CHAPTER 1

Chapter 1 Opener: Courtesy of the Adler Planetarium. Hand-colored engraving by Johann Bayer (1572-1625). **Fig. 1-1:** Dennis Milon/Photo by Allan E. Morton. **Fig. 1-2:** National Optical Astronomy Observatories. **Figs. 1-4a, 1-19a-b:** Arnim D. Hummel Planetarium, EKU. **Fig. 1-5:** Harvard College Observatory. **Fig. 1-18:** The Granger Collection, New York.

CHAPTER 2

Chapter 2 Opener: Courtesy of The Collection of Historical Scientific Instruments, Harvard University. **Figure 2-1:** The Bettman Archives. **Figs. 2-12, 2-14:** The Granger Collection, New York. **Fig. 2-13:** Photo Researchers, Inc. **Fig. H2-1:** By permission of the Houghton Library, Harvard University.

CHAPTER 3

Chapter 3 Opener: NASA. **Fig. 3-1:** The Granger Collection, New York. **Fig. 3-2:** Werner Sabo. **Fig. 3-4:** Lowell Observatory Photograph. **Figs. H3-1, 3-8:** The Granger Collection, New York. **Fig. C3-2:** NASA. **Fig. 3-17:** Author. **Fig. 3-23:** AIP Niels Bohr Library.

CHAPTER 4

Chapter 4 Opener: Charles H. Phillips/Smithsonian Books. **Fig. 4-2:** David Parker/Photo Researchers, Inc. **Figs. 4-6, 4-7:** Author. **P. 104:** D. Golimowski. **Fig. 4-11:** Deutches Museum, Munich. **Fig. H4-1:** AIP Niels Bohr Library, Margrethe Bohr Collection. **Fig. 4-18a:** Author. **Fig. 4-18b:** Education Development Center.

CHAPTER 5

Chapter 5 Opener: NOAO/I. Gatley and R. Probst. **Fig. 5-3a:** Author. **Fig. 5-8:** Author. **Fig. 5-10:** Author. **Fig. 5-11a:** Harvard College Observatory. **Fig. 5-12:** George C. Atamian. **Fig. 5-15:** Yerkes Observatory Photo. **Fig. 5-16b:** Author. **Fig. 5-17a:** © California Institute of Technology, 1979. **Fig. 5-18:** European Southern Observatory. **Fig. 5-19:** California Association for Research in Astronomy. **Fig. 5-20:** © 1993 Roger Ressmeyer-Starlight. **Fig. 5-21:** © 1993 Roger Ressmeyer-Starlight. **Fig. 5-22:** National Optical Astronomy Observatories. **Fig. 5-23:** NASA. **Fig. 5-24:** NASA/Galileo Project. **Fig. 5-25:** University of Texas McDonald Observatory. **Fig. 5-26:** Courtesy, NRAO/AUI. **Figs. 5-27a-b:** The Arecibo Observatory is part of the National Astronomy and Ionosphere Center which is operated by Cornell University under a cooperative agreement with the National Science Foundation. **Fig. 5-33:** Doug Johnson/Science Source, Photo Researchers, Inc. **Fig. 5-34:** NASA, Smithsonian Astrophysical Observatory. **Fig. 5-35:** STS-82 Crew/HST/NASA.

CHAPTER 6

Chapter 6 Opener: Celestron International. **Fig. 6-2:** By permission of Johnny Hart and Creator Syndicate. **Figs. 6-4a-b, 6-7:** Author. **Fig. 6-8:** Dennis Milon/Photo by Dennis Trail. **Figs. 6-10:** Lick Observatory. **Fig. 6-14b:** Adapted from Sky and Telescope Magazine. **Fig. 6-16:** Photo by Jim Rouse of the 8-16-89 lunar eclipse. All three exposures were 30 second on Ektachrome 400, using a 4-inch flourite refractor. Exposures were made at the beginning, middle, and end of totality. **Fig. 6-20:** Alex S. York. **Fig. 6-21b:** NASA. **Fig. 6-23a:** National Optical Astronomy Observatories. **Fig. 6-23b:** © Hans Vehrenberg, Hansen Planetarium. **Fig. 6-26b:** Nancy Rudger/The TASS /Sovfoto. **Figs. C6-1a-b:** NASA. **Fig. C6-2:** National Optical Astronomy Observatories. **Fig. 6-31:** Photo Researchers, Inc. **Fig. EA6-1:** NASA. **Fig. MO6-2, 6-37a-b:** Lick Observatory. **Fig. 6-38:** NASA. **Fig. C6-3:** NASA/Galileo Imaging Team. **Fig. 6-41:** Lick Observatory. **Fig. 6-43:** Sterne Photography. **Fig. 6-44:** NASA/Galileo Imaging Team. **Fig. A6-2:** Author.

CHAPTER 7

Chapter 7 Opener: NASA. **Fig. 7-2:** Author. **Fig. H7-1:** Yerkes. **Fig. 7-14:** Royal Observatory, Edinburgh/AATB/Science Photo Library. **Fig. 7-15:** D. Berry/STSCI AUL. **Fig. 7-16:** NASA. **Fig. 7-17:** Coronagraphic image of the ß Pictoris circumstellar disk in red light. Courtesy of D.A. Golimowski, S. T. Durrance, and M. Clampin.

CHAPTER 8

Chapter 8 Opener: HST/ NASA. **Figs. 8-3, 8-5, 8-6, 8-7:** NASA. **Fig. 8-11:** Courtesy Jet Propulsion Laboratory, California Institute of Technology/NASA. **Fig. ME8-2:** NASA. **Fig. 8-13:** TASS/Sovfoto. **Figs. 8-14a-b, 8-15a-b, 8-16, 8-17, VE8-2:** NASA/JPL. **Fig. 8-20a and b:** S. Larson, University of Arizona. **Fig. 8-20c:** D. Crisp/WFPC2 Team/NASA. **Fig. 8-21:** Worlds in Comparison, Astronomical Society of the Pacific. **Fig. 8-23:** Lowell Observatory Photograph. **Fig. 8-24a:** NASA/JPL. **Fig. 8-24b:** Astronomical Society of the Pacific. **Fig. 8-26:** NASA. **Fig. 8-27a:** Astronomical Society of the Pacific. **Fig. C8-2:** NASA /JPL. **Figs. 8-28, 8-29:** NASA. **Fig. 8-30:** Jeffrey S. Kargel, (U.S. Geological Survey). **Fig. 8-31:** NASA/Johnson Space Center. **Fig. 8-32:** D. Crisp/WFPC2 Team/NASA. **Fig. 8-34:** IMP Team/NASA/JPL. **Fig. H8-1:** NASA/JPL. **Fig. MA8-2:** U.S. Geological Survey, Flagstaff, AZ.

CHAPTER 9

Chapter 9 Opener: NASA/JPL. **Fig. 9-1:** © California Institute of Technology 1965. **Figs. 9-2, 9-3, 9-4, 9-5, C9-1:** NASA. **Fig. 9-9:** Astronomical Society of the Pacific. **Fig. 9-10:** NASA/JPL. **Fig. 9-11:** NASA. **Fig. 9-12:** Space Telescope Science Institute. **Figs. 9-13, 9-14, 9-15, 9-16, JU9-2:** NASA. **Fig. 9-17:** NASA. **Fig. 9-18a-f:** Lowell Observatory Photographs. **Figs. 9-21, 9-22a-d, 9-23a-b, 9-24, SA9-2, 9-26:** NASA. **Figs. UR9-2, 9-28a-b, 9-29:** NASA/JPL. **Figs. 9-33a-b, 9-34, 9-35, 9-36, C9-2, C9-3:** NASA. **Fig. C9-4:** NASA/JPL. **Figs. 9-38, NE9-2, 9-39:** NASA.

CHAPTER 10

Chapter 10 Opener: Tim Schroder. **Figs. 10-1a-b:** Lowell Observatory Photograph. **Fig. 10-2a:** U.S. Naval Observatory Photograph. **Figs. 10-2b-c:** NASA. **Fig. 10-4:** Yerkes. **Fig. 10-5:** NASA Galileo Imaging Team. **Fig. PL10-2:** A. Stern (SwRI), M. Buie (Lowell), NASA, ESA. **Fig. 10-6:** Author. **Fig. 10-8:** Dennis Milon/Photo by George East. **Fig. 10-9:** Courtesy of The Astronomer magazine. **Fig. 10-10:** Photo by Nick James. **Fig. 10-11:** J.A. DeYoung (USNO), 61-cm Telescope, Washington, DC. **Fig. 10-11b:** Copyright Max-Planck-Institute für Aeronomie, Courtesy H.U. Keller. **Fig. 10-14:** Palomar Observatory Photograph. **Figs. 10-17: top and bottom:** David Jewitt (Institute for Astronomy, U. Hawaii) and Jane Luu (U.C. Berkeley). **Fig. 10-19:** Dennis Milon. **Fig. 10-20:** NOAO (Kitt Peak Observatory). **Figs. 10-24a-b:** Fotosmith. **Fig. 10-24c:** Field Museum of Natural History: #39663. **Fig. C10-1:** John Bortle, W.R. Brooks Observatory, Stormville, NY. **Fig. 10-25:** Meteor Crater Enterprises, Inc., Flagstaff, Arizona. **Fig. 10-26:** Harvard-Smithsonian Center for Astrophysics, 60 Garden St. Cambridge, MA 02138.

CHAPTER 11

Chapter 11 Opener: NASA. **Fig. 11-1:** Owen Franken, Stock Boston. **Fig. 11-3:** Sterne Photograph. **Fig. 11-3b:** The Observatory of the Carnegie Institution of Washington. **Fig. 11-14:** Raymond Davis, Jr., Brookhaven National Laboratory. **Fig. 11-15:** Courtesy Marshall Spaceflight Center. **Figs. 11-18, C11-2, 11-21:** National Optical Astronomy Observatories. **Fig. 11-22:** National Center for Atmospheric Research/University Corporation for Atmospheric Research/National Science Foundation. **Fig. 11-23:** Photo by High Altitude Observatory, National Center for Atmospheric Research. The National Center for Atmospheric Research is sponsored by The National Science Foundation. **Fig. 11-24:** NASA. **Figs. 11-25, 11-26:** NASA/MSFC/Hathaway. **Figs. 11-27a-b:** National Optical Astronomy Observatories.

CHAPTER 12

Chapter 12 Opener: Institute for Astronomy, University of Hawaii. **Fig. 12-1:** Arnim D. Hummel Planetarium, EKU. **Figs. 12-8a-b:** Yerkes Observatory Photo. **Fig. 12-10:** University of Michigan Astronomy Department. **Fig. 12-12:** Photograph by David Malin, © The Anglo-Australian Telescope Board. **Fig. 12-13:** Harvard College Observatory. **Fig. 12-14:** University of Michigan Observatories. **Fig. 12-25a:** George C. Atamian. **Fig. 12-26:** Swathmore College Observatory. **Fig. 12-27:** Lick Observatory. **Figs. 12-39, 12-40:** Harvard College Observatory.

CHAPTER 13

Chapter 13 Opener: "Two Armed Instability of Rotating Polytropic Star" Principle Investigators: Durisen, Richard; Yang, Shelby; Grabhorn, Robert; Department of Astronomy, Indiana University. Visualization: Yost, Jeffrey; NCSA. **Fig. 13-1a:** Martin C. Germano. **Fig. 13-1b:** Copyright © Royal Observatory, Edinburgh, and Anglo-Australian Telescope Board. **Fig. 13-2:** NASA. **Fig. 13-5a:** National Optical Astronomy Observatories. **Fig. 13-5b:** C.R. O'Dell, S.K and Wong (Rice University), NASA. **Fig. 13-7:** Copyright © Royal Observatory, Edinburgh, and Anglo-Australian Telescope Board. **Fig. 13-9:** Courtesy Anglo-Australian Observatory. **Fig. 13-10:** Jeff Hester and Paul Scowen (Arizona State University), NASA. **Fig. 13-11:** National Optical Astronomy Observatories. **Fig. C13-1:** Courtesy Anglo-Australian Observatory. **Fig. 13-13:** NASA. **Fig. 13-15:** National Optical Astronomy Observatories. **Fig. 13-16:** R. Mundt, Max Lanck, Institut fur Astronomie; Photo obtained at the Calar Alto-3.5 m telescope in Spain. **Fig. 13-17a:** Jack Newton. **Fig. 13-17b:** Yerkes Observatory Photograph. **Fig. 13-18:** U.S. Naval Observatory.

CHAPTER 14
Chapter 14 Opener: J.P. Harrington and K.J. Borkowski (University of Maryland), NASA. **Fig. 14.1b:** S. Kulkarni © Caltech, D. Golimowski (JHU), NASA. **Fig. 14-15a:** National Optical Astronomy Observatories. **Fig. 14-15b:** Courtesy Anglo-Australian Observatory. **Fig. 14-16a:** Celestion International. **Fig. 14-18a:** © California Institute of Technology 1965. **Fig. 14-18b:** H. Bond (STScI), NASA. **Fig. 14-20b:** Arnim D. Hummel Planetarium, EKU. **Figs. 14-21, 14-25:** Lick Observatory. **Fig. 14-27:** © Copyright 1980, Royal Observatory, Edinburgh.

CHAPTER 15
Chapter 15 Opener: Courtesy of Adler Planetarium. Hand-colored engravings by Johann Bayer (1572-1625). **Fig. 15-1:** A. Dupree (CfA), NASA. **Fig. 15-5a: top and bottom:** Jack Newton. **Fig. 15-6:** Caltech/David Malin/Jay Pasachoff. **Fig. 15-7a:** Jack Newton. **Fig. 15-7b:** NASA. **Figs. 15-8a, 15-8b:** Chris Floyd. **Fig. 15-9:** University of Cambridge, Mullard Radio Astronomy Observatory. With Compliments of Professor Antony Hewish. **Fig C15-0:** Jeff Hester and Paul Scowen (Arizona State University)/NASA. **Fig. 15-19:** Giffith Observatory, Painting by Lois Cohen.

CHAPTER 16
Chapter 16 Opener: The National Radio Astronomy Observatory, operated by Associated Universities, Inc. under contract with the National Science Foundation. Acknowledgment: Farhad Yusef-Zadeh, Mark R. Morris, Don R. Chance. **Fig. 16-1:** Photo by Akira Fujii. **Fig. 16-3b:** Arnim D. Hummel Planetarium, EKU. **Fig. 16-4a:** National Optical Astronomy Observatories. **Figs. 16-6, 16-7:** Yerkes Observatory Photograph. **Fig. 16-8:** National Optical Astronomy Observatories. **Fig. 16-17a:** © 1980 Anglo-Australian Telescope Board. **Fig. 16-17b:** Courtesy Anglo-Australian Observatory. **Fig. C16-1:** Photo by Kim Zussman, Thousand Oaks, CA. **Fig. 16-21a:** NOAO/N. A. Sharp. **Fig. 16-21b:** The Observatories of the Carnegie Institution of Washington. **Fig. 16-22:** NASA.

CHAPTER 17
Chapter 17 Opener: R. Williams/Hubble Deep Field Team (STScI), NASA. **Fig. 17-1:** Palomar Observatory. **Figs. 17-2, 17-3:** Courtesy Anglo-Australian Observatory. **Fig. 17-4a:** National Optical Astronomy Observatories. **Fig. 17-4b:** Joe Liddell/Astronomy. **Fig. 17-4c:** Kim Zussman, Thousand Oaks, CA. **Fig. 17-5a-c:** Palomar Observatory Photograph. **Fig. 17-6a:** Barney Magrath/Science Photo Library. **Fig. 17-6b:** Courtesy Anglo-Australian Observatory. **Fig. 17-7:** Courtesy Anglo-Australian Observatory. **Fig. 17-9a:** © Copyright 1959 by California Institute of Technology and the Carnegie Institution of Washington. **Fig. 17-9b:** Hale Observatories. **Fig. 17-13:** Palomar Observatory Photograph. **Fig. 17-18:** Royal Observatory, Edinburgh/AATB/Science Photo Library. **Fig. 17-20a:** © 1980 Anglo-Australian Telescope Board. **Fig. 17-20b:** Courtesy NRAO/AUI. Acknowledgements: J.O. Burns, E. J. Schreier, E. D. Feigelson. **Fig. 17-21a:** Courtesy Anglo-Australian Observatory. **Fig. 17-21b:** Tod R. Lauer/NASA. **Fig. 17-24:** Palomar Observatory Photograph. **Fig. 17-26a:** NRAO/AUI. **Fig. 17-26b:** K. Ratnatunga (JHU), NASA. **Fig. 17-30a:** Courtesy NRAO/AUI. Acknowledgement: R.A. Perley, J.W. Dreher, J.J. Cowan. **Fig. 17-30b:** Courtesy NRAO/AUI. Acknowledgement: E. Fomalont, R. Ekers, K. Ebneter, W. van Breugel. **Fig 17-c:** NASA. **Fig. 17-31:** K. Gebhardt (University of Michigan), Tod Lauer (NOAO). **Figs. 17-33, 34:** National Optical Astronomy Observatories.

CHAPTER 18
Chapter 18 Opener: © 1980 Anglo-Australian Telescope Board. **Fig. 18-9:** Photri. **Fig. 18-11:** Courtesy of AT&T. **Fig. 18-18:** NASA. **Fig. C18-2:** © Ian Berry/Magnum. **Fig. 18-19:** NASA/Goddard Space Flight Center. **Fig. 18-20:** R.J.E. Peebles. **Figs. 18-21, 18-22:** NASA/Goddard Space Flight Center.

End of book star charts:
Griffith Observer, Griffith Observatory, Los Angeles.

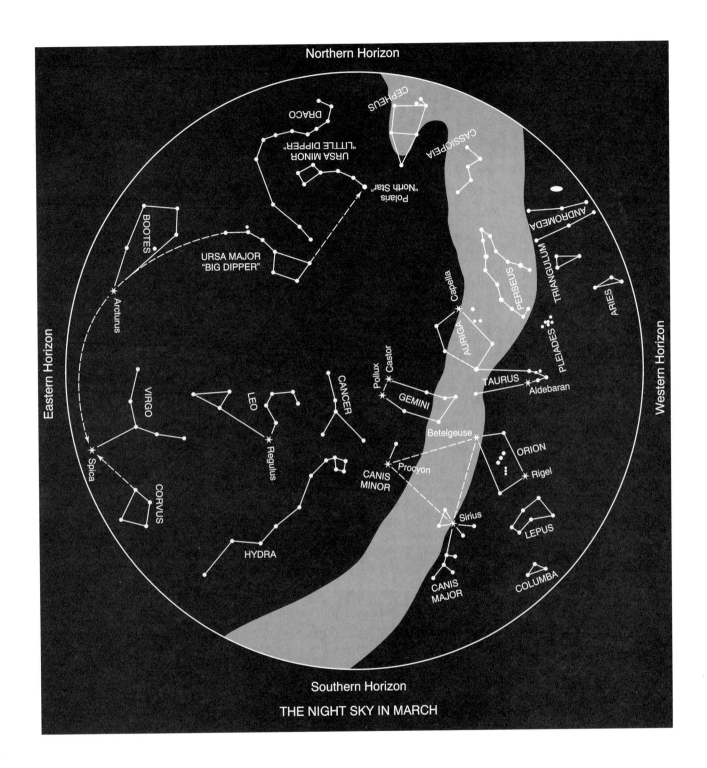

To use: Hold chart vertically and turn it so the direction you are facing shows at the bottom.

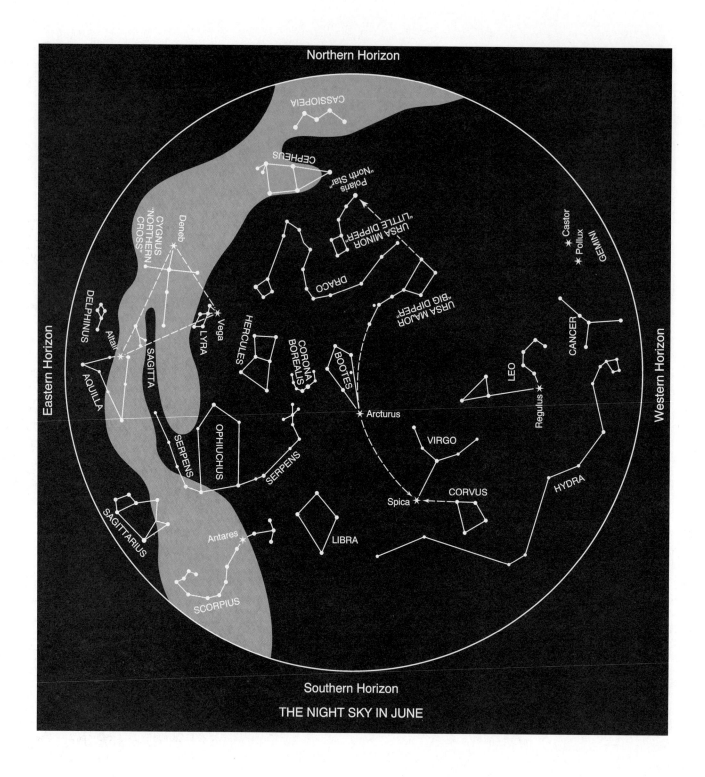

To use: Hold chart vertically and turn it so the direction you are facing shows at the bottom.

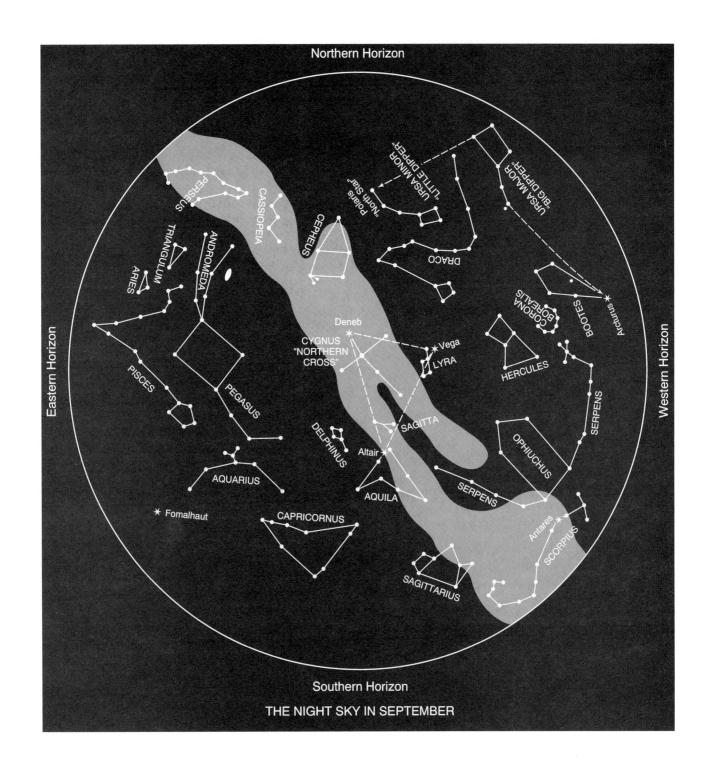

To use: Hold chart vertically and turn it so the direction you are facing shows at the bottom.

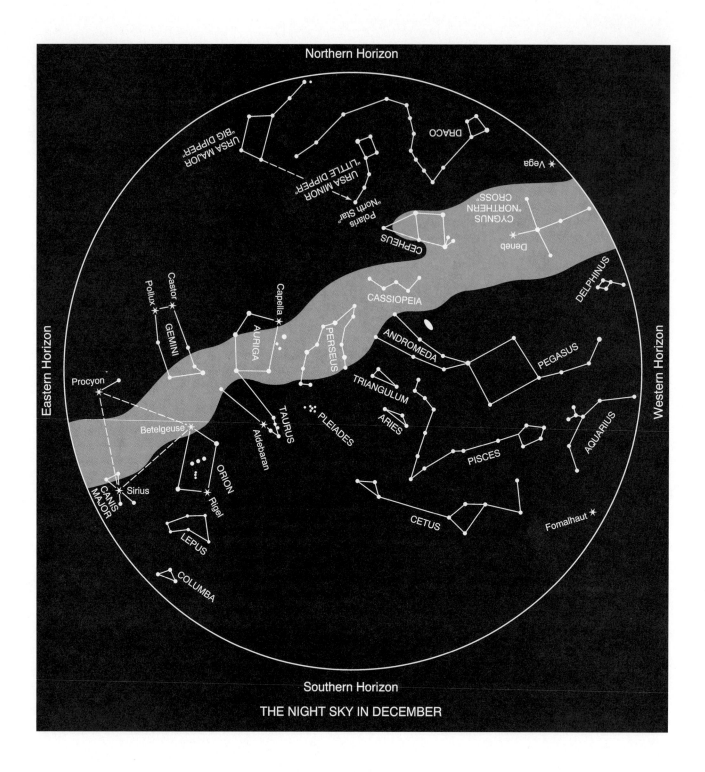

To use: Hold chart vertically and turn it so the direction you are facing shows at the bottom.